D1601624

Lecture Notes in Physics

Springer
Berlin
Heidelberg
New York
Barcelona
Hong Kong
London
Milan
Paris
Tokyo

Editorial Policy

The series *Lecture Notes in Physics* (LNP), founded in 1969, reports new developments in physics research and teaching -- quickly, informally but with a high quality. Manuscripts to be considered for publication are topical volumes consisting of a limited number of contributions, carefully edited and closely related to each other. Each contribution should contain at least partly original and previously unpublished material, be written in a clear, pedagogical style and aimed at a broader readership, especially graduate students and nonspecialist researchers wishing to familiarize themselves with the topic concerned. For this reason, traditional proceedings cannot be considered for this series though volumes to appear in this series are often based on material presented at conferences, workshops and schools (in exceptional cases the original papers and/or those not included in the printed book may be added on an accompanying CD ROM, together with the abstracts of posters and other material suitable for publication, e.g. large tables, colour pictures, program codes, etc.).

Acceptance

A project can only be accepted tentatively for publication, by both the editorial board and the publisher, following thorough examination of the material submitted. The book proposal sent to the publisher should consist at least of a preliminary table of contents outlining the structure of the book together with abstracts of all contributions to be included.
Final acceptance is issued by the series editor in charge, in consultation with the publisher, only after receiving the complete manuscript. Final acceptance, possibly requiring minor corrections, usually follows the tentative acceptance unless the final manuscript differs significantly from expectations (project outline). In particular, the series editors are entitled to reject individual contributions if they do not meet the high quality standards of this series. The final manuscript must be camera-ready, and should include both an informative introduction and a sufficiently detailed subject index.

Contractual Aspects

Publication in LNP is free of charge. There is no formal contract, no royalties are paid, and no bulk orders are required, although special discounts are offered in this case. The volume editors receive jointly 30 free copies for their personal use and are entitled, as are the contributing authors, to purchase Springer books at a reduced rate. The publisher secures the copyright for each volume. As a rule, no reprints of individual contributions can be supplied.

Manuscript Submission

The manuscript in its final and approved version must be submitted in camera-ready form. The corresponding electronic source files are also required for the production process, in particular the online version. Technical assistance in compiling the final manuscript can be provided by the publisher's production editor(s), especially with regard to the publisher's own Latex macro package which has been specially designed for this series.

Online Version/ LNP Homepage

LNP homepage (list of available titles, aims and scope, editorial contacts etc.):
http://www.springer.de/phys/books/lnpp/

LNP online (abstracts, full-texts, subscriptions etc.):
http://link.springer.de/series/lnpp/

N. J. Balmforth A. Provenzale (Eds.)

Geomorphological Fluid Mechanics

Springer

Editors

N. J. Balmforth
University of California at Santa Cruz
School of Engineering
1156 High St.
Santa Cruz, CA 95064, USA

A. Provenzale
Istituto di Scienze dell'Atmosfera e del Clima, CNR
Corso Fiume 4
10133 Torino, Italy

Cover Picture:A photograph of desert sand ripples (by Balmforth et al. p.3).

Library of Congress Cataloging-in-Publication Data applied for.

Die Deutsche Bibliothek - CIP-Einheitsaufnahme

Geomorphological fluid mechanics / N. J. Balmforth ; A. Provenzale (ed.). - Berlin ; Heidelberg ; New York ; Barcelona ; Hong Kong ; London ; Milan ; Paris ; Tokyo : Springer, 2001
(Lecture notes in physics ; Vol. 582)
(Physics and astronomy online library)
ISBN 3-540-42968-9

ISSN 0075-8450
ISBN 3-540-42968-9 Springer-Verlag Berlin Heidelberg New York

Springer-Verlag Berlin Heidelberg New York
a member of BertelsmannSpringer Science+Business Media GmbH

http://www.springer.de

Typesetting: Camera-ready by the authors/editors
Camera-data conversion by Steingraeber Satztechnik GmbH Heidelberg
Cover design: *design & production*, Heidelberg

Printed on acid-free paper
SPIN: 10857564 54/3141/du - 5 4 3 2 1 0

Preface

This volume contains the lecture notes from the Gran Combin Summer School, held in Saint Oyen, Aosta, Italy, in the second half of June 2000. The lectures, loosely connected through the broad heading of "Geomorphological Fluid Mechanics," explored a variety of topics involving the fluid flows encountered in geomorphology and in related geological problems. Specific topics included lava and mud flows, ice dynamics, snow avalanches, river and coastal morphodynamics, and landscape formation. The aim of the school was to unite these topics using their common mathematical and physical language. The lecture notes have four parts, each with a particular theme: Fundamentals, Hot, Cold and Dirty. In these parts, we include chapters of a more basic nature (Fundamentals), chapters relevant to magma and lava (Hot), to ice flow (Cold), and to fluids in which the transport of suspended particles is critical (Dirty). Each division has its own opening chapter that gives a brief introduction to each theme.

The directors of the course "Geomorphological Fluid Mechanics,"were Neil J. Balmforth (University of California, Santa Cruz, USA) and Antonello Provenzale (Istituto di Cosmogeofisica, CNR, Torino, Italy), who also acted as scientific editors for these lecture notes. Jost von Hardenberg (Istituto di Cosmogeofisica, CNR, Torino, Italy) was the scientific secretary and Laura Roma was the administrative assistant of the school. Costanza Piccolo was our industrious technical editor who collected together all the notes and painstakingly edited them; without her assistance, the chapters may never have seen the light of day.

The main lecturers of the school were Ross Griffiths (Australian National University), Kolumban Hutter (University of Darmstadt), Chiang C. Mei (Massachusetts Institute of Technology), Gary Parker (University of Minnesota), Giovanni Seminara (University of Genova), Terence R. Smith (University of California at Santa Barbara), Jack Whitehead (Woods Hole Oceanographic Institution) and Andy Woods (University of Cambridge). Special lectures were also given by Christophe Ancey (CNRS, Grenoble), Augusto Biancotti (University of Torino), Richard Craster (Imperial College, London), Andrew Fowler (University of Oxford), Stuart B. Savage (McGill University, Montreal), and John Wettlaufer (University of Washington).

The Gran Combin Summer School is a joint enterprise of the French CNRS and Italian CNR, whose general theme is "Fundamental Problems in Geophysical and Astrophysical Fluid Dynamics."Local organization and funding each year is provided by the "Istituto di Cosmogeofisica"(CNR, Torino, Italy), by the

Groupement de Recherche “Mécanique Fondamentale des Fluides Géophysiques et Astrophysiques”(CNRS, France) and by the Laboratoire de Meteorologie Dynamique, ENS-CNRS, Paris. Support for the summer school also comes from the APT Gran St. Bernard, Valle d'Aosta (Italy), the Regional Government of Valle d'Aosta, and the Comunitá Montana “Grand Combin.”

Woods Hole,
August 2001

Neil J. Balmforth
Antonello Provenzale

Contents

Part I Fundamentals: Methods, Materials and Metaphors

1 The Language of Pattern and Form
N.J. Balmforth, A. Provenzale, J.A. Whitehead 3

2 Geophysical Aspects of Non-Newtonian Fluid Mechanics
N.J. Balmforth, R.V. Craster 34

3 Introduction to Rheology and Application to Geophysics
C. Ancey 52

4 Granular Material Theories Revisited
Y. Wang, K. Hutter 79

Part II Hot

5 Earth's Surface Morphology and Convection in the Mantle
R.W. Griffiths, J.A. Whitehead 111

6 Morphological Instabilities in Flows with Cooling, Freezing or Dissolution
J.A. Whitehead, R.W. Griffiths 138

7 Shallow Lava Theory
N.J. Balmforth, A.S. Burbidge, R.V. Craster 164

8 Explosive Volcanic Eruptions
A.W. Woods 188

Part III Cold

9 The Dynamics of Snow and Ice Masses
J.S. Wettlaufer 211

10 Response of Italian Glaciers to Climatic Variations
A. Biancotti, M. Motta 218

11 Asymptotic Theories of Ice Sheets and Ice Shelves
D.R. Baral, K. Hutter 227

12 Aspects of Iceberg Deterioration and Drift
S.B. Savage 279

13 Snow Avalanches
C. Ancey 319

14 Dense Granular Avalanches: Mathematical Description and Experimental Validation
Y.-C. Tai, K. Hutter, J.M.N.T. Gray 339

Part IV Dirty

15 Patterns of Dirt
N.J. Balmforth, A. Provenzale 369

16 Invitation to Sediment Transport
G. Seminara 394

17 Types of Aeolian Sand Dunes and Their Formation
H. Tsoar 403

18 Dunes and Drumlins
A.C. Fowler 430

19 Estuarine Patterns: An Introduction to Their Morphology and Mechanics
G. Seminara, S. Lanzoni, M. Bolla Pittaluga, L. Solari 455

20 Longshore Bars and Bragg Resonance
C.C. Mei, T. Hara, J. Yu 500

21 Debris Flows and Related Phenomena
C. Ancey 528

22 Mud Flow – Slow and Fast
C.C. Mei, K.-F. Liu, M. Yuhi 548

Index 579

List of Contributors

Christophe Ancey
Cemagref, Unité Erosion Torrentielle, Neige et Avalanches,
Domaine Universitaire, 38402
Saint-Martin-d'Hères Cedex, France

Neil J. Balmforth
Department of Applied Mathematics and Statistics,
School of Engineering,
University of California at Santa Cruz, CA 95064, USA

Dambaru R. Baral
Institute of Mechanics,
Darmstadt University of Technology,
Hochschulstr. 1, 64289 Darmstadt, Germany

Augusto Biancotti
Dipartimento di Scienze della Terra,
Universitá di Torino,
Torino, Italy

Michele Bolla Pittaluga
Dipartimento di Ingegneria Ambientale, Universitá di Genova,
Via Montallegro 1, 16145 Genova, Italy

Adam S. Burbidge
School of Chemical Engineering,
University of Birmingham,
Edgbaston, Birmingham, B15 2TT, UK

Richard V. Craster
Department of Mathematics,
Imperial College of Science,
Technology and Medicine,
London, SW7 2BZ, UK

Andrew Fowler
Mathematical Institute,
Oxford University,
24-29 St. Giles', Oxford OX1 3LB, UK

J.M.N.T. Gray
Department of Mathematics,
University of Manchester, Manchester M13 9PL, UK

Ross W. Griffiths
Research School of Earth Sciences,
The Australian National University,
Canberra, 0200 ACT, Australia

Tetsu Hara
Graduate School of Oceanography,
University of Rhode Island, Kingston, RI 02881, USA

Kolumban Hutter
Institute of Mechanics,
Darmstadt University of Technology,
Hochschulstr. 1, 64289 Darmstadt, Germany

Stefano Lanzoni
Dipartimento di Ingegneria Idraulica, Marittima e Geotecnica,
Universitá di Padova,
Via Loredan 20, 35131 Padova, Italy

Ko-fei Liu
Department of Civil Engineering,
National Taiwan University, Taipei,
Taiwan, Republic of China

Chiang C. Mei
Department of Civil & Environmental Engineering,
Massachusetts Institute of Technology,
Cambridge, MA 02139, USA

Michele Motta
Dipartimento di Scienze della Terra,
Universitá di Torino,
Torino, Italy

Antonello Provenzale
Istituto di Cosmogeofisica,
Corso Fiume 4, 10133 Torino, Italy;
and ISI Foundation,
V.le Settimio Severo 65, 10133 Torino,
Italy

Stuart B. Savage
McGill University, Montreal,
Quebec H3A 2K6, Canada

Giovanni Seminara
Dipartimento di Ingegneria Ambientale, Università di Genova,
Via Montallegro 1, 16145 Genova,
Italy

Luca Solari
Dipartimento di Ingegneria Civile,
Universitá di Firenze,
Via S. Marta, 50135 Firenze, Italy

Yih-Chin Tai
Institute of Mechanics,
Darmstadt University of Technology,
Hochschulstr. 1, 64289 Darmstadt,
Germany

Haim Tsoar
Department of Geography and
Environmental Development,
Ben-Gurion University of the Negev,
Beer-Sheva 84105, Israel

Yongqi Wang
Institute of Mechanics,
Darmstadt University of Technology,
Hochschulstr. 1, 64289 Darmstadt,
Germany

John A. Whitehead
Woods Hole Oceanographic Institution,
Woods Hole, MA 02543, USA

John S. Wettlaufer
Applied Physics Laboratory and
Department of Physics,
University of Washington, Seattle,
WA 98105-5640, USA

Andrew W. Woods
BP Institute, University of Cambridge,
Cambridge, UK

Jie Yu
Division of Earth and Ocean Sciences,
Duke University, Box 90227,
Durham, NC 27708-0227, USA

Masatoshi Yuhi
Department of Civil Engineering,
Kanazawa University, Kanazawa,
Ishikawa, 920-8667, Japan

Part I

Fundamentals: Methods, Materials and Metaphors

1 The Language of Pattern and Form

N.J. Balmforth[1], A. Provenzale[2], and J.A. Whitehead[3]

[1] Department of Applied Mathematics and Statistics, School of Engineering, University of California at Santa Cruz, CA 95064, USA
[2] Istituto di Cosmogeofisica, Corso Fiume 4, 10133 Torino, Italy; and ISI Foundation, V.le Settimio Severo 65, 10133 Torino, Italy
[3] Woods Hole Oceanographic Institution, Woods Hole, MA 02543, USA

1.1 Introduction

Geology and geomorphology deal with some of the most striking patterns of Nature. From mountain ranges and mid-ocean ridges, to river networks and sand dunes, there is a whole family of forms, structures, and shapes that demand rationalization as well as mathematical description. In the various chapters of this volume, many of these patterns will be explored and discussed, and attempts will be made to both unravel the mathematical reasons for their very existence and to describe their dynamics in quantitative terms. In this introductory chapter, we discuss some of the methods that can be adopted in the study of patterns, and use the specific examples of *convection* – an evergreen classic in nonlinear fluid dynamics – and of the formation of *aeolian ripples* – another phenomenon that strikes the imagination of anybody who has been travelling in a sand desert.

The first observation is that, in many instances, patterns form because of a simple linear instability. In the first example we treat in this chapter, a fluid layer heated from below, the instability is one in which a heated fluid parcel rises upwards and displaces the ambient fluid downwards. In response to geometrical constraints and detailed physical effects, the simple instability often chooses its own structure, or spatial pattern – certain motions are "easiest" or most favourable, leading to a preferred pattern. For convection, fluid motions can create an aesthetic network of rolls, squares or hexagons, which has partly lead to its frequent portrayal in works on pattern formation. In other situations, however, there can be several patterns that are roughly equally preferred, and the system is "frustrated" in its passage to a final state. This is also what one sees in convection, where there can be a pronounced competition between, for example, a planform of hexagons and one of rolls. Some of the goals of "pattern theory" are to understand how the system sorts itself out in these situations. Other goals are to predict the precise pattern itself, which must be an attracting solution of the governing equations, and not all that easy to construct.

The wide spectrum of different patterns and systems in which they arise makes a theoretical exploration appear to be both complicated and very specific to each situation. Indeed, each pattern has its own unique elements, and is studied by employing specific methodologies that reflect both the tradition of the disciplines involved and the idiosyncracies of the explorers. However, from the mathematical perspective there are usually common, connecting threads:

First, linear instability theory is a general approach that invariably follows the same lines (the decomposition of a disturbance of infinitesimal amplitude into a set of normal modes and the subsequent computation of their growth or decay rates) whatever the problem. Further, the onset of an instability is a "bifurcation," and we may usefully think of the transition using ideas from dynamical systems theory. In fact, mathematically, the most common bifurcations aways occur in certain standard ways. According to dynamical systems theory, each of the standards has a specific underlying mathematical description – an "amplitude equation," or a set of them, for the unstable normal modes. The main challenge, then, is to determine which of the standard bifurcations takes place, and to reduce the governing equations to the relevant set of amplitude equations.

Practically, this goal is achieved by positioning oneself at the point of bifurcation, and performing some asymptotic expansion; we give an example for convection below. In doing so, the details of the particular problem all become subsumed into the specific values of a set of constants (the coefficients or parameters of the amplitude equations). This is why, but for those special parameter values, quite different patterns can be described by the same mathematics. Of course, to visualize the emerging pattern we must reconstruct the entire solution from the mode amplitudes, and this is where all the details of the particular problem matter. Nevertheless, the underlying mathematical description provides a tool to understand pattern formation in general situations, which is why this approach has become so popular in recent years and spawned a generation of roving nonlinear dynamicists in search of problems to apply their technology.

There is one severe limitation with the reductive idea: the derivation and validity of the amplitude equations is restricted to conditions near the brink of instability – the patterns must be just able to form. What this signifies is that the growth of the instability, the creation of the pattern, is a slow process, and all of the other complicating processes that could occur in the system do so and subside beforehand. Mathematically, we filter these complications out, leaving a relatively simple description of the slow pattern formation. Unfortunately, these conditions do not last once we move away from the onset of instability, and the filtered complications invariably return to thwart the reduction scheme. The further from the initial bifurcation, the more complicated the dynamics usually becomes, producing rich spatio-temporal dynamics and even turbulence. Describing those situations is a far harder problem. In fact, out of these very complicated states, other kinds of patterns can emerge, such as the coherent structures of turbulent fluids (the most notable, perhaps, being Jupiter's Great Red Spot). Because we no longer have a simple equilibrium state, the coherent structures that punctuate a turbulent background have an origin that is much less clear than that of the patterns at the onset of the instability, and the mathematics is correspondingly more complicated (and much remains to be understood).

As it turns out, even the theory of patterns close to the bifurcation point is often not very simple. It is simpler than what we started with, but, often, also requires a great deal of labour to understand (in "spatially extended" systems, pattern equations can be partial differential equations with complicated chaotic solutions). This aspect, coupled with the need to go beyond the small window near onset, typically means that the reductive style of approach – the detection of a mildly unstable pattern, and the reduction of the governing equations to an amplitude equation – is not always helpful. Worse still, for many problems in geology and geomorphology, the governing equations themselves are either practically intractable, or not even known; our example of the sand ripples provides an illustration.

All this dirty washing highlights how the usual geophysical problem is far more complicated than we would truly like to admit when we embark on a mathematical description of the forming patterns. Our mathematical tools are largely fashioned for a physical situation that is more like a controlled laboratory experiment; our description of the convection problem follows largely such lines. To advance further we need to make more and more less refined approximations, or simply resort to big numerical simulations. The former pathway eventually leads into the realm of crude phenomenological models, which our consideration of viscous conduits and sand ripples also illustrates. The approach of full-scale numerical simulation has an important place in the exploration of many problems, but this introduction is meant to offer some technology for mathematical modelling, and so we have little more to say about simulation, other than it can be done and is important.

1.2 Convection

Convection is a classic problem in geophysical fluid mechanics. In its incarnation as the Rayleigh–Bénard problem, it models the thermal convection of a differentially heated fluid layer, and can be studied from both the theoretical and experimental perspectives. In its simplest form, convection occurs when the fluid is contained between two plates maintained at different temperatures and the lower plate is heated sufficiently so that small horizontal inhomogeneities in the fluid density induce the warm fluid near the bottom to float up buoyantly and displace the denser fluid above. Nevertheless, convection can occur in different geometries (like a sphere), over extensive regions (such as the Earth's mantle), and when there are other competing physical effects (embedded magnetic fields, internal heat sources, suspended or sedimenting particles, and so on). Consequently, this problem has many applications in geophysics. Moreover, it is a paradigm of a system that develops patterns and turbulent fluid motions. Several reviews of theory and experiment are available [1–6].

A key physical variable for convection is the degree of differential heating. A convenient, dimensionless measure of this quantity is the so-called Rayleigh number,

$$R = \frac{g\alpha D^3(T_2 - T_1)}{\nu\kappa}\,, \tag{1.1}$$

defined in terms of the temperature difference between the two plates, $(T_2 - T_1)$, the fluid depth, D, the gravitational acceleration, g, the coefficient of thermal expansion, α, the kinematic viscosity of the fluid, ν, and l its thermal conductivity, κ. The Rayleigh number is one of the single most important quantities in convection theory because it provides a crude yardstick for what we expect the fluid to do in any given situation. For example, in the laboratory, for small R, heat is transferred between the plates purely by conduction and there is no motion in the fluid. But when the Rayleigh number exceeds a critical value, the fluid begins to overturn and steady convection patterns occur that take the form of a network of convection cells. The shape of the cells varies depending on the fluid properties and lateral shape of the fluid layer, but cells with the form of hexagons or rolls are common (see Fig. 1.1 – in the second panel rolls are gradually replacing hexagons from one side). With a further increase of R the cellular pattern changes and becomes progressively more complex (see Figs. 1.2 and 1.3). Eventually, the cells become time-dependent, and when R is sufficiently large, the convective pattern degrades and the fluid can become turbulent.

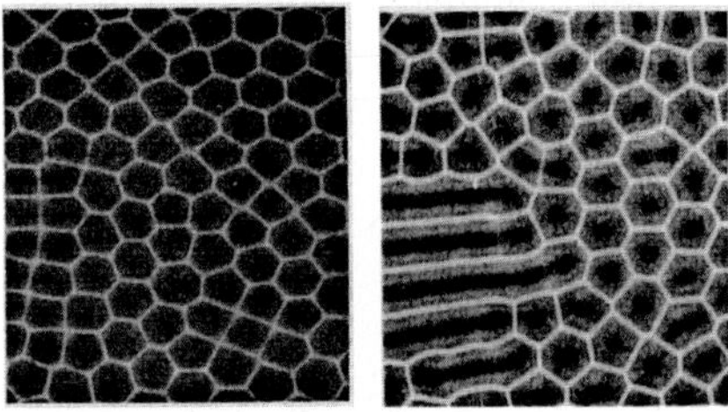

Fig. 1.1. Shadowgraphs (plan view) of hexagonal/polygonal planforms. These are slowly being replaced by rolls. In these and subsequent figures, white lines or spots indicate cold, dense material while dark lines or spots indicate warm, buoyant material

In the turbulent regime in the laboratory, no regular patterns are evident. However, large-scale flows can still persist, and sweep fluid around over the scale of the container. Moreover, much of the temperature variation becomes confined to boundary layers adjacent to the plates. The reduction in scale over these boundary layers enhances the dissipative effects of thermal conduction and viscosity, with the results that motion in these regions is more laminar than in the interior of the fluid. However, these layers are themselves in an unsteady balance: The natural tendency of the boundary layers to thicken by thermal diffusion is held in check by the eruption of localized plumes from them that become swept and mixed into the turbulent interior.

Besides turbulent convection in the laboratory, there are many geophysical flows in which the Rayleigh number is even higher (such as convection in the Earth's core). In these systems, the popular conception is that the driving of the fluid by the differential heating is so high that the motion is turbulent even inside any boundary layers. The molecular values of the conductivity and viscosity are then claimed to become irrelevant in determining the amount of heat transported by convection. This leads to a characteristic asymptotic scaling of the degree of

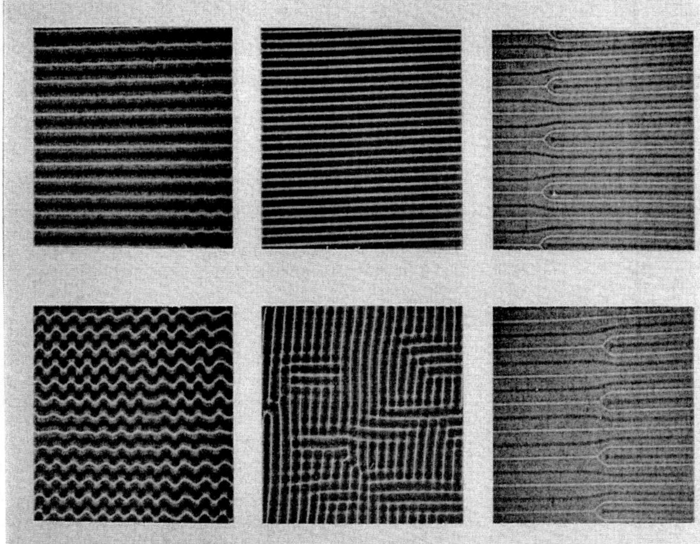

Fig. 1.2. Experimental observations of the breakdown of convection rolls. On the left is a zig-zag instability. In the middle is a cross-roll instability. On the right is a pinch

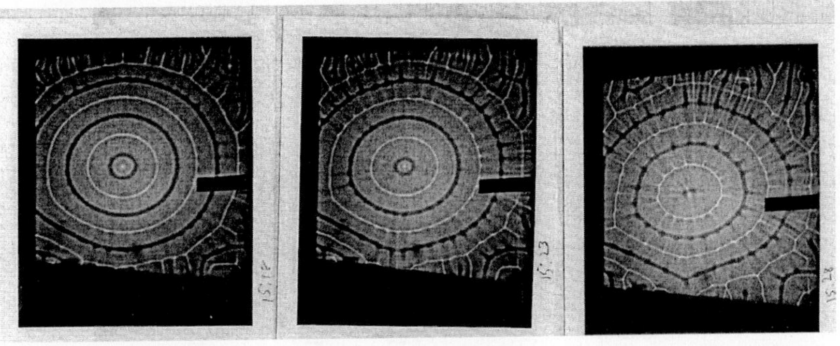

Fig. 1.3. More shadowgraphs showing "bimodal flow," which is roughly a combination of rolls with different orientation

heat transfer with Rayleigh number (the infamous $R^{1/2}$ scaling), which is based essentially on dimensional analysis [2].

As described in the chapters to come, a specific case relevant to landscapes is convection in the Earth's mantle. In this instance, the Rayleigh number is not especially high (compared to the core, for example) and fluid motions are somewhat less vigorous. However, one key difference with many other instances of convection, is that the fluid (magma) has a very high viscosity relative to the conductivity. It is usual to measure the relative importance of heat conduction and viscosity using the Prandtl number, $\sigma = \nu/\kappa$. In the Earth's mantle, $\sigma \gg 1$, and the resulting "high Prandtl number" convection can differ significantly in its form from that encountered in fluids like air and water for which σ is much smaller [3]. In particular, a common vision is that the high viscosity impedes fluid motion so much that the fluid interior contains only a relatively weak flow. Instead, because heat flows out from the underlying Earth's core and is produced internally in the lower mantle by radioactive decay, temperature gradients build up high enough at the base of the mantle to create large-scale plumes that rise up from the lower boundary and ascend through the entire fluid layer. These plumes subsequently collide with the Earth's crust, spread laterally, and then, after cooling, descend as "subducting" slabs.

What we describe next is a far cry from this image of upwelling plumes, and subducting slabs. Indeed, we retire from the high Rayleigh number problem completely, and think only of the inception of convection itself in an idealized problem (a plane fluid layer between two plates held at different temperatures). The reason is partly because we aim to be pedagogical, but there is also a deeper theme – as implied by our introductory discussion, the inception of convection is the only physical regime in which we can operate by mathematical reduction of the governing equations and truly believe the results. This is not to say we cannot proceed further, simply that greater and greater doubts enter as we try to advance beyond this point.

1.2.1 Mathematical Formulation

To deal with convection problems, we often make use of the "Boussinesq approximation." This is basically an assumption about scale separation that simplifies the equations (specifically the buoyancy force and equation of state) and is suitable for systems in which the fluid layer is relatively shallow (much less than the natural scale height). We follow such a route also in this chapter, and so we begin not from the full Navier–Stokes equations for a fluid, but from a slightly abridged version of them:

$$\frac{\partial \mathbf{u}}{\partial t} + (\mathbf{u} \cdot \nabla)\,\mathbf{u} = -\frac{1}{\rho}\nabla p + \nu \nabla^2 \mathbf{u} + \mathbf{g}\alpha T \tag{1.2}$$

and

$$\nabla \cdot \mathbf{u} = 0\,, \tag{1.3}$$

where $\mathbf{u}$ is velocity, p is pressure, T is the temperature and ρ is density. We also need a heat equation,

$$\frac{\partial T}{\partial t} + (\mathbf{u} \cdot \nabla)\,T = \kappa \nabla^2 T\,. \tag{1.4}$$

In the following, we also make the assumption of two-dimensionality. That is, we take the fluid flow to be uniform along one of the horizontal directions, which we call y. The other horizontal direction is x, while z points vertically upwards, perpendicularly to the two bounding plates located at $z = 0$ and $z = D$. The divergence-free condition in (1.3) can be automatically taken into account on defining a streamfunction: $u = -\partial\psi/\partial z$ and $w = \partial\psi/\partial x$. The pressure can then be eliminated from the momentum equation to give the vorticity equation,

$$\nabla^2\psi_t + [\psi, \nabla^2\psi] = \nu\nabla^4\psi + g\alpha T_x \ , \tag{1.5}$$

where $[f, g] \equiv f_x g_z - f_z g_x$ is the two-dimensional Jacobian operator (and subscripts indicate partial derivatives).

For boundary conditions, we follow the normal practice of adopting $\psi = \nabla^2\psi = 0$ on $z = 0$ and D, $T = T_1$ on $z = 0$ and $T = T_2$ on $z = D$. These conditions imply there is no flow through the boundary vertical plates which are stress free and held at the temperatures T_1 and T_2. In the state of no motion ($\psi = 0$), the temperature field has the conduction solution, $T(z) = T_1 + (T_2 - T_1)z/D$.

The equations derived above can be nondimensionalized to pave the way for subsequent analysis. We choose

$$(x, z) = D(\tilde{x}, \tilde{z}) \ , \qquad t = \frac{D^2}{\kappa}\tilde{t} \ , \qquad \psi = \kappa\tilde{\psi} \ , \qquad T = T_1 + (T_2 - T_1)\left(\frac{z}{D} + \theta\right) \ ,$$

and then rewrite the equations in terms of the dimensionless variables:

$$\frac{1}{\sigma}\nabla^2\psi_t + \frac{1}{\sigma}[\psi, \nabla^2\psi] = \nabla^4\psi + R\theta_x \tag{1.6}$$

and

$$\theta_t + [\psi, \theta] - \psi_x = \nabla^2\theta \ , \tag{1.7}$$

where, as defined above, σ is the Prandtl number and R is the Rayleigh number, and we have dropped the tilde decoration.

As a final preparation, we move to the large Prandtl number limit relevant to convection in magma in the Earth's interior. The terms on the left-hand side of (1.6) can then be dropped leaving the simpler system,

$$\nabla^4\psi = -R\theta_x \qquad \text{and} \qquad \theta_t + [\psi, \theta] - \psi_x = \nabla^2\theta \ , \tag{1.8}$$

that we explore next.

1.2.2 Convective Instability

To determine the linear stability of the basic, motionless state, we assume the dependence, $\psi_x, \theta \propto \sin(n\pi z)\, \mathrm{e}^{ikx+\lambda t}$, with n denoting the order of the vertical "overtone," k denoting the horizontal wavenumber, and drop all nonlinear terms. Then, (1.8) reduce to:

$$(k^2 + n^2\pi^2)^2\psi_x = k^2 R\theta \qquad \text{and} \qquad (\lambda + k^2 + n^2\pi^2)\theta = \psi_x \ , \tag{1.9}$$

which imply that

$$\lambda = \frac{k^2 R - (k^2 + n^2\pi^2)^3}{(k^2 + n^2\pi^2)^2} \ . \tag{1.10}$$

Provided $(k^2 + n^2\pi^2)^3 > k^2 R$, small-amplitude perturbations decay, reflecting how thermal conduction damps motion sufficiently to stabilize the fluid layer. But when $(k^2+n^2\pi^2)^3 < k^2 R$, buoyancy forces overcome the damping and there is an unstable linear mode. Thus, there is a special Rayleigh number,

$$R = R_0(k) = \frac{(n^2\pi^2 + k^2)^3}{k^2} , \tag{1.11}$$

beyond which convective rolls with wavenumber k and vertical order n will overturn. The critical Rayleigh number R_{crit} is the minimum value of $R_0(k)$ over all possible wavenumbers k and order n, and is the Rayleigh number at which overturning first begins. Clearly, the most important vertical overtone is that with $n = 1$. Then, if all horizontal wavenumbers are permitted (when the layer is infinite), a simple calculation shows that

$$R_{\mathrm{crit}} = \frac{27\pi^4}{4} , \tag{1.12}$$

and the marginal wavenumber (that is, the wavenumber of the first roll to overturn) is $k = k_{\mathrm{crit}} = \pi/\sqrt{2}$. The critical Rayleigh number can be different if the values of the horizontal wavenumber are restricted (when the layer has finite horizontal size). Different boundary conditions on the horizontal plates (such as no-slip conditions) lead to different values of R_{crit} and k_{crit}, but the overall ideas are the same.

1.2.3 Weakly Nonlinear Convective Rolls

Consider now a finite fluid layer in which the first convective roll to overturn has $n = 1$ and wavenumber k (so implicitly, we assume that the horizontal boundary conditions are periodic). Just beyond the threshold of instability, $R = R_0(k)$, there is a slightly unstable convective roll with wavenumber k. At these temperature differences, the convective roll develops slowly because the instability is weak. Moreover, in overturning, the roll convects heat and slightly depresses the destabilizing temperature gradient. As a result, the overturning suppresses the instability. That is, the convective instability saturates. We may mathematically formulate this saturation process in terms of weakly nonlinear theory [7].

This theory is basically an asymptotic development of the problem which is valid in a physical regime surrounding the stability threshold $R = R_0(k)$. We introduce a small parameter ϵ to measure the range of this validity, and organize the asymptotic expansion using ϵ:

$$\psi = \epsilon\psi_1 + \epsilon^2\psi_2 + \ldots \qquad \theta = \epsilon\theta_1 + \epsilon^2\theta_2 + \ldots \qquad T = \epsilon^2 t \qquad R = R_0 + \epsilon^2 R_2 . \tag{1.13}$$

The scaling of the amplitude of the convective roll ensures weakly nonlinear motions. The rescaling of time reflects that the temporal development is slow. The parameter, R_2, can be taken to have either sign so that we may explore both sides of the stability boundary (but is otherwise not necessary since ϵ estimates the proximity to the stability boundary).

We introduce the sequences into the governing equations, and gather together all terms of order ϵ, then ϵ^2 and so forth. Next we solve each set of equations separately. At leading order ($O(\epsilon)$), we find

$$\nabla^4\psi_{1x} + R_0\theta_{1xx} = 0 \qquad \text{and} \qquad \psi_{1x} + \nabla^2\theta_1 = 0\,. \tag{1.14}$$

We solve these equations by choosing

$$\begin{pmatrix}\psi_{1x}\\ \theta_1\end{pmatrix} = [A(T)\mathrm{e}^{\mathrm{i}kx} + c.c.]\begin{pmatrix}1\\ (k^2+\pi^2)^{-1}\end{pmatrix}\sin z\,, \tag{1.15}$$

where $A(T)$ is the amplitude of the neutrally stable mode which is not yet known, and *c.c.* denotes complex conjugate.

At second order ($O(\epsilon^2)$), we have

$$\nabla^4\psi_{2x} + R_0\theta_{2xx} = 0 \qquad \text{and} \qquad \psi_{2x} + \nabla^2\theta_2 = \frac{2\pi}{(k^2+\pi^2)}|A|^2\sin 2\pi z\,. \tag{1.16}$$

We solve these equations by taking

$$\psi_2 = 0 \qquad \theta_2 = -\frac{|A|^2}{2\pi(k^2+\pi^2)}\sin 2\pi z\,. \tag{1.17}$$

Physically, this solution represents the modification to the mean temperature gradient caused by the average effect of the convective rolls; this feedback is crucial to the nonlinear saturation.

Finally, at third order, we obtain

$$\nabla^4\psi_{3x} + R_0\theta_{3xx} = -R_2\theta_{1xx} \qquad \text{and} \qquad \psi_{3x} + \nabla^2\theta_3 = \theta_{1T} + \psi_{1x}\theta_{2z}\,. \tag{1.18}$$

Or,

$$\begin{aligned}\nabla^6\theta_3 - R_0\theta_{3xx} = (k^2+\pi^2)\left(A_T - \frac{k^2R_2A}{(k^2+\pi^2)^2} + \frac{1}{2}|A|^2A\right)\mathrm{e}^{\mathrm{i}kx}\sin\pi z\\ -\frac{(k^2+9\pi^2)^2}{2(k^2+\pi^2)}|A|^2A\sin 3\pi z + c.c.\end{aligned} \tag{1.19}$$

The various inhomogeneous terms in this equation indicate that there should be a number of particular solutions for θ_3. The inhomogeneous term with dependence $\mathrm{e}^{\mathrm{i}kx}\sin 3\pi z$ (or its complex conjugate) generates a particular solution with the same spatial structure and can be found straightforwardly; this term is not especially important. However, the other inhomogeneous terms with dependence $\mathrm{e}^{\pm\mathrm{i}kx}\sin\pi z$ are problematic: we cannot find particular solutions with the same spatial structure by simply substituting into the equation. In fact, standard techniques for solving equations of this form would actually produce solutions with dependences of, say, $x\mathrm{e}^{\mathrm{i}kx}\sin\pi z$ or $z\mathrm{e}^{\mathrm{i}kx}\sin\pi z$, which violate the boundary conditions. This problem arises, of course, because the difficult inhomogeneous terms have the same structure as the original linear mode – that is, there is a "resonance" because these inhomogeneous terms have the form of homogeneous solutions. To surmount the problem, our only option is to impose a condition on the amplitude, $A(T)$, that ensures that the problematic inhomogeneous terms in (1.19) all cancel. This "solvability condition" assures that we find a consistent solution at this order of the asymptotic expansion, a procedure also called "taking the Fredholm Alternative."[1] When we adopt the procedure here, we find the amplitude equation,

$$A_T = \frac{k^2 R_2}{(k^2+\pi^2)^2} A - \frac{1}{2}|A|^2 A \,. \tag{1.20}$$

Equation (1.20) is known as a Landau equation. Here the equation has real coefficients. This means that we can simplify the equation by setting $A(T) = a(T)e^{i\phi}$ where ϕ is a constant phase, and write

$$a_T = \frac{k^2 R_2}{(k^2+\pi^2)^2} a - \frac{1}{2} a^3 \,. \tag{1.21}$$

This equation has fixed points, where $a_T = 0$, given by $a = a_0$ with

$$a_0 \left[\frac{k^2 R_2}{(k^2+\pi^2)^2} - \frac{1}{2} a_0^2 \right] = 0 \,. \tag{1.22}$$

That is, $a_0 = 0$, or

$$a_0 = \pm \frac{\sqrt{2k^2 R_2}}{(k^2+\pi^2)} \,. \tag{1.23}$$

The first solution, $a_0 = 0$, represents the state of no motion. The other two fixed points only exist provided $R_2 > 0$; that is, if the system is linearly unstable. In that circumstance, the solutions (1.23) represent states of finite-amplitude convection. It is not difficult to solve (1.21) explicitly and show that the system, if unstable, always converges to one of these finite-amplitude states (that is, they are the two attractors of the system). The bifurcation that occurs as R_2 passes through zero is a "supercritical pitchfork" bifurcation, in which two new stable equilibrium states appear by emerging from the original one, that itself becomes unstable for $R_2 > 0$. The two different finite-amplitude states correspond to rolls with the two possible senses of rotation (clockwise and anti-clockwise).

At this point, a short detour into bifurcation theory is appropriate. A bifurcation is in general found at a critical value of one of the control parameters of the system (in the above example, the Rayleigh number). At this critical parameter value, new things happen and the system changes its qualitative behavior. For example, a new equilibrium appears, or disappears, or an equilibrium point changes its stability. The bifurcation relevant to our convection problem is illustrated in Fig. 1.4.

[1] In general, to take the Fredholm alternative, we take an inner product of the equation with the adjoint of the resonant homogeneous solution, which is not necessarily the same as forcing the problematic inhomogeneous terms to cancel exactly. This signifies that the general solvability condition contains a bunch of integrals involving the various homogeneous solutions (the eigenmodes) and their adjoints.

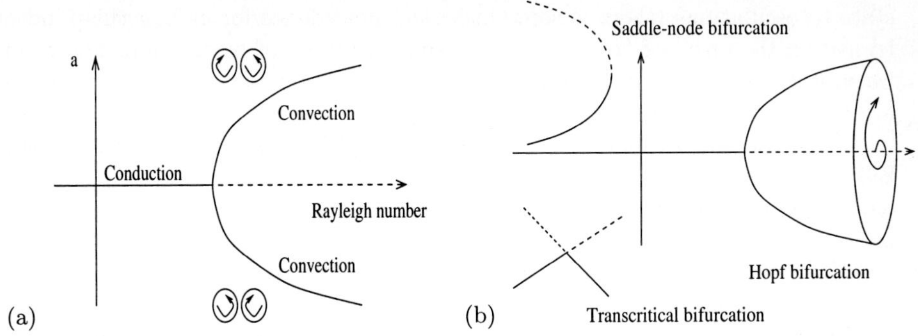

Fig. 1.4. (**a**) Bifurcation diagram for the onset of convection. At the critical Rayleigh number, there is a pitchfork bifurcation in which two new solutions appear; these are convective rolls with opposite senses of rotation. Panel (**b**) shows a sketch of the saddle-node, transcritical and Hopf bifurcations – some of the other commonly encountered bifurcations. Solid lines show stable solutions; unstable equilibria are indicated by dashed lines

The pitchfork is one of the more common bifurcations we encounter, and it is called supercritical when the two new states are stable. The other possibility – that the new equilibria are unstable – is called a subcritical pitchfork, and arises when the nonlinear term in the Landau equation has the opposite sign. The subcritical case harbours many problems for weakly nonlinear theory because the equation in that case has no bounded solutions – the amplitude of an unstable mode simply grows without saturation. Worse still, if the mode is stable (so we are below the threshold of instability), a big enough kick sends the system again onto a diverging solution. In other words, the stable system is unstable to finite-amplitude perturbations even when the equilibrium point is linearly stable. Of course, this is a feature of the asymptotic amplitude equation, and not necessarily the behaviour of the original equations. In fact, one would hope fervently that it is not, and solutions to the governing equations remain bounded and physically plausible. The lesson is that in the subcritical case, there is some physics missing from the weakly nonlinear theory, namely a saturation mechanism that limits the mode amplitude at a level beyond the scale accounted for in the asymptotic scheme. In other words, the transition in system behaviour is harder than expected. Fortunately, for the convection problem considered here, the bifurcation is supercritical and all is well.

Other standard bifurcations are the saddle-node, transcritical and Hopf bifurcations (Fig. 1.4). In the first, a stable equilbrium collides with an unstable equilibrium, and both disappear; what happens beyond is determined by what other attractors exist in the system. For the transcritical bifurcation, two equilibria cross and exchange stability. In the Hopf bifurcation, a stable equilibrium becomes unstable because an oscillating solution – a limit cycle – is born. In this type of bifurcation, the system changes its longterm behavior from a stationary state to oscillations (there is some historical precedence for calling this kind of transition the onset of "overstability," see e.g. [1]). Hopf bifurcations can again be supercritical or subcritical (leading to saturated, low-amplitude oscillations, or to an oscillatory growth out of the asymptotic regime).

There are many other types of bifurcations. Here we just note that all this talk about bifurcations and criticality comes from dynamical systems theory, which one could regard as the theory creating the necessary tools to uncover and categorize the underlying mathematical description of emerging patterns. The Landau equation is by far the most commonly encountered descriptor of this kind, and has come up in many different contexts such as for bar formation in rivers – see [8]. This reflects the universality of this equation and the particular type of transition to instability associated with it, namely a supercritical pitchfork bifurcation. Thus, one should not be surprised by the appearance of the Landau equation in the study of many types of weakly nonlinear patterns.

1.2.4 Extended Systems

To return to convection, the theory we have just described applies only to simple, periodic roll solutions, and so we are unable to describe any richer dynamics, such as the competing patterns shown in Figs. 1.1–1.3. These require the third spatial dimension, which allows more unstable linear modes. When we add these extra modes to the weakly nonlinear analysis, we get more amplitude equations and the idea (hope) is that there are various different kinds of fixed points that correspond to rolls, hexagons or squares, and the evolution in phase space of the amplitude equations allows us to decide which are preferred. Much work has been done in this direction for Rayleigh–Bénard convection, and many of the details and dynamics of convective patterns have been understood in this way.

Another limitation of the theory presented above is that we assumed the fluid layer to be periodic. What happens when all wavenumbers are allowed? In this case, there can be effects of spatial propagation, which allows travelling, nonlinear structures to form (as in the invading rolls shown in Fig. 1.1 – more generally, there can be solitary waves, shocks, fronts, pulses and so on). These objects are one of the most interesting and commonly encountered entities in extended nonlinear systems, and so we spend some time exploring how one fits these into a weakly nonlinear description.

To allow the solution to develop horizontally we must look for an amplitude equation that describes both a temporal growth and a spatial variation. But we cannot allow the original kind of spatial variations in x, since then we would be unable to simplify the governing equations. The way forward is to recognise that, near the onset of convection, rolls develop both on long timescales and also on large lengthscales. Consider again the growth rate λ for an infinite layer:

$$\lambda = \frac{k^2[R - R_0(k)]}{(k^2+\pi^2)^2} \to \frac{2}{9\pi^2}\left[R - \frac{27\pi^4}{4} - 36\pi^2\left(k - \frac{\pi}{\sqrt{2}}\right)^2\right] , \tag{1.24}$$

near onset and the critical wavenumber (onset occurs at $R = 27\pi^4/4$ and $k = \pi/\sqrt{2}$ with $n = 1$). The origin of the scaling of the long timescale T is evident from this relation: we require $\lambda \propto (R - R_0) = \epsilon^2 R_2$, and so $\lambda t \propto \epsilon^2 t = T$. On allowing for spatial variations, we see there is also a wavenumber dependence near criticality. Evidently, to preserve the asymptotic scalings, we should set $(k - \pi/\sqrt{2}) = \epsilon K$, a small wavenumber corresponding to a long lengthscale, $X = \epsilon x$. We modify the asymptotic scheme accordingly:

$$\psi = \epsilon\psi_1 + \ldots \quad \theta = \epsilon\theta_1 + \ldots \quad R = R_0 + \epsilon^2 R_2 \quad \partial_t \to \epsilon^2 \partial_T \quad \partial_x \to \partial_x + \epsilon\partial_X \; . \tag{1.25}$$

Note that we must still allow for the order unity spatial dependence (the original x) because we still need to capture the shape of the rolls. This is now the scheme of Newell and Whitehead [9] and Segel [10] for weakly nonlinear, finite bandwidth convection.

We now proceed as before; we omit the details except for the important differences. At leading order we find the linear mode solution, with unknown amplitude A. This time we take this amplitude to have both slow time and long space dependence: $A = A(X,T)$. No changes arise at second order, but there are some new inhomogeneous terms in the equations at order ϵ^3. To eliminate the problematic terms (take the Fredholm alternative; enforce a solvability condition), we set

$$A_T = \frac{2R_2}{9\pi^2} A - \frac{1}{2}|A|^2 A + 8A_{XX} \; . \tag{1.26}$$

This amplitude equation is commonly called the the Ginzburg–Landau equation. In the problem at hand, the coefficients of the equation are all real (giving the "real GL" equation – an equation that has been extensively studied in problems of phase separation in condensed matter physics). For other systems (such as thermohaline convection [11], or bar instability in river flow [12]) the coefficients can be complex, in which case the system is refered to as the "complex GL" equation.

For the real GL equation, only the amplitude of A matters, and after suitable rescalings ($|A| = a(2/3\pi)\sqrt{R_2}$, $X \to 6\pi X/\sqrt{R_2}$ and $T \to (9\pi^2/2R_2)T$, assuming R_2 is positive) we then have

$$a_T = a - a^3 + a_{XX} \; . \tag{1.27}$$

The equation has the spatially homogeneous solutions, $a = 0$ and $a = \pm 1$. The equilibrium $a = 0$ is unstable, but the finite amplitude states are stable. One of the most distinctive features of the real Ginzburg–Landau equation is how its solutions develop from low-amplitude initial conditions near the unstable homogeneous phase. One such example is shown in Fig. 1.5.

Initially, the system diverges exponentially quickly from the low-amplitude initial condition because of the linear instability. In other words, the system tries to form rolls everywhere. However, a problem soon appears in this headlong rush to the steady state: depending on the precise details of the initial condition, the system attempts to form rolls that rotate one way in some places, and rolls with the opposite sense of rotation elsewhere. The regions of different rolls then jam up against one another and sharp gradients develop in the amplitude $a(X,T)$ in

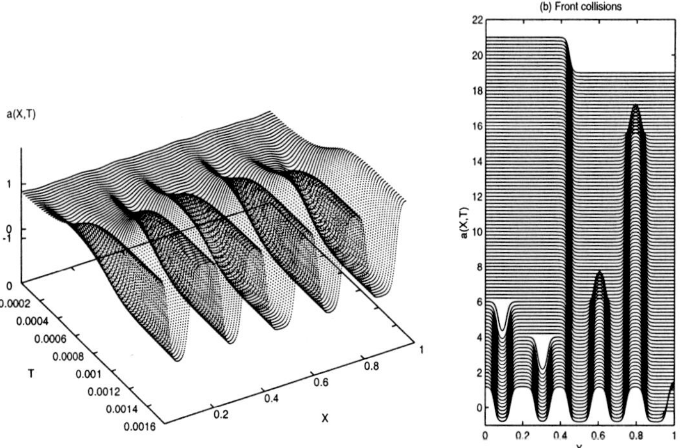

Fig. 1.5. Evolution in the real Ginzburg–Landau equation, solved on a finite domain with no-flux boundary conditions, $a_X = 0$. First panel shows the initial evolution; the second pair show longer timescale evolution

the interfaces separating regions with $a > 0$ and $a < 0$. This produces a state in which there are regions with $a \sim -1$ and other regions with $a \sim 1$, separated by steep steps, or *fronts*. The final solution in the first panel of Fig. 1.5 is of this form. It is not, however, stationary: The fronts drift on a much longer timescale. The more closely spaced pairs approach one another and collide. The collisions gradually annihilate the fronts until only a single front remains. In other words, the dynamics is one of a rundown process (indeed, (1.27) has a Lyapunov functional that explicitly requires such a rundown).

Although the steep steps connecting the two homogeneous phases drift over a longer timescale, they are almost steady. Approximately, the fronts are given by

$$a_{XX} + a - a^3 \approx 0 \, . \tag{1.28}$$

This is a nonlinear oscillator equation in which our spatial coordinate plays the role of time. The fixed points are just the spatially homogeneous equilibrium solutions. Moreover, there is an orbit with $a \to -1$ to the left and $a \to 1$ to the right, given by

$$a(X) = F(X) = \tanh(X/\sqrt{2}) \, , \tag{1.29}$$

which describes an upward step. Similarly, there is a downward step with $a(X) = -F(X)$. In dynamical systems language, these particular solutions are the separatrices or "heteroclinic orbits" of the phase portrait of the oscillator – see Fig. 1.6.

The fronts are examples of *spatially localized solutions*. In order to construct such localized solutions, their tails must become flat. That is, $a \to$ constant, as $X \to \pm\infty$, which means that the tails must correspond to the fixed points of

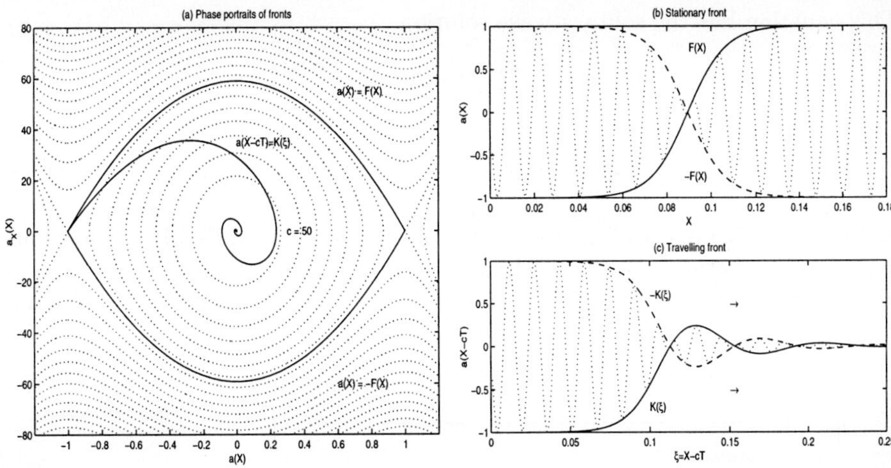

Fig. 1.6. Heteroclinic orbits of the oscillator equation (1.27). Panel (**a**) shows the orbits as phase portaits on the (a, a_X)–plane. Panel (**b**) shows the structure of the stationary front ($c = 0$), and a travelling front ($c = 50$). In both cases, the dotted lines show the actual structure of the reconstructed temperature field

the oscillator in (1.28), which are also the homogeneous states. This is why the front solutions correspond to separatrices or heteroclinic orbits.

The fronts of (1.29) are not the only types of localized solutions to the real Ginzburg–Landau equation. In fact, there is an entire family of *moving* fronts for which $a = a(X - cT)$, where c is a travelling wave speed. These fronts are solutions to the damped oscillator equation,

$$a_{\xi\xi} + ca_\xi + a(1 - a^2) = 0 \; , \tag{1.30}$$

where $\xi = X - cT$ is a *travelling-wave coordinate.* The fronts connect the stable homogeneous phases, $a = \pm 1$ to the unstable phase $a = 0$, as shown in Fig. 1.6, and describe the rapid evaporation of the unstable phase through the propagation of fronts into it. In the convective problem, these fronts describe the invasion of convection cells into motionless conducting regions.

For arbitrary complex coefficients, the Ginzburg–Landau system can generate chaotic solutions, which has attracted many researchers to explore the equation as an analogue model for turbulence. However, the equation is one-dimensional and not quantitatively similar to real turbulence, so it is largely a metaphor. Actually, even this metaphor is pretty complicated and difficult to understand, which epitomizes how we fall severely short of developing a true analogue model for turbulent systems.

Another important limit of the complex Ginzburg–Landau equation is when the coefficients are purely imaginary. In this case, the equation reduces to the cubic Schrödinger equation:

$$\mathrm{i}A_T = A_{XX} + 2|A|^2 A \; . \tag{1.31}$$

This equation has the solitary-wave solution,

$$A = \mathrm{i}k\mathrm{e}^{-\mathrm{i}(\Phi-\Phi_0)}\mathrm{sech}k(X - VT) , \tag{1.32}$$

where

$$\Phi = \frac{1}{2}VX - \left(\frac{1}{4}V^2 - k^2\right)T , \tag{1.33}$$

and k, V and Φ_0 are constants. One such solution is shown in Fig. 1.7; it describes a localized packet or pulse of travelling waves. As a phase portrait on the (A, A_X) plane, this is a "homoclinic" orbit connecting the state $A = 0$ to itself (see panel (b)).

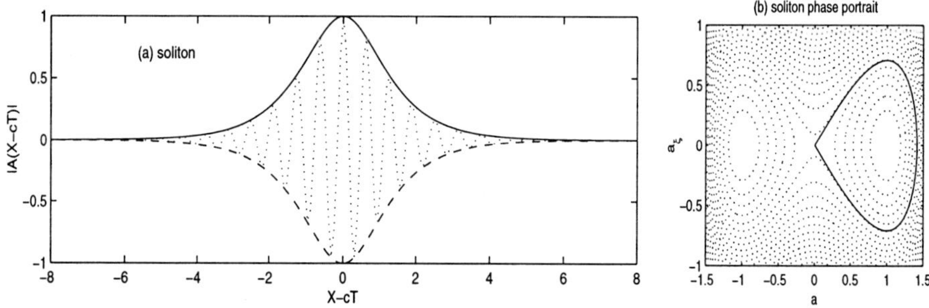

Fig. 1.7. Soliton structure in the cubic Shrödinger equation. Panel (**a**) shows the soliton shape; the dotted lines show the wave packet of travelling convective rolls that the soliton approximates. In panel (**b**) we use the transformation $A = ka(\xi)\exp(-\mathrm{i}\Phi)$, with $\xi = X - VT$, to write an oscillator equation for a, $a_{\xi\xi} = k^2 a(1 - a^2)$, and then draw the soliton solution on the phase plane (the homoclinic orbit connecting $a = 0$ to itself)

The cubic Schrödinger equation is an example of an integrable system and its solutions can be studied using inverse scattering techniques [13]. In fact, the solution in (1.32) is a soliton. The Inverse Scattering Transform is unusually powerful in that it allows us to generate multiple solitary-wave equilibria and consider soliton dynamics within the framework of an exact theory. In most situations we are not this fortunate, but these integrable systems offer us a glimpse of some of the properties of nonlinear systems.

1.3 Asymptotics, Galerkin Approximation and Conceptual Models

The weakly nonlinear amplitude expansion derives an equation for the behaviour of the fluid just beyond the bifurcation to instability. The applicability of the theory is limited to the narrow region surrounding the stability boundary, which also means that the dynamics captured by the amplitude equation is very restricted. Importantly, the Landau and Ginzburg–Landau equations describe the onset of steady convective rolls, and nothing more. To try to capture the richer dynamics of the fluid we must go beyond the weakly nonlinear description described above, but how may we do this? One option is the Galerkin procedure, which we now sketch out. The procedure is far less mathematically rigorous than the weakly nonlinear theory, so we will be correspondingly more woolly.

The key effect uncovered by the weakly nonlinear theory is the suppression of the background temperature gradient created by the convective heat flux associated with an overturning roll. We can build a theory including both the linear mode and the suppression of the mean temperature gradient by assuming that

$$\psi = \frac{2Rk^2 a(t)}{(k^2+\pi^2)^3} \sin \pi z \cos kx \tag{1.34}$$

and

$$\theta = \frac{2a(t)}{k^2+\pi^2} \sin \pi z \cos kx + \frac{b(t)}{2\pi(k^2+\pi^2)} \sin 2\pi z\ , \tag{1.35}$$

where $a(t)$ and $b(t)$ are unknown (real) amplitudes to be determined. The precise form of this *ansatz* is guided by the theory above. The *ansatz* is, of course, not a solution to the equations. For example, it is easy to see that the nonlinear terms in the equations, if the solution is given by this form, create terms with dependences of $\cos kx \sin 3\pi z$, which will never cancel exactly to zero. Instead, we must assume that these terms are always small and can be neglected (one can mathematically formulate these statement a little better, and this is what "Galerkin projection" is all about). Proceeding in this way – by introducing the *ansatz* (1.35) into the governing equations and dropping "unnecessary" terms – we derive the system:

$$\dot{a} = \lambda a + \frac{Rk^2}{2(k^2+\pi^2)^3} ab \qquad \text{and} \qquad \dot{b} = -4\pi^2 b - \frac{2Rk^2}{(k^2+\pi^2)^3} a^2\ , \tag{1.36}$$

where λ is given by (1.10).

This is a truncated model system that captures the physics we put into it via the *ansatz*. It is truncated because we dropped lots of nonlinear terms without any real justification. However, there are some pleasing virtues in the reduced model. First, if we drop the nonlinear terms, we find $\dot{a} = \lambda a$ and $\dot{b} = -4\pi^2 b$. The first of these is our old friend, the linear mode. The other is another linear mode, this time a decaying thermal diffusion mode (it is a second vertical overtone with $n = 2$ and zero horizontal wavenumber). In other words, we recover aspects of the linear problem.

Second, consider the case relevant to the weakly nonlinear problem: $Rk^2 \approx (k^2+\pi^2)^3$ or λ small. In this instance, we expect the system to evolve slowly. However, the diffusion mode is strongly damped (the damping rate, $4\pi^2$, is order one and therefore relatively large). Thus we anticipate that this mode rushes to a local equilibrium in which $\dot{b} \approx 0$, or

$$4\pi^2 b \approx \frac{2Rk^2}{(k^2+\pi^2)^3} a^2 . \tag{1.37}$$

If we substitute this approximation into the equation for a, and approximate R, we recover the weakly nonlinear theory.

We illustrate the evolution of the truncated system in Fig. 1.8. As one might have expected, when unstable, the system always evolves to a state of steady convection from an arbitrary initial condition. Moreover, when λ is small, the evolution proceeds through two distinct steps. First there is a relaxation of the thermal mode onto the rough equilibrium (1.37), and then a more gradual growth and saturation of the unstable mode. The curve denoting the rough equilibrium is a "slow manifold" for the reduced system, since when it gets there, evolution proceeds relatively slowly. The dynamics that is captured by the model is therefore little different from the Landau description obtained earlier. However, we can add more variables and structure into the Galerkin ansatz to enrich the dynamics, as is often done to arrive at the celebrated Lorenz equations with chaotic solutions.

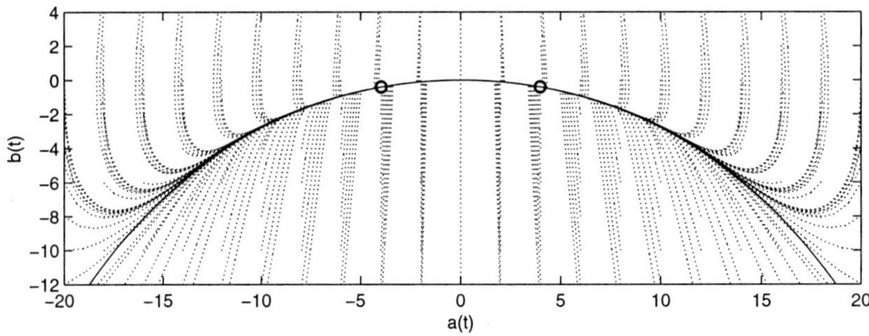

Fig. 1.8. Sample phase portrait of the Galerkin truncation. The dotted lines show a set of trajectories of the system; the solid curve shows the slow manifold along which b is slaved to a, so $b = b(a)$. The circles show the stable equilibria

The relaxation to the local equilibrium in Fig. 1.8 illustrates nicely why weakly nonlinear theory works. This theory filters out the fast relaxation and leaves only the more gradual modal growth. In fact, it filters *all* the other stable linear modes, not just our thermal mode $b(t)$. The resulting amplitude equation is essentially a projection onto the slower manifold. In the language of dynamical systems theory, this geometrical structure is the "centre manifold," and the procedure of weakly nonlinear theory is called "centre-manifold reduction," when the mathematics is made formal and rigorous.

The more adaptable Galerkin procedure has no rigorous foundation in dynamical systems theory. It is based more on physical arguments of plausibility.

One must *assume* that the solution takes a certain form. The power in the procedure is in the simpler reduced model that it furnishes, but one should always be cautious about its accuracy since the approximations involved in the Galerkin projection are uncontrolled. Although these statements seem rather out of place in this book on geomorpholgical fluid mechanics, we make them because they are actually very relevant. The reason is that, as we shall see in many of the upcoming chapters, when we seek to model mathematically the phenomena of interest, approximations are necessary because the governing equations are usually too hard to solve. Sometimes, as with the lubrication theory used for slowly flowing lava and glaciers, the approximation amounts to an asymptotic expansion, in which case, the errors are known. On other occasions, as in the depth-averaged models used for relatively fast debris and mud flows, one assumes a certain structure for the flow, and this is essentially a Galerkin projection that attempts to extend the lubrication analysis (as we used it here to extend the weakly nonlinear theory).

Unfortunately, on occasions, even Galerkin is not enough to allow us access to the physical regime that might be of interest. That is, there may be no plausible approximations evident, or maybe the number of approximations needed gets far out of hand. Instead, another kind of modelling is needed. Perhaps this is full-scale numerical simulation. But this approach is always restricted by the power of computations. Indeed, current technology precludes us from taking this approach too far without entering into situations in which we do not fully resolve the solutions and cannot make parameterical studies to gauge their robustness. Nevertheless, this approach is popular, and has been very successful in furthering our understanding of turbulent and geophysical fluids.

There is, however, another direction that we can also exploit. This direction avoids the governing equations and big computers altogether. The idea is that there is a process or phenomenon that is of interest, and we need to determine whether a particular physical effect is capable of explaining it. Then, to test out the hypotheses, we write down a "conceptual" or "phenomenological" model. In other words, we throw away asymptotic arguments and attempt to write a mathematical metaphor of the physical process. The idea is somewhat similar to Galerkin projection, although Galerkin usually proceeds from the governing equations by making uncontrolled approximations. The reduced phenomenological or conceptual models can have different degrees of sophistication and contain as much physics as one tries to incorporate. Moreover, the models themselves need not even be theoretical – analogue experimental models can serve equally well if not better. In future chapters, the reader should recognize many of these conceptual models. Below we give examples relevant to convection in the earth's mantle and to aeolian sand ripples.

1.4 Solitary Waves in Conduits

In this section, we show an example of how solitary wave solutions can naturally emerge in the dynamics of geophysical systems. We mentioned earlier how, in a popular image, convection in the Earth's mantle produces rising plumes. The plumes rise upwards and drain fluid from below. They leave behind them a trailing conduit which, under favourable circumstances, can remain open and continue to drain fluid. This conduit is roughly a uniform cylinder. If the source of the upcoming fluid is not steady, a disturbance in the source flux travels up the conduit as a nonlinear travelling structure or structures. In particular, a source with sporadic fluctuations can produce solitary waves [14,15].

Given that the conduits themselves are supposed to form from a vigorous convective instability, and pierce a complicated background flow, it is impossible to proceed from the governing equations in the reductive fashion outlined above in order to construct a model for these objects. How can we proceed in a situation like this? We outline two complementary approaches, which illustrate this modelling style.

First, it is not necessary to proceed purely theoretically – experiments can also provide analogue models of the phenomenon of interest. In fact, using soluble fluids (to eliminate surface tension) with different properties one can quite easily create distinct conduits of up-going fluid. One way to do this is to use fluids with different density, in which case if we confine the light fluid beneath the heavier one, the lower fluid will rise upward, buoyantly displacing the heavier fluid (see, for example, the chapters on mantle convection). With a suitable experimental set up, this Rayleigh–Taylor instability can be engineered to create a single upward plume with a trailing conduit. Moreover, by injecting fresh light fluid at the bottom we can vary the source flux in a controlled way. Fig. 1.9 shows photographs of isolated disturbances propagating up a conduit produced in this fashion using a mixture of syrup and water as a light fluid, and pure syrup as the heavy fluid.

The disturbances generated in this way appear to have almost permanent form, meaning they preserve their identity as they traverse the conduit. In other words, they are close approximations of solitary waves. Even more surprising is that multiple solitary waves can interact in a fashion very reminiscent of the dynamics of solitons [16,17]. This is what is illustrated in the right-hand panels of Fig. 1.9 – two structures approach one another, collide, and then separate with only a slight change in form (injected die serves to highlight the initially lower solitary wave, and then to reveal the mass exchange between the objects).

The second approach is the theoretical one. Here, however, instead of attempting to derive a reduced model from the governing equations, we opt for writing the model down from phenomenological or plausibility arguments alone. For the conduit problem, we aim to describe propagating structures, so we need a fair degree of sophistication (however, we also throw far more away). Let z denote a height variable in what we pretend is a conduit of slowly moving, immiscible fluid; we need a partial differential equation in z and time, t. The main physical variables will be the cross-sectional area of the conduit, $A(z,t)$, and the local mass flux, $Q(z,t)$.

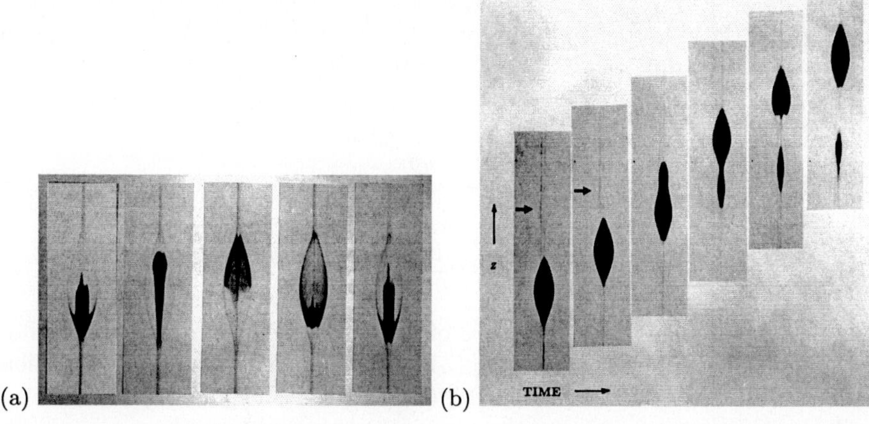

Fig. 1.9. (**a**) Sequential photographs (with equal time intervals) of a solitary wave with a small amount of dye. The circulation of dyed fluid in closed streamlines is illustrated. The conduit and most of the wave contains clear fluid which is scarcely distinguishable. The camera was traveling upward with the waves to take these photographs. (**b**) A solitary wave containing dyed fluid overtakes a small wave without fluid that is initially invisible. Dyed material is injected into the leading wave which then becomes larger and leaves the depleted wave behind

Now, for an incompressible immiscible fluid in the conduit, mass must conserved, and so

$$\frac{\partial A}{\partial t} + \frac{\partial Q}{\partial z} = 0 \,. \tag{1.38}$$

If we assume that the mass flux changes due to pressure variations, $P(z,t)$, which are compensated by viscous forces within the conduit, then we can write an equation of the form,

$$Q = -\frac{A^2}{8\pi\mu_i}\frac{\partial P}{\partial z} \,, \tag{1.39}$$

where μ_i is the dynamic viscosity of the conduit fluid. This form assumes that the fluid moves up the conduit with a vertical velocity profile that is parabolic in the radial coordinate. Finally, the pressure variation is assumed to be caused by buoyancy and temporal changes in cross-sectional area, which induce flow and therefore pressure variations outside the conduit:

$$P = -(\rho_e - \rho_i)gz + \frac{\mu_e}{A}\frac{\partial A}{\partial t} \,, \tag{1.40}$$

where μ_e is the dynamic viscosity of the external fluid.

Now we nondimensionalize:

$$\tilde{A} = A/A_o \qquad \tilde{z} = z/L \qquad \tilde{Q} = Q/Q_o \qquad \tilde{t} = t/T \,, \tag{1.41}$$

where

$$L = \left(\frac{\mu_e A_o}{8\pi\mu_i}\right)^{1/2} \quad \text{and} \quad T = \frac{1}{g(\rho_e - \rho_i)} \left(\frac{8\pi\mu_i\mu_e}{A_o}\right)^{1/2} . \tag{1.42}$$

The dimensionless equations are then,

$$\frac{\partial A}{\partial t} + \frac{\partial Q}{\partial z} = 0 \tag{1.43}$$

$$Q = A^2 \left[1 + \frac{\partial}{\partial z}\left(\frac{1}{A}\frac{\partial Q}{\partial z}\right)\right] , \tag{1.44}$$

where, in our by-now standard abuse of notations, we have dropped the tilde. In the limit of small-wave amplitude ($|A| \ll 1$), one can reduce these equations to the Korteweg–de-Vries equation [18] (another soliton equation).

Travelling-wave solutions to the above equations are given by

$$\frac{\mathrm{d}A}{\mathrm{d}\xi} = \pm \frac{A}{c^{1/2}}[1 + c - 2cA^{-1} - 2\ln A - (1-c)A^{-2}]^{1/2} , \tag{1.45}$$

where $A = A(\xi)$ and $\xi = z - ct$, which can be reduced to quadrature. The equation has a fixed point $A = 1$, corresponding to the undisturbed conduit, and so there can be families of solitary wave solutions with different speeds, c, that approach $A = 1$ as $\xi \to \pm\infty$. There is a second fixed point at $A = A_m$ and $\mathrm{d}A/\mathrm{d}\xi = 0$, which corresponds to the peak of the solitary wave, and provides a relation between wave speed and maximal conduit area:

$$c = (2A_m^2 \ln A_m - A_m^2 + 1)/(A_m - 1)^2 . \tag{1.46}$$

As shown in Fig. 1.10, the two branches of (1.45) define two curves on the $(A, \mathrm{d}A/\mathrm{d}\xi)$ plane; provided $c > 2$, we can construct the solitary wave by joining together the segments of these two curves that connect the two fixed points at $A = 1$ and $A = A_m$. For low amplitude waves, $A_m \to 1$, $c \to 2$, which is the phase speed of long linear waves. In the other, large-amplitude limit, $A_m \gg 1$, and it can be shown that the solitary wave shape is the Gaussian, $A \approx \exp(-\xi^2/2c)$ [17].

Given our initial assumptions, we can write the dimensionless velocity profile in the conduit in the form,

$$u = 2A\left(1 - \frac{r^2}{A}\right)\left[1 + \frac{\partial}{\partial z}\left(\frac{1}{A}\frac{\partial Q}{\partial z}\right)\right] \tag{1.47}$$

for $0 \le r \le A^{1/2}$, where r is the radial coordinate. This is the parabolic vertical velocity profile, and simplifes further,

$$u = \frac{2}{A}\left(1 - \frac{r^2}{A}\right)(cA + 1 - c) , \tag{1.48}$$

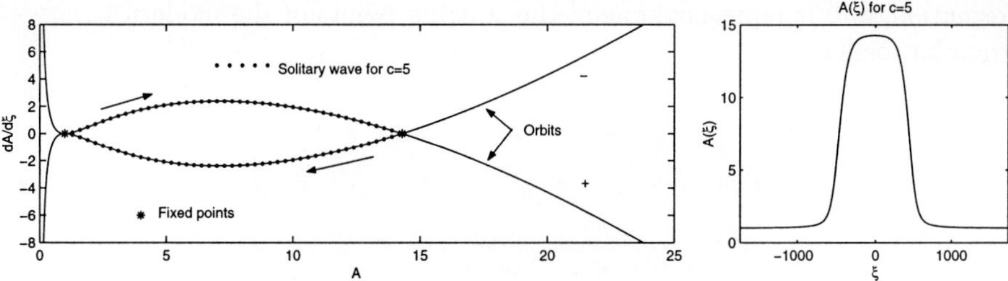

Fig. 1.10. The solitary wave on the $(A, \mathrm{d}A/\mathrm{d}\xi)$ plane and plotted against $\xi = x - ct$ for $c = 5$

giving the maximal speed,

$$u_m = 2A_m^{-1}(cA_m + 1 - c)\ . \tag{1.49}$$

It is readily seen that $u_m/c > 1$ for all $A_m > 1$, i.e. fluid in the wave moves faster than the wave itself. In contrast, ahead and behind the wave, $A = 1$ and $u = 2$, which is less than u_m. Thus the solitary waves must contain trapped, recirculating fluid. This is precisely what the injected die reveals in Fig. 1.9. Further comparisons between the model and experiments can be found in [16,17].

1.5 Blown by Wind

Convection is a splendid example both for the patterns that it creates, and the ease by which we can go through the linear stability theory and weakly nonlinear development of the governing equations (the Navier–Stokes equations in Boussinesq approximation). Few other systems are this submissive. In fact, the governing equations for many fluids relevant to geomorphology and geology are either intractable or unknown, as we shall see in chapters to follow. Most notable for their difficulty are the equations for granular media. Thus, now that we finally arrive at the problem of sand ripples, we do not even have a solid foundation on which to build our study. Instead, we must base the whole exploration on a phenomenological model. Whilst this is less appealing than beginning from accepted governing equations, provided the model is well-posed, we can continue as before and use the mathematical tools of stability and weakly nonlinear theory we have already developed to explore the problem. An illustration of this philosophy is one of our goals in this section.

Aeolian ripples are commonly found in sand deserts, often atop dunes, and on sandy beaches. An example of the patterns formed by these structures is shown in Fig. 1.11. The physical mechanism responsible for the formation of ripples is thought to be the action of the wind on loose sand. A classic reference to ripples and dunes is the book "The Physics of Blown Sand and Desert Dunes" by R.A. Bagnold [19]. This book describes in detail the process of sand saltation (the hopping of grains above the sand surface) that is responsible for ripple formation, and it represents one of the starting points of the modern theory of granular media.

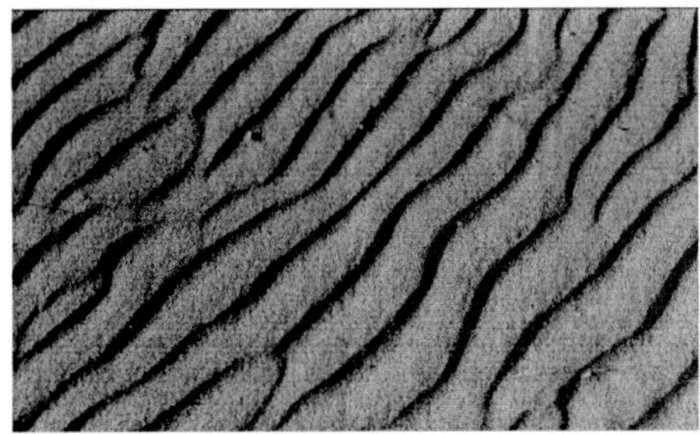

Fig. 1.11. A photograph of desert sand ripples

1.5.1 Saltation and Reptation

The essential idea is that when the wind strength is above some threshold, grains are displaced by the direct action of the wind, and are lifted into the air (in a process analogous to the lifting of sediment grains under water, that is described in some detail in the fourth part of this volume). Even for strong winds, however, sand grains are too heavy to stay suspended and return to the ground (unlike what happens for sand in water). During their flight, the grains reach a velocity that is approximately that of the wind, and upon their impact with the surface, they impart their energy and momentum to the sand, and eject other grains. For sufficiently large wind velocities, the bombardment by sand grains accelerated by the wind generates a cascade process, and an entire population of saltating grains hopping on the sand surface emerges. During strong winds, this layer of saltating grains can reach a thickness of more than a meter.

Experimental results [20–22] indicate that the cascading bombardment process generates a bimodal population of moving grains: One part of the population is formed by grains that are ejected with large energy; these reach higher elevations and are directly accelerated by the wind. These high-energy grains bombard the surface elsewhere and maintain the cascade process. The second population consists of grains that are ejected with low energy, and stay close to the sand surface. These "crawling" grains compose what is called the "reptating" population.

Following Anderson [23], we build a heuristic model of the sand transport process based on the conservation law,

$$(1-\lambda_p)\rho_p \frac{\partial \zeta}{\partial t} = -\frac{\partial Q}{\partial x} \tag{1.50}$$

where $\zeta(x,t)$ is the height of the sand surface, ρ_p is the density of a sand grain, λ_p is the porosity of the bed (typically, $\lambda_p \approx 0.35$), and $Q(x,t)$ is the sand flux.

Equation (1.50) is known, in fluvial geomorphology, as the Exner equation, and it is discussed further in the fourth part of this volume.

The flux, Q, is the sum of the flux of saltating grains, Q_s, and the flux of reptating grains, Q_r. Saltating grains are accelerated by the wind to speeds close to the windspeed, and follow a ballistic path to their next impact with the ground. If their arced trajectory is a long one, it seems plausible that the angle at which the grains descend back to the bed is dictated largely by the wind speed, and that the flux of saltating grains is fairly uniform. Hence we assume that Q_s is approximately independent of x, and we do not consider it further in the Exner equation (though we should remember that Q_s is the driving force of the reptation flux). If reptating grains undergo a small jump of length $\bar{a}$, the flux $Q_r(x,t)$ at point x is proportional to the number of reptating grains that have been ejected between the point $x-\bar{a}$ and the point x:

$$Q_r(x,t) = m_p \int_{x-\bar{a}}^{x} N_{ej}(x',t)\mathrm{d}x' \qquad (1.51)$$

where $N_{ej}(x,t)$ is the number density of reptating grains ejected per unit area of bed and unit time, and m_p is the mass of each grain.

Now, the number density N_{ej} of ejected reptating grains must be given by the number density of impacting saltating grains, N_{im}. As a simple model, we assume that $N_{ej} = n_r N_{im}$ where n_r is the average number of reptating particles ejected by a single impact of a saltating grain. Finally, because the saltating flux is constant, and characterized by the fixed angle ϕ at which the grains descend back to the ground, the number density of impacting grains changes only because of variations in the slope of the bed. Based on geometrical considerations, we obtain [19,23]:

$$N_{im} = N_{im}^0 \left[1 + \frac{\tan\theta}{\tan\phi}\right]\cos\theta = N_{im}^0 \frac{1+\zeta_x \cot\phi}{\sqrt{1+\zeta_x^2}} , \qquad (1.52)$$

where N_{im}^0 is the number density of impacting grains on an horizontal surface and θ is the inclination of the bed ($\theta > 0$ when the bed dips upwind). Provided this slope is small, we can expand the above expression and obtain

$$N_{im}(x) \approx N_{im}^0 \left[1 + \zeta_x \cot\phi\right] . \qquad (1.53)$$

By inserting this expression in (1.50), we obtain

$$\frac{\partial\zeta}{\partial t} = -\beta \frac{\partial}{\partial x}\left[\zeta(x) - \zeta(x-\bar{a})\right] , \qquad (1.54)$$

where $\beta = m_p n_r N_{im}^0 \cot\phi / [\rho_p(1-\lambda_p)]$.

Equation (1.54) is our phenomenological equation that we hope governs ripple formation. Indeed, there is a linear instability in the model that we uncover by the usual procedures: Let $\zeta \propto \exp[ik(x-ct)]$. By inserting this solution into (1.54), we obtain the dispersion relation,

$$c = \beta\left[1 - \exp(-ik\bar{a})\right] . \qquad (1.55)$$

Thus there is an instability with growth rate, $\sigma = k\text{Im}(c) = \beta k \sin(k\bar{a})$. The ripples move downwind with the phase speed $\text{Re}(c) = \beta[1 - \cos(k\bar{a})]$. The first peak in the ripple's growth rate, at $k\bar{a} = \pi/2$, corresponds to ripples with wavelength $\lambda = 4\bar{a}$, which coincides with that expected from qualitative arguments [23]. Note that the origin of this instability resides in the non-locality of the flux of transported grains (i.e. the presence of a space integral in the expression for the sediment flux Q_r – the instability disappears for $\bar{a} = 0$). A similar situation is encountered when dealing with simplified models of fluvial ripples and dunes, as discussed in the fourth part of this book.

Evidently the simple model has the serious drawback that the growth rate diverges for $k \to \infty$, indicating that it is not well-posed. Partly responsible is the hypothesis that the reptation length is constant. Anderson showed that the unpleasant behavior can be cured by making the hypothesis that the reptation length is not a fixed number, but it is a random variable, a, sampled from values α, $-\infty < \alpha < \infty$, using a probability distribution $p(\alpha)$. The spatially varying part of the reptation flux then becomes

$$Q_r(x,t) = m_p n_r \int_{-\infty}^{\infty} [N_{im}(x) - N_{im}(x-\alpha)] p(\alpha) \mathrm{d}\alpha \ . \tag{1.56}$$

The simpler, ill-posed model is recovered with the choice $p(\alpha) = \delta(\bar{a} - \alpha)$. On linearization, one then finds the dispersion relation,

$$c = \beta\,[1 - \hat{p}(k)] \ , \tag{1.57}$$

where $\hat{p}(k)$ is the Fourier transform of $p(\alpha)$. Provided that $\hat{p}(k)$ decreases at large wavenumber, the growth rate remains finite for $k \gg 1$. For example, if

$$p(\alpha) = \begin{cases} \alpha\lambda^2 \mathrm{e}^{-\lambda\alpha} & \alpha \geq 0 \\ 0 & \alpha < 0 \ , \end{cases} \tag{1.58}$$

where λ is a parameter, we find

$$c = \frac{\beta k^2}{\lambda^2 + k^2} + \frac{2\mathrm{i}k\beta\lambda^3}{(\lambda^2 + k^2)^2} \ . \tag{1.59}$$

This predicts that the growth rate vanishes for $k \to \infty$, although all wavenumbers remain unstable.

An alternative regularization comes from the observation that the flux of reptating particles depends on the local bed slope [24]: Reptating particles have a harder time in climbing up a positive slope than in rolling down an incline. This can be accounted for by writing

$$Q_r(x,t) = m_p n_r \int_{-\infty}^{\infty} [N_{im}(x) - N_{im}(x-\alpha)] p(\alpha) \mathrm{d}\alpha - \mu_0 \zeta_x \ , \tag{1.60}$$

where μ_0 is a parameter weighting the sensitivity of the flux to the bed slope. The stability analysis now results in

$$c = \beta\,[1 - \hat{p}(k)] - \mathrm{i}\mu k \ , \tag{1.61}$$

where $\mu = \mu_0/[\rho_p(1-\lambda_p)]$. For example,

$$c_r = \begin{cases} \beta(1-\cos k\bar{a}) \\ \beta k^2/(\lambda^2+k^2) \end{cases} \qquad \sigma = \begin{cases} \beta k \sin k\bar{a} - \mu k^2 \\ 2\beta k^2\lambda^3/(\lambda^2+k^2)^2 - \mu k^2 \end{cases}, \tag{1.62}$$

for the two models considered above. The slope-dependence of the flux therefore stabilizes short-wavelength perturbations. Moreover, it is clear that the destabilizing cascade (as measured by β) must now overcome the stabilizing influence of slope-induced reptation (estimated by μ) in order to form ripples. In other words, by varying the relative strength of these two effects, we can control the instability and tune the system to be near a marginally stable state. That state is given by $\beta\bar{a} = \mu$ for both models (given that $\bar{a} = \int \alpha p(\alpha) \mathrm{d}\alpha = 2/\lambda$ for the second model), at which point *long* waves with $k \ll 1$ are on the brink of instability. This sets the stage for further analysis of the problem, and so (1.60) is a convenient point of departure for our theory of sand ripples.

1.5.2 A Minimal Model

Given the formulation of a well-posed model, we now nondimensionalize to streamline the formulae, eliminate distracting constants and isolate the important parameters of the problem. We introduce the non-dimensional variables,

$$\tilde{x} = x/\bar{a} \ , \ \ \tilde{\alpha} = \alpha/\bar{a} \ , \ \ \tilde{\zeta} = \zeta/\bar{a} \ , \ \ \tilde{t} = \beta t/\bar{a} \ , \ \ \tilde{p}(\tilde{\alpha}) = \bar{a}p(\alpha) \ , \tag{1.63}$$

where $\bar{a}$ is the average reptation length, and we obtain the nondimensional evolution equation

$$\frac{\partial \zeta}{\partial t} = \int_{-\infty}^{\infty} [\mathcal{F}(x-\alpha) - \mathcal{F}(x)] p(\alpha) \mathrm{d}\alpha + \kappa \zeta_{xx} \tag{1.64}$$

where we have again dropped the tilde decoration,

$$\mathcal{F}(x) = \frac{\tan\phi + \zeta_x}{\sqrt{1+\zeta_x^2}} \ , \tag{1.65}$$

and $\kappa = \mu/(\beta\bar{a})$.

This equation is still quite complicated, but reduces further near the onset of instability. To achieve this reduction, we exploit the long-wave character of the instability near onset: We set $\zeta(x,t) = \epsilon\zeta(X,t)$, where $X = \epsilon x$ or $\partial_x = \epsilon\partial_X$. In this case,

$$\mathcal{F}(x-\alpha) \to \mathcal{F}(X-\epsilon\alpha) \approx \mathcal{F}(X) - \epsilon\alpha\mathcal{F}_X(X) + \frac{\epsilon^2\alpha^2}{2}\mathcal{F}_{XX}(X) - \frac{\epsilon^3\alpha^3}{6}\mathcal{F}_{XXX}(X) + \ldots \ , \tag{1.66}$$

and so we may write

$$\zeta_t = -\epsilon \left[1 - \frac{1}{2}\epsilon\overline{a^2}\partial_X + \frac{1}{6}\epsilon^2\overline{a^3}\partial_X^2\right] \mathcal{F}_X + \kappa\epsilon^2\zeta_{XX} + O(\epsilon^4) \tag{1.67}$$

where $\overline{a^p} = \int \alpha^p p(\alpha) d\alpha$. Moreover,

$$\mathcal{F}(x) \approx \epsilon \zeta_X - \frac{1}{2}\epsilon^2 \zeta_X^2 \tan\phi - \frac{1}{2}\epsilon^3 \zeta_X^3 + O(\epsilon^4) . \tag{1.68}$$

Thence,

$$\zeta_t = -\epsilon^2(1-\kappa)\zeta_{XX} + \frac{1}{2}\epsilon^3 \overline{a^2} \zeta_{XXX} - \frac{1}{6}\epsilon^4 \overline{a^3} \zeta_{XXXX} + \frac{1}{2}\epsilon^3 \tan\phi (\zeta_X^2)_X + \frac{1}{2}\epsilon^4 (\zeta_X^3)_X - \frac{1}{4}\epsilon^4 \overline{a^2} \tan\phi (\zeta_X^2)_{XX} + O(\epsilon^5) . \tag{1.69}$$

Finally, by introducing $u = \zeta_X$ and some rescaling we arrive at the approximation,

$$u_t = -u_{\xi\xi} + \gamma u_{\xi\xi\xi} - u_{\xi\xi\xi\xi} + \delta(u^2)_{\xi\xi} + (u^3)_{\xi\xi} - \varpi(u^2)_{\xi\xi\xi} , \tag{1.70}$$

where ξ is the rescaled spatial coordinate and γ, δ and ϖ are parameters.

The linear instability is present in (1.70) through the diffusion term which has a negative sign. Thus the ripple instability can be rationalized in terms of negative diffusion. It should be noted that the model (1.70) is not strictly an asymptotic one, since we mix orders, but we do at least have an idea of the size of the neglected terms. Also, if we ignore the dispersive terms, and set $\gamma = \varpi = 0$, we arrive at a version of the so-called Cahn–Hilliard model – another canonical equation that describes many long-wave instabilities. Like the real Ginzburg–Landau equation, the Cahn–Hilliard equation has a Lyapunov functional and describes the formation and coarsening of front patterns. However, the dispersive terms of (1.70) remove the Lyapunov functional, and so one cannot anticipate a coarsening dynamics for the ripple model without solving the equation. In fact, over a fairly wide parameter range, the forming patterns do coarsen, as illustrated in Fig. 1.12, which shows a numerical solution of the equation. The dispersive terms also introduce a strong drift in the solution.

It is no accident that the minimal model in (1.70) has the form of a generalized Cahn–Hilliard equation. The long-wave analysis naturally reduces the original equations to this form; the various terms are the lowest orders of the power series expansion in ϵ. A key difference is that the Cahn–Hilliard model has the additional symmetry $x \to -x$, which removes the two dispersive terms in (1.70). Nevertheless, the symmetries of the original governing equations are responsible for the appearance of each kind of term in (1.70) and the absence of some other kinds (such as a linear term, u_ξ, or nonlinear terms without derivatives like u^2 – these are forbidden by the original symmetries). Any set of governing equations with the same symmetries should lead to the model (1.70) on long-wave expansion. Thus provided we have the right symmetries, (1.70) is actually a far better model than we might originally appreciate – although the current derivation is based on what we admit is a flawed metaphor for sand dynamics, the long-wave expansion of better governing equations would produce the same model. Indeed, a different approach altogether would have been to write down the symmetries of the original system, argue for a long-wave instability, and

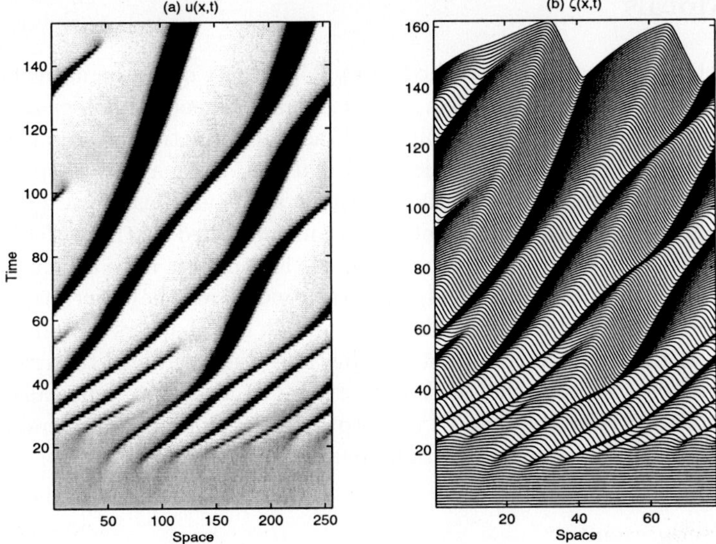

Fig. 1.12. A numerical solution of the minimal model (1.70), for $\gamma = 1$, $\delta = 3$ and $\varpi = 1/2$. The domain is periodic with size 25π. Panel (**a**) shows $u(\xi, t)$ as a density on the (ξ, t) plane, and panel (**b**) shows the corresponding solution for ζ, defined by $u = \zeta_\xi$, as a progression of snapshots. The code for this computation was kindly provided by Yuan-Nan Young (Northwestern University)

simply write down (1.70). Symmetry arguments of this kind, because they can be used to avoid a complicated asymptotic expansion, are popular in nonlinear dynamics.

Given these arguments, it is perhaps not so surprising that the model (1.70) is also similar to some other equations derived for sand ripples [25–29]. Most of these latter models [26–29] base their description on an extension of the so-called BCRE equations for the dynamics of granular sandpiles [30]. What is appealing here is that the model is obtained by introducing very little new physics over the approaches of Bagnold and Anderson.

After all these heuristic approximations and rather uncontrolled expansions, we may of course wonder whether (1.70) has anything to do with real aeolian ripples. Numerical integrations indicate a qualitative resemblance between the solution of the equation and the general aspect of aeolian ripples. To obtain serious progress, however, one now needs a quantitative comparison between the predictions of these simplified models and the results of laboratory experiments and field measurements, as well as an extension of the models to take into account the spatial two-dimensionality of real aeolian ripples and the presence of sand grains with different size. Here, we are content with the fact that one can derive a simple phenomenological model of ripple dynamics, based on a blend of physical intuition and mathematical manipulations.

1.6 Morals

In this Introduction, we have rushed through a variety of mathematical methods that can be used to build models of geological and geomorphological patterns. Some of these methods are rigorous, such as the derivation of amplitude equations in proximity of a supercritical bifurcations, but have correspondingly a limited range of applicability. Other methods are heuristic, and are based on a mixture of phenomenological intuition and uncontrolled mathematical approximations. Most of the methods mentioned in this chapter have a long history of successes and failures, and provide a starting point for the description of complex geological and geomorphological systems. In the chapters to come, we shall see many of these methods in action. Sometimes they will lead to great success, sometimes not. In any event, it is worth trying to use them, especially because there is not much else that can be done to obtain a theoretical understanding of the systems we are dealing with. Bon voyage.

Acknowledgements

Antonello Provenzale thanks L. Prigozhin for discussions.

References

1. S. Chandrasekhar: *Hydrodynamic and Hydromagnetic Stability* (Oxford University Press, Oxford 1961) 643 pp.
2. J.S. Turner: *Buoyancy Effects in Fluids* (Cambridge University Press, New York 1973) 367 pp.
3. G. Schubert: Ann. Rev. Fluid Mech. **24**, 395–397 (1992)
4. M.C. Cross, P.C. Hohenberg: Rev. Mod. Phys. **65**(3), 851–1112 (1993)
5. J.R. de Bruyer, E. Bodenschatz, S.W. Morris, S.P. Trainoff, Y. Hu, et al.: Rev. Sci. Inst. **67**(6), 2043–2067 (1996)
6. E. Bodenschatz, W. Pesch, G. Ahlers: Ann. Revs. Fluid Mech. **32**, 709–778 (2000)
7. W.V.R. Malkus, G. Veronis: J. Fluid Mech. **4**, 225–260 (1958)
8. M. Colombini, G. Seminara, M. Tubino: J. Fluid Mech. **181**, 213–232 (1987)
9. A.C. Newell, J.A. Whitehead: J. Fluid Mech. **28**, 279–303 (1969)
10. L.A. Segel: J. Fluid Mech. **38**, 203–224 (1969)
11. C.S. Bretherton, E.A. Spiegel: Phys Lett. A **96**, 152 (1983)
12. R. Schielen, A. Doelman, H.E. de Swart: J. Fluid Mech. **252**, 325–356 (1993)
13. M.J. Ablowitz, H. Segur: *Solitons and the Inverse Scattering Transform. SIAM Studies in Applied Mathematics* (SIAM, Philadelphia 1981)
14. D.R. Scott, D.J. Stevenson, J.A. Whitehead, Jr.: Nature, **319**, 759–761 (1986)
15. P. Olson, U. Christensen: J. Geophys. Res. **91**(B), 6367 (1986)
16. J.A. Whitehead: Am. J. Phys. **55**(11), 998–1003 (1987)
17. K.R. Helfrich, J.A. Whitehead: Geophys. Astrophys. Fluid Dyn. **51**, 35–52 (1990)
18. J.A. Whitehead, K.R. Helfrich: Geophys. Res. Letts. **13**, 545–546 (1986)
19. R.A. Bagnold: *The Physics of Blown Sand and Desert Dunes* (Chapman and Hall, London 1941)
20. D.A. Rumpel: Sedimentology **32**, 267–275 (1985)

21. B.B. Willetts, M.A. Rice: 'Inter-saltation collisions'. In: *Proc. Int. workshop on the physics of blown sand, Dept. of theoretical statistics, Unversity of Aarhus* **1**, 83–100 (1986)
22. J. Ungar, P.K. Haff: Sedimentology **34**, 289–299 (1987)
23. R.S. Anderson: Sedimentology **34**, 943–956 (1987)
24. H. Nishimori, N. Ouchi: Phys. Rev. Lett. **71**, 197 (1993)
25. R. Hoyle, A. Woods: Phys. Rev. E **56**, 6861 (1997)
26. O. Terzidis, P. Claudin, J.-P. Bouchaud: Eur. Physics Journal B **5**, 245 (1998)
27. A. Valance, F. Rioual: Eur. Physics Journal B **10**, 543 (1999)
28. L. Prigozhin: Phys. Rev. E, **60**, 729–733 (1999)
29. Z. Csahok, C. Misbah, F. Rioual, A. Valance. Dynamics of aeolian sand ripples. Preprint cond-mat/0001336 (2000)
30. J.-P. Bouchaud, M.E. Cates, J.R. Prakash, S.F. Edwards: J. Phys. France I **4**, 1383 (1994)

2 Geophysical Aspects of Non-Newtonian Fluid Mechanics

N.J. Balmforth[1] and R.V. Craster[2]

[1] Department of Applied Mathematics and Statistics, School of Engineering, University of California at Santa Cruz, CA 95064, USA
[2] Department of Mathematics, Imperial College of Science, Technology and Medicine, London, SW7 2BZ, UK

2.1 Introduction

Non-Newtonian fluid mechanics is a vast subject that has several journals partly, or primarily, dedicated to its investigation (Journal of Non-Newtonian Fluid Mechanics, Rheologica Acta, Journal of Fluid Mechanics, Journal of Rheology, amongst others). It is an area of active research, both for industrial fluid problems and for applications elsewhere, notably geophysically motivated issues such as the flow of lava and ice, mud slides, snow avalanches and debris flows. The main motivation for this research activity is that, apart from some annoyingly common fluids such as air and water, virtually no fluid is actually Newtonian (that is, having a simple linear relation between stress and strain-rate characterized by a constant viscosity). Several textbooks are useful sources of information; for example, [1–3] are standard texts giving mathematical and engineering perspectives upon the subject. In these lecture notes, Ancey's chapter on rheology (Chap. 3) gives further introduction.

Non-Newtonian fluids arise in virtually every environment. Typical examples within our own bodies are blood and mucus. Other familiar examples are lava, snow, suspensions of clay, mud slurries, toothpaste, tomato ketchup, paints, molten rubber and emulsions. Chemical engineers, and engineers in general, are faced with the (often considerable) practical difficulties of modelling a variety of industrial processes involving the flow of some of these materials. Consequently, much theory has been developed with this in mind, and our aim in this review is to guide the reader through some of the developments and to indicate how and where this theory might be used in the geophysical contexts.

2.2 Microstructure and Macroscopic Fluid Phenomena

Most non-Newtonian fluids are characterized by an underlying microstructure that is primarily responsible for creating the macroscopic properties of the fluid. For example, a variety of non-Newtonian fluids are particulate suspensions – Newtonian solvents, such as water, that contain particles of another material. The microstructure that develops in such suspensions arises from particle–particle or particle–solvent interactions; these are often of electrostatic or chemical origin.

A common example of such a suspension is a slurry of kaolin (clay) in water. Kaolin particles roughly take the form of flat rectangular plates with different electrostatic charges on the faces and on the sides; their physical size is of the order of a micron. In static fluid, the plates stack together like a giant house of cards. This structure becomes so extensive that the electrostatic forces that hold the structure together engender a macroscopic effect, namely the microstructure is able to provide a certain amount of resistance to fluid flow [4].

Of course, the image of the kaolin structure within the slurry as a giant house of cards is a gross idealization. Undoubtedly, the kaolin forms an inhomogeneous, defective structure with a variety of length scales. Nevertheless, the important idea is that microstructure can lead to macroscopic observable effects on the flow of the fluid. For the kaolin slurry, we anticipate that microstructure adds to the resistance to flow provided the shearing (rate of deformation) is not too great. However, once the fluid is flowing and shearing over relatively long scales, the microstructure must disintegrate – the house of cards collapses. Thus, for greater shearing (larger rates of deformation), the fluid begins to flow more easily. This macroscopic, non-Newtonian effect of "shear thinning" is well documented and a key effect in suspension mechanics. The crudest model of the phenomenon is to make the fluid viscosity a decreasing function of the rate of strain. In this simple departure from the regular fluid behaviour, one then makes the shear stress a nonlinear function of the strain rate. This is an example of a "constitutive law"; we elaborate further on such laws soon, but first we continue with a brief discussion of other non-Newtonian effects.

If the concentration of kaolin is sufficiently high, the microstructure can provide so considerable a resistance to deformation that material does not flow at all until a certain amount of stress is exerted on the fluid. At smaller stresses, the fluid behaves like an elastic solid, and simply returns to its original state if the applied stress is removed. Above the critical stress, the "yield stress", the material begins to flow. Materials exhibiting yield behaviour are said to behave plastically, and when they flow viscously after yield, the terminology viscoplastic is often used.

The kaolin–water slurry is what one might call a "pure" form of mud. But, when the mud is less pure, and contains numerous embedded particles, grains or boulders with widely varying sizes (as in most geophysical conditions), the clay particles still form microstructure, with the attendant macroscopic effects. Hence muds are a classic example of a geophysical viscoplastic fluid. But there are also other geophysical materials with microstructure. For example, snow flakes, through a process of partial melting and refreezing, act to form a static coherent structure; this is relevant when considering avalanches, see also Chap. 13. And lava has a microstructure of bubbles and silicate crystals suspended within a hot viscous solvent.

Shear thinning and yield stresses are common effects in particle suspensions, but they are not the only type of non-Newtonian behaviour we can encounter. Another type of behaviour arises in polymeric fluids. Here, the fluid is laced with high molecular weight deformable molecules (polymers), whose length can be so long that the collective effect of the deformations of individual molecules affects the flow. Notably, because polymers coil and entangle themselves and their neighbours through weak molecular interactions (such as hydrogen bonding), they provide an effective elastic force that resists flow deformations which separate, straighten and stretch them. Moreover, because the forces produced by molecular rearrangements depend on their original orientations, polymeric fluids can also display significant memory dependence; that is, the fluid "remembers" the way in which it has been deformed. The macroscopic consequence is that the fluid can display highly elastic effects, such as the recoil of the fluid back into a container after it has begun to pour out of it.

Some of the effects of such "viscoelasticity" can be rather weird and surprising, and in all discussion of such fluids it is customary to mention a few examples: The Weissenberg effects [5] include die swell [6,7], wherein fluid emerges from a pipe and then undergoes a subsequent and sudden radial expansion downstream, and rod climbing, where the free surface of a rotating fluid rises up around the rod forcing it into motion (the surface of a Newtonian fluid would be depressed there). In the flow of a viscoelastic liquid down an open channel, the free surface bulges slightly to create a rounded fluid profile [8]. Viscoelastic flow past a bubble [9] leads to a distinct cusp at the rear stagnation point due to a long filament of highly stretched polymers in the bubble wake.

An important point that one should take from this discussion is that non-Newtonian fluid effects can be varied and unusual. As a result, the literature on non-Newtonian fluid mechanics contains many models of suspensions and polymeric fluids, each adding or encapsulating some observed effect. Unfortunately many of these models are designed with precisely one set of effects in mind and none adequately deal with the general non-Newtonian fluid. Consequently, because non-Newtonian effects all typically stem in some way from the underlying fluid microstructure, one should keep the microscopic physics in mind whilst negotiating one's way through the minefield of rheological models to which we now give some introduction.

2.3 Governing Equations

To begin, we must first describe the continuum approximation that underlies the models to be discussed here. This continuum approximation assumes that the dimensions of the flow fields we are considering, with lengthscale L, are far greater than the lengthscale of the microstructure of the fluid l; that is, $L \gg l$. Given this continuum hypothesis we can derive the governing equations for a fluid using conservation of mass and examining the rate of change of momentum within a volume of fluid with lengthscale L. If the fluid is incompressible, mass conservation yields

$$\nabla \cdot \mathbf{u} = 0 \, , \tag{2.1}$$

where $\mathbf{u}$ denotes the Eulerian velocity field (here we shall only consider incompressible fluids). Conservation of momentum leads us to

$$\varrho \frac{\mathrm{D}\mathbf{u}}{\mathrm{Dt}} = \nabla \cdot \boldsymbol{\sigma} + \mathbf{F} , \tag{2.2}$$

where the fluid density is ϱ, the convective derivative is $\mathrm{D/Dt} \equiv \partial/\partial \mathrm{t} + \mathbf{u} \cdot \nabla$, the stress tensor is $\boldsymbol{\sigma} \equiv \{\sigma_{ij}\}$, and $\mathbf{F}$ denotes a body force, such as gravity. For incompressible fluids, the stress tensor is conveniently split into an isotropic piece $-p\mathbf{I}$, where p is the pressure field, and a remainder, here denoted by $\boldsymbol{\tau} \equiv \{\tau_{ij}\}$, called the deviatoric stress tensor. Thus,

$$\boldsymbol{\sigma} = -p\mathbf{I} + \boldsymbol{\tau} \qquad \text{or} \qquad \sigma_{ij} = -p\delta_{ij} + \tau_{ij} , \tag{2.3}$$

and the momentum equation becomes

$$\varrho \frac{\mathrm{D}\mathbf{u}}{\mathrm{Dt}} = -\nabla p + \nabla \cdot \boldsymbol{\tau} + \mathbf{F} . \tag{2.4}$$

So far, apart from the continuum hypothesis, and for brevity and practicality assuming incompressibility, we have not made any statement about the fluid itself; mass conservation and the momentum equation are valid for all fluids. Thus the development so far parallels that of a Newtonian fluid, much as can be found in textbooks such as [10].

To produce a closed model, we must further specify how the deviatoric stress tensor τ_{ij} is related to the properties of the fluid. Many non-Newtonian fluid models do this by relating the deviatoric stress to the rate-of-strain tensor, $\dot{\gamma}_{ij}$, here defined as

$$\dot{\boldsymbol{\gamma}} = \nabla \mathbf{u} + (\nabla \mathbf{u})^T \qquad \text{or} \qquad \dot{\gamma}_{ij} = \frac{\partial u_i}{\partial x_j} + \frac{\partial u_j}{\partial x_i} ; \tag{2.5}$$

where the superscript T denotes the transpose (some other authors use a minor variation with an extra factor of 1/2). Further variables are also sometimes included, such as the strain tensor γ_{ij} (which arises in linear elasticity), temperature, pressure, or particulate concentration. The relationship between τ_{ij}, $\dot{\gamma}_{ij}$ and any other variables is the *constitutive relation* of the fluid, and closes the set of governing equations. This relation is the key ingredient to non-Newtonian fluid models and contains all of the fluid microphysics; unsurprisingly, the constitutive law can be extremely complicated. Indeed, there is considerable freedom in deciding how the fluid behaves due to changes in its deformation (the instantaneous strain, strain rates or strain history), or the behaviour due to its surroundings (such as temperature or pressure).

If the fluid is temperature-dependent and in a situation where the temperature can change, as is often the case for ice or lava flows, then we also require an energy equation. This equation describes, for example, how mechanical energy is converted by molecular friction into heat. Such frictional heating is often negligible in many fluid problems – after all we do not heat cups of coffee by stirring them. But in ice flows, this effect can be important (see Chap. 11). Of much more importance in general fluid problems, however, is that a change in temperature can affect the fluid microstructure. This may give rise to magnitudes of variation in macroscopic material properties. Indeed, many fluids are Newtonian at fixed temperature, but have viscosities that are dramatically affected by temperature changes, as spreading golden syrup upon hot toast will demonstrate.

The energy equation is:

$$\varrho c \frac{\mathrm{DT}}{\mathrm{Dt}} = \frac{1}{2}\tau_{ij}\dot{\gamma}_{ij} + \nabla \cdot (\mathcal{K}\nabla T) \,. \tag{2.6}$$

The parameters c and $\mathcal{K}$ are the specific heat (at constant pressure or volume, as the fluid is incompressible) and conductivity. In deriving this equation we have assumed that the thermal expansion coefficient for the fluid is negligible, and we have ignored other energy sources or sinks, such as from plastic or elastic work, or from inelastic collisions between particles within the microstructure. The energy equation describes how the temperature field evolves in the fluid as a result of advection, diffusion and frictional heating. Such thermal evolution subsequently affects fluid microstructure and, thence, material properties. In turn, this modifies the fluid flow according to the constitutive law.

2.4 Constitutive Models

Newtonian fluids are characterized by an isotropic microstructure of passive spherical molecules that do not chemically interact with one another. The constitutive law is particularly simple: the deviatoric stress is linearly proportional to the rate of strain and the coefficient of proportionality is the viscosity, μ. Thus

$$\tau_{ij} = \mu \, \dot{\gamma}_{ij} \,,$$

and (2.2) reduces to the more familiar Navier–Stokes equation,

$$\varrho \frac{\mathbf{Du}}{\mathrm{Dt}} = -\nabla p + \mu \, \nabla^2 \mathbf{u} + \mathbf{F} \,.$$

For non-Newtonian fluids the constitutive relations can be much more complicated and must be built to reflect the macroscopic properties engendered by the fluid microstructure. There are several ways in which one goes about this construction; here we mention four different styles.

The first kind of approach is theoretical and "kinetic": one assembles a model of the molecular anatomy of the fluid and then builds a kinetic theory for the fluid microstructure. Sometimes, this goes by way of an investigation of the flow around a single idealized model polymer, or emulsion droplet, and then the generation of the appropriate constitutive equation for a dilute suspension via an averaging procedure [11]. But other routes are also possible, including the representation of the fluid microstructure as a regular lattice or network of interacting elements [12]. These theories furnish a fluid model directly from the input microscopic physics, and in an idealized world would be the most sensible approach. Unfortunately, such kinetic approaches have only recently become possible, and even then only for very simple fluids. Moreover, the mathematics behind them is often based upon physical approximations rather than asymptotic analysis. The problem is that it is currently technically impossible to build a kinetic theory for anything more than a very simple range of molecular models. For example, a popular model in visco-elasticity is a perfect network of identical elastic rods. But real fluids never conform to the idealizations necessary in order to fabricate kinetic theories, and even the simplest of such theories can lead to constitutive laws with very convoluted forms. Nevertheless, much progress has been made in the recent non-Newtonian fluid literature in this direction.

A second style of approach is purely phenomenological: one simply writes down a convenient model equation that represents how one imagines the fluid microstructure to affect the flow. Historically, this type of approach was the first used in non-Newtonian fluid mechanics. For example, Maxwell's model of a viscoelastic fluid was largely phenomenological – the stresses have a "fading memory" of the strain rates, which models the relaxation of the fluid to applied deformation at a molecular level.

The third approach was taken somewhat after the first phenomenological models and is largely an attempt to improve on them. The phenomenological theories provided a set of simple constitutive relations that at times did not possess some of the symmetries of the fluid. For example, the original Maxwell model was not "objective" when written in three dimensions, meaning that it took different forms in different frames of reference (see later). The third approach was therefore to write down the simplest kinds of constitutive models that possessed the same symmetries as the fluid. Thus Oldroyd wrote down a general constitutive model for a linear visco-elastic fluid model. This "Oldroyd-8" model contains a set of free parameters and has been claimed to work well in several situations. Moreover, several kinetic theories have also eventually led to the same kinds of models.

The difficulty in proceeding theoretically to furnish the constitutive law has led to a very popular fourth approach which is practical, but empirical. One performs various experiments upon the fluid using, for example, a viscometer, and then postulates a plausible stress strain-rate relation. Experiments for non-Newtonian fluids are not necessarily easy to perform [6] and a considerable amount of effort is sometimes required to neatly design experiments that isolate a particular factor. This empirical approach focusses on the macroscopic behaviour of the fluid and to a large extent simply takes the fluid microstructure for granted. Needless to say, the empirical models that one derives in this way are dangerous in that they are derived for specific experimental conditions and are not necessarily suitable once one changes those conditions. However, given some non-Newtonian fluid with a complicated and possibly unknown microstructure, the empirical approach is often the most expedient way forward.

This discussion should illustrate to the reader how non-Newtonian fluid mechanics has a certain schizophrenic aspect to it. On the one hand, the theory is mathematically complicated and furnishes unwieldy constitutive laws. And on the other, there is a pragmatic approach that provides workable, but potentially unreliable, models. Below we give some examples of the forbidden fruit of the marriage of the two.....

2.5 Generalized Newtonian Models

Generalized Newtonian fluid models assume a fairly simple constitutive relation in which one modifies the linear relationship between the stresses and the strain rates by making the constant of proportionality, the viscosity, a prescribed function of strain-rate, temperature or particulate concentration. Thus,

$$\tau_{ij} = \mu(\dot{\gamma}, T, \phi)\, \dot{\gamma}_{ij} \qquad \text{(Generalized Newtonian model)}\ , \tag{2.7}$$

where we use τ and $\dot{\gamma}$ to represent the second invariants of the stress and strain rate,

$$\tau = \sqrt{\tau_{ij}\tau_{ij}/2}\ , \quad \dot{\gamma} = \sqrt{\dot{\gamma}_{ij}\dot{\gamma}_{ij}/2}\ ; \tag{2.8}$$

($\dot{\gamma}$ can be thought of as a measure of the magnitude of the deformation rate) and ϕ is the particle concentration.

2.5.1 Power-law Fluids and the Herschel–Bulkley Model

A popular example of this kind of model is the power law fluid:

$$\mu(\dot{\gamma}) = K\dot{\gamma}^{n-1} \qquad \text{(power law model)}\ . \tag{2.9}$$

This viscosity function has two parameters, the consistency K and the index n. If $n = 1$ we revert to Newtonian behaviour and the consistency is just the viscosity. If $n < 1$, the effective viscosity decreases with the amount of deformation. Thus this models the disintegration of fluid structure under shear, the shear thinning effect mentioned earlier.

Conversely, if $n > 1$, the viscosity increases with the amount of shearing, which implies that the fluid microstructure is build up by the fluid motion. This kind of effect can occur if the molecules of the microstructure can bind together on contact; during increasing flow these molecules can come into contact more regularly and thus larger structures are created. Examples of such "shear thickening" materials are corn flour (which is used to thicken soup) and highly concentrated suspensions. The latter show shear thickening due to dilatancy [3]: at low shear rates the particles are closely packed together and a small amount of fluid lubricates the flow of particles. But, at higher shear rates, the close packing is disrupted and the material expands (dilates) and there is no longer enough fluid to lubricate particle–particle interactions. The resistance to flow then increases substantially.

The empirical power law model is a useful fit to the observed data, and can often provide good quantitative results over many decades of the shear rate.

However, it does not capture the effects of yield stress. Probably the most popular model that incorporates both shear thinning or thickening and a yield stress is the Herschel–Bulkley model [13]:

$$\begin{array}{ll} \tau_{ij} = \left(K\dot{\gamma}^{n-1} + \tau_p/\dot{\gamma}\right)\dot{\gamma}_{ij} & \text{for } \tau \geq \tau_p \\ \dot{\gamma}_{ij} = 0 & \text{for } \tau < \tau_p \end{array} \quad \text{(Herschel – Bulkley model)} . \tag{2.10}$$

The new parameter τ_p that we introduce is the yield stress. This formula also contains an even simpler model, the Bingham fluid, which is given by (2.10) with $n = 1$. For this model, the fluid flows as a Newtonian fluid, with strain rate proportional to the difference between the applied and yield stresses, once it has yielded. With $n \neq 1$, the Herschel–Bulkley model allows also for shear thinning or thickening beyond yielding. Two recent review articles upon yield stress phenomena are [14] and [15] and the model often appears in geophysical models, as illustrated in several other chapters in this volume.

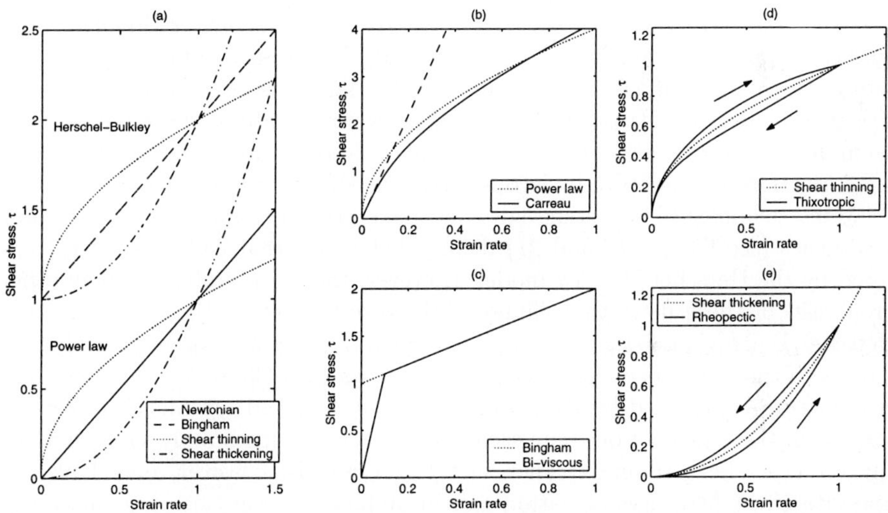

Fig. 2.1. Non-Newtonian fluid models. A sketch of the constitutive models for a variety of rheological models. In (**a**) we show the power-law and Herschel–Bulkley models. Three curves are shown in each case, displaying shear-thinning and shear-thickening flow curves. The Bingham fluid and a Newtonian fluid are also shown. In panel (**b**) we display the Carreau model, $\mu(\dot{\gamma}) = \mu_\infty + (\mu_0 - \mu_\infty)/[1+(\lambda\dot{\gamma})^2]^{(1-n)/2}$ (μ_0, μ_∞, λ and n are constants), which regularizes the infinite viscosity of the shear-thinning power-law fluid at zero strain rate. In (**c**) we show the bi-viscous regularization of the Bingham model, which allows flow for all strain rates. Panels (**d**) and (**e**) show thixotropic and rheopectic hysteresis curves. The scales are arbitrary

2.5.2 Variants and Deviants

There are many other empirical equations that provide stress-strain-rate relations within the generalized Newtonian framework, although the power law, Bingham and Herschel–Bulkley models are those most widely used; an illustration showing these models is in Fig. 2.1. However, this is not to say that they are uniformly accepted. Indeed, there is much discussion in the recent literature over whether these models are physically plausible. For example, the shear-thinning power law fluid predicts an infinite viscosity at zero strain rate. Even the concept of a yield stress has received much recent criticism, with evidence presented to suggest that most materials weakly yield or creep near zero strain rate [15]. Moreover, from a mathematical perspective, the discontinuous surface defined by the yield condition, $\tau = \tau_p$, introduces several undesirable features into the non-Newtonian fluid model, mainly because this surface is difficult to track accurately. Such criticisms have fuelled the introduction of further models that go some way to avoid the problems (see [3] p. 14, and [1]). For example, the Carreau model regularizes the infinite viscosity of the shear-thinning power-law fluid (see Fig. 2.1). And various regularizations of the Herschel–Bulkley or Bingham fluid modify the constitutive law so that, for $\dot{\gamma} \to 0$, the stress abruptly decreases to zero in the manner of a Newtonian fluid with a large viscosity. The latter regularizations allows flow to occur even at very low strain rates and are particularly useful for numerical work, [16–18]. A popular, although not necessarily optimal, regularization is to adopt a biviscous model, as shown in Fig. 2.1.

Many geophysical materials such as muds [19,20], debris flows and snow avalanches (see Chaps. 13 and 21) display behaviour that can be crudely captured by the Herschel–Bulkley model. However, there are probably many other properties of these flows that cannot [21]. Nevertheless, at the very least, the Herschel–Bulkley model can be used as the starting point for more elaborate models. This model has also been used for lavas (see Chap. 7). Here, the microstructure is provided by a combination of bubbles and crystals. Bubbles deform with the fluid motion; numerical computations with bubbly viscous fluids suggest that shear thinning can result [22]. Crystals, however, may have the opposite effect [23]: crystallization can be induced by the shearing motion of the fluid and so microstructure can be build up in a shear thickening fashion. Both effects may compete in lava, and which dominates depends on the ambient conditions.

2.5.3 Temperature Dependence

Many materials have strongly temperature-dependent microstructure. For generalized Newtonian fluids, the most common way of accounting for this dependence is to make the viscosity a function of temperature. A popular choice is an exponential, Arrhenius, dependence:

$$\mu(T) = \mu_* \exp(Q/RT) \tag{2.11}$$

where μ_* is the viscosity value evaluated at some reference temperature, Q is the activation energy and R is the universal gas constant. Sometimes it is more convenient to use the approximation,

$$\mu(T) = \mu_* \exp[-\tilde{G}(T - T_a)] , \tag{2.12}$$

where T_a and $\tilde{G}$ are two more prescribed constants. Provided the temperature variation is relatively small, (2.12) can be considered as an approximation to (2.11); in some other contexts, this is referred to as the Frank–Kamenetski approximation. Exponential forms for the temperature dependence are commonly used for lavas [23–27], laboratory materials used to model magma and lava (such as wax, paraffin and corn syrup [28,29]), muds [30–32], and ice sheets [33].

Some fluids display both strong temperature dependence and other non-Newtonian effects, like shear thinning or yield behaviour. Lava and ice are two such materials. Within those subjects there have been attempts to generate empirical models incorporating all these features. Typically, they proceed by simply combining the earlier models. For example, one particular model that has found a niche of geophysical importance is Glen's Law [34,35] for the flow of ice. It has the stress-strain-rate relation,

$$\mu(\dot{\gamma}, T) = \exp(Q/nRT)\dot{\gamma}^{(n-1)/n}, \quad n \sim 3 , \tag{2.13}$$

and combines an Arrhenius temperature dependence with shear thinning. Typically the constitutive law is written in terms of the second invariant of the stress, rather than the strain rate, for reasons of algebraic ease in subsequent analysis. However, despite the wide usage of this law, there is significant disagreement between measurements taken in various laboratory experiments and from actual ice flows [36]. Part of the reason for this disagreement seems to be that ice relaxes under stress only over long times, and this relaxation has not been correctly taken into account in most measurements.

2.5.4 Concentration Dependence

Another issue that often arises in fluid suspensions is how the microstructural effects depend upon the particle concentration, ϕ. For Newtonian fluids, the Einstein relation was deduced to give the viscosity correction due to a dilute suspension of rigid spheres within a solvent of viscosity μ_0:

$$\mu = \mu_0 \left(1 + \frac{5}{2}\phi\right) . \tag{2.14}$$

Strictly speaking, this model is only suitable if the suspension is very dilute. A simple resummation of (2.14) that attempts to extend the formula to much larger concentrations is the Einstein–Roscoe relation:

$$\mu = \mu_0 \left(1 - \frac{\phi}{\phi_m}\right)^{-\alpha} . \tag{2.15}$$

The quantity ϕ_m is a maximum packing fraction beyond which the suspension cannot flow; for a suspension of solid spheres, $\phi_m \approx 0.68$, but this quantity depends on the shape of the particles and how they organize themselves into a lattice structure. Experiments with concentrated non-colloidal suspensions [37] suggest that a good empirical fit is achieved if $\alpha \approx 1.82$. Other related models are reviewed in [38]. Similar approximations have been developed for lava, where one argues that the role of the suspended particles is played by silicate crystals [39], and in temperate ice (a binary mixture of ice and water at the melting temperature), where the concentration does not refer to particles at all, but to the water content [40].

Particle concentration also affects the yield stress in viscoplastic fluids [14], and so we need another formula for $\tau_p(\phi)$ in the constitutive law. In geophysical contexts, the combined effect of concentration dependence on viscosity and yield stress may be important for lava (because crystallization occurs when the temperature falls) and for some debris flows.

Given that the fluid properties depend on particle concentration, one should also add an equation that determines ϕ. In some situations, it may be possible to treat the concentration as though it were homogeneous; then ϕ is simply a parameter. However, the origin of many effects observed in suspensions can be traced to the appearance of an inhomogeneous particle distribution. A notable example that plagues chemical engineers is wall slip. Many rheometers operate by creating a shear flow inside the fluid by rotating the walls containing the material. Often it is observed that high shear layers build up near these walls in which the particle concentration is depleted. Because the fluid is then relatively dilute in these region, and they are frequently extremely thin, they act like lubricating "slip" layers. As a result, the direct measurements taken with the instrument can be in error.

Another example that may be of geophysical relevance is viscous resuspension. The observation here is that particles in a shearing suspension tend to migrate away from regions with relatively large shear. This migration provides an uplift in flows over plates that can oppose and even dominate the natural tendency to sediment [41].

To deal with concentration variations, we need a conservation equation for ϕ. One relevant to viscous resuspension is [42] :

$$\frac{\mathrm{D}\phi}{\mathrm{D}t} + \nabla \cdot [J_c + J_\mu] = 0 \tag{2.16}$$

$$J_c = -K_c a^2 \phi \nabla(\phi \dot{\gamma}) \,, \quad J_\mu = -K_\mu \dot{\gamma} \phi^2 \frac{a^2}{\mu} \frac{\mathrm{d}\mu}{\mathrm{d}\phi} \nabla \phi \,. \tag{2.17}$$

Here J_c and J_μ are the fluxes due to particle collisions and spatially varying viscosity; the particular forms quoted are given by heuristic arguments in [42]. The parameters K_c and K_μ are constants determined experimentally and a is the particle radius.

In lava, particle diffusion and migration may be unimportant for silicate crystals. However, crystals form when the temperature decreases, and so one should add sources and sinks associated with the phase change of solidification. Moreover, as in ice, the crystal structure may form anisotropically and with a broad distribution of sizes. The particle concentration ϕ in lava could equally well be considered to be the concentration of bubbles or dissolved volatiles. We mentioned earlier the effect of bubbles, but volatiles add chemical effects that can also modify microstructure (for example, OH^- ions are observed to inhibit polymerization of silicon–oxygen bonds). Furthermore, as temperature and pressure changes, the bubble and volatile content can also change, with one being converted to the other. Overall, this makes the modelling of lava an extremely challenging problem.

2.5.5 Hysteresis

There are complicating issues that the generalized Newtonian models do not capture. One often overlooked issue is hysteresis. As described above, for a static viscoplastic material there is a microstructure that prevents flow until the yield stress is exceeded. Once flowing the structure is gradually broken down with increasing shear, and this gradual attrition of the microstructure leads to non-linear stress strain-rate behaviour. The reverse situation, in which the strain-rate is decreased until the structure reforms, is conceptually identical. However, there is no pressing reason why structure should reform in the same way that it disintegrates; in practice some hysteresis occurs. As a result the stress-strain-rate relation is not identical when the same material is measured with increasing or decreasing strain-rates. That is, the "up-curves" and "down-curves" on the $\dot{\gamma}$–τ plane are different.

The most common types of hysteretic curves are illustrated in the final two panels of Fig. 2.1. The "thixotropic" fluid is shear thinning, and microstructure disintegrates due to the flow of the fluid. Thus the viscosity decreases during the experiment. The "rheopectic" fluid is shear thickening and structure builds up during the experiment. Both thixotropic and rheopectic behaviour have been observed in lavas [23]; thixotropy may be associated with the effects of bubbles, whereas shear-induced crystallization may be responsible for the rheopexy.

We illustrate hysteresis with some rheological measurements for a kaolin–water slurry and a celacol (Methyl–Cellulose) solution. The data is taken with a TI Instruments CSL 500 controlled-stress, cone-and-plate rheometer (6 cm, 2 degree measurement geometry). The results are shown in Fig. 2.2; this also shows the Herschel–Bulkley models that were used to fit the data. Hysteresis is certainly evident for the kaolin slurry. There are also some sharp changes in the up-curves that are possibly indicative of wall slip in the cone and plate device. The extreme example of celacol shows a material that behaves viscoplastically at first, but the destruction of the microstructure is permanent, and on decreasing the applied stress the material behaves viscously.

Another form of hysterisis occurs if the yield strength is itself time dependent, with a distinct gellation timescale. In this case, the structure that creates the yield strength takes time to form. Thus the material may have different yield strengths dependent upon when we choose to disturb it or bring it to rest [43].

Table 2.1. The properties of the experimental materials; the ratios are kaolin:water on a weight basis. Also shown are the parameters of the Herschel–Bulkley model from down-curves of stress sweeps with virgin material (see Fig. 2.2) using 6 cm 2 degree plate with CSL 500 Carrimed. †The Celacol data is taken from the up-curve

Material	0.6:1	0.8:1	1:1	1.2:1	Syrup	Celacol†
Density ϱ (g/cm^3)	1.1	1.2	1.33	1.47	1.0	1.0
Yield stress τ_p (dyne/cm^2)	20	130.0	500.0	1320.0	0.0	0.0
Consistency K (units)	61	240	408	946	690	28.5
Index n	0.5	0.75	0.54	0.42	1.0	0.08

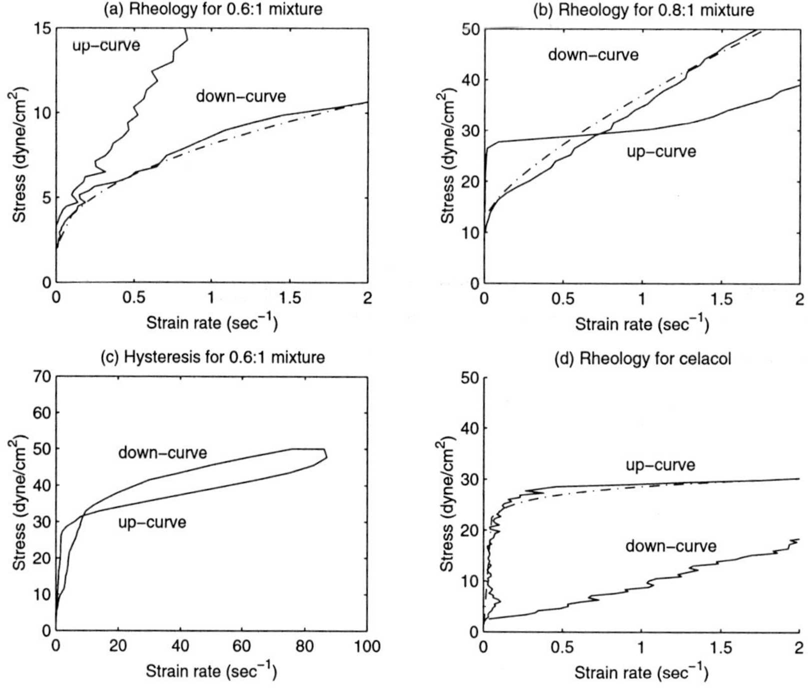

Fig. 2.2. The rheological data collected using a controlled stress sweep. Panels (**a**) and (**b**) show the stress strain-rate curves for the 0.6:1 and 0.8:1 kaolin–water mixtures. The up- and down-curves relate to whether the data was collected whilst the applied stress was increasing or decreasing; the dot-dash lines show the Herschel–Bulkley fit using the parameters of Table 2.1. Panel (**c**) shows the 0.6:1 data over a substantially extended range of strain-rates. The rheology of the celacol solution is shown in panel (**d**)

2.6 Viscoelasticity

Under some circumstances a material will exhibit both elastic and viscous behaviour; in response to some applied shear many materials show initially viscous behaviour and then 'relax' to elastic behaviour. The generalized Newtonian fluid model does not incorporate any elastic effects whatsoever, and so is inappropriate for such flows. Instead, it is usually necessary to introduce the strains as well as strain rates into the constitutive law. This is apparent from the form the constitutive law must take in the extreme limits: an incompressible linear elastic material has the stress is proportional to the strain, whereas a Newtonian fluid has the stress proportional to the rate of strain. Thus, for a general viscoelastic fluid, the constitutive law takes the form of an evolution equation.

The appearance of time evolution terms in the rheology relation reflects the relaxational character of the fluid stresses, and leads to the notion of a characteristic relaxation timescale. Many rheological measurement devices for viscoelastic fluids are designed with this in mind. One standard experiment is to apply instantaneously a shear at the surface of a sample material. If the material is linearly elastic the resulting stress is zero before the application of the shear, and constant immediately afterwards. On the other hand, if the material is a Newtonian fluid, the stress is infinite at the instant the stress is applied, but thereafter is zero. Thus elastic and viscous responses are markedly different, and many real materials have elements of both types of response. A viscoelastic material will have an initially large stress due to the viscous component, but the stress then decreases over the relaxation time to a constant value arising due to the intrinsic elasticity.

If we assume that the relation between the deviatoric stress and the strain rates is purely linear, then a general constitutive law can be stated:

$$\tau_{ij} = \int_{-\infty}^{t} G(t-\tau)\dot{\gamma}_{ij}(\tau)\mathrm{d}\tau \ . \tag{2.18}$$

Here, $G(t)$ is called the relaxation function, and builds in the elastic and viscous behaviour. Implicitly, the shape of the function $G(t)$ determines the characteristic relaxation timescale (or timescales if there are more than one).

The relaxation time is important because it characterizes whether viscoelasticity is likely to be important within an experimental or observational timescale. For example, we might consider the continents upon the earth's surface as solid over a timescale based upon the human lifespan, but upon a geological timescale they could be considered as a viscous, or viscoelastic, fluid. Many fluids, particularly those in industrial situations containing polymers or emulsion droplets, exhibit both elastic and viscous responses on an experimental or observational timescale.

For a Newtonian fluid, $G(t) = \mu\delta(t)$ and relaxation is immediate. For a linear elastic material, $G(t) = \mu H(t)$. If we denote the relaxation time by λ then the simplest viscoelastic model, the Maxwell model, has $G(t) = \mu \exp(-t/\lambda)/\lambda$ and the integral relation above can be recast in the form of a differential constitutive relation,

$$\tau_{ij} + \lambda \dot{\tau}_{ij} = \mu \dot{\gamma} . \tag{2.19}$$

Much can be achieved with this simple extension to the Newtonian constitutive model, and in many circumstances, particularly if one wishes to investigate whether viscoelasticity can be important, this linear theory suffices. Extensions to multiple relaxation times with a sequence of relaxation functions are also straightforward.

Unfortunately, the Maxwell model (2.19) has at least one major failing – it is not frame indifferent (objective). That is, if we change to a moving coordinate frame the equations also change. Since we are concerned with material behaviour this should not occur. One crude, effective and ad-hoc cure is to replace the time derivatives in (2.19) with more complicated operators that build in the convection, rotation and stretching of the fluid motion. These operators, called either Oldroyd or Jaumann derivatives, render the equations frame indifferent; in usual tensor notation, the Oldroyd (upper convected) derivative, $\overset{\triangledown}{\mathbf{b}}$, for a tensor $\mathbf{b}$ is

$$\overset{\triangledown}{\mathbf{b}} = \frac{\mathrm{D}\mathbf{b}}{\mathrm{D}t} - \mathbf{b}\cdot(\nabla\mathbf{u}) - (\nabla\mathbf{u})^T\cdot\mathbf{b} \quad \text{or} \quad \overset{\triangledown}{b}_{ij} = \dot{b}_{ij} + u_k b_{ij,k} - u_{j,k}b_{ki} - u_{i,k}b_{kj} . \tag{2.20}$$

These derivatives involve the local fluid motion, and so substantially complicate the constitutive law, and therefore computations using them.

Although we introduce these derivatives as a mathematical device to improve the linear model, one can also obtain these derivatives by working with dilute suspensions and low Reynold's number hydrodynamics – the kinetic approach mentioned earlier. By studying the fluid motion around a single elastic sphere, emulsion droplet, or a dumbbell connected with an elastic spring, and then analyzing the force exerted by the droplet upon the fluid, one can construct constitutive relations. Rather pleasingly these also involve Oldroyd, or Jaumann, derivatives and so the apparently crude mathematical fix has some physical basis. Further details of this approach can be found in [44] or [1].

A popular, more refined version of the Maxwell model is the so-called Oldroyd-B model; a simplification of his Oldroyd-8 model. The Oldroyd-B model takes account of the stresses due to both the Newtonian solvent and the polymeric constituents:

$$\boldsymbol{\tau} = \boldsymbol{\tau}_s + \boldsymbol{\tau}_p . \tag{2.21}$$

The total viscosity μ is also written as the sum of solvent and polymeric viscosities, $\mu = \mu_s + \mu_p$. Thus, if $\eta = \mu_s/(\mu_s + \mu_p)$, the stress is written as

$$\boldsymbol{\tau} = \mu[\eta\dot{\boldsymbol{\gamma}} + (1-\eta)\mathbf{a}] . \tag{2.22}$$

The constitutive equation for the extra stress tensor $\mathbf{a}$ takes the form,

$$\mathbf{a} + \lambda\overset{\triangledown}{\mathbf{a}} = \dot{\boldsymbol{\gamma}} , \tag{2.23}$$

where λ is the polymer relaxation time. There are several problems with the Oldroyd-B model [45], which suggest that it should not be used indiscriminately to model viscoelastic flows. On the other hand, this model gives a reasonable description for some flows of dilute polymeric suspensions in highly viscous solvents with a single characteristic relaxation time ("Boger fluids" – [46]), and has been used extensively in attempting to characterize and interpret fluid flows [47,48].

One might imagine that because viscoelasticity is commonly engendered by dissolved polymers, there are few geophysical fluids which behave in this fashion. In fact, somewhat surprisingly, lava has been observed to show some viscoelastic non-Newtonian effects. For example, the Weissenberg effect (rod climbing) was observed in some laboratory experiments, and upward bulges have been seen on lava flows on Mount Etna [23]. Also, prolonged time-dependent relaxational effects are seen in measurements of density, pressure and sound speed [49]; relaxation times range from seconds to weeks.

2.7 Concluding Remarks

In this chapter we have given a brief overview of some phenomena and rheological models of non-Newtonian fluid mechanics. However, this is a notoriously involved subject, mainly due to the wide range of often complex and sometimes unexpected behaviours that real fluids and fluid-like materials exhibit. We can only hope to scratch the surface of the subject here, provide references to allow the interested reader to delve further into the subject, and draw together the underlying theory required in later chapters.

It is also important to appreciate the limitations of the models we have described. Indeed, this subject is not like Newtonian fluid mechanics where the Navier–Stokes equation is uniformly accepted; there is still much debate over which constitutive models are appropriate for different materials, and this is particularly prevalent for viscoelastic fluids. The generalized Newtonian models that seem easiest to use are empirical, and the explanation for the experimentally observed behaviour is based upon heuristic microstructural arguments. However, the models are essentially curve fits to observed data that have a convenient mathematical form. Some of the viscoelastic models have a sounder physical foundation, but they are typically far more complicated and are often designed with a specific phenomenon in mind and fail to incorporate the behaviour one wishes to model. None the less, many models exist with a spectrum of degrees of sophistication that build in both physical behaviour and mathematical niceties.

Despite all of these efforts much remains to be understood for non-Newtonian flows in general. Later chapters on debris flows, ice, snow avalanches and lava highlight aspects of the behaviours we have discussed in this chapter: yield stress, shear thinning, temperature dependence and particle concentration dependence. These chapters also describe the current modelling difficulties that remain. For example, the Bingham and Herschel–Bulkley models have had some success for concentrated mud flows containing fine particles [50,51], but have been less successful for flows containing larger particles [21]. Debris flows (Chap. 21) incorporate a range of particle sizes, that at one extreme may be so significant that we violate the continuum approximation. The detailed failure of the Herschel–Bulkley model in these cases is due to several effects. The model does not allow for fluid motion relative to solid debris, it does not incorporate energy dissipation for the solid boulders and grains interacting, or for the way that such large objects can slide or roll along the base of the flow. None the less for primarily shear-dominated flows of concentrated suspensions of fine particles, Bingham-like models can provide good predicative and quantitative information. Indeed, in a later chapter we shall adopt the Herschel–Bulkley model to analyse some isothermal viscoplastic lava flows.

Lastly, we have focussed exclusively on fluids in this chapter. Yet some geophysical materials ought probably not to be treated as fluids at all. For example, the bubbly magma that rises through the conduits within volcanos (see Chap. 8) is much closer to being a foam, and dry landslides and avalanches and some debris flows [52] are fully fledged granular media (see Chap. 4).

Acknowledgements

The financial support of an EPSRC Advanced Fellowship is gratefully acknowledged by RVC. The authors also thank Adam Burbidge for useful interactions.

References

1. R.B. Bird, R.C. Armstrong, O. Hassager: *Dynamics of polymeric liquids, Vol. 1: Fluid dynamics* (Wiley, New York 1977)
2. A.S. Lodge: *Elastic Liquids* (Academic Press, New York 1964)
3. R.I. Tanner: *Engineering Rheology* (Clarendon Press, Oxford 1985)
4. D.H. Everett: *Basic principles of colloid science* (Royal Society of Chemistry, London 1988)
5. K. Weissenberg: Nature **159**, 310 (1947)
6. K. Walters: *Rheometry* (Chapman Hall, London 1975)
7. D.D. Joseph, J.E. Matta, K.P. Chen: J. Non-Newtonian Fluid Mech. **24**, 31 (1987)
8. M.V. Keentok, A.G. Georgescu, A.A. Sherwood, R.I. Tanner: J. Non-Newt. Fluid Mech. **6**, 303 (1980)
9. O. Hassager: Nature **279**, 402 (1979)
10. G.K. Batchelor: *An introduction to Fluid Mechanics* (Cambridge University Press, Cambridge 1967)
11. J.G. Oldroyd: Proc. Roy. Soc. Lond. A **232**, 567 (1955)
12. M. Doi, S.F. Edwards: *The theory of polymer dynamics* (Oxford University Press, Oxford 1986)
13. W.H. Herschel, R. Bulkley: Am. Soc. Testing Mater. **26**, 621 (1923)
14. Q.D. N'Guyen, D.V. Boger: Ann. Rev. Fluid Mech. **24**, 47 (1992)
15. H.A. Barnes: J. Non-Newtonian Fluid Mech. **81**, 133 (1999)
16. J.P. Dent, T.E. Lang: Ann. Glaciology **4**, 42 (1983)
17. I.C. Walton, S.H. Bittleston: J. Fluid Mech. **222**, 39 (1991)
18. A.N. Beris, J.A. Tsamopoulos, R.C. Armstrong, R.A. Brown: J. Fluid Mech. **158**, 219 (1985)
19. P. Coussot: *Mudflow rheology and dynamics* (IAHR Monograph Series, Balkema 1997)

20. K.F. Liu, C.C. Mei: J. Fluid Mech. **207**, 505 (1989)
21. R.M. Iverson: Reviews of Geophysics **35**, 245 (1997)
22. M. Manga, J. Castro, K.V. Cashman, M. Loewenberg: J. Volcan. Geotherm. Res. **87**, 15 (1998)
23. H. Pinkerton, G. Norton: J. Volcan. Geotherm. Res. **68**, 307 (1995)
24. H.R. Shaw: J. Petrology **10**, 510 (1969)
25. A.R. McBirney, T. Murase: Ann. Rev. Earth Planet. Sci. **12**, 337 (1984)
26. D.K. Chester, A.M. Duncan, J.E. Guest, C.R.J. Kilburn: *Mount Etna: The anatomy of a volcano* (Chapman Hall, London 1985)
27. F.J. Spera, A. Borgia, J. Strimple: J. Geophys. Res. **93**, 10273 (1988)
28. J.H. Fink, R.W. Griffiths: J. Fluid Mech. **221**, 485 (1990)
29. M.V. Stasiuk, C. Jaupart, R.S.J. Sparks: Geology **21**, 335 (1993)
30. M.R. Annis: J. Petrol. Technol. **19**, 1074 (1967)
31. H.J. Alderman, A. Gavignet, D. Guillot, G.C. Maitland: SPE (1988), paper 18035
32. B.J. Briscoe, P.F. Luckham, S.R. Ren: Phil. Trans. R. Soc. Lond. A **348**, 179 (1994)
33. K. Hutter: *Theoretical Glaciology* (D. Reidel, Dordrecht 1983)
34. W.S.B. Paterson: *The physics of glaciers* (Pergamon, Oxford 1969)
35. J.W. Glen: Proc. Roy. Soc. Lond. A **228**, 519 (1955)
36. R.L. Hooke: Rev. Geophys. Space Phys. **19**, 664 (1981)
37. I.M. Kreiger: Adv. Colloid Interface Sci. **3**, 111 (1972)
38. P.M. Adler, A. Nadim, H. Brenner: Adv. Chem. Eng. **15**, 1 (1990)
39. H. Pinkerton, R.J. Stevenson: J. Volcan. Geotherm. Res. **53**, 47 (1992)
40. K. Hutter, H. Blatter, M. Funk: J. Geophys. Res. **93**, 12205 (1988)
41. D. Leighton, A. Acrivos: J. Fluid Mech. **181**, 415 (1987)
42. R.J. Phillips, R.C. Armstrong, R.A. Brown, A.L. Graham, J.R. Abbott: Phys. Fluids A **4**, 29 (1992)
43. P. Coussot, under review (unpublished)
44. J.M. Rallison: Ann. Rev. Fluid Mech. **16**, 45 (1984)
45. J.M. Rallison, E.J. Hinch: J. Non-Newt. Fluid Mech. **29**, 37 (1988)
46. D.V. Boger: J. Non-Newt. Fluid Mech. **3**, 87 (1977)
47. C.W. Butler, M.B. Bush: Rheol. Acta **28**, 294 (1989)
48. K.P. Jackson, K. Walters, R.W. Williams: J. Non-Newt. Fluid Mech. **14**, 173 (1984)
49. S. Webb: Rev. Geophys. **35**, 191 (1997)
50. P. Coussot: J. Hydr. Res. **32**, 535 (1994)
51. X. Huang, M.H. Garcia: J. Fluid Mech. **374**, 305 (1998)
52. R.M. Iverson, M.E. Reid, R.G. LaHusen: Ann. Rev. Earth Planet Sci. **25**, 85 (1997)

3 Introduction to Rheology and Application to Geophysics

C. Ancey

Cemagref, unité Erosion Torrentielle, Neige et Avalanches, Domaine Universitaire, 38402 Saint-Martin-d'Hères Cedex, France

3.1 Introduction

This chapter gives an overview of the major current issues in rheology through a series of different problems of particular relevance to geophysics. For each topic considered here, we will outline the key elements and point the reader to ward the most helpful references and authoritative works. The reader is also referred to available books introducing rheology [1,2] for a more complete presentation and to the tutorial written by Middleton and Wilcock on mechanical and rheological applications in geophysics [3]. This chapter will focus on materials encountered by geophysicists (mud, snow, magma, etc.), although in most cases we will consider only suspensions of particles within an interstitial fluid without loss of generality. Other complex fluids such as polymeric liquids are rarely encountered in geophysics.

The mere description of what the term rheology embraces in terms of scientific areas is not easy. Roughly speaking, rheology distinguishes different areas and offshoots such as the following:

- *Rheometry.* The term "rheometry" is usually used to refer to a group of experimental techniques for investigating the rheological behavior of materials. It is of great importance in determining the constitutive equation of a fluid or in assessing the relevance of any proposed constitutive law. Most of the textbooks on rheology deal with rheometry. The books by Coleman, Markovitz, and Noll [4], Walters [5] and by Bird, Armstrong, and Hassager [6] provide a complete introduction to the viscometric theory used in rheometry for inferring the constitutive equation. Coussot and Ancey's book [7] gives practical information concerning rheometrical measurements with natural fluids. Though primarily devoted to food processing engineering, Steffe's book presents a detailed description of rheological measurements; a free sample is available on the web [8]. In Sect. 3.2, we will review the different techniques that are suitable to studying natural fluids. Emphasis is given both to describing the methods and the major experimental problems encountered with natural fluids.
- *Continuum mechanics.* The formulation of constitutive equations is probably the early goal of rheology. At the beginning of the 20th century, the non-Newtonian character of many fluids of practical interest motivated Professor Bingham to coin the term *rheology* and to define it as the study of the

deformation and flow of matter. The development of a convenient mathematical framework occupied the attention of rheologists for a long time after the Second World War. At that time, theoreticians such as Coleman, Markovitz, Noll, Oldroyd, Reiner, Toupin, Truesdell, etc. sought to express rheological behavior through equations relating suitable variables and parameters representing the deformation and stress states. This gave rise to a large number of studies on the foundations of continuum mechanics [6]. Nowadays the work of these pioneers is pursued through the examination of new problems such as the treatment of multiphase systems or the development of nonlocal field theories. For examples of current developments and applications to geophysics, the reader may consult papers by Hutter and coworkers on the thermodynamically consistent continuum treatment of soil–water systems [9,10], the book by Vardoulakis and Sulem on soil failure [11], and Bedford and Dumheller's review on suspensions [12]. A cursory glance at the literature on theoretical rheology may give the reader the impression that all this literature is merely an overly sophisticated mathematical description of the matter with little practical interest. In fact, excessive refinements in the tensorial expression of constitutive equations lead to prohibitive detail and thus substantially limit their utility or predictive capabilities. This probably explains why there is currently little research on this topic. Such limitations should not prevent the reader (and especially the newcomer) from studying the textbooks in theoretical rheology, notably to acquire the basic principles involved in formulating constitutive equations. Two simple problems related to these principles will be presented in Sect. 3.3 to illustrate the importance of an appropriate tensorial formulation of constitutive equations.

- *Rheophysics.* For many complex fluids of practical importance, bulk behavior is not easily outlined using a continuum approach. It may be useful to first examine what happens at a microscopic scale and then infer the bulk properties using an appropriate averaging process. Kinetic theories give a common example for gases [13] or polymeric liquids [6], which infer the constitutive equations by averaging all the pair interactions between particles. Such an approach is called *microrheology* or *rheophysics.* Here we prefer to use the latter term to emphasize that the formulation of constitutive equations is guided by a physical understanding of the origins of bulk behavior. Recent developments in geophysics are based on using kinetic theories to model bed load transport [14], floating broken ice fields [15], and rockfall and granular debris flows [16]. It is implicitly recognized that thoroughly modeling the microstructure would require prohibitive detail, especially for natural fluids. It follows that a compromise is generally sought between capturing the detailed physics at the particle level and providing applicable constitutive equations. Using dimensionless groups and approximating constitutive equations are commonly used operations for that purpose. In Sect. 3.4, we will consider suspensions of rigid particles within a Newtonian fluid to exemplify the different tools used in rheophysics. Typical examples of such fluids in a geophysical context include magma and mud. Chapters 4 and 14 provide further examples of rheophysical treatments with granular flows.

Other aspects of rheology, such as complex flow modeling and computational rheology, are not addressed in this introductory chapter. Chapter 2 in this book introduces the reader to the main rheological properties (viscoplasticity, time-dependent behaviour, etc.) encountered in geophysics. The reader is referred to examples of application to geophysical problems that are given in other chapters, notably Chap. 7 for lava flows, Chap. 13 for snow avalanches, Chaps. 22 and 21 for mud and debris flows.

3.2 Rheometry

At the very beginning, the term *rheometry* referred to a set of standard techniques for measuring shear viscosity. Then, with the rapid increase of interest in non-Newtonian fluids, other techniques for measuring the normal stresses and the elongational viscosity were developed. Nowadays, rheometry is usually understood as the area encompassing any technique which involves measuring mechanical or rheological properties of a material. This includes visualization techniques (such as photoelasticimetry for displaying stress distribution within a sheared material) or nonstandard methods (such as the slump test for evaluating the yield stress of a viscoplastic material). In most cases for applications in geophysics, shear viscosity is the primary variable characterizing the behavior of a fluid. Thus in the following, we will mainly address this issue, leaving aside all the problems related to the measurement of elongational viscosity. Likewise, the description of the most relevant procedures in rheometric measurement is not addressed here. We will first begin by outlining the main geometries used in rheometry. The principles underlying the viscometric treatment will be exposed in a simple case (flow down an inclined plane). Then, we will examine the most common problems encountered in rheometry. We will finish this section by providing a few examples of rheometric measurements, which can be obtained without a laboratory rheometer.

3.2.1 Standard viscometers

The basic principle of rheometry is to perform simple experiments where the flow characteristics such as the shear stress distribution and the velocity profile are known in advance and can be imposed. Under these conditions, it is possible to infer the *flow curve*, that is, the variation of the shear stress as a function of the shear rate, from measurements of flow quantities such as torque and the rotational velocity for a rotational viscometer. In fact, despite its apparent simplicity, putting this principle into practice for natural fluids raises many issues that we will discuss below. Most rheometers rely on the achievement of curvilinear (viscometric) flow [4]. The simplest curvilinear flow is the simple shear flow achieved by shearing a fluid between two plates in a way similar to Newton's experiment depicted in Sect. 3.3. But, in practice many problems (fluid recirculation, end effect, etc.) arise, which preclude using such a shearing box to obtain accurate measurements. Another simple configuration consists of an inclined plane or channel. To exemplify the viscometric approach, we will show how some flow properties such as the *discharge equation* (variation of the fluid discharge as a function of the flow depth) can be used to infer the constitutive equation characteristics. We consider a gravity-driven free-surface flow in a steady uniform regime down an inclined channel. The plane is tilted at an inclination θ to the horizontal. We use the Cartesian coordinate system of origin 0 and of basis $\boldsymbol{e}_x$, $\boldsymbol{e}_y$, $\boldsymbol{e}_z$, as depicted in Fig. 3.1.

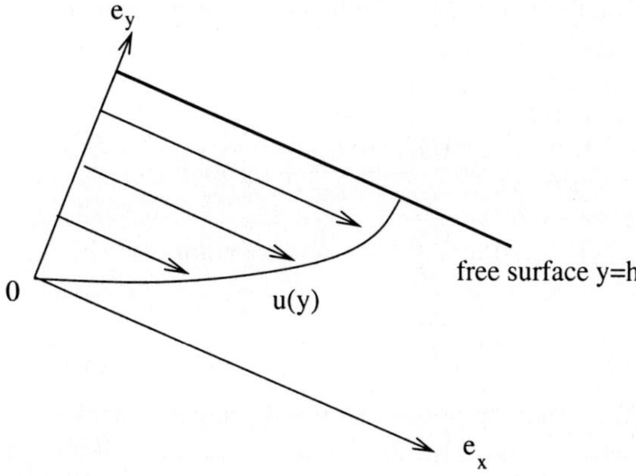

Fig. 3.1. Definition sketch for steady uniform flow

The velocity field $\boldsymbol{u}$ only depends on the coordinate y and takes the following form: $u_x = u(y)$, $u_y = 0$, and $u_z = 0$, where u is a function of y to be determined. Accordingly, the strain-rate tensor $\dot{\boldsymbol{\gamma}} = (\nabla \boldsymbol{u} + {}^t\nabla \boldsymbol{u})/2$ has the following components in the coordinate system:

$$\dot{\boldsymbol{\gamma}} = \frac{\dot{\gamma}}{2} \begin{bmatrix} 0 & 1 & 0 \\ 1 & 0 & 0 \\ 0 & 0 & 0 \end{bmatrix} , \tag{3.1}$$

where the shear rate $\dot{\gamma}$ is defined as a function of the coordinate y and implicitly of the inclination θ: $\dot{\gamma}(y) = (\partial u / \partial y)_\theta$. The momentum balance can be written as:

$$\varrho \frac{\mathrm{d}\boldsymbol{u}}{\mathrm{d}t} = \varrho \boldsymbol{g} + \nabla . \boldsymbol{\sigma} , \tag{3.2}$$

where ϱ and $\boldsymbol{g}$ respectively denote the local material density and gravitational acceleration. We assume that there is no slip at the bottom: $u(y) = 0$. Furthermore, we assume that there is no interaction between the free surface and the ambient fluid above except the pressure exerted by the ambient fluid. Notably, we ignore surface tension effects on the free surface. Without restriction, the stress tensor can be written as the sum of a pressure term p and a deviatoric term called the extra-stress tensor $\boldsymbol{s}$ (see also Sect. 3.3) [2,4]: $\boldsymbol{\sigma} = -p\mathbf{1} + \boldsymbol{s}$. For a homogeneous and isotropic simple fluid, the extra-stress tensor depends on the strain rate only: $\boldsymbol{s} = G(\dot{\boldsymbol{\gamma}})$, where G is a tensor-valued isotropic functional. In the present case, it is straightforward to show that the stress tensor must have the form:

$$\boldsymbol{\sigma} = -p\mathbf{1} + \begin{bmatrix} s_{xx} & s_{xy} & 0 \\ s_{xy} & s_{yy} & 0 \\ 0 & 0 & s_{zz} \end{bmatrix} . \tag{3.3}$$

Thus, the stress tensor is fully characterized by three functions: the shear stress $\tau = \sigma_{xy} = s_{xy}$, and the normal stress differences: $N_1 = s_{xx} - s_{yy}$ and $N_2 = s_{yy} - s_{zz}$, called the first and second normal stress differences, respectively. Since for steady flows acceleration vanishes and the components of $\boldsymbol{s}$ only depend on y, the equations of motion (3.2) reduce to:

$$0 = \frac{\partial s_{xy}}{\partial y} - \frac{\partial p}{\partial x} + \varrho g \sin\theta , \tag{3.4}$$

$$0 = \frac{\partial s_{yy}}{\partial y} - \frac{\partial p}{\partial y} - \varrho g \cos\theta , \tag{3.5}$$

$$0 = \frac{\partial p}{\partial z} . \tag{3.6}$$

It follows from (3.6) that the pressure p is independent of z. Accordingly, integrating (3.5) between y and h implies that p must be written: $p(x,y) - p(x,h) = s_{yy}(y) - s_{yy}(h) + \varrho g(h-y)\cos\theta$. It is possible to express (3.4) in the following form:

$$\frac{\partial}{\partial y}\left(s_{xy} + \varrho\, g\, y \sin\theta\right) = \frac{\partial p(x,h)}{\partial x} . \tag{3.7}$$

Equation (3.7) has a solution only if both terms of this equation are equal to a function of z, which we denote $b(z)$. Moreover, (3.6) implies that $b(z)$ is actually independent of z; thus, in the following we will note: $b(z) = b$. The solutions to (3.7) are: $p\,(x,h) = bx + c$ and $s_{xy}(h) - s_{xy}(y) - \varrho\, g\, y \sin\theta = b(h-y)$, where c is a constant, which we will determine. To that end, let us consider the free surface. It is reasonable and usual to assume that the ambient fluid friction is negligible. The stress continuity at the interface implies that the ambient fluid pressure p_0 exerted on an elementary surface at $y = h$ (oriented by $\boldsymbol{e}_y$) must equal the stress exerted by the fluid. Henceforth, the boundary conditions at the free surface may be expressed as: $-p_0\boldsymbol{e}_y = \boldsymbol{\sigma}\boldsymbol{e}_y$, which implies in turn that: $s_{xy}(h) = 0$ and $p_0 = p(x,h) - s_{yy}(h)$. Comparing these equations to former forms leads to $b = 0$ and $c = p_0 + s_{yy}(h)$. Accordingly, we obtain for the shear and normal stress distributions:

$$\tau = \varrho g(h-y)\sin\theta , \tag{3.8}$$

$$\sigma_{yy} = s_{yy} - (p - p_0) = -\varrho g(h-y)\cos\theta . \tag{3.9}$$

The shear and normal stress profiles are determined regardless of the form of the constitutive equation. For simple fluids, the shear stress is a one-to-one function of the shear rate: $\tau = f(\dot{\gamma})$. Using the shear stress distribution (3.8) and the inverse function f^{-1}, we find: $\dot{\gamma} = f^{-1}(\tau)$. A double integration leads to the flow rate (per unit width):

$$q = \int_0^h \mathrm{d}y \int_0^y f^{-1}(\tau(\xi))\,\mathrm{d}\xi \,. \tag{3.10}$$

Taking the partial derivative of q with respect to h, we obtain:

$$\dot{\gamma} = f^{-1}(\tau(h)) = \frac{1}{h}\left(\frac{\partial q}{\partial h}\right)_\theta \,. \tag{3.11}$$

This relation allows us to directly use a channel as a rheometer. The other normal components of the stress tensor cannot be easily measured. The curvature of the free surface of a channeled flow may give some indication of the first normal stress difference. Let us imagine the case where it is not equal to zero. Substituting the normal component s_{yy} by $s_{yy} = s_{xx} - N_1$ in (3.5), then integrating, we find:

$$s_{xx} = p + \varrho g y \cos\theta + N_1 + d \,, \tag{3.12}$$

where d is a constant. Imagine that a flow section is isolated from the rest of the flow and the adjacent parts are removed. In order to hold the free surface flat (it will be given by the equation $y = h$, $\forall z$), the normal component σ_{xx} must vary and balance the variations of N_1 due to the presence of the sidewalls (for a given depth, the shear rate is higher in the vicinity of the wall than in the center). But at the free surface, the boundary condition forces the normal stress σ_{xx} to vanish and the free surface to bulge out. To first order, the free surface equation is:

$$-\varrho g y \cos\theta = N_1 + d + \mathcal{O}(y) \,. \tag{3.13}$$

If the first normal stress difference vanishes, the boundary condition $-p_0\boldsymbol{e}_y = \boldsymbol{\sigma}\boldsymbol{e}_y$ is automatically satisfied and the free surface is flat. In the case where the first normal stress difference does not depend on the shear rate, there is no curvature of the shear free surface. The observation of the free surface may be seen as a practical test to examine the existence and sign of the first normal stress difference and to quantify it by measuring both the velocity profile at the free surface and the free-surface equation. Computation of the shear-stress function and normal stress differences is very similar for other types of viscometers. Figure 3.2 reports the corresponding functions for the most common viscometers. All these techniques are robust and provide accurate measurements for classic fluids, with uncertainty usually less than 2%. For geophysical fluids, many problems of various types may arise.

First, the viscometric treatment relies on the crucial assumption that the extra-stress tensor is a one-to-one function of the strain-rate tensor only (class of simple fluids). Many classes of material studied in geophysics are not in fact homogeneous, isotropic, or merely expressible in the form $\boldsymbol{\sigma} = -p\mathbf{1} + \boldsymbol{s}(\dot{\boldsymbol{\gamma}})$. For instance, for materials with time-dependent properties (thixotropic materials, viscoelastic materials), the constitutive equation can be expressed in the form $\boldsymbol{\sigma} = -p\mathbf{1} + \boldsymbol{s}(\dot{\boldsymbol{\gamma}})$ only for a steady state. Another example is provided by granular

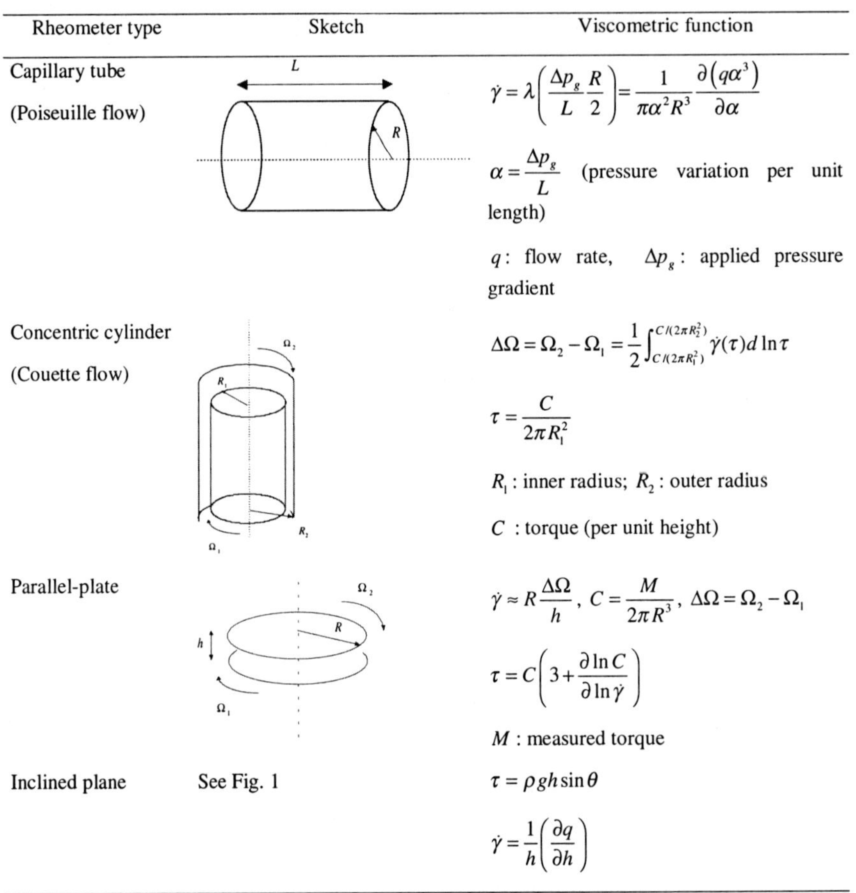

Rheometer type	Sketch	Viscometric function
Capillary tube (Poiseuille flow)		$\dot{\gamma}=\lambda\left(\frac{\Delta p_g}{L}\frac{R}{2}\right)=\frac{1}{\pi\alpha^2R^3}\frac{\partial\left(q\alpha^3\right)}{\partial\alpha}$ $\alpha=\frac{\Delta p_g}{L}$ (pressure variation per unit length) q: flow rate, Δp_g: applied pressure gradient
Concentric cylinder (Couette flow)		$\Delta\Omega=\Omega_2-\Omega_1=\frac{1}{2}\int_{C/(2\pi R_1^2)}^{C/(2\pi R_2^2)}\dot{\gamma}(\tau)d\ln\tau$ $\tau=\frac{C}{2\pi R_1^2}$ R_1 : inner radius; R_2 : outer radius C : torque (per unit height)
Parallel-plate		$\dot{\gamma}\approx R\frac{\Delta\Omega}{h}$, $C=\frac{M}{2\pi R^3}$, $\Delta\Omega=\Omega_2-\Omega_1$ $\tau=C\left(3+\frac{\partial\ln C}{\partial\ln\dot{\gamma}}\right)$ M : measured torque
Inclined plane	See Fig. 1	$\tau=\rho gh\sin\theta$ $\dot{\gamma}=\frac{1}{h}\left(\frac{\partial q}{\partial h}\right)$

Fig. 3.2. Chief geometries used in rheometry

flows. In this case, when applied to experimental data obtained by studying dry granular flows down an inclined channel [17], the viscometric treatment leads to the conclusion that the flow curve should be a decreasing function of the shear rate in violation of a stability criterion imposing that the flow curve be an increasing function. Although such a decrease in the flow curve cannot be directly interpreted in terms of a constitutive equation, it provides interesting rheological information that can be explained on the basis of microstructural theories [18].

Second, for most viscometers, computing the shear rate from experimental data can raise serious problems. A major source of uncertainty is that in most viscometric procedures the shear rate is expressed as a derivative – for instance $\partial q/\partial h$ in (3.11) – which must be estimated from experimental data. To do so, different procedures are available but they do not always provide the same results, especially when data are noisy [19]. A typical example of these problems is given by the concentric-cylinder rheometer (or Couette rheometer). The shear rate is inferred from the rotational velocity Ω and the torque (per unit depth) C using the following relationship:

$$\Omega = -\frac{1}{2}\int_{C/(2\pi R_1^2)}^{C/(2\pi R_2^2)} \dot{\gamma}(\tau)\,\mathrm{d}(\ln\tau)\ . \tag{3.14}$$

When the gap between the two cylinders is narrow, it is possible to approximate the shear rate as: $\dot{\gamma} = R_1\Omega/(R_2 - R_1) + \mathcal{O}(1 - R_2/R_1)$. However, such a geometry is not very suitable to studying natural fluids (slipping, size effects, etc.) and usually a large gap is preferred. For large gaps, one of the most common approximations is attributed to Krieger who proposed for Newtonian and power-law fluids [20,21]:

$$\dot{\gamma} = \frac{2\Omega(1+\alpha)}{1-\beta^f} f \tag{3.15}$$

with $f = \mathrm{d}\ln\Omega/\mathrm{d}\ln C$; $\alpha = f'f^{-2}\chi_1(-f\log\beta)$; $\chi_1(x) = x(x\mathrm{e}^x - 2\mathrm{e}^x + x + 2)(\mathrm{e}^x - 1)^{-2}/2$, $\beta = (R_1/R_2)^2$. However, this method can give poor results with yield stress fluids, especially if it is partially sheared within the gap. In this case, Nguyen and Boger [22] have proposed using $\dot{\gamma} = 2\Omega\,\mathrm{d}\ln\Omega/\mathrm{d}\ln C$. In their treatment of debris suspensions, Coussot and Piau [23] used an alternative consisting of an expansion into a power series of (3.15). They obtained: $\dot{\gamma} = 2\Omega\sum_{n=0}^{\infty} f\left(\beta^n C/(2\pi R_1^2)\right)$. For methods of this kind, computing the shear rate requires specifying the type of constitutive equation in advance. Furthermore, depending on the procedure chosen, uncertainty on the final results may be as high as 20% or more for natural fluids. Recently, a more effective and practical method of solving the inverse problem has been proposed [24,25]: the procedure based on Tikhonov regularization does not require the algebraic form of the $\tau - \dot{\gamma}$ curve to be prespecified and has the advantage of filtering out noise. The only viscometer that poses no problem in converting experimental data into a $\tau - \dot{\gamma}$ curve is the parallel-plate rheometer. In this case, the shear rate distribution is imposed by the operator: $\dot{\gamma} = \Omega R/h$. But such a relationship holds provided centrifugal forces are negligible compared to the second normal stress difference: $\varrho R^2 w^2 \ll N_2$, where w is the orthoradial component of the velocity. Such an effect can be detected experimentally either by observing secondary flows or by noticing that doubling both the gap and the rotational velocity (thus keeping the shear rate constant) produces a significant variation in the measured torque.

Third, any rheometer is subjected to end effects, which have to be corrected or taken into account in the computation of the flow curve. For instance, end effects in a channel are due to the finite length of the channel as well as the sidewalls, both producing potentially significant variations in the flow depth. Likewise, in a Couette rheometer, the measured torque includes a contribution due to the shearing over the bottom surface of the bob. Such a contribution is substantially reduced using a bob with a hole hollowed on the bottom surface so that air is trapped when the bob is immersed in the fluid. But this can be inefficient for natural fluids, such as debris suspensions, and in this case, the bottom contribution to the resulting torque must be directly assessed using the method proposed by Barnes and Carnali [26]. For a parallel-plate rheometer, the fluid surface at the peripheral free surface may bulge out or creep, inducing a significant variation in the measured torque, possibly varying with time. Furthermore, many natural fluids encountered in geophysics are suspensions with a large size distribution. The size of the rheometer should be determined such that its typical size (e.g. the gap in a rotational viscometer) is much larger than the largest particle size. For instance, for debris flows, this involves using large-sized rheometers [23,27].

Last, many disturbing effects may arise. They often reflect the influence of the microstructure. For instance, for a particle suspension, especially made up of nonbuoyant particles, sedimentation and migration of particles can significantly alter the stress distribution and thus the measured torque. Likewise, for concentrated pastes, a fracture inside the sheared sample may sometimes be observed, usually resulting from a localization of shear within a thin layer. Other disturbing effects are experimental problems pertaining to the rheometer type. For instance, when using a rotational viscometer with a smooth metallic shearing surface, wall slip can occur. Apart from effects resulting from microstructural changes, which are a part of the problem to study, it is sometimes possible to reduce disturbing effects or to account for them in the flow-curve computation. For instance, to limit wall slip, the shearing surfaces can be roughened. Another strategy involves measuring the slipping velocity directly and then computing an effective shear rate. Still another possibility requires using the same rheometer with different sizes, as first proposed by Mooney for the capillary rheometer.

All the above issues show that, for complex fluids (the general case for natural fluids studied in geophysics), rheometry is far from being an ensemble of simple and ready-for-use techniques. On the contrary, investigating the rheological properties of a natural material generally requires many trials using different rheometers and procedures. In some cases, visualization techniques (such as nuclear magnetic resonance imagery, transparent interstitial fluid and tools, birefringence techniques) may be helpful to monitor microstructure changes. Most of the commercialized rheometers are now controlled by a PC-type computer, both controlling the measurements and providing automatic procedures for computing the flow curve. Such procedures should be reserved for materials whose rheological behavior is well known, and consequently are of limited interest for natural fluids.

3.2.2 What Can Be Done Without a Rheometer?

In the laboratory, it is frequently impossible to investigate the rheological properties of a natural fluid using a rheometer. For instance, with snow or magma, such tests are almost always impractical. For debris suspensions, it is usually impossible to carry out measurements with the complete range of particle size. This has motivated researchers to developed approximate rheometric procedures and to investigate the relations between field observations and rheological properties. For instance, given the sole objective of determining the yield stress, the semiempirical method referred to as a *slump test* can provide an estimate of the yield stress for a viscoplastic material. This method involves filling a cylinder with the material to be tested, lifting the cylinder off and allowing the material to flow under its own weight. The profile of the final mound of material as well as the difference δ between the initial and final heights is linked to the yield stress. Pashias and Boger [28] have found:

$$\frac{\delta}{h} = 1 - 2\frac{\tau_c}{\varrho g h}\left[1 - \ln\left(2\frac{\tau_c}{\varrho g h}\right)\right] , \tag{3.16}$$

where h is the cylinder height, ϱ the material density. Close examination of experimental data published by Pashias and Boger shows a deviation from the theoretical curve for yield stress values in excess of approximately $0.15\varrho g h$. For yield stress values lower than $0.15\varrho g h$ (or for $\delta/h > 0.4$), uncertainty was less than 10% for their tests. The explanation of the deviation for higher yield stress values lies perhaps in the weakness of the assumption on the elastoplastic behavior for very cohesive materials. Coussot, Proust and Ancey [29] developed an alternative approach based on an interpretation of the deposit shape. They showed that the free surface profile (the relationship between the material height y and the distance from the edge x) depends on the yield stress only. On a flat horizontal surface, the free surface profile has the following expression:

$$\frac{\varrho g y}{\tau_c} = \sqrt{2\frac{\varrho g x}{\tau_c}} . \tag{3.17}$$

Comparisons between rheological data deduced from a parallel plate rheometer and free surface profile measurements showed an acceptable agreement for fine mud suspensions and debris flow materials. Uncertainty was less than 20%, within the boundaries of acceptable uncertainty for rheometrical measurement. The major restriction in the use of (3.17) stems from the long-wave approximation, which implies that the mound height must far outweigh the extension of the deposit: $h - \delta \gg \tau_c/(\varrho g)$. The method proposed by Coussot et al. [29] can be extended to different rheologies and boundary conditions. In the field, such a method applied to levee profiles of debris flow can provide estimates of the bulk yield stress provided that the assumption of viscoplastic behavior holds.

Observing and interpreting natural deposits may provide interesting information either on the flow conditions or rheological features of the materials involved [30]. For instance, laboratory experiments performed by Pouliquen with granular flows have shown that the flow features (e.g. the mean velocity) of a dry granular free-surface unconfined flow can be related to the final thickness of the deposit [31]. Although fully developed in the laboratory, such a method should be applicable to natural events involving granular flows. More evidence of the interplay between the deposit shape, the flow conditions, and the rheological features is given by the height difference of two lateral levees deposited by a debris flow in a bending track [32].

3.3 The Contribution of Continuum Mechanics

In 1687, Isaac Newton proposed that "the resistance which arises from the lack of slipperiness of the parts of the liquid, other things being equal, is proportional to the velocity with which the parts of the liquid are separated from one another" [33]. This forms the basic statement behind the theory of Newtonian fluid mechanics. Translated into modern scientific terms, this sentence means that the resistance to flow (per unit area) τ is proportional to the velocity gradient U/h:

$$\tau = \mu \frac{U}{h}, \tag{3.18}$$

where U is the relative velocity with which the upper plate moves and h is the thickness of fluid separating the two plates (see Fig. 3.3). μ is a coefficient intrinsic to the material, which is termed *viscosity*. This relationship is of great practical importance for many reasons. It is the simplest way of expressing the constitutive equation for a fluid (linear behavior) and it provides a convenient experimental method for measuring the constitutive parameter μ by measuring the shear stress exerted by the fluid on the upper plate moving with a velocity U (or conversely by measuring the velocity when a given tangential force is applied to the upper plate).

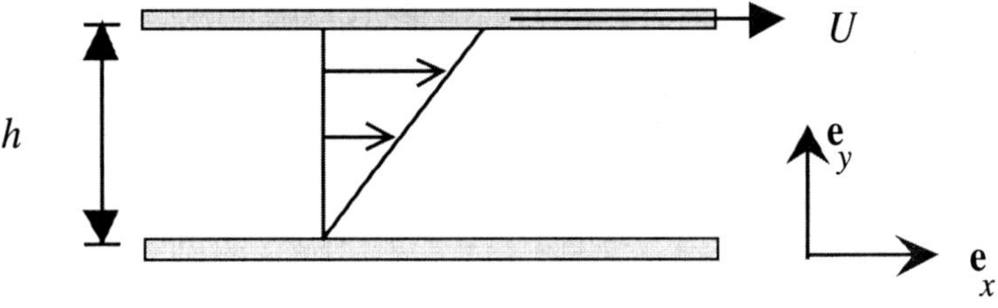

Fig. 3.3. Illustration of a fluid sheared by a moving upper plate

In 1904, Trouton did experiments on mineral pitch involving stretching the fluid with a given velocity [34]. Figure 3.4 depicts the principle of this experiment. The fluid undergoes a uniaxial elongation achieved with a constant elongation rate $\dot{\alpha}$, defined as the relative deformation rate: $\dot{\alpha} = \dot{l}/l$, where l is the fluid sample length. For his experiments, Trouton found a linear relationship between the applied force per unit area σ and the elongation rate:

$$\sigma = \mu_e \alpha = \mu_e \frac{1}{l} \frac{\mathrm{d}l}{\mathrm{d}t}. \tag{3.19}$$

This relationship was structurally very similar to the one proposed by Newton but it introduced a new material parameter, which is now called *Trouton viscosity*. This constitutive parameter was found to be three times greater than the Newtonian viscosity inferred from steady simple-shear experiments: $\mu_e = 3\mu$.

At first glance, this result is both comforting since behavior is still linear (the resulting stress varies linearly with the applied strain rate) and disturbing since the value of the linearity coefficient depends on the type of experiment. In fact, Trouton's result does not lead to a paradox if we are careful to express the constitutive parameter in a tensorial form rather than a purely scalar form.

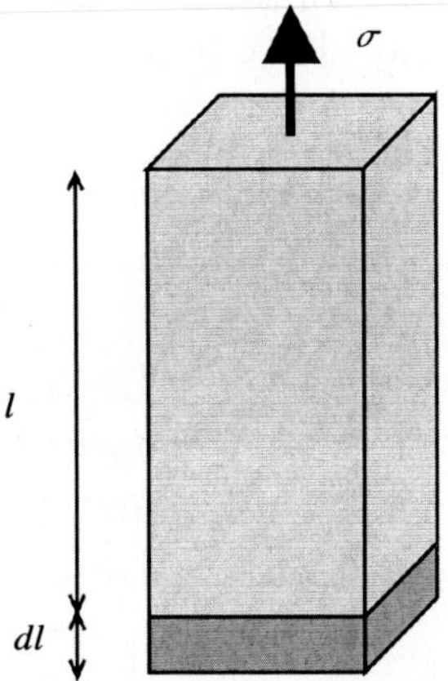

Fig. 3.4. Typical deformation of a material experiencing a normal stress σ

This was achieved by Navier and Stokes, who independently developed a consistent three-dimensional theory for Newtonian viscous fluids. For a simple fluid, the stress tensor $\boldsymbol{\sigma}$ can be cast in the following form:

$$\boldsymbol{\sigma} = -p\mathbf{1} + \boldsymbol{s} \tag{3.20}$$

where p is called the *fluid pressure* and $\boldsymbol{s}$ is the extra-stress tensor representing the stresses resulting from a relative motion within the fluid. It is also called the *deviatoric stress tensor* since it represents the departure from equilibrium. The pressure p is defined as (minus) the average of the three normal stresses $p = -\mathrm{tr}\,\boldsymbol{\sigma}/3$. This also implies that $\mathrm{tr}\,\boldsymbol{s} = 0$. The pressure used in (3.20) is analogous to the static fluid-pressure in the sense that it is a measure of the local intensity of the squeezing of the fluid. Contrary to the situation for fluids at rest, the connection between this purely mechanical definition and the term pressure used in thermodynamics is not simple. For a Newtonian viscous fluid, the Navier–Stokes equation postulates that the extra-stress tensor is linearly linked to the strain rate tensor $\dot{\boldsymbol{\gamma}} = (\nabla\boldsymbol{u} + {}^t\nabla\boldsymbol{u})/2$ (where $\boldsymbol{u}$ is the local fluid velocity):

$$\boldsymbol{s} = 2\eta\dot{\boldsymbol{\gamma}} \tag{3.21}$$

where η is called the Newtonian viscosity. It is worth noticing that the constitutive equation is expressed as a relationship between the extra-stress tensor and the local properties of the fluid, which are assumed to depend only on the instantaneous distribution of velocity (more precisely, on the departure from uniformity of that distribution). There are many arguments from continuum mechanics and analysis of molecular transport of momentum in fluids, which show that the local velocity gradient $\nabla \boldsymbol{u}$ is the parameter of the flow field with most relevance to the deviatoric stress (see [37]). On the contrary, the pressure is not a constitutive parameter of the moving fluid. When the fluid is compressible, the pressure p can be inferred from the free energy, but it is indeterminate for incompressible Newtonian fluids. If we return to the previous experiments, we infer from the momentum equation that the velocity field is linear : $\boldsymbol{u} = U\boldsymbol{e}_x y/h$. We easily infer that the shear rate is: $\dot{\gamma} = \partial u/\partial y = U/h$ and then comparing (3.21) to (3.18) leads to: $\eta = \mu$. Thus, the Newtonian viscosity corresponds to the simple shear viscosity. In the case of a uniaxial elongation, the components of the strain-rate tensor are:

$$\dot{\boldsymbol{\gamma}} = \begin{bmatrix} \dot{\alpha} & 0 & 0 \\ 0 & -\dot{\alpha}/2 & 0 \\ 0 & 0 & -\dot{\alpha}/2 \end{bmatrix} . \tag{3.22}$$

At the same time, the stress tensor can be written as:

$$\boldsymbol{\sigma} = \begin{bmatrix} \sigma & 0 & 0 \\ 0 & 0 & 0 \\ 0 & 0 & 0 \end{bmatrix} . \tag{3.23}$$

Comparing (3.20), (3.22), and (3.23) leads to: $p = -\eta\dot{\alpha}$ and $\sigma = 3\eta\dot{\alpha}$, that is: $\mu_e = 3\eta$, confirming that the Trouton elongational viscosity is three times greater than the viscosity. It turns out that Trouton's and Newton's experiments reflect the same constitutive behavior. This example shows the importance of an appropriate tensorial form for expressing the stress tensor. In the present case, the tensorial form (3.21) may be seen as a simple generalization of the simple shear expression (3.18).

In many cases, most of the available information on the rheological behavior of a material is inferred from simple shear experiments (see Sect. 3.2). But, contrary to the Newtonian (linear) case, the tensorial form cannot be merely and easily generalized from the scalar expression fitted to experimental data. First, building a three-dimensional expression of the stress tensor involves respecting a certain number of formulation principles. These principles simply express the idea that the material properties of a fluid should be independent of the observer or frame of reference (principle of material objectivity) and the behavior of a material element depends only on the previous history of that element and not on the state of neighboring elements [6]. Then it is often necessary to provide extra information or rules to build a convenient expression for the constitutive equation. To illustrate this, we shall consider a simple example: the Bingham equation (see also Chaps. 2 and 22). When a fluid exhibits viscoplastic properties, we usually fit experimental data with a Bingham equation as a first approximation [35,36,38]:

$$\dot{\gamma} > 0 \Rightarrow \tau = \tau_c + K\dot{\gamma} . \tag{3.24}$$

Equation (3.24) means that for shear stresses in excess of a critical value, called the *yield stress*, the shear stress is a linear function of the shear rate. Conversely when $\tau \leq \tau_c$ there is no shear within the fluid ($\dot{\gamma} = 0$). The question arises as to how the scalar expression can be transformed into a tensorial form. The usual but not the only way is to consider a process, called *plastic rule*, as the key process of yielding. A plastic rule includes two ingredients. First, it postulates the existence of a surface in the stress space $(\sigma_1, \sigma_2, \sigma_3)$ delimiting two possible mechanical states of a material element (σ_i denotes a principal stress, that is an eigenvalue of the stress tensor) as depicted in Fig. 3.5. The surface is referred to as the *yield surface* and is usually represented by an equation in the form $f(\sigma_1, \sigma_2, \sigma_3) = 0$. When $f < 0$, behavior is generally assumed to be elastic or rigid. When $f = 0$, the material yields. Second it is assumed that, after yielding, the strain-rate is directly proportional to the surplus of stress, that is, the distance between the point the representing the stress state and the yield surface. Translated into mathematical terms, this leads to write: $\dot{\gamma} = \lambda \nabla f$ with λ a proportionality coefficient (Lagrangian multiplier).

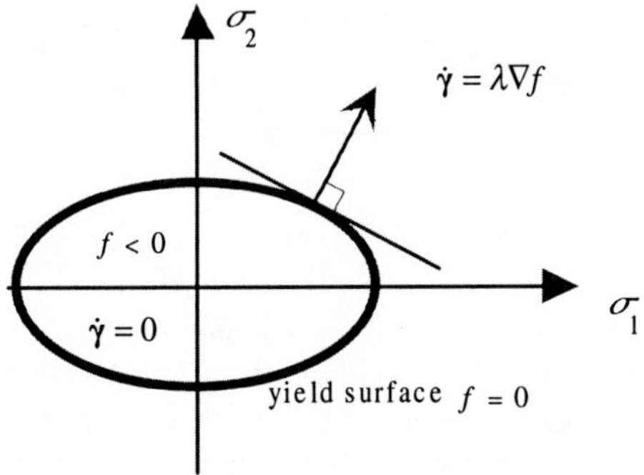

Fig. 3.5. Yield surface delimiting two domains

How must the yield function f be built to satisfy the principle of material objectivity? For f to be independent of the frame, it must be expressed not as a function of the components of the stress but as a function of its invariants. An *invariant* is a quantity that does not depend on the frame in which it is expressed. For instance, it is well known that the determinant of a tensor is an invariant. In contrast with tensor invariants used in mathematics without physical meaning, it is usual in mechanics to use specific forms for the invariants of the stress tensor: they are defined in such a way that they can be used as the coordinates of the point representing the stress state M in the stress space (see Fig. 3.6). The first invariant $I_1 = \operatorname{tr} \boldsymbol{\sigma} = \sigma_1 + \sigma_2 + \sigma_3$ represents the *mean stress* multiplied by 3 ($|\boldsymbol{OP}| = I_1/3$ in Fig. 3.6), the second invariant $I_2 = (\operatorname{tr}^2\boldsymbol{\sigma} - \operatorname{tr}\boldsymbol{\sigma}^2)/2 = -\operatorname{tr}(\boldsymbol{s}^2)/2$ can be interpreted as the deviation of a stress state from the mean stress state ($|\boldsymbol{PM}|^2 = -2I_2$ in Fig. 3.6) and is accordingly called the *stress deviator*. The third invariant $I_3 = -\operatorname{tr} \boldsymbol{s}^3/6$ reflects the angle in the deviatoric plane made by the direction $\boldsymbol{PM}$ with respect to the projection of σ_1-axis and is sometimes called the *phase* ($\cos^2 3\varphi = I_3^2/I_2^3$ in Fig. 3.6).

If the material is an isotropic and homogenous fluid, the yield function f is expected to be independent of the mean pressure and the third invariant (for reasons analogous to those given above for explaining the form of the constitutive equation). Thus we have $f(\sigma_1, \sigma_2, \sigma_3) = f(I_2)$. In plasticity, the simplest yield criterion is the von Mises criterion, asserting that yield occurs whenever the deviator exceeds a critical value (whose root gives the yield stress): $f(I_2) = \sqrt{-I_2} - \tau_c$. As depicted in Fig. 3.6, the resulting yield surface is a cylinder of radius τ_c centered around an axis $\sigma_1 = \sigma_2 = \sigma_3$. (If we draw the yield surface in the extra-stress space, we obtain a sphere of radius $\sqrt{2}\tau_c$.)

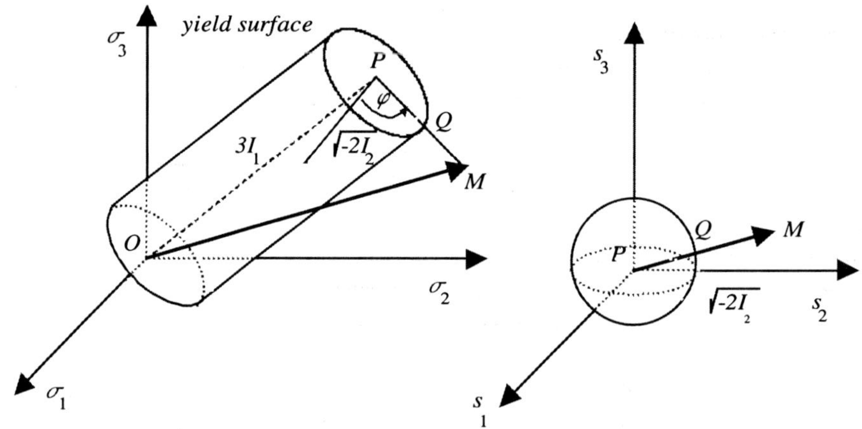

Fig. 3.6. On the left, the yield surface in the stress space when the von Mises criterion is selected as yield function. A stress state is characterized by its three principal stresses and thus can be reported in the stress space. The three invariants of the stress tensor can be interpreted in terms of coordinates

Once the stress state is outside the cylinder defined by the yield surface, a flow occurs within the material. As stated above, it is assumed that the strain rate is proportional to the surplus of stress. This leads to the expression:

$$\dot{\boldsymbol{\gamma}} = \lambda \partial f / \partial \boldsymbol{s} = \lambda \left(\sqrt{I_2} - \tau_c \right) \frac{\boldsymbol{s}}{\sqrt{I_2}} \, . \tag{3.25}$$

For convenience, we define the proportionality coefficient as: $\lambda^{-1} = 2\eta$. It is generally more usual to express the constitutive equation in the converse form $\boldsymbol{s}(\dot{\boldsymbol{\gamma}})$. To that end, we express the second invariant of the strain rate tensor J_2 as $J_2 = -\mathrm{tr}(\dot{\boldsymbol{\gamma}}^2)/2 = \left[\lambda\left(\sqrt{-I_2} - \tau_c\right)\right]^2$. Then we deduce:

$$\dot{\boldsymbol{\gamma}} = 0 \Leftrightarrow \sqrt{-I_2} \leq \tau_c \, , \tag{3.26}$$

$$\dot{\gamma} \neq 0 \Leftrightarrow \boldsymbol{\sigma} = -p\mathbf{1} + \left(2\eta + \frac{\tau_c}{\sqrt{-J_2}}\right)\dot{\boldsymbol{\gamma}}\,, \tag{3.27}$$

which is the usual form of the Bingham constitutive equation. It is worth noting that contrary to the Newtonian case, the general tensorial expression (3.26)–(3.27) cannot not easily be extrapolated from the steady simple-shear equation (3.24).

3.4 Rheophysics

The rheophysical approach seeks to derive the bulk properties by examining what may happen at the microscopic scale. Generally the bulk stress tensor is computed by averaging the local stresses. Accurate computation has been achieved in a certain number of simple cases. Kinetic theories for gases, polymers, and granular media (rapidly sheared) are typical examples. In most cases for fluids involved in geophysics, computations are so much more complex that analytical results cannot be provided. One can, however, benefit from this approach either by building approximate rheological models or by finding convenient scalings for the key variables describing bulk behavior. Typical examples include all the treatments focusing on the rheology of concentrated suspensions. To begin with, we will outline the principles used in deriving the bulk constitutive equations. This will lead to introducing important concepts such as the *pair distribution function*, the *averaging operator*, *particle interactions*, and *evolution equations.* We will examine these different notions through the example of Newtonian suspensions with no loss of generality since they can be encountered with a similar meaning in other theories such as the kinetic theories for granular flows [39]. Then we will examine how it is possible to simplify the constitutive equation to obtain approximate equations. The last subsection will demonstrate the advantages of dimensional analysis combined with a microstructural analysis of particle interactions in deriving appropriate scalings for experimental data and theoretical results.

3.4.1 Definition of the Bulk Stress Tensor and Selected Applications

One of the key questions in rheophysics is to establish the way in which bulk behavior can be deduced from the microstructure properties. For suspensions, this is generally achieved by averaging the local stress and particle interactions. As all the issues around the most appropriate averaging procedure are still being debated, here we restrict our attention to the approach followed by Batchelor and many subsequent authors. The reader interested in further information on averaging is referred to specific papers [41–48,40].

In the following, we consider a suspension of rigid spherical particles of radius a within an incompressible Newtonian fluid with viscosity η. Particles are assumed to be identical and neutrally buoyant. The solid fraction ϕ is defined as the ratio of the solid volume to the total volume. In a fundamental paper,

Batchelor showed that the bulk stress is the sum of a fluid contribution and a particle contribution [49]:

$$\bar{\boldsymbol{\sigma}} = \bar{\boldsymbol{\sigma}}^{(f)} + \bar{\boldsymbol{\sigma}}^{(p)} , \tag{3.28}$$

where the fluid part can be written as

$$\bar{\boldsymbol{\sigma}}^{(f)} = 2\eta\bar{\dot{\boldsymbol{\gamma}}} - \langle p_f \rangle \mathbf{1} - \varrho_f < \boldsymbol{u}' \otimes \boldsymbol{u}' > , \tag{3.29}$$

where $\bar{\dot{\boldsymbol{\gamma}}}$ denotes the averaged strain-rate tensor, $\langle p_f \rangle$ is the mean interstitial fluid pressure, ϱ_f is the fluid density, $\boldsymbol{u}'$ refers to velocity fluctuations, and $\otimes$ is the tensor product. We use brackets and the bar symbol to represent ensemble and volume-averaged quantities respectively. The ensemble average of a quantity $f(\boldsymbol{r}, t)$ at position $\boldsymbol{r}$ and time t, is computed by performing a large number of experiments ("realizations"), with the same macroscopic initial and boundary conditions, and measuring f at $\boldsymbol{r}$ at the same time relative to the beginning of each experiment. The average of these realizations forms the ensemble average. To do such a computation, we have to record the configuration C^N of N particles (specified by their positions, linear, and angular velocities) contained in a volume V. After calculating the probability $P(C^N, t)$ of observing a given configuration C^N at time t, we can define the ensemble average as $< f(\boldsymbol{r}, t) >= \int P(C^N, t) f(\boldsymbol{x}, t; C^N) \mathrm{d}C^N$. Such a definition is not very practical since it implies to specify the positions and velocities of all the particles contained in V. A strategy to bypass this difficulty is to focus on a single particle ("test particle") and examine how other particles are distributed with respect to this particle. This leads to introduce the *pair distribution function* P_2, which is the probability of finding a particle located at $\boldsymbol{y}$ when the centre of the test particle is simultaneously in $\boldsymbol{x}$. Formulated in mathematical terms, this leads to write the ensemble average of $f(\boldsymbol{r}, t)$ as:

$$< f(\boldsymbol{r}, t) >= \int_{C^2} P_2(t; \boldsymbol{x}, \boldsymbol{y}) f^{(2)}(\boldsymbol{x}, t; C^2) \mathrm{d}\boldsymbol{x} \mathrm{d}\boldsymbol{y} \approx \int_{C^2} P_2(t; \boldsymbol{x}, \boldsymbol{y}) f(\boldsymbol{x}, t) \mathrm{d}\boldsymbol{x} \mathrm{d}\boldsymbol{y} \tag{3.30}$$

where $f^{(2)}$ denotes the conditional averaged function when the position of two spheres is fixed. It is usually assumed that the conditional averaged function $f^{(2)}$ can be merely replaced by f. For dilute suspensions, apart from systems governed by fluctuations (critical phase transition), such an assumption is generally sound but remains to be proven for concentrated suspensions. The ensemble average is conceptually very convenient since it offers a sound statistical description of suspensions and it has the advantage that the operations of differentiation and ensemble averaging commute. However, its use is restricted by the poor knowledge that we may have on the distribution of particles in the suspension. An alternative is to use a volume average, that is, to average the quantity f over a control volume V, whose length scale must be large compared to the average distance between particles but small with respect to a distance over which the average of the property at hand varies appreciably. According we define the volume-averaged quantity $\bar{f}$ as $\bar{f}(\boldsymbol{r}, t) = \int_V f(\boldsymbol{x}, t) \mathrm{d}\boldsymbol{x} / V$.

In parallel to the fluid contribution, it is possible to obtain a generic expression of the particle contribution [40]:

$$\bar{\boldsymbol{\sigma}}^{(p)} = \bar{\boldsymbol{\sigma}}^{(p)}_{\text{surface}} - \frac{1}{2} J_p < \boldsymbol{\Omega}' \otimes \boldsymbol{\Omega}' > - \varrho_p < \boldsymbol{u}' \otimes \boldsymbol{u}' > \tag{3.31}$$

where $\bar{\boldsymbol{\sigma}}^{(p)}_{\text{surface}}$ denotes the contribution due to forces exerted on the particle surface, $\boldsymbol{\Omega}'$ the fluctuations of angular velocity of particles, and J_p the inertia moment. It can be shown that the surface contribution $\bar{\boldsymbol{\sigma}}^{(p)}_{\text{surface}}$ reflects the effects of local forces at the particle level and may be deduced by averaging the local forces [40]:

$$\bar{\boldsymbol{\sigma}}^{(p)}_{\text{surface}} = \frac{a}{V} \sum_{m=1}^{N} \int_{A_p^{(m)}} \boldsymbol{\sigma k} \otimes \boldsymbol{k} \mathrm{d}\boldsymbol{k} = a\, n \left\langle \boldsymbol{\sigma k} \otimes \boldsymbol{k} \right\rangle \tag{3.32}$$

where $\boldsymbol{\sigma k}$ is the local stress acting on the particle surface ($\boldsymbol{\sigma k}\mathrm{d}\boldsymbol{k}$ is sometimes referred to as the contact force), $\boldsymbol{k}$ is the outward normal at the contact point, $\mathrm{d}\boldsymbol{k}$ the angle around $\boldsymbol{k}$, n is the number density ($n = \phi/(4\pi a^3/3)$). In the first equality in (3.32), we use a volume average of all contact forces acting on the surface $A_p^{(m)}$ of N beads included in a control volume V. The second equality is a simple translation of the first one in terms of ensemble average, which is more usual in kinetic theories or homogenization techniques.

To compute the two contributions, we have to introduce further ingredients. In particular, information on the particle distribution and the forces acting on particles is needed. In fact these two elements are tightly connected. It can be easily shown by first taking $f = 1$ in (3.30), then calculating the total time derivative that the pair distribution function satisfies an evolution equation called the Smoluchowski equation:

$$\frac{\partial P_2}{\partial t} + \nabla_x . P_2 \boldsymbol{U}_x^{(2)} + \nabla_r . P_2 \boldsymbol{U}_r^{(2)} = 0 \tag{3.33}$$

where $\boldsymbol{U}_x^{(2)}$ and $\boldsymbol{U}_r^{(2)}$ are the conditionally averaged velocity and relative velocity between the two particles located at $\boldsymbol{x}$ and $\boldsymbol{x} + \boldsymbol{r}$. From a general point of view, these two velocities depend on the interparticle forces $\boldsymbol{F}^{(\text{hyd})}$, the Brownian motion, etc., which in turn depend on the imposed velocity gradient $\dot{\gamma}$. There is no for-all-purpose solution to this equation, but several particular applications have been completely or partially explored. The simplest application of this theory is to consider suspensions sufficiently dilute for the hydrodynamic interplay between two particles to be negligible. In this case, if the Reynolds particle number $Re_p = 2\varrho a \left|\boldsymbol{U}\right| / \eta$ (with $\boldsymbol{U}$ the particle velocity relative to the fluid) comes close to zero, the hydrodynamic force that the particle undergoes is the Stokes force: $\boldsymbol{F}^{(\text{hyd})} = 6\pi\eta a \boldsymbol{U}$ [37]. (This force is inferred from the so-called Stokes equation, that is, the Navier–Stokes equation in which the inertial terms have been neglected since $Re_p \to 0$: $\mu \nabla^2 \boldsymbol{u} = \nabla p_f$.) Both the disturbances in the fluid velocity and fluid stress fields can be inferred from Stokes problem. At a point $\boldsymbol{x}$ from the particle center, the disturbance in the fluid stress due to the slow motion of the particle can be expressed as: $\boldsymbol{\sigma}^{(f)} = -\boldsymbol{x}.\boldsymbol{f}/(4\pi|\boldsymbol{x}|^3)\mathbf{1} + \eta(\nabla\boldsymbol{u} +^t \nabla\boldsymbol{u})$, where $\boldsymbol{u} = (1+\boldsymbol{x}\boldsymbol{x}/|\boldsymbol{x}|^2).\boldsymbol{f}/(8\pi\eta|\boldsymbol{x}|)$ is the disturbance in the velocity field and $\boldsymbol{f}$ a constant such that $\int \boldsymbol{\sigma}^{(f)}\boldsymbol{k}\,\mathrm{d}\boldsymbol{k} = \boldsymbol{F}^{(\mathrm{hyd})}$ [37,50]. Using (3.32) with $P_2 = 1$ (assumption of dilute suspensions), we deduce that the bulk stress tensor can be expressed as:

$$\bar{\boldsymbol{\sigma}} = -\langle p_f \rangle \mathbf{1} + 2\eta \left(1 + \frac{5}{2}\phi\right) \bar{\dot{\boldsymbol{\gamma}}} \,. \tag{3.34}$$

Thus the well-known Einstein relationship for the effective viscosity of a dilute suspension is obtained: $\eta_{eq}/\eta = 1+2.5\phi+\mathcal{O}(\phi)$, holding for solid fractions lower than 2%. This method has been progressively extended to take further interactions into account. Batchelor and Green [51,52] provided the pair distribution function and the disturbances in the velocity and pressure fields when the solid concentration is increased so that the velocity and pressure caused by the motion of a particle is significantly influenced by the presence of another particle. This leads to modifying the Einstein equation as follows: $\eta_{eq} = \eta+2.5\phi+7.6\phi^2+o(\phi^2)$. Subsequently, the Brownian force [53], colloidal forces [54], the effect of solid fraction [55,56], and the particle surface roughness [57] have been included in the bulk stress computation.

3.4.2 Approximate Models

Because of the complexity of the dynamics of multiparticle interactions, rigorous microstructural theories generally do not provide analytical results. For instance, no analytical constitutive equation is available to predict the bulk behavior of Newtonian suspensions or granular flows at high solid fractions. A common way of overcoming this difficulty is to approximate the pair distribution function and the particle interaction expressions. This leads to a wide range of approximate models, whose applicability compensates for the introduction of *ad hoc* approximations. It is worth noting that numerical simulations of particle dynamics are increasingly used as an intermediate step between the theoretical models and the approximate equations. Typical examples include the treatment performed by Zhang and Rauenzahn [46,58] for granular flows and by Phan Thien [59,60] for concentrated viscous suspensions. Here, to exemplify the derivation of approximate models, we present the reasoning for deriving the bulk viscosity (see also [40,61]). The first step is to specify the approximate pair distribution function. This is usually done by considering a given configuration of particles (generally assumed to be cubic) and by assuming that the face-to-face distance between particles (ξ) is fixed on average and related to the solid fraction as follows:

$$\frac{\xi}{a} = 2\frac{\varsigma}{1-\varsigma}\,, \qquad \text{with } \varsigma = 1 - \sqrt[3]{\frac{\phi}{\phi_m}}\,, \tag{3.35}$$

where ϕ_m is the maximum random solid concentration ($\phi_m \approx 0.635$ for unimodal suspensions of spherical particles). The pair distribution function may thus be written as:

$$P_2(\boldsymbol{k})|_{r=\xi} = \sum_{i=1}^{n_c} \delta(\boldsymbol{k} - \boldsymbol{k}_i) \ , \tag{3.36}$$

where δ is the Dirac function, $\boldsymbol{k}_i$ denotes the directions of the neighboring particle centers in the considered configuration with respect to the test-particle center, n_c the coordination number (number of indirect contacts). The lubrication force between two spheres can be divided into three contributions: a squeezing contribution, a shearing contribution, and a term due to the rotation of spheres. It can be shown that, in a steady state, the squeezing contribution is to leading order [62]:

$$\boldsymbol{F}_{\text{sq}} = \frac{3\pi}{2}\eta\frac{a^2}{\xi}\boldsymbol{c}_n \ , \tag{3.37}$$

where $\boldsymbol{c}_n$ is the normal component of the relative particle velocity $\boldsymbol{c}$. The force due to shearing motion can be written to first order: $\boldsymbol{F}_{\text{sh}} = \pi\eta a \ln(\xi/a)\, \boldsymbol{c}_t$ (with $\boldsymbol{c}_t$ the tangential component of the relative particle velocity) and the force due to the rotation of particles is: $\boldsymbol{F}_{\text{rot}} = 2\pi\eta a^2 \ln(\xi/a)\ \boldsymbol{k}\times\boldsymbol{\Omega}$. These two contributions are of the same order and their magnitudes increase as $\ln(\xi/a)$. Consequently, for concentrated suspensions, to leading order in ξ/a, they are negligible compared to the squeezing force. All the above expressions tend toward infinity when the gap becomes extremely small, which would preclude any direct contact. The squeezing contribution can be evaluated by incorporating (3.37) into (3.32):

$$\boldsymbol{\sigma}_{\text{sq}}^{(p)} = \frac{3\pi}{2}\frac{a^3}{\xi}\mu n_d \langle \boldsymbol{c}_n \otimes \boldsymbol{k}\rangle \ . \tag{3.38}$$

The relative velocity is computed as the average velocity imposed by the bulk flow:

$$\boldsymbol{c} \approx 2a\bar{\boldsymbol{L}}\boldsymbol{k} - 2a < \boldsymbol{\Omega} > \times\boldsymbol{k} = 2a(\bar{\dot{\boldsymbol{\gamma}}}\boldsymbol{k} - (< \boldsymbol{\Omega} > - \bar{\boldsymbol{\omega}}) \times \boldsymbol{k}) \ , \tag{3.39}$$

where $\bar{\boldsymbol{L}} = \nabla\bar{\boldsymbol{u}}$ denotes the bulk velocity gradient, $\bar{\boldsymbol{\omega}}$ is the curl of $\bar{\boldsymbol{L}}$, and $\bar{\dot{\boldsymbol{\gamma}}}$ is the symmetric part of $\bar{\boldsymbol{L}}$. It follows that the squeezing velocity can be written:

$$\boldsymbol{c}_n = 2a(\bar{\dot{\boldsymbol{\gamma}}} : \boldsymbol{k}\otimes\boldsymbol{k})\boldsymbol{k} \ . \tag{3.40}$$

The contribution due to the squeezing motion is directly deduced from (37):

$$\boldsymbol{\sigma}_{\text{sq}}^{(p)} = \frac{9}{4}\frac{a}{\xi}\eta\phi(\bar{\dot{\boldsymbol{\gamma}}} : \boldsymbol{k}_i\otimes\boldsymbol{k}_i)\, \boldsymbol{k}_i\otimes\boldsymbol{k}_i \ . \tag{3.41}$$

It should be noted that the Newtonian character of bulk stress is dictated by the symmetry of the directions $\boldsymbol{k}_i$ with respect to the principal directions of the strain-rate tensor. Let us consider a simple shear flow. If we assume that (i) the particle configuration is cubic, (ii) its privileged axes coincide with the principal axes of the strain-rate tensor, (iii) the predominant action is due to squeezing, then we can deduce that the bulk viscosity varies as:

$$\eta_{\text{eq}} = \alpha\frac{a}{\xi}\eta \ , \tag{3.42}$$

with $\alpha = 9\phi/4$. Thus it is shown that the bulk viscosity of a concentrated suspension should tend towards infinity when the solid concentration comes closer to its upper limit ϕ_m.

The main drawback in the derivation of approximate models lies in the speculative character of many assumptions. As pointed out by different authors [63,64], the mean-field approach presented here suffers a great deal from questionable approximations. Among others, it is obvious from (3.40)–(3.41) that the resulting bulk stress tensor depends to a large extent on the particle arrangement, the face-to-face distance between particles, and the velocity field. For instance, using different methods or assumptions, most authors have obtained a bulk viscosity whose expression is structurally similar to (3.42), but sometimes with a different value for α. For instance, using a similar approach, Goddard [65] found $\alpha = 3\phi/8$ while van den Brule and Jongshaap arrived at $\alpha = 9\phi/4$ [61]. Using an energy-based method, Frankel and Acrivos obtained $\alpha = 9/4$ [66]. Sengun and Probstein [67] inferred a more complicated expression from energy considerations but, asymptotically for solid concentrations near the maximum concentration, they found a comparable expression for the bulk viscosity, with $\alpha \approx 3\pi/4$, close to the value determined by Frankel and Acrivos. On the basis of energy and kinematic considerations, Marrucci and Denn [64] argued that coefficient α is not constant and must vary as $\alpha \propto \ln(a/\xi)$ in the worst case. Likewise, Adler et al. [63] put forward that averaging the different configurations through which the particle arrangement passes does indeed smooth the singularity $1/\varsigma$ and consequently the bulk viscosity does not diverge when the solid concentration tends to its maximum.

It is worth noting that approximate models can be built using empirical reasoning without any recourse to a detailed analysis of particle interactions. A typical example in the area of suspensions is given by Krieger and Dougherty's model [68]. The authors assumed that within a suspension of non-Brownian, noncolloidal particles, a particle sees a homogeneous fluid surrounding it, whose viscosity depends only on the solid fraction and the interstitial fluid viscosity. This is obviously a crude assumption since this particle is more influenced by nearby particles than by more distant particles. Using dimensional analysis (see below), it may be shown that the bulk viscosity is of the form: $\eta_{\mathrm{eq}} = \eta f(\phi)$. The bulk viscosity can be computed by assuming that one first introduces a solid fraction ϕ_1, then a solid fraction ϕ_2 so that the resulting solid concentration is ϕ. For doing so, we must choose ϕ_2 such that it satisfies: $\phi_2 = (\phi - \phi_1)/(1 - \phi_1)$. Finally we must have: $f(\phi_1)f(\phi_2) = f(\phi)$, which must hold whatever the solid fractions. It can be shown that the only function obeying such an equality is of the form: $f(\phi) = (1 - \phi)^{-\beta}$. Experimentally, β has been generally estimated at approximately 2. Krieger and Dougherty's expression has been modified to represent experimental data over as wide a range of solid concentrations as possible:

$$\frac{\eta_{\mathrm{eq}}}{\eta} = \left(1 - \frac{\phi}{\phi_m}\right)^{-[\eta]\phi_m}, \tag{3.43}$$

where $[\eta] = \lim_{\phi \to 0} (\eta_{\text{eq}} - \eta)/(\eta\phi) = 2.5$ is called the *intrinsic viscosity*. Such a relation matches the Einstein expression at low solid fractions. Many expressions with a form similar to (3.43) have been proposed to take further phenomena (aggregating of particles [69], shear-thinning, colloidal effects, polydispersity [70,71], etc.) into account. A common element in several models is to consider that the maximum solid concentration is not constant but is rather a shear-rate-dependent function since it should reflect changes in the microstructure. For instance, in order to make an allowance for viscoplastic behavior, Wildemuth and Williams [73,72] have assumed that the maximum solid fraction relaxes with shear stress from a lower value ϕ_0 to an upper bound ϕ_∞:

$$\frac{1}{\phi_m} = \frac{1}{\phi_0} - \left(\frac{1}{\phi_0} - \frac{1}{\phi_m}\right) f(\tau) \tag{3.44}$$

where $f(\tau) = (1 + A\tau^{-m})^{-1}$, A and m are two constants intrinsic to the material. This also implies that such a suspension (with $\phi_0 \le \phi \le \phi_\infty$) exhibits a yield stress:

$$\tau_c(\phi) = \left[A\left(\frac{\phi/\phi_0 - 1}{1 - \phi_m/\phi_\infty}\right)\right]^{1/m} . \tag{3.45}$$

It should be noted that in the model and experiments presented by Wildemuth and Williams, the yield appearance reflects either colloidal effects or structural changes in the particle arrangement (jamming, friction between coarse particles) or both of them.

In contrast, Sengun and Probstein [67] proposed different arguments to explain the viscoplastic behavior observed in their investigations on the viscosity of coal slurries (with particle size typically ranging from 0.4 µm to 300 µm). Their explanation consists of two approximations. First, as it is the interstitial phase, the dispersion resulting from the mixing of fine colloidal particles and water imparts most of its rheological properties to the entire suspension. Secondly, the coarse fraction is assumed to act independently of the fine fraction and to enhance the bulk viscosity. They introduced a *net viscosity* η_{nr} of a bimodal slurry as the product of the fine relative viscosity η_{fr} and the coarse relative viscosity η_{cr}. The fine relative viscosity is defined as the ratio of the apparent viscosity of the fine-particle suspension to the viscosity of the interstitial fluid: $\eta_{\text{fr}} = \eta_f/\eta_0$. The coarse relative viscosity is defined as the ratio of the apparent viscosity of the coarse-particle slurry to the viscosity of the fine-particle suspension: $\eta_{\text{cr}} = \eta_c/\eta_f$. The two relative viscosities depend on the solid concentrations and a series of generalized Péclet numbers. For the coarse-particle suspensions, all the generalized Péclet numbers are much greater than unity. Using a dimensional analysis, Sengun and Probstein deduced that the coarse relative viscosity cannot depend on the shear rate. In contrast, bulk behavior in fine-particle suspensions is governed by colloidal particles and thus at least one of the generalized Péclet numbers is of the order of unity, implying that the fine relative viscosity is shear-dependent. Sengun and Probstein's experiments on viscosity of coal slurries confirmed the reliability of this concept [67]. Plotting $\log \eta_{\text{nr}}$ and $\log \eta_{\text{fr}}$

against $\log \dot{\gamma}$, they found that over a wide range of concentrations, the curves were parallel and their distance was equal to $\log \eta_{\mathrm{cr}}$ (see Fig. 3.7). However, for solid concentrations in the coarse fraction exceeding 0.35, they observed a significant departure from parallelism which they ascribed to nonuniformity in the shear rate distribution within the bulk due to squeezing effects between coarse particles.

Generally, all these empirical models successfully provide an estimation of bulk viscosity over a wide range of solid fraction, as shown in Fig. 3.8, provided that the maximum solid concentration has been correctly evaluated. In practice, for natural fluids such as debris suspensions, this evaluation may be problematic and lead to a large uncertainty in computing bulk viscosity.

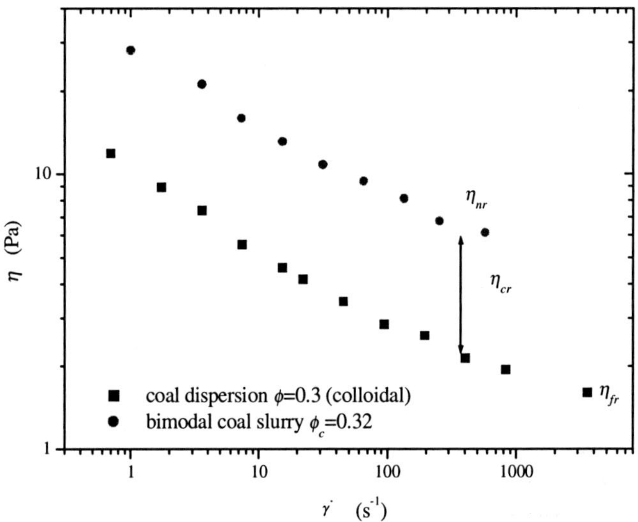

Fig. 3.7. Variation of the bulk viscosity of coal slurry as a function of the shear rate. The bulk viscosity curve is parallel to the curve obtained with the fine fraction. After [67]

3.4.3 Contribution of Dimensional Analysis

Expressing bulk behavior in terms of dimensionless groups is a practical and usual way of identifying the most relevant variables and delineating flow regimes. A certain number of studies have so far focused on suspensions of rigid spherical particles within a Newtonian fluid with a narrow size distribution [7,54,76,78]. In this case, a suspension of noninteracting particles is characterized by eight variables: (i) for particles, the density ϱ_p, the radius a, and the solid volume concentration ϕ; (ii) for the interstitial fluid, the viscosity η and the density ϱ_f; (iii) for the conditions imposed during an experiment, the temperature T, the

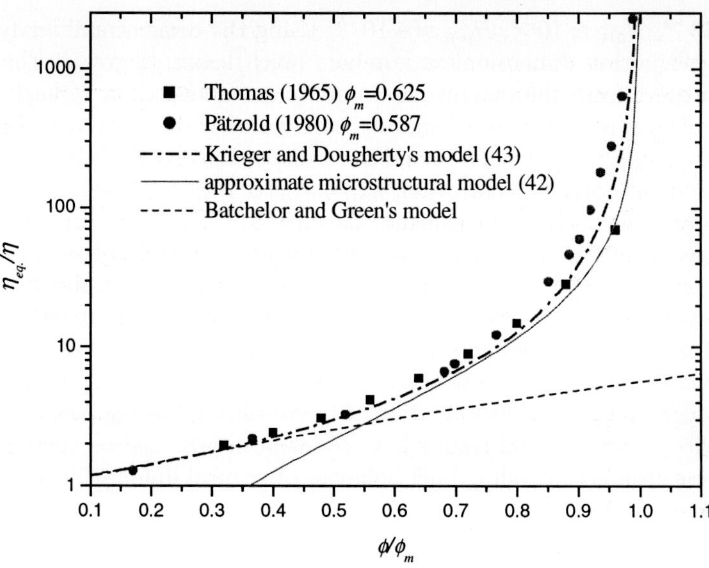

Fig. 3.8. Variation in the bulk viscosity as a function of the reduced fraction. Typical data obtained by Thomas [74] and Pätzold [75] are reported

shear rate $\dot{\gamma}$ (or equivalently the shear stress τ), and the experiment duration t_{exp}. According to the principles of dimensionless analysis, the bulk viscosity can be expressed as a function of $8-3=5$ dimensionless groups. The following numbers are preferentially formed: the solid fraction ϕ, the *Reynolds particle number* $Re = (2a)^2\dot{\gamma}/\eta$ reflecting fluid inertia at the particle scale, the *Péclet number* $Pe = 6\pi\dot{\gamma}a^3\eta/(kT)$ (where k refers to the Boltzmann constant) defined as the ratio of viscous forces to Brownian forces, the *Deborah number* expressed as the ratio of a particle relaxation time t_p to the typical time of the experiment $De = t_p/t_{\text{exp}}$ (depending on the particle size, the particle relaxation can be linked to the Brownian diffusion time $t_p = 6\pi a^3\eta(kT)^{-1}$ or the Stokes relaxation time $t_p = 2a^2\varrho_p(9\eta)^{-1}$), the *Stokes number* $St = 2\varrho_p Re/(9\varrho_f)^{-1}$ defined as the ratio of a particle relaxation time to a fluid characteristic time. If the particles are colloidal, van der Waals' attraction and electrostatic repulsion must be taken into account, giving rise to two dimensionless groups: an attraction number $N_{\text{att}} = \eta a^3\dot{\gamma}/A$, where A is the Hamacker constant of the colloidal particles, and a repulsion number $N_{\text{rep}} = \eta a^2\dot{\gamma}/(\varepsilon\psi_0^2)$, where ε is the fluid permittivity and ψ_0 the surface potential. As examples, taking $a = 0.5\,\text{mm}$, $\dot{\gamma} = 1\,\text{s}^{-1}$, $\eta = 10^{-3}\,\text{Pa.s}$, $t_{exp} = 10\,\text{s}$, $T = 293\,\text{K}$, $\varrho_p = 2500\,\text{kg/m}^3$ for a suspension of coarse particles slowly sheared (typically a suspension of particles in a water-glycerol solution), we find: $Re = 10^{-3}$, $Pe = 580\ 10^6$, $St = 5\ 10^{-4}$, $De = 10^{-2}$. Taking $a = 0.5\,\mu\text{m}$, $A \approx 10^{-20}\,\text{J}$, $\varepsilon = 7\ 10^{-10}\,\text{C}^2\text{J}^{-1}\text{m}^{-1}$, $\psi_0 \approx 100\,\text{mV}$, $\varrho_p = 2650\,\text{kg/m}^3$ for a suspension of colloidal particles slowly sheared (typically a water–kaolin dispersion), we find: $Re = 10^{-9}$, $Pe = 0.6$, $St = 6\ 10^{-10}$,

$De = 6\ 10^{-2}$, $N_{\mathrm{att}} \approx 10^{-2}$, $N_{\mathrm{rep}} \approx 4\ 10^{-5}$. Using the dimensional analysis principles (i.e. ignoring dimensionless numbers much lesser or greater than unity) [79], we expect from the magnitude orders found above that, typically for the viscosity of a coarse-particle suspension, bulk viscosity depends on the solid concentration mainly: $\eta_{\mathrm{eq}}/\eta = f(\phi)$, and for a dispersion, it depends on the Péclet number and the solid concentration: $\eta_{\mathrm{eq}}/\eta = f(\phi, Pe)$. Such scalings have been successfully compared to experimental data [80,81]. The main problem encountered in geophysics is that fluids generally involve a wide range of size particles and different types of particle interaction. For instance, typically for a debris flow, the particle size ranges from $1\,\mu\mathrm{m}$ to more than $1\,\mathrm{m}$ and particle interactions can include colloidal effects, collisional, frictional, lubricated contacts, etc. Thus the large number of physical parameters intervening in the problem makes any thorough and general examination of the resulting flow regimes intricate. To our knowledge, only partial results have so far been provided on the relevant dimensionless groups controlling bulk behavior of natural fluids [7] (see also Chap. 21).

References

1. H.A. Barnes, J.F. Hutton, K. Walters: *An introduction to rheology* (Elsevier, Amsterdam 1997)
2. R.I. Tanner: *Engineering Rheology* (Clarendon Press, Oxford 1988)
3. G.V. Middleton, P.R. Wilcock: *Mechanics in the Earth and Environmental Sciences* (Cambridge University Press, Cambridge 1994)
4. B.D. Coleman, H. Markowitz, W. Noll: *Viscometric flows of non-Newtonian fluids* (Springer, Berlin 1966)
5. K. Walters: *Rheometry* (Chapman and Hall, London 1975)
6. R.B. Bird, R.C. Armstrong, O. Hassager: *Dynamics of polymeric liquids* (John Wiley & Sons, New York 1987)
7. P. Coussot, C. Ancey: *Rhéophysique des pâtes et des suspensions* (EDP Sciences, Les Ulis 1999) (in French)
8. J.F. Steffe: *Rheological methods in food process engineering* (Freeman Press, East Langing, USA 1996). Free sample available at: http://www.egr.msu.edu/~steffe/freebook/offer.html
9. Y. Wang, K. Hutter: Granular Matter **1**, 163 (1999)
10. K. Hutter, B. Svendsen, D. Rickenmann: Continuum Mech. Therm. **8**, 1 (1996)
11. I. Vardoulakis, J. Sulem: *Bifurcation Analysis in Geomechanics* (Blackie Academic & Professional, Glasgow 1995)
12. A. Bedford, D.S. Dumheller: Int. J. Eng. Sci. **21**, 863 (1983)
13. S. Chapman, T.G. Cowling: *The mathematical theory of nonuniform gases* (Cambrige University Press, Cambrige 1970)
14. J.T. Jenkins, H.M. Hanes: J. Fluid Mech. **370**, 29 (1998)
15. S.B. Savage: 'Marginal ice zone dynamics modelled by computer simulations involving floe collisions'. In: *Mobile particulate systems, Carghese, 1994*, ed. by E. Guazelli and L. Oger (Kluwer Academic Publishers, Dordrecht 1995); S.B. Savage, G.B. Crocker, M. Sayed, T. Carrieres, Cold Reg. Sci. Technol. **31**, 163 (2000)
16. S.B. Savage: 'Flow of granular materials'. In: *Theoretical and Applied Mechanics*, ed. by P. Germain, J.-M. Piau, D. Caillerie (Elsevier, Amsterdam 1989)

17. C. Ancey, P. Coussot, P. Evesque: Mech. Cohes. Frict. Mat. **1**, 385 (1996)
18. C. Ancey, P. Evesque: Phys. Rev. E **62**, 8349 (2000)
19. A. Borgia, F.J. Spera: J. Rheol. **34**, 117 (1990)
20. T.M.T. Yang, I.M. Krieger: J. Rheol. **22**, 413 (1978)
21. I.M. Krieger: Trans. Soc. Rheol. **12**, 5 (1968)
22. Q.D. Nguyen, D.V. Boger: Ann. Rev. Fluid Mech. **24**, 47 (1992)
23. P. Coussot, J.M. Piau: Rheol. Acta **39**, 105 (1995)
24. Y.L. Yeow, W.C. Ko, P.P.P. Tang: J. Rheol. **44**, 1335 (2000)
25. Y.T. Nguyen, T.D. Vu, H.K. Wong, Y.L. Yeow: J. Non-Newtonian Fluid Mech. **87**, 103 (1999)
26. H.A. Barnes, J.O. Carnali: J. Rheol. **34**, 851 (1990)
27. C.J. Phillips, T.R.H. Davies: Geomorphology **4**, 101 (1991)
28. N. Pashias, D.V. Boger: J. Rheol. **40**, 1179 (1996)
29. P. Coussot, S. Proust, C. Ancey: J. Non-Newtonian Fluid Mech. **66**, 55 (1996)
30. R.W. Griffith: Ann. Rev. Fluid Mech. **32**, 477 (2000)
31. O. Pouliquen: Phys. Fluids **11**, 542 (1999)
32. A.M. Johnson, J.R. Rodine: 'Debris flow'. In: *Slope Instability*, ed. by D. Brundsen, D.B. Prior (John Wiley & Sons, Chichester 1984)
33. I. Newton: *Philosophiae naturalis principia mathematica.* English translations of this historical text are available, e.g. F. Cajori: *Sir Isaac Newton's mathematical principles of natural philosophy and his system of the world* (University of California Press, Berkeley 1962)
34. F.T. Trouton, E.S. Andrews: Phil. Mag. **7**, 347 (1904). F.T. Trouton: Proc. Roy. Soc. London Ser. A **77**, 426 (1906)
35. R.B. Bird, G.C. Dai, B.J. Yarusso: Rev. Chem. Eng. **1**, 1 (1983)
36. J.D. Sherwood, D. Durban: J. Non-Newtonian Fluid Mech. **77**, 155 (1998)
37. G.K. Batchelor: *An introduction to fluid mechanics* (Cambridge University Press, Cambridge 1967)
38. R. Hill: *Mathematical theory of plasticity* (Oxford University Press, Oxford 1950)
39. C.S. Campbell: Ann. Rev. Fluid Mech. **22**, 57 (1990)
40. C. Ancey, P. Coussot, P. Evesque: J. Rheol. **43**, 1673 (1999)
41. G.K. Batchelor: Ann. Rev. Fluid Mech. **6**, 227 (1974)
42. R. Herczynski, I. Pienkowska: Ann. Rev. Fluid Mech. **12**, 237 (1980)
43. R.J.J. Jongschaap, D. Doeksen: Rheol. Acta **22**, 4 (1983)
44. M. Lätzel, S. Luding, H.J. Herrmann: Granular Matter **2**, 123 (2000)
45. D.Z. Zhang, A. Prosperetti: Int. J. Multiphase Flow **23**, 425 (1997)
46. D.Z. Zhang, R.M. Rauenzahn: J. Rheol. **41**, 1275 (1997)
47. B. Cambou, P. Dubujet, F. Emeriault, F. Sidoroff: Eur. J. Mech. A/Solids **14**, 255 (1995)
48. Y.A. Buyevich, I.N. Shchelchkova: Prog. Aero. Sci. **18**, 121 (1978)
49. G.K. Batchelor: J. Fluid Mech. **41**, 545 (1970)
50. S. Kim, S.J. Karrila: *Microhydrodynamics: principles and selected applications* (Butterworth-Heinemann, Stoneham 1991)
51. G.K. Batchelor, J.T. Green: J. Fluid Mech. **56**, 401 (1972)
52. G.K. Batchelor, J.T. Green: J. Fluid Mech. **56**, 375 (1972)
53. G.K. Batchelor: J. Fluid Mech. **83**, 97 (1977)
54. W.B. Russel, D.A. Saville, W.R. Schowalter: *Colloidal dispersions* (Cambridge University Press, Cambridge 1995)
55. J.F. Brady, J.F. Morris: J. Fluid Mech. **348**, 103 (1997)
56. R.A. Lionberger, W.B. Russel: J. Rheol. **41**, 399 (1997)

57. H.J. Wilson, R.H. Davis: J. Fluid Mech. **421**, 339 (2000)
58. D.Z. Zhang, R.M. Rauenzahn: J. Rheol. **44**, 1019 (2000)
59. N. Phan-Thien, X.-J. Fan, B.C. Khoo: Rheol. Acta **38**, 297 (1999)
60. N. Phan-Thien: J. Rheol. **39**, 679 (1995)
61. B.H.A.A. van den Brule, R.J.J. Jongshaap: J. Stat. Phys. **62**, 1225 (1991)
62. R.G. Cox: Int. J. Multiphase Flow **1**, 343 (1974)
63. P.M. Adler, M. Zuzovski, H. Brenner: Int. J. Multiphase Flow **11**, 387 (1985)
64. G. Marrucci, M. Denn: Rheol. Acta **24**, 317 (1985)
65. J.D. Goddard: J. Non-Newtonian Fluid Mech. **2**, 169 (1977)
66. N.A. Frankel, A. Acrivos: Chem. Eng. Sci. **22**, 847 (1967)
67. M.Z. Sengun, R.F. Probstein: Rheol. Acta **28**, 382 (1989)
68. I.M. Krieger, T.J. Dougherty: Trans. Soc. Rheol. **3**, 137 (1959)
69. C. Tsenoglou: J. Rheol. **34**, 15 (1990)
70. C. Chang, R.L. Powell: J. Rheol. **38**, 85 (1994)
71. P. Gondret, L. Petit: J. Rheol. **41**, 1261 (1997)
72. C.R. Wildemuth, M.C. Williams: Rheol. Acta **23**, 627 (1984)
73. C.R. Wildemuth, M.C. Williams: Rheol. Acta **24**, 75 (1985)
74. D.G. Thomas: J. Colloid Sci. **20**, 267 (1965)
75. R. Pätzold: Rheol. Acta **19**, 322 (1980)
76. I.M. Krieger: Adv. Colloid. Interface Sci. **3**, 111 (1972)
77. P. Coussot, C. Ancey: Phys. Rev. E **59**, 4445 (1999)
78. A.I. Jomha, A. Merrington, L.V. Woodcock, H.A. Barnes, A. Lips: Powder Tech. **65**, 343 (1990)
79. G.I. Barenblatt: *Scaling, self-similarity, and intermediate asymptotics* (Cambridge University Press, Cambridge 1996)
80. W.B. Russel: J. Rheol. **24**, 287 (1980)
81. J.C. van der Werff, V.F. de Kruif: J. Rheol. **33**, 421 (1989)

4 Granular Material Theories Revisited

Y. Wang and K. Hutter

Institute of Mechanics, Darmstadt University of Technology
Hochschulstr. 1, 64289 Darmstadt, Germany

4.1 Introduction

A granular material is a collection of a large number of discrete solid particles with interstices filled with a fluid or a gas. If the interstitial fluid plays an insignificant role in the transportation of momentum, flows of such materials can be considered as dispersed single-phase flows. In other occasions, when the mass of the interstitial fluid is comparable to that of the solids the interactions between the fluid and solid phases are significant, the motion of the fluid can then provide the driving force for the flow of the solid phase. The dynamical behaviour of these materials can be very complex; its description involves aspects of traditional fluid mechanics, plasticity theory, soil mechanics and rheology.

4.1.1 Some Distinctive Features of Granular Materials

Granular materials exhibit a number of distinctive features not shared by "ordinary" solids, fluids and gases. In fact, depending on the externally applied mechanisms they behave somewhat like solids or fluids or gases. Furthermore, their behaviour can in a given process change from, say being fluid-like to suddenly solid-like, often repeatedly, so that an *intermittent* reaction results from a driving mechanism that may strictly be continuous. Distinctive features of granular materials are the following:

Dilatancy. Deformations in a granular body are always accompanied by volume changes and can in principle be easily understood. If an array of identical spherical grains at closest packing is subjected to a load so as to cause a shear deformation, then from pure geometrical considerations that particles must ride one over another it follows that an increase in volume of the bulk material will occur, see Fig. 4.1. This property was termed *dilatancy* by Reynolds in 1885. Despite this fact, there are many phenomena for the description of which this non-volume preserving – a kind of compressibility – need not be accounted for. For instance, in a granular avalanche followed from initiation to runout, the mass expands at the onset of its motion and it contracts at the moment of settling, but while rapidly moving down the mountain flank variations of the volume are generally small.

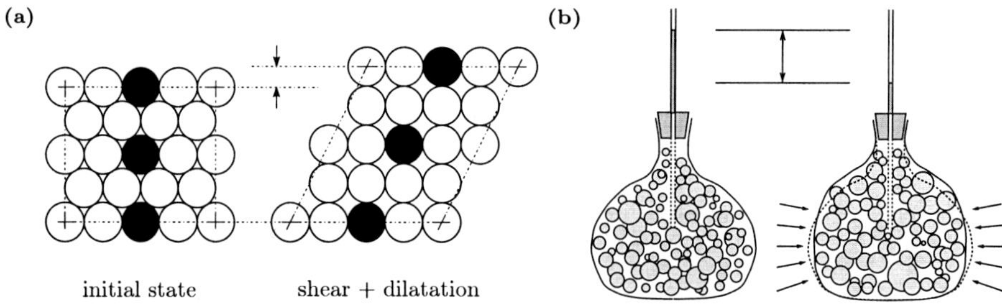

Fig. 4.1. Explaining dilatancy: (**a**) Identical circular disks in closest packing and when they have been sheared. The layer thickness must increase if such shear deformations are possible. (**b**) A rubber bellows, filled with a granular material of densest packing and sealed with a plug and pore space filled with water, of which the filling is made visible by the liquid level in the pipette. Outside pressure deforms the content, also by shear; the water level in the pipette falls as a result of the pore space extension

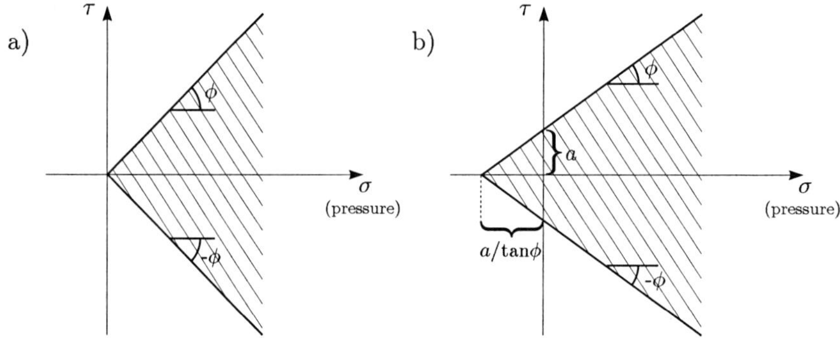

Fig. 4.2. Mohr–Coulomb yield criterion ($\sigma > 0$ as a pressure): If the point representing normal and shear tractions, (σ, τ), at an interior surface element lies in the shaded area then that element can be in equilibrium without deformation, if it lies on the limiting straight lines, then yielding occurs. (**a**) holds for a material without, (**b**) with cohesion

Near-Coulomb Behaviour. When grains are poured on a rough horizontal plane from a fixed source point they pile up in a heap (triangular in 2d and circular cone in 3d). The surface angle θ, called *angle of repose* is that limiting angle below which the heap stays unchanged at rest and above which surface grains move down as avalanches to reconstitute the limiting angle.

The behaviour inside the material is analogous and described by the *Mohr–Coulomb yield criterion*, which states that yielding will occur on a plane element at an interior point, when the shear, τ, and normal, σ, tractions acting on the plane element are related by (see Fig. 4.2a)

$$|\tau| = (\tan\phi)\sigma\ , \quad \sigma > 0 \quad \text{as a pressure}\ . \tag{4.1}$$

ϕ is the static *internal angle of friction*, and it is generally assumed that $\phi = \theta$. Typical values are from 25° to 40°. The law (4.1) ignores *cohesion* which is of

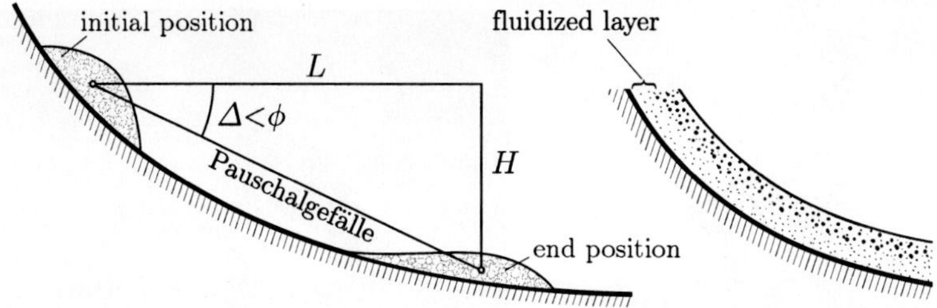

Fig. 4.3. Definition of the "Pauschalgefälle": It is the angle $\triangle$ between a line connecting the centres of mass in the initial and end positions and the horizontal. It is smaller than the internal angle of friction ϕ, because during motion a fluidized layer is formed close to the bed

significance e.g. when sand is wet. With cohesion the law (4.1) is replaced by (see Fig. 4.2b)

$$|\tau| = a + (\tan\phi)\sigma \ , \quad \sigma > -\frac{a}{\tan\phi} \ . \tag{4.2}$$

Many physical phenomena, whether they are of static or dynamic nature, can be very well described by employing the above Mohr–Coulomb material behaviour. Recent continuum-mechanical developments, however, generally involve Coulomb behaviour in special limiting situations but treat the sand, porous material or snow as a material with elastic, viscous and/or plastic behaviour.

Fluidization. It is well known that avalanches (of snow or gravel or rock) travel very large distances, generally much larger than one would expect on the basis that the loss in potential energy from initiation to runout is balanced by the work done due to basal sliding (evaluated with the given internal angle of friction, ϕ), see Fig. 4.3. Several postulates have been proposed (hover-craft action at the base, melting of rock, fluidization aided by the presence of fine dust). The most acceptable explanation is that in a very thin layer immediately above the sliding surface the strong shearing gives rise to enhanced collisions of the particles, leading to an increase of the mean particle distance and thus reducing the effective friction angle. One way to handle this situation is to ignore the thickness of the boundary layer and to introduce a basal Mohr–Coulomb type friction law with a *bed friction* angle $\delta < \phi$. Alternatively, one may resolve the boundary layer with a theory that accounts for the dilatation due to the particle collisions.

Liquefaction by Seismic Waves. The devastating earth quakes in Niigata (1992) and Kobe (1995) left behind a number of large residential buildings, erected in saturated soil, which sunk into the ground and are now tilted, but otherwise practically left intact, see Fig. 4.4, right. This phenomenon can be

Fig. 4.4. Left: Glacier moraines in Tuyk Valley, Alaarcha basin North Tien Shan, Kirgizstan. The moraine to the left has been displaced by a (probably seismically induced) debris flow. (Courtesy Dr. Vladimir Aizen, University of California at Santa Barbara). **Above**: Overturned buildings. Photo probably taken after the devastating earthquake in Kobe or Niigata. The overturning is the result of the liquefaction of the soil at the passage of the seismic wave. (Courtesy Prof. Dr. D. Kolymbas, Innsbruck)

explained as follows: Before the passage of the earthquake the building was in equilibrium with the buoyancy forces and the shear stresses that were established between the water saturated soil and the base of the building. When the seismic wave was passing the shear stresses were suddenly released, the weight of the building and the buoyancy force out of equilibrium, so that a motion could set in. Liquefaction phenomena are also in action when the soil in a slope suddenly becomes unstable and moves catastrophically downhill (Fig. 4.4, left). The reason is often heavy rainfall so that the soil is quickly becoming soaked with water. In such debris and mud flow events water plays a significant role; it follows that theories of dry granular materials are likely not appropriate for their description.

The phenomenon discussed next is so significant in geological flows that we reserve its own subsection to it: *Particle size separation* or *particle size segregation.* It is seen almost everywhere in granular deposits, and its phenomenology is understood but the theoretical state of its description is still fairly meagre.

4.1.2 Particle Size Segregation

It is a common experience for everyone who wishes to mix different types of particles that it is very difficult to achieve homogeneous mixing of several sorts of grains, whereas it is, in general fairly easy to achieve homogeneous mixing with miscible fluids. Conversely, moving a spoon in a jar of well mixed dry-freezed coffee shows that the large coffee grains will rise to the surface. Factors that can

Fig. 4.5. (**a**) Sketch of a profile from a deposit of a pyroclastic flow due to the volcanic eruption of Mount St. Helens, 12 June 1980. The profile is taken from a position about 6.7 km north of the crater and 1 km southwest of the Spirit Lake. One complete "flow unit" is shown that is under- and overlain by other flow units. The profile indicates a clear reverse grading in which larger grains are at the upper portions of the flow unit, while smaller grains are in its lower parts. Each flow unit corresponds to the passage of a pyroclastic flow (Courtesy of S. Straub [113]). (**b**) Debris flow deposit form a disastrous flow event on 31 July–1 Aug. 1996 in Taiwan. In the picture the road in the front has been cleared. It demonstrates particle size separation. The free surface of the deposit is covered by large bolders, whilst the lower part consists of the fine material

give rise to separation are differences in size, density and shape and differences in resilience (i.e. interaction forces) during impact.

Avalanche-, Debris- and Pyroclastic Flow Deposits. Such particle size separations are often observed in avalanche-, debris- and pyroclastic flow deposits. In dynamical systems of such flows one generally observes that the large particles move to the front and to the top surface whilst small particles accumulate at the bottom and in the rear part of the avalanche. In deposits of pyroclastic flows due to volcanic eruptions or in marine deposits in the depository zone of turbidity currents the following is encountered. Deposition profiles show a repetitive occurrence of flow units with the dust particles at the bottom and particle size increasing as one moves higher up until a level is reached where a new flow unit commences. Each flow unit corresponds to the passage of an avalanche, and obviously it is characterised by *reverse* (or *inverse*) grading, see Fig. 4.5a. This same structure of inverse grading can also be observed in deposits of debris flows. Often a rather thin "skin" of larger particles covers the top, whilst the main part of the body is occupied by the smaller size components, see Fig. 4.5b.

The phenomenon of inverse grading can relatively easily be understood, if one assumes that gravity plays a significant role in explaining it. A simple mathematical model was presented by Savage and Lun [105] that allows the quantification of the process of gravity separation of fine from coarse spherical particles during the shearing flow of an initially randomly mixed material down an inclined chute. The model is restricted to shear flow of a cohesionless granular material which consists of spherical particles of two sizes. During the shearing motion the particles experience continued rearrangements. These rearrangements are assumed to be *random*. The shearing motion is supposed to be the macroscopic manifestation of a large number of such microstate realizations, and the laws of statistics are supposed to hold in order to derive the properties of the macrosystem. At any instance, there will be a distribution of void spaces. If a void space at a certain depth is sufficiently large, then a particle from a position immediately above can fall into it when it moves past this void space in the shearing motion. For a given realization of the solid concentration, the probability of finding a hole that a small particle can fall into is obviously larger than the probability of finding a hole that a large particle can fall into. This will lead to a tendency for particles to segregate out, with fines at the bottom and coarse ones at the top. This mechanism is, of course, orientation dependent due to the action of gravity. This mechanism is called the *random fluctuating sieve mechanism.*

It is clear that this gravity-induced size-dependent void filling mechanism is insufficient to explain the phenomenon of inverse grading, because there exists a net mass flux perpendicular to the direction of the shearing motion towards the bed. A second mechanism must therefore exist for the transfer of particles from one position to another which gives rise to a counter flow so as to accommodate for the mass loss in the transverse direction of motion that would otherwise exist. Savage and Lun [105] propose that, as a result of the fluctuating contact forces on an individual particle, that there can occur force imbalances such that a particle is *squeezed* out of its own position into a position above or below. This mechanism cannot be gravity driven nor be size dependent. It must be as large as need be to compensate the mass flux towards the non-movable bed. This proposal is called the *squeeze expulsion mechanism.*

Sand-Piles and Rotating Drums. Inverse grading can easily be made experimentally visible in laboratory experiments.

In the first experiment, consider two vertical glass plates forming a narrow gap; together with a basal plate and side walls they form a plane "silo". This space is filled from a single central point source at the upper edge with a binary mixture of cohesionless (white) sugar crystals of 0.5 mm nominal diameter and (dark) spherical iron powder with mean diameter 0.34 mm. Although material is continuously deposited at the top of the pile it does not immediately flow down the faces because of the differences between the static and dynamic angles of friction. Once the static friction angle is exceeded the avalanche flows down the face of the pile and forms a roll-wave, as shown in Fig. 4.6a, in which the kinetic sieving takes place. As the avalanche reaches either the base or the wall of the silo it is rapidly brought to rest by an upslope moving shock wave, as shown in Fig. 4.6b. These upslope moving shock waves freeze the particle size distribution into the deposited granular material and thus preserve the pattern formed during the avalanche motion. Successive and alternating avalanche releases on both faces

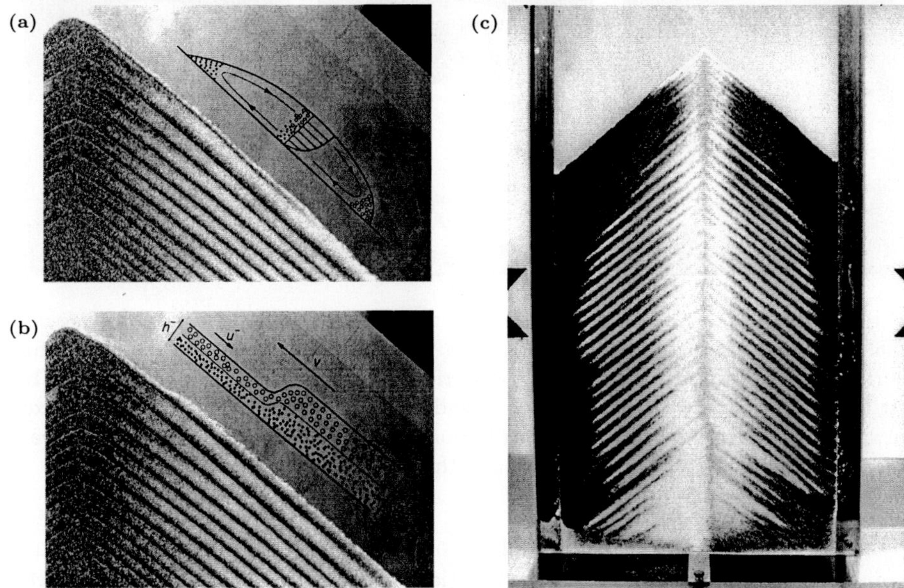

Fig. 4.6. Sand-piles formed by pouring a mixture of white large and dark small particles between the slit of two parallel glass plates. (**a**) Photograph and schematic diagram of a granular avalanche in a typical roll wave configuration. An inverse-graded particle size distribution rapidly develops in which the large (white) particles overlie the small (dark) particles forming a stripe. Velocity shear through the avalanche thickness then transports the larger (white) particles to the front. (**b**) Photograph and schematic diagram of the upward propagating dispersed shock wave. The material below the shock is at or near rest, whilst the grains above the shock are flowing rapidly downslope. (**c**) Photograph of the final deposition of the material as a pine tree type sand pile. As opposed to panels (**a**) and (**b**), the larger particles are here dark and the small ones white. (From [34])

of the triangular pile build up a sequence of such layers giving rise to a pine tree pattern as shown in Fig. 4.6c. It should also be mentioned that there is a tendency for the upslope propagating shock wave to destabilize the granular material on the opposite face of the pile as it reaches the centre, so that avalanches tend to form first on one side and then on the other.

In the second experiment the same granular mixture is contained within the small gap between two disks (diameter 25 cm) with a free surface that lies above the centre as shown in Fig. 4.7; particle size segregation may then occur if the disk is rotated about its centre. To emphasize the pattern formation, the disk is laid horizontally and gently shaken so that all the small particles fall to the bottom. Once gently returned to the vertical, one side of the disk is completely white whilst the other is completely dark. When the disk is rotated at constant rate (110 seconds per revolution), intermittent avalanches are formed at the free surface. The intermittency again stems from the difference between static and

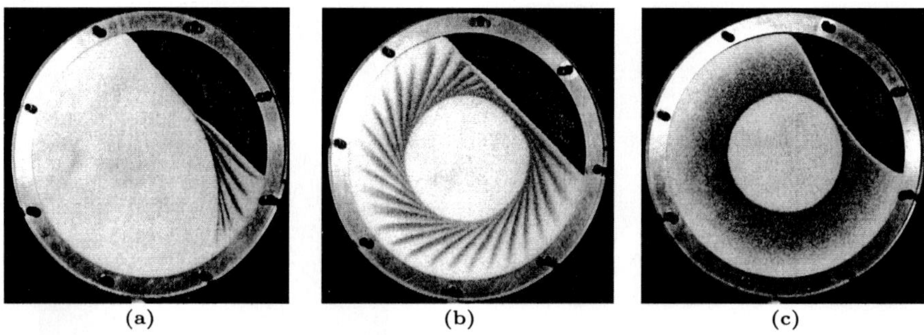

Fig. 4.7. At small rotation rates intermittent avalanche release in a thin rotating disk filled with a granular mixture leads to the formation of stripes tangent to the free surface (**a**), which are then rotated and buried to form a Catherine wheel effect (**b**). At faster rotation rates a quasi-steady flow develops in which the free surface is fixed in space and there is a continuous distribution of particle sizes outside the central core (**c**). The large particles are now white and the smaller ones dark. (From [34])

dynamic internal friction angles. The central circular core of material remains completely undisturbed by the slow rotation of the drum [77,81]. Each avalanche release sorts the material, forming a stripe, which is frozen into the deposit by the shock wave and subsequently rotated and buried in the undisturbed material below the free surface. Subsequent releases create a sequence of stripes (Fig. 4.7a) tangent to the central core, which create a Catherine wheel effect (Fig. 4.7b).

At faster rotation rates ($<$ 20 seconds per revolution) the intermittency of the avalanche ceases, the shock waves and the stripes disappear and a steady-state flow regime dominates, Fig. 4.7c. The material is continuously released on the upper side and continuously deposited on the lower side of the concave free surface and is transported between the two positions by a quasi-steady avalanche in which kinetic sieving takes place. Since the smaller particles are concentrated at the bottom of the avalanche they are the first to get deposited on the lower half of the free surface and a new pattern develops in which the central core is undisturbed, and there is a continuous distribution of grain sizes outside the central core, starting with a high concentration of small particles near the core and ending with a high concentration of large particles near the outer wall as shown in Fig. 4.7c.

4.1.3 Structure of Theories

Structurally, granular materials are described by three different theoretical concepts:

- **Molecular Dynamics**: One models the granular material as an assemblage of a large number of rigid bodies interacting with one another. Postulates are introduced to describe the interaction between the particles at the contact points and Euler's equations of the motion of each finite dimensional body are formulated and solved. This discrete particle method is applied to a large number of bodies in 2d and 3d flow configurations, see e.g. [11,12,23,42,72,97,119–121].

- **Statistical Mechanics**: In the limit as the number of particles becomes infinitely large this method is replaced by a statistical approach, in which moments of a Boltzmann type equation are used. Interactions between the individual bodies are expressed by the collision operator, here accounting for the loss of energy under collision. The number of moments taken defines the complexity of the theory, which is now continuous for fields that are statistical averages of fields exhibiting large fluctuations on the microscale, see e.g. [39,50,53–55,73,104].
- **Continuum Mechanical Models**: These are purely phenomenological descriptions and are restricted to macroscopic length scales that extend over many particle diameters. Closure conditions are based on common rules of rational thermodynamics and may account for microstructural effects, see e.g. [1,2,31,91,114,122,123,126].

In the next sections we shall illustrate accounts on all three of these concepts.

4.2 Single-Phase Theories

In many flows involving granular materials, the interstitial fluid plays an insignificant role in the transportation of momentum, and thus flows of such materials can often be considered as dispersed single-phase rather than multi-phase flows. Rockfalls, landslides and flow avalanches of snow, but also pipe flows of grains and pills in the food and pharmaceutical industry are examples of this sort. For dry materials (i.e. granules which are *suspended* in a gas of negligible density) there are three mechanisms that contribute to the generation of stress:

(1) dry Coulomb-type rubbing friction,
(2) transport of momentum by particle translation between contacts,
(3) dispersive momentum transport by collisional interactions.

In general, all three mechanisms are effective, however, there are flow regimes in which only a single one plays a dominant role. For instance, at high solid concentrations and low shear rates, the particles will be in close contact; as a result the stresses are of the *quasi-static, rate-independent Coulomb-type*. On the other hand, at very low concentrations and high shear rates the particles are likely to be in contact a very short time, and mean free paths are large as compared to the particle diameter. The transport of momentum by particles is significant and the bulk material will in some way behave *like a dilute gas*. When concentrations and shear rates are large, momentum transfer occurs as a result of collisional interactions, since the void spaces are too small to permit essential particle transport between collisions. This is referred to as the *grain inertia regime*.

In the following we will introduce various methods to describe the flows of granular materials.

4.2.1 Molecular Dynamics

The molecular dynamics approach to simulate the dynamics of granular materials is a direct numerical simulation of particle–particle interaction [119]. The calculations are typically carried out for a fixed number of spherical particles (or in two dimensions circular disks) that are usually bounded on the four sides by stationary or periodic boundaries. Initially, the particles have assigned random velocities. The numerical method involves explicit integration of Newton's law of motion. The collision between the spheres is often assumed to be elastic; slip could be allowed during contact as well as frictional resistance. While dealing with spherical particles it is unnecessary to know all the Euler angles to determine the subsequent motion of the particles; however the method could in principle be extended to simulations involving non-spherical particles in which case the initial orientations of the individual particles would be necessary[1].

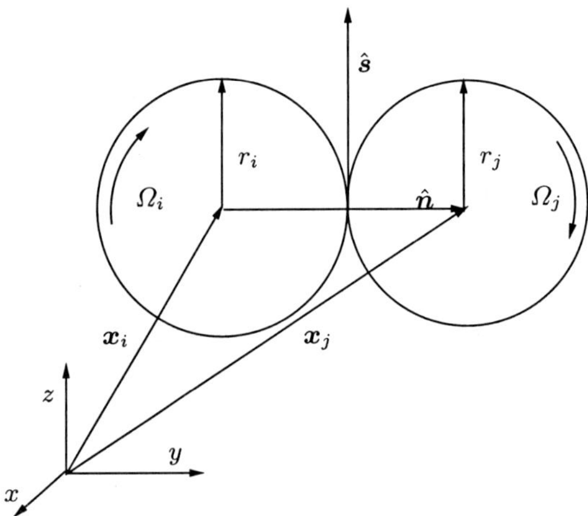

Fig. 4.8. Sketch of particle–particle contacts

Newton's second law of motion reads for a system of grains

$$\boldsymbol{F}_i = m_i \frac{\mathrm{d}^2 \boldsymbol{x}_i}{\mathrm{d}t^2} , \tag{4.3}$$

where $\boldsymbol{F}_i$ denotes the force acting on particle i and $\boldsymbol{x}_i$ stands for the position vector of the i-th particle. The crucial point and the physical input that enters

[1] For a non-spherical body the angular momentum equation is described in a frame fixed with the body: the Euler equations. The degrees of freedom are the so-called Euler angles; they describe at each instance the orientation of the body relative to a fixed coordinate system. Since a sphere is symmetric relative to any plane through its center only one equation is needed to describe its rotation. This leads to a substantial reduction of the complexity of the numerical integrations of many particle systems.

into the simulations are the forces. Besides external forces like gravity, the only forces are the ones acting during particle–particle and particle–wall collisions.

Whenever two spherical particles are closer to one another than the sum of their radii ($d := r_i + r_j$) they interact via normal and shear forces. In the normal direction $\hat{\boldsymbol{n}}$, given by the line connecting the two centers of mass of the colliding particles (see Fig. 4.8), the first contribution exerted on particle i comes from an *elastic restoring force*

$$F^i_{\text{elastic}} = -k_n \left(r_i + r_j - (\boldsymbol{x}_i - \boldsymbol{x}_j) \cdot \hat{\boldsymbol{n}}\right)^\alpha \,, \tag{4.4}$$

where k_n denotes Young's modulus. For $\alpha = 1$, (4.4) corresponds to Hooke's law, and for $\alpha = 1.5$ the Hertzian contact force due to slightly deformable spheres is investigated [43,68]. The second contribution acting in the direction of $\hat{\boldsymbol{n}}$ is a *dissipative force* proportional to the relative particle velocity in the normal direction

$$F^i_{\text{diss}} = -\gamma_n m_{\text{eff}} (\dot{\boldsymbol{x}}_i - \dot{\boldsymbol{x}}_j) \cdot \hat{\boldsymbol{n}} \,, \quad \text{with } m_{\text{eff}} := \frac{m_i m_j}{m_i + m_j} \,, \tag{4.5}$$

where γ_n stands for a phenomenological friction coefficient.

The force in the tangential direction $\hat{\boldsymbol{s}}$ is somewhat more artifical. The first contribution is a *viscous friction force*, assumed to be proportional to the relative velocity difference of the surfaces of the particles in this tangential direction,

$$F^i_{\text{viscous}} = -\gamma_s m_{\text{eff}} \left[(\dot{\boldsymbol{x}}_i - \dot{\boldsymbol{x}}_j) \cdot \hat{\boldsymbol{s}} + r_i \Omega_i + r_j \Omega_j \right] \,, \tag{4.6}$$

where Ω_i represents the angular velocity of the i-th particle (clockwise rotation is chosen as positive). To account for real static friction, which might lead to stable, static arches, a virtual spring is put at the point of contact inception during a collision [69,98]. This leads to a *static friction force*, neglecting rotation,

$$F^i_{\text{static}} = -k_s \int (\dot{\boldsymbol{x}}_i - \dot{\boldsymbol{x}}_j) \cdot \hat{\boldsymbol{s}} \, \mathrm{d}t \,, \tag{4.7}$$

in which integration is over the contact time. Due to the Coulomb yield criterion (4.2), the magnitude of the shear force exerted on particle i, F^i_{shear}, is given by

$$|F^i_{\text{shear}}| = \min \left(\left| F^i_{\text{viscous}} + F^i_{\text{static}} \right|, a + \mu \left| F^i_{\text{elastic}} + F^i_{\text{diss}} \right| \right) \,, \text{ with } \mu = \tan\phi \,, \tag{4.8}$$

where ϕ is the static internal angle of friction, while a is the cohesion. Different authors used different approaches in modeling the collisional forces (4.4)–(4.7) and criterion (4.8) in whole or in part. Despite the differences, however, they obtained many reasonable numerical results [24,28,69,72,92,97,117,118,121,128].

Since all forces, except the external forces like gravity, only act during particle contacts, one must keep track of all the collisions during numerical simulations. A naive implementation would check at each time step for all N particles whether they are in contact with any of the other $N-1$ particles. This is very inefficient and hardly feasible for system sizes of more than a few thousand particles. In order to achieve as effective numerical simulations as possible, different numerical methods have been suggested, see e.g. [3,10,80,95]. Numerical results obtained by the molecular dynamic simulations exhibit fair to good agreement with observations. For example, in a simulation of plane Couette flow by Thompson and Grest [117] a Hooke type elastic force model with 750 soft particles with equal radii is employed. A plug-like motion of the core and a thickness of the boundary shear layers of 6 to 12 particle diameters is found, which shows good agreement with experimental results [41]. Surprisingly, the shear stress did not show a quadratic dependence on the mean shear rate as expected [5] but became constant for large shear rates; this was explained by the dilatancy in the steady-state regime. Besides, *segregation by size* or *mass* can also be modelled by the use of molecular dynamics [40,89,96].

Such direct simulations can hardly handle real practical problems involving hundreds of thousands, in fact millions of particles, in which the interactions between any particle and its neighbours are far from simple. Nonetheless, they do provide useful insight into the formulation of theories, much the same as experimental results. This is demonstrated in the various computer codes that have been developed and are in use (see the review of Campbell [11] and Savage [101]).

Above, most molecular dynamic simulations have been performed for dry granular materials. The stickiness due to the humidity of the surrounding air may make it necessary to account for the *cohesion*. In such cases and when the viscous nature of the surrounding fluid is large, the interstitial fluid is significant. In such cases the Navier–Stokes equations can be used to model the fluid phase. What remains is to adequately incorporate the interactions between the grains and the fluid.

4.2.2 Statistical Mechanics

Procedures of statistical mechanics along the lines of molecular dynamics of dense gases or liquids have also been developed. The important results are the evolution equations for the density, the velocity and the granular temperature. Included in these expressions are also explicit formulas for the constitutive quantities. The statistical theory approaches that have been used are extensions of ideas of Brownian motion, Grad's thirteen moment method [53,54], Bhatnagar–Gross–Krook relaxation model [6,74] and Enskog dense gas theory [9,56,73,104].

A three-dimensional constitutive equation derived from the kinetic theory of a rarefied gas was proposed for the behaviour of dry granular flows in the *inertia* regime. The basic model was first presented by Savage and Jeffrey [104] then corrected and improved by Jenkins and Savage [56] and Lun et al. [73]. This model involves all sophisticated mathematical tools used in the kinetic theory. Its two fundamental assumptions are that momentum transfers via collisions prevail and that only binary collisions occur. This implies that the granular phase must be sufficiently dispersed which, for dry granular flows subject to gravity, seems possible only for very rapid motions. Some simpler models considered other types of energy dissipation [79,88,111,112] occurring specifically during collisions, but the collision process was still considered to be predominant. Unfortunately, most flows of practical interest probably fall into the intermediate regime where both frictional contacts and particle–particle collisions are significant. For example the experimental results of Johnson et al. [61] and Ancey et al. [4] clearly demonstrated that dry granular flows exhibit complex properties and that their behaviour cannot in general be described with the statistical theory alone. With a view to providing a more realistic approach capable of predicting both slow and rapid granular flows, Johnson and Jackson [60] proposed adding a Coulomb frictional yield stress term to the stress found in the Lun et al. [73] development. Nevertheless a complete, theoretical treatment of slow or moderately rapid, channelled, granular flows taking into account both friction and collisions has yet to be performed.

Some detailed reviews about the statistical mechanics approach have been presented by Hutter and Rajagopal [48].

4.2.3 Continuum Mechanical Models without Additional Balance Laws

Higher Order Closure Models. Many experimental results show that dry granular materials exhibit non-Newtonian behaviour [5,41,100,107,110]. On the basis of these observations purely mechanical theories for the grain inertia regime were designed; they made use of the balance laws of mass and momentum and constitutive equations for the stress tensor. The balance equations of mass and linear momentum read as follows

$$\frac{\mathrm{d}\rho}{\mathrm{d}t} + \rho \,\mathrm{div}\, \boldsymbol{v} = 0 \,, \quad \rho\frac{\mathrm{d}\boldsymbol{v}}{\mathrm{d}t} = \mathrm{div}\, \boldsymbol{T} + \rho \boldsymbol{b} \,, \quad \text{with } \frac{\mathrm{d}(\cdot)}{\mathrm{d}t} = \frac{\partial(\cdot)}{\partial t} + \boldsymbol{v}\cdot \mathrm{grad} \,, \tag{4.9}$$

where $\boldsymbol{v}$ is the velocity, $\boldsymbol{T}$ the Cauchy stress tensor, $\boldsymbol{b}$ the specific body force and ρ is the density, also given as $\rho = \gamma\nu$, with the granular true density γ and the grain volume fraction ν. For incompressible grains, the balance of mass $(4.9)_1$ can also be rewritten as

$$\frac{\mathrm{d}\nu}{\mathrm{d}t} + \nu \,\mathrm{div}\, \boldsymbol{v} = 0 \,. \tag{4.10}$$

In the early purely mechanical phenomenological theories [30,31,51,79,87] and [91,100,110], the Cauchy stress $\boldsymbol{T}$ is of the form

$$\boldsymbol{T} = \boldsymbol{f}(\nu, \mathrm{grad}\, \nu, \boldsymbol{D}) \,, \tag{4.11}$$

where $\boldsymbol{D} = \mathrm{sym}\,\mathrm{grad}\, \boldsymbol{v}$ is the stretching tensor. It is not physically obvious why the stresses should depend upon the gradients of ν, however (see [100]). It should also be emphasized that models of the form (4.11) have limited applicability: dense slow to moderately fast flows of granular materials, but by the same token models that arise from kinetic theory approaches also have equally limited applicability, namely very rapid flows. A thorough discussion of models of the class (4.11) can be found in [76].

The most general isotropic representation for the stress given by (4.11) is

$$T = a_0 I + a_1 D + a_2 D^2 + a_3 \operatorname{grad}\nu \otimes \operatorname{grad}\nu + a_4 \operatorname{sym}(\operatorname{grad}\nu\,(D \otimes \operatorname{grad}\nu)) + a_5 \operatorname{sym}\left(\operatorname{grad}\nu\left(D^2 \otimes \operatorname{grad}\nu\right)\right) , \tag{4.12}$$

where the coefficients $a_1, a_2, \ldots, a_5$ are functions of

$$a_i = \tilde{a}_i\left(\nu, I_D, II_D, III_D, \operatorname{grad}\nu\cdot\operatorname{grad}\nu, \operatorname{grad}\nu\cdot D\operatorname{grad}\nu, \operatorname{grad}\nu\cdot D^2\operatorname{grad}\nu\right)$$

with the principal invariants

$$I_D = \operatorname{tr}D\ , \quad II_D = \tfrac{1}{2}\left((\operatorname{tr}D)^2 - \operatorname{tr}(D^2)\right)\ , \quad III_D = \det D\ . \tag{4.13}$$

However, this representation is not very useful here. The early models were therefore essentially generalizations of Reiner–Rivlin fluids and had the structure

$$T = \alpha_1 I + \alpha_2 D + \alpha_3 D^2\ , \tag{4.14}$$

where α_i, $i = 1, 2, 3$ are functions of the principal invariants of D and ν. However, these models are fraught with internal inconsistencies, which has succinctly been discussed in [108]. They cannot exhibit all normal stress differences in simple shear flow, while those whose stress depends on both $\operatorname{grad}\nu$ and D can do so.

Here, we shall now introduce a model belonging to the class (4.11) that is appropriate for flows of such granular materials under non-isothermal conditions. In the model of interest, the Cauchy stress T takes the form [93]

$$T = (\beta_0(\nu) + \beta_1(\nu)\operatorname{tr}(\operatorname{grad}\nu\otimes\operatorname{grad}\nu) + \beta_2(\nu)\operatorname{tr}D)\, I + \beta_3(\nu) D + \beta_4(\nu)\,(\operatorname{grad}\nu\otimes\operatorname{grad}\nu)\ . \tag{4.15}$$

One can arrive at a model with the precise structure shown in (4.15) from a kinetic theory approach based on Enskog's dense gas theory [76]. Amongst the many specific continuum models, (4.15) has been used to study a variety of problems: Massoudi [75] describes flows in problems involving fluidization, and later Johnson et al. [57–59] use the same model in flows involving suspensions of particles in fluids, in various geometries. Another related study is that of the flow of granular materials down a vertical pipe due to the action of gravity, by Gudhe et al. [37]. They carried out a detailed parametric study of the problem and found that for a certain range of parameters, the predictions of the theory agree quite well with the experimental results of Savage [100]. Rajagopal et al. [94] have also studied the flow of a granular material using a similarity transformation $v = u(y)\hat{\imath}$, $\nu = \nu(y)$ and showed that for a range of values of the material parameters, the equations admit non-unique solutions, one in which the volume fraction increases monotonically from the free surface to the bottom plane, and the other in which it decreases monotonically from the free surface to the plane.

Constitutive Postulates for Rapid Flow of Cohesionless Materials. The constitutive equation of a suspension in sufficiently rapid flows where collisions are predominant was proposed by Bagnold [5]. Indeed, in the so-called *inertia* regime, Bagnold assumed that momentum transfer was due to elastic collisions between particles of parallel layers in relative motion. The momentum transmitted through each collision is proportional to the relative velocity between the two colliding particles which for one-dimensional shear is proportional to the shear rate $\gamma = \mathrm{d}u/\mathrm{d}y$, because they belong to two adjacent layers in relative motion. For the same reason the collision frequency is also proportional to γ. The shear stress originating in collisions is therefore proportional to γ^2. The momentum transfer process also yields a normal stress proportional to shear stress.

These trends were found to be in agreement with some experimental results [5,107]. Bagnold's experimental results in his annular cell viscometer [5], in which identical, rigid, neutrally buoyant spheres in a Newtonian fluid under shear were considered, have corroborated that the dispersive pressure and the shear stress depend quadratically upon shear rate, if the dynamic friction angle is independent of shear rate. This strong rate dependence differs sharply from simple Newtonian behaviour. The constitutive relation for the stress tensor $\boldsymbol{T}$ in a general three-dimensional case is postulated as suggested in (4.14) for the Reiner–Rivlin fluid with coefficient functions α_i $(i = 1, 2, 3)$ such that Bagnold's shear cell results for the shear and normal stresses are reproduced as closely as possible. The constitutive relation achieving this is not unique. The following proposed relations satisfy this requirement:

$$
\begin{aligned}
&\text{McTigue [78]:} && \boldsymbol{T} = f(\rho)\left(\sqrt{II_{\boldsymbol{D}}}\boldsymbol{D} - A\boldsymbol{D}^2\right) , \\
&\text{Savage [100]:} && \boldsymbol{T} = f(\rho)\left(-A II_{\boldsymbol{D}}\boldsymbol{I} + \sqrt{II_{\boldsymbol{D}}}\boldsymbol{D}\right) , \\
&\text{Jenkins and Cowin [52]:} && \boldsymbol{T} = f(\rho)\left(\sqrt{II_{\boldsymbol{D}}}\boldsymbol{D} - (1/2)A\boldsymbol{D}^{\triangle}\right) ,
\end{aligned}
\qquad (4.16)
$$

where $II_{\boldsymbol{D}}$ is the second invariant of the tensor $\boldsymbol{D}$ and

$$
\boldsymbol{D}^{\triangle} := \dot{\boldsymbol{D}} + \boldsymbol{D}\,\boldsymbol{L} + \boldsymbol{L}^T\boldsymbol{D} , \qquad \boldsymbol{L} = \operatorname{grad}\boldsymbol{v}
$$

and $f(\rho)$ and A are supposed to be known from parameterisations. These stress-stretching relationships are rate dependent[2] and do only represent the dynamic portion of the stress tensor. They can describe rapid flows without cohesion. For a slow flow of granular material, a quasi-static, rate-independent part must be added to it.

Constitutive Postulates with Cohesion. In general terms, the flows of granular materials exhibit *plastic* as well as *viscous* behaviour. Soil under quasi-static loads exhibits plastic behaviour, while rapid shearing of a debris flow is predominantly viscous. Examples of purely viscous behaviour are the Newtonian fluids,

[2] A stress stretching relation is called rate dependent if a replacement of $\boldsymbol{D}$ by $\lambda\boldsymbol{D}$, $\lambda \neq 0$, yields a functional relation for $\boldsymbol{T}$ that depends on λ. If the emerging relationship should be independent of λ, it is called rate independent. The versor $\hat{\boldsymbol{D}}$ defined in (4.18) is rate independent. Rate dependent is also called viscous, while rate independent is called plastic.

the Bagnold fluid and essentially all constitutive models derived from statistical mechanics. Purely plastic behaviour is exhibited by all Mohr–Coulomb models (see [25]). Simple viscoplastic models are the Bingham and Herschel–Bulkley bodies (see [14] and [17] as well as Chap. 2).

The usual approach taken is that, for a viscoplastic shear flow in a gravitational field, one might represent the total stresses as the linear sum of a rate-independent dry friction part describing the quasi-static flow regime plus a rate dependent *viscous* part covering the grain inertia regime.

As a result of the cohesion a granular mass sliding down an inclined plane develops a vigorously active fluidized layer only very close to the bed, whereas the layer on top of this is more or less passively riding with the lower layer. If the fluidized bed is very thin in comparison to the passive layer it may be ignored altogether and incorporated in the basal boundary condition. This is done by Savage and Hutter [102,103]. This theory has been widely used to describe the two-dimensional (later extended to three-dimensional) motion of a finite mass avalanche over a rough inclined slope [33,35,36,44–46,49,65,116]. These applications show a good agreement with experimental results. For details see Chap. 14.

If in a granular flow the sheared and the passive layers are of comparable thickness, then a more detailed analysis is required. This is the combined shear-plug-flow regime. Norem et al. [85] introduce an extension of the Criminale–Ericksen–Filbey fluid [22] as a model for rapid shear flow of a granular material and demonstrate that their proposed constitutive relation fitted the experimental data of Savage and Sayed [107] well and was, in steady shear flow with free surface, capable of having both a shear-deformation and a plug-flow regime. Their constitutive relation has the form

$$\boldsymbol{T} = -p\boldsymbol{I} + 2\mu\hat{\boldsymbol{D}} + 2\eta\boldsymbol{D} + (2\Psi_1 + 4\Psi_2)\boldsymbol{D}^2 + \Psi_1\boldsymbol{A} \,, \tag{4.17}$$

where $\hat{\boldsymbol{D}}$ is called a *versor* [29] and $\boldsymbol{A}$ is the second *Rivlin–Ericksen–tensor*, expressed by

$$\hat{\boldsymbol{D}} = \boldsymbol{D}/\sqrt{\mathrm{tr}\boldsymbol{D}^2} \,, \qquad \boldsymbol{A} = \dot{\boldsymbol{D}} - \boldsymbol{W}\boldsymbol{D} + \boldsymbol{D}\boldsymbol{W} \,, \tag{4.18}$$

in which $\boldsymbol{W} = \mathrm{skw}\,\mathrm{grad}\,\boldsymbol{v}$, and μ, η, Ψ_1 and Ψ_2 are phenomenological coefficients, where η is a viscosity and Ψ_1 and Ψ_2 are viscometric functions. In a steady simple shear flow, they account for primary and secondary normal stress effects. The coefficient μ is a *plastic modulus*, assumed by Norem et al. [85] in the form

$$\mu = \frac{1}{\sqrt{2}}\left(a + \beta p_e^k\right) \,, \qquad \beta = \tan\phi \,, \tag{4.19}$$

where a, β and k are constants while p_e is their *effective pressure*, which will simply be the pressure p, as any pore pressure is ignored. To find the interpretation of the plastic modulus, consider a pure shear deformation $\partial u/\partial y = \gamma(y)$, $v = w = 0$ in the limit as $\gamma \to 0$. Then, (4.17) implies the shear stress

$$\tau = a + (\tan\phi)p^k \,, \tag{4.20}$$

which, for $k = 1$, corresponds to a Coulomb friction model. Thus, a is the cohesion and ϕ the internal friction angle. Thus, (4.17) combines a dry Coulomb plastic behaviour with a non-Newtonian viscous behaviour.

Another three-dimensional formulation of the constitutive equation for viscoplastic fluids takes the form [18,27]

$$\begin{cases} \boldsymbol{D} = \boldsymbol{0}\,, & \sqrt{|II_{\boldsymbol{T}}|} < \tau_c\,, \\ \boldsymbol{T} = \sqrt{2}\,\tau_c \hat{\boldsymbol{D}} + f(II_{\boldsymbol{D}})\boldsymbol{D}\,, & \sqrt{|II_{\boldsymbol{T}}|} \geq \tau_c\,, \end{cases} \tag{4.21}$$

where τ_c is the yield stress, $II_{\boldsymbol{T}} = \frac{1}{2}\left((\mathrm{tr}\boldsymbol{T})^2 - \mathrm{tr}(\boldsymbol{T}^2)\right)$ is the second invariant of $\boldsymbol{T}$ and $f(II_{\boldsymbol{D}})$ a positive continuous function of $II_{\boldsymbol{D}}$.

Special cases are the *Bingham fluid* ($f = \mu_B =$ constant) and *Herschel–Bulkley fluid* ($f = 2^n K/(\sqrt{|II_{\boldsymbol{D}}|})^{1-n}$), in which K and n are positive parameters [13,15,19,84,109]. Another model that fits into the same class is

$$\boldsymbol{T} = \begin{cases} \rho\nu_1 \boldsymbol{D}\,, & \sqrt{|II_{\boldsymbol{T}}|} < \tau_c\,, \\ \sqrt{2}\,\tau_c\left(1 - \dfrac{\nu}{\nu_1}\right)\hat{\boldsymbol{D}} + \rho\nu\boldsymbol{D}\,, & \sqrt{|II_{\boldsymbol{T}}|} \geq \tau_c\,, \end{cases} \tag{4.22}$$

in which τ_c is the yield stress and ν, ν_1 are positive constants. Liu and Mei [71] proposed this model for debris flows and were able to separate with it regions of weak and strong shearings. They also analysed with it the linear and nonlinear stability of shear flows and the formation of *roll waves* in free surface gravity flows.

Yet another proposal is a combination using (4.16) and (4.21) as follows

$$\begin{cases} \boldsymbol{D} = \boldsymbol{0}\,, & \sqrt{|II_{\boldsymbol{T}}|} < \tau_c\,, \\ \boldsymbol{T} = \sqrt{2}\,\tau_c \hat{\boldsymbol{D}} + \boldsymbol{T}_{(4.16)}(\boldsymbol{D})\,, & \sqrt{|II_{\boldsymbol{T}}|} \geq \tau_c\,, \end{cases} \tag{4.23}$$

in which $\boldsymbol{T}_{(4.16)}(\boldsymbol{D})$ is one of the stress expressions in (4.16). For steady shear all expressions (4.23) reduce to the one-dimensional constitutive relation for the shear stress τ, proposed by Julien and Lan [62]

$$\begin{cases} \dfrac{\mathrm{d}u}{\mathrm{d}y} = 0\,, & \tau < \tau_c\,, \\ \tau = \tau_c + \mu_d\left(\dfrac{\mathrm{d}u}{\mathrm{d}y}\right) + \mu_c\left(\dfrac{\mathrm{d}u}{\mathrm{d}y}\right)^2, & \tau \geq \tau_c\,, \end{cases} \tag{4.24}$$

where μ_d denotes dynamic viscosity, μ_c a dispersive and turbulence parameter, and τ_c is the yield stress. In this constitutive relation the shear stress contains the yield stress, a linear and a quadratic rate-dependent part. Using (4.24), Julien and Lan successfully simulated various experimental data obtained by Govier et al. [32], Savage and McKeown [106] and Bagnold [5].

4.2.4 Constitutive Theories with Additional Balance Laws

The main difference between a classical theory of solid or fluid bodies and the theory of granular or porous materials is connected with the existence of the pore space that affects the kinematics of the material at the macroscopic scale encompassing several microstructural granular elements. An additional variable is introduced – the porosity or its complement, the solid volume fraction – to describe the distribution of the grain volume fraction in the total microscopic control volume. In the construction of a theoretical model the classical balance laws of mass, momenta and energy together with constitutive relations for the internal (or free) energy, stress tensor and heat flux vector still form the "back bone"of a theoretical formulation of granular materials, but these laws must now be supplemented by relations describing the evolution of the grain volume fraction ν (or porosity $1-\nu$). A number of models of this class have been proposed; they can be divided into two classes:

- additional constitutive relations are introduced,
- additional field equations in the form of either evolution equations or balance equations are proposed.

The simplest models of the first class are those proposed by Bowen [8] and Sampaio and Williams [99]. These authors write $\rho = \gamma\nu$, where ρ is the density in the granular assemblage, γ the true mass density of the granules and ν the solid volume fraction as before. For constant [3] γ, the material of which the grains are made is density preserving and the emerging theory is formally analogous to a classical compressible mixture. This theory has been demonstrated to be flawed especially when it is applied to dynamical and relaxation processes, see [7].

The most commonly used model within the second class seems to be that proposed by Goodman and Cowin [30,31]. It makes also use of $\rho = \gamma\nu$, but does not a priori impose the constancy of γ. This enlarges the number of field variables by one, the volume fraction ν, and entails in compensation the introduction of an additional field equation. It is called the *balance law of equilibrated forces* and is analogous to the classical balance equation of linear momentum and motivated by a variational analysis [20].

The field equations of this theory are given by

$$
\begin{aligned}
\mathcal{R} &:= (\gamma\nu)^{\cdot} + \gamma\nu \operatorname{div} \boldsymbol{v} = 0 \, , \\
\boldsymbol{\mathcal{M}} &:= \gamma\nu\dot{\boldsymbol{v}} - \operatorname{div} \boldsymbol{T} - \gamma\nu\boldsymbol{b} = \boldsymbol{0} \, , \\
\mathcal{N} &:= \gamma\nu k\ddot{\nu} - \operatorname{div} \boldsymbol{h} - \gamma\nu f = 0 \, , \\
\mathcal{E} &:= \gamma\nu\dot{\varepsilon} - \boldsymbol{T}\cdot\boldsymbol{D} - \boldsymbol{h}\cdot\operatorname{grad}\dot{\nu} + \gamma\nu f\dot{\nu} + \operatorname{div}\boldsymbol{q} - \gamma\nu r = 0 \, ,
\end{aligned}
\tag{4.25}
$$

where $(\bullet)^{\cdot}$ is the material time derivative defined in (4.9). $\mathcal{R} = 0$ and $\boldsymbol{\mathcal{M}} = \boldsymbol{0}$ are the balance laws of mass and linear momentum (and balance of moment of

[3] This is a special constitutive relation; more general would be, if γ would depend on other variables as for instance the pressure, volume fraction, etc.

momentum is identically satisfied by the requirement that $\boldsymbol{T}$ is symmetric). The third equation in (4.25), $\mathcal{N} = 0$, is new, has scalar structure and contains new fields: $k =$ constant is called the coefficient of *equilibrated inertia*; $\boldsymbol{h}$ and f are the *equilibrated stress vector* and *intrinsic equilibrated body force*, respectively. This equation comprises the *equilibrated force balance* and is supposed to model the microstructural force systems operative in granular materials. The conservation of energy $(4.25)_4$, $\mathcal{E} = 0$, differs from the traditional statements. It does not only balance internal energy ε, heat flux $\boldsymbol{q}$, stress power $\boldsymbol{T} \cdot \boldsymbol{D}$ and radiation r, it also contains two additional terms, $-\boldsymbol{h} \cdot \operatorname{grad} \dot{\nu} + \gamma \nu f \dot{\nu}$, interpretable as the power of working of the equilibrated stress and intrinsic equilibrated force.

Equations (4.25) constitute six scalar equations for e.g. γ, ν, $\boldsymbol{v}$ and θ. These are called the independent fields. The remaining variables arising in (4.25), namely[4]

$$\mathcal{C} := \{\boldsymbol{T}, \boldsymbol{h}, \boldsymbol{q}, f, \varepsilon, \eta, \boldsymbol{\phi}\} \, , \tag{4.26}$$

must then be functionally related to these independent fields. The form of this dependence defines the constitutive class and a popular choice is

$$\mathcal{S} := \{\nu, \operatorname{grad} \nu, \dot{\nu}, \gamma, \theta, \operatorname{grad} \theta, \boldsymbol{D}\} \; . \tag{4.27}$$

$\mathcal{S}$ is called the state space and $\mathcal{C} = \hat{\mathcal{C}}(\mathcal{S})$ defines the material model. The assumed dependence on the rate of deformation tensor implies a fluid-like behaviour, it is suitable for describing rapid granular shear flows. In another study, Nunziato and Cowin [86] considered a slightly different theory in which the dependence on the rate of deformation tensor was suppressed, but the dependence upon the deformation gradient (among other things) were retained. These assumptions imply a solid-like behaviour, and it is thought that this type of model is appropriate for quasi-static motions of porous solid materials and pressed powders. A combination of both is, of course, equally possible.

The postulation $\mathcal{C} = \hat{\mathcal{C}}(\mathcal{S})$ is not arbitrary as it must be in conformity with the second law of thermodynamics. The latter is commonly expressed as a balance of entropy

$$\Pi = \gamma \nu \dot{\eta} + \operatorname{div} \boldsymbol{\phi} - \gamma \nu s \geq 0 \, , \tag{4.28}$$

where η, $\boldsymbol{\phi}$, s and Π are the specific entropy, its flux, supply and production densities, respectively. The first two are equally constitutive quantities. The inequality (4.28) must be satisfied for all processes satisfying the balance laws (4.25) and constitutive relations $\mathcal{C} = \hat{\mathcal{C}}(\mathcal{S})$. This requirement constrains the form of the constitutive relations. There are several approaches available in the literature how this reduction of the constitutive relations is executed; they differ not only in details but also in fundamental physical assumptions, thus corresponding to

[4] Note that (4.26) contains two additional fields, not arising in (4.25): η the specific entropy and $\boldsymbol{\phi}$, the entropy flux. Their occurrence will become apparent later.

different second laws of thermodynamics[5]. Results obtained with the entropy principle of Müller [83] depend on the functional form of the Helmholtz free energy

$$\psi = \varepsilon - \theta\eta \,.$$

If ψ does not depend upon $\dot{\nu}$, $\psi \neq \hat{\psi}(\bullet, \dot{\nu})$, then it can be shown that

$$\begin{aligned} &\psi = \hat{\psi}(\nu, \operatorname{grad}\nu \cdot \operatorname{grad}\nu, \gamma, \theta) \,, \qquad && \phi(\mathcal{S}) = \boldsymbol{q}(\mathcal{S})/\theta \,, \\ &\boldsymbol{h} = \mathcal{A}\operatorname{grad}\nu \quad \text{with } \mathcal{A} = 2\gamma\nu \frac{\partial\psi}{\partial(\operatorname{grad}\nu\cdot\operatorname{grad}\nu)} \,. \end{aligned} \tag{4.29}$$

Thus, the free energy depends only on subspace $\mathcal{E} := \mathcal{S} \setminus \{\boldsymbol{D}\}$ of the state space $\mathcal{S}$, and $\boldsymbol{h}$ is given once ψ is specified. Moreover, the entropy flux assumes the classical relationship $\boldsymbol{q}/\theta$, where θ is the absolute temperature. The constitutive relations for the stress tensor $\boldsymbol{T}$, heat flux vector $\boldsymbol{q}$ and intrinsic equilibrated body force f can be decomposed into equilibrium, $(\cdot)^E$, and non-equilibrium, $(\cdot)^D$, parts

$$\boldsymbol{T} = \boldsymbol{T}^E + \boldsymbol{T}^D \,, \quad \boldsymbol{q} = \boldsymbol{q}^E + \boldsymbol{q}^D \,, \quad f = f^E + f^D \,, \tag{4.30}$$

of which the former are given by

$$\boldsymbol{T}^E = -\nu p \boldsymbol{I} - \mathcal{A}\operatorname{grad}\nu \otimes \operatorname{grad}\nu \,, \quad \boldsymbol{q}^E = \boldsymbol{0} \,, \quad f^E = \frac{p-\beta}{\gamma\nu} \,, \tag{4.31}$$

with

$$p = \gamma^2 \frac{\partial\psi}{\partial\gamma} \,, \qquad \beta = \gamma\nu\frac{\partial\psi}{\partial\nu} \,, \tag{4.32}$$

whilst the latter are general relations of the form $\mathcal{C} = \hat{\mathcal{C}}(\mathcal{S})$ such that they vanish in the equilibrium state: $\boldsymbol{T}^D(\mathcal{E}) = \boldsymbol{0}$, $\boldsymbol{q}^D(\mathcal{E}) = \boldsymbol{0}$ and $f^D(\mathcal{E}) = 0$. If, for the dynamic parts a linear theory in the non-equilibrium variables is considered, then

$$\boldsymbol{q}^D = -\kappa \operatorname{grad}\theta \,, \quad \boldsymbol{T}^D = \xi\dot{\nu}\boldsymbol{I} + \lambda(\operatorname{tr}\boldsymbol{D})\boldsymbol{I} + 2\mu\boldsymbol{D} \,, \quad f^D = -\zeta\dot{\nu} - \delta\operatorname{tr}\boldsymbol{D} \,, \tag{4.33}$$

where

$$\kappa \geq 0 \,, \quad \mu \geq 0 \,, \quad \lambda + \frac{2}{3}\mu \geq 0 \,, \quad \xi \geq 0 \,, \quad \zeta \geq 0 \,, \quad \delta \geq 0 \,. \tag{4.34}$$

[5] One approach is that of Coleman–Noll, using the Clausius–Duhem inequality [16]; another is the entropy principle of Müller [82,83] and Liu [70] which is more general. A comparison of the two methods for the granular material with the constitutive class (4.27) is given in [124], with details contained in [122]. Results obtained with the two approaches are different under dynamic conditions, but the same in thermodynamic equilibrium.

Some typical results calculated for the granular volume fraction, dimensionless velocity for a gravity-flow problem down a inclined plate are shown in Fig. 4.9 for various values of the dimensionless granular flow thickness $\bar{L}$. If the layer thickness is small (e.g. only several grain diameters), the shear can extend from the bottom to the free surface, which behaves much like an incompressible fluid, and the volume fraction experiences only a small change across the depth, whereas for thicker grain flow the flow structure is far from an incompressible fluid, in which in a large region near the free surface of the grain flow is similar to that of a plug flow, with a nearly constant velocity and less changed volume fraction, the shear layers close to the bottom, where dilatation has occurred, may be very thin.

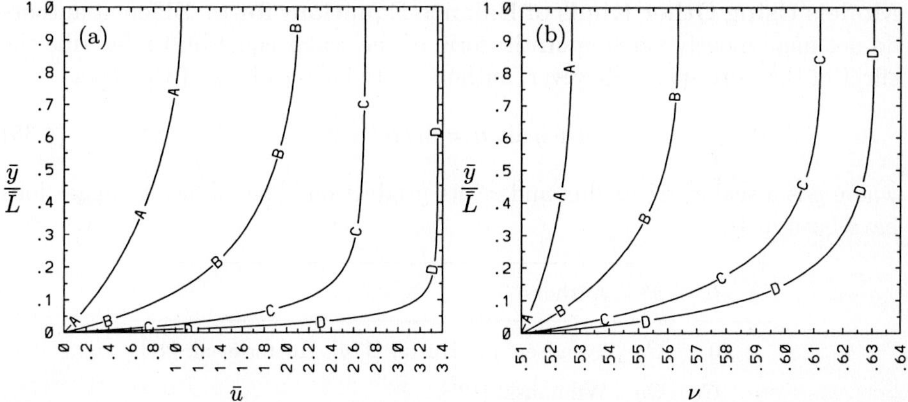

Fig. 4.9. Non-dimensional velocity profiles (**a**) and volume fraction profiles (**b**) of a steady granular gravity flow down a rough inclined plate for various values of the dimensionless flow thickness $\bar{L}$ indicating the ratio of the flow thickness to an internal length scale (e.g. the grain diameter): $\bar{L} = 5\,(A)$; $10\,(B)$; $20\,(C)$ and $30\,(D)$. The inner free energy ψ and the viscosity μ are chosen according to [90,91] and [100]. For details see [122]

The results of the gravity shear flow problem indicate that the internal length scale naturally provided by this model makes it possible to achieve blunt velocity profiles in gravity driven shear flows. This is a property that is experimentally corroborated. Nonetheless, the model is doubtful and the question of its applicability to real granular materials has not yet been adequately answered. First, as known from Bagnold's work [5], $\boldsymbol{T}^D$ should show quadratic dependence on the shear rate. This problem can be rectified for instance by taking a dissipative stress representation according to (4.16). Second, equilibrium results are ill behaving, since nontrivial gradients in the volume fraction are needed in order for nontrivial solutions to exist. Cowin and Nunziato [21] give examples. Third, at very small flow rates, rate independent behaviour would be expected because in this range of deformation dissipation is based on particle-to-particle rubbing. All these effects can, in principle, be incorporated, but the theoretical formulation is very complicated.

4.3 Other Models Based on Additional Balance Laws – Discussion

It was pointed out in the last section that microstructural effects in a granular porous materials can be accounted for by introducing the solid volume fraction ν as an additional field. The equilibrated force balance was the field equation postulated to model its redistribution. There are other equations which have been proposed to model microstructure properties, and here we briefly list these.

Models Using Other Kinds of Balance Equations for ν. Different authors do not unanimously agree upon the form of the scalar equation to describe the effect of the pore space. However, authors write balance laws of the form

$$\dot{g} + g \operatorname{div} \boldsymbol{v} = \operatorname{div} \boldsymbol{h}^g + \pi^g \tag{4.35}$$

where g is a scalar, $\boldsymbol{h}^g$ its flux and π^g its production. The following suggestions have been made:

g	$\boldsymbol{h}^g$	π^g	Authors
ν	0	Cq	Svendsen and Hutter [114], Hutter et al. [47]
ν	Cq	Cq	Wilmanski [125]
γ	Cq	Cq	Bluhm et al. [7]

$Cq \mathrel{\hat{=}}$ Constitutive quantity

In the first and second approach a balance of solid volume fraction (or porosity) is formulated with and without flux term $\boldsymbol{h}^g$, in the third approach this balance law is postulated for the true density. The energy equation in these proposals is not altered by adding power of workings due to $\boldsymbol{h}^g$ and π^g as was done in the Goodman–Cowin approach. Because $\rho = \gamma\nu$ the structures of the formulations by Bluhm and Wilmanski are not really different; the first formulation in the above table, however, yields different resulting theories, since the flux term $\boldsymbol{h}^g$ is set to zero ab initio. It may also be no big surprise that the essential differences in comparison to the Goodman–Cowin approach will show in thermomechanically coupled processes, since the essential differences arise in the energy equation.

The questions touched upon above are still open and form topics of today's research in granular media. A comparison of the thermodynamic formulation of the Goodman–Cowin, Svendsen–Hutter and Wilmanski approaches has been given by Kirchner [63], summarized in [64], but typical flow problems (e.g. gravity driven shear) have not yet been compared.

Model with an Additional Balance Equation for an Internal Variable Describing Internal Friction. The behaviour of a granular material at low volume fraction and high (kinetic) energy, i.e. in the grain inertia regime of Bagnold [5], appears to be modelled quite well by the application of Enskog's theory of dense kinetic gases (e.g. [53,73]). Unfortunately, the extension of this approach to high grain density and moderate to low energy, where frictional interactions become dominant, appears intractable (see discussion in [48]). Simple Mohr–Coulomb-type phenomenological continuum models for the quasi-static frictional behaviour of granular materials enjoy wide use in the continuum modelling of granular flows [33,102]. Among the generalizations of the Mohr–Coulomb idea one finds the viscoplastic model of Goddard [29], and more recently, the hypoplastic approach of Kolymbas [38,66,67,127]. The formulation of all these models has been based up to this point on statistical mechanical and/or direct phenomenological considerations. For materials whose quasi-static behaviour is governed by internal friction, e.g. dry granular materials, the process of internal friction is represented phenomenologically with the help of a second-order, symmetric-tensor-valued internal variable $\boldsymbol{Z}$. This variable is by interpretation associated with the effective *contact* stress in the granular material. In the phenomenological setting, $\boldsymbol{Z}$ is modelled constitutively by an incremental relation of the form

$$\overset{\triangledown}{\boldsymbol{Z}} = \boldsymbol{\Phi} \,, \tag{4.36}$$

where $\overset{\triangledown}{\boldsymbol{Z}}$ is an objective time (e.g. Jaumann) derivative of $\boldsymbol{Z}$ and $\boldsymbol{\Phi}$ represents the constitutive part of (4.36). A choice recovering the hypoplastic material would be $\boldsymbol{\Phi} = \alpha \boldsymbol{T}$, where α is a constant. Svendsen et al. [115] provide a thermodynamic formulation for a granular continuum incorporating such quasi-static frictional behaviour and Kirchner [63] extends it to a formulation of the Goodman–Cowin type as presented in Sect. 4.2.4.

Other Effects Modelling Internal Structure. In the above the solid volume fraction has been used as the kinematic variable describing the internal structure of the pore space. Despite the mathematical complexity, this is a very simple description of the porosity. It does not account for pore space *anisotropy* or *tortuosity* which would require additional internal variables of tensorial complexity. Furthermore other internal effects such as *fragmentation* and *abrasion* have not been touched upon so far. The former is the fracturing of grains into parts of comparable size, the latter is a grain-surface smoothening or roughening.

There is limited progress on these subjects. Kirchner [63] has modelled abrasion by a scalar variable a, having the dimension of a length and being a measure of surface roughness of the grains. A balance law of the Goodman–Cowin type is formulated for its time rate of change, $\dot{a}$, and a thermodynamic theory is formulated to describe its effects macroscopically. First shear flow computations show that roughness inhomogeneities may trigger localization phenomena. Induced anisotropies will develop in rapid flows of nonspherical grains such as rise or lenticular peas. Theoretical formulations accounting for such effects are under way (H. Ehrentraut: personal communication). Concepts are very much like those used in rheology to describe the flow of suspensions of long chain polymers.

How can particle size segregation be handled? One approach is mixture theory which is not the subject of this article. The particles within a certain range of nominal diameter will define one of the constituents of the mixture and the segregation will be described by the different motions of these constituents. Such an approach is taken by Kirchner [63] on the basis that each constituent is described by a set of equations of the Goodman–Cowin type. Fragmentation could also be described by a model of such a structure, but has not yet been done to our knowledge. The disadvantage of this approach is that the particles in such a mixture have discrete sizes; a better model would allow for a continuous distribution of particle size. How this should be done is not yet clear, even though first steps have been done in a formulation of *mixture with continuous diversity*, see [26].

4.4 Concluding Remarks

From a continuum mechanical point of view, in order to close the system of field equations, constitutive equations must be introduced. These equations connect certain mechanical or thermodynamic quantities via material-dependent coefficients which are determined by test observations. Furthermore, if new variables are introduced, e.g. volume fraction as a measure of the microstructure to obtain a model for a *smeared* continuum and describe behaviours such as dilatancy or/and an internal variable for internal friction, additional equations must be added to close the system of equations. It is, however, difficult to gain additional field equations since the new variables, such as volume fraction or internal friction concern quantities of the microscale for which balance or constitutive equations are not formulated and homogenizations to the macroscopic level of the mixture theory are not performable. Therefore, much effort must be devoted to overcome this lack of knowledge. This effort starts by introducing further evolution equations for additional variables. This procedure solves – from the mathematical point of view – the closure problem.

In the last decades great strides were taken in the formulation of constitutive equations for dry and saturated granular materials. However, many of them are rather complicated. Without any doubt, those constitutive equations may closely describe the stress-deformation relations of the special mechanical behaviour of materials. However, in many cases this has only been achieved by introducing many parameters and neglecting requirements due to mechanical and thermodynamic principles. Those constitutive equations may, however, be so complicated in explicit boundary- and initial-value problems to make their use rather difficult. Additionally, entrainment and deposition processes, ignored in this analysis may overshadow the role of the constitutive relation of stress. Therefore, the future goal should be to formulate relatively simple constitutive equations, but to account properly for boundary conditions.

There is as yet still no rigorous theory, which can describe all essential features of the flow of granular materials in the geophysical context. Depending on the application, the shear stresses result from different mechanisms; in the static case they are due to from dry interparticle friction and particle interlocking whereas in the shear-flow case particles override other particles and inertia associated with interparticle collisions becomes more important. In other words, the static and flowing cases may be regarded as two different states, rather like (metaphorically speaking) a solid and a liquid when viewed on the microscopic scale. There may be no smooth transition from one state to the other as the strain rate $\boldsymbol{D} \to \mathbf{0}$ and thus a constitutive equation suitable for flowing materials need not necessarily be appropriate to describe the state of the static equilibrium. Perhaps special constitutive relations should be separately developed that are only suitable to special flow regimes.

Acknowledgement

Financial support by the German Research Foundation (DFG) through the special collaborative research project SFB 298 is gratefully acknowledged. We thank the editors for their review for an earlier manuscript.

References

1. G. Ahmadi: Int. J. Non-linear Mechanics **15**, 251–262 (1980)
2. G. Ahmadi: Acta Mech. **44**, 299–317 (1982)
3. M.P. Allen, D.J. Tildesley: *Computer Simulation of Liquids* (Oxford University Press, Oxford 1987)
4. C. Ancey, P. Coussot, P. Evesque: Mech. Cohesive-Frictional Mat. **1**, 385–403 (1996)
5. R.A. Bagnold: Proc. R. Soc. London **A225**, 49–63 (1954)
6. P.L. Bhatnagar, P. Gross, M. Krook: Physical Reviews **94**, 5511–5536 (1954)
7. J. Bluhm, R. De Boer, K. Wilmanski: Mech. Res. Comm. **22**, 171–180 (1995)
8. R.M. Bowen: Int. J. Engng. Sci. **18**, 1129–1148 (1980)
9. E.J. Boyle, M. Massoudi: Kinetic theories of granular media with applications to fluided beds. U. S. Department of Energy, Technical Note, DOE/METC-89-4088 (1989)
10. V. Buchholtz, T. Pöschel: Physica A **202**, 390 (1994)
11. C.S. Campbell: Ann. Rev. Fluid Mech. **22**, 57–92 (1990)
12. C.S. Campbell, C.E. Brennen: J. Fluid Mech. **151**, 167–188 (1985)
13. R.P. Chhabra, P.H.T. Uhlherr: 'Static equilibrium and motion of spheres in viscoplastic liquids'. In: *Encyclopedia of Fluid Mechanics*, ed. by N.P. Cheremisinoff (Gulf Publishing Coo., Houston 1988) Chapter 21, Vol. 7
14. C.L. Chen: Geol. Soc. Am. Rev. Eng. Geol. Vol. VII, 13–29 (1987)
15. C.L. Chen, C.H. Ling: J. Eng. Mech. **122**, 469–480 (1996)
16. B.D. Coleman, W. Noll: Arch. Rat. Mech. Anal. **13**, 167–178 (1963)
17. P. Coussot: J. Hydr. Res. **32**(4), 535–559 (1994)
18. P. Coussot: *Mudflow Rheology and Dynamics.* (Balkema, Rotterdam 1997)
19. P. Coussot, J.M. Piau: Rheol. Acta **33**, 175–184 (1994)

20. S.C. Cowin, M.A. Goodman: Zeitschrift für Angewandte Mathematik and Mechanik **56**, 281–286 (1976)
21. S.C. Cowin, J.W. Nunziato: Int. J. Engng. Sci. **19**, 993–1008 (1981)
22. W.O. Criminale, J.L. Ericksen, G.L. Filbey: Arch. Rational Mech. Anal. **1**, 410–417 (1958)
23. P.A. Cundall, O.D.L. Strack: Gèotechnique **29**(1), 47–65 (1979)
24. C.M. Dury, G.H. Ristow: J. Phys. I France **7**, 737 (1997)
25. W. Ehlers: 'Constitutive equations for granular materials in geomechanical context'. In: *Continuum Mechanics in Environmental Sciences and Geophysics*, ed. by K. Hutter (Springer Verlag, Heidelberg, New York 1993) pp. 313–402
26. S.H. Faria: Continuum Mech. Thermodyn. in press (2001)
27. A.G. Fredrickson: *Principles and Applications of Rheology.* (Prentice-Hall, Englewood Cliffs 1964)
28. J.A.C. Gallas, H.J. Herrmann, S. Sokolowski: J. Phys. II France **2**, 1389 (1992)
29. J.D. Goddard: Acta Mechanica **63**, 3–13 (1986)
30. M.A. Goodman, S.C. Cowin: J. Fluid Mech. **45**, 321–339 (1971)
31. M.A. Goodman, S.C. Cowin: Arch. Rat. Mech. and Anal. **44**, 249–266 (1972)
32. G.W. Govier, C.A. Shook, E.O. Lilge: Trans. Can. Insti. Mining and Met. **60**, 147–154 (1957)
33. J.M.N.T. Gray, M. Wieland, K. Hutter: Proc. R. Soc. Lond. A**455**, 1841–1874 (1999)
34. J.M.N.T. Gray, K. Hutter: Continuum Mech. Thermodyn **9**, 341–345 (1997)
35. R. Greve, K. Hutter: Phil. Trans. R. Soc. London A**342**, 573–604 (1993)
36. R. Greve, T. Koch, K. Hutter: Proc. R. Soc. London A**445**, 399–413 (1994)
37. R. Gudhe, R.C. Yalamanchili, M. Massoudi: The flow of granular materials in a pipe: numerical solutions. Rec. Adv. Mech. Strcted. Continua, AMD-160/MD-41, 41–53 (1993)
38. G. Gudehus: Soils and Foundations **36**, 1–12 (1996)
39. P.K. Haff: J. Fluid Mech. **134**, 401–430 (1983)
40. P.K. Haff, B.T. Werner: Powder Technol. **48**, 239 (1987)
41. D.M. Hanes, D.L. Inman: J. Fluid Mech. **150**, 357–380 (1985)
42. H.J. Herrmann, S. Luding: Continuum Mech. Thermodyn. **10**, 1–48 (1998)
43. H. Hertz: J. für die reine und angew. Math. **92**, 136 (1882)
44. K. Hutter: Acta Mechanica (Suppl.) **1**, 167–181 (1991)
45. K. Hutter: 'Avalanche Dynamics'. In: *Hydrology of Disasters*, ed. by V.P. Singh (Kluwer Academic Publ., Dordrecht–Boston–London 1996) pp. 317–394
46. K. Hutter, R. Greve: J. Glaciology **39**, 357–372 (1993)
47. K. Hutter, L. Laloui, L. Vulliet: Mech. Cohesive-Frictional Mat. **4**(4), 295–338 (1999)
48. K. Hutter, K.R. Rajagopal: Continuum Mech. Thermodyn. **6**, 81–139 (1994)
49. K. Hutter, T. Koch: Phil. Trans. R. Soc. London, A**334**, 93–138 (1991)
50. H. Hwang, K. Hutter: Continuum Mech. Thermodyn. **7**, 357–384 (1995)
51. J.T. Jenkins: 'Balance laws and constitutive relations for rapid flows of granular materials'. In: *Proc. Army Research Office Workshop on Constitutive Relations*, ed. by J. Chandra, R. Srivastava (Philadephia, 1987)
52. J.T. Jenkins, S.C. Cowin: 'Theories for flowing granular materials'. In: *The Joint ASME-CSME Appl. Mech. Fluid Engng. and Bioengng. Conf.*, AMD-Vol. 31, pp. 79–89 (1979)
53. J.T. Jenkins, M.W. Richman: Arch. Rat. Mech. Anal. **87**, 355–377 (1985a)
54. J.T. Jenkins, M.W. Richman: Phys. Fluids **28**, 3485–3494 (1985b)

55. J.T. Jenkins, M.W. Richman: J. Fluid Mech. **171**, 53–69 (1986)
56. J.T. Jenkins, S.B. Savage: J. Fluid Mech. **130**, 186–202 (1983)
57. G. Johnson, M. Massoudi, K.R. Rajagopal: Chem. Engng. Sci. **46**(7), 1713–1723 (1991)
58. G. Johnson, M. Massoudi, K.R. Rajagopal: Int. J. Engng. Sci. **29**, 649–661 (1991)
59. G. Johnson, M. Massoudi, K.R. Rajagopal: Rec. Adv. Mech. Strctd. Cont. AMD **117**, 97–105 (1991)
60. P.C. Johnson, R. Jackson: J. Fluid Mech. **130**, 187–202 (1987)
61. P.C. Johnson, P. Nott, R. Jackson: J. Fluid Mech. **210**, 501–535 (1990)
62. P.Y. Julien, Y. Lan: J. Hydraulic Eng. **117**(3), 346–353 (1991)
63. N. Kirchner: Thermodynamics of Structured Granular Media. Ph.D. Thesis, Institute of Mechanics, Darmstadt University of Technology (2001)
64. N. Kirchner, K. Hutter: 'Thermodynamic modelling of granular continua exhibiting quasi-static frictonal behaviour with abrasion'. In: *Mathematical Models of granular and porous materials.* ed. by G. Capriz, G. Ghionna, P. Giovine (Modelling and Simulation in Science, Engineering and Technology Series, Birkhaeuser 2001)
65. T. Koch, R. Greve, K. Hutter: Proc. R. Soc. London A**445**, 415–435 (1994)
66. D. Kolymbas: Ing. Arch. **61**, 143–151 (1991)
67. D. Kolymbas, W. Wu: 'Introduction to hypoplasticity'. In: *Modern Approaches to Plasticity*, ed. by D. Kolymbas (Elsevier, 1993) pp. 213–223
68. L.D. Landau, E.M. Lifschitz: *Elastizitätstheorie* (Akademie-Verlag, Berlin 1989)
69. J. Lee, H.J. Herrmann: J. Phys. A: Math. Gen. **26**, 373 (1993)
70. I-Shih Liu: Arch. Rat. Mech. and Anal. **46**, 131–148 (1972)
71. K.F. Liu, C.C. Mei: Physics of Fluids **A6**(8), 2577–2590 (1994)
72. S. Luding: *Die Physik kohäsionsloser granularere Medien.* Habilitation (Logos Verlag, Berlin 1997) pp. 1–197
73. C.K.K. Lun, S.B. Savage, D.J. Jeffrey, N. Chepurniy: J. Fluid Mech. **140**, 233–256 (1984)
74. D.N. Ma, G. Ahmadi: Power Tech. **56**, 191 (1988)
75. M. Massoudi: Application of mixture theory to fluidized beds. Ph.D. Thesis, University of Pittsburgh (1986)
76. M. Massoudi, E.J. Boyle: A review of theories for flowing granular materials with applications to fluidized beds and solids transport. *U.S. Department of energy Report, DOE/PETC/TR-91/8* (1991)
77. J.J. McCarthy, J.E. Wolf, T. Shinbrot, G. Metcalfe: AICHE **42**, 3351–3363 (1996)
78. D.F. McTigue: A nonlinear continuum theory for flowing granular materials. Ph.D. Thesis, Dept. of Geol., Stanford University (1979)
79. D.F. McTigue: J. Appl. Mech. **49**, 291–296 (1982)
80. S. Melin: Int. J. Mod. Phys. C **4**, 1103 (1993)
81. G. Metcalfe, T. Shinbrot, J.J. McCarthy, J.M. Ottino: Nature **374**, 39–41 (1995)
82. I. Müller: Arch. Rat. Mech. and Anal. **40** (1971)
83. I. Müller: *Thermodynamics* (Pitmann, Boston 1985)
84. Q.D. Nguyen, D.V. Boger: J. Rheol. **27**, 321–349 (1983)
85. H. Norem, F. Irgens, B.A. Schieldrop: 'A continuum model for calculating snow avalanches'. In: *Avalanche Formation, Movement and Effects*, ed. by B. Salm, H. Gubler (IAHS publ. Noangemessen. 126, 1987) pp. 363–379
86. J.W. Nunziato, S.C. Cowin: Arch. Rotational Mech. Anal. **72**, 175–201 (1979)
87. J.W. Nunziato, S.L. Passman, J.P. Thomas: J. Rheol. **24**, 395–420 (1980)
88. S. Ogawa, A. Unemura, N. Oshima: J. Appl. Math. Phys. **31**, 483–493 (1980)

89. T. Ohtsuki, Y. Takemoto, T. Hata, S. Kawai, A. Hayashi: Int. J. Mod. Phys. B **7**, 1865-1872 (1993)
90. S.L. Passman, J.W. Nunziato, P.B. Bailey: J. Rheol. **30**, 167–192 (1986)
91. S.L. Passman, J.W. Nunziato, P.B. Bailey, J.P. Thomas: J. Eng. Mech. Division, ASCE **106**, 773–783 (1980)
92. T. Pöschel: J. Phys. II France **3**, 27 (1993)
93. K.R. Rajagopal, M. Massoudi: A method for measuring material moduli of granular materials: flow in an orthogonal rheometer. *Topical Report U, Department of Energy. DOE/PETC/TR-90/3* (1990)
94. K.R. Rajagopal, W.C. Troy, M. Massoudi: Eur. J. Mech., B/Fluids **11**, 265–276 (1992)
95. G.H. Ristow: Int. J. Mod. Phys. C **3**, 1281–1293 (1992)
96. G.H. Ristow: Europhys. Lett. **28**, 97–101 (1994a)
97. G.H. Ristow: Flow properties of granular materials in three-dimensional geometries. Habilitation, Philipps-Universität Marburg, pp. 1–111 (1998)
98. G.H. Ristow, H.J. Herrmann: Phys. Rev. E **50**, R5–R8 (1994)
99. R.S. Sampaio, W.O. Williams: J. de Mechanique **18**, 19–45 (1979)
100. S.B. Savage: J. Fluid Mech. **92**, 53–96 (1979)
101. S.B. Savage: 'Granular flows down rough inclines – review and extension'. In: *Mech. of Granular Materials: New Models and Constitutive Relations*, ed. by J.T. Jenkins, M. Satake (Elsevier 1983) pp. 261–282
102. S.B. Savage, K. Hutter: J. Fluid Mech. **199**, 177–215 (1989)
103. S.B. Savage, K. Hutter: Acta Mech. **86**, 201–223 (1991)
104. S.B. Savage, D.J. Jeffrey: J. Fluid Mech. **110**, 255–272 (1981)
105. S.B. Savage, C.K.K. Lun: J. Fluid Mech. **189**, 311–335 (1988)
106. S.B. Savage, S. McKeown: J. Fluid Mech. **127**, 453–472 (1983)
107. S.B. Savage, M. Sayed: J. Fluid Mech. **142**, 391–430 (1984)
108. T. Scheiwiller, K. Hutter: 'Lawinendynamik, Übersicht über Experimente und theoretische Modelle von Fließ- und Staublawinen'. *Mitteilung N0. 58 der Versuchsanstalt für Wasserbau, Hydrologie und Glaziologie an der ETH Zürich*, pp. 166 (1982)
109. W.R. Schowalter: *Mechanics of non-Newtonian fluids* (Pergamon Press, Oxford 1978)
110. M. Shahinpoor, S.P. Lin: Acta Mechanica **42**, 183–196 (1982)
111. M. Shahinpoor, J.S.S. Siah: J. Non-Newt. Fluid Mech. **9**, 147–156 (1981)
112. H. Shen, N.L. Ackerman: J. Eng. Mech. Div. **108**, 748–763 (1982)
113. S. Straub: Schnelles granular Fließen in subaerischen pyroklastischen Strömen. PhD thesis, Bayerische Julius-Maximilians-Universität Würzburg (1994)
114. B. Svendsen, K. Hutter: Int. J. Engng Sci. **33**, 2021–2054 (1995)
115. B. Svendsen, K. Hutter, L. Laloui: Continuum Mech. Thermodyn. **11**, 263–275 (1999)
116. Y.C. Tai: Dynamics of Granular Avalanches and their Simulations with Shock-Capturing and Front-Tracking Numerical Schemes. Ph.D. Thesis, Darmstadt University of Technology, 146 p (2000)
117. P.A. Thompson, G.S. Grest: Phys. Rev. Lett. **67**, 1751 (1991)
118. Y. Tsuji, T. Tanaka, T. Ishida: Powder Technol. **71**, 239 (1992)
119. O.R. Walton, R.L. Braun: J. Rheology **30**, 949–980 (1986)
120. O.R. Walton, R.L. Braun, R.G. Mallon, D.M. Cervelli: 'Particle-dynamics calculations of gravity flow of inelastic, frictional spheres'. In: *Micromechanics of Granular Materials*, ed. by M. Satake, J.T. Jenkins (Elsevier 1987) pp. 153–162

121. O.R. Walton, H. Kim, A. Rosato: 'Micro-structure and stress difference in shearing flows of granular materials'. In: *Proc. ASCE Eng. Mech. Div. Conf.* (Columbus, Ohio, May 19–21, 1991)
122. Y. Wang, K. Hutter: Particulate Science and Technology **17**, 97–124 (1999)
123. Y. Wang, K. Hutter: Granular Matter **1**(4), 163–181 (1999)
124. Y. Wang, K. Hutter: Arch. Mech. **51**(5), 605–632 (1999c)
125. K. Wilmanski: Arch. Mech. **48**, 591–628 (1996)
126. K. Wilmanski: *The thermodynamical model of compressible porous materials with the balance equation of porosity.* Preprint No. 310, ed. by Weierstraß-Institute für Angewandte Analysis und Stochastik, Berlin (1997)
127. W. Wu, E. Bauer, D. Kolymbas: Mech. Mat. **24**, 45–69 (1996)
128. Y. Zhang, C.S. Campbell: J. Fluid Mech. **237**, 541 (1992)

Part II

Hot

5 Earth's Surface Morphology and Convection in the Mantle

R.W. Griffiths[1] and J.A. Whitehead[2]

[1] Research School of Earth Sciences, The Australian National University, Canberra 0200 ACT, Australia
[2] Woods Hole Oceanographic Institution, Woods Hole, MA 02543, USA

5.1 Introduction

It is now generally agreed that the Earth's solid mantle is undergoing thermal convection. Much of the evidence for this conclusion is derived from geological and geophysical observations of the Earth's surface, its relative horizontal motions and its topography. Direct consequences of the mantle flow include plate tectonics, which refers to the relative motions of the continents, spreading of the sea-floor, creation of new crust and mid-ocean ridges at spreading centres, and subduction at ocean trenches, along with associated phenomena such as mountain building and volcanism. The motion of the mantle over geological time scales is driven by gravity acting on density differences, which result from loss of heat from the Earth's surface and, to a lesser extent, from transfer of heat from the Earth's core to the mantle. Mantle convection phenomena are reviewed here in the context of geomorphology because they are responsible for producing much of the large-scale topography (horizontally $> 10\,\mathrm{km}$) of the Earth's surface. This topography, in turn, imposes strong influences on the atmosphere and ocean circulation patterns, affects precipitation, and provides the base on which erosion and sedimentation processes act. The surface transport processes can also couple back to mantle flow and topography through redistribution of loading on the mantle.

We briefly introduce the nature of the mantle, the behaviour of convection at large Rayleigh numbers, and the mantle's expected response to boundary heat fluxes. We then outline convective instability of a boundary layer, several forms of large-amplitude plume flows, and the formation and subduction of oceanic lithosphere plates. We conclude with a discussion of the surface topographic expressions of these phenomena. These phenomena are discussed in the context of two main notions: 1) we paint a picture of the mantle as a convecting viscous fluid in which heat lost from the Earth's core drives blobs and continuous streams of fluid to ascend from the core-mantle boundary to the surface as plumes, where they create isolated morphological features such as island chains and flood basalt plateaux; 2) the plumes, however, are relatively minor in the heat budget of the mantle and they ascend through much larger-scale convective flows driven by the cooling of the lithosphere. The surface cooling produces subducting slabs that plummet downward and morphological features such as deep ocean trenches and mid-ocean ridges. The article is concerned with a few of the dynamical phenomena in the mantle rather than with the geological evidence but it should be recognised that the dynamical modelling discussed here must go hand in hand with a range of observational evidence.

5.2 Some Basic Assumptions and Deductions

Before treating a number of specific problems in thermal convection we consider three basic concepts which, explicitly or implicitly, enter into every physically realistic discussion of convection in the mantle.

5.2.1 The Rheology of the Mantle

Seismic and petrological evidence indicates that the bulk of the mantle is a crystalline solid. However, imposed stresses can produce irreversible deformation or creep. The two 'flow' mechanisms considered most relevant to the mantle are 'diffusion creep', in which the strain rate is proportional to the stress; and 'dislocation creep', in which the strain rate is proportional to a higher power of the stress [20]. Both these behaviours allow arbitrarily large strains, so that solids with these properties have no long-term strength. This ensures that in both cases an "effective viscosity" can be defined for mantle materials on geological timescales (although this "viscosity" depends on the average stress level, if dislocation creep is appropriate). Hence the mantle is treated as a viscous fluid in analytical and numerical models of mantle convection, and laboratory experiments directly relevant for the understanding of mantle dynamics (i.e. properly scaled to duplicate the dynamics of the Earth) can be carried out with linear viscous fluids.

Regardless of the details of the rheology, the effective viscosity is strongly temperature-dependent. Assuming diffusion creep is the mechanism by which deformation is accommodated, the viscosity η will be of the form

$$\eta = \eta_0 \exp(AT_M/T) \ , \tag{5.1}$$

where T_M is the melting temperature and A the activation energy. For a mantle of olivine, $A = 30$ at the pressures of interest and $\eta_0 = 10^5\,\mathrm{Pa\,s}$ ([63] and summarized by [61]). For $\eta = 10^{22}\,\mathrm{Pa\,s}$ (a mean value to order of magnitude inferred from postglacial uplift) [38] $T = 0.77\,T_M$ and η changes by an order of magnitude as T/T_M changes by only about 5%. We will see below that this strong dependence of η on temperature ensures that it adjusts to a value which depends on the presence of mantle convection. That is, the value of this material property is determined, through the temperature and within wide bounds set by the microscopic mechanics of the mantle material, by the dynamics and motions of the mantle. This conclusion contrasts with the view that whether or not mantle convection occurs is predetermined by the viscosity. The viscosity will, of course, vary from place to place within the convection system according to the temperature variation, and further studies have considered the additional effects of a probable pressure-dependence of the viscosity [16].

5.2.2 Thermal Forcing and the Inevitability of Convection in the Mantle

The Earth's mantle is bounded above by the oceans and atmosphere, and below by the outer core of liquid metal. The mantle is losing heat from the surface at a rate of approximately 3.5×10^{13} W, mostly through the oceanic crust where the fluxes are between $40\,\mathrm{mW\,m^{-2}}$ through old crust and $100\,\mathrm{mW\,m^{-2}}$ through young crust. This heat originates largely from the radioactive decay of elements distributed throughout the mantle (so called 'internal heating'), with a small but significant component (estimated to be approximately 10% of the total surface heat loss, [17]) entering from the core. The latter flux represents a cooling of the core through geological time and is expected to provide the driving force for the geodynamo (through both thermal and compositional convection, the latter resulting from the cooling and consequent solidification of components of the outer core on to a growing solid inner core [5]).

Following the argument put forward originally by Tozer [62], and restated by Stevenson and Turner [61], we consider the behaviour of the mantle when subjected to a purely vertical temperature gradient, and begin by assuming that the physical properties are uniform. The stability of such a fluid layer, heated from below or cooled from above, is a classic problem in fluid mechanics and we quote only the basic results. The onset of convection in this simplest approximation is governed entirely by the Rayleigh number, Ra, which is essentially the ratio of the driving force (due to thermal buoyancy and influenced by diffusion of heat) to the retarding force (due to diffusion of momentum by viscous stresses). For a fluid layer of depth H, with constant kinematic viscosity $\nu = \eta/\varrho$ and thermal diffusivity κ,

$$Ra = \frac{g\alpha\beta H^4}{\nu\kappa} , \tag{5.2}$$

where g is the acceleration due to gravity, α is the coefficient of thermal expansion, and β is the difference between the actual overall temperature gradient (from top to bottom boundary) and the adiabatic temperature gradient. If Ra exceeds a critical value, Ra_c, of about 10^3 (the exact value depending on the boundary conditions) then convection will occur. For internal heating at a prescribed flux and cooling from the top boundary the relevant Rayleigh number can still be defined as in (5.2), except that β is now the (horizontally averaged) superadiabatic temperature gradient that would be required for a conductive steady state given the imposed rate of heat generation.

Rather than trying to evaluate Ra in the Earth using the poorly known present values of the physical properties (β being a particularly large source of uncertainty), the inevitability of mantle convection can be demonstrated by an idealized thermal evolution calculation based on the strong temperature dependence of viscosity (5.1). Consider again a horizontal layer of thickness H, but now containing a uniformly distributed energy source, representing heating due to decay of radioactive elements. The bottom boundary is supposed to be insulated, and the top temperature is fixed at $T = 0\,°\mathrm{C}$. At time $t = 0$, we suppose that the temperature $T = T_0$ everywhere and that subsequently, but before convection occurs, the temperature distribution obeys the diffusion equation (with a source term included) – that is, the heat generated is transported only by conduction.

As discussed in more detail by Stevenson and Turner [61], the scale and conductivity of the Earth are such that the heat generated cannot escape by conduction alone in the age t_E of the Earth. The diffusion lengthscale $l \approx (\kappa t)^{1/2}$ is a few hundred kilometres when $t = t_E$, so that a small body could lose most of its heat by conduction as it is generated. However, the much larger model Earth heats up, developing a temperature profile which is fixed at the surface but with increasing temperature and temperature gradient at all depths. As T increases, η given by (5.1) rapidly decreases, and a time is inevitably reached when Ra over some depth interval exceeds the critical value for convection to occur, virtually whatever the magnitude of the temperature gradient. The subsequent behaviour is for all regions eventually to become convective (except possibly the outermost highly viscous layer, which is a boundary layer and will be discussed in more detail below). This follows from the fact that any non-convecting region must continue to heat up, because conduction is too small to remove the heat generated, and so it must achieve a progressively lower viscosity until it takes part in the convection. Given the large depth of the mantle and the expected values of the constants in (5.1) and (5.2), a small enough viscosity is achieved at subsolidus temperatures for convective heat transport to become possible before melting occurs at any depth.

5.2.3 Boundary Layers in Convection at High Rayleigh Numbers

The above argument concentrates on the initiation of convection in the interior of a progressively heated mantle. It is clear that the eventual steady state must have a much larger heat transport than can be achieved by conduction, and that the corresponding Rayleigh number will be much greater than the critical value.

Two other points are useful in understanding the finite amplitude flow in the earth's mantle. The viscosity ν is very large, effectively infinite, relative to the thermal diffusivity κ (i.e. infinite Prandtl number $Pr = \nu/\kappa$), and so the viscous response to a perturbation is instantaneous relative to the thermal response. Secondly, for large Rayleigh numbers the convective heat transport is much more important than conductive heat transport over most of the depth (the ratio $uH/\kappa \approx 10^3$, where u is a typical flow velocity such as that of the tectonic plates). Conduction remains important, however, in thin boundary layers through which heat is transported to and from the interior, and which in fact determine the magnitude of the flux which must be carried by the convection in the interior.

Some fundamental predictions can be made on the basis of dimensional reasoning, as follows. Suppose that the flux does depend only on the material properties and on conditions very near the boundaries, i.e. that it is independent of the total depth H. It follows from their definitions that the Nusselt number Nu, the ratio of the actual heat flux to the purely conductive flux down a linear (super-adiabatic) temperature gradient between the two boundaries, and the

Rayleigh number are related by:

$$Nu = cRa^{1/3} , \tag{5.3}$$

since this is the only form which gives a flux independent of H. The constant $c \approx 0.1$ but can depend on the boundary conditions. A phenomenological theory due to Howard [36] suggests that the conductive boundary layer is inherently unsteady, with cold (or hot) material breaking away intermittently. The mean thickness δ of the boundary layer is such that the Rayleigh number based on δ, Ra_δ say, is just critical ($\approx 10^3$). Thus

$$Nu = H/\delta = (Ra/Ra_\delta)^{1/3} = 0.1Ra^{1/3} , \tag{5.4}$$

in reasonable agreement with experiments [65] using large-Prandtl number fluids. Expressions (5.2–5.4) have been written with the Bénard problem in mind (i.e. with ΔT the temperature difference between the two boundaries and an equal heat flux passing through both boundaries). However, they apply equally well to the more general case in which the heat flux through the top boundary is equal to the sum of the bottom flux and internal heat generation by radioactive decay. In the limit of zero bottom flux, ΔT becomes the temperature drop across the upper thermal boundary layer alone, and (5.3) remains valid.

The expression (5.4) allows one to make crude estimates of Ra and η for the mantle. Using a (poorly constrained) temperature of 3500 °C at the base of the mantle [4], at a depth $H = 3000$ km, an estimate of the overall temperature gradient through the mantle is 1.2 K km^{-1}. The measured temperature gradient near the Earth's surface is of order 20 K km^{-1}. Thus the conducting upper boundary layer, the lithosphere, is very thin compared to H and (5.4) implies that $Nu > 10$, hence $Ra > 10^6$. Inserting the depth and other properties [1] in (5.2) we deduce that the average viscosity is less than $\eta \approx 6 \times 10^{22}$ Pa s. The average viscosity is thus determined by the heat flux and the efficiency of mantle convection. These conclusions, which are based on the assumption of uniform material properties, provide a first approximation to the mantle. As will be seen below, there will be quantitative differences resulting from the temperature- and pressure-dependence of viscosity and other material properties, but the basic conclusions remain unchanged.

The above very robust general arguments show that the existence of a heat flux through a boundary of a convecting region inevitably implies that there will be an unstable conductive boundary layer. However, the two boundary layers at the top and bottom of the Earth's mantle are very different. Because of the strong temperature-dependence of viscosity the upper cold boundary layer will be stiff, and this property will affect the horizontal dimensions of the plates

[1] The values substituted into (5.2) are $\alpha = 3 \times 10^{-5}\,\mathrm{K}^{-1}$, $\kappa = 10^{-6}\,\mathrm{m}^2\mathrm{s}^{-1}$, $\varrho = 3 \times 10^3\,\mathrm{kg\,m}^{-3}$ and $\beta = 0.9\,\mathrm{K\,km}^{-1}$. Remember that by definition β is the difference between the overall temperature gradient over the whole depth, with or without convection, and the adiabatic gradient of 0.3 K km^{-1}. In the convecting region the gradient will of course be much closer to the adiabatic value.

and the behaviour of subducting slabs (Fig.5.1). If the plates are able to move and sink sufficiently rapidly, as is apparently the case for the present oceanic lithosphere, then they represent the unstable boundary layer. On the other hand, it is possible that the surface layer could be so viscous (or strong) that it is stable and does not take part in the underlying convection, instead forming a thick stagnant lid which supresses heat transport, as suggested to be the case on Venus over the past 500 Myr [54,59]. The behaviour in systems with very viscous, non-convecting upper boundary layers (a problem that is relevant also in the dynamics of cooling magma chambers) has been addressed through laboratory experiments by Davaille & Jaupart [10,11].

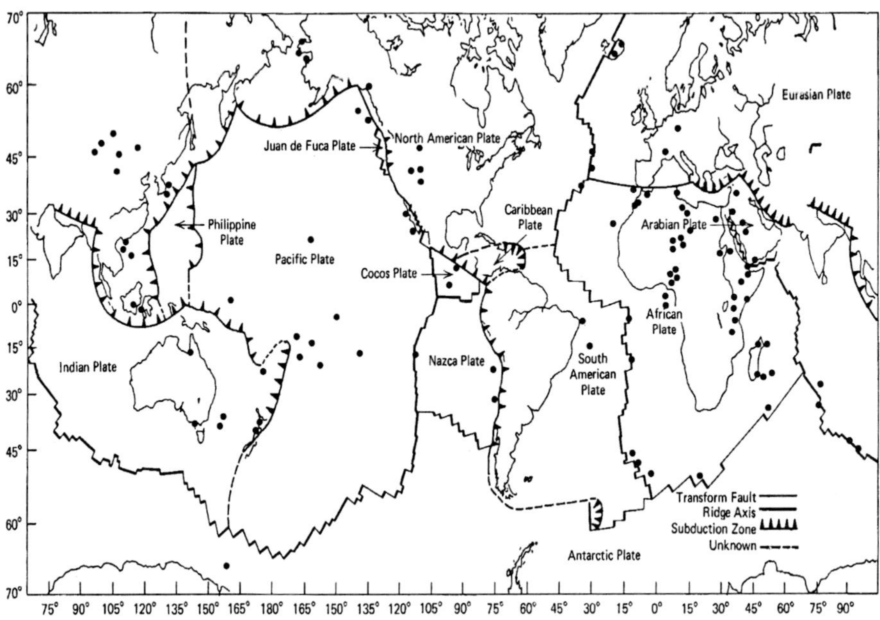

Fig. 5.1. A rendition of the major active boundaries of tectonic plates on Earth, showing the mid-ocean ridges (at divergent boundaries) and subduction zones (at convergent boundaries). Also shown are many of the known "hotspot" plumes that create tracks of volcanism across the moving surface plates. (Adapted from [64])

Since the Earth as a whole, including the core, is cooling, there will be a heat flux out of the core and into the base of the mantle, estimated to be of the order of 10% of the Earth's total surface heat flux (see review by Davies and Richards [17]). The resulting boundary layer of hot, less dense and less viscous material behaves quite differently from the plates produced by surface cooling and may give rise to upwelling plumes (as discussed below). In addition, if there are any internal density interfaces in the mantle separating distinct convecting layers, then boundary layers must form on each side of such interfaces.

Important questions to be answered are: what aspects of the mantle motion lead to topography at its top and bottom boundaries, and can material arising at one boundary layer deliver a sufficient buoyancy force or thermal anomaly to the opposite boundary such that it generates topography by virtue of the buoyant support or the production and eruption of melts? The answer to the latter is clearly 'yes' in the case of upwelling plumes, which are believed to be the cause of surface phenomena such as chains of intraplate volcanos [21], uplift of the seafloor surrounding hotspots by the order of 1000 m, and eruptions of flood basalts sequences 10 km deep and covering millions of square kilometres [53]. It is also clear that temperature (density) differences within the surface boundary layer itself produce surface topography (noteably an increase in ocean depth with distance from the spreading centres due to conductive cooling of the lithosphere). At the opposite boundary, the sinking of lithospheric plates may potentially affect the dynamical processes at the core-mantle boundary if they are able to penetrate to sufficient depths.

5.3 Upwelling Thermals and Plumes

We now turn to a discussion of models of specific convective processes in the mantle, starting from the core-mantle boundary (CMB) and working upwards. First we need to consider the implications of a heat flux through the CMB itself. It is also useful to keep in mind the application of these same concepts to convection arising at an internal interface, heated from below.

5.3.1 The Initiation of Convection at the Base of the Mantle

There is a large density difference between the core and the mantle. The best estimates of the temperatures of the outer core and the lowermost mantle (the latter from extrapolation of the upper mantle temperature adiabatically to the CMB), indicate that there is also a large temperature difference (approximately 1300 K; [4]), so that there is a conductive heat flux from the core to the base of the mantle. This temperature drop must occur across a thermal boundary layer. There is direct seismic evidence for a spatially inhomogeneous boundary layer, the so-called D″ layer, above the CMB, which in places is a few hundred kilometers thick [37]. Although there may be significant compositional differences within the D″ layer, it is likely that it also contains the thermal boundary layer.

Because of the strong temperature-dependence of the effective viscosity, there will be a gradient of viscosity through this boundary layer at the bottom of the mantle, with a minimum at the CMB. This reduced viscosity will enhance the flow of the boundary-layer material into any region which has begun to break away from the boundary and convect upwards. An analysis of this lateral flow [60], assuming steady conditions, showed that it will be concentrated in a rheological boundary layer which is much thinner than the thermal boundary layer, and that the lateral flow can be replaced by a slow subsidence of the overlying mantle. Davies [14] combined heat flux estimates with this theory to deduce the thicknesses of the two boundary layers. As a result of the viscosity variation, the temperature of the rising plume material will be strongly weighted towards the highest temperature in the thermal boundary-layer. However, Griffiths & Campbell [28] noted that the temperature of the plume source material may be much less than that of the core since a thin gravitationally stable conductive layer may persist between a partially miscible or reactive mantle and the much denser core. Such a dense stable layer (not to be confused with either the unstable boundary layer or the D″ layer) will support a large temperature drop without taking part in the boundary layer convection.

In this picture each plume draws boundary layer material from a horizontal area determined only by the separation of unstable convective events. Presumably, if plumes are too far apart, perturbations on the boundary layer between grow to large amplitude and a new plume develops. There is as yet no prediction of this separation distance for large amplitude motions, and hence no prediction of the mean heat and buoyancy flux in each plume. However, we do anticipate from theoretical stability arguments and a variety of experiments, some described here, that the mean separation of plumes will be related to the depth of the boundary layer and not to the overall depth of the convecting layer.

A relevant model here is the so-called Rayleigh–Taylor instability of a thin horizontal layer of fluid beneath a deep fluid of larger density and viscosity. In contrast to convective instability, the effects of heat conduction are removed, the layer depth is prescribed and each layer is uniform. However, the result gives a first estimate of the role of the viscosity contrast and of the horizontal length scale for instability of a convective boundary layer. Figure 5.2 shows a laboratory experiment that exhibits a Rayleigh–Taylor instability, which is a candidate model of instability of the hot boundary layer at the base of the mantle. A lower viscosity layer of dyed fluid lies under a clear deep immiscible fluid of much greater viscosity in a transparent tank. After being left overnight the tank is rapidly inverted and the results photographed. Four or five regularly spaced protrusions were observed shortly after inversion. Within the confines of the box, the protrusions arranged themselves quite uniformly throughout the tank. The dyed fluid had developed long waves which allowed it to buoyantly pass through a clear fluid of much greater viscosity. The wavelength was almost 10 times the depth of the thin layer. The wavelength of maximum growth rate and the exponential time constant for growth have been theoretically and numerically predicted for a number of geometries and boundary conditions for problems like this [1–3,8,9,44–46], [47–50,56,70]. Demonstration experiments with putty and non-Newtonian fluids have been extensively photographed and compared to geological formations by Nettleton [41], Parker and McDowell [43] and Ramberg [44,45,49]. There was no intercomparison between the laboratory experiments and theory owing to the unknown rheology of the laboratory materials.

In general, if we have two layers of viscous fluid they obey the equations

$$\nabla \cdot \tilde{u} = 0 \,, \tag{5.5}$$

$$(\partial/\partial t - \nu\nabla^2)\,\tilde{u} = -(1/\rho)\nabla p \,. \tag{5.6}$$

Fig. 5.2. A photograph of a laboratory experiment in which a thin bottom layer of low-viscosity (dyed) fluid is placed beneath a deep upper layer of more viscous fluid and denser clear silicon oil. Viscosity ratio is 43. The bottom layer is initially 5 mm deep and the tank is $18.5\,\mathrm{cm}^2$

Here $\tilde{u}$ is the velocity vector of the fluid, ν is the kinematic viscosity, ρ is density of the mantle and p is the deviation from hydrostatic pressure. These equations can be expected to be valid only for a system in which inertia of the fluid is negligible so that $U_{\max}L/\nu \ll 1$, where L is the largest length scale in the problem (in this case it is either the depth of the layer, the wavelength of a perturbation, or $(\nu^2/g)^{1/3}$, where g is gravity). Since fluid velocity in a viscous medium would be proportional to $g\Delta\rho L^2/\rho\nu_1$, this criterion is easily met in solid Earth geophysics for all the length scales above. We can immediately write down a class of general solutions to these equations in two regions that correspond to the deep mantle and the thin bottom boundary layer respectively. By taking the curl of (5.6) and using (5.5), the equation for the vertical component of velocity w is

$$(\partial/\partial t - \nu\nabla^2)\,\nabla^2 w = 0\,. \tag{5.7}$$

This equation can be applied in each region. At the boundaries corresponding to the Earth's surface and the core, $z = h_1, h_2$, we apply zero disturbance boundary conditions. For example, the conditions of zero normal velocity and either zero tangential velocity, or zero tangential stress might be applied. Thus no external forces are driving the fluid at the boundaries. The general expressions for velocity are

$$w_1 = \left[A\mathrm{e}^{kh_1} + B\mathrm{e}^{-kh_1} + C\mathrm{e}^{q_1h_1} + D\mathrm{e}^{-q_1h_1}\right] f(x,y)\mathrm{e}^{nt}, \tag{5.8a}$$

$$w_2 = \left[E\mathrm{e}^{kh_2} + F\mathrm{e}^{-kh_2} + G\mathrm{e}^{q_2h_2} + H\mathrm{e}^{-q_2h_2}\right] f(x,y)\mathrm{e}^{nt}, \tag{5.8b}$$

where $q_1 = [k^2 + (n/\nu_1)]^{1/2}$, $q_2 = [k^2 + (n/\nu_2)]^{1/2}$ and $\partial^2 f/\partial x^2 + \partial^2 f/\partial y^2 = k^2 f$. To this order, the analysis admits a multiplicity of solutions, each one's growth rate depending on a two-dimensional wave number vector on the horizontal plane. This degeneracy is reduced by finite amplitude effects (Sect. 5.3.2). At the interface, horizontal velocities u, v, vertical velocity w, tangential stresses, and normal stress must be matched. The linearized expressions of these matching conditions are

$$w_1 = w_2 \, , \tag{5.9}$$

$$\frac{\partial w_1}{\partial z} = \frac{\partial w_2}{\partial z} \, , \tag{5.10}$$

$$\eta_1 \left(\frac{\partial^2}{\partial z^2} + k^2 \right) w_1 = \eta_2 \left(\frac{\partial^2}{\partial z^2} + k^2 \right) w_2 \, , \tag{5.11}$$

$$\left[\rho_1 \frac{\partial}{\partial t} - \eta_1 \left(\frac{\partial^2}{\partial z^2} - k^2 \right) \right] \frac{\partial w_1}{\partial z} + 2k^2 \eta_1 \frac{\partial w_1}{\partial z} =$$

$$\left[\rho_2 \frac{\partial}{\partial t} - \eta_2 \left(\frac{\partial^2}{\partial z^2} - k^2 \right) \right] \frac{\partial w_1}{\partial z} + 2k^2 \eta_2 \frac{\partial w_2}{\partial z} + k^2 g (\rho_2 - \rho_1) \hat{z} \, . \tag{5.12}$$

Equation (5.12) is a balance of normal stress, where the interface is slightly distorted by an amount $\hat{z}(x, y, t) = z - h$ so that a buoyancy force is produced.

The interface is swept along with the fluid so that

$$\frac{\partial \hat{z}}{\partial t} + u \frac{\partial \hat{z}}{\partial x} + v \frac{\partial \hat{z}}{\partial y} + w \frac{\partial \hat{z}}{\partial z} = 0 \, . \tag{5.13}$$

For small distortions (5.13) can be expanded in a Taylor series

$$\frac{\partial \hat{z}}{\partial t} + u \frac{\partial \hat{z}}{\partial x} + v \frac{\partial \hat{z}}{\partial y} = w(h) + \hat{z} \frac{\partial w}{\partial z} + \ldots \, , \tag{5.14}$$

where velocities and their derivatives are evaluated at the point $z - h = \hat{z} = 0$. For arbitrarily small $\hat{z}$, (5.14) reduces to

$$\frac{\partial \hat{z}}{\partial t} = w(h) \, . \tag{5.15}$$

Using the solutions given by (5.8a, 5.8b) in (5.9)–(5.12) and using (5.15) we obtain eight linear homogeneous equations for the eight constants. The determinant of these eight equations must be zero.

The limit in which one layer is both thinner and of lower viscosity than the other is particularly relevant to the geophysical context. The wavelength λ of fastest growth is

$$\lambda = 4.6 \, d \, \varepsilon^{1/3} \, , \tag{5.16}$$

and the growth rate is

$$\sigma = 0.232 \left(\frac{g' d}{\nu_2} \right) \varepsilon^{1/3} \, , \tag{5.17}$$

where $\varepsilon = \nu_2/\nu_1$, ν_1 is kinematic viscosity of the thin layer that is of depth d, ν_2 is viscosity of the infinitely deep fluid above it, and $g' = g\Delta\rho/\rho$ is called the reduced gravity.

The scaling law for a thin layer of relatively low viscosity fluid whereby the wavelength is proportional to the viscosity ratio to the one-third power is very commonly found, although it is not completely universal. The physical interpretation of the 1/3 power law is that it is more efficient for the low-viscosity fluid to flow large lateral distances up a gradual slope, and to accumulate in massive diapers, than it is to push straight up through the stiff material with shorter wavelength perturbations. This aspect of the dynamics will be illustrated through the use of a scaling argument here; the complete mathematical derivations are available in the original papers.

Assuming long wavelength compared to depth of the fluid, for a small disturbance the force balance in the thin layer is between the lateral pressure difference p and the viscous drag along the thin sheet, so

$$\frac{p}{\lambda} = \mu_1 \frac{u}{d^2} \,. \tag{5.18}$$

In the deep fluid above it the force balance is between the pressure, buoyancy and drag from the vertical deformation of the interface so that

$$\frac{p}{\lambda} = \frac{g'\eta}{\lambda} + \mu_2 \frac{w}{\lambda^2} \,. \tag{5.19}$$

This combines with (5.18) to give

$$\mu_1 \frac{u}{d^2} - \mu_2 \frac{w}{\lambda^2} = \frac{g'\eta}{\lambda} \,. \tag{5.20}$$

Continuity (conservation of volume flux) is

$$\frac{u}{\lambda} + \frac{w}{d} = 0 \tag{5.21}$$

and the kinematics of the interface is linearized so that

$$\frac{\mathrm{d}\eta}{\mathrm{d}t} = w \,. \tag{5.22}$$

Growth will be exponential as $w = w_0 \mathrm{e}^{\sigma t}$, where

$$\sigma = \frac{g'}{\nu_2} \left[\frac{\lambda}{\lambda^3/(\varepsilon d)^3 + 1} \right] \,. \tag{5.23}$$

Maximum growth rate occurs at

$$\frac{\lambda}{d} = 1.26\, \varepsilon^{1/3} \tag{5.24}$$

and the maximum growth rate is

$$\sigma = \frac{0.42\, g' d}{\nu_2} \varepsilon^{1/3} \,. \tag{5.25}$$

Thus the wavelength of fastest growth for a thin layer of smaller viscosity is much greater than the depth of the thin layer, but is unrelated to the depth of the overlying very deep layer. The low viscosity fluid accumulates in pockets that push their way into the more viscous fluid assisted by their relatively large volume. One might expect that the hot lower boundary layer of large Rayleigh number convection might exhibit blobs of fluid such as this, with the wavelength again determined by the boundary layer thickness and viscosity contrast as given by (5.16) or (5.24), and not by the overall depth of the convecting layer. We leave predictions for the mantle until Sect. 5.4.

We next consider the large-amplitude structures that arise from boundary layer instability. Experiments in which the buoyancy is due to temperature differences inevitably include the effects of heat conduction and have identified three basic forms of flow that may occur once convection has begun. First, a blob may become detached from the source boundary layer to form an isolated 'thermal'. Thermals are common in both experiments and numerical solutions of very viscous high-Rayleigh number convection with uniform viscosity, forming when flow sweeps away the feeding conduit or when nearby instabilities on the boundary layer remove the supply of heat. However, it is not clear whether they form in convection with large viscosity variations. Second, the boundary layer instability may lead to an initial transient flow in which the convection forms a mushroom-shaped 'starting plume' consisting of a large head and narrow tail, the later acting as a conduit through which fluid continues to be supplied to the head. Third, once a starting plume reaches the opposite boundary a more steady conduit flow may persist. The starting plumes and conduits are likely to be the dominant forms of motion when the viscosity is strongly temperature-dependent. Solutions for each of these forms of convective flow are summarised below.

5.3.2 Isolated "Thermals"

When a volume of buoyant fluid breaks away from the boundary, the resulting structure is known as a 'thermal', because of the superficial resemblance to the turbulent atmospheric thermals sought by birds and gliding enthusiasts to provide lift. During ascent of a 'thermal' heat can spread and warm up the surrounding cooler material (by conduction in the case of the extremely viscous mantle). However, the warmed material also becomes buoyant and begins to take part in the convection, with the result that the heat is not lost from the convecting region.

Consider first for comparison the case of a bubble of fluid for which the buoyancy is a consequence of an essentially non-diffusive property as in the Rayleigh–Taylor problem above (or a compositional difference in the mantle). In this case the volume of the less dense fluid remains constant. It can be shown that the bubble will become spherical and that the velocity of rise U for a bubble of volume V and diameter D is given by Stokes law:

$$U = \left(\frac{B}{2\pi D \eta_m} \right) f\left(\eta/\eta_m\right) , \tag{5.26}$$

where $B = g\Delta\varrho V$ is the total buoyancy, η_m is the viscosity far from the bubble and $\Delta\varrho$ is the density difference. The factor $f = 1$ when the ratio of viscosities inside and outside the bubble is small and the outer viscosity always has the dominant effect on the rate of ascent.

Theory and laboratory experiments show that the balance of buoyancy and drag (5.26) applies to the rise of thermals in which the density difference $\Delta\varrho = \varrho_m \alpha \Delta T$ (where ϱ_m is the environment density and α is again the coefficient of thermal expansion) is due to a temperature difference ΔT, despite the effects of the conduction of heat [24,25]. Assuming that no heat is lost from a thermal during its ascent, and that α is constant, conservation of heat implies that the buoyancy B (where in this case $B = g\alpha \int \Delta T \mathrm{d}V$) is conserved. As heat diffuses outwards into a thin boundary layer of thickness $\delta \approx (\kappa D/U)^{1/2}$ around the thermal, the newly heated layer becomes buoyant (and less viscous) and is drawn into the moving region, so increasing its volume V. The inward volume flux due to this process of 'thermal entrainment' is of order $\mathrm{d}V/\mathrm{d}t \approx UD\delta$ and the overall flow is characterized by a Rayleigh number $Ra_T = B/\kappa\nu_m$, where $\nu_m = \eta_m/\varrho_m$ is the kinematic viscosity of the environment.

A solution for self-similar flow can be derived using the above entrainment flux, conservation of buoyancy and the velocity (5.26) [24,25]. We predict the diameter D and height of rise z (above a virtual source at the point $z = 0$, where $D = 0$ and $t = 0$) as functions of time t:

$$D = C Ra_T^{1/4} (\kappa t)^{1/2} , \tag{5.27}$$

and

$$z = (f/\pi C) Ra_T^{3/4} (\kappa t)^{1/2} , \tag{5.28}$$

where C is a similarity constant of order unity. The value of C can in principle be predicted using numerical simulations capable of resolving details of the flow within the boundary layer [15,22]. However, it has only been evaluated from experiments (see below). Combining (5.27) and (5.28) shows that the diameter increases linearly with height,

$$D = 2\varepsilon z , \tag{5.29}$$

with a half-angle of spread $\varepsilon = (\pi C^2/2f) Ra_T^{-1/2}$ which is smaller for larger Rayleigh numbers. Hence the thermal enlarges less before reaching a given height for a larger temperature difference or smaller outer viscosity. The requirement that $\delta \ll D$ implies that the analysis applies to cases where $Ra_T \gg 1$. In addition to calculating the size and rate of ascent of a thermal, the above solution can also be used to calculate the shape of particle paths in the fluid, determine which fluid parcels will be entrained, and find the shapes of passive dye markers placed in the flow (Fig. 5.3).

Experiments in which known volumes of heated viscous oil were injected into a cooler environment of the same oil [24] showed that the behaviour was well described by (5.27)–(5.29). Fitting both (5.28) and (5.29) to the data, the similarity constant was found to be $C = 1.0 \pm 0.4$. This laboratory value of C will

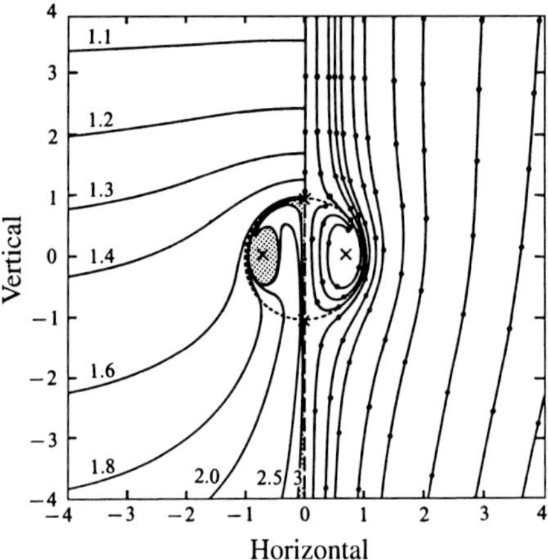

Fig. 5.3. The particle paths (*right*) and deformation of material surfaces (*left*) near a thermal with Rayleigh number $Ra_T \approx 1736$, relative to a frame of reference that is expanding with the diameter of the thermal. All fluid initially lying in a cone above the thermal (and bounded by a dividing streamsurface) is eventually heated and entrained into the thermal. The material surfaces illustrate the large vertical displacement of surrounding fluid that does not form a part of the warm thermal (From [28])

also be applicable to thermals in the mantle, provided the underlying assumptions are satisfied, and allows predictions of ascent speed and plume properties. The predicted shapes into which passive tracers are moulded by the flow compare well with those found in experiments [25]. For example, at $Ra_T > 200$ the internal circulation forms a torus into which all the material originally in the thermal is eventually advected (Fig. 5.3). Since the entrainment process relies on conduction, it is clear that the heat is distributed more widely through the surrounding spherical volume. However, the details of the temperature distribution, which are expected to be of lesser significance for the overall evolution of the flow, are not given by the above solution.

5.3.3 Starting Plumes

When a steady flux of buoyancy is suddenly supplied at the base of a region of viscous fluid (by heating the boundary or by injecting hotter fluid), it produces a nearly spherical volume of buoyant fluid that grows slowly until it becomes large enough to leave the boundary. As the spherical volume rises it remains attached to the source by a cylindrical conduit through which buoyant fluid continues to flow, so increasing the buoyancy and volume of the plume 'head' [70]. When the plume fluid has a relatively low viscosity, flow in the conduit (i.e. the hot plume tail) can be rapid, whereas the motion of the 'head' remains slow as a result of the larger outer viscosity.

Analysis of a starting plume driven by thermal buoyancy [28] involves only a simple modification of the theoretical treatment for isolated thermals. Conduction round the head again leads to warming and entrainment of surrounding fluid but we must this time take into account the increasing buoyancy in the head with time due to the source flux and the increase in volume due to both the source flux and entrainment. We define an average temperature anomaly for the plume head and, as a result of entrainment and cooling, this will be less than the temperature of the source fluid arriving at the top of the plume head through the axial conduit. The evolution of the plume head is governed by the heat conservation relation

$$V \Delta T = q_0 \Delta T_0 (t - t_0) , \tag{5.30}$$

where V is the volume, q_0 is the source volume flux and ΔT_0 is the source temperature anomaly. Along with (5.30), we have the momentum equation (5.26) and the head volume

$$\frac{\mathrm{d}V}{\mathrm{d}t} \approx q_0 + UD\delta . \tag{5.31}$$

For large times, when entrainment has become important, the solutions for the diameter D, velocity U and temperature anomaly ΔT of the head have the asymptotic forms

$$D \approx z^{3/5} , \qquad U \approx z^{1/5} , \qquad \Delta T / \Delta T_0 \approx z^{-1} , \tag{5.32}$$

where the constants of proportionality are functions of the plume Rayleigh number, defined in this case by $Ra_p = g\alpha \Delta T_0 q_0^3 / \kappa^4 \nu_m$ [27]. Note that in deriving the above solution we do not need to make any specific assumptions about the form of the profiles of velocity or temperature either in the feeding conduit or in the plume head. The essential assumption is that these profiles remain similar as the flow develops; use of a mean temperature does not require an assumption that the temperature is constant across the plume. But in each of the relations such as (5.32) there is a similarity constant which depends on the real profiles, and which has been evaluated experimentally (see below). In principle it could also be found through finite-element numerical models such as those of Davies [15].

Photographs of a hot starting plume in the interior of a laboratory tank are shown in Fig. 5.4. This plume was produced by injecting hot, dyed syrup at a steady rate into the same (but cold and very viscous) syrup [28]. There was little cooling of the fluid flowing up the conduit until it arrived at the forward stagnation point of the rising head, where it met the resistance of the overlying fluid. There it spread laterally and axisymmetrically as a sheet, facilitating a more efficient heat transfer to a boundary layer in the surrounding fluid, which henceforth became part of the plume head. After the head had ascended a large distance a continuous axisymmetric spiral of dyed material extended inward to a toroidal focus.

Although experiments with continuously fed plumes have gone some way towards determining the coefficient C [23], there are still considerable uncertainties

Fig. 5.4. Photographs of a laboratory starting plume after it has left the source region, enlarges both by continued addition from the source through the conduit and by entrainment, and eventually spreads beneath the free surface. The hot source fluid is dyed. The temperature distribution is not seen. This is one of a series of experiments designed to test the theoretical similarity solution and it illustrates the predicted nature of newly forming mantle plumes. However, for scaling to the mantle it is necessary to use the theory referred to in Sect. 5.3 [28]

in its value (which relates to the rate of incorporation of external fluid into the rising plume head). Departures from self-similarity during the plume ascent in the experiments (due to viscosity changes in the head, a finite volume in the conduit, temporal changes in the head shape, and side-wall effects, all of which are neglected in the simple formulation given here) make it difficult to determine the coefficient to better than a factor of two. However, when the experiments are compared with the solution after small correction terms are included in (5.26, 5.30, 5.31), the result is consistent with that for detached thermals ($C \approx 2$) and is robust enough to allow some firm predictions to be made about the scale and ascent rate of plume heads in the mantle (summarised in Sect. 5.4). The quantitative application to the mantle is also consistent with a range of geophysical data and has been supported by more recent numerical modelling results [12,14,22]. Modifications of the plume behaviour in a mantle of power-law rheology have been computed [67]. These show that plume heads may ascend more rapidly than predicted for a Newtonian mantle, and reach farther into the base of the lithosphere, but that entrainment and head size are not greatly changed.

The distribution of source fluid in the plume head, as seen in Fig. 5.4, does not indicate the temperature distribution, which we can safely assume will be much more smoothly distributed through the bulk of the plume head as a result of the nature of the 'thermal entrainment' process and continued dispersion within the head. The axial conduit and the radial outflow near the top of the head will be almost as hot as the source, and there may be some small remnant temperature maximum near the toroidal focus, whereas the remainder of the source and entrained material in the head will be significantly cooler. Davies [15] and Farnetani & Richards [22] have computed temperature distributions which confirm these ideas, in particular the conclusion that there are only small temperature gradients everywhere, except around the axial conduit and the horizontal outflow at the top of the head. They show that the coupling of advection and conduction is so effective at re-distributing heat between source and entrained material that there is only a small temperature maximum near the toroidal focus of the flow. The hot outflow layer at the top of the plume head and the axial conduit appear as the dominant features in the temperature plots.

5.3.4 Long-lived Plumes

After the head of an (isolated) new plume reaches the top of the layer through which it is rising, and if the source flux is constant, the flow in the trailing conduit delivering material from the source tends toward a steady state. If the surrounding fluid is otherwise at rest, the conduit will be vertical and axisymmetric.

A similarity solution [39] for flow in the vicinity of a steady conduit shows a very strong tendency for the vertical velocity within the conduit to be confined to a thin low-viscosity core, along with a radial balance between the horizontal diffusion of heat out of the conduit and a slow inflow driven by a radial pressure gradient (low pressure in the hotter, lower density conduit). This in effect produces an insulating sheath around the conduit flow. Another similarity solution by Hauri et al. [34] incorporated a wide range of effects of temperature and shear stress on viscosity along with depth-dependent viscosity and thermal expansivity. For a wide range of plausible rheologies, and for buoyancy fluxes of 10^3–10^5 N s^{-1} (see Sect. 5.4.1), vertical velocities in the conduit predicted by this solution range over 0.030–100 m yr^{-1} and conduit radii range over 30–250 km. The extent of dilution by entrainment of surrounding mantle into the conduit flow ranges from under 5% to over 90%, with small buoyancy flux associated with the most entrainment. Most of the entrained material originates from the lower half of the layer traversed by the conduit.

Since mantle plumes are expected to be produced by only a small fraction of the heat flow at the top of the mantle, it is natural to expect the plumes to be strongly influenced by mantle motion driven by movement of tectonic plates and deep subduction (see Sect. 5.5). When there is a larger scale systematic motion in the surroundings, such as a superimposed horizontal shear flow, the conduit will be bent over in the direction of the horizontal flow. The relationship between the shear velocity and the tilt can be simplified to a relation in terms of a vector addition of the horizontal advection velocity and the vertical Stokes velocity of a sphere. For example, with a linear shear profile (in which the horizontal velocity varies by u_0 over a depth h) and a conduit of fixed diameter D, the formula for deflection $x(z)$ is [52]

$$\frac{x}{h} = \frac{u_0}{2kU}\left(\frac{z}{h}\right)^2 , \tag{5.33}$$

where U is given by (5.26) for a sphere with $f = 1$. Laboratory results give $k = 0.54$ [52]. This kinematic theory gives the tilt at the top of the sheared region as

$$\frac{\mathrm{d}z}{\mathrm{d}x} = \frac{kU}{u_0} \tag{5.34}$$

because the conduit is rotated by an amount that depends on the Stokes rise time of a given conduit element through the depth of the shear zone. If the shear is intense enough to rotate the conduit by more than about 55 degrees from the vertical before it rises out of the region with shear, it will develop an instability [31,52,57]. With instability present, a conduit that is fed steadily at its base will not expel the material steadily at the top of the shear zone, but will develop a chain of new plume heads [69], each rising to a different spot on the surface.

The most obvious application is to the Hawaii-Emperor seamount chain, which not only has been actively producing volcanos for more than 80 Myr, but also experienced a change in plate motion about 40 Myr ago. During this change of motion, the trend of the hot spot track on the Earth's surface changed direction and produced a bend in its path with a radius of less than 200 km, which implies that the upwelling conduit had a small horizontal deviation from the plume source to the surface [31]. Using $u_0 = 0.1\,\mathrm{m\ yr}^{-1}$, and $U = 0.05\,\mathrm{m/yr}$, corresponding to a density difference of $50\,\mathrm{kg\,m}^{-3}$, and a (somewhat arbitrarily chosen) conduit radius of 70 km in (5.33)–(5.34) gives a lateral deflection almost exactly equal to the depth of the mantle (which is unknown for the mantle), and the conduit is tilted to 64 degrees. This does not fit the above observation of the abruptness of the bend in the hotspot track unless the shear zone is less than 200 km deep. However, a mantle flow underlain by a return flow gives considerably smaller deflections which do satisfy the observations [31].

Plume conduit solutions based on measured plume buoyancy fluxes (see Sect. 5.4.1), rather than an assumed conduit radius, and which allow for effects of tilting on entrainment have also been constructed [29]. These too predict small horizontal deflections of conduits carrying the relatively large buoyancy flux of the Hawaiian plume. Hence tilt angles for Hawaii are expected to be below the critical angle. Thus strong plumes are expected to experience relatively small deflection and inject a steady flow of mantle material to the base of the lithospheric plates. Weaker plume conduits will be deflected more and thus break up more readily, setting the lower limit to the buoyancy flux that will generate significant effects at the surface. Deep shear zones tend to produce more lateral deflection and thus a greater tilt angle. Shear zones concentrated near the surface tend to produce less lateral deflection and a smaller tilt angle. However, with continuous variation of properties, and with two or three dimensional mantle convection, these simple models must be considered a starting hypothesis at best.

Another consequence of the tilting of plume conduits in a mantle 'wind' is enhanced entrainment of the surroundings. While hot material flows upward along the conduit, each part of an inclined thermal-plume conduit must be rising through and continually displacing its surroundings upward if it is to maintain a steady shape, and therefore must contain a circulation in planes normal to the axis of the conduit; the quasi-two-dimensional equivalent of that shown in Fig. 5.3 for the axisymmetric plume head. Surrounding material is again heated in a boundary layer round the rising cylindrical region and is drawn into it, so increasing the volume flux in the conduit. The source material is concentrated into two cores, leaving a central strip which is relatively free of source fluid [53]. The solutions referred to above for the thermal conduit flow [29] predict that entrainment has a much greater effect on bent-over plumes when the temperature-dependent viscosity in the plume is allowed to increase with distance from the source due to entrainment and cooling, since in order to cope with the imposed buoyancy flux the diameter of the conduit must then increase (with height) as the plume cools. The solution also predicts that, as for starting plumes, the behaviour is a function of the plume Rayleigh number $Ra_p = g\alpha\Delta T_0 q_0^3/\kappa^4\nu_m$, where q_0 is the source volume flux. Plumes with larger buoyancy fluxes will be less tilted and entrain a volume flux from the surroundings that is smaller relative to the source volume flux (i.e. they will be less diluted).

5.3.5 Surface Uplift

Experiments have also been used to help predict the surface topography generated by the arrival of a plume head beneath the surface of a convecting fluid or the continued upwelling of a plume conduit. For example, capacitance [42] and optical interferometric observations [31] give the maximum surface uplift and the maximum rate of uplift over a rising spherical diapir

$$h_{\max} = 0.27(\Delta\varrho/\varrho_m)D_0 \ , \tag{5.35a}$$

$$v_{\max} = 0.16(\Delta\varrho/\varrho_m)U_0 \ , \tag{5.35b}$$

where $\Delta\varrho$ is the density anomaly, ϱ_m is the surrounding density and D_0 and U_0 are the diameter and velocity of the diapir when it is far from the surface. The diapir or plume head spreads horizontally as it approaches closer than one radius from the surface, but the rate of spreading decreases as $D \sim t^{1/5}$, so that further spreading is very slow after the horizontal radius has doubled [30]. This result is in agreement with theoretical scaling laws for the radial spreading of a low-viscosity blob into a high-viscosity fluid. The surface reaches a maximum uplift after which it slowly subsides and the width of the surface swell increases. During this collapse a thin layer of more dense outer fluid remains above the top of the plume and thins (according to a $t^{-1/2}$ law) to around $0.1D_0$ when the plume diameter has doubled. In the laboratory experiments, this layer becomes gravitationally unstable at a time $Ut/D_0 \approx 10$ after maximum uplift is reached, and overturning with the underlying plume fluid leads to either axisymmetric convective flow or to irregular three-dimensional convection. This smaller scale of convection is able to more rapidly continue the release of potential energy and enable the plume fluid to penetrate closer to the surface.

Similar behaviour is expected where a continuing conduit flow impinges beneath migrating oceanic crust. In this case, uplift of the surface occurs upstream of and above the upwelling flow in the conduit, where buoyancy is being added to the mantle beneath the crust, leading to the seafloor swell. Subsidence of the swell and its associated volcanic island chain occurs with age downstream of the conduit location, where horizontal spreading flow of the hot plume material beneath the lithosphere proceeds to slowly redistribute the buoyancy across an increasingly broad region.

5.4 Mantle Plumes and Surface Topography

In order to apply the theoretical, computational and laboratory results discussed above to predict plume velocities and sizes in the mantle, one first needs to make realistic estimates from geophysical data of the material properties (in particular the viscosity), and also of the temperature anomaly and heat flux at the source.

5.4.1 Plume Fluxes from Hotspot Tracks

One of the most important inputs to quantitative predictions for plumes in the mantle is the boundary condition on temperature or heat flux, or both, at the base. The plume heat flux $F_H = q_0 \varrho c_p \Delta T_0$ (or, more precisely, the plume buoyancy flux $F_B = g\alpha F_H / c_p$, where c_p is the specific heat capacity and q_0 is again the source volume flux, as in (5.30)) has the primary control on the plume flow. The temperature anomaly plays a lesser role through its influence on the viscosity difference and partial melting. The range of plume fluxes to be found in a convecting fluid with temperature-dependent viscosity is not yet understood: it will be related to the plume spacing. However, we can understand individual plumes by considering a single plume in isolation from other plumes and lateral boundaries, and specifying both a source temperature anomaly and buoyancy flux[3] based on geophysical constraints.

We begin by applying the results for Rayleigh–Taylor instability of a thin layer to the bottom boundary layer of the mantle in order to estimate the separation of plumes. Assuming the unstable layer is 50 to 100 km thick, and a viscosity contrast $\varepsilon \approx 10^3$ the wavelength (5.16) is of order 600–1200 km. This wavelength is consistent with the separation of volcanic hotspots within the Pacific Plate, where the separation is significantly smaller than the depth of the mantle. The result indicates a strong effect of the viscosity contrast. The analysis

[3] In most computer models of a convecting layer, only one of these is imposed, since the other is then determined by the coupling of conduction and convection of the bottom boundary layer. In numerical experiments [15,22] this was done by applying a temperature anomaly over a finite area of the bottom boundary. In the laboratory experiments described above the temperature anomaly and source mass flux are prescribed.

is expected to apply to the onset of instability on the boundary layer between existing plumes, where new plumes may be initiated. However, it should be remembered that the linear stability analysis does not necessarily predict well the separation of well-established plume conduits or the effects of large-scale convective flow driven by sinking of the cold surface boundary layer, which may advect plumes into regions of convergence of the large scale flow.

Estimates of the plume buoyancy and heat fluxes in the mantle have been made using observations of the surface effects of long-lived plumes in oceanic settings, where the crust exerts relatively little masking compared to that of thicker continental crust [12,58]. The size of the hotspot swell can be combined with the velocity of the plate over the hotspot to obtain the rate of production of anomalous topography, from which we infer the buoyancy flux carried by the plume. For example, the existence of the Hawaiian swell, about 1000 km wide and 1 km high, propagating across the Pacific plate at about 100 mm yr^{-1}, implies a buoyancy (or mass-deficit flux) in the plume of 7.3×10^3 kg s^{-1}. This mass-deficit flux is actually $\alpha \Delta T Q$, where Q and ΔT are the mass flux and temperature difference at any depth. This is related to the more physically meaningful buoyancy and heat fluxes through $F_B = g(\alpha \Delta T Q)$, and $F_H = (c_p/\alpha)(\alpha \Delta T Q)$, respectively. For the Hawaiian plume $F_B \approx 8 \times 10^4$ N s^{-1} and $F_H \approx 3 \times 10^{11}$ W. The distribution of plume mass-deficit fluxes (calculated by Sleep [58] and Davies [12] for 35 oceanic hotspots) therefore imply the buoyancy fluxes as plotted in Fig. 5.5 [32]. Although there are many uncertainties in such estimates, they do indicate that plumes carry a range of fluxes, and that the distribution is (logarithmically) centred about 10^4 N s^{-1}.

An estimate of the volume or mass flux requires independent knowledge of the temperature anomaly, which is usually obtained from the petrology of erupted melts. However, the mass flux is not a conserved quantity in that it, like the temperature, may vary with height along a plume. Nor is the mass flux well-defined: on the one hand, the mass flux of *hot material* near the top of the plume (the flux that is relevant to melt production) for the Hawaiian example becomes $Q_{\text{top}} = F_B/(g\alpha\Delta T) \approx 3 \times 10^5$ kg s^{-1} (assuming an average temperature anomaly of 100 C and no large scale shearing); on the other hand, the movement of the lithospheric plate over the plume implies, as we have already explained, that upper mantle is continuously being displaced by the plume and that there must be a vertical mass flux in the cooler surroundings. The upward mass flux relevant to overall motion and stirring in the mantle is then made up of both the slow broad motion of the surroundings (associated with the Stokes ascent of the plume conduit in the presence of plate migration), and the Poiseuille-like pipe flow (at relatively large velocities) of low-viscosity material upward through the narrow conduit. Unpublished experiments (by RG) with stirring when a plume conduit passes through a larger-scale overturning convection cell will not be discussed here but show how the former of these two transport components may be dominant and lead to a large vertical transport of the surroundings up toward the surface, as well as to disturbance of particle paths in the large scale cell.

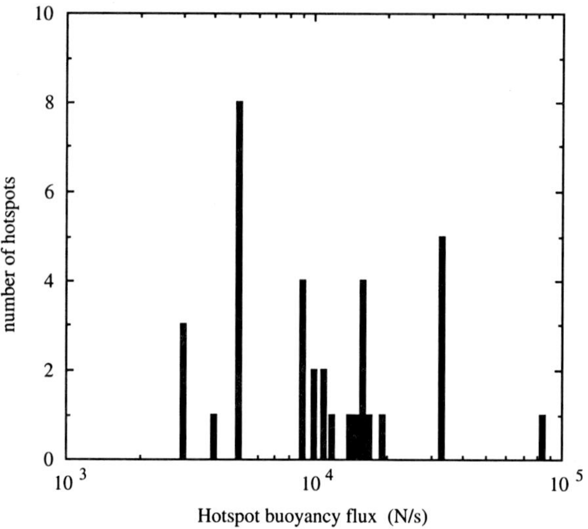

Fig. 5.5. Plume buoyancy fluxes, adapted from Sleep's [58] estimates of the fluxes for 35 oceanic hotspots [29,32]. The buoyancy flux is Sleep's 'mass exchange flux' mutiplied by g

5.4.2 New Plumes and Flood Basalts

Predictions can be made for new plume heads by assuming that the rate of supply of buoyancy from the source boundary layer during this early stage in the life of a plume falls in the same range as the buoyancy fluxes derived for currently active hotspot tracks. In that case a mantle viscosity of 10^{22} Pa s implies that heads will grow as large as 400–600 km in diameter at the core–mantle boundary before their ascent speed is large enough to cause them to break away. Application of the complete form of (5.32) [28] to the ensuing motions leads to the prediction of a further doubling of the diameter (and an increase of volume by an order of magnitude) as the plume heads ascend through 2800 km. Thus plume heads that reach the lithosphere while still receiving a constant influx from their source region are predicted to be extremely large: 800–1200 km in diameter. They will also have incorporated a volume of lower mantle material comparable with the total volume supplied from the source, though the ratio of these two volumes depends on the source flux. The head size, however, is insensitive to the flux. The diameter D is instead dependent primarily on the mantle viscosity ($D \approx \eta_m^{1/5}$).

As a plume head approaches the upper boundary (the free surface in the laboratory tank, or the stiff lithosphere of the Earth) it must flatten and spread. Thus a spherical head, predicted to be of order 1000 km in diameter, should produce a pancake-shaped thermal anomaly about 2000 km across at the base of the lithosphere. It should be remembered that the dimension given by the model is the diameter of the equivalent sphere that would contain the plume head buoyancy at an average temperature, and that some of the head will be much cooler while the top of the head will contain the hottest material supplied from the source via the conduit. Similar head sizes and ascent times are predicted by numerical experiments simulating mantle conditions [15,22], and the chronology, tectonics and geochemistry of flood basalt provinces, believed to be attributable to plume heads, are consistent with the 1000–2000 km scale. Furthermore, continental flood volcanism is known to be characterised by a sudden onset, with most of the magmas erupted within a short period of 1 to 3 Myr [53] and over a roughly equant region 2000–2500 km across, followed by slow subsidence. Use of these comparisons to argue in the opposite direction provides evidence that plume heads responsible for the major flood basalts had dimensions consistent only with an origin deep in the lower mantle and therefore most probably in the thermal boundary layer at the CMB.

5.5 The Upper Boundary Layer

We turn now to the cooled upper boundary, and the generation of the primary motions of mantle convection. The total heat flux at the Earth's surface, apparently an order of magnitude greater than the flux carried by hot plumes, is largely due to loss of the heat generated by radioactive decay. That is, the mantle may be regarded as a layer of viscous fluid, largely internally heated, and cooled from above.

Early notions about mantle convection regarded plate tectonics as the surface reaction to an underlying pattern of convection occurring especially in the upper mantle. This view required the plates to be dragged along by a faster motion beneath. When the observations (Fig. 5.1) were compared with laboratory studies with this picture in mind, it was puzzling that the inferred convection cells are so much wider than their depth (often presumed to be that of the upper mantle), and this led to many investigations of the effect of variable fluid properties and different boundary conditions on the aspect ratio. A more consistent view is that the buoyancy forces acting on the colder, denser plates are the primary driving mechanism of convection, at least under the present tectonic regime, so subduction and descent of lithospheric slabs is an active part of convection, not a reaction to it. The plates are the upper thermal boundary layer. Those earlier questions about the horizontal scale are then readily answered by 1) considering that the convection may penetrate the full (2900 km) depth of the mantle and 2) noting that the strength of the lithosphere (which can yield and break only at stresses greater than a few hundred MPa) can inhibit the initiation of subduction and thus increase the horizontal scale of convection cells. In this view cold material can break away from the upper boundary only at plate boundaries where one plate may slide under the other. Near mid-ocean ridges there is a compensating, passive ascending flow – this upwelling limb of the convection is not a hot active plume. Thus much of the structure of convection in the mantle is organized by the pattern of the plates, though the prediction of the criteria for formation and size of these plates, and for the initiation of subduction, remains a major theoretical challenge.

The most significant topographic characteristics of plate convection are the deep ocean trenches (up to 5000 m below the mean sea floor) at the subduction zones and the mid-ocean ridges (standing 3000 m above the plate spreading centres). The trenches are clearly the effect of the presence below the surface of a larger mass of cold dense lithosphere (this negative buoyancy will pull down the surface even in the absence of motion) as well as the downward motion of the slab. The topographic high above the spreading centres and its steady fall-off with distance from the ridge have been shown to be a simple consequence of conductive cooling of the oceanic lithosphere and thermal contraction while remaining in isostatic balance with the whole of the seafloor [18]. The plate motion, driven by surface cooling and subduction, produces a pressure gradient that "pulls" warm mantle material up to the surface at the spreading centre. Close to the ocean floor this material proceeds to cool and hence increases its density as it moves away from the centre. This produces an additional ocean depth Δd which can be found from the buoyancy balance $g\Delta\varrho\Delta d = g\varrho_m\alpha\Delta T z$, where $z = 2(\kappa t)^{1/2}$ is the conductive thickness of the lithosphere, κ is the thermal diffusivity of the lithosphere, t is the age of any section of the lithosphere, ϱ_m is the density of the mantle, ΔT is the average temperature of the lithosphere relative to the surface and $\Delta\varrho$ is the density difference between the mantle and seawater. The result is $\Delta d = 2\alpha\Delta T(\varrho_m/\Delta\varrho)(\kappa t)^{1/2}$, which predicts about 3000 m relief between old seafloor (100 Myr old and 100 km thick) and the ridge crest (using $\alpha = 3 \times 10^{-5}\,°\mathrm{C}^{-1}$, $\varrho_m = 3,300\,\mathrm{kg\,m^{-3}}$, $\Delta T = 650\,°\mathrm{C}$). This square root of seafloor age relation explains most of the measured topography, which is therefore consistent with a predominantly internally-heated mantle undergoing convection due to surface cooling, and with a passive upwelling at normal mantle temperature under the ridge.

5.6 Synopsis

The dominant large-scale morphology of the earth's surface is a direct consequence of the dynamics of thermal convection in the Earth's solid mantle. The two major topographic features of the ocean floor, ocean trenches and mid-ocean ridges, represent active boundaries of the tectonics plates where the upper thermal boundary layer of the mantle convection system is foundering and sinking into the interior or just beginning its thermal development, respectively. The topography is dynamic, being produced by buoyancy differences within the lithosphere. We have not discussed continents and their major mountain belts, however these too are directly produced by mantle convection through plate collisions (either of continent with continent, such as the formation of the Himalaya by the collision of the Indian sub-continent with the Eurasian continent and the formation of the European Alps by the collision of Africa with Europe, or of oceanic plate and continent, as in the Sierra Ranges and Andes of the Americas). Looking at more isolated structures, but still at a large scale, continental flood basalt provinces (such as the Deccan Traps of India) and oceanic plateaux (such as the Ontong-Java plateau in the Pacific), each of order 1000–2000 km across and 1000–2000 m high, and volcanic ('hotspot') tracks such as ocean island chains and their accompanying, broader seafloor swells, are all thought to be generated by the buoyancy of hot upwelling plumes. These most probably ascending from the core-mantle boundary. Plumes are also thought to have been the cause of continental rifting and the initiation of the opening of new ocean basins [35,68]. The role of mantle convection in the generation of surface morphology has meant that knowledge of the surface topography has provided important evidence about the way in which the mantle works. This topography strongly influences ocean and atmosphere circulation patterns. It also acts as major drainage highs, which are eroded during and after their formation, and drainage lows, which capture vast amounts of sediment.

References

1. H. Berner, H. Ramber, O. Stephansson: Tectonophysics **15**, 197 (1972)
2. M.A. Biot, H. Ode: Geophysics **30**, 213 (1965)
3. M.A. Biot: Geophysics **30**, 153 (1966)
4. R. Boehler: Nature **363**, 534 (1993)
5. B.A. Buffet, H.E. Huppert, J.R. Lister, A.W. Woods: J. Geophys. Res. **101**, 7989 (1996)
6. I.H. Campbell, R.W. Griffiths: Earth Planet. Sci. Lett. **99**, 79 (1990)
7. I.H. Campbell, R.W. Griffiths: J. Geol. **92**, 497 (1992)
8. S. Chandrasekhar: *Hydrodynamics and Hydromagnetic Stability* (Oxford University Press, New York 1961)
9. Z.F. Danes: Geophysics **29**, 414 (1964)
10. A. Davaille, C. Jaupart: J. Fluid Mech. **253**, 141 (1993)
11. A. Davaille, C. Jaupart: J. Geophys. Res. **99**, 19,853 (1994)
12. G.F. Davies: J. Geophys. Res. **93**, 10,467 (1988)
13. G.F. Davies: J. Geophys. Res. **93**, 10,451 (1988)
14. G.F. Davies: Geophys. J. Int. **115**, 132 (1993)
15. G.F. Davies: Earth Planet. Sci. Lett **133**, 507 (1995)
16. G.F. Davies, M. Gurnis: Geophys. J. Roy. Astr. Soc. **85**, 523 (1986)
17. G.F. Davies, M.A. Richards: Geophys. J. Geol. **100**, 151 (1992)
18. G.F. Davies: *Dynamic Earth: Plates, Plumes and Mantle Convection* (Cambridge University Press, Cambridge 1999)
19. M.B. Dobrin: Eos Trans. AGU **22**, 528 (1941)
20. M.R. Drury, J.D. FitzGerald: 'Mantle Rheology: Insights from laboratory studies of deformation and phase transformation'. In: *The Earth's Mantle: Composition, Structure and Evolution*, ed. by I. Jackson (Cambridge University Press, New York 1998) pp. 503–559
21. R.A. Duncan, M.A. Richards: Rev. Geophys. **29**, 31 (1991)
22. C. Farnetani, M.A. Richards: Earth Planet. Sci. Lett. **136**, 251 (1995)
23. M.F. Fitzpatrick: The dynamics of viscous thermal plumes in the Earth's mantle. Honours Thesis, The Australian National University (1991)
24. R.W. Griffiths: J. Fluid Mech. **166**, 115 (1986)
25. R.W. Griffiths: J. Fluid Mech. **166**, 139 (1986)

26. R.W. Griffiths: Earth Planet. Sci. Lett. **78**, 435 (1986)
27. R.W. Griffiths: Phys. Fluids, A **3**, 1233 (1991)
28. R.W. Griffiths, I.H. Campbell: Earth Planet. Sci. Lett. **99**, 66 (1990)
29. R.W. Griffiths, I.H. Campbell: Earth Planet. Sci. Lett. **103**, 214 (1991)
30. R.W. Griffiths, I.H. Campbell: J. Geophys. Res. **96**, 18, 295 (1991)
31. R.W. Griffiths, M.A. Richards: Geophys. Res. Lett. **16**, 437 (1989)
32. R.W. Griffiths, J.S. Turner: 'Understanding mantle dynamics through mathematical models and laboratory experiments'. In: *The Earth's Mantle: Composition, Structure and Evolution*, ed. by I. Jackson (Cambridge University Press, New York 1998) pp. 191–227
33. R.W. Griffiths, M. Gurnis, G. Eitelberg: Geophys. J. **96**, 1 (1989)
34. E.H. Hauri, J.A. Whitehead, S.R. Hart: J. Geophys. Res. **99**, 24275 (1994)
35. R.I. Hill, I.H. Campbell, G.F. Davies, R.W. Griffiths: Science **256**, 186 (1992)
36. L.N. Howard: 'Convection at high Rayleigh number'. In:*Proc. 11th Int. Congress Applied Mechanics, Münich 1964*, ed. by H. Görtler (Springer-Verlag, Berlin) pp. 1109–1115
37. B.L.N. Kennett, R. van der Hilst: 'Seismic structure of the mantle: From subduction zone to craton'. In: *The Earth's Mantle: Composition, Structure and Evolution*, ed. by I. Jackson (Cambridge University Press, New York 1998) pp. 381–404
38. K. Lambeck, P. Johnston: 'The viscosity of the mantle: Evidence from analyses of glacial-rebound phenomena'. In: *The Earth's Mantle: Composition, Structure and Evolution*, ed. by I. Jackson (Cambridge University Press, New York 1998) pp. 461–502
39. D.E. Loper, F.D. Stacey: Phys. Earth Planet. Interiors **33**, 304 (1983)
40. W.J. Morgan: Nature **230**, 42 (1971)
41. L.L. Nettleton: Amer. Ass. Petrol. Geol. Bull. **18**, 1175 (1934)
42. P.L. Olson, I.S. Nam: J. Geophys. Res. **91**, 7181 (1986)
43. T.J. Parker, A.N. McDowell: Amer. Ass. Petrol. Geol. Bull. **39**, 2384 (1955)
44. H. Ramberg: Bull. Geol. Inst. Univ. Uppsala **42**, 1 (1963)
45. H. Ramberg: Geophys. J. **14**, 307 (1967)
46. H. Ramberg: Phys. Earth Planet. Interiors **1**, 63 (1968)
47. H. Ramberg: Phys. Earth Planet. Interiors **1**, 427 (1968)
48. H. Ramberg: Phys. Earth Planet. Interiors **5**, 45 (1968)
49. H. Ramberg: Geol. J. Spec. Issue **2**, 261 (1970)
50. Lord Rayleigh: Scientific papers, ii (Cambridge University Press, Cambridge 1900), 200–207
51. M.A. Richards, R.A. Duncan, V.E. Courtillot: Science **246**, 103 (1989)
52. M.A. Richards, R.W. Griffiths: Geophys. J. **94**, 367 (1988)
53. M.A. Richards, R.W. Griffiths: Nature **342**, 900 (1988)
54. G.G. Schaber, R.G. Strom, H.J. Moore, L.A. Soderblom, R.L. Kirk, D.J. Dawson, L.R. Gaddis, J.M. Boyce, J. Russell: J. Geophys. Res. **97**, 13,257 (1992)
55. J.G. Sclater, J. Francheteau: Geophys. J. Royal. Astron. Soc. **20**, 509 (1970)
56. F. Selig: Geophysics **30**, 633 (1965)
57. J.N. Skilbeck, J.A. Whitehead: Nature **272**, 499 (1978)
58. N.H. Sleep: J. Geophys. Res. **95**, 6715 (1990)
59. V.S. Solomatov, L.N. Moresi: J. Geophys. Res. **101**, 4737 (1996)
60. F.W. Stacey, D.E. Loper: Phys. Earth Planet. Interiors **33**, 45 (1983)
61. D.J. Stevenson, J.S. Turner: 'Fluid models of mantle convection.' In: *The Earth: Its origin, structure and evolution*, ed. by M.W. McElhinny (Academic Press, London 1979) pp.227–263

62. D.C. Tozer: Phil. Trans. R. Soc. A **258**, 252 (1965)
63. D.L. Turcotte, E.R. Oxburgh: Ann. Rev. Fluid Mech. **4**, 252 (1972)
64. D.L. Turcotte, G. Schubert: *Geodynamics applications of continuum physics to geological problems* (John Wiley and Son, New York 1982)
65. J.S. Turner: *Buoyancy Effects in Fluids* (Cambridge University Press, Cambridge 1973)
66. J.S. Turner: Earth Planet. Sci. Lett. **17**, 369 (1973)
67. R. Weinberg, Y.Y. Podladchikov: J. Structural Geol. **17**, 1183 (1995)
68. R. White, D. McKenzie: J. Geophys. Res. **94**, 7685 (1989)
69. J.A. Whitehead: Ann. Rev. Fluid Mech. **20**, 369 (1988)
70. J.A. Whitehead, P.S. Luther: J. Geophys. Res. **80**, 705 (1975)

6 Morphological Instabilities in Flows with Cooling, Freezing or Dissolution

J.A. Whitehead[1] and R.W. Griffiths[2]

[1] Woods Hole Oceanographic Institution, Woods Hole, MA 02543, USA
[2] Research School of Earth Sciences, The Australian National University, Canberra 0200 ACT, Australia

6.1 Introduction

The Earth's crust is shaped by a wide range of fluid flows and their characteristic instabilities. Here we consider the flow of silicate melts, either within the crust or as surface lava flows, and the way in which these flows are affected by variable viscosity due to cooling or by a yield strength resulting from solidification. These effects invariably lead to non-uniform or three-dimensional flow patterns, particularly fingering and channelisation. In the case of solidifying free-surface flows there is, in addition, a range of three-dimensional surface structures or deformation styles depending on flow conditions. Parallels can be drawn with channeling instabilities that occur in either the dissolution of a porous matrix or precipitation reactions within a matrix during the percolation of an interstitial fluid.

The flow of magma through the upper-most solid mantle and crust is fundamental to the formation and evolution of the crust, which is formed and reworked through geological time by the rise of melts towards the surface. The melts, whether granitic or basaltic, do not ascend in a uniform and steady fashion but are instead influenced by heat loss to the relatively cool surrounding rock and consequent variation in viscosity. They rise through channels or, in some cases, as diapirs. The channels, at least sills and dykes in the upper crust, where temperature differences are larger, tend also to evolve from two-dimensional slots to more focused three-dimensional channels [6,24]. Thus volcanic eruptions tend to occur not through a uniform percolation of melt, nor through uniform two-dimensional flow from a dyke, but from localised vents. We therefore begin with a discussion of the fingering instability that causes the flow of cooling melts in two-dimensional sills or dykes to become focused into three-dimensional channels. We then turn in Sect. 6.3 to the case of channelling of the flow of aqueous solution through a soluble porous medium.

Much of the Earth's surface, indeed much of the mass of the planet's crust, was at some time laid down by lava flows, whether submarine (from mid-ocean ridges, seamounts or plume-related volcanic hotspots) or sub-aerial (from volcanism above subduction arcs or hotspots) (see Chap. 5). Hence we devote some of this article to simple models of lava flows and the instabilities that shape them, noting two factors in particular. First, that lava is not a simple Newtonian liquid but is generally a mix of silicate liquid, crystals and gas bubbles with complicated rheology that may vary from close to Newtonian in the case of the hotter basaltic eruptions (e.g. Hawaiian channel flows) to a rheology having a large yield strength in the case of the high crystal content andesitic lavas of arc volcanoes (such as those of Japan and Indonesia). Second, that surface heat fluxes from the lava flows are generally large enough to cause rapid quenching of a thin surface layer [22], whereas the slower process of crystallization leads eventually to complete solidification of the flow. The behaviour of lava flows, their structure, rate of flow front advance and instabilities varies according to the properties of the erupted magma, the effusion rate, the ground topography over which it flows and the rate of heat loss (determined primarily by the environment) [22,42,47]. The flow front eventually comes to a halt as a result of cooling or ground topography.

In Sect. 6.4 we analyse the spreading of a high yield strength material under gravity as a free-surface gravity current on a sloping plane, but ignoring effects of cooling. In Sect. 6.5 we discuss dimensional analysis and laboratory experiments with cooling and freezing gravity currents on a horizontal base which demonstrate many of the instabilities and morphological characteristics of real flows. These show some similarities with the fingering instability due to viscosity increase of Sect. 6.1, but are substantially different in that the cooling is confined to a very thin surface boundary layer and in that a wider range of behaviour is observed. We then turn in Sect. 6.6 to a discussion of experiments with freezing flows on a slope. A more extensive review of work on cooling and solidifying free-surface flows can be found in [22].

6.2 Viscous Fingering Instabilities

It is well-known that the intrusion of a less viscous fluid into a two-dimensional slot (or a porous medium) containing a more viscous fluid leads to inter-fingering of the two fluids as a result of the Saffman–Taylor instability [43]. This is an isothermal phenomenon but relies on the viscosity contrast across the advancing front. A more complex case, but one which is more relevant to the flow of hot silicate melts in the earth's crust, is that in which the flow involves only one fluid but the viscosity is a function of temperature, so that the dynamics of flow become strongly coupled with the heat flow.

Following [48] we consider a flow in a narrow slot with walls at temperature $T_W(z)$ and which is fed from below by hotter fluid at temperature T_H from a two-dimensional chamber as sketched in Fig. 6.1.

The bottom of the chamber is fed by an initially uniform volumetric flux Q propelled by pressure P. The slot gap width is d, the depth of the slot in the z direction is L and the slot is infinite in the lateral (x) direction. Reynolds number is small, so the flow is governed by a balance between viscous drag and pressure. Following [23,52] the velocity is two dimensional and tangential to the slot walls, so $\mathbf{u}' = (u', w')$ where u' is lateral velocity and w' is out of the chamber. Primes on velocity, pressure, and viscosity denote dimensional quantities. Corrections due to three dimensional variations in temperature and velocity are developed in

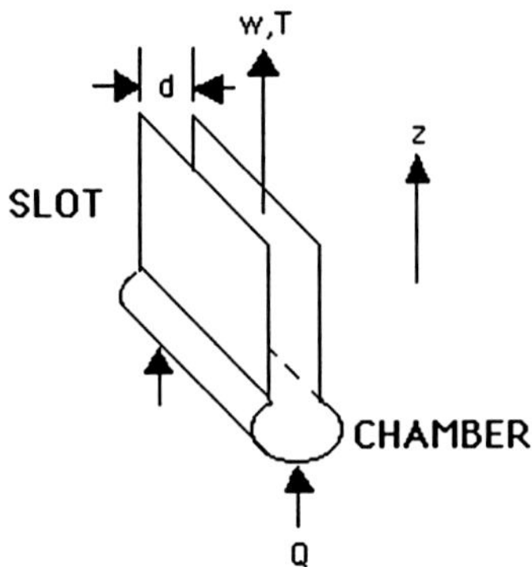

Fig. 6.1. Sketch of the idealized system for finger instability. Fluid comes in at the bottom of a chamber. It flows through a slot whose sidewalls are cooler than the fluid

[51], but here we make a simple narrow gap approximation. If the dimensionless number $\alpha = g\rho\alpha'(T_H - T_W)L/P$ is small, where ρ is density in the chamber and α' is coefficient of thermal expansion, effects of gravity g are negligible [23]. The basic equations are

$$\frac{12\mu'(T)}{d^2}\mathbf{u}' = -\nabla p' \,, \tag{6.1}$$

$$\nabla \cdot \mathbf{u}' = 0 \,, \tag{6.2}$$

and

$$\frac{\partial T}{\partial t} + \mathbf{u}' \cdot \nabla T = -\frac{\pi^2 \kappa}{d^2} T \,. \tag{6.3}$$

Following [23], we investigate two-dimensional flow in the slot. We use a model where the temperature of the walls is uniform and viscosity obeys the law $\mu' = \mu_H \exp[\lambda'(T_H - T)]$, which in nondimensional form is $\mu = \exp[\lambda(1-\theta)]$ where μ_H and λ' are constants. This introduces the dimensionless quantities $\mu = \mu'/\mu_H$ and $\theta = (T - T_W)/(T_H - T_W)$, and the dimensionless number $\lambda = \lambda'(T_H - T_W)$. In the scaling we also use L for a length scale, L/U for a time scale, and $U = Pd^2/12L\mu_H$ for a velocity scale. The dimensionless streamfunction defined by $(u, w) = (-\psi_z, \psi_x)$ (primes are dropped for dimensionless flow variables) obeys the scaled vorticity equation

$$\nabla^2 \psi = -\frac{\mathrm{d}\ln\mu}{\mathrm{d}\theta} \nabla\psi \cdot \nabla\theta \,. \tag{6.4}$$

In the heat equation we neglect lateral conduction of heat in the plane of the slot compared to conduction from the fluid to the walls, leaving

$$\frac{\partial \theta}{\partial t} + J(\psi, \theta) = -\delta\theta \,. \tag{6.5}$$

Here, the second dimensionless number in the problem is $\delta = 12\pi^2\kappa\mu_H L^2/Pd^4$. This is the inverse of a modified Peclet number $Ud^2/\kappa L$ [48]. Physically, δ is the ratio of the time that it takes for hot fluid to traverse the slot length divided by the time scale for the fluid temperature to respond to changes in wall temperature. Typically, the usual Peclet number $Pe = UL/\kappa$ will be quite large. For example, for a fissure with $U = 1\,\mathrm{m\,s^{-1}}$, $l = 10^3\,\mathrm{m}$ and $\kappa = 10^6\,\mathrm{m^2\,s^{-1}}$, $Pe = 10^9$. However, the ratio d^2/L^2 is typically small and may span wide ranges from 10^{-6} to 10^{-12}, reflecting a variation in expected dike widths from $10^{-3}\,\mathrm{m}$ to $1\,\mathrm{m}$. Thus the modified Peclet number δ can span a wide range of values both greater and less than one. The dynamic boundary condition at the flow exit is $u = -\partial\psi/\partial z = 0$ at $z = 1$. At the entrance, the condition is $\theta = 1$.

The basic steady-state temperature and viscosity fields are readily found as a function of imposed flow w_0 and correspond to a multi-valued pressure drop for $\lambda > 3.03$ (as shown in Fig. 6.2). Inspection shows that for sufficiently large λ there are values of w_0 that lead to a decreasing magnitude of the pressure gradient for increasing w_0. This happens because viscosity is small for the warmer temperatures at higher flow rates and is larger for the cooler temperatures at slower flow rates. For a balance between pressure gradient and steady flow, the only recourse is for uniform flow to break down to spatially uneven flow. The feedback mechanism, whereby a change in flow rate makes an inverse change in flow resistance, is the essential factor that causes instability.

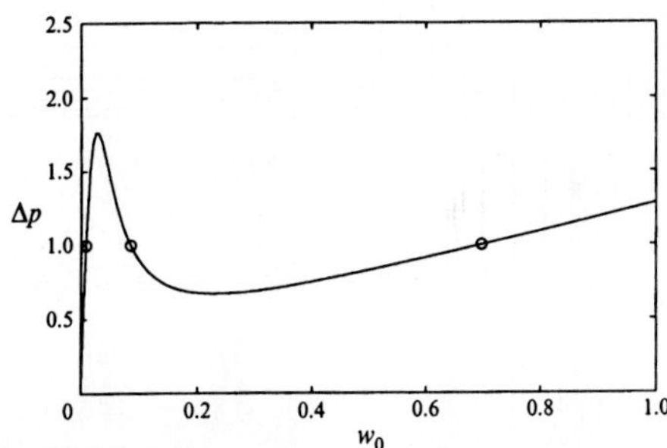

Fig. 6.2. Pressure difference across the slot as a function of steady flow rate. In this example $\lambda = 5$ and $\delta = 0.1$. The three circles show that there are three values of flow rate for the same value of pressure difference. At the intermediate circle the flow can be expected to be unstable since a faster flow makes less resistance

A linear stability analysis [23] reveals that, if the basic flow is set in the region where pressure decreases with increasing flow, the flow becomes channelized. The wavelength of most rapid growth is sensitive to the source conditions used (Fig. 6.3). With constant source flux the wavenumber of most rapid growth is of order one. With uniform source pressure the fastest growth is found for zero wavenumber.

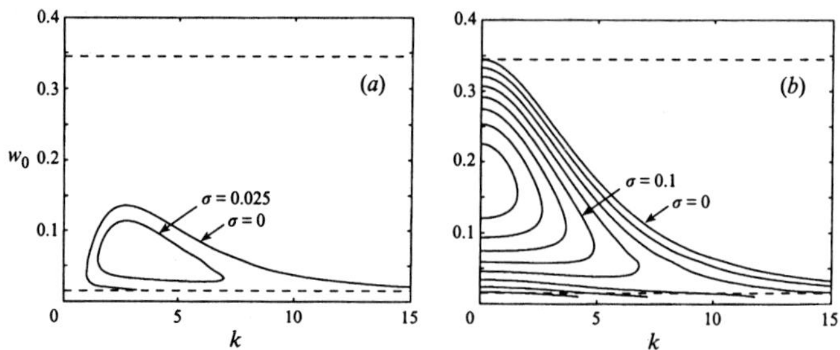

Fig. 6.3. Growth rate as a function of wavenumber for two boundary conditions with $\lambda = 5$. The values of w_0 are varied rather than λ because the latter is not single valued. On the left a constant source flux is imposed into the slot and on the right a constant pressure is imposed

Numerical studies of the evolution of the temperature and streamfunction field (Fig. 6.4) illustrate the formation of fingers, which represent channels of enhanced flow. Similar channels have been viewed in syrup flowing between two walls with one wall highly cooled [52].

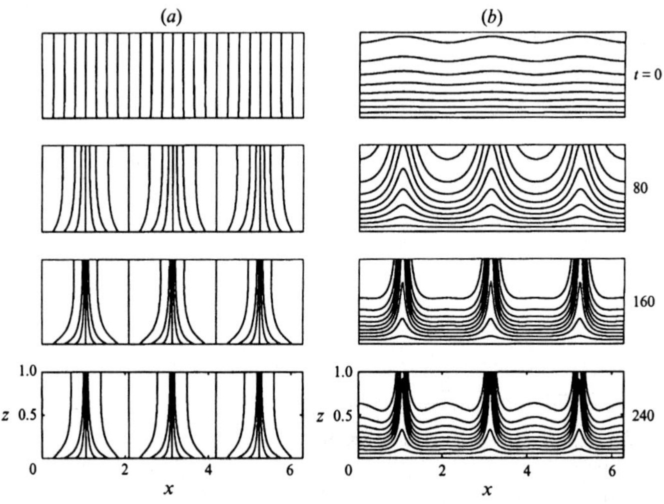

Fig. 6.4. Isotherms (*right*) and streamfunction (*left*) at different times for the finger instability. In this case $\lambda = 5$, and $w_0 = 0.0425$ ($\delta = 0.075$). The initial condition was the linear solution for fastest growth

Experiments with liquid paraffin, and described in the following paragraphs, demonstrate the transition from uniform flow to fingering flow as time progresses. The apparatus consisted of a thick square aluminum plate placed horizontally and levelled carefully in a pan of ice water. Only the underside of the plate was in contact with the water. The temperature of the ice water was estimated to be approximately 5 ° C in contact with the plate, since the ice floated only around the edges of the plate. A 11 mm thick square plexiglas plate 4.6 mm on a side was clamped over the aluminum with spacers between the aluminum and plexiglas leaving a narrow gap of 2.4 ± 0.7 mm. A hole drilled in the center of the plexiglas was connected by a heated hose to a reservoir containing melted paraffin. As a run commenced, paraffin was delivered to the hole at a known constant rate (5.5 ml/s) by gravity feed. The paraffin initially began to spread out in a growing pattern that was close to perfectly circular. After 16 seconds there was a rapid growth of radial finger-like bulges (Fig. 6.5a) with round tips. Ten or twelve fingers grew within four seconds but many of them stopped growing during the next four seconds. The only change in the pattern subsequently was that four fingers reached the edge of the tank, the rest froze. At this stage oil soluble dye was injected into the paraffin source and it was observed that most flux was into the two largest fingers. Forty eight seconds later the flow was through only one finger (Fig. 6.5b), fed by a single channel, in a flow pattern that then continued indefinitely, with little apparent change.

Assuming that both the lid and the aluminum plate cool the paraffin as it flows along the slot, the thermal time constant for the initial paraffin flow in the gap is of order $h^2/4\kappa = 14.5\,\mathrm{s}$ ($\kappa = 0.001\,\mathrm{cm^2\,s^{-1}}$), which is similar to the observed time to instability of 16 seconds. Furthermore, the final channel (of width approximately 1.5 cm) carried the full source flux from the source to the edge of the plate at a relatively large velocity of around $15\,\mathrm{cm\,s^{-1}}$. Hence the fluid at this stage spent less than 2 s in the slot, a time that is short compared to

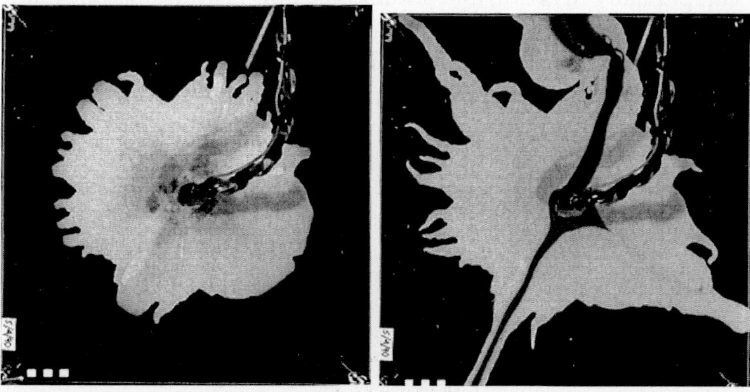

Fig. 6.5. The evolution of paraffin flowing with constant volume flux through a cooled annular slot from a small source. (**a**) Numerous fingers have broken out at 20 seconds from an intrusion that was circular at a time of 12 seconds. (**b**) At 92 seconds dark dye reveals that the fingers have all stopped except for one channel. The dark dye was placed in the fluid earlier and it indicates a previous time when there were two channels. Extended fingers also reveal that there was a time with four channel flow

the time required for it to cool. In this manner the fluid adopted both a long and a short length scale for the final flow. The long distance between the remaining active channels led to faster and hence warmer flows in the remaining channels. At the same time each channel became narrow enough to continue to support rapid flow. Thus large channel spacing combined with narrow channels allowed transit of fluid parcels without them becoming too cold.

Our aim here was to demonstrate the simple concept that the system establishes, through instability, a flow pattern that allows fluid to escape from the slot before it cools enough to greatly increase its viscosity. It was found that instability ocurred on the time scale for diffusion of heat through the width of the slot, and it took on a large length scale. These results are probably relevant to the flow of magma through conduits in the Earth's crust [6]. Later in this chapter we will see that free-surface gravity currents too, such as lava flows, are unstable due to cooling, but that instability occurs on a very much shorter time scale during which only a very thin superficial boundary layer is cooled and solidified. Again, under some conditions, long narrow channels containing high flow velocity are formed.

6.3 Dissolution Instability and Channelised Flow in Permeable Media

A related problem, which in addition involves latent heat effects, is the melting of a permeable matrix (or freezing of melt) during percolation in a porous medium. We consider, as an example, the particular case of a liquid flowing uniformly through a porous material where it encounters a material of lower permeability which it can partially dissolve, and assume the flow is at a constant temperature. Dissolution increases the permeability of the solid matrix. A similar fingering instability that again leads, at finite amplitude, to the channelisation of flow occurs in the percolation of a solvent, such as a hot hydrothermal solution, through a partially soluble permeable matrix.

Following [32], the system to be analysed is sketched in Fig. 6.6.

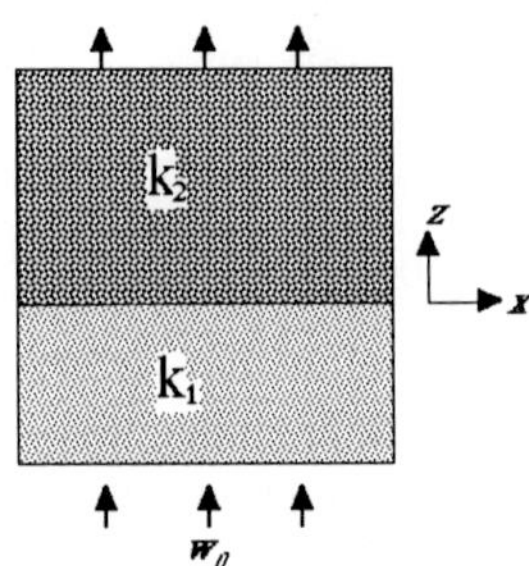

Fig. 6.6. Sketch of the initial state for the model of liquid flowing upward through a porous material which it can dissolve

Let the interface lie at

$$\eta(x,t) = z \,, \tag{6.6}$$

and the equation for the dissolution of the interface be

$$\frac{\partial \eta(x,t)}{\partial t} = \gamma w \,. \tag{6.7}$$

In this model, dissolution causes the interface to migrate at a rate proportional to the amount of fluid flowing through the interface. We assume that dissolution at the interface immediately increases the permeability from the initial value above the interface to the higher permeability of the material below the initial position of the interface.

For small times we let the total vertical Darcy velocity at the interface w consist of a steady uniform component plus a small perturbation, and describe the interface as a horizontal plane plus a small deviation:

$$w = w_0 + w'(x,t) \,, \tag{6.8}$$

$$\eta = \eta_0(t) + \eta'(x,t) \,. \tag{6.9}$$

The steady flow and the mean interface position are given by:

$$\frac{\partial \eta_0}{\partial t} = \gamma w_0 \,, \tag{6.10}$$

$$\eta_0 = \gamma w_0 t = z \,, \tag{6.11}$$

whereas equations governing the perturbations throughout the porous materials are given by Darcy's law:

$$\frac{k_n}{\mu}\frac{\partial p_n}{\partial x} = -u_n \,, \tag{6.12}$$

$$\frac{k_n}{\mu}\frac{\partial p_n}{\partial z} = -w_n \,, \tag{6.13}$$

with velocity components

$$u_n = \frac{\partial \psi_n}{\partial z} \,, \tag{6.14}$$

$$w_n = -\frac{\partial \psi_n}{\partial x} \,, \tag{6.15}$$

where the streamfunctions ψ_n satisfy

$$\nabla^2 \psi_n = 0 \,. \tag{6.16}$$

Here, μ is the dynamic viscosity, k_n is the matrix permeability and subscripts $n = 1, 2$ denote the region upstream or downstream of the interface.

For a perturbation to the interface of the form

$$\eta' = N(t)\cos(lx) \tag{6.17}$$

and, assuming firstly, that $w_0 \gg w'_n$, secondly, that we start with $t = 0$ so the unperturbed interface is at $z = 0$, and thirdly that $N(t)$ is smaller than the length scale l^{-1}, we integrate (6.13) for w upward from a plane normal to the bottom of the perturbed interface at $z = -N$ to a plane normal to the top of the perturbed interface $z = N$:

$$\begin{aligned} p_2(N) - p_1(-N) &= -\frac{\mu}{k_2}\int_{\eta'}^{N} w_0 \mathrm{d}z - \frac{\mu}{k_1}\int_{-N}^{\eta'} w_0 \mathrm{d}z \\ &= -\mu w_0 \left\{ N\left[\frac{1}{k_2} + \frac{1}{k_1}\right] + N\cos(lx)\left[\frac{1}{k_1} - \frac{1}{k_2}\right]\right\} . \end{aligned} \tag{6.18}$$

A contribution to the above calculation for pressure from perturbation velocity w' is of order $w'N$ and is assumed to be negligible. Equation (6.18) divided by $2N$ is the average vertical pressure drop in this region. Of primary interest is the laterally varying component of pressure, which by symmetry, we take to be zero at the origin, so that

$$\left.\frac{\partial p_2}{\partial x}\right|_N = -\left.\frac{\partial p_1}{\partial x}\right|_{-N} \tag{6.19}$$

and laterally varying pressure at $z = N$ is

$$\frac{\partial p_2}{\partial x} = \frac{\mu w_0 N l}{2}\left[\frac{1}{k_1} - \frac{1}{k_2}\right]\sin(lx) . \tag{6.20}$$

Since N is indefinitely small, this condition will be assumed to apply at $z = 0$.

Using (6.12) and the streamfunction definition

$$u_2 = \frac{\partial \psi_2}{\partial z} = -\frac{k_2}{\mu}\frac{\partial p_2}{\partial x} = -\frac{w_0 N l k_2}{2}\left[\frac{1}{k_1} - \frac{1}{k_2}\right]\sin(lx) . \tag{6.21}$$

A solution to (6.16) is

$$\psi_2 = A(t)\sin(lx)\mathrm{e}^{-lz} , \tag{6.22}$$

which with (6.21) at $z = 0$ produces

$$A(t) = k_2 w_0 \left[\frac{1}{k_1} - \frac{1}{k_2}\right] N(t) . \tag{6.23}$$

Using (6.7), (6.8), (6.10), (6.15), (6.17), and (6.23) results in

$$\frac{\partial N}{\partial t} = -\gamma l w_0 \left[\frac{k_2}{k_1} - 1\right] N , \tag{6.24}$$

which has an exponentially growing solution for $k_1 > k_2$ (upstream permeability greater than downstream). Equation (6.24) shows that the larger the wavenumber l the faster the growth rate. Therefore, very small length scale perturbations grow most rapidly.

The stability of an initially planar interface and the subsequent spatial distributions of permeability, porosity, solute concentration and water composition were studied by Ortoleva et al. [38,39]. They found that the planar interface is unstable with the fastest growing wavelength determined by matrix size, initial modal amount of reactive mineral in the rock, initial porosity, the composition and the velocity of the inlet fluid. In essence their result simplifies to the wavelength of fastest growth being proportional to thickness of the front, which is determined by the effective solute diffusivity divided by the fluid Darcy velocity. Since the effective diffusivity (due to mechanical dispersion) of a solute flowing through a porous material is proportional to the Darcy velocity times grain size, this reduces to the simple result that finger width is proportional to grain size. The purpose here was to present the essentials of such an instability in as simple form as possible. Hence many of the elements included in the original analysis, such as grain size, porosity variation, and change in volume of the solute, were neglected.

The more complete theories of Ortoleva et al. [38,39] have shown that diffusion processes limit the magnitude of the fastest growing wavenumber, so that fastest growth is scaled by grain size. The first wavelengths to appear in experiments with water percolating through salt agree with this [38]. They are small but larger than grain size, which contrasts with the very long wavelength favored in the viscous fingering case of Sect. 6.2. For longer times the short wavelength distortions attained a finite size and stopped their exponential growth, while longer wavelengths continued to grow. Flows produce drainage channels that exhibit both coalescence and branching [32,49].

Finger instability may also be encountered in reaction–dissolution effects upon the migration of melt. In order to describe the migration of melts to produce magmas it is necessary to add the process of compaction (the driving of the fluid by gravity acting on a viscous deformable solid matrix of different density). For example, mid-ocean ridge basalts (MORB) are produced through pressure-dependent melting (the melting temperature decreases with decreasing pressure) coupled with compaction-driven flow. Melt accumulates around grain boundaries and is squeezed upwards by the slowly deforming denser mantle crystals. The composition of the mantle is such that rising mantle material undergoes partial melting as it reaches lower pressures. Hence melt percolating upward will be out of chemical equilibrium with the remaining matrix of mantle crystals and will produce additional reaction that in general will reduce the permeability. In the absence of compaction [1], the porous flow of a reacting fluid through a soluble matrix with gradually changing solubility has growing finger instabilities over the entire range of Damkohler numbers $Da = l/L_{\text{eq}}$. Here $L_{\text{eq}} = \phi_0 w_0 \rho_f / R_{\text{eff}}$ is the distance that a perturbation in chemical concentration will travel before becoming chemically equilibrated, and thus it is the product of velocity and reaction time. (In this formula, ϕ_0 is the porosity, ρ_f is density of the melt, R_{eff} is the reaction rate of the melt in contact with a crystal matrix, and the other symbols are as above.) However, in a compacting matrix not all reacting flows are unstable [1]. The criterion for instability is that $Da > 1/Cn$, where Cn

measures the effects of the change in solubility over one compaction length h_{comp} multiplied by the ratio of compaction length to matrix depth. A compaction length is the distance a change in matrix porosity can migrate in a compacting flow before decaying. The stability criterion is also written as $\beta' h^2_{\rm comp}/L_{\rm eq} > 1$, where $\beta' h_{\rm comp}$ is the change in solubility over one compaction length.

Finite amplitude effects of instability have been observed in laboratory studies without matrix compaction [32,49] and also in a number of computer studies [2,33,45]. In numerical studies the channels that break out as the result of instability typically branch and coalesce again. Branching is also seen in the related problem of Saffman–Taylor instability and may be simply a function of the degree to which the stability criterion is exceeded. However, the thermal channels in cooled laboratory viscous flows observed to date do not exhibit such branching, and the exact causes of channel branching and coalescence are poorly understood.

6.4 The Shapes of Free-Surface Yield-Strength Flows on a Slope

Some lava flows, especially those having relatively high silica content, tend to be erupted with high crystal and vesicle fractions that give them a highly non-Newtonian rheology. This rheology is most simply characterised by the addition of a large yield stress to the viscous stress in the stress-strain rate relation. Therefore, a very useful (but highly simplified) flow to understand is the isothermal flow of a yield-strength material as it is slowly extruded onto a sloping plane from a localised source (or vent). This is the next step beyond an analysis of the radial spreading of a viscoplastic fluid from a source on a horizontal plane [3,5], where the flow is characterised (apart from significant viscous stresses in a small neighbourhood of the source) by a static, or quasi-equilibrium, balance between gravity and yield strength at any distance from the axial source. A simple parabolic height profile with radius results, and the dome remains axisymmetric as it expands over the horizontal base, apart from a set of orthogonal spiral glide planes on which the material achieves the deformation that is necessary for it to spread radially.

Early realisation that the levee banks created by long basalt flows implied non-Newtonian flow led Hulme [26] to consider the unconfined motion of a Bingham fluid of yield stress σ_0 down a slope. He considered long flows and assumed that all quantities are independent of distance x down-slope. Near the edges of a flow its depth $h(y)$ becomes small and the lateral flow was assumed to cease when the cross-slope pressure gradient is balanced by the basal yield stress, as expressed in

$$\frac{\partial P}{\partial y} = \rho g \frac{\partial h}{\partial y} = \frac{\sigma_0}{h} \, . \tag{6.25}$$

This model is based on the assumption that the fluid does not deform anywhere but at its base, where the pressure is greatest and (in order for the fluid to have reached its current shape) equal to the yield stress. The solution to

(6.25), originally obtained in the context of icesheet dynamics [37], with $h = 0$ at $y = W$, is simply

$$h^2 = \frac{2\sigma_0}{\rho g}(W - y) , \tag{6.26}$$

which implies that the central height $H = h(0)$ and the half-width W are always related by $H = C(\sigma_0 W/\rho g)^{1/2}$, where $C = \sqrt{2}$. If the flow depth is assumed to be constant in the down-slope direction at any value of y, then motion requires $\rho g h(y) \sin\beta > \sigma_0$, where β is the slope of the base from the horizontal. Hence there is a critical depth

$$h_s = \frac{\sigma_0}{\rho g \sin\beta} \tag{6.27}$$

below which there will be no down-slope motion. Substituting this depth into the cross-slope balance (6.26) gives the width of the region of stationary fluid along the edge of the flow:

$$w_s = \frac{\sigma_0}{2\rho g \sin^2\beta} = \frac{h_s}{2\sin\beta}. \tag{6.28}$$

Between these two stationary regions there is free visco-plastic flow down-slope, which Hulme approximated as the two-dimensional flow between a parallel stress-free surface and no-slip bottom plane, leading to the depth-averaged velocity [26,44]:

$$u = \frac{\rho g \sin\beta h_s^2}{3\eta}\left[\left(\frac{h}{h_s}\right)^3 - \frac{3}{2}\left(\frac{h}{h_s}\right)^2 + \frac{1}{2}\right] . \tag{6.29}$$

An error in Hulme's analysis is that, for the cross-slope motion to cease, it is necessary to consider more than the cross-slope component alone of the basal stress: the total stress $\sigma = \rho g h[\sin^2\beta + (\partial h/\partial y)^2]^{1/2}$ at the base (where the down-slope thickness gradient might be neglected) must become equal to the yield stress.

Laboratory experiments with kaolin–water slurry on a slope [26] revealed the presence of stationary levees bounding long down-slope flows. The height of the levees was consistent with the formula (6.27), which was then applied to lava flows to find yield strengths for various flows (of order 10^3 Pa for low silica contents to 10^5 Pa for higher silica contents) from the height of levees (5–30 m) and the underlying topographic slope. This much of the behaviour of long flows, and particularly the observed levees, can therefore be explained in terms of isothermal flows having a yield strength. The levee-derived correlation between silica content and strength for terrestrial flows, along with remote measurements of levee heights, were even used to estimate compositions of lunar flows. In detail, real flow levees are formed of cooled flow-front or surface material pushed aside by the advancing flow front, so that only the levees are required to have a yield strength and the flow is not of uniform rheology. However, the principle and the application of (6.27) are unchanged.

A more difficult problem is posed by domes of extremely 'stiff', high-silica content, lavas erupted on to slopes; these are not the very long flows (or relatively low-viscosity basalt) which motivated the previous work [21,22]. Instead, the challenge is to predict the fully three-dimensional shapes, including the extent of up-slope flow from the vent. A solution for the analogous problem of Newtonian viscous flows was given by Lister [35]. The solution can also be found for the three-dimensional case in the limit of slow flow or high yield strength (i.e. when $B \to \infty$, where $B = \sigma_0/\eta\dot{\varepsilon} = \sigma_0 h_s/\mu U$, with U a velocity scale and μ the viscosity, is the Bingham number comparing yield stress to viscous stresses). In this case, the extruded material causes a force that exceeds the yield stress. This produces a flow that terminates as a new balance between yield stress and hydrostatic forces is produced. Elements of this solution were given by Coussot et al. [8,9], who thought the complete solution would be non-unique, but the unique solution for emplacement from a small source was obtained by Osmond & Griffiths [40]. This solution is summarised here and gives the three-dimensional shape for static finite volumes. It also gives the final width of very long down-slope flows (i.e. for large volumes), which turns out to be independent of the viscosity.

We assume $H/L \ll 1$, a hydrostatic gradient in the vertical and a static balance between gravity and yield stress (as in (6.25) but this time in the plane parallel with the base slope) and that the total stress at the base is equal to the yield stress. We readily obtain an equation for thickness $h(x, y)$ normal to the base [40]:

$$\left(\frac{\partial h}{\partial x} - \tan\beta\right)^2 + \left(\frac{\partial h}{\partial y}\right)^2 = \left(\frac{\sigma_0}{\rho g h \cos\alpha}\right)^2 . \qquad (6.30)$$

The axes x and y lie in the plane of the sloping base, the z axis normal to the base. Assuming symmetry about the down-slope (x) axis through the source implies $[\partial h/\partial y]_{y=0} = 0$ (except at $x = 0$, where $\partial h/\partial y$ must be discontinuous in order to force radial flow from the vent). Then (6.30) can be solved for the thickness profile $h(x, 0)$. Scaling thickness h by h_s (6.27) and distance x parallel to the sloping base by $h_s/\sin\beta$ gives the thickness profiles (with all quantities now dimensionless)

$$\begin{aligned} x &= h - H + \ln \mid (1-h)/(1-H) \mid, \; x \geq 0 , \\ x &= h - H + \ln \mid (1+h)/(1+H) \mid, \; x \leq 0 , \end{aligned} \qquad (6.31)$$

on the down-slope ($x > 0$) and up-slope ($x < 0$) sides, where $h = H$ at $x = 0$. The leading edges of the dome are found at $h = 0$ and from (6.31) we have

$$x_d = -H - \ln \mid 1 - H \mid , \qquad x_u = -H + \mid 1 + H \mid \qquad (6.32)$$

(downslope and up-slope respectively) or a total dimensionless flow length $L = -\ln \mid 1 - H^2 \mid$. The cross-slope thickness profile of the dome can be approximated by neglecting $\partial h/\partial x$ in (6.30) in the region of maximum width (down-slope from $x = 0$). The dimensionless maximum width is given by $W \approx 2(1 - \sqrt{1 - H^2})$. It tends to be more useful to describe these flows in terms of their volume V at any time, where V is the extruded volume normalised by the volume scale: $\sigma_0^3/[(\rho g)^3 \sin^5 \beta]$; the dome is not much influenced by the topography for $V \ll 1$

but strongly influenced and displaced somewhat down-slope from the vent for $V > 1$. When $V \gg 1$ (and the thickness H tends to 1) the down-slope length of the dome tends to infinity. This reflects the fact that the critical thickness h_s (6.27) is the maximum dome thickness that can be supported on the slope in a static balance; larger thickness would imply a basal region of dynamic viscous flow. For $V \ll 1$ (i.e. $H \ll 1$ as a result of small volume, large yield strength, small slope or reduced gravity) the dome is not influenced by the base slope, is close to axisymmetric and (6.31) approaches the quadratic profile (6.26).

In order to obtain the complete locus of the dome perimeter and contour plots of flow thickness as a function of H (or of V) equation (6.30) was solved numerically. Sample solutions are shown in Fig. 6.7 for the flow depth contours at several values of the dimensionless volume. Miyamoto & Sasaki [36] treat a similar problem through numerical simulation.

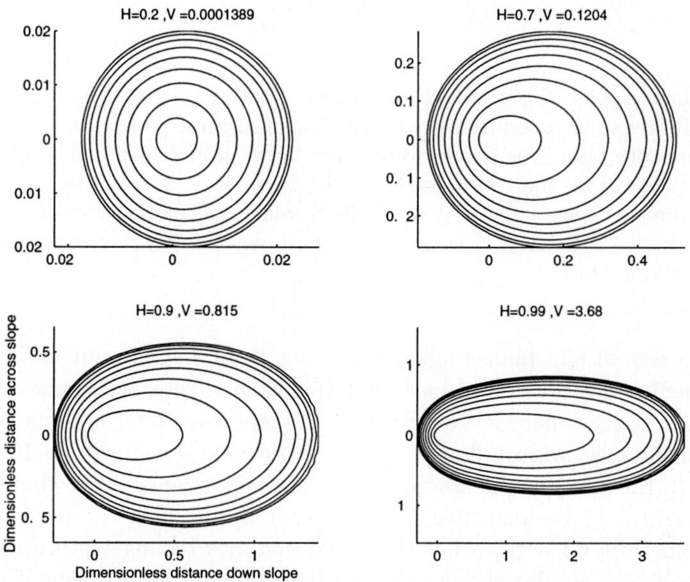

Fig. 6.7. Contour plots of flow depth obtained from the numerical solutions (6.30) for yield strength flow on a slope. Solutions are shown for four values of the dimensionless central thickness H (or dimensionless volume V). Note the change in scale on the axes between the various plots. From [40]

The solutions compare well with isothermal experiments with slurries of kaolin in polyethylene glycol wax as well as kaolin in water, both on a sloping base (Fig. 6.8). In experiments with $V \approx 0.1$ ($H \approx 0.7$), the flow margin begins to depart noticeably from circular and the down-slope length is more than twice the up-slope length from $x = 0$. For $V \approx 1.5$ ($H \approx 0.95$), the down-slope length is eight times greater than the upslope length and nearly twice the full

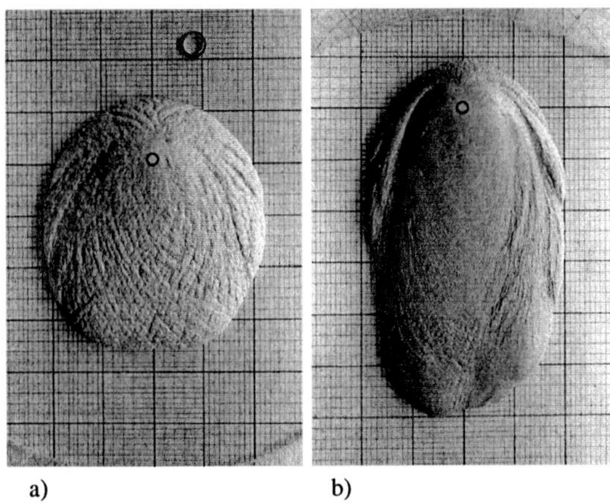

Fig. 6.8. Photographs of isothermal laboratory flows of Bingham fluid on a planar slope. The fluid was extruded in many small volume increments from a 1 cm diameter hole in the smooth base. The domes were static between increments. The black circle shows the location of the source. (**a**) kaolin/water slurry, slope $\beta = 12°$, volume $900\,\mathrm{cm}^3$, $\sigma_0 = 92\,\mathrm{Pa}$, dimensionless volume $V \approx 0.8$; (**b**) kaolin/PEG slurry, $\beta = 18°$, $1000\,\mathrm{cm}^3$, $\sigma_0 = 84\,\mathrm{Pa}$, $V \approx 12$. 'Leeves' develop for $V > 10$, when further spreading is largely down-slope. (From [40])

width. Two sets of slip planes again curve out from the summit as in Blake's [5] experiments on a horizontal base, but this time they are asymmetric in the x-direction. The stationary levees of Hulme [26] are seen to form along the edges of the down-slope laboratory flow at very large flow elongations ($V > 10, H \to 1$; Fig. 6.8b). In the analysis the assumption of a static balance everywhere implies that the origin is to be identified with the vent from which the fluid was supplied, and there is no implication that fluid volumes having histories different from this will take similar shapes: the static shape will be different if the base slope is changed after the volume is emplaced, or if the flow is rapid and partly viscous for a time before taking on the quasi-static shape controlled by yield strength.

One conclusion to be drawn from these observations is that complex flow structure is not always the result of instability. The levee structures along the edge of the flow form simply because the edges must be shallow and therefore cannot move down-slope in the presence of the yield strength. A down-slope channel is formed, somewhat like those that form in the cooled slot (Sect. 6.2) as a consequence of increased viscosity (or solidification). However, the detailed texture of the surface and the structure of the intersecting yield surfaces have not been addressed. In real lava flows, there are additional complications due to the effects of cooling and varying rheology, which may greatly alter some flows while leaving others, presumeably those emplaced relatively rapidly, less affected.

6.5 Instabilities of Solidifying Free-Surface Gravity Currents

In the case of the spreading of a fluid on a plane under gravity the flow has a free surface and hence additional freedom to respond to cooling and rheology variation. Isothermal gravity currents, both turbulent and very viscous, have been studied extensively. However, the more complicated (and much richer) problem of cooling gravity currents having temperature-dependent rheology or solidification has barely been touched. Recent work in this area has been motivated by modelling of lava flows, but is also relevant to some industrial problems such as the flow of molten metals.

There are a number of styles and morphologies of lava flow, each presumably reflecting a different dynamical flow regime. Some examples are shown in Fig. 6.9. The style of flow is contolled by factors such as: lava composition and rheology, eruption temperature, effusion rate, base topography and whether it flows under water, air or a vacuum [12]. Initial heat loss is predominantly by radiation under the relatively thin atmospheres of Earth and Mars (Earth's atmosphere has sufficient heat capacity that convection provides a comparable flux only after the surface temperature has fallen to less than $\approx 200\,^\circ$ C [18,22]). In contrast, radiation is less important than convection for temperatures less than $900\,^\circ$ C under the dense atmosphere of Venus (where $T_a \approx 450\,^\circ$ C). For eruptions under water radiation is always negligible relative to the very rapid convective transport [19].

The effects of cooling will depend on the rate of spreading of the flow relative to the rate of cooling through a temperature range sufficient to cause rheological change. It is therefore logical to first consider the role of a dimensionless parameter expressing this ratio. A comparison of the conductive transport of heat within the lava to the advection of heat with the flow reveals that for almost all lava flows the Peclet number $Pe = UH/\kappa$ (U a velocity scale, H a flow depth scale and κ the thermal diffusivity) is very large, ranging from 10^2 to 10^5 for slow-growing lava domes through to $> 10^6$ for fast channelized basalts flows. Thus cooling is confined to a very thin boundary layer at the free surface and, if mixing is absent, the interior remains largely isothermal. In this thin boundary layer regime surface solidification commences at a distance $d_s \approx ut_s$ from the vent, where u is the surface velocity and t_s is the time taken for the surface temperature T_c to cool from the vent temperature T_e to the solidification temperature T_s. Scaling distance by H and velocity by a suitable scale U one can define the dimensionless parameter [12,20]

$$\Psi = \frac{Ut_s}{H} = \frac{t_s}{t_A}\,, \tag{6.33}$$

where $t_A = H/U$ is the time scale for lateral flow through a distance H. Equivalently, $d_s/H \approx \Psi$. Here the velocity scale U and the advection timescale T_a will

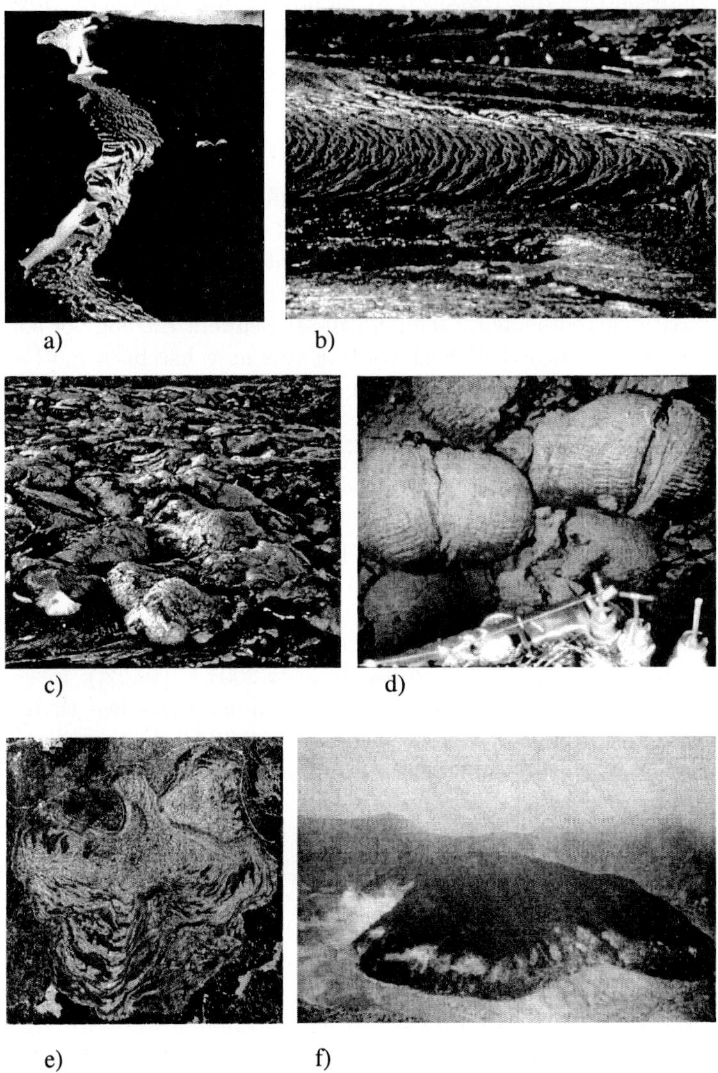

Fig. 6.9. Examples of some lava flow forms: (**a**) channelised basalt flow from Kilauea Volcano, Hawaii (flow channel roughly 20 m wide); (**b**) and (**c**) 'ropy pahoehoe' and 'toey paheohoe', respectively from Kilauea ('ropes' have wavelength $\approx$ 20 cm, 'toes' are typically 30 cm across); (**d**) submarine pillow basalts, each approximately 1 m across, on the East Pacific Rise; (**e**) Little Glass Mountain rhyolite flow, northern California, showing flow around an obstacle on a gentle slope and transverse surface ridges $\approx$ 5 m in height (image 2.8 km across); (**f**) a lava dome 850 m across and 130 m high in the crater of La Soufrière, St Vincent, 1979. Photographs courtesy of J.H. Fink and R. Embley

depend upon the governing dynamics of the flow, as indicated below. The value of t_s depends on the surface heat flux and the dimensionless temperature of solidification $\Theta_s = (T_s - T_a)/(T_e - T_a)$, the proximity of the eruption temperature T_e to the solidification temperature T_s [12,20]. This time scale (and the surface temperature T_c) must be obtained from a heat transfer calculation, accounting for radiation and convection from the surface [18,19] matched to the conductive heat flux within the lava. The surface solidification time was found to be of the order of 0.1 s for submarine lavas, 100 s for sub-aerial basaltic lavas (on Earth and Mars), and approximately 60 s for the cooler highly-silicic lavas under air. Note that the parameter Ψ is defined for extrusions of constant volume flux Q in terms of the advective time scale t_A appropriate to the corresponding *isothermal* Newtonian gravity currents. A similar parameter Ψ_B can be defined, again by (6.33), when the flow is plastic [21].

For the Newtonian case (and point source) a global velocity scale $U = Q/H^2 \approx (\rho g Q/\eta)^{1/2}$ and depth scale $H = (Q\eta/\rho g)^{1/4}$ are found for the isothermal flow [30] and give $t_A = (\eta/\rho g)^{3/4} Q^{-1/4}$, so that the dimensionless solidification time becomes

$$\Psi = \left(\frac{\rho g}{\eta}\right)^{3/4} Q^{1/4}\, t_s\,. \tag{6.34}$$

For the plastic case $U = Q(\rho g/\sigma_0)^2$ and $H = \sigma_0/\rho g$ [21], and these lead to

$$\Psi_B = \left(\frac{\rho g}{\sigma_0}\right)^{3} Q\, t_s\,. \tag{6.35}$$

These definitions represent an attempt at describing a flow in a global sense, recognising that the advection velocity at a given radius can vary with time (as the depth changes or the flow becomes non-axisymmetric) and depends on distance from the vent. Thus there remains scope for time-dependence of the effects of solidification within a flow having a fixed value of Ψ. Of course, variations of source volume flux lead to changes in Ψ and this is explicit in (6.34, 6.35).

At distances from the vent greater than d_s the layer of solid crust will thicken in a manner that, again, can be calculated by coupling conduction in the lava to the surface heat flux through the surface temperature $T_c(t)$ [18,19]. Note that in terms of the external dimensionless parameters, Ψ and Ψ_B provide a general indication of whether the solid crust thickens quickly or slowly relative to the lateral motion. These parameters are more relevant to the thickness of the rheological boundary layer than is Pe, at least at early times, since the latter relates only to the thickness of the thermal boundary layer (given by $\delta_T \approx (\Psi/Pe^{1/2})H$ at the location of the onset of solidification), and the thermal boundary layer is not directly related to the presence or thickness of crust.

Laboratory analog experiments serve to test the hypothesis that the primary effects of cooling and solidification for slow laminar flows are captured by differences in the parameter Ψ. The experiments used viscous polyethylene glycol (PEG) wax, which freezes at a convenient temperature of 18–19 °C, extruded from a small circular (or narrow linear) vent under cold water onto a horizontal or sloping base [12,13]. The cold water gave rise to a sufficiently large turbulent convective heat flux and solidification times comparable to horizontal advection times. The results revealed a sequence of distinct flow regimes (Fig. 6.10) and these correlated with intervals of Ψ [13].

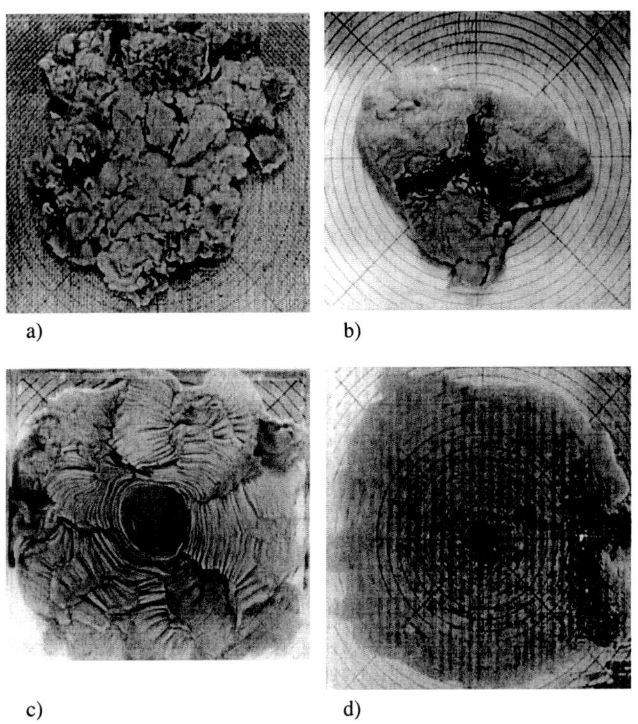

Fig. 6.10. Examples of solidifying gravity currents showing four flow types in laboratory experiments with polyethylene glycol wax flowing over a horizontal floor. The Newtonian liquid was extruded from a small hole onto the base of a tank of cold water. Some of the surface subsequently solidified. (**a**) 'pillow' growth at $\Psi = 0.11$; (**b**) 'rifting' flow with separating rigid surface plates at $\Psi = 2.7$; (**c**) 'folded' flow at $\Psi = 3.0$; (**d**) largely axisymmetric flow with weak cooling and solid confined to 'levees' along the flow front at $\Psi = 7.3$ (these values of Ψ have been corrected for a previous numerical mistake: all values reported in [12] must be divided by $10^{2/3}$)

At $\Psi < 0.7$, where cooling is rapid or extrusion is slow, the flow was fully encased in solid and spread through many small bulbous outgrowths reminiscent of submarine lava 'pillows'; at $0.7 < \Psi < 2.5$ thick solid extended over most of the surface and formed rigid plates separated by divergent 'rifts', complete with transform faults, where solid continued to accrete onto the plates; at $2.5 < \Psi <$

6 solid became more widely distributed (except over the vent) but was thin and tended to buckle or fold, forming many small transverse ridges and ropy structure; at $6 < \Psi < 16$ crust was seen only around the margins of the flows, where it formed 'levees'; and at $\Psi > 16$ no solid crust formed before the flow front reached the side walls of the container (the values of Ψ given here are smaller, by a factor $10^{2/3}$, than those originally reported in [12] because an incorrect value for the water viscosity was originally used). In addition, for $\Psi < 6$, the flows ceased to spread when the source flux was turned off, indicating control of spreading by the strength of the solid (i.e. a balance of buoyancy and crust yield strength). The forms of surface deformation and flow morphology observed are similar to some of the main characteristics found on basaltic (low viscosity) lava flows and traditionally used to categorise them. In particular, they include submarine 'pillow basalts', submarine jumbled plates, sub-aerial ropy pahoehoe and sheet pahoehoe flows where "pahoehoe" refers to a smooth glassy surface.

Experiments similar to the wax studies above but using instead a kaolin-PEG slurry [21], which has both the freezing temperature of the PEG and a yield strength, reveal a different sequence of morphologies (Fig. 6.11). Hence the rheology of the interior fluid plays a role in controlling the forms of flow and deformation, even though the rate of solidification relative to advection, expressed in Ψ_B, again determined which of a sequence of morphologies occurred. At $\Psi_B > 15$ (fast extrusion and slow cooling) the slurry spread axisymmetrically almost as if there were no cooling; at $0.9 < \Psi_B < 15$ there were strong rigid plates over most of the surface and later upward extrusion of ridges with smooth striated sides; at $0.12 < \Psi_B < 0.9$ the flow commenced as a set of four to six (most often five) radially moving lobes having a weak tendency to spiral. Under rapid cooling or very slow effusion, $\Psi_B < 0.12$, the lobes were more like vertical spines and were extruded upward only from the vicinity of the source. In these experiments the transitions between regimes were more gradual than those for the viscous fluid. These morphologies strongly resemble qualitative characteristics of many highly-silicic lava domes [14].

There has been no adequate theoretical description of the above cooling and solidifying flows and the various instabilities that lead to asymmetric spreading and irregular structure. Only a gravitational instability in a density stratified lava dome [16] and the surface buckling instability [11,15] have been analysed. There is good agreement between the wavelengths of observed folds (both on 'ropy pahoehoe' and on the laboratory wax flows) and that predicted for the buckling of layers of differing viscosity or yield strength subjected to a compressive stress [4].

6.6 Freezing Flows down a Slope

Experiments with solidifying gravity currents have also shown that the effects of a sloping base are important, leading both to a flow elongation down-slope and to greater channelisation of the flow (Fig. 6.12). The down-slope flow can be channelised by solidified edges in the levee and surface folding regimes [17].

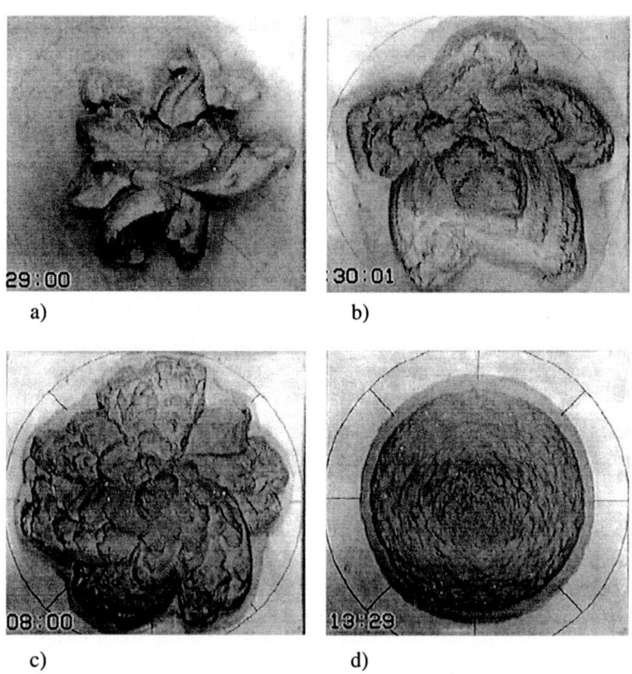

Fig. 6.11. Solidifying flows of a Bingham fluid using a slurry of kaolin in PEG. Apart from the fluid rheology the experiments were similar to those of Fig. 8.7. (**a**) a 'spiny' extrusion at $\Psi_B = 0.09$; (**b**) a lobate extrusion showing a typical 5-lobe pattern at $\Psi_B = 0.79$; (**c**) a flow without distinct lobes but surfaced by solid plates with curving segments, $\Psi_B = 1.3$; (**d**) an axisymmetric flow almost unaffected by cooling at $\Psi_B = 30$. (Heaviest grid lines are 5 cm apart; from [14])

Hence we have flows that form their own channels, somewhat similar to the large-amplitude flow following fingering in a cooled slot. In contrast to the formation of narrow and fast flow in a small and decreasing number of channels in the cooled slot which implies increasing pressure drop for an imposed volume flux, the free-surface flow can only draw on the gravitational head (the height of the flow) for its forcing. However, in free surface flows there are other mechanisms that allow flow to continue. In particular, the formation of solid crust can lead to covered lava tubes at smaller Ψ and these tend to increase the insulation of the flow against cooling and thereby further enhance the distance the flow can travel without cooling so much that it solidifies. However, the regimes identified in terms of surface deformation and overall morphology (aside from the down-

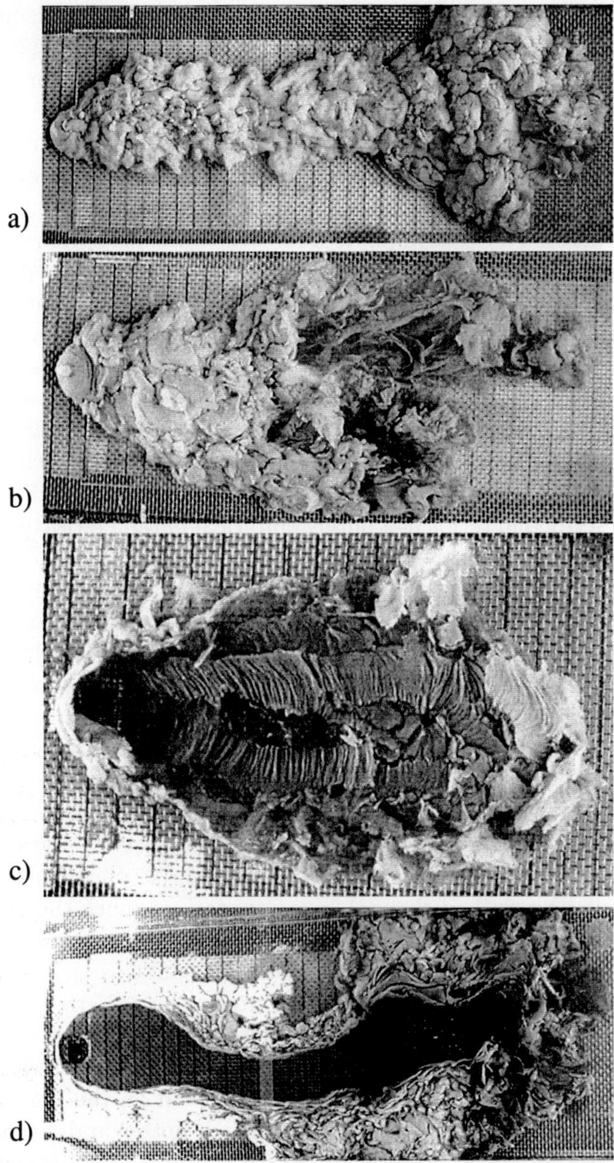

Fig. 6.12. Laboratory experiments with PEG wax flowing from a small source on a planar slope under cold water. The base slopes downward to the right and is covered with mesh to make a rough floor. (**a**) 'pillow' flow; (**b**) 'rifting' flow; (**c**) 'folded' flow; (**d**) 'leveed' flow. (The tank is 30 cm wide; from [17]). (c) and (d) are similar to ropy pahoehoe and long channelised flows observed in Hawaiian lava flows

slope elongation) are not much different from those on a horizontal base, apart from a shift of the regime transitions to smaller values of Ψ [17].

Turning to observations of long basalt flows that extend for many kilometres from their vent (see e.g. [7]), the flow behaviour again reflects, albeit in ways that remain poorly understood, differing vent fluxes, eruption duration, underlying topography and whether they flow under air or water. Field evidence indicates that surface cooling leads to the formation of a glassy crust while internal mixing in these moderate Reynolds number flows can cause disruption or entrainment of crust, cooling and rheological changes in the interior. The development of levees removes mass from the advancing flow front and represents formation of a flow-defined channel, whereas the solidification of crust to form lava tubes respresents a major change in the cooling rate. There are so many processes involved that, in past attempts to model these flows, some processes are approximated by empirical parameterizations. A key factor which has proved particularly difficult to model in a predictive manner is the effect of cooling, which depends on the amount, and rate of disruption, of cooled surface crust. The disruption of crust has been described in terms of a purely empirical parameterisation of the fraction of the surface representing exposed incandescent fluid from the flow interior [10]. Conditions for the disruption and mixing of surface crust under stresses imposed by the underlying flow, and conditions for stable crust, also are not known for either laminar or turbulent flows, yet they determine the distance down-channel at which vertically-mixed flow gives way to stratified flow, the onset of a thickening surface layer [34] and the formation of lava tubes [31,41].

The aim is again to predict factors such as the rate of cooling with distance downstream, flow thickness, the speed of advance of the flow front, changes in flow regime, and the final length of a flow as a function of erupted volume. Given the complexities of long lava flows both simple theoretical results and complex parameterised computational models will be valuable. Significantly, long cooling flows without a prescribed channel have not received much theoretical attention. In this case flow may spread across-slope, form levees or branch (as in numerical experiments with complex distributary systems [36]).

An additional process that can contribute to the pattern of lava flow is melting and thermal erosion of the base underlying a flow. For example, thermal erosion due to melting of underlying sediments or rock by basaltic lava flows was investigated as the cause of sinuous rilles observed on the moon [25,27]. Theoretical modelling [28,29,46] suggests that much hotter and low viscosity melts, called komatiites, which erupted on Earth some 2.7 billion years ago, flowed for large distances as turbulent currents. These would have had high cooling rates under seawater and could have produced thermal erosion 10–100 km from their sources. The extent of melting may have led to significant contamination of the flow by the assimilated melt. Further analysis [50] indicates that erosion is strongly dependent on the nature of the base material, with hydrous sediment being fluidised by vaporised seawater and strongly eroded, whereas relatively little erosion is predicted to occur for consolidated anhydrous sediment.

6.7 Conclusions

The dynamics of flows involving cooling and temperature-dependent viscosity, a yield strength, freezing, melting, or dissolution pose many challenges. These flows can form complex shapes and flow patterns, but they can also evolve toward simpler active flow patterns such as a single channel. We have introduced several simplified models which illustrate the nature of some of the underlying flow instabilities. Along with laboratory experiments these models help to explain many characteristics of geological flows. These models also serve as a basis of comparison for more complex models.

Given that magmas in the upper crust and lavas erupted on to the surface have temperatures up to 1200 ° C above those of their new environment, but less than only 200 ° C above their solidus, it is not surprising that the effects of heat loss can be large. The thermal effects and consequent rheological change (or, in the extreme, solidification) often lead to the onset of a larger viscosity or a yield strength in cooled portions of the flow. This influences the overall flow depth and average spreading rate. In the case of free-surface flows, a yield stress of a surface boundary layer is generally responsible for eventually halting the advance of the flow front. The thermal effects and rheological heterogeneity also lead to a range of complexity and instabilities such as flow fingering, lobes, branching, channelisation, and the formation of surface deformation structures such as folds, 'ropes', rifts and faults. Laboratory analog experiments have been invaluable in relating these instabilities to flow conditions, especially the rate of cooling relative to the rate of flow, and base slope. However, many processes remain poorly understood and lacking a theoretical description. In the case of free-surface flows, difficulties are introduced by a moving free surface that is also the boundary at which the thermal and rheological changes tend to be strongly concentrated, and where flow instabilities arise. In the case of melting, dissolution or reaction in permeable media, theoretical difficulties are introduced by large changes in the permeability, which feed back strongly to the flow structure.

References

1. E. Aharonov, J.A. Whitehead, P.B. Kelemen, M. Spiegelman: J. Geophys. Res. **100**, 20,433 (1995)
2. E. Aharonov, M. Spiegelman, P.B. Kelemen: J. Geophys. Res. **102** 14, 821 (1997)
3. N.J. Balmforth, A.S. Burbidge, R.V. Craster, J. Salzig, A. Shen: J. Fluid Mech. **403**, 37 (2000)
4. M.A. Biot: Bull. Geol. Soc. Am. **72**, 1595 (1961)
5. S. Blake: 'Emplacement Mechanisms and Hazard Implications'. In: *Lava Flows and Domes*, ed. by J.H. Fink (IAVCEI Proc. in Volcanology, vol. 2, 1990) pp. 88–128
6. P.M. Bruce, H.E. Huppert: Nature **342**, 665–667 (1989)
7. K.V. Cashman, H. Pinkerton, P.J. Stephenson: J. Geophys. Res. **103** 27281–9 (1998)
8. P. Coussot, S. Proust: J. Geophys. Res. **101**, 25217 (1996)

9. P. Coussot, S. Proust, C. Ancey: J. Non-Newtonian Fluid Mech. **66**, 55 (1996)
10. J. Crisp, S. Baloga: J. Geophys. Res. **95**, 1255 (1990)
11. J.H. Fink, R.C. Fletcher: J. Volcanol. Geotherm. Res. **4**, 151 (1978)
12. J.H. Fink, R.W. Griffiths: J. Fluid. Mech. **221**, 485 (1990)
13. J.H. Fink, R.W. Griffiths: J. Volcanol. Geotherm. Res. **54**, 19 (1992)
14. J.H. Fink, R.W. Griffiths: J. Geophys. Res. **103**, 527 (1998)
15. J.H. Fink: Geology **8**, 250 (1980a)
16. J.H. Fink: Tectonophys. **66**, 147 (1980b)
17. T.K.P. Gregg, J.H. Fink: J. Volcanol. Geothermal Res. **86**, 145 (2000)
18. R.W. Griffiths, J.H. Fink: J. Geophys. Res. **97**, 19739 (1992)
19. R.W. Griffiths, J.H. Fink: J. Geophys. Res. **97**, 19729 (1992)
20. R.W. Griffiths, J.H. Fink: J. Fluid Mech. **252**, 667 (1993)
21. R.W. Griffiths, J.H. Fink: J. Fluid Mech. **347**, 13 (1997)
22. R.W. Griffiths: Ann. Rev. Fluid Mech. **32**, 477 (2000)
23. K. Helfrich: J. Fluid Mech. **305**, 219 (1995)
24. C.J. Hughes: *Igneous Petrology* (Elsevier Scientific Publishing Company, New York 1982)
25. G. Hulme: Mod. Geol. **4**, 107 (1973)
26. G. Hulme: Geophys. J. Roy. Astr. Soc. **39**, 361 (1974)
27. G. Hulme: Geophys. Surv. **5**, 245 (1982)
28. H.E. Huppert, R.S.J. Sparks, J.S. Turner, N.T. Arndt: Nature **309**, 19 (1984)
29. H.E. Huppert, R.S.J. Sparks: J. Petrol. **26**, 694 (1985)
30. H.E. Huppert: J. Fluid Mech. **121**, 43 (1982)
31. J.P. Kauahikaua, K.V. Cashman, T.N. Mattox, K. Hon, C.C. Heliker, M.T. Mangan, C.R. Thornber: J. Geophys. Res. **103**, 27303 (1998)
32. P.B. Kelemen, J.A. Whitehead, E. Aharonov, K. Jordahl: J. Geophys. Res. **100**, 475 (1995)
33. P.B. Kelemen, E. Aharonov: 'Periodic formation of magma fractures and generation of layered gabbros in the lower crust beneath ocean spreading centres.' In: *Faulting and Magmatism at Mid-Ocean Ridges*, ed. by W.R. Buck, P.T. Delaney, J.A. Karson, Y. Lagabrielle (Geophysical Monograph 106, American Geophysical Union, Washington DC 1998) pp. 267–280
34. C.R.J. Kilburn: 'Lava crust ʻaʻa flow lengthening and the pahoehoe–ʻaʻa transition', ed. by C.R.J. Kilburn. In: *Active Lavas* (UCL Press, London 1993) pp. 263–280
35. J.R. Lister: J Fluid Mech. **242**, 631 (1993)
36. H. Miyamoto, S. Sasaki: J. Geophys. Res. **103**, 27489 (1998)
37. J.F. Nye: J. Glaciol. **2**, 82 (1952)
38. P. Ortoleva, E. Merino, C. Moore, J. Chadam: American Journal of Science **287**, 979 (1987)
39. P. Ortoleva, J. Chadam, E. Merino, A. Sen: American Journal of Science **287**, 1008 (1987)
40. D.I. Osmond, R.W. Griffiths: J. Fluid Mech. (in press 2001)
41. D.W. Peterson, R.T. Holcomb, R.I. Tilling, R.L. Christiansen: Bull. Volcanol. **56**, 343 (1994)
42. H. Pinkerton: Endeavour **11**, 73 (1987)
43. P.G. Saffman, G.I. Taylor: Proc. Roy. Soc. A **245**, 312 (1958)
44. A.H.P. Skelland: *Non-Newtonian flow and heat transfer* (Wiley, New York 1967)
45. M. Spiegelman, P.B. Kelemen, E. Aharonov: J. Geophys. Res. (submitted 2000)
46. J.S. Turner, H.E. Huppert, R.S.J. Sparks: J. Petrol. **27**, 397 (1986)

47. G.P.L. Walker: Phil. Trans. R. Soc. A **274**, 107 (1952)
48. J.A. Whitehead, K.R. Helfrich: G. Geophys. Res. B3 **96**, 4145 (1991)
49. J.A. Whitehead, P. Kelemen: 'Fluid and thermal dissolution instabilities in magmatic systems'. In: *Magmatic Systems*, ed. by M.P. Ryan (Academic Press, New York, 1994) pp. 355–379
50. D.A. Williams, R.C. Kerr, C.M. Lesher: J. Geophys. Res. **103**, 27533 (1998)
51. J. Wylie, J.R. Lister: J. Fluid Mech. **305**, 329 (1995)
52. J. Wylie, K.R. Helfrich, B. Dade, J.R. Lister, J.F. Salzig: Bull. Volcanol. **60**, 432 (1999)

7 Shallow Lava Theory

N.J. Balmforth[1], A.S. Burbidge[2], and R.V. Craster[3]

[1] Department of Applied Mathematics and Statistics, School of Engineering, University of California at Santa Cruz, CA 95064, USA
[2] School of Chemical Engineering, University of Birmingham, Edgbaston, Birmingham, B15 2TT, U.K.
[3] Department of Mathematics, Imperial College of Science, Technology and Medicine, London, SW7 2BZ, UK

7.1 Introduction

In Chap. 2, we mentioned that lava was a non-Newtonian fluid, and discussed a variety of state-of-the-art constitutive laws that crudely model some of the properties of such fluids. In the current chapter, we go further in this direction and describe more developments of a theoretical model for lava flows. Lava flows have recently been the subject of a review by Griffiths [1] (see also Chap. 6). Our aim here is to illustrate the use of viscoplastic rheological models in this problem.

Viscoplastic fluid models are appropriate because silicic lava contains large quantities of silicate crystals that provide a significant yield stress and crystallize with temperature to produce highly temperature-dependent material properties. Many lava formations are built from this material. For example, silicic lava forms the bulk of the lava domes that emerged after eruptions on Katmai and Mount St. Helens, and which are shown in Figs. 7.1 and 7.2. These structures were gradually built up by the slow effusion of lava from a smaller vent. Other lava flows contain less silicates, such as the basaltic lavas of Mount Etna and Hawaii. These lavas generally have both a smaller yield stress and viscosity, with the result that they flow much more easily and create morphology more like that of rivers, see Fig. 7.3.

Although we have models from non-Newtonian fluid mechanics at our disposal for roughly describing some of the rheology of lava, it is still a formidable task to solve the resulting governing equations – we have a non-isothermal, three-dimensional evolving fluid flow with a free surface and strongly varying material properties. Though this does not rule out full-scale numerical simulation as an option, it does mean that such an approach is far from straightforward. Moreover, because we do not completely understand all the input physics, one can justifiably question the usefulness of embarking on such a difficult exercise. Fortunately, lava flows are often relatively shallow, and as in other fields of geophysical fluid dynamics, one is tempted to exploit this attribute to build simpler theoretical models describing the phenomena. The construction amounts to an asymptotic expansion of the governing equations, and furnishes a "shallow-lava theory." This is entirely analogous to theories developed for avalanches, ice, mud

Fig. 7.1. The Novarupta dome that formed after the 1912 Katmai eruption in Alaska; the dome has diameter 800 ft and is 200 ft high. This photograph is courtesy of the USGS/Cascades Volcano Observatory and further details regarding this dome and others can be found at http://vulcan.wr.usgs.gov/home.html

Fig. 7.2. The lava dome inside the crater of Mount St. Helens. Photographs courtesy of USGS

and debris flows, as described in other chapters in this volume. Here we describe elements of a shallow-lava theory.

Theoretical modelling of this kind can be complemented by laboratory experiments: extrusions in the laboratory with fluids that act as analogues of lava provide a controllable visualization of the important fluid mechanics. The most commonly used analogue fluids for isothermal flows are kaolin–water slurries [2,3], which, as we saw in earlier chapters, are approximately Herschel–Bulkley fluids. Later in this chapter we describe some experiments with such slurries. These experiments nicely demonstrate some of the fluid dynamical effects present in lava flows and which can be understood with the theory. Moreover, detailed comparisons verify that, in the simpler isothermal limit, the theory compares quantitatively with laboratory analogues. Non-isothermal experiments have also been conducted using wax, syrup and slurries of wax and kaolin [4–6] – as de-

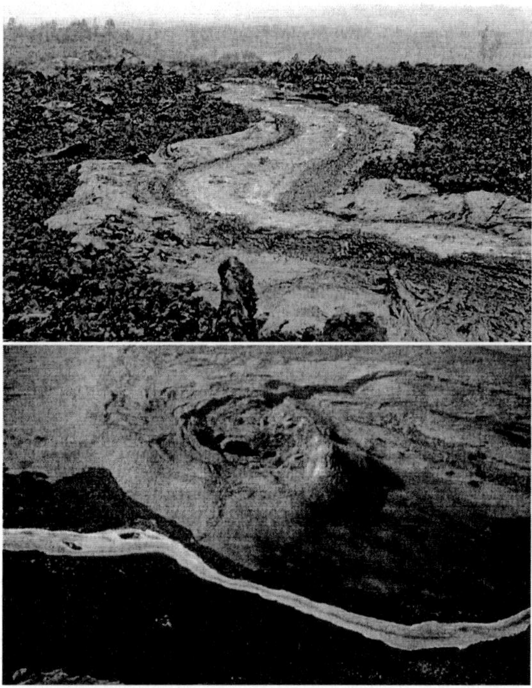

Fig. 7.3. Two Hawaiian lava flows. Photographs courtesy of USGS

scribed in Chap. 6, these experiments have many common morphological features with real lava formations.

We open our discussion with the derivation of the shallow-lava theory for axisymmetrical, cooling lava domes. Our main aim is to summarize the equations that one needs to solve for cooling domes; this theory is also relevant in some entirely different subjects, such as spreading non-isothermal fluids in chemical engineering [7]. But when we deal with explicit solutions and experimental comparisons, we retire to the simpler isothermal limit. After discussing isothermal domes, we switch problems and turn to isothermal lava flows on slopes. The mathematical formulation is much the same, and we focus on a specific geological issue – the creation of "levees" bordering downslope flows.

7.2 Mathematical Formulation

7.2.1 Governing Equations in Axisymmetrical Geometry

Our vision of the problem (Fig. 7.4) is one in which there is a vent centred at the origin of a cylindrical polar coordinate system (r, θ, z). The material (lava or analogue fluid) is extruded through the vent and then spreads out laterally over a horizontal plate located on the plane $z = 0$. Assuming axisymmetry and incompressibility, the governing equations consist of conservation of momentum,

$$\rho\left(u_t + uu_r + wu_z\right) = -p_r + \partial_r\tau_{rr} + \partial_z\tau_{rz} + \frac{1}{r}\tau_{rr} \tag{7.1}$$

and

$$\rho\left(w_t + uw_r + ww_z\right) = -p_z - \rho g + \partial_r\tau_{zr} + \partial_z\tau_{zz} + \frac{1}{r}\tau_{rz}\;, \tag{7.2}$$

continuity,

$$\frac{1}{r}\partial_r(ru) + w_z = 0\;, \tag{7.3}$$

and the heat equation,

$$\rho c_p\left(T_t + uT_r + wT_z\right) = \frac{1}{2}\tau_{ij}\dot{\gamma}_{ij} + \mathcal{K}\left[\frac{1}{r}\partial_r(rT_r) + T_{zz}\right] + \mathcal{S}\;. \tag{7.4}$$

In these equations, the fluid motions are described by the velocity field $(u, 0, w)$, the pressure p, density ρ, and temperature T. Also, g is gravity, c_p is specific heat, $\mathcal{K}$ is the thermal conductivity, and $\mathcal{S}$ denotes any latent heat release on solidification or crystallization in the material. The subscripts (r, z) denote partial derivatives, except in the case of the stress components, τ_{ij}, and then we use the notation ∂_r, and so on. The material variables, c_p, ρ and $\mathcal{K}$, could, in principle, be temperature dependent, but for simplicity we treat them as constants.

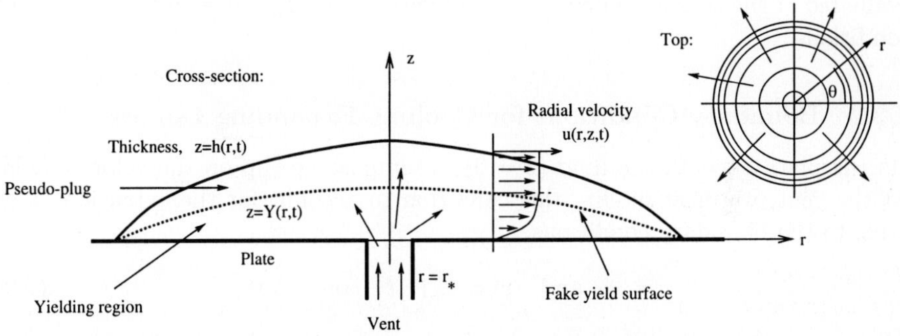

Fig. 7.4. Sketch of an expanding dome

For the cooling problem of interest here, the main source of latent heat release is through a gradual process of crystallization: Lava is a cocktail of different minerals, each crystallizing at a temperature that depends upon the composition. As a result, lava solidifies not at a single temperature, but over a range bounded by the "liquidus" and "solidus" temperatures (where the material is completely fluid and a crystalline solid respectively). The crystal content (expressed as a volumetric fraction, ϕ) varies throughout this range, and the resulting fluid structure can be very complex. Here, we ignore the complicated physical details of the solidification process and opt for a simple model in which the crystal content is a known function of the temperature alone: $\phi = \phi(T)$. Then,

$$\mathcal{S} = \rho\mathcal{L}(\phi_t + u\phi_r + w\phi_z) \equiv \rho\mathcal{L}\phi_T(T_t + uT_r + wT_z) , \tag{7.5}$$

where $\mathcal{L}$ is the latent heat of crystallization and $\phi_T = \mathrm{d}\phi/\mathrm{d}T$.

As discussed in Chap. 2, we adopt the Herschel–Bulkley model for the rheology of the fluid:

$$\tau_{ij} = \left(K\dot{\gamma}^{n-1} + \frac{\tau_p}{\dot{\gamma}}\right)\dot{\gamma}_{ij} \qquad \text{for } \tau \geq \tau_p \tag{7.6}$$

and

$$\dot{\gamma}_{ij} = 0 \qquad \text{for } \tau < \tau_p , \tag{7.7}$$

where τ_p is the yield stress, K is the consistency and n the power-law parameter. Also required are the second invariants of the stress and strain rate:

$$\tau = \sqrt{\tau_{ij}\tau_{ij}/2} , \quad \dot{\gamma} = \sqrt{\dot{\gamma}_{ij}\dot{\gamma}_{ij}/2} . \tag{7.8}$$

We allow for the temperature and crystal dependence of the material by allowing the consistency and yield stress to vary: $K \to K(\phi, T)$ and $\tau_p \to \tau_p(\phi, T)$. We leave the precise dependences arbitrary, but sensible choices include the Arrhenius law and the Einstein–Roscoe relation (Chap. 2). To derive a thin layer model, we prescribe $K(\phi_a, T_a) = K_*$ and $\tau_p(\phi_a, T_a) = \tau_{p*}$ as the values evaluated at the crystal content, ϕ_a, and temperature, T_a, relevant to the ambient conditions.

7.2.2 Boundary Conditions for Cooling, Expanding Domes

On the plate beneath the fluid ($z = 0$), we impose no-slip on the velocity field. At the vent, we must modify this condition to account for the extrusion. This leads to the boundary conditions,

$$u = 0 \quad \text{and} \quad w = w_s(r, t) \quad \text{on } z = 0 , \tag{7.9}$$

where $w_s(r, t)$ is the vertical velocity of material exiting the vent. For simplicity, we also prescribe the heat flux on $z = 0$:

$$\mathcal{K}T_z = \rho(c_p - \phi_T\mathcal{L})(T - T_e)w_s \qquad \text{on } z = 0 , \tag{7.10}$$

where T_e is the "eruption" temperature. This means that the plate is insulating away from the vent, but the arrival of hot fluid generates an incoming heat flux.

The surface of the dome, $z = h(r, t)$, is stress-free, and so

$$h_t + uh_r = w \tag{7.11}$$

and

$$\begin{pmatrix} \tau_{rr} - p & \tau_{rz} \\ \tau_{zr} & \tau_{zz} - p \end{pmatrix}_{z=h} \begin{pmatrix} -h_r \\ 1 \end{pmatrix} = \begin{pmatrix} 0 \\ 0 \end{pmatrix} . \tag{7.12}$$

The thermal boundary condition incorporates surface cooling:

$$\mathcal{K}\mathbf{n} \cdot \nabla T = -F(T) , \tag{7.13}$$

where $\mathbf{n}$ is the outward pointing normal. Various forms are possible for $F(T)$, depending on the specific physical conditions. The simplest model is Newton's law of cooling: $F(T) = a(T - T_a)$, where a is a constant. For lava, if the dominant heat loss is through thermal radiation, the Stefan–Boltzmann black-body law is appropriate, although forced convection of heat by wind can also be appreciable [8]. For the experimental slurries, domes are cooled by both conduction and convection in overlying water, each characterized by some functional form for $F(T)$ [6].

7.2.3 Thin-layer Theory

The full governing equations compose a system of coupled partial differential equations with an evolving free boundary. One could embark upon a heavy numerical simulation using, for example, finite element calculations. However, given the relatively thin profiles of lava domes we are also primed for an asymptotic reduction using thin-layer theory. The aim of the theory is to reduce the complexity of the equations, whilst still retaining the most important physics.

To perform the analysis, it is first expedient to non-dimensionalize the equations as follows: we take H, a characteristic thickness of the fluid layer, as the dimension of the vertical coordinate, and L as a horizontal length-scale. We measure the velocities, u and w, by V and HV/L respectively, and time by L/V. Then we set

$$r = L\tilde{r} , \quad z = H\tilde{z} , \quad u = V\tilde{u} , \quad w = (VH/L)\tilde{w} , \tag{7.14}$$

$$t = (L/V)\tilde{t} \quad h = H\tilde{h} \quad \text{and} \quad p = \rho g H \, \tilde{p} ; \tag{7.15}$$

the tilde decoration denotes the non-dimensional variables. The temperature field is non-dimensionalized using the temperature drop between eruption and ambient temperature:

$$T = T_a + (T_e - T_a)\Theta \equiv T_a + \Delta T \Theta . \tag{7.16}$$

Now, given our non-dimensional units, we may measure the stresses by the quantity, $\rho g H^2/L$. However, units for the stresses can also be given based on the constants of the constitutive model. As a result, there is a relationship amongst the various units that we may choose to have the form,

$$V^n = \frac{\rho g H^{n+2}}{K_* L} . \tag{7.17}$$

This relation also reflects a balance of terms in the momentum equations (the horizontal pressure gradient with the force from the vertical shear stress) which is standard in "lubrication theory".

Thin-layer theory proceeds by introducing the scalings above into the governing equations and then taking the limit, $H/L = \epsilon \to 0$, with a number of non-dimensional numbers held fixed [9]. To leading order, the governing equations become

$$p_r - \partial_z \tau_{rz} = 0 , \qquad p_z + 1 = 0 , \tag{7.18}$$

$$\frac{1}{r}\partial_r(ru) + w_z = 0 \tag{7.19}$$

and

$$\Theta_t + u\Theta_r + w\Theta_z = \kappa\Theta_{zz} , \tag{7.20}$$

where

$$\kappa = \left(1 - \frac{\phi_T \mathcal{L}}{c_p}\right)^{-1} \frac{\mathcal{K} L}{\rho c_p V H^2}$$

is a dimensionless, effective diffusivity (an inverse Peclet number) depending on temperature. The acceleration terms disappear from the momentum equations because the Reynolds number can be taken to be small (the flow is typically laminar), and viscous heating can be ignored for lava and most laboratory analogue fluids (the "Brinkman number" is small). The crucial parameter in the energy equation is κ: If $\kappa \gg 1$, the diffusive term is dominant in the energy equation, and further asymptotic simplification follows [9]. This limit corresponds to rapid heat diffusion, and in the lava literature this is sometimes called the thermally mixed limit. However, for lava and many analogue materials, κ is order one, and heat diffuses relatively slowly. In this circumstance, we are faced with dealing with the heat equation as a partial differential equation at leading order.

The rescaling of the constitutive equation leads to

$$\tau_{rz} = \frac{1}{\dot\gamma}\left[\mathcal{A}(\Theta)\dot\gamma^n + \mathcal{B}(\Theta)\right] u_z \qquad \text{for } \mathcal{B}(\Theta) < \tau , \tag{7.21}$$

$$u_z = 0 \quad \text{for } \mathcal{B}(\Theta) > \tau , \tag{7.22}$$

where

$$\tau \equiv |\tau_{rz}| \qquad \text{and} \qquad \dot\gamma = |u_z| , \tag{7.23}$$

and, given that $\phi = \phi(\Theta)$,

$$\mathcal{A}(\Theta) = \frac{K(\phi, T)}{K_*} , \qquad \mathcal{B}(\Theta) = B\frac{\tau_p(\phi, T)}{\tau_{p*}} \qquad \text{and} \qquad B = \frac{\tau_{p*} L}{\rho g H^2} . \tag{7.24}$$

The "Bingham number", B, is a dimensionless measure of the yield stress.

The boundary conditions become

$$u = 0 , \quad w = w_s , \quad \kappa\Theta_z = (\Theta - 1)w_s \qquad \text{on } z = 0 , \tag{7.25}$$

and

$$h_t + uh_r = w , \qquad \tau_{rz} = p = 0 , \qquad \Theta_z = -\alpha\Theta \qquad \text{on } z = h(r,t) , \tag{7.26}$$

where α denotes the non-dimensional "cooling law",

$$\alpha(\Theta) = \frac{H}{\mathcal{K}\Delta T} F(T) . \tag{7.27}$$

The momentum equations (7.18) can be integrated once:

$$p = h - z , \qquad \tau_{rz} = -h_r(h - z) , \qquad \tau = (h - z)|h_r| . \tag{7.28}$$

The magnitude of the shear stress is measured by τ. This decreases from a maximal value of $\tau(r, z = 0, t) = h|h_r|$ on the base of the fluid, to zero on the stress-free surface. If $h|h_r| < \mathcal{B}[\Theta(r, 0, t)]$, the fluid is not stressed sufficiently to yield anywhere over its depth, and the dome is stationary. But, when $h|h_r| > \mathcal{B}[\Theta(r, 0, t)]$, the fluid near the base of the dome must yield and flow. In this case, because τ decreases with z to zero, there is a surface, $z = Y(r, t)$, on which $\tau = \mathcal{B}$, given by

$$Y(r, t) = h + \frac{\mathcal{B}[\Theta(r, Y, t)]}{h_r} . \tag{7.29}$$

Above this surface, the stress apparently fall beneath the yield stress, and so the fluid is predicted to flow like an unyielded, rigid "plug" with $u_z = 0$. This result is surprising given that the dome is expanding – such spreading flows are divergent, and so the fluid cannot be truly rigid. In the past this apparent contradiction has mistakenly led to the belief that lubrication-style analyses of the sort described here are not self-consistent. The mistake is to identify the flow in $z > Y(r, t)$ as a true "plug flow" – a more refined asymptotic analysis shows that this region is actually weakly yielding [10], and sufficiently so to account for the spreading of the dome. A better terminology is to refer to the weakly yielding region as a "pseudo-plug." (One sees this feature also in Chap. 22).

Equations (7.21) and (7.28) can now be combined into

$$u_z = \begin{cases} -h_r^{(1-n)/n}[\mathcal{A}(\Theta)]^{-1/n}(Y - z)^{1/n} h_r & z < Y(r, t) \\ 0 & z \geq Y(r, t) \end{cases} \tag{7.30}$$

(at least to leading order), which means that the flow is approximately parabolic in the lower, yielding region and constant in the pseudo-plug (see the definition sketch in Fig. 7.4).

We next integrate the continuity equation (7.19) in z, using the boundary conditions at the surface and base, to obtain an evolution equation for the height $h(r, t)$:

$$h_t + \frac{1}{r}\partial_r (r\mathcal{U}) = w_s , \tag{7.31}$$

where

$$\mathcal{U}(r, t) = \int_0^h u \mathrm{d}z = \sigma h^2 |h_r|^{1/n} \int_0^\eta \frac{(1 - \zeta)(\eta - \zeta)^{1/n}}{[\mathcal{A}(\Theta)]^{1/n}} \mathrm{d}\zeta , \tag{7.32}$$

with $\sigma = \mathrm{sgn}(h_r)$ and $\eta = Y/h$ ($\zeta \equiv z/h$). Because η depends on temperature, we cannot integrate this equation without solving the energy equation (7.20), and we cannot evolve that equation without knowing $h(r,t)$ and the velocity field. Thus, our shallow-lava theory now grinds to a halt analytically, leaving a coupled, integro-differential system for $h(r,t)$ and $\Theta(r,z,t)$. Though this system is still rather complicated, it is simpler than the original governing equations.

7.3 Isothermal Domes

7.3.1 Shallow Isothermal Domes

The shallow-lava theory simplifies significantly if the temperature dependence of the fluid drops out of the problem. Such is the case if the fluid did not have time to cool, or cooled to the ambient temperature immediately. Then we may omit the heat equation and set $\mathcal{A}=1$ and $\mathcal{B}=B$, leaving only a single evolution equation for the height field:

$$h_t + \frac{1}{r}\partial_r\left(r\mathcal{U}\right) = w_s\ , \qquad \mathcal{U} = -\frac{n\eta h^2(1+2n-n\eta)}{(n+1)(2n+1)}|hh_r\eta|^{1/n}\mathrm{sgn}(h_r)\ , \quad (7.33)$$

where $\eta = \mathrm{Max}(1-B/|hh_r|, 0)$.

Representative solutions of these equations are shown in Fig. 7.5 for $n=1$ and an influx given by $w_s = 0.1\ \mathrm{Max}(r_*^2 - r^2, 0)$, where $r_* = 0.15$ is the dimensionless vent radius. We also pre-wet the plate beneath the dome (by taking initial conditions with $h(r,t)$ small but everywhere finite) in order to avoid mathematical complications associated with contact lines at the rim of the dome. Figure 7.5 shows the height and yield surfaces for three values of the Bingham number B. Newtonian-like domes (with $B \ll 1$, as in panel (a)) spread laterally much further than yield-stress-dominated domes (with $B \sim 0.1$ or larger, as in panel (c)); the latter rise to greater heights due to the conspiracy between the viscous and yield stress.[1]

For Newtonian domes ($B=0$), $\eta=1$, and one can find a similarity solution to the thin-layer equations for point sources [11]. This solution predicts that $R(t) \sim t^{1/2}$ and $h(0,t) \sim t^0$, which also follow directly from dimensional scaling analysis of the full governing equations [4].

In yield-stress dominated domes, only a thin fluid layer near the base yields. Hence, $\eta \to 0$, giving $h \approx -B/h_r$, and then

$$h = \begin{cases} \sqrt{2B(R-r)} & r < R \\ 0 \quad r > R\,. \end{cases} \qquad (7.34)$$

[1] Formally speaking, the thin-layer theory is not valid at the vent, where $h_r \to 0$, and at the rim, where radial gradients become as sharp as vertical ones. The condition $h_r(0,t)=0$ also leads to the curious behaviour of the apparent yield surfaces in Fig. 7.5 near $r=0$. Neither problem is especially important to the overall evolution of the dome. A similar difficulty arises in shallow-ice theory, and a later chapter by Hutter is partly motivated by them.

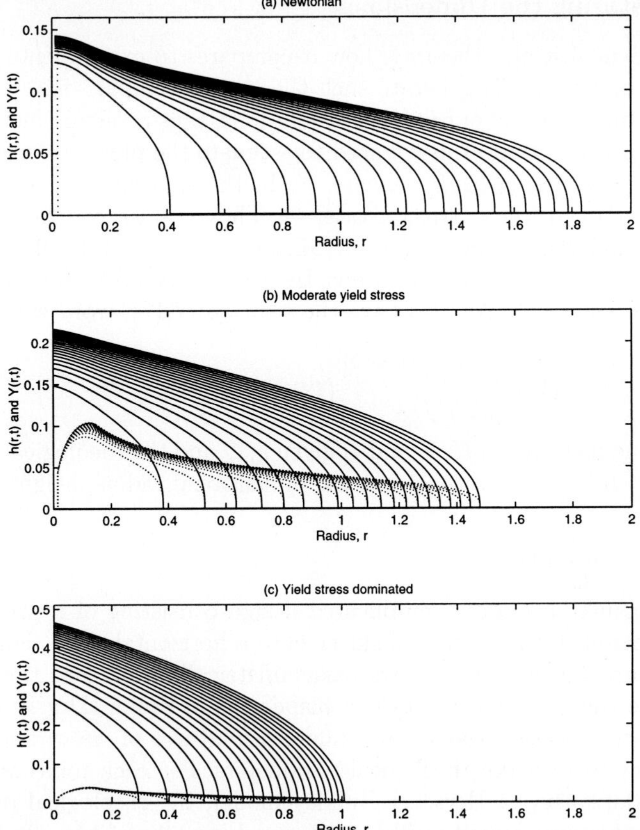

Fig. 7.5. Evolution of the height field, $h(r,t)$, together with the "yield" surface, $Y(r,t)$ (shown by the dotted lines) for various values of B; snapshots of the solution are shown every 500 time units. In panel (**a**), $B = 10^{-5}$ and the dome is effectively Newtonian. For panel (**b**) $B = 0.01$, and the dome in panel (**c**) with $B = 0.1$ is dominated by the yield stress

The time rate of change of the radius, $R(t)$, is dictated by the mass conservation law,

$$\frac{\mathrm{d}}{\mathrm{d}t}\int_0^\infty h(r,t)r\mathrm{d}r = \int_0^\infty w_s(r,t)r\mathrm{d}r \, . \tag{7.35}$$

With a constant inflow rate, Q,

$$R(t) = \frac{1}{(2B)^{1/5}}\left(\frac{15Qt}{8\pi}\right)^{2/5} , \qquad h(0,t) = (2BR)^{1/2} \, , \tag{7.36}$$

an asymptotic result also deduced by Nye [12].

7.3.2 Restoring the Dimensions

One convenient test of a theory is how it compares to experiments designed as laboratory analogues. To generate such theoretical comparisons, we must first restore the dimensions in our numerical solutions, thus reversing our earlier non-dimensionalization. To do this we need to estimate the physical length scales, L and H, and the characteristic velocity, V. In the experiments, we fix the vent radius, R_*, and set the extrusion rate, Q. These values can be compared to the dimensional vent radius, $r_* L = 0.15L$, and extrusion rate, $0.05\pi r_*^4 LHV \approx 8\times 10^{-5} LHV$, used in the computations. Hence, $L = 1\,\mathrm{cm}$ and $HV = 1.26\times 10^6 Q$ (with Q in mks units). We also have the relation (7.17), from which it follows that

$$H = \left(\frac{KL}{\rho g}\right)^{1/(n+2)} \left(1.26 \times 10^6 Q\right)^{n/(n+2)} . \tag{7.37}$$

This allows us to compute H, V and B given ρ, Q and the rheological parameters of the fluid, and thereby reconstruct the dimensional radius, height and time.

7.3.3 Experiments

The experiments have an uncomplicated design consisting of a piston that extrudes a controlled volume flux of slurry onto a horizontal plate. For the domes that then form (which were always axisymmetrical), we record the radius and height above the vent. The slurry is a suspension of kaolin (Dry Branch Kaolin Company) in de-ionized water, and different mixtures of water and kaolin are used in order to vary the rheological parameters. For each mixture, we fit the rheological data using a Herschel–Bulkley model; the rheological properties of the slurries are summarized in Table 2.1 and Fig. 2.2 of Chap. 2. A variety of (time-independent) flow rates, Q, is also used; we quote results for the fastest and slowest of these ($0.18\,\mathrm{cm}^3/\mathrm{s}$ and $0.54\,\mathrm{cm}^3/\mathrm{s}$).

The heights and radii are shown versus time in Fig. 7.6 and 7.7 for kaolin–water slurries mixed in the ratio 0.6:1 and 0.8:1 by weight. The theoretical curves from the shallow lava theory are added for comparison, and are in fair agreement. The dome heights compare least favourably, but this should be tempered by the fact that there were some experimental difficulties in taking this measurement. The figures also show the asymptotic result for large yield stress (Nye's theory), which overestimates the radii and underestimates the heights (Blake [3] uses an empirical correction to account for this error). More paste-like materials with kaolin to water ratios of 1:1 and 1.2:1 were modelled with similar accuracy by the thin-layer theory. Typically these had larger Bingham numbers, ~ 0.19, and were also adequately modelled using Nye's solution.

In Chap. 2 we mentioned that kaolin slurries show some hysteresis in their stress-strain-rate relations, suggesting that the fluid microstructure does not reform in the same way as it is destroyed. In the extrusion experiments, the microstructure disintegrates as the fluid is pushed up the vent, and then reforms as the dome spreads and the stresses gradually decline. This means that the "down-curve" is most suitable for modelling the experiments. We illustrate

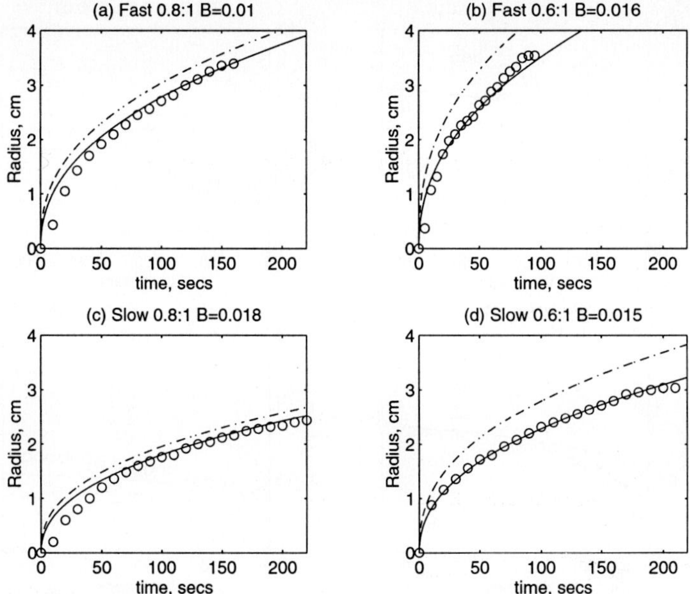

Fig. 7.6. Experimental and theoretical comparisons of dome radii for kaolin–water domes. The solid line gives the result found from shallow lava theory, the circles are the experimental data and the dot-dash line is Nye's result for large B. The values of B are estimated as described in Sect. 7.3.2

the importance of this choice by using another material, Celacol (Courtaulds), also described in Chap. 2, that shows pronounced hysteresis. The comparison between theory and experiment for this material is shown in Fig. 7.8. The two are in agreement only if rheological data from the down curve are used to fit the parameters of the Herschel–Bulkley model; use of the data from the "up-curve" leads to significant disagreement. Evidently, the best model for Celacol would be one accounting for hysteresis, but if we insist on using a model like Herschel–Bulkley we should exercise care in interpreting the rheological data.

7.4 Flows on Inclined Planes

7.4.1 Shallow Flow Dynamics

If the plate beneath the fluid is inclined, the circular symmetry of an expanding dome is broken. Instead, the fluid slumps downslope, leading to elliptical domes for low inclinations, and fully fledged channel flows on larger slopes. We now turn to a theoretical consideration of these structures, again specializing to isothermal conditions.

To generalize the theory it is first helpful to consider a new, Cartesian coordinate system, (x, y, z), in which $z = 0$ again coincides with the base of the

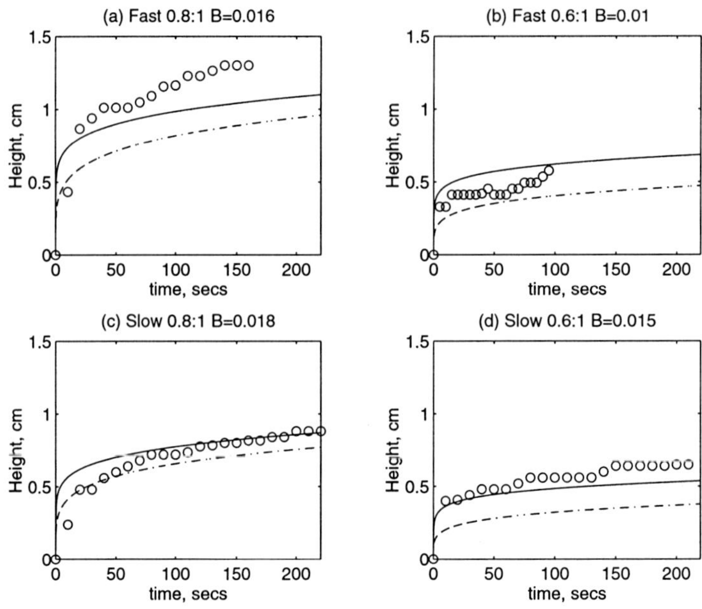

Fig. 7.7. Experimental and theoretical comparison of dome heights for kaolin–water domes. The solid line gives the result found from shallow lava theory, the circles are the experimental data and the dot-dash line is Nye's asymptotic, large B, result

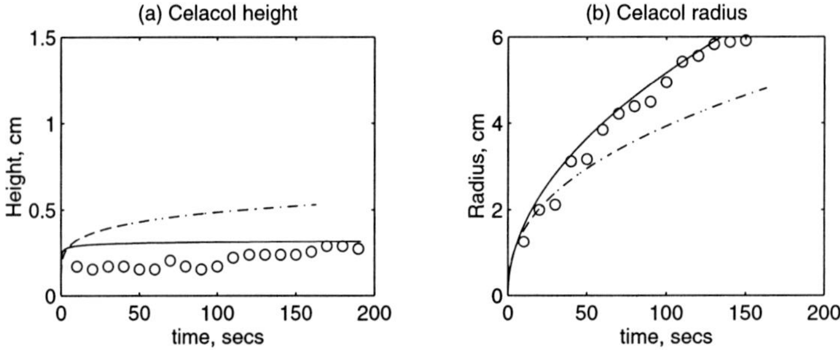

Fig. 7.8. Experimental and theoretical comparison of dome evolution for a Celacol dome. Panel (**a**) shows height measurements and in panel (**b**) we compare radii measurements. The circles show the experimental data. The solid lines are the numerical results using the rheological data from the down curves, and the dash-dotted lines show the corresponding results using the up-curves

fluid, which is now on an inclined plane. As shown in Fig. 7.9, we also take the coordinate x to lie in the downslope direction, and ϕ to be the angle of the plane's inclination from the horizontal. From here we could again write down the governing equations, non-dimensionalize, expand asymptotically, and finally arrive at a relevant thin-layer model [10] (see also Chap. 22). We will not go through the details here, and instead offer some simple arguments that indicate how we should generalize (7.33).

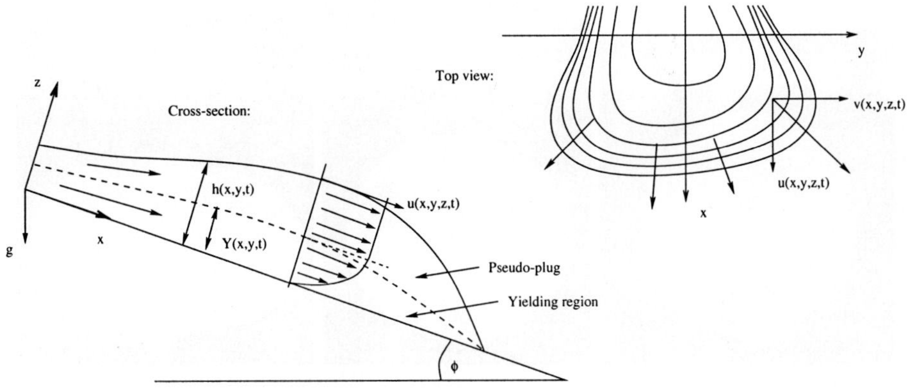

Fig. 7.9. Sketch of a flow on an inclined plane. ϕ is the angle of inclination

The key feature of the thin-fluid dynamics is that thickness variations drive a flow that is down-gradient with respect to the height field: $\mathcal{U} \sim -|h_r|^{1/n-1}h_r$. A natural generalization is therefore to introduce an analogous depth-integrated lateral velocity, $\mathcal{V}$, and take $(\mathcal{U}, \mathcal{V}) \sim -s^{1/n-1}(h_x, h_y)$, where $s = \sqrt{h_x^2 + h_y^2}$ is the mean surface gradient. This naturally accounts for the shape of the fluid, but not the background slope, which also forces flow in the x–direction. To take account of the slope we make the replacement $h_x \to h_x - S$, where $S = \epsilon^{-1} \tan\phi$ is a measure of the slope relative to the fluid's typical aspect ratio (assumed to be order one, so that the slope must be sufficiently gentle). Thence, with other dependences as before,

$$h_t + \mathcal{U}_x + \mathcal{V}_y = w_s \,, \qquad \begin{pmatrix} \mathcal{U} \\ \mathcal{V} \end{pmatrix} = \frac{n\eta h^2 (s\eta h)^{1/n}(1 + 2n - n\eta)}{s(n+1)(2n+1)} \begin{pmatrix} S - h_x \\ -h_y \end{pmatrix} \,, \tag{7.38}$$

with

$$\eta = \mathrm{Max}\left(1 - \frac{B}{hs}, 0\right) \qquad \text{and} \qquad s = \sqrt{(S - h_x)^2 + h_y^2} \tag{7.39}$$

(once again, w_s denotes the extrusion speed above any vents and $Y = h\eta$ is the fake yield surface), which also results from a proper expansion (see Chap. 22).

7.4.2 Inclined Domes

Experiments illustrating how domes slump downhill and lose symmetry are shown in Fig. 7.10. The slurry used in these experiments is a mixture of water and "joint compound" (a commonly available, kaolin-based material). Some more careful experiments with a true kaolin slurry are shown in Chap. 6 and explored further in [15].

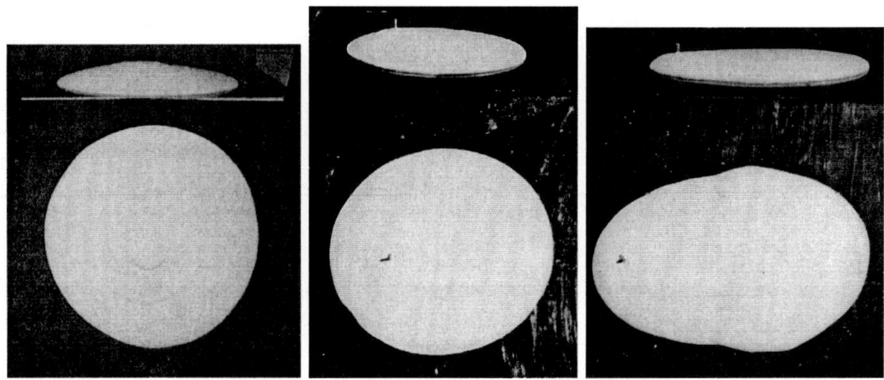

Fig. 7.10. Slumped domes on inclined planes. Shown are domes on slopes with inclinations of roughly 0, 10 and 20 degrees. The slurry, a mixture of water and "joint compound" (a kaolin-based material that is commonly available at hardware shops), is fed onto the plane through a narrow tube from a reservoir held just above. An inclined mirror at the top of each pictures gives a side view of the domes. For the second two domes, a marker indicates the position of the feeder from the reservoir

To compare with experimental images like these we solve the thin-layer equations numerically. As for symmetrical domes we take $w_s = 0.1\ \mathrm{Max}(0.15^2 - r^2, 0)$, and use a numerical scheme: VLUGR2 [16]. One such computation is shown in Fig. 7.11. This shows a dome with $B = 0.01$ and $n = 1$ on a slope with $S = 0.5$. As indicated by the fake yield surfaces, this dome is not far from being Newtonian. The yield stress has most effect upstream of the vent where the fluid becomes almost stationary over longer times. The overall appearance of the dome is similar to the experimental pictures.

When $\eta \to 0$ (large B), and the dome is dominated by yield stress, the thin-layer model simplifies substantially. From the condition, $\eta \approx 0$, we obtain the nonlinear first-order partial differential equation,

$$(S - h_x)^2 + h_y^2 = B^2/h^2 . \tag{7.40}$$

This simpler equation determines the structure of domes that are either dominated by the yield-stress, or slump to rest at the termination of an extrusion. We can solve the equation using Charpit's method [14]: First we scale the equations

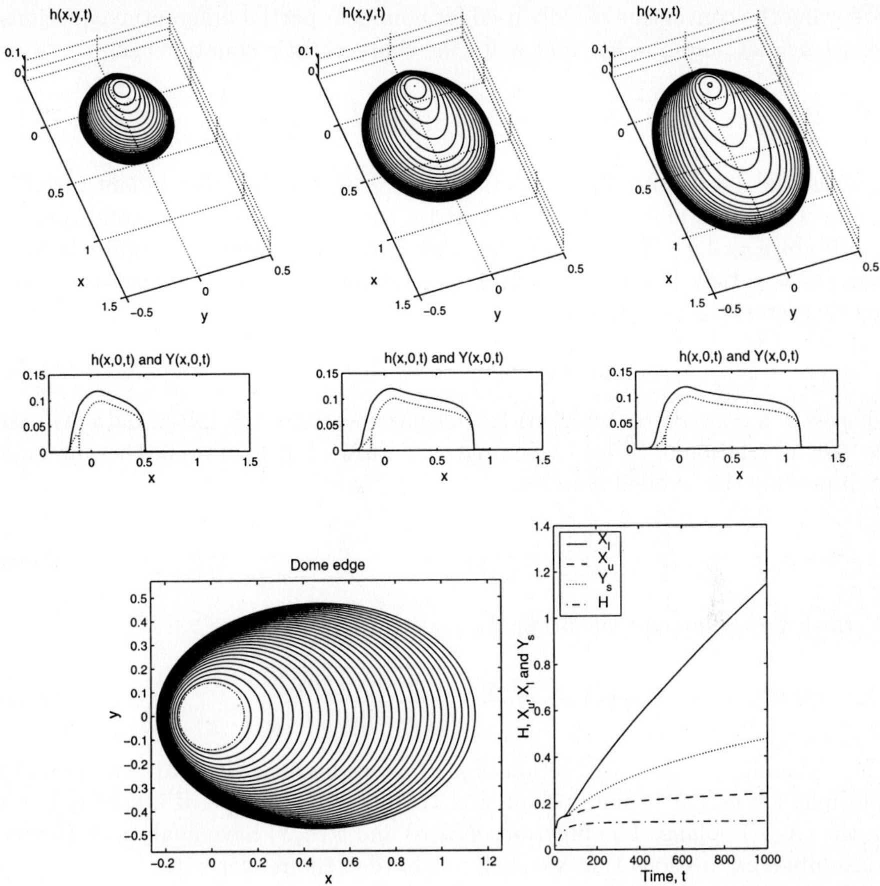

Fig. 7.11. Slumped domes on inclined planes, computed numerically using the thin-layer model. The top row of pictures show three snapshots of the domes at times 333, 666 and 1000. Directly below are the corresponding height profiles and yield surfaces along the midsection ($y = 0$). The lower panels show a sequence of curves showing the dome's edge (the curves show the edge every 6.66 time units), and the evolution of the cross-stream half-thickness (Y_s), the downslope and upslope lengths (X_l and X_u), and the maximum height (H)

to eliminate some distracting constants; set

$$X = Sx/H\ , \quad Y = Sy/H\ , \quad h = Hu \quad \text{and} \quad b = B/(HS)\ , \tag{7.41}$$

where H is the dome height at $x = y = 0$. Then,

$$(1 - u_X)^2 + u_Y^2 = \frac{b^2}{u^2}\ . \tag{7.42}$$

Following the convention usually used for nonlinear partial differential equations, we set $p = u_X$ and $q = u_Y$, and write the characteristic equations as:

$$\dot{X} = 2(1-p)\ , \quad \dot{Y} = -2q\ , \quad \dot{u} = 2p(1-p)-2q^2\ , \quad \text{and} \quad \frac{\dot{p}}{p} = \frac{\dot{q}}{q} = \frac{2b^2}{u^3}\ , \tag{7.43}$$

in which the dot denotes differentiation with respect to the independent variable, τ, the coordinate along each characteristic curve. Suitable initial conditions at $\tau = 0$ are $u = 1$ at $X = Y = 0$, together with parameterized conditions for p and q that satisfy (7.40) at $u = 1$. Two relations follow straightforwardly from the characteristic equations:

$$q = ap \quad \text{and} \quad aX - Y = 2a\tau\ , \tag{7.44}$$

where a is a constant of integration that parameterizes the initial data. We use the second relation of (7.44) to eliminate $\tau = (aX-Y)/2a$. Two further integrals then provide the implicit solution,

$$X = -\int_u^1 \frac{[1-p(\hat{u})]\mathrm{d}\hat{u}}{p(\hat{u})[1-p(\hat{u})(1+a^2)]}\ , \quad Y = a\int_u^1 \frac{\mathrm{d}\hat{u}}{[1-p(\hat{u})(1+a^2)]}\ , \tag{7.45}$$

in which we can exploit the original equation (7.42) to write

$$p(u) = \frac{u \pm \sqrt{b^2(1+a^2) - u^2a^2}}{u(1+a^2)}\ , \tag{7.46}$$

The ambiguity in the construction of $p(u)$ arises because there are two possible solutions for $u(X,Y)$, one upslope and the other downslope of a special curve on the (X,Y)–plane. The functions $\Phi(u,a)$ and $\Psi(u,a)$ have analytical, though convoluted, expressions that we shall not burden the reader with.

Explicit solutions follow for $a = 0$ and $a \gg 1$: For $a = 0$, $Y = 0$ and

$$X = \begin{cases} u - 1 + b\log[(u-b)/(1-b)], & X > 0 \\ u - 1 - b\log[(u+b)/(1+b)], & X < 0, \end{cases} \tag{7.47}$$

as in [15] (see also Chap. 6 and [17]). For $a \gg 1$,

$$X = -1 + u - \frac{b}{2}\log\left[\frac{(b-1)(b+u)}{(b+1)(b-u)}\right] \tag{7.48}$$

and

$$Y = \pm\left(\sqrt{b^2-u^2} - \sqrt{b^2-1}\right)\ . \tag{7.49}$$

This second second curve is the junction dividing the two pieces of the solution for $u(X,Y)$.

Sample solutions are shown in Fig. 7.12. As $b \to 1$, the domes are increasingly slumped ($b \propto S^{-1}$, so a decrease in b corresponds to an increase of the slope). The limiting solution for $b \to 1$ is shown in Fig. 7.13. The solution does not work if $b < 1$, indicating that the dome is no longer able to support itself against gravity.

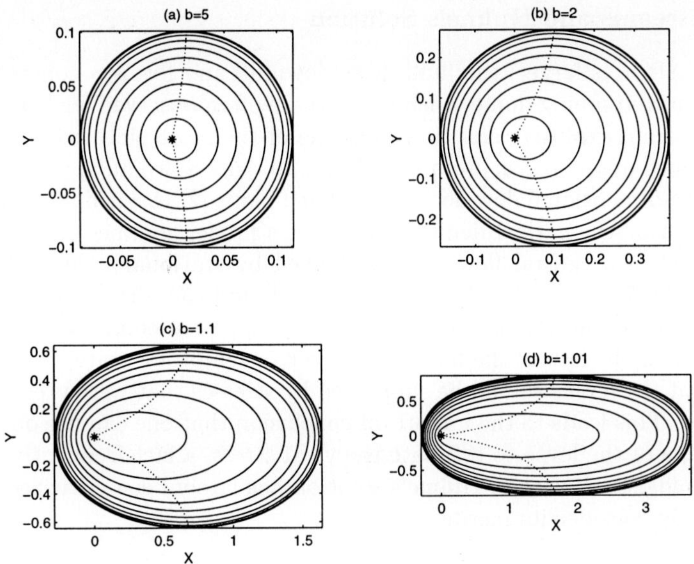

Fig. 7.12. Contours of constant height for yield-stress dominated domes on sloping planes. The dotted curves show the junction between the two parts of the solution

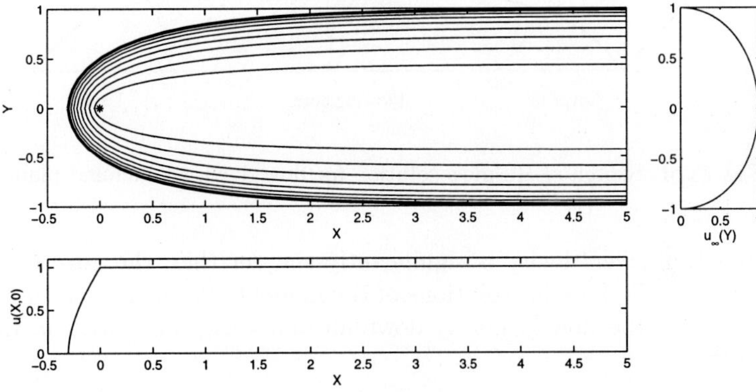

Fig. 7.13. Contours of constant height for the limiting shape as $b \to 1$. Also shown is the thickness profile along the $Y = 0$ axis, given by $u = 1$ for $X > 0$ and the second relation of (7.47), and the thickness profile in Y far downstream ($X \gg 1$), which is a semicircle

7.4.3 Streams and Hulme's Solution

When the slope is larger, the fluid flows downhill and forms a coherent stream rather than a slumping dome. Such streams are often observed for basaltic lavas, and, aside from explosive eruptions, are one of the most pictured volcanic phenomena. Such observations prompted Hulme to write down an approximate solution for one-dimensional flow down an inclined plane [2]. Argued more from plausibility than mathematical deduction, Hulme's model assumes that the stream is composed of a central flowing core flanked by stationary "levees", as illustrated in Fig. 7.14; the flow is purely downhill, and there is no variation in the downslope direction. By assuming that the levees were composed of fluid that naturally came to rest as the flow settled to its asymptotic state, Hulme further argued that the levees should be supported by stresses that were precisely at the yield value. This leads to the important conclusion that one can use observations of the shape of the levee to estimate the yield stress, a fact frequently exploited by volcanologists. However, Hulme's solution has lately been criticized [18], and so we briefly consider its merits.

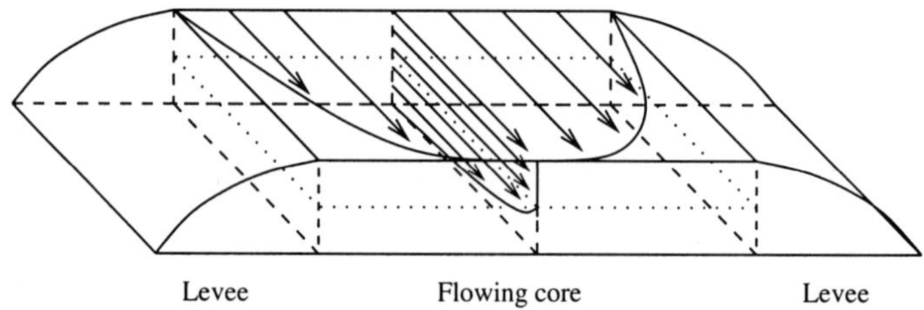

Fig. 7.14. Sketch of Hulme's solution for flow down an inclined plane

Because thin-layer theory is significantly simpler than the governing equations, one can easily look for solutions of the model that correspond to Hulme's. If we insist that the flow is purely downhill and varies only with y, then one concludes that

$$\mathcal{V} = h_y = 0\,, \quad h = 1\,, \quad \eta = 1 - \frac{B}{S} \quad \text{and} \quad s = S\,, \tag{7.50}$$

in regions where the fluid yields. That is, a uniform flow. In the levee, on the other hand, the fluid is on the brink of yielding, which implies that $\eta = 0$, or

$$B = h\sqrt{S^2 + h_y^2} \geq SH_l\,, \tag{7.51}$$

where H_l is the maximal thickness of the levee. Unless $H_l < 1$ (and the levee is shallower than the core), this contradicts the flowing solution which requires that $\eta = 1 - B/S > 0$. In other words, one cannot connect the central flowing channel with the stationary levee. This difficulty is not simply a problem with thin-layer theory – even if one begins from the governing fluid equations, Hulme's construction still appears to be impossible for the same reasons.

The problem with Hulme's construction is that it uses a one-dimensional version of the Bingham fluid model, and consequently does not have the correct, two-dimensional yield criterion: Hulme shapes the levee according to Nye's solution, $B = |hh_y|$, which predicts a parabolic profile, rather than (7.51) (which indicates the profile far downstream is semicircular – see Fig. 7.13). In other words, one assumes that the lateral structure of the levee is the same as the shape the fluid would take on a flat plane – the yielding induced by the downstream flow is ignored. When one takes the extra degree of yield into account, one is led to the inescapable conclusion that, if the levees are to remain as thick as the main channel, the shape of the levees forces lateral flow.

A simple demonstration that the downstream flow affects the lateral shape is afforded by the following experiment with kaolin slurry: We allow a corridor of fluid to slump laterally to a static equilibrium on a horizontal plate. We then tilt the plate in the direction of the central axis of the fluid to create a channel flow. As shown in Fig. 7.15, the main effect of the flow is to allow further lateral spreading of the fluid (except near the upper end of the column, where the thickness remains roughly the same, but some of the fluid drains away). Note also the creation of streamwise flow dependence.

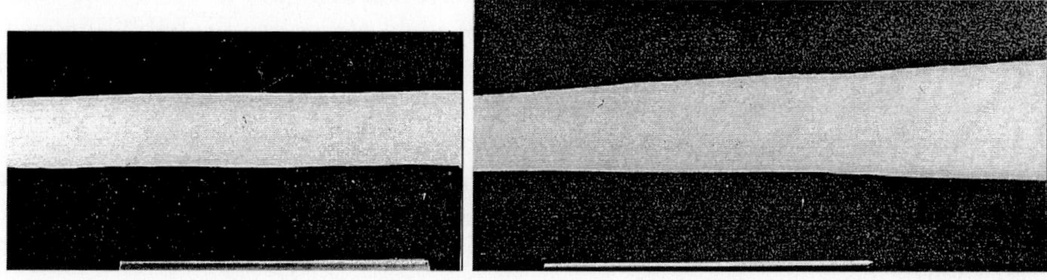

Fig. 7.15. Photographs from an experiment in which a column of viscoplastic fluid (a mixture of water and joint compound) was first allowed to slump to rest on a horizontal plane (first panel). The plane was then tilted at an angle of roughly 25 degrees and as fluid flowed downhill, the column spread laterally and downstream (second panel). The two photographs have the same scale

Because of the theoretical difficulties with Hulme's model, Coussot & Proust suggested another solution taking account of the true yield criterion: Let the flow be independent of time, but not of the streamwise coordinate, x. The thin-layer model equations then become

$$[\mathcal{F}(S - h_x)]_x = [\mathcal{F}h_y]_y \,, \qquad \mathcal{F} = \frac{n\eta h^2 (s\eta h)^{1/n}(1 + 2n - n\eta)}{s(n+1)(2n+1)} \,, \tag{7.52}$$

with η and s as before, and assuming no vertical mass flux (the source of fluid is upstream). Because there is now streamwise variation, the free surface is not necessarily flat, and Coussot & Proust [18] construct solutions with some correspondence with experiments (see also [19]). Actually, Coussot & Proust do not use the full equation (7.52), but an approximation obtained by neglecting h_x and h_y in comparison with S. Thus s becomes S and we arrive at the parabolic equation $S\mathcal{F}_x = [\mathcal{F}h_y]_y$, with $\mathcal{F}$ as above and $\eta = \text{Max}(1 - B/Sh, 0)$. This approximation cannot be accurate at the edges of the stream where the gradients of $h(x, y)$ diverge, but these regions are also where the thin-layer model breaks down.

The analytical solution pictured in Fig. 7.13 shows similar features to Coussot & Proust's downstream spreading flows. Further numerical computations are shown in Fig. 7.16. In this case, with $B = 0.06$ and $S = 2$, the fluid immediately slumps downhill without forming a dome, and creates a gradually widening stream.

A laboratory illustration of a stream flow is shown in Fig. 7.17. This inclined flow has well-defined levees bordering the flow but also spreads laterally with distance downstream. Similar features can be seen in Osmond & Griffiths's domes (Chap. 6 and [15]). Thus, although Hulme's "solution" is not actually a solution of the equations, the image is not entirely wrong: the stationary levees supported by stresses at the yield value do exist – Hulme's precise construction is invalid because the flow spreads downstream and the levees are not shaped according to Nye's solution. Thus Hulme's image is qualitatively correct, if not quantitatively.

Although the final conclusion is that Hulme's solution is in error, the ramifications in geology regarding estimations of yield stress are probably inconsequential: Rheological measurements of lava are exceptionally difficult because of its extreme temperature, and it is probably fair to say that actual values of the yield stress are not known to within orders of magnitude. Hence, although the correction to the shape of the levee given the slope S will certainly change the inferred value of the yield stress by an order one amount, this is insignificant in comparison to other rheological uncertainty. Osmond & Griffiths [15] discuss further how to infer yield stresses given the proper yield condition.

7.5 Concluding Remarks

The purpose of a thin-layer theory is to reduce the full, governing fluid mechanical equations to a more manageable form. For non-isothermal lava flows, because heat conduction occurs relatively slowly, the thin layer theory remains fairly complicated, as in shallow-ice theories. However, the reduced equations contain all the relevant physics in a concise form, and filter out any complicating, but inessential details. The analogue experiments for isothermal domes show that thin-layer theory is accurate over a wide range of extrusion rates and rheological parameters. Some related experiments, with fluid flowing down an inclined plane (and performed to test similar theory for mud flow), show a comparable degree of agreement [13]. Hence shallow-fluid theory appears to be a useful route to take whilst modelling geophysical viscoplastic fluids.

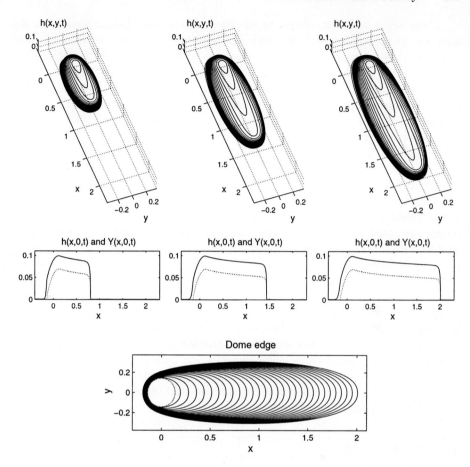

Fig. 7.16. A flow down an inclined planes, computed numerically using the thin-layer model. The top row of pictures show three snapshots of the domes at times 333, 666 and 1000. Directly below are the corresponding height profiles and yield surfaces along the midsection ($y = 0$). The lower panel shows a sequence of curves showing the dome's edge (the curves show the edge every 6.66 time units)

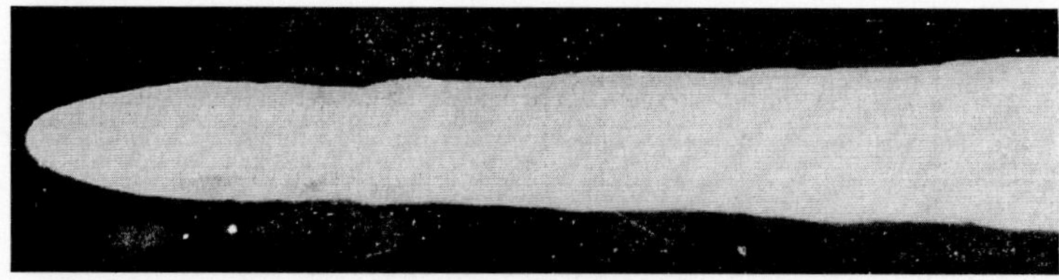

Fig. 7.17. Photographs from an experiment in which viscoplastic fluid (a mixture of water and joint compound) was extruded onto a sloping plane (inclined by roughly 30 degrees). Shown is the final shape after the extrusion was terminated and the fluid allowed to come to rest on the plane

There are several extensions to the theory that must be pursued for a description of lava flows, and the subject is rich with problems to examine. For example, it is essential to solve the non-isothermal problem outlined at the beginning of this article. But there are several other issues that we have not dwelt upon here, such as the detailed mechanics at the edge of the fluid where the flow over-rides the substrate. Notably, the analogue experiments should also be taken further, and theoretical computations should be compared systematically with non-isothermal extrusions.

Beyond the formulation and testing of a shallow-lava theory lies the applications to real geological problems. The shallow-lava theory provides a computationally convenient tool to analyze lava flows in geological settings. For example, one might wish to predict the direction of a lava flow over complex terrain and assess possible hazards. Alternatively, the goal may be to predict which lava domes are most likely to allow the internal build up of hot gas, which could lead to structural failure, explosions and pyroclastic flows (Chap. 8). Of course, we already have some answers to such questions, based on cruder theoretical models (such as approximating the lava as an isothermal viscous fluid) or qualitative arguments (for example, if the lava has a yield stress, a dome can sustain large internal pressures, thus trapping gas within the dome). The real question is whether our more quantitative modelling has any advantage over these simpler arguments, given the uncertainties, idealizations and approximations in the theory. We cannot answer this particular question until we have fully formulated a usable non-isothermal shallow-lava theory, but the hope is that the theory will significantly improve our predictive capabilities, and place the modelling of lava flows on a more solid foundation.

Acknowledgements

The financial support of an EPSRC Advanced Fellowship is gratefully acknowledged by RVC. This work was supported by the National Science Foundation (Grant No. DMS0072521).

References

1. R.W. Griffiths: Ann. Rev. Fluid Mech. **32**, 477 (2000)
2. G. Hulme: Geophys. J. Roy. Astron. Soc. **39**, 361 (1974)
3. S. Blake: 'Viscoplastic models of lava domes'. In: *Lava flows and domes: emplacement mechanisms and hazard implications*, ed. by J.H. Fink, pp. 88–128. IAVCEI Proc. in Volcanology, vol. 2, Springer–Verlag (1990)
4. R.W. Griffiths, J.H. Fink: J. Fluid Mech. **252**, 667 (1993)
5. M.V. Stasiuk, C. Jaupart, R.S.J. Sparks: Geology **21**, 335 (1993)
6. R.W. Griffiths, J.H. Fink: J. Fluid Mech. **347**, 13 (1997)
7. B. Reisfeild, S.G. Bankohh, S.H. Davis: J. Appl. Phys. **70**, 5258 and 5267 (1991)
8. A. Neri: J. Volcan. Geotherm. Res. **81**, 215 (1998)
9. N.J. Balmforth, R.V. Craster: J. Fluid Mech. **422**, 225 (2000)
10. N.J. Balmforth, R.V. Craster: J. Non-Newtonian Fluid Mech. **84**, 65 (1999)

11. H.E. Huppert: J. Fluid Mech. **121**, 43 (1982)
12. J.F. Nye: J. Glaciology **2**, 82 (1952)
13. K.F. Liu, C.C. Mei: J. Fluid Mech. **207**, 505 (1989)
14. I.N. Sneddon: *Elements of Partial Differential Equations* (McGraw–Hill 1957)
15. D.I. Osmond, R. Griffiths: J. Fluid Mech. (in press 2001)
16. J.G. Blom, R.A. Trompert, J.G. Verwer: ACM Trans. Math. Software **22**, 302 (1996)
17. P. Coussot, S. Proust, C. Ancey: J. Non-Newtonian Fluid Mech. **66**, 55 (1996)
18. P. Coussot, S. Proust: J. Geophys. Res. **101**, 25217 (1996)
19. S.D.R. Wilson, S.L. Burgess: J. Non-Newtonian Fluid Mech. **79**, 77 (1998)

8 Explosive Volcanic Eruptions

A.W. Woods

BP Institute, University of Cambridge, Cambridge, England

8.1 Introduction

During explosive volcanic eruptions, up to 10^{14} kg of volcanic ash may be erupted from a vent forming violent ash flows or towering eruption columns. This massive amount of material is subsequently deposited on the ground, with much of the coarser fraction of the flow being deposited within a few hundred kilometres of the volcanic edifice. This may lead to a substantial regional change to the topography, with ash flow deposits being tens to hundreds of metres deep and air-fall deposits being several metres deep. In addition, the eruption of such a large mass of material from a volcanic edifice may lead to collapse of the crust above the sub-surface magma reservoir. This leads to the formation of calderas which are large depressions in the surface topography, often extending tens of kilometres in diameter and being several hundred metres deep. In summary, explosive volcanic eruptions can produce major changes in surface topography owing to the very powerful transport and redistribution of mass. In this contribution, we aim to develop quantitative models to predict the dynamics and deposition patterns of this erupted material, and where possible we compare this with field data.

To set the scene and understand the scale of these phenomena, it is worth examining some of the key parameters which are involved in explosive volcanic eruptions. During explosive volcanic eruptions, dense liquid magma, which is stored in a crustal reservoir, at typical depths of 5–10 km, rises to the surface and issues from a vent as a fragmented mixture of volcanic ash, pumice and gas. This mixture may erupt at speeds of 100–300 m/s, with pressures in the range 1–100 Pa and with density in the range 1–100 kg/m^3. Following eruption from the vent, the mixture may rise as high as 30–40 km into the atmosphere as a buoyant convective plume, or if it remains denser than the atmosphere, it will form a dense, turbulent ash flow. These flows spread over the ground with speeds as large as 100–200 m/s. Buoyant eruption columns shed large particles which fall out to the ground in the vicinity of the volcanic vent, while the smaller particles are carried high into the atmosphere, and may then spread over many hundreds or thousands of kilometers before falling back to the ground. Dense ash flows, in contrast, spread many 10's of km over the ground, leaving a carpet of ash and pumice particles which becomes progressively finer grained with distance from the vent. In some cases, as the flow sediments particles and entrains and mixes air, the residual material in the flow becomes less dense than the air and lifts off the ground to form an ascending plume. The resulting cloud of fine ash can then disperse the fine-grained material over a much wider area.

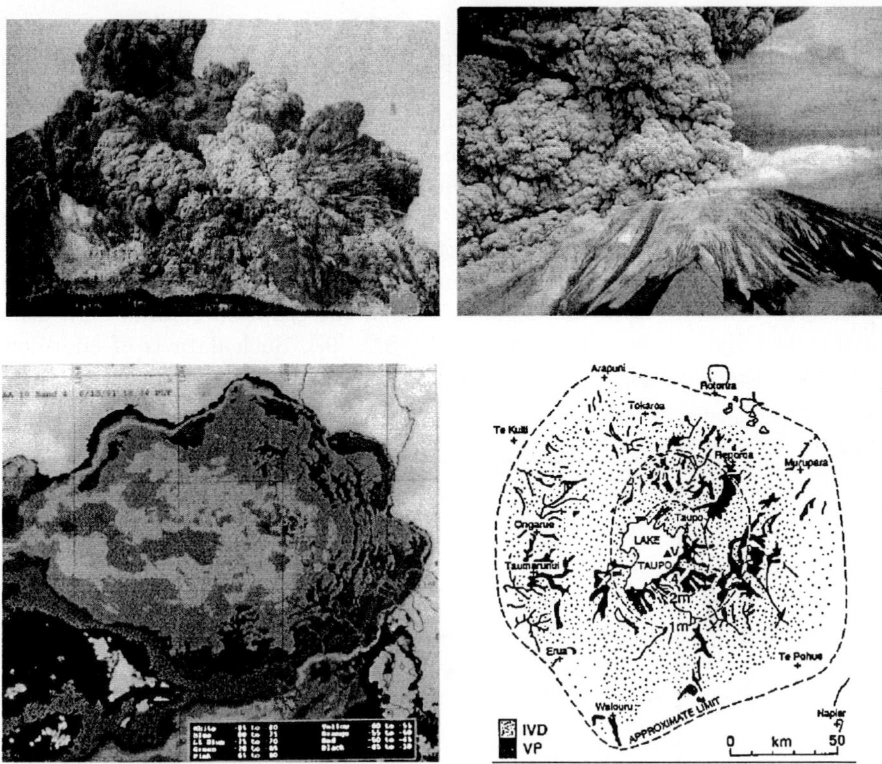

Fig. 8.1. (**a**) Picture of the spreading ash flow during the initial blast of Mount St Helens, May 18 1980. (**b**) Picture of the subsequent momentum jet issuing from the volcanic vent at Mount St Helens (**c**) Picture of the umbrella cloud which formed at a height of 34 km and spread over 1000 km. (**d**) Figure illustrating the radial extent of the Taupo ash flow deposit

Several recent historical eruptions have provided much important new data and observational evidence about the processes involved in explosive volcanism. These include the May 18 1980 eruption of Mount St Helens. In this eruption, an initial blast caused one flank of the volcano to fail, producing a dense flow which spread over 15 km from the volcano before lifting off and rising 15–20 km into the atmosphere (Fig. 8.1a). This initial blast was followed by a more sustained eruption, in which a steady eruption column ascended directly above the volcanic vent (Fig. 8.1b). As the eruption continued, the height of this steady column gradually decreased [7]. Other important eruptions, which have contributed to our understanding of the dynamics of explosive volcanic eruptions, include the 1991 eruption of Mt. Pinatubo, in the Phillipines. In this eruption, a buoyant eruption column was formed and ascended over 35 km into the atmosphere, where it formed a neutrally buoyant cloud which spread radially about 1000 km (Fig. 8.1c). The fine ash in this cloud was then carried around the globe by zonal winds. The 1994–2000 eruption of Soufriere Hills volcano, Montserrat involved the growth of a highly viscous lava dome at a rate of about 2–5 m^3/s. This dome was charged with gas at high pressure, and as it periodically failed it shed small, dense ash flows around the flanks of the volcano. The eruption of Taupo volcano, New Zealand, about 3500 BP, produced an enormous ash flow deposit, which extended over 80 km radially from the volcanic centre (Fig. 8.1d). This deposit has been extensively studied and shows a gradual waning of grain size and thickness of the deposit with distance. This data is extremely useful for comparing with dynamical models of the flow. Many other specific eruptions have been documented and studied, and in a number of cases, field data can be used to calculate both the height of rise of the eruption column and, using independent data, the eruption rate (Fig. 8.2, [9]). Such data is of enormous value in testing models, as we shall see later in the text.

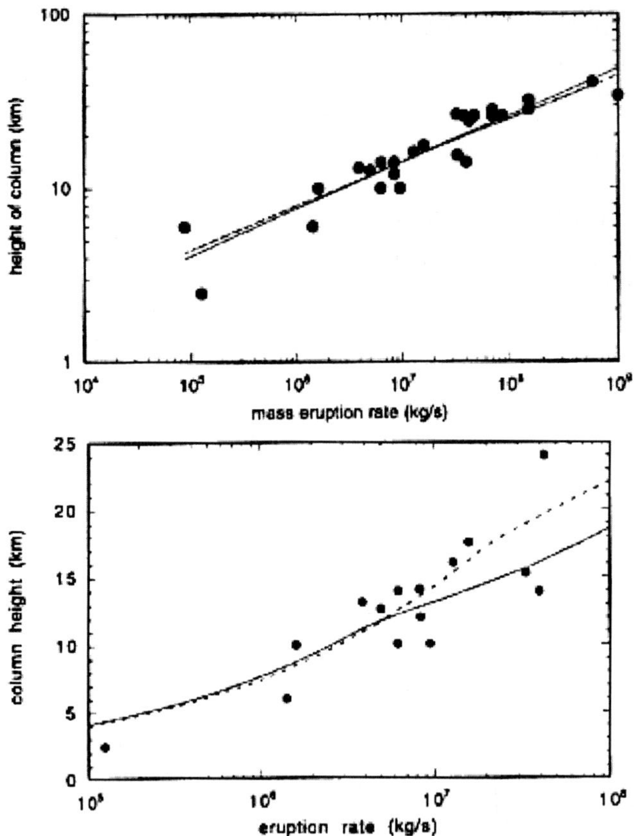

Fig. 8.2. Relationship between the height of rise of the eruption column and the eruption mass flux

The controls on the style and intensity and magnitude of a volcanic eruption are multifold, but perhaps the key parameters are: (i) the dissolved volatile gas content, primarily composed of water, and the viscosity/rheology of the magma; (ii) the geometry of the conduit from the magma reservoir to the surface; (iii) and the depth/size of this reservoir [1]. At the high pressures of a magma reservoir, of order 10^7–10^8 Pa, the volatiles typically remain in solution in the melt. However, as the mixture ascends towards the surface, the pressure falls and volatile gases (primarily water vapour) are exsolved. This decreases the density of the mixture and causes the flow to accelerate towards the surface, owing to the exsolution and expansion of the gas. In addition, the exsolution of volatile gases leads to an increase in the viscosity of the residual magma. These effects set the stage for an explosive eruption of the magma stored in the reservoir. Eruption will occur if the material is able to decompress, with the volatile gases expanding and the mixture accelerating to the surface. In order for this to occur, a conduit or flow path to the surface is required. Once such a conduit has formed, owing to the fracturing of the crust overlying the magma reservoir, an eruption can commence.

Initially, the eruption occurs along a planar fracture or dike which intersects the surface. However, as the eruption develops, the conduit to the surface tends to localise spatially, becoming more cylindrical. This occurs possibly due to thermal or mechanical instability: the faster moving magma remains hot, as there is less time for heat transfer to the cold surrounding walls before the magma reaches the surface, and therefore the continuing flow is able to erode the conduit walls, leading to an increased flux. Any slower moving magma tends to cool more, becomes more viscous and hence the flux decreases locally [2]. The larger the conduit, the greater the flow rate, and hence the more intense the eruption. Mechanical as well as thermal erosion may become important as the eruption continues, and this can lead to an increasing eruption rate with time given a uniform source of magma. However, eruptions also evolve with time owing to changes in the magma properties in the reservoir. The reservoir may be stratified or layered, as for example was the case during the AD79 eruption of Mt. Vesuvius, and the change in magma properties may lead to abrupt changes in the eruption style (and hence nature of the deposits) from column forming to flow forming activity.

Although the processes described in the previous paragraph provide key source conditions which ultimately determine the style of an eruption, here we primarily focus on the dynamics of the erupted material once it issues from the volcanic vent into the atmosphere. However, first we mention the key features of models of magma ascent from a magma reservoir to the surface, in order to illustrate the origin of the source conditions for the models of the flow above the surface. We then move on to describe eruption column dynamics and this naturally leads to the ideas of column collapse and ash flow propagation.

8.2 Ascent of Magma to the Surface

To model the ascent of magma to the surface, we require a simple description of the physical properties of magma. Although magma is a highly complex, multiphase material, we can model the dissolved volatile content of the melt using Henry's Law,

$$n_s = s\,P^{1/2}\;, \tag{8.1}$$

which is a good approximation for silicic magma, where s the solubility coefficient, which is of order $4 \times 10^{-6}\,\mathrm{Pa}^{1/2}$ [9]. The viscosity of the residual melt typically increases by about an order of magnitude for a change in volatile mass fraction of about 0.01, and so we take

$$\mu = \mu_o 10^{100(n_o - n_s)} f(\phi/\phi_o)\;, \tag{8.2}$$

where $n = n_o - n_s$ is the exsolved gas mass, f is a function of the volume fraction ratio ϕ/ϕ_o, and subscript o denotes the initial value in the magma reservoir beneath the conduit. This dependence of the viscosity of the bubble–magma mixture on the bubble volume fraction may be modelled by a law of the form

$$f(\phi/\phi_o) = (1 - \phi/\phi_o)^{-5/2}\;. \tag{8.3}$$

These physical properties are used in a model of flow along the conduit from the reservoir to the surface. The detailed geometry and dimensions of the conduit are poorly constrained from observations, and it is likely to be strongly influenced by both the regional stresses as well as the properties of the rock, which are often heterogeneous, layered and fractured. Therefore, for simplicity, we assume a regular geometry for the conduit (Fig. 8.3), but recognise that this is a rich area for future research. Since the conduit is long, of order 5–10 km, and relatively narrow, of diameter 10–100 m, we expect that the pressure varies primarily along the conduit. In that case, the flow along the conduit, averaged over the cross-section, is governed by a law of the form

$$\rho u \frac{\mathrm{d}u}{\mathrm{d}z} = -\frac{\mathrm{d}p}{\mathrm{d}z} - \rho g - f u\;, \tag{8.4}$$

where

$$f = \frac{\mu}{12 d^2} + \frac{\rho C_d u}{d}\;, \tag{8.5}$$

where f is a friction factor [11] for the conduit flow whose magnitude depends on the Reynolds number of the flow, with typical value of the turbulent drag coefficient being $C_d \approx 0.01$. For low Reynolds number flow, the viscous friction dominates, whereas for higher speed, high Reynolds number flow, which applies primarily after the viscous liquid has disrupted or fragmented, the turbulent drag law is the dominant source of frictional resistance. This dynamical law is coupled with an equation for mass conservation, which accounts for the compressibility of the mixture

$$\rho u A = Q\;, \tag{8.6}$$

and the equation for the density of the mixture

$$\rho = \left(\frac{1-n}{\rho_m} + \frac{nRT}{P} \right)^{-1} , \tag{8.7}$$

where ρ_m is the magma density and R is the gas constant, T the gas temperature and P the pressure, and n is the exsolved gas content, defined in terms of the total mass fraction of volatiles and the mass of volatiles in solution, $n = n_o - n_s$.

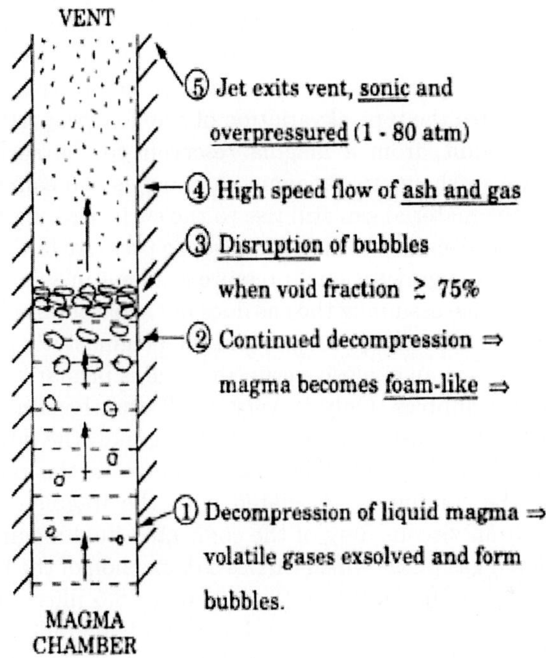
VENT
⑤ Jet exits vent, sonic and overpressured (1 - 80 atm)
④ High speed flow of ash and gas
③ Disruption of bubbles when void fraction ≳ 75%
② Continued decompression ⇒ magma becomes foam-like ⇒
① Decompression of liquid magma ⇒ volatile gases exsolved and form bubbles.
MAGMA CHAMBER

Fig. 8.3. Schematic illustrating the flow along the conduit as used for developing the flow model

In order to solve these equations for a steady flux Q along the conduit, originating from a magma reservoir at known pressure and depth, additional boundary conditions are required at the surface, where the flow issues into the environment. Either the pressure equals atmospheric, and the flow is sub-sonic or the pressure is in excess of atmospheric and the flow is sonic or choked at the vent, with the speed of sound being given by $(\mathrm{d}p/\mathrm{d}\rho)^{1/2}$. In the present context, this speed of sound corresponds to the speed of pressure waves in the isothermal, bubbly mixture, where it is assumed that the compressibility originates from the compression of the gas phase. Choked flow corresponds to the situation in which the flow speed matches this speed of pressure waves.

One complication to this picture is that once the gas volume fraction reaches a critical value, the liquid films of melt surrounding the bubbles begin to disrupt and fracture, rather than flowing as a liquid. Ultimately, the mixture undergoes a transition from being a bubbly liquid, with the liquid being the continuous phase, to being a gas dispersion, with the gas being the continuous phase. At this transition, the viscosity of the mixture decreases dramatically towards that of a particle-laden gas. For simplicity, we take this point of transition to occur when the gas volume fraction has value 0.7. However, there are more complex models of the transition based on arguments about the balance between the fracture strength of the material and the viscous stresses of an expanding bubble, but owing to the complication of the range of bubble sizes present in the mixture, and hence of the critical decompression rate at which liquid films rupture, we take a fixed value of 0.7 in this model. For larger volume fractions, we effectively set $\mu = 0$ in (8.5) which parameterises the total frictional stress.

Figure 8.4 illustrates the typical variation of the flow properties as a function of height in the conduit, from a magma reservoir to the surface. Note that even if the pressure in the source magma chamber is the same as that of the surrounding rock, the material can still rise to the surface owing to the release of bubbles and hence the decrease of the magmastatic pressure in the conduit. Here, magmastatic denotes the gravitational pressure gradient in a column of stagnant magma and exsolved gas, assuming the gas does not separate from the liquid. The difference between the lithostatic pressure and the magmastatic pressure is one of the main driving forces in an explosive eruption, and is primarily associated with the exsolution of gas bubbles. Only in very small eruptions are the conditions at the vent subsonic; for most realistic eruption regimes, the flow is choked and highly overpressured at the vent. The sharp change in the rate of change of properties within the conduit is associated with the fragmentation transition and the evolution from viscous drag of the connected fluid magma, to turbulent drag of the connected gas phase which carries a dispersion of the liquid fragments above the fragmentation horizon (see (8.5)). Figure 8.5 illustrates calculations of the flow rate in a circular conduit as a function of the radius of the conduit, for magma volatile contents in the range 1–5 wt%, corresponding to a mass fraction of 0.01–0.05. The largest known eruptions, with eruption rates of order 10^9–10^{10} kg/s, require conduit radii of about 50–100 m, implying very substantial erosion of the walls of the conduit during the initial stages of the eruption. There is some limited field evidence from exposures of old volcanic conduits which are tens of metres in dimension and which were associated with relatively mild eruptions, which support the order of magnitude of these predictions (e.g. [6]).

Predictions of vent velocity and flow rate enable us to relate the different possible flow regimes above the surface to the different source conditions. Since the flow issues from the vent at an elevated pressure compared to the atmosphere, there is an initial stage of decompression. One important observation is that typically the scale of the flow above the surface is much larger than the length-scale over which the flow decompresses. As a result we can model the decompression of the flow which occurs directly above the surface, using a localised model based on the conservation of mass and momentum. This then gives the actual source conditions for the eruption column model [14].

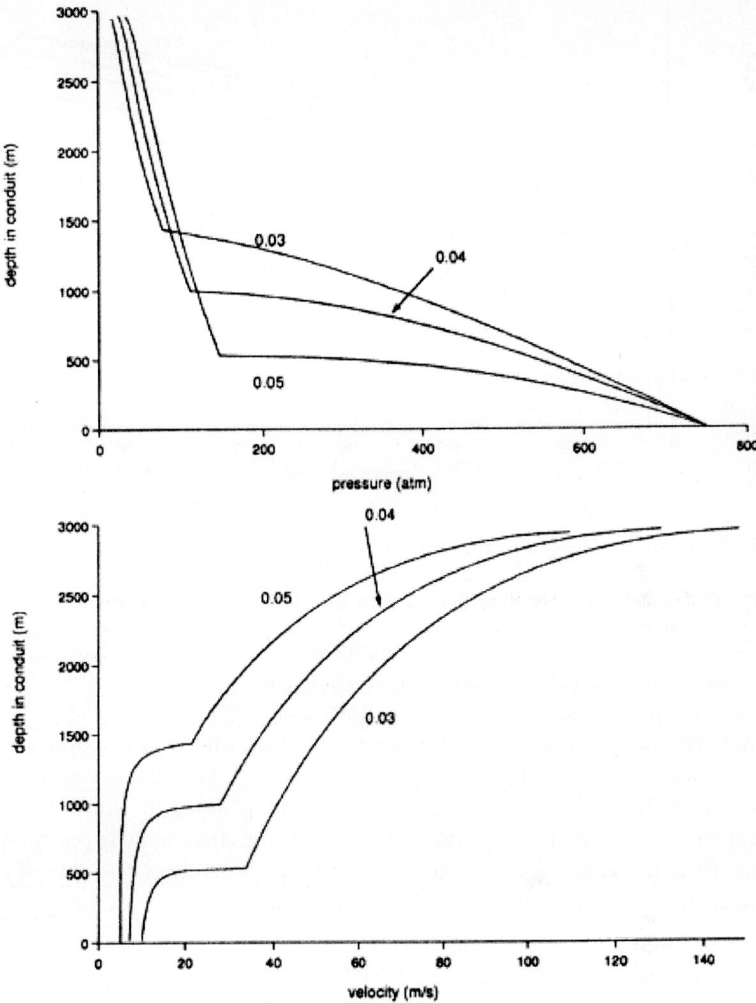

Fig. 8.4. Variation of the velocity and pressure in the ascending mixture of magma and volatiles as it ascends the conduit. Curves are shown for magma with initial dissolved water mass fractions 0.03, 0.04 and 0.05. The chamber is assumed to be 3 km below the surface and the conduit radius is 20 m

8.3 Eruption Column Models

The dense, high speed mixture of ash and gas issuing from the volcanic vent into the atmosphere produces a complex, time-dependent turbulent flow. With flow speeds up to 100–200 m/s, and typical length scales (radii) of the column in the range 100–1000 m, the time-scale for eddy turnover may be as much as 5–10 s. In contrast, the time-scale of an eruption is of order hours or, in the longest cases, even a few days. Therefore, in order to capture the main properties of the

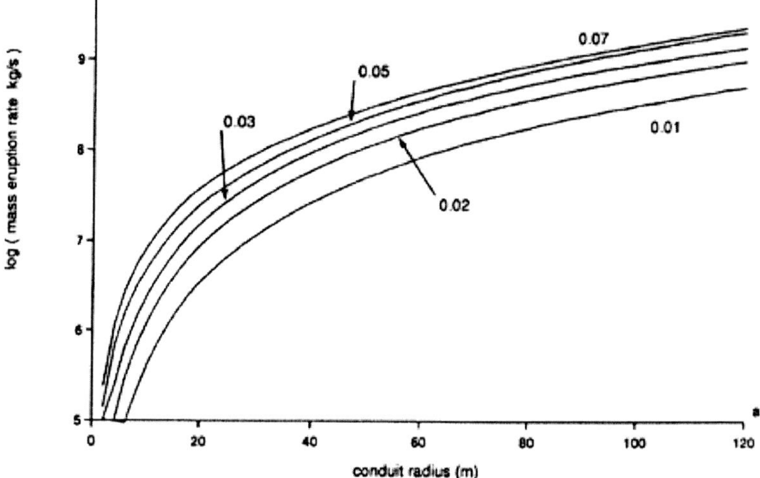

Fig. 8.5. Flow rate at the top of the conduit as a function of the conduit radius. Curves are shown for magma with initial dissolved water mass fractions 0.01, 0.02, 0.03, 0.05 and 0.07. The magma reservoir is assumed to be 3 km below the surface

flow, we average over a time-scale longer than the typical eddy turnover time, and develop a quasi-steady model of the flow, as illustrated in Fig. 8.6. We work with the horizontally averaged properties of the flow, and in particular the mass, momentum and energy fluxes. We denote the average velocity across the plume as $\overline{u}$. We draw from models of turbulent buoyant plumes in which the rate of turbulent mixing between the plume fluid and the environment is parameterised in terms of the mean upward velocity of the plume [8], so that the rate of change of the mass flux with height in the plume is given by

$$\frac{\mathrm{d}Q}{\mathrm{d}z} = 2\pi\,\epsilon\,\overline{u}\,\rho_a b\ , \tag{8.8}$$

where ρ denotes density, with ρ_a the ambient density, $Q = \int_o^\infty 2\pi u r \rho \mathrm{d}r = \pi\overline{\rho}\,\overline{u}\,b^2$ and ϵ is the entrainment coefficient, which has been measured experimentally to have value of order 0.1, and b is the effective radius of the column.

The momentum flux

$$M = \int_0^\infty 2\pi\, u^2 r\rho \mathrm{d}r = \pi b^2 \overline{\rho}\,\overline{u}^2$$

evolves as a result of the buoyancy force on the column

$$\frac{\mathrm{d}M}{\mathrm{d}z} = g(\overline{\rho} - \rho_a)b^2 \tag{8.9}$$

where g is the acceleration due to gravity. Finally, the steady flow energy equation has the form

$$\frac{\mathrm{d}[Q(u^2/2 + gz + C_p\theta)]}{\mathrm{d}z} = 2\pi\epsilon\overline{u}b(gz + C_p\theta_a)\ , \tag{8.10}$$

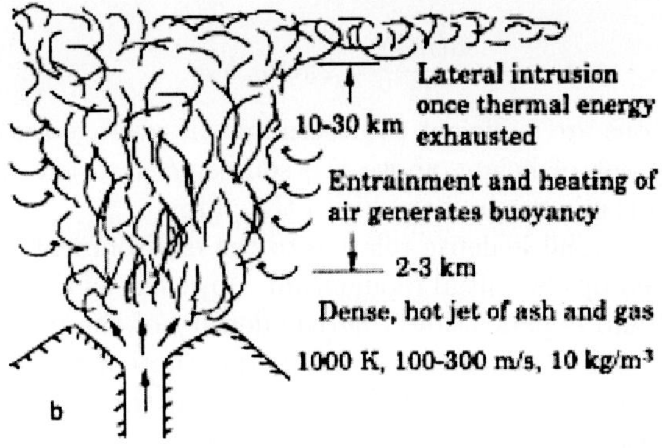

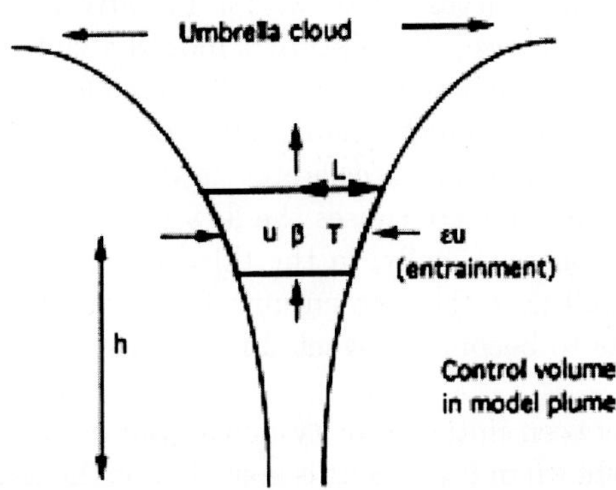

Fig. 8.6. Cartoon of a time-dependent turbulent eruption column and of the time-averaged picture being used in the development of the quantitative model. In the figure the bulk density of the material in the eruption column is denoted as β and this corresponds to $\overline{\rho}$ in the model. The distance above the ground is denoted by h and this corresponds to z in the model, the column radius is denoted as L which corresponds to b in the model

where the term on the right hand side denotes the enthalpy of the material entrained into the column, and θ denotes temperature, with θ_a the ambient temperature, and C_p the specific heat of the mixture at constant pressure. The model is completed by the equation for the density of the mixture,

$$\rho = \left(\frac{nR\theta}{P} + \frac{1-n}{\rho_m} \right) , \tag{8.11}$$

where R is the gas constant for the water vapour, with value 462 J/kg/K, and P is the pressure in the mixture, and an equation for the conservation of the gas flux

$$\frac{\mathrm{d}(nQ)}{\mathrm{d}z} = 2\pi\epsilon\overline{u}\rho_a b\,. \qquad (8.12)$$

These five equations are sufficient to determine the averaged quantities b, $\overline{u}$, $\overline{\rho}$, n and θ as functions of height above the source. R is calculated as the mass average for the volcanic gas and the air in the plume.

Initially, the material is dense relative to the environment, and so the mixture ascends driven by the initial momentum. However, as the flow entrains air, the air is heated and expands, lowering the density of the mixture. Eventually the bulk density of the mixture may fall below that of the air. In this case, a buoyant eruption column develops. In contrast, if the upward momentum of the mixture decreases to zero while the mixture is still dense, then a collapsing fountain develops around the volcano, shedding dense ash flows. This transition in behaviour may be seen in Fig. 8.7 which illustrates the variation of the velocity and density of the mixture as a function of height in the column for three different flow regimes. First, curves are shown for a relatively fast moving jet issuing from the vent. In that case, the mixture is indeed able to entrain sufficient air to become buoyant. In the second case, with a smaller initial velocity, the mixture velocity decreases to a much smaller value before the mixture becomes buoyant. The buoyant material then accelerates upwards for some distance until the stratification of the atmosphere causes the flow to come to rest and intrude laterally into the atmosphere. Finally, in the third calculation of Fig. 8.7, the initial velocity is so small that the momentum of the mixture falls to zero before the material is able to become buoyant. In that case, a collapsing fountain develops.

The relationship between initial velocity and initial mass flux at which column collapse occurs is shown in Fig. 8.8. It is seen that as the mass flux increases, the minimum velocity required for collapse also increases. This is because the entrainment process occurs around the periphery of the column, and as the mass flux increases, the mass of air required to generate a buoyant column increases with the area of the column, while the mass of air entrained into the column only increases with the radius of the column. The figure also illustrates that there is only a relatively weak dependence between the collapse threshold and the volatile content of the magma, owing to the very much larger mass of air entrained into the column once it rises through the atmosphere.

The height of rise of the eruption column, assuming it does become buoyant, depends largely on (i) the initial flux of the energy which is available for conversion to potential energy through entrainment and heating of air, and (ii) the stratification of the atmosphere, which tends to suppress the ascent of the buoyant mixture. Dimensional analysis may be used to determine the scaling for the height of rise of the column, following the classical models of the ascent of a buoyant plume through a stratified environment

$$H = 5B^{1/4}N^{-3/4}\,, \qquad (8.13)$$

where N is the Brunt–Vaiasala frequency of the atmosphere, $N^2 = -g\mathrm{d}(\ln\rho)/\mathrm{d}z$, and where B is the source buoyancy flux defined as $B = Qg(\rho_a - \rho)/\rho_a$. Figure 8.9

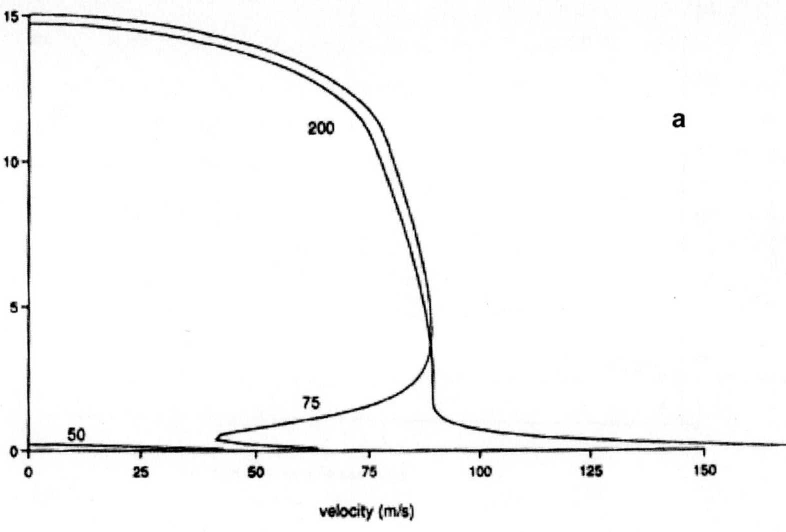

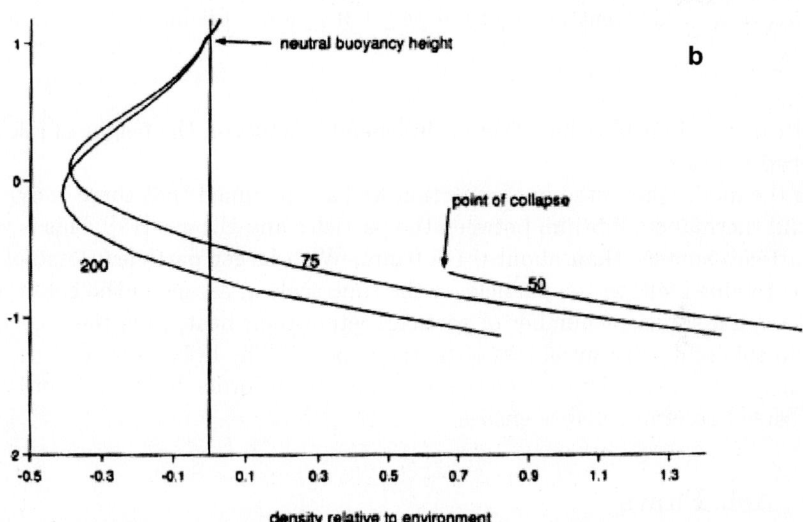

Fig. 8.7. Variation of the flow velocity and density as a function of the height in the eruption column. The three curves refer to three different initial velocities, 50, 75 and 200 m/s for magma with an initial dissolved water content of 3 wt%

compares the prediction of this simple scaling for the height of rise of the column, with the full prediction from the numerical solution of the quantitative model shown above. Also shown on the graph are a series of data points which denote

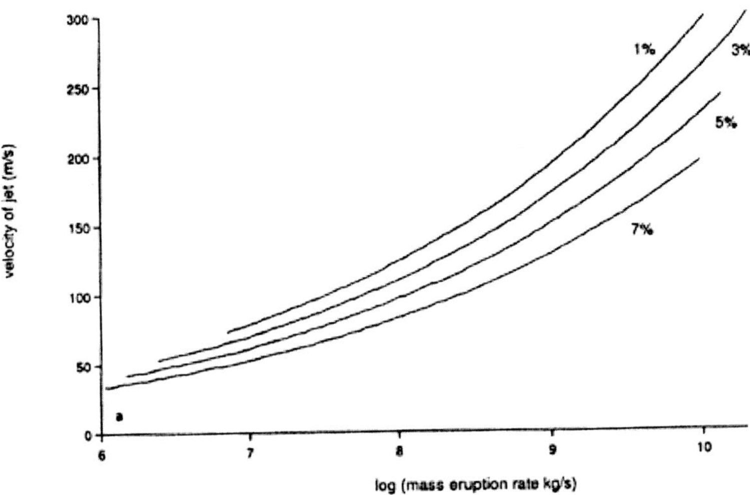

Fig. 8.8. Variation of the critical initial velocity of the erupting material such that the eruption column is just able to develop. Values are shown as a function of the mass eruption rate, for different values of the weight % of water originally dissolved in the magma

historical eruptions for which there is independent data on the height of rise and the eruption rate.

In the model presented in this section we have assumed that there is dynamical and thermal equilibrium between the particles and the gas [15]. This is valid for particles smaller than about 0.1–1.0 mm. With larger particles, some of the heat is retained within the particle on the time-scale of ascent in the column. In this case, if a sufficient number of particles retain their heat, then the ascending column suffers a substantial loss of thermal energy. The bulk density of the column may not fall below that of the environment on order to become buoyant, and instead column collapse ensues.

8.4 Ash Flows

We can develop a similar model to describe the propagation of ash flows. We assume the mixture propagates as a dilute suspension of particles and gas, which is sufficiently thin that the variation in pressure over the depth of the flow is small relative to the hydrostat. As a simplification we assume that the flow is highly turbulent and use depth averaged quantities, which is equivalent to assuming implicitly that the turbulent fluctuations are sufficiently rapid to mix the flow. We allow for sedimentation from the base of the flow using the Einstein sedimentation model, in which we assume that particles fall into the base of the flow and are not subsequently resuspended. Thus the rate of loss of particles

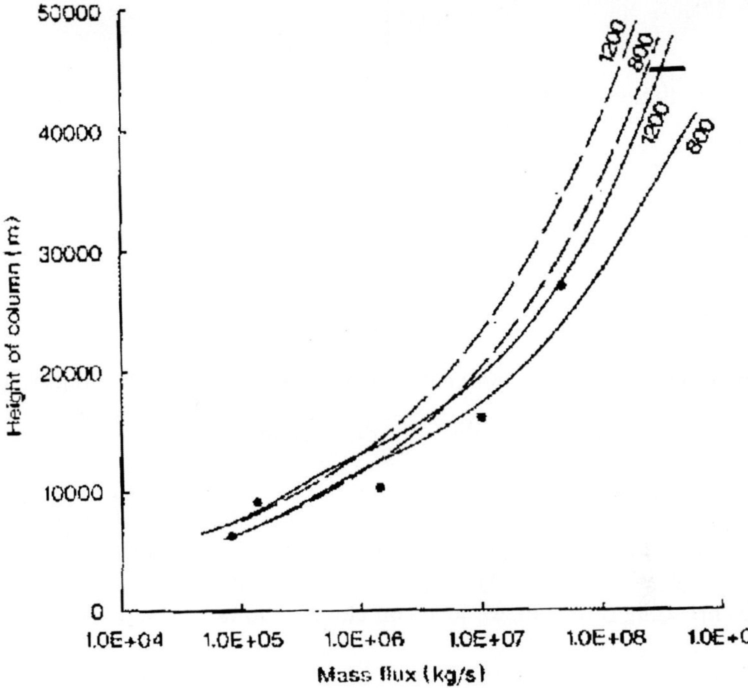

Fig. 8.9. Height of rise of an eruption column as a function of the eruption mass flux, as calculated by the theoretical model for both a mid-latitude (*solid*) and tropical atmosphere (*dotted*). The mid-latitude standard atmosphere has the tropopause at 11 km while the tropical atmosphere has the tropopause at 15 km. Solid circles denote historical data for comparison

is proportional to the fall speed divided by the depth of the flow times the concentration of the flow. As well as sedimenting particles, such flows are likely to entrain air as they propagate. Since the flow is horizontal, any mixing with the air is limited by the potential energy required to movde the less dense ambient downwards into the flow. This is expressed in terms of the Richardson number of the flow, $Ri = (\rho - \rho_a)hg/\rho u^2$, with a fast, thin flow having more energy for entrainment of the relatively light overlying air than a deep, slow flow.

Given these physical processes, a vertically averaged model of the motion of a steady, isothermal, one-dimensional ash flow in a channel (Fig. 8.10) follows the mass conservation relation

$$\frac{\mathrm{d}(\rho u h)}{\mathrm{d}x} = \epsilon(Ri)\rho_a u - v_s C \ , \tag{8.14}$$

where C is the particle volume concentration, h the flow depth, v_s the particle fall speed and u the current speed. Also, the particle flux conservation satisfies the relation

$$\frac{\mathrm{d}(\rho u h C)}{\mathrm{d}x} = -v_s C \,. \tag{8.15}$$

Also the gas flux conservation has the form

$$\frac{\mathrm{d}(n\rho u h)}{\mathrm{d}x} = \epsilon(Ri)\rho_a u \,, \tag{8.16}$$

where the gas mass fraction in the flow, n is given by

$$n = \left(1 - \frac{\rho_s C}{\rho}\right) . \tag{8.17}$$

On near horizontal terrain, momentum conservation has the leading order form

$$\rho u \frac{\mathrm{d}u}{\mathrm{d}x} = -g \frac{\mathrm{d}(\rho - \rho_a)h}{\mathrm{d}x} \,. \tag{8.18}$$

These equations may be solved, together with an empirical relation for the entrainment coefficient as a function of the Richardson number [3], to describe the motion of ash flows [3]. Analogous equations may also be formulated for a radially spreading flow, but we focus on channel flows in the present chapter.

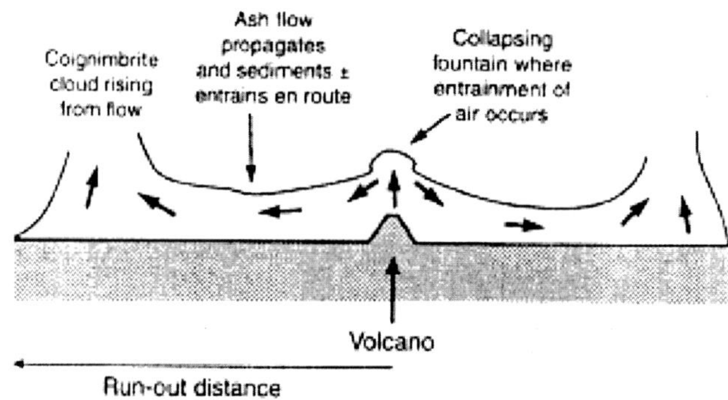

Fig. 8.10. Schematic of a one-dimensional ash flow as used in the model developed in Sect. 8.4

Numerical solutions of the equations are shown in Fig. 8.11, illustrating the typical form of propagation of an ash flow. In these solutions, we impose the initial velocity and depth of the current, as well as the initial gas mass fraction and temperature. Two different kinds of behaviour may be recognised depending on whether the initial velocity of the current is sub or supercritical relative to the speed of gravity waves, $(g'h)^{1/2}$. The supercritical currents have relatively

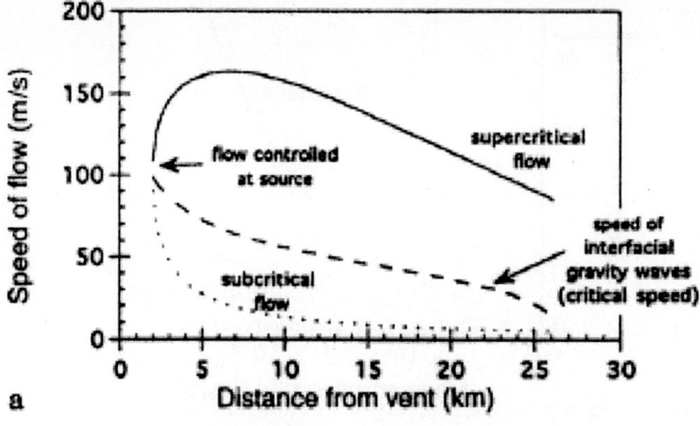

Fig. 8.11. Variation of the velocity of an ash flow as a function of the initial conditions. Curves may be seen for flows which are both subcritical and supercritical

small Ri and hence entrain air at a significant rate. As a result, the flow becomes progressively less dense and eventually the density equals that of the environment at which point the current lifts off forming an ash cloud. For sub-critical flows, there is much less entrainment of air, and the currents propagate further before becoming less dense than the air through sedimentation of particles. Figure 8.12 illustrates the difference in runout distance of both sub and supercritical currents as a function of the eruption flux, for currents with mean particle size 3 mm. The particle size is important in this calculation as it controls the fall speed of particles. It is interesting to note that the initial stages of a flow, or the propagation of a shortlived flow may also be modelled using a similar approach, except that an additional boundary condition at the head of the current is then also needed [5,13].

This model of ash flow propagation provides a plausible physical explanation of the tremendous mobility of large ash flows, and also their ability to scale substantial topographic obstacles far from the source of the flow. The latter effect is most easily understood by noting that as the flow propagates and sediments particles, it becomes progressively less dense. As a result of this decreasing density, less work against gravity is required for the flow to scale topography of given elevation. Figure 8.13 illustrates the height of a ridge which a typical flow can scale as a function of distance from the source [16], adapted from classical ideas of hydraulics, but including the evolution of the flow density with distance from the source.

8.5 Analogue Laboratory Models

The models of the physics which control some of the key physical processes involved in explosive volcanic eruptions may be tested using analogue laboratory

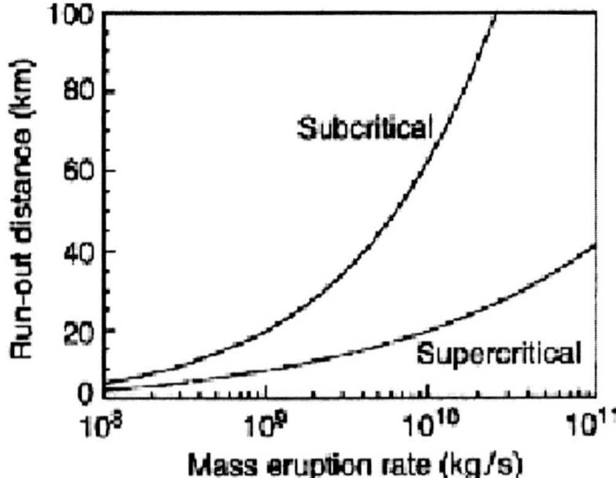

Fig. 8.12. Variation of total run-out distance as a function of the mass eruption rate for both sub and super critical ash flows. The difference in run-out arises owing to the difference in entrainment rate between sub and supercritical flows. In both cases, the mean particle size was set to be 1 mm, and the mass fraction of water in the magma was 0.01

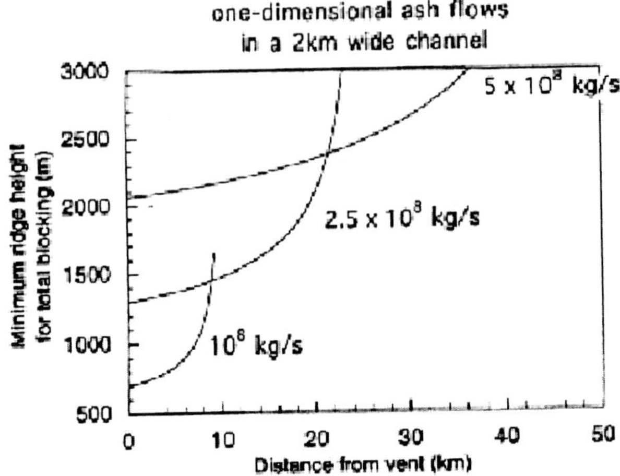

Fig. 8.13. Height of a long ridge in the path of the flow which an ash flow may scale, calculated from the channel ash flow model, assuming that the ridge inclination is small. The different curves refer to eruptions with the different initial mass fluxes of 10^8, 2.5×10^8 and 5×10^8 kg/s

experiments. Here we review some of the experimental work aimed at testing models of eruption column formation and collapse, and also models of ash flow propagation and the associated particle sedimentation.

In order to simulate the dynamics of an eruption column in which the buoyancy of a plume evolves with height and may even change sign, two approaches may be adopted. First, plumes of fresh, particle laden water may be injected into a tank of saline water, to simulate the effects of particle fallout. Second, plumes of MEG (Methanol and ethylene glycol) may be used – on mixing with water, the density of methanol varies in a non-linear fashion, and so the density of a MEG plume may change sign relative to that of the ambient water. This may be used to simulate changes in density which arise owing to the heating of the entrained air.

With particle laden plumes, a series of phenomena may arise as the density of the injected fluid changes as a result of particle sedimentation. If a relatively buoyant mixture of fresh water and particles is injected at the base of the tank, this will rise through the ambient fluid to form a plume. The plume may then rise to the top of the tank, where it spreads laterally and sediments its particles as in an eruption column. However, as the plume rises through the tank, it entrains the saline water, and therefore at some height its density may in fact fall below that of the ambient fluid, owing to the presence of particles. As this happens, the plume fluid will slow down and come to rest, with particles then falling back to the base of the reservoir. This leads to an increase in the buoyancy of the residual, relatively fresh plume fluid, which is then able to rise upwards, until the continuing flow has carried sufficient particles upwards that the motion is again arrested. In this way, an oscillatory motion develops from a steady source of particle-laden fresh water ([13]; Fig. 8.14). In contrast, by injecting a relatively dense mixture of particles and fresh water at the base of the tank, a collapsing fountain forms, and spreads laterally around the source. However, as the flow spreads and particles sediment from the flow, the residual fresh-water particle suspension becomes less dense than the overlying fresh water, and separates from the base of the tank to form a so-called coignimbrite ash cloud above the flow [4].

With MEG plumes, the density of the MEG increases on mixing with water, and so an initially buoyant mixture may become dense on mixing with water. To simulate a volcanic eruption, a relatively buoyant jet of MEG is injected downwards from a source at the top of a tank. As this mixes, it may then become dense and forms a descending plume, thereby simulating the transition in buoyancy of an eruption column. However, if the initial momentum of the plume is too small, then the motion will be arrested before the plume becomes dense, and in that case, the flow will rise to the surface of the tank, and spread laterally. As it spreads, it continues mixing and will eventually become denser than the water in the tank. At this stage it separates from the upper boundary and sinks through the fluid as with a coignimbrite ash cloud [17]. The critical momentum flux required to produce a buoyant plume, rather than the inverted collapsing fountain may be predicted using a model which is analogous to that of section 2, but replacing the steady flow energy equation with a relation which specifies the MEG mass fraction in the plume. This may then be coupled with a model of the density as a function of MEG content. Together, these relations lead to the prediction of the critical flow rate for plume formation rather than

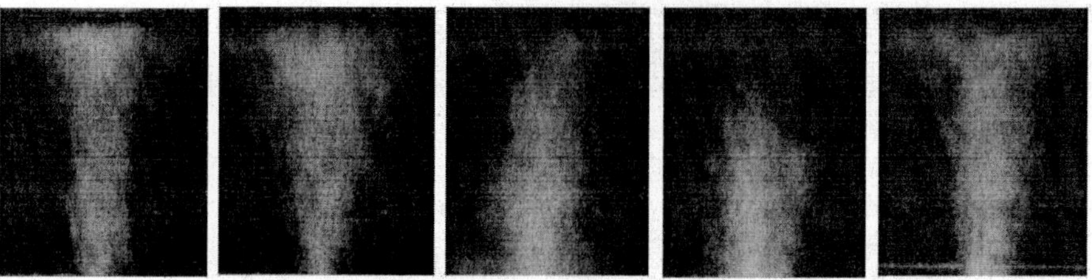

Fig. 8.14. Series of photographs which illustrate the oscillation of a particle-laden plume of fresh water rising through a tank of saline water

fountain collapse. Figure 8.15 illustrates the comparison between the model and theory, which supports the modelling approach used herein (see [17]).

In order to model the motion of ash flows experimentally, a similar series of experiments may be performed, using dense, particle laden suspensions of water, but now advancing along the lower boundary of an experimental tank. The deposition of particles from such a flow may be determined by measuring the deposit in the experimental system. For a steady flow, the deposit is expected to thin exponentially with distance from the source, and this has been confirmed (e.g. [16]).

8.6 CO_2-Charged Lake Eruptions

One closely related process is the overturn of CO_2-charged crater lakes, such as occured at Lake Nyos in 1986. As CO_2 percolates through the crust from degassing magma and into deep confined lake water, the CO_2 may become dissolved in the water, since the solubility of CO_2 in water increases with pressure. As a result, the lake becomes charged with CO_2. If the water at the base of the lake becomes saturated with CO_2, then any further input of CO_2 will lead to exsolution of bubbles and formation of an ascending plume of bubbles. This ascending plume will entrain lake water, which is carried upwards, decompresses and may itself then become saturated. If this occurs, then more bubbles are released from the water, and a non-linear runaway process ensues, with a turbulent bubble plume being produced. However, if the water higher in the lake is unsaturated in dissolved CO_2, then the CO_2 gas may dissolve back into the water, thereby removing the source of buoyancy in the flow. This may lead to the arrest of the upward motion of the plume. The flow may be modelled in a similar fashion to that of an eruption column, but the thermodynamic model for the density of the plume is replaced with a model of the conservation of CO_2 in the plume and an expression for the density of the bubble-water mixture. This varies with bubble content and gas density. Woods and Phillips [18] have modelled this process using these conservation laws and find that for sufficiently small mass fluxes of CO_2 charging the base of the lake, then the motion is arrested and the ascending plume of CO_2 intrudes into the lower part of the lake. However, for larger fluxes of CO_2, the plume becomes more vigorous and is able to ascend to the lake surface, leading to a new degassing of CO_2 from the lake.

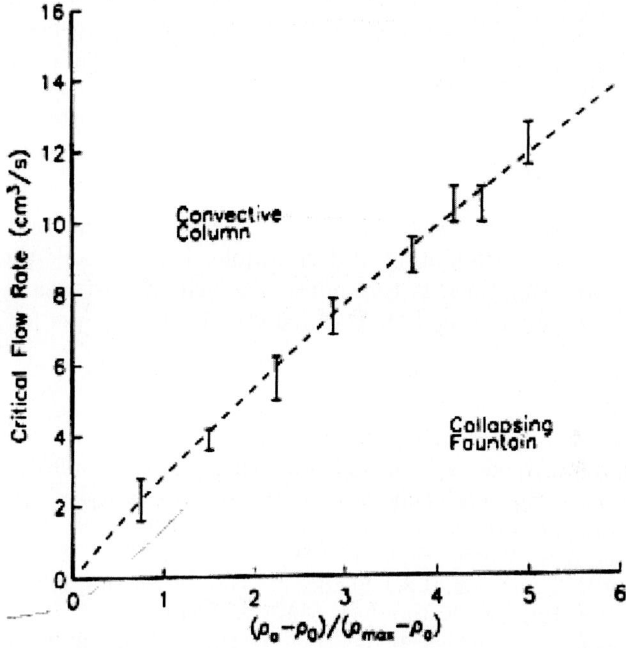

Fig. 8.15. Comparison of the experimentally measured and theoretical prediction of the critical flow rate above which a dense plume develops in the MEG experiments. The critical flow rate is shown as a function of the magnitude of the initial negative buoyancy of the source fluid

8.7 Discussion

In these lectures we have developed some simple modelling approaches to capture the dominant dynamics of explosive volcanic eruptions, focussing on the dispersal of ash and fragmented magma on the Earth's surface. The modelling has identified conditions under which collapsing fountains develop to form high speed, turbulent ash flows, and under which turbulent buoyant plumes develop and rise several tens of kilometres into the atmosphere. We have also shown how dimensional scaling arguments lead to accurate predictions of the height of rise of volcanic columns. A series of analogue laboratory models have also been discussed, and used to test and validate the theoretical models. These phenomena can have an important impact on the local topography around a volcano, especially in extremely massive eruptions which can deposit 10^{14}–10^{15} kg of material onto the Earth's surface. The models described for ash flows indicate that such massive eruptions may disperse the erupting material over distances of order 100 km, with deposits tens of metres deep near the source, thereby have a major geomorphological impact. Air-fall deposits from towering eruption columns lead to a more widely dispersed ash blanket extending over 100's or even 1000's of kilometres. Although this may have less impact on the surface morphology, it has a more immediate impact on terrain, especially through agriculture and building stability.

Acknowledgements

I would like to thank Neil and Antonello for arranging this summer school, which was extremely stimulating and enjoyable. I also thank them for their patience while awaiting these lecture notes. The help of Costanza is also greatly appreciated, especially in preparing the final version of the paper for publication.

References

1. S. Bower, A.W. Woods: J. Volc. and Geoth. Res. **73**, 19–32 (1996)
2. P. Bruce, H.E. Huppert: 'Solidification and melting in dykes'. In: *Ryan, MP, Magma Transport and Storage* (Wiley, New York 1989) pp. 87–101
3. M.I. Bursik, A.W. Woods: Bull. Volc. **58**, 175–193 (1996)
4. S.N. Carey, H. Sigurdsson, R.S.J. Sparks: J. Geophys. Res. **93**, 15314–15328 (1988)
5. B. Dade, H.E. Huppert: Nature **381**, 509–512 (1996)
6. J. Eichelberger, C. Carrigan, H. Westrich, R. Price: Nature **323**, 598–602 (1986)
7. P. Lipman, D. Mullineaux: USGS Prof. Pap. 1250. The 1980 eruptions of Mount St Helens, Washington, USA (1981)
8. B. Morton, G.I. Taylor, J.S. Turner: Proc. Roy. Soc. A256, 1–23 (1956)
9. R.S.J. Sparks, M.I. Bursik, S.N. Carey, J. Gilbert, J. Glaze, H. Sigurdsson, A.W. Woods: *Volcanic Plumes* (Wiley, New York 1997)
10. G. Veitch, A.W. Woods: J. Geophys. Res. **105**, 2829–2842 (2000)
11. C.N.J. Wilson: Phil. Trans. Roy. Soc. Lond. A314, 229–310 (1985)
12. L. Wilson, R.S.J. Sparks, G.P.L. Walker: Geophy. J. Roy. Astron. Soc. **63**, 117–148 (1980)
13. A.W. Woods: Phil. Trans. Roy. Soc. Lond. A358, 1705–1724 (2000)
14. A.W. Woods, S.N. Bower: Earth Plan. Sci. Lett. **131**, 189–205 (1995)
15. A.W. Woods, M.I. Bursik: Bull. Volcanol. **53**, 559–570 (1991)
16. A.W. Woods, M.I. Bursik, A. Kurbatov: Bull. Volc. **60**, 38–51 (1998)
17. A.W. Woods, C.P. Caulfield: J. Geophys. Res. **97**, 6699–6712 (1992)
18. A.W. Woods, J.C. Phillips: J. Volc. Geoth. Res. **92**, 259–270 (1999)

Part III

Cold

9 The Dynamics of Snow and Ice Masses

J.S. Wettlaufer

Applied Physics Laboratory and Department of Physics, University of Washington, Seattle, WA 98105-5640, USA

9.1 Ice: Land, Sea and Air

On Earth today we enjoy a relatively comfortable climate, which is a fortunate consequence of the present extent of the global ice cover. Although more than two-thirds of the surface of Earth is covered by water, it is the water to ice conversion, and vice versa, that makes an important fraction of the globe habitable today. Hence, changes in the global scale dynamics of the ice cover capture scientific and public interest principally because of their role in global warming and ice-age events. It is in this sense that ice is the ultimate geomorphological fluid mechanic.

Field observations [1] and modeling studies [19] of past and "future" climates teach us that the ice cover is an extremely sensitive geophysical variable. Among other things, the eccentricity, obliquity and precession index of Earth's orbit, the optical depth of the atmosphere, and the storage of heat in the oceans underlie the present tropical-to-polar difference in mean surface temperature of approximately 50° C. Because water freezes near the middle of this range, the suggestion of advancing or retreating ice extent is not hard to grasp. Indeed, our contemporary polar oceans undergo dramatic seasonal variations in their sea-ice covers, amounting to approximately 18 million square kilometers in the Antarctic and 8 million square kilometers in the Arctic, where a perennial ice cover persists. The swift ice streams of West Antarctica are believed to modulate sea-level by influencing the storage of relatively slow inland (upstream) ice [2]. These contemporary observations give strength to the notion of rapid ice motion with consequences for all of Earth's inhabitants [3].

Most of what we study concerning the dynamics of the present ice cover involves our interest in understanding how, and how fast, circumstances might change. We study the past, as far back as 420,000 years, principally through the analysis of ice cores from the great ice sheets [1], for they trap in their polycrystalline matrix particulate and chemical clues concerning the history of the state of the Earth's past environments. Deriving a truly quantitative understanding of these environments constitutes a challenging inverse problem, for a host of post-depositional dynamical processes can act to redistribute climate proxies.

An ice sheet is maintained by the deposition of snow on its surface. The nucleation and growth of snow in the atmosphere occurs under chemical and dynamical conditions that mirror important aspects of climate. Our common experience tells us that a meter or so of snow can form a relatively loose aggregate of granular material, a fact in strong evidence during an avalanche. As snow accumulates on an ice sheet or glacier it is compressed into "firn", which is less dense than ice and more dense than snow. In this layer, typically tens of meters thick, some atmospheric gases are displaced by the advection of seasonal meltwater or vapor transport through the connected network of air pockets that separate individual ice grains. Eventually, deeper down, the air pockets are sealed and the interfaces between the grains confine impurities. The persistence of unfrozen liquid separating these grain boundaries is a basic aspect of the phase behaviour of all polycrystalline material called *premelting* [24], [25]. But at the cold temperatures that persist near the surface of the polar ice sheets, such liquid is likely to be present only in small quantities. Hence, because of the extremely slow solid-state diffusion through single ice grains, the impurities are normally considered to be "frozen" in place. However, at higher temperatures and solute concentrations, premelted liquid at grain boundaries provides an alternative route for diffusive transport.

A proper accounting of the role of premelted liquid in the bulk diffusive properties of ice reveals that the interaction between compositional diffusion and the phase relationships determining the fraction of unfrozen liquid causes advection of the bulk-compositional signal towards warmer regions while maintaining its spatial integrity [17]. We can illustrate the basic effect of how the migration of such a climate signal evolves by modeling the diffusion of a single impurity such as H_2SO_4. What theory predicts is that, under conditions representative of those encountered in the Eemian interglacial ices of central Greenland, impurity fluctuations may be separated from ice of the same age by as much as half a meter, which is a distance comparable to the thickness of the apparent sudden-cooling events detected in Eemian ices from the GRIP core [17]. Moreover, when considering the premelting-enhanced diffusion of two species, we find that features of their evolution can mimic what has been ascribed to irreversible chemical reactions. The theory should help guide the analysis of existing and incipient deep ice cores. To the uninitiated it may seem paradoxical that a quantitiative understanding of the past global climate hinges on an understanding of the basic microscopics of ice, but it an unavoidable fact, and one that studies of future climates must also come to grips with.

The future is dealt with principally through the speculative viewing glass of prognostic global and regional models, which are initialized using various aspects of the present and past record. Although such models emerge out of the rostrum of geophysical fluid dynamics, they sacrifice the "rigor" of process studies to construct "realism" on the large scale. Large scale models incorporate a plethora of approaches which include various degrees of "realistic physics", in order to study the sensitivity of a prediction to particularly well known feedbacks. Because of the sensitivity of the polar regions in model simulations and the recent changes in the Arctic climate [11], [18], air/sea/ice interactions are at the forefront of efforts to understand how the past climate has and can influence the future. During approximately the last decade a dramatic weakening of the central Arctic basin sea-level pressure, coupled with the European subarctic low pressure cell, ascribed to a positive phase of the North Atlantic Oscillation [6], has driven changes in polar atmospheric circulation [23]. These phenomena are part of a larger scale circulation pattern, called the Arctic Oscillation [22], with intraseasonal, interannual, and interdecadal time scales, interpreted as modulations in the strength of the polar vortex. Over the same time scales there have been large scale increases in surface atmospheric temperature and the temperature and salinity of the upper Arctic Ocean, the areal extent of sea ice has decreased in the Arctic and increased in the Antarctic, and large areas of permafrost have thawed [12].

These observed phenomena are difficult to model with entirely prognostic methods that do not restore forcing to deep-ocean climatology and/or observed atmospheric temperatures (e.g. [27]). The thermodynamic state of the atmosphere and the ocean are more readily observed than that of the sea ice that separates the two, and yet it is the change in the area of sea ice that drives ice-albedo feedback. Sea ice grows and melts in response to atmospheric and oceanic changes, it is redistributed by wind stress at its surface and it undergoes deformation in which ice of one thickness becomes ice of another [20]. The first thermodynamic models of sea ice developed for climate studies were one-dimensional and hence avoided the complication of deformation [10]. It was recognized then that the two-phase, two-component nature of the sea ice matrix, through its influence on the thermophysical properties of the layer, strongly influenced the agreement between model predictions and field observations. We now understand that a complete treatment of the thermodynamic properties of undeformed sea ice involves taking account of both diffusion and convection *within* the layer [26], and yet one of the principal simplifications of the original thermodynamic sea ice model [10] is to ascribe the same constant value of salinity to all ice thicknesses no matter what their particular history. Moreover, present day numerical models are not able to make an accurate accounting of the space-time variation in the distribution of sea ice thickness, but progress is being made [27], and when the observed trends and changes can be predicted reliably, the future may reveal itself. An enduring question concerns just which ostensibly small scale thermodynamic phenomena can be approximated to achieve a practical, computable scheme for making such predictions. Increasing computational power is not always the solution to the problem, for we are often in a position to make a "simplification" with the aim of enhancing computability, when in fact "simplification" is simply a synonym for ignorance of the underlying process. It is only years later, after that simplification becomes part of the fabric of the modeling, that the physics is rooted out.

There are other ways of predicting change. So called low-order models can focus our attention on the dominant balances in a system thereby providing insight on issues such as how sea ice responds to changes in poleward heat transport or ocean temperature [21]. Some variations, such as wind speed, temperature and atmospheric CO_2 content are predicted to be fast, whereas others, such as ice mass, are believed to respond more slowly, indicating a separation of time scales in the climate system [19]. Such a separation is again distinguished in studies of ice variations on daily, seasonal, decadal, millennial,...periods. The spatial extent of geophysical ice forms, and the characteristic length scales of processes controlling them, also provide natural divisions of ice research.

9.2 Ice Flow: As Clear as Mud

The essence of modern dynamical glaciology emerged out of basic considerations of the flow of ice as a problem in plasticity theory [14], [13], and we now understand that the thermal history of a parcel of glacier ice can be influenced by diffusion, advection and strain heating, and the temperature itself can modify chemical and isotopic signatures. Ice flows under its own weight and, because it builds up on, and creates, irregular landforms, understanding the dynamics of large snow and ice masses constitutes a formidable task [15]. Even along ice divides, which are the preferred drilling sites for ice cores because of their relatively simple flow pattern, reconstructing thermal and mechanical states is ultimately a computational undertaking [9].

In the present day, ice movements constitute one of the most important geomorphological sculptors in colder regions of the Earth, and in the past they were responsible for the landscape we presently observe in vast portions of the globe that are ice free. The considerations employed to great advantage in understanding sediment dynamics in rivers, lava and mud flow, dune formation, and the dynamics of ice masses, constitute a unifying theme for this volume, tying together these seemingly distinct flow processes that shape features of the Earth's surface.

It is the aspect ratio of the important features of these diverse flows that underlies their mathematical similarities. This is displayed in the generally shallow nature of glaciers and ice sheets, with depth to width/length ratios of about 10^{-3} [15], [7], [4]. A kind of minimal model of a glacier is depicted in Fig. 9.1. The glacier has thickness h, much smaller than its length and width, overlying

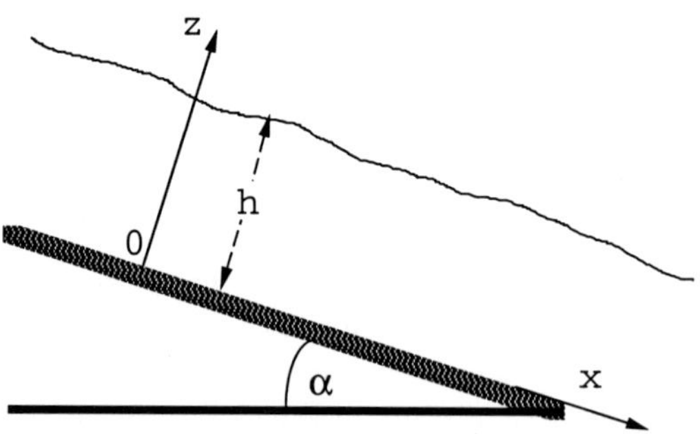

Fig. 9.1. Schematic of a parallel-sided slab [15]

a sticky inclined bed of slope α. In real glaciers, there is a difference between the surface and the bed slopes, but the essential behavior is not changed when this difference is small, except near margins, where the assumption tends to fail. The aspect ratio allows the neglect of longitudinal stresses, an assumption which breaks down in the vicinity of ice divides where the surface slope and basal shear stress vanish.

Traditionally, when employing these assumptions, dynamical glaciologists have referred to this treatment as the "parallel-sided slab" model (e.g. [15]), but as these phenomena drew other scientists into the field it became known as the *shallow-ice approximation* in analogy with the widely used *shallow-water equations* of geophysical fluid dynamics [16]. In the latter case, the essential notion is that the average fluid depth, which is the length scale characterizing the vertical scales of motion, is much smaller than the scale over which horizontal motions are important. Additionally, this approximation assumes that the density is constant in the layer and that the fluid is incompressible thereby decoupling the dynamics from the thermodynamics. The qualitatively similar approach of the shallow-ice approximation is used to study the surface profiles of ice sheets, to detect ice thickness changes and to reconstruct past ice sheets.

Ice sheets are rarely in steady state and hence mass conservation demands that

$$h_t + q_x = \mathcal{S} + \mathcal{B}\,, \tag{9.1}$$

where q is the ice flux and $\mathcal{S}$ and $\mathcal{B}$ are the mass per unit area per unit time being added or removed from the surface and bed of the glacier. Typically, the mass balance is dominated by the surface term. As in the case of *lubrication theory* of thin viscous flows, or in the evolution of dunes, the specific dynamics of the system resides in an understanding of the driving fluxes. Ice is often treated as a "viscous" fluid, but in fact it is a non-newtonian material with a power law rheology which for simple shear takes the form

$$\dot{\epsilon}_{xz} = A{\tau_{xz}}^3\,, \tag{9.2}$$

wherein $\dot{\epsilon}_{xz}$ is the strain rate and τ_{xz} is the stress, A is an Arrhenius factor modelling the temperature dependent creep of ice (treated as a constant here), and the exponent 3 derives from Glen's experiment [5]. The ice flows under its own weight with a component of force parallel to the bed $\rho g h \sin\alpha$ which is balanced by the resisting stress at the bed. At a given depth in the ice the shear stress is $\tau_{xz} = \rho g(h-z)\sin\alpha$ and therefore, because $2\,\dot{\epsilon}_{xz} \equiv u_z + w_x$ and $w_x \approx 0$, we have

$$u(z) = u_s - \frac{A}{2}(\rho g h \sin\alpha)^3 (h-z)^4\,, \tag{9.3}$$

where u_s is the surface velocity. Because the ice is frozen to the bed, the ice flux q can be written in terms of the depth averaged velocity $\bar{u}$ as

$$q = h\bar{u} = \frac{2A}{5}(\rho g h \sin\alpha)^3 h^5\,, \tag{9.4}$$

and hence, neglecting transverse strain rate, the simplest shallow-ice model can be written

$$h_t + (h\bar{u})_x = \mathcal{S} + \mathcal{B}\,, \tag{9.5}$$

describing how the longitudinal strain rate varies with mass balance and thinning or thickening; a sudden increase in snow fall in the accumulation zone must result is a change in the ice thickness and a flux down glacier. Such simple models show that the dynamics of glaciers and ice sheets have the signature of climate variations encoded in them. Other examples wherein models of the form of (9.5) arise appear throughout the volume; resulting from the combination of conservation laws and shallow configurations.

9.3 Drumlins, Glaciers, Icebergs and Avalanches

In the section of the volume that follows, the authors discuss processes that span many of the space and time scales of active interest. Kolumban Hutter focuses on theories used to describe the dynamics of ice sheets and shelves such as the shallow-ice approximation. Such approaches develop the asymptotic limits that provide insight into the essential dynamics and often simplify the calculational efforts required to make long term predictions. When an ice sheet retreats it leaves geomorphological clues of its past existence in a region. One such clue takes the form of a long ridge, or oval-shaped hill, called a "drumlin" by glaciologists. Andrew Fowler presents a theory of drumlin formation in which these landforms emerge out of an instability coupling the interaction between the ice flow and the properties of the till below it. He also employs a shallow-ice methodology. Much of the hype associated with global warming is fueled by the periodic calving of gigantic icebergs from Antarctic ice shelves. An active area of present research concerns the armada of icebergs believed to have been calved from the Laurentide Ice Sheet during the last deglaciation [8]. The sinking of the Titanic focused our attention on the practical aspects of icebergs – their drift and deterioration – which is the topic of the chapter by Stuart Savage. A deeply aesthetic aspect of the hydrological cycle is the freezing of atmospheric water vapor to form snowflakes. The microscopics of this commonplace event have a plethora of macroscopic consequences [24] ranging from stratospheric processes to the ski slopes. Avalanche disasters in the Alps have a long history and yet a quantitative understanding is in its infancy. Much of the research, as described by Christophe Ancey, focuses on empirical correlations and modeling the dynamics like that of a debris flow. A great deal of damage can be mitigated, but in the long term a reliable predictive methodology is necessary.

As the present day snow and ice masses wax and wane, we think of the oscillations and their stability. Could an understanding of the qualitative mechanisms associated with the bifurcations of past climates, analogous to the patterns emerging out of convection in a layer of fluid, be sufficient to make progress in understanding the future? Nevertheless, the present is peppered with ice problems of pressing importance; problems suggested by studies of the past, the need to understand practical issues in the present and to make our best guess at what is ahead of us.

Acknowledgements

Support for my participation in the school from the US National Science Foundation, the *Istituto di Cosmogeofisica* – CNR (Torino, Italy), and the Groupement de Recherche *Mécanique Fondamentale des Fluides Géophysiques et Astrophysiques* (CNRS, France), is gratefully acknowledged. Richard Alley, Alan Rempel, George Veronis, Ed Waddington and Grae Worster kindly read or discussed various versions of this overview and helped to improve it. Caustic comments by the organizers of the school made my participation all the more enjoyable.

References

1. R.B. Alley: Proc. Nat. Acad. Sci. **97**, 1331 (2000)
2. S. Anandakrishnan, D.D. Blankenship, R.B. Alley, P.L. Stoffa: Nature **394**, 62 (1998)
3. M.I. Budyko: *Climate and Life* (Academic Press, New York 1974)
4. A.C. Fowler: *Mathematical models in the applied sciences* (C. U. P., Cambridge 1997)
5. J.W. Glen: Proc. Roy. Soc. A **207**, 519 (1955)
6. J.W. Hurrell: Science **269**, 676 (1995)
7. K. Hutter: Theoretical Glaciology (D. Reidel, Dordrecht)
8. D.R. MacAyeal: Paleoceanography **8**, 775 (1993)
9. S.J. Marshall, K.M. Cuffey: Earth Planet. Sci. Lett. **179**, 73 (2000)
10. G.A. Maykut, N. Untersteiner: J. Geophys. Res. **76**, 1550 (1971)
11. M.G. McPhee, T.P. Stanton, J.H. Morison, D.G. Martinson: Geophys. Res. Lett. **25**, 1720 (1998)
12. J.H. Morison, K. Aagaard, M. Steele: Arctic **53**, 359 (2000)
13. J.F. Nye: Proc. Roy. Soc. Lond. **207**, 554 (1951)
14. E. Orowon: J. Glaciol. **1**, 231 (1949)
15. W.S.B. Paterson: *The Physics of Glaciers*, 3rd ed. (Pergamon, Oxford 1994)
16. J. Pedlosky: *Geophysical Fluid Dynamics*, 2nd ed. (Springer, New York 1987)
17. A.W. Rempel, E.D. Waddington, J.S. Wettlaufer, M.G. Worster: Nature **411**, 568 (2001)
18. D.A. Rothrock, Y. Yu, G.A. Maykut: Geophys. Res. Lett. **26**, 3469 (1999)
19. B. Saltzman, H. Hu, R.J. Oglesby: Dyn. Atmos. & Oceans **27**, 619 (1998)
20. A.S. Thorndike: J. Geophys. Res. **97**, 12601 (1992)
21. A.S. Thorndike: in *Ice Physics and the Natural Environment* NATO ASI, Series 1, Vol. 56 (ed. by J.S. Wettlaufer, J.G. Dash, N. Untersteiner) 169–184 (Springer–Verlag, Berlin 1999)
22. D.W.J. Thompson, J.M. Wallace: Geophys. Res. Lett. **25**, 1297 (1998)
23. J.E. Walsh, W.L. Chapman, T.L. Shy: J. Climate **9**, 480 (1996)
24. J.S. Wettlaufer: Phil. Trans. Roy. Soc. A **357**, 3403 (1999)
25. J.S. Wettlaufer, J.G. Dash: Sci. American **282**, 56 (2000)
26. J.S. Wettlaufer, M.G. Worster, H.E. Huppert: J. Geophys. Res. **105**, 1123 (2000)
27. J. Zhang, D.A. Rothrock, M. Steele: J. Climate **13**, 3099 (2000)

10 Response of Italian Glaciers to Climatic Variations

A. Biancotti and M. Motta

Dipartimento di Scienze della Terra, Università di Torino, Italy

10.1 Introduction

The glaciers of the southern alpine slope have been investigated using a variety of different methods. The most common ones are:

a) glacial inventories, which consist of an overall description, the geographical location, the classification, and the area and volumetric measurements. Four such inventories, all focussed on the Italian Alps, were carried out in 1925, 1958, 1976 and in 1989 [1,2]. The first one is the result of the consultation of I.G.M. topographic maps, the second comes from measurements in the field, and the latter two come from aerophotogrammetric observations. The data base for the last one was provided by Italy Flight 1988;

b) annual glacial campaigns, during which snout variations of the ablating tongue are measured. The operation, carried out by land surveys, refers to a sample of about 15% of the total population of glaciers, and it includes almost all the glacial bodies of major dimension and importance [3];

c) mass balances, with the calculation of the volumetric variation of the glacier in time and its role in the local climatic evolution. Only a few sample glaciers, located in different parts of the Alpine range, are taken into consideration.

10.2 Glacial Inventories

The inventory of 1989 (Table 10.1) includes 706 glacial bodies with a surface larger than 0.05 km^2 (a dimension limit recommended by the *World Glaciers Inventory*). Figure 10.1 shows the distribution according to the historical division of the Alps. In addition to these bodies, there are others with a surface area less than 0.05 km^2, mostly of the type called *glacionevato*,[1] and a collection of glaciers that were extinguished during the past 100 years, but whose history is known. From the comparison with the inventory of 1958, one concludes that the number of minor and extinguished glaciers is increasing, while that of glaciers with a surface area exceeding 0.05 km^2 is decreasing. Moreover, the total glacial area has shrunk by 8.2%. It should be noted that the comparison is only suggestive, given the different methods adopted for the two inventories.

[1] *A glacionevato* is "a more or less homogeneous and compact mass, formed by snow and/or ice, of different extension and form, that lasts for two or more years and is not in motion" [4]. It differs from a glacier because the latter moves slowly [5].

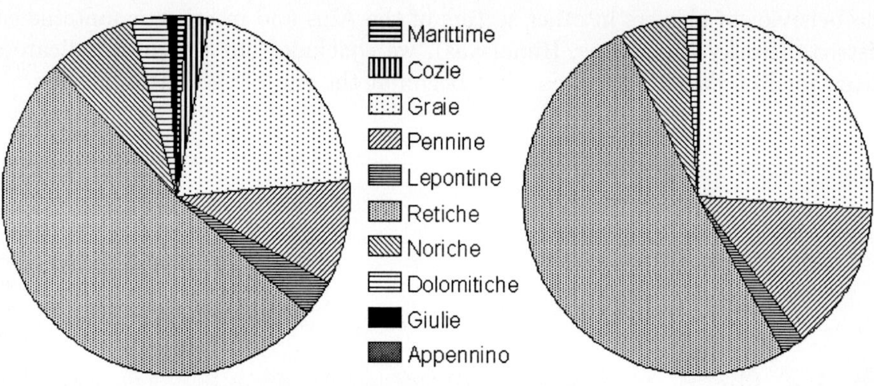

Fig. 10.1. Percent distribution of glaciers in the historical segments of the Italian Alps, according to number (*left*) and surface (*right*). Data from the inventory of 1989

Table 10.1. Variation of the number and surface of glacial bodies in the Italian Alps. The comparison refers to the data of the inventory of 1967–68 (*normal typeface*) and to those of the inventory of 1989 (**bold**)

Glacial bodies	Exting-uished	Surface less than 0.05 km^2	Surface exceeding 0.05 km^2	Total	Total area in km^2
Western Alps	41 → **90**	15 → **36**	307 → **272**	363 → **398**	237.6 → **201.1**
Central Alps	108 → **117**	35 → **28**	220 → **243**	363 → **388**	161.0 → **163.3**
Eastern Alps	52 → **100**	40 → **36**	217 → **191**	309 → **327**	126.4 → **117.5**
Apennines	–	0 → **1**	1 → **0**	1 → **1**	0.06 → **0.02**
Total	201 → **307**	90 → **101**	745 → **706**	1036 → **1114**	525.0 → **481.9**

10.3 Glacial Campaigns

The graph in Fig. 10.2 shows the glacial snout variations, expressed as a percentage of advancing glaciers, for the sample measured from 1925 to 1998. The series contains an interruption of four years, from 1943 to 1946. The data show that, until 1965, only about 10% of the measured glaciers were advancing, while the great majority of glacial bodies showed a regression. Between 1966 and 1980 there were large annual variations, but in general the percentage of advancing glaciers increased significantly, including about 80% of the observed bodies. In the last twenty years, there was again an increase of the number of regressing glacial tongues, that led to values comparable with those measured in the period before 1965. Based on these measurements, and on the comparison with the behavior of glaciers in other sectors of the Alps and in other mountaineous districts (Rocky Mountains, Himalayas), we conclude that the data indicate a clear trend of glacial regression, that began at the end of the Little Ice Age.

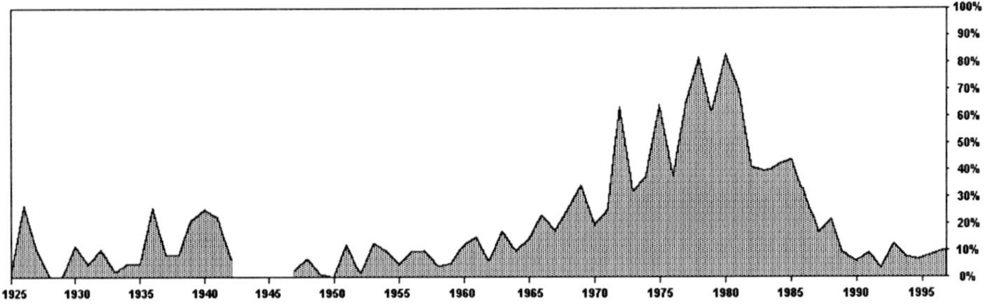

Fig. 10.2. Percentage of advancing glaciers from 1925 to 1997 (data from [3])

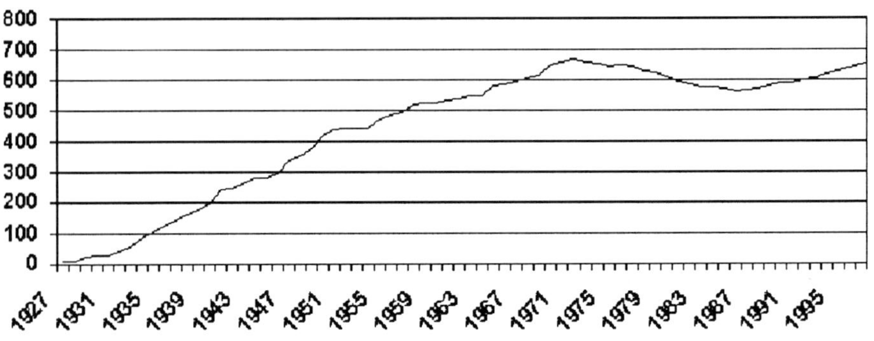

Fig. 10.3. Annual average snout variations of the Lys glacier from 1927 to 1997. On the y-axis is the distance (in m) from the initial position. The only significant advance was in the years 1972–1986

The only exception to this trend was an advance that lasted from the beginning of the 1970's until the middle of the 1980's. This exceptional behaviour is well illustrated by the snout variations of two glaciers: the Lys glacier in the Aosta Valley (Fig. 10.3), and the northern glacier of Locce in the Ossola Valley (Table 10.2).

In Figs. 10.4 and 10.5 we show the snout variations in the period 1991–1997 of some of the most typical Italian glaciers. An overall regression is evident. This regression, which is drawn as though it were a continuous smooth curve, can in reality be either regular or irregular, as when the tongue is subject to big

Table 10.2. Snout variations of the Northern Glacier of Locce (data analysed by Mazza, [6])

Northern Gl. of Locce	
1915–1934	About −200 m
1934–1968	About −250 m
1968–1979	Stationary?
1979–1982	About +30 m
1982–1985	Advance?
1985–1992	About −20 m
1992–1993	−7 m
1993–1994	−4 m
1994–1995	+0.5 m
1995–1996	−7 m
1996–1997	−6 m

and sudden fall phenomena. The latter situation occurred for the Forni glacier (Fig. 10.5), whose snout, in the hydrologic years 1993–94, suddenly regressed by about one hundred meters. This regression is five times larger than that observed for any other glacier. The upper glacier of Coolidge behaved similarly: it collapsed almost completely, on the northern side of the Monviso mountain, on July 6, 1989 [7]. With the persisting tendency to regression, the probability of an increase of traumatic collapse events in the Alps grows. As a consequence, there is also an increase in the risk of accidents in the defrosted areas, which are more and more frequently visited by tourists.

Snout variations can be used as climatic indicators only if they are compared with one another, and are analysed in such a way as to eliminate irregularities due to variations in local conditions. To compensate for such irregularities, it is possible to express the trend of snout variations by using standardized time series, in which the raw data, X_t, are replaced by normalized data, $X'_t = (X_t - \hat{X})/\sigma$, where $\hat{X}$ is the mean of the signal X_t and σ^2 is its variance. With this normalization (also called standardization), the data sets for different glaciers become more homogeneous, although there still remain intrinsic variations due to differences in altitude and orientation of the individual glacial bodies. Clearly, the standardized data indicate more accurately the actual climate variations in the area under study. However, some anomalies persist (Fig. 10.6), such as the ones of the Money and Tribolazione glaciers (signal BV1, central branch), which are due to irregularities of these specific glaciers.

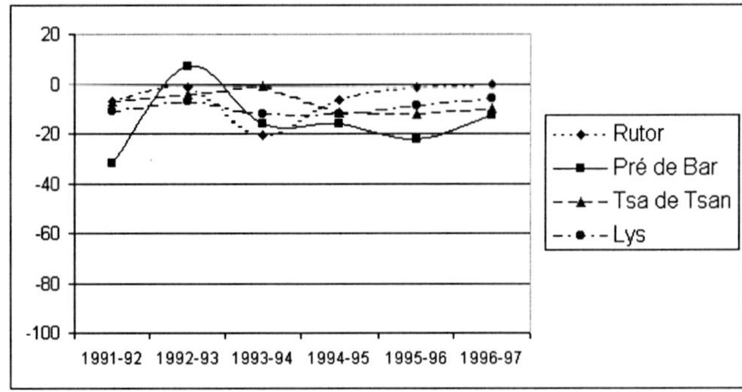

Fig. 10.4. Relative annual snout variations from 1991 to 1997, measured in meters, for four large valley glaciers in the Aosta Valley

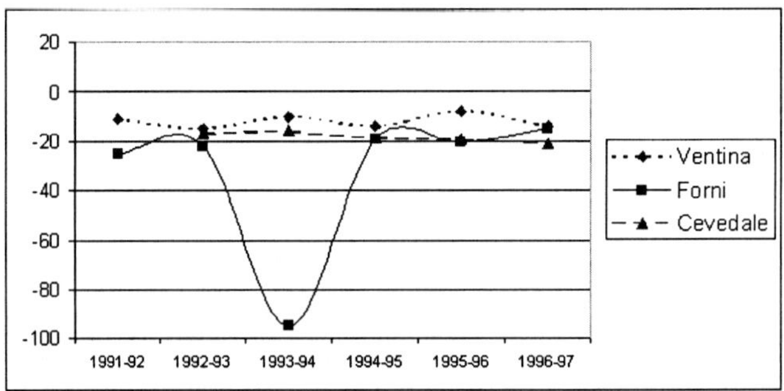

Fig. 10.5. Relative annual snout variations from 1991 to 1997, expressed in meters, for three large valley glaciers in the Central Alps

Other exceptional behaviours were noticed in glaciers near to extinction.[2] Many small glaciers of the Italian Alps (at least 106 from 1959 to 1989, as shown from the comparison of the corresponding inventories), have extinguished. The process has often consisted of the arrest of the ice flux, caused by the reduction of the glacial body because of climatic changes. To be considered a glacier, a body of ice must necessarily be moving; however, when its dimensions are too small, the weight of the overhanging mass is not enough to trigger the ice flux, the body stops and must be considered extinguished as a glacier. In the years immediately before extinction, snouts have been observed to advance relatively quickly because melting leads to a reduction in the friction over the underlying bed, and so the whole mass slides towards the valley (Fig. 10.7). It is obvious that the snout advance recorded in such a case must not be understood as an index in favour of glacial advance.

[2] A glacier is extinguished when it is not "a large, slowly moving accumulation of ice and firn resulting from consolidation and transformation of atmospheric precipitation provided its perennial balance is positive" [5].

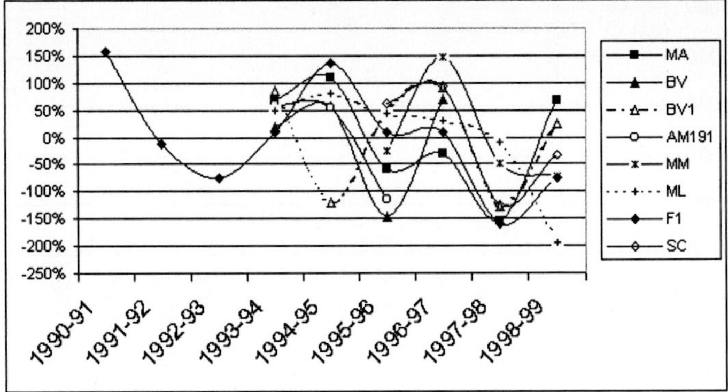

Fig. 10.6. Trend of snout variation of the head glaciers in Valnontey (Aosta Valley) expressed through a series of standardised variables. Measures surveyed at signals MA (Gl. of Grand Croux), BV and BV1 (Gl. of Tribolazione), AM191 (Gl. of Coupè of Money), MM (Gl. of Dzasset), ML (Gl. of Money), F1 and SC (Gl. of Lauson)

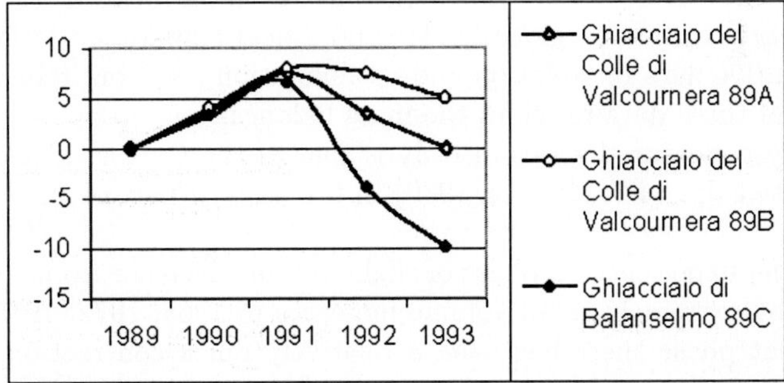

Fig. 10.7. Accumulated snout variations of the Colle of Valcournera and Fontanella glaciers, expressed in meters, just before their transformation to motionless dead ice masses. It is worth highlighting the glacial advance recorded in the years 1989–1991, before the regression that has led to the extinction of these two glaciers in 1993. The advance has occurred by a shift of the whole mass, and it was the same for the two signals of the Valcournera glacier. The signal 89A was measured in front of the central part of the glacier snout of Colle of Valcournera, and the signal 89B was obtained in front of the hydrographic left side of the glacier. The signal 89C was obtained in front of the left side of the snout of the Balanselmo glacier

These advances, that take place just before the interruption of the glacial flux, do not occur only for whole glacial bodies, turning them from glaciers into *glacionevati*, but also occur for the snout of glacial tongues that move independently (from a dynamical point of view) from the main mass of an active glacier [8]. This process explains some of the data indicating positive snout variations for glaciers undergoing a significant contraction phase, but not yet extinguished.

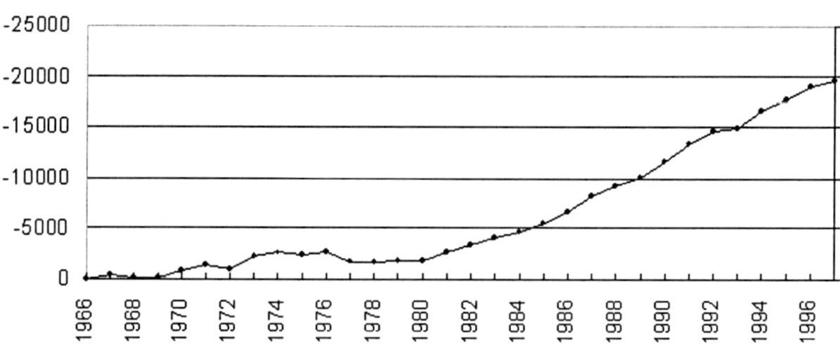

Fig. 10.8. Net glacial balance, in mm of water equivalent (WE), calculated for the Careser glacier. Data reported by Zanon [9] and included in glacial campaigns for the Triveneto sector

10.4 Mass Balances

As previously discussed, snout variations have a climatic value on alpine scale only if they are considered globally. To study local climatic variations, the annual quantitative data on ablation and accumulation are more reliable, and the comparison of these data provides the mass balance.

The longest measured series in Italy is that of the Careser glacier, a glacial body of limited size (4.241 km^2 in 1989) in the eastern part of the alpine range (Fig. 10.8).

This glacier experienced a phase of slight volume decrease from 1966 to 1980, interrupted by extremely small volume increases in 1968, 1972, 1975 and 1977. Following that phase there has been a relatively quick contraction, due to an average increase in the altitude of persisting snow, which has been above the maximum height of the glacier during the most recent summers [9].

The biggest glacial body (10.195 km^2 in 1989) for which the annual mass balance has been measured is the Lys glacier (Table 10.3). The data of this balance appear to be consistent with the measurements of snout variations of the same glacial body, although the variation of volume seems to change from year to year much more than the snout position.

10.5 Morphologic Variations Associated with the Regression

A glacier does not represent an isolated body in the scenery of high alpine valleys. On the contrary, it is the centre around which the dynamics of the overhanging sides develop as well as the centre of the territory used by humans. Significant glacial variations cause noticeable changes in the landscape. This can be seen for example in the glaciers of Val Ferret, such as the Triolet glacier, located in a deep valley descending from the Grandes Jorasses (4206 m above sea level). The outlet of the deep valley is obstructed by the high snout and the side moraines, which

Table 10.3. Mass balance of the Lys glacier (data from Michele Motta, unpublished)

	Accumulation	Ablation	Net balance
1993	$2.5 \cdot 10^9$ kg	$3.9 \cdot 10^9$ kg	$-1.4 \cdot 10^9$ kg
	212 ± 42 mm WE	330 ± 42 mm WE	-118 mm WE
1994	–	$7.6 \cdot 10^9$ kg	–
		682 ± 50 mm WE	–
1995	$4.0 \cdot 10^9$ kg	$3.5 \cdot 10^9$ kg	$+0.5 \cdot 10^9$ kg
	367 ± 183 mm WE	323 ± 5 mm WE	$+44$ mm WE
1996	$9.8 \cdot 10^9$ kg	$3.6 \cdot 10^9$ kg	$+6,2 \cdot 10^9$ kg
	900 ± 46 mm WE	334 ± 4 mm WE	$+566$ mm WE
1997	$2.7 \cdot 10^9$ kg	$16.0 \cdot 10^9$ kg	$-13.2 \cdot 10^9$ kg
	249 ± 46 mm WE	1465 ± 4 mm WE	-1216 mm WE

were built during the advance in the Little Ice Age (mid nineteenth Century). As in almost every alpine valley, these structures represent the more evident morphological elements. At its bottom, this deep valley is characterized by glacial moulding and it has been, for long time, subject only to the action of other mountain geomorphological processes (running waters, landslides, avalanches). On the top, the glacier has shrunk over the past few decades, and its moulding action appears to be just finished. The vegetation is scarce or absent, and steep piles of debris prevail, where masses of "dead ice," isolated and currently motionless reminders of the glacier, are common. The continuous settling of these unstable areas makes them difficult to manage and impossible to use for human beings. These areas do indeed represent a significant source of danger for the valleys below: Heavy rains can trigger landslide phenomena of such dimension to reach the built-up areas. In the last decade, events of this kind often happened in the valley of the Gura Glacier, just over the built-up area of Forno Alpi Graie, Italy. Dangerous collapses can happen also in active glaciers. In the case of Triolet glacier, for example, the withdrawing snout hangs unstably on the upper edge of steep rocky walls.

Less evident but possibly deeper changes affect the mountain sides over the glacier. Once mainly snowy, these areas are now constituted by bare rock that suffers from frost-defrost cycles contributing to its progressive disintegration. This leads to a significant increase in fragment and block detachments from the mountains overhanging the glacier. These changes indicate the need for a complete revision of the use of the glacial areas by tourists and mountaineers. The old itineraries to the top, that followed channels or snowy slopes and were both less steep and easier to climb, are becoming increasingly dangerous and often impassable. The indications given by old route descriptions and the lack of updated information can therefore increase the frequency of accidents. In the Triolet basin, for example, the local tour operators (guides, hut managers) have been forced to completely revise the trail maps, advising against old itineraries and arranging for new routes entirely on rock.

10.6 Conclusions

The data discussed here confirm the current persistence of a defrost phase that started around 1980, and has involved the whole alpine range. Local climatic variations do not change significantly this picture, as they would under less critical conditions. Most of the positive snout variations recorded during the current period of defrost are due to peculiarities of the glacial dynamics and do not derive from local microclimatic variations.

The phase of local advance or stationarity, observed for several glacial snouts from 1966 to 1980, and confirmed by the mass balance of the Careser glacier, represents the only interruption to the global negative trend that was recorded since the end of the nineteenth Century and that restarted in 1980, continuing until today.

More in detail, the snout withdrawal of the Lys Glacier (Fig. 10.3) measured during the last few years appears to be weaker now than before 1966, suggesting that the current phase of defrost could be less intense than the one recorded before the interruption of 1966–1980. Other data, however (such as the percentage of advancing snouts), display a weaker negative trend in the first half of the twentieth Century than in the last twenty years. It is therefore reasonable to conclude that the current phase of defrost is a natural continuation of that recorded from the end of the nineteenth Century to the years around 1966.

References

1. F. Porro: Boll. C.A.I., XLII, n. 75, 309–322, Torino (1925)
2. C.N.R.-C.G.I. *Catasto dei ghiacciai italiani.* 4 volumes, C.G.I., Torino (1959–1961)
3. The reports of the **annual glacial campaigns** by the Comitato Glaciologico Italiano were published in the *Bollettino del Comitato Glaciologico Italiano*, printed in Torino by the same institution. This magazine was also called, starting from the third series, "Geografia Fisica e Dinamica Quaternaria"
4. F. Secchieri: Geogr. Fis. Din. Quat. **8** (2), 156–165, Torino (1985)
5. V.M. Kotlyakov, N.A. Smolyarova: *Elsevier's Dictionary of Glaciology* (Elsevier, Amsterdam 1990)
6. A. Mazza: Geogr. Fis. Din. Quat. **21**, 233–244, Torino (1998)
7. F. Dutto, F. Godone, G. Mortara: Rev. Géogr. Alp. **79**, 7–18, Lyon (1991)
8. L. Motta, M. Motta: Annalen der Met. **30**, 295–298, Offenbach am Main (1994)
9. G. Zanon: Geogr. Fis. Din. Quat. **15**, 215–219, Torino (1992)

11 Asymptotic Theories of Ice Sheets and Ice Shelves

D.R. Baral and K. Hutter

Institute of Mechanics, Darmstadt University of Technology, Hochschulstr. 1, 64289 Darmstadt, Germany

11.1 Introduction

11.1.1 Motivation

In climate dynamics of the Globe the atmosphere, hydrosphere and cryosphere interplay with one another with various different time scales, typically from years to several millennia. Ice sheets and ice shelves, which are the grounded and floating components of the large ice masses such as Greenland and Antarctica and the former Fennoscandinavian and Laurentide ice sheets are those subsystems of the geosphere, which respond to and interplay with climate variations with periods of 10^3 to 10^5 years. 100000 years ago the amount of water bound in solid ice was so large that the ocean surface was about 120–150 m below its present level; alternatively, the complete melting of the Greenland ice sheet or Antarctica under a future Greenhouse scenario would raise the ocean surface by approximately 7 and 65 m, respectively. Because the socio-economic impact of the sea level rise due to an increase of the mean temperature of the Earth's surface is immense, it is absolutely vital that the nourishment and wastage of the large ice masses are properly understood and transformed into sea level status. This requires careful computation of the flow, phase change mechanisms as well as geometric evolution of such ice masses.

Ice shelves, because they are floating, do not contribute to sea level rise when disintegrating, but they hold the inland ice in its position; this stabilising mechanism is lost when the ice shelves melt away (which one presently believes might happen catastrophically in Antarctica). Much more inland ice will likely flow thereby into the ocean and then contribute to sea level rise, another reason why ice sheet – ice shelf dynamics is so crucial in future climate scenarios.

There is, however, also a further perhaps more technical reason why ice sheet flow needs to be computed very accurately. This is dating of ice in ice sheets at large depths. Ice cores have been and are being drilled both in Greenland (GRIP, GISP, NORTH-GRIP) and Antarctica (VOSTOK, DOME C, EPICA) for reconstruction of the past climate over 500000 years from CO^2, O^2, D^2 etc. isotope compositions trapped in the ice specimens of the cores. The climate reconstruction requires knowledge of the depth-ice age correlation within such ice cores amounting to determining the trajectories traced by the ice particles through time.

Geometrically, ice sheets and ice shelves are shallow objects, with large horizontal extent as compared to their thickness. Likewise, horizontal velocities are basically large and vertical velocity components much smaller, and topographic variations in the horizontal extent are small. This suggests to introduce an aspect ratio $\epsilon =$ "typical thickness to typical horizontal extent" as a scaling parameter and to seek the mathematical description in terms of perturbation expansions of ϵ . The lowest order approximations, in which all terms of order ϵ^m, $m \geq 1$ are dropped in the governing equations, are called the *shallow ice approximation* (SIA) for ice sheets and the *shallow shelf approximation* (SSA) for ice shelves. These are zeroth order approximations and possess their inadequacies – they are poor approximations e.g. exactly where ice cores are best drilled and therefore must be improved, as we shall see to second order, $\mathcal{O}(\epsilon^2)$, for ice sheets and to first order for ice shelves, $\mathcal{O}(\epsilon)$, making the determination of the flow, temperature, geometry and ice age as functions of space and time a challenging problem of mathematical glaciology.

11.1.2 A Descriptive View of Ice Sheet and Ice Shelf Flows

Ice sheets are large ice masses resting on the solid Earth; they are thus bounded by the atmosphere along the free surface and by the rock bed on which they rest, see Fig. 11.1. They grow or shrink according to as mass is added by solid precipitation or subtracted by melting (and evaporation), and their thermal regime is governed by the geothermal heat provided by the interior of the Earth and the surface temperature described by the atmospheric temperature at the surface of the ice sheet. The flow of such an ice sheet is much like the flow of honey on the breakfast plate, only slower. Physically an essential difference is that ice sheets may gain and loose mass from above or below so that steady state configurations do exist if the accumulated mass equals the ablated mass, whereas this is not so with the honey on the breakfast plate. The flow in such a very viscous fluid is driven by gravity, and strong vertical shear develops, that vertical profiles of horizontal velocities have belly-type shape with strong shear close to the rockbed and almost none at the free surface. Of special significance are the margin neighbourhoods, where the free surface touches ground, the free surface and the vicinity of the summit, a dome or ice divide. It is plausible to suppose that the flow is towards the sides from the position of the dome, roughly in the direction of steepest descent of the free surface. This can easily be demonstrated by experiment with honey on the breakfast plate. Thus at the dome, where the tangent plane to the free surface is horizontal, the horizontal velocity components must vanish, and there can only be a vertical velocity. The dome location separates the ice flow from one direction to the other and therefore is a true ice divide.

This behaviour is essentially described by the SIA mentioned above, i.e. the zeroth order approximation in the aspect ratio ϵ of the thermomechanically coupled dynamical equations of creeping flow of ice under the external action of gravity. In fluid mechanics this approximation is also known as the "thin film theory"; analysis shows that the SIA is capable of modelling ice sheet flow adequately everywhere except in three local regions, in a marginal region, in a

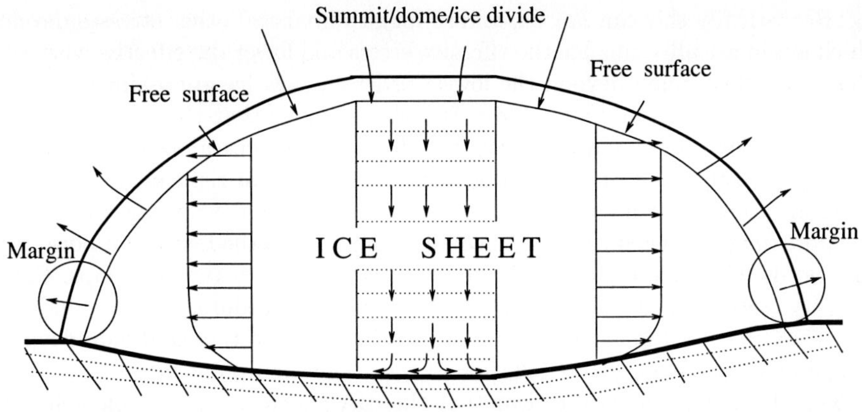

Fig. 11.1. Sketch of an ice sheet with vertical scale exaggerated 100–1000 times. The arrows at the free surface symbolize the accumulation and ablation of ice mass via precipitation and melting. The ice velocity in the vicinity of a dome or ice divide is primarily vertical with a stagnation point at the base. Away from domes it is largely horizontal with large shear at the base and practically none close to the surface

boundary layer close to the free surface and in the vicinity of ice divides. Unfortunately, ice domes are the preferred locations where ice cores are drilled, since the coring equipment is subject to least shearing, and obviously, the surface near boundary layer is most easily accessible to observation. The reasons for the inaccuracy of the SIA in the mentioned subregions are as follows:

Close to the margins the surface slopes are large – in some cases very close to vertical, implying that the shallowness assumption is no longer satisfied. Hence, the full equations would have to be solved, but fortunately, the region is so small and the effects remain so local that in any numerical scheme of global ice sheet dynamics these effects are of subgrid size, and the problem can formally be dismissed. The difficulties arising close to the free surface and in the vicinity of ice divides are of a quite different nature. They are based on the fact that ice behaves as a strongly nonlinear viscous fluid with an effective viscosity that is infinitely large at zero strain rate or at least very large[1]. Thus, in the upper parts of the ice sheet, perhaps 10 to 20% below the surface where the shearing is practically zero (see the velocity profile in Fig. 11.1) the effective viscosity is very (infinitely) large making the free surface very (infinitely) stiff against this shear-

[1] In plane flow the thin-film or SIA approximation of a power law fluid operates with a stress-strain rate relationship of the form $\dot{\gamma} = A\tau|\tau|^{n-1}$ or $\tau = A^{-n}\dot{\gamma}|\dot{\gamma}|^{1/n-1}$, where $\dot{\gamma}$ is the horizontal shear strain rate (shearing) and τ the shear stress and A a constant. Thus, except for a constant factor, $|\dot{\gamma}|^{1/n-1}$ is the viscosity which for $n > 1$ becomes infinitely large when $\dot{\gamma} \to 0$. This is what is meant by infinite stiffness; it is not restricted to power laws but can be avoided by using a "finite viscosity law" as we shall see.

ing. Realistically this can not be, and so longitudinal and other stresses develop which automatically enhance the effective stress and lower the effective viscosity. This is an effect going beyond the lowest order shallow ice approximation.

In the vicinity of an ice divide it is rather obvious that the prerequisites of the shallow ice approximation must fail. Here, the velocities are nearly vertical from the free surface to the bed, diverting to the left and right close to the bed in a stagnation-point flow manner. A formal singularity develops in the shallow ice approximation when the no-slip condition applies at the bed and a power law rheology is used, the reason again being infinite viscosity that develops. Finite viscosity laws again regularize the formal solution procedure, but this does not alter the fact that the constructed SIA solution close to the ice divide deviates a large amount from the true solution of the stagnation-point flow.

This description makes it clear that the SIA generates reasonable solutions everywhere except in the three mentioned regions. By using a finite viscosity law the lowest order SIA can be regularized such that it becomes uniformly valid, and an extension of a perturbation expansion to second order in ϵ will improve on this SIA. We shall demonstrate in these lecture notes how the SIA is constructed and in which way we improve it to eliminate the mentioned inadequacies.

Ice shelves are large ice masses which are floating on the ocean on most of their parts, are fed by the ice from the inland ice sheet and may also occasionally touch the ocean bottom, Fig. 11.2. They are connected to the inland ice sheet at the grounding line where the discharging ice from the ice sheet becomes afloat. The ice velocity is primarily horizontal and towards the front, where the ice sheet looses its mass through calving into icebergs. In plan view they may be bounded by mountain flanks, for instance when they move through a fjord-like narrow bay. When they reach the ocean bottom, part of their weight is carried by the solid ground, (the other by buoyancy [1]), generating basal frictional resistance and enhanced free surface elevations. If this interaction through the ocean bottom is weak the surface elevation above the very flat ice-shelf surface is small, say less than 50 m, forming what is called *ice rumples.* The flow of the ice is slowed as a result of the existence of these rumples, but the ice continues to go through them, being perhaps somewhat diverted, but not to the extent that rumples would be dynamically separated from the remaining ice shelf. In contrast, *ice rises* are regions of an ice shelf of which a substantial part rests on the ground. This results in higher surface elevations of the ice rises above the remaining ice-shelf surface, perhaps 200 m or more. The flow of the ice-shelf ice is basically around the ice rises, and the flow within the ice rises is reminiscent of that of ice sheets, along the direction of steepest descent of their surface topography. In other words, the dynamics of ice rises are disconnected from those of shelves.

Ice shelves tribute their existence to the strong mass flux of inland ice through the grounding line, defined as a transition zone between an inland ice-sheet and an ice shelf, but also precipitation from the atmosphere. At the ice–ocean interface ice may melt or ocean water may freeze and the frazil ice suspended in the cold ocean water may also contribute via accretion to the addition of marine ice from the bottom. As a result, ice shelves consist of two sorts of ice, *meteoric ice* which is ice from fallen snow, fresh water ice from inland, and *marine ice*, which is frozen ocean water. In these lecture notes we will limit attention to locations where no marine ice exists close to the ice–ocean interface; in so doing we will considerably simplify the mathematics (and restrict the physical applicability). Nevertheless the essentials of the improved approximations of the so-called shallow shelf approximation (SSA) will become apparent. The complete theory is still under construction.

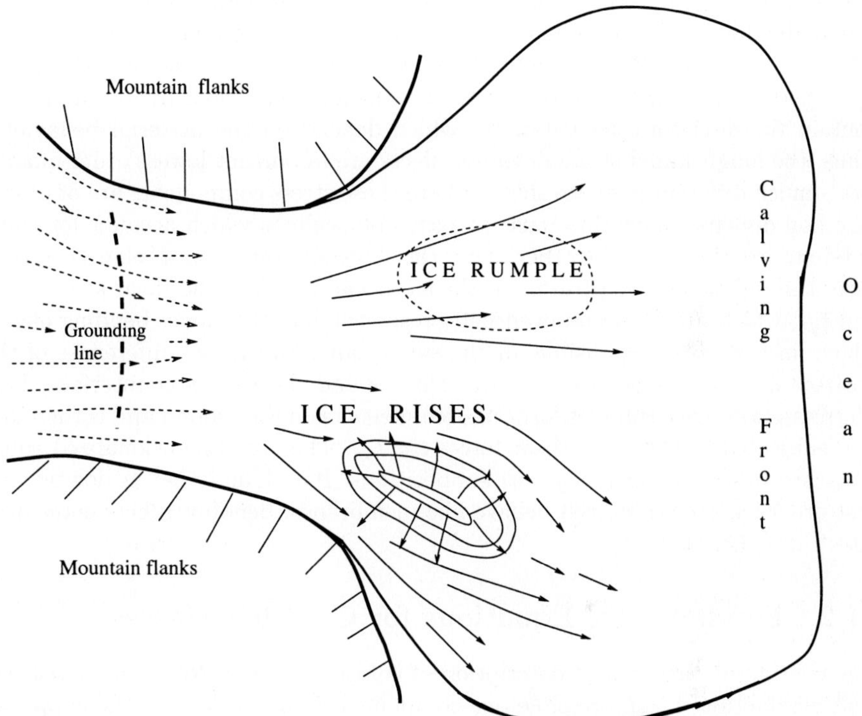

Fig. 11.2. Plan view of an ice shelf. Such ice shelves are bounded by the grounding line, the side flanks of the rising topography and the free side margins (front)

To write down the general three-dimensional equations for such a model is rather complex; it involves establishing the balance equations of mass, momentum and energy as well as the kinematic surface conditions and dynamic jump conditions at the free surface, the ice ocean interface, the grounding line, the side flanks of the rising topography and the free side margins (front). Among the kinematic boundary conditions the evolution equation of the ice front is important, because it involves a dynamically significant variable, the calving rate, i.e. the amount of ice mass lost per unit time along the ice front. This variable must be parameterized, and its parameterization describes an important component of the temporal response of ice shelves to climate variations. This is still one of the open problems of glaciology today.

Shallowness, i.e. small depth in comparison to the horizontal extent and primarily horizontal motion, is the distinctive feature of ice shelves. The aspect ratio ϵ is again the parameter motivating a perturbation solution procedure; however, the stress scalings must be quite different from those of ice sheets. Indeed, ice shelves are essentially a skin or membrane – despite their thickness of 200–1500 m – covering the ocean. So their deformations are essentially due to membrane forces, tensions and in-plane shear, which generate in-plane velocities which should largely be independent of the vertical coordinate where-ever the ice shelf is floating. These membrane forces are made up of the normal stresses t_{xx}, t_{yy} and the shear stresses t_{xy}, if x, y are the horizontal coordinates, and these contain the deviatoric contributions which depend on the material behaviour. Thus, the longitudinal stress deviator effects are important lowest order quantities – much different from ice sheets where these stress components are of $\mathcal{O}(\epsilon^2)$. The non-dimensionalization must be based on scalings which account for these different orderings. On the other hand, the thickness variation in plan view activates through the water pressure at the ice–ocean interface the shear stresses t_{xz} and t_{yz}, which are $\mathcal{O}(\epsilon)$ and eventually responsible for the weak bending effects which, in turn, are responsible for the small, but measurable, differences of the horizontal velocities between the free surface and the ice–water interface. Furthermore, where ice rumples form, the ice shelf sits on the ground and is therefore also subjected to the same shear forces t_{xz}, t_{yz}. The non-dimensionalized equations are based on these a priori suppositions. It is found that to lowest and first order ice shelves indeed behave like membranes; bending effects enter only afterwards, i.e. at $\mathcal{O}(\epsilon^2)$.

11.2 Fundamental Equations for Cold Ice Masses

The continuum mechanical description of cold ice is a non-Newtonian, viscous, heat conducting, incompressible one-component fluid. Neglecting the supply of solar radiation in the energy equation (which is justified, since the radiation depth is only a cm), we write the balance equations of mass, momentum and energy as

$$\begin{aligned} \operatorname{div} \boldsymbol{v} &= \operatorname{tr} \boldsymbol{D} = 0 \,, \\ \rho \dot{\boldsymbol{v}} &= -\operatorname{grad} p + \operatorname{div} \boldsymbol{t}^D + \rho \boldsymbol{g} \,, \\ \rho \dot{u} &= -\operatorname{div} \boldsymbol{q} + \operatorname{tr} (\boldsymbol{t}^D \boldsymbol{D}) \,, \end{aligned} \tag{11.1}$$

where the stress tensor $\boldsymbol{t} = -p\mathbf{1} + \boldsymbol{t}^D$ is decomposed into the constraint pressure p and the viscous extra stress $\boldsymbol{t}^D$. Moreover, $\boldsymbol{v}$ is the velocity vector, ρ the density, p the pressure, $\boldsymbol{t}^D$ the stress deviator, $\boldsymbol{g}$ the acceleration due to gravity, u the internal energy, $\boldsymbol{q}$ the heat flux and $\boldsymbol{D} = \operatorname{sym}(\operatorname{grad} \boldsymbol{v})$ the stretching or strain rate tensor. These balance equations must be complemented by the material equations which are given as

$$\boldsymbol{D} = A(T')f(\sigma)\,\boldsymbol{t}^D \,, \quad \dot{u} = c(T)\,\dot{T} \,, \quad \boldsymbol{q} = -\kappa(T) \operatorname{grad} T \,, \tag{11.2}$$

where c and κ are functions of the temperature T, known respectively, as the specific heat and thermal conductivity [2]. A is the rate factor, a function of the homologous temperature $T' = T - T_M$, defined here as the difference between the true and the melting temperature T_M; f is a creep response function that depends on the effective shear stress $\sigma = \left(\frac{1}{2}\mathrm{tr}\,(\boldsymbol{t}^D)^2\right)^{1/2} = \left(II_{\boldsymbol{t}^D}\right)^{1/2}$, the root of the second invariant of the stress deviator. Parameterizations for f and A are

$$
\begin{aligned}
f(\sigma) &= \sigma^{n-1} + \frac{1}{\eta}\,, \quad n \geq 1 \quad (\text{usually} \quad n = 3)\,, \\
A(T') &= A_0 \exp\left(-\frac{Q}{R(T_M + T')}\right)\,,
\end{aligned}
\tag{11.4}
$$

where η is a dimensionless viscosity, $\eta = \infty$ characterizing an infinite viscosity law; Q is the activation energy, R the universal gas constant and

$$
T_M = T_0 - \beta \frac{p}{\rho g} \tag{11.5}
$$

the melting temperature; β is the Clausius–Clapeyron constant, T_0 the melting temperature at normal pressure and p the pressure deviating from it. Upon substituting the material equations in the energy balance equation, we obtain the evolution equation for the temperature in terms of convection, conduction and dissipation,

$$
\rho c \dot{T} = \rho c \left\{ \frac{\partial T}{\partial t} + \underbrace{(\mathrm{grad}\, T) \cdot \boldsymbol{v}}_{convection} \right\} = \underbrace{\mathrm{div}\,(\kappa\, \mathrm{grad}\, T)}_{conduction} + \underbrace{2A(T') f(\sigma) \sigma^2}_{dissipation}\,. \tag{11.6}
$$

This equation together with the first two balance equations (11.1) and the constitutive relation for the stress deviator (11.2) constitute the field equations for cold ice.

The above field equations must be complemented by *boundary conditions* at the free surface, the ice–rock interface and, at the floating portion, at the ice–water interface. These comprise kinematic as well as dynamic statements, relating to the geometry and the fluxes, respectively.

At the *free surface* $z = h(x, y, t)$ the following conditions must hold:

$$
\frac{1}{N_s}\frac{\partial h}{\partial t} + \boldsymbol{v} \cdot \boldsymbol{n}_s = a_s^{\perp}(\boldsymbol{x}, t)\,, \quad \boldsymbol{t}\boldsymbol{n}_s = \boldsymbol{0}\,, \quad T = T_s(\boldsymbol{x}, t)\,, \tag{11.7}
$$

[2] An inverse relationship to $(11.2)_1$ is when the stress deviator is expressed in terms of the stretching tensor $\boldsymbol{t}^D = \tilde{B}(T')\tilde{g}(d/A(T'))\boldsymbol{D}$ where $d := \left(\frac{1}{2}\mathrm{tr}\boldsymbol{D}^2\right)^{1/2}$, $\tilde{B}(T') = 1/A(T')$ and $\tilde{g}(\xi) = 1/f([\mathfrak{F}(\xi)]^{-1})$ and $\mathfrak{F}^{-1}$ is the inverse of $\mathfrak{F}(\xi) = f(\xi)\xi$. One usually absorbs the $A(T')$-dependence in the argument of $\tilde{g}$ into $\tilde{B}(T')$ with a new B and then writes

$$
\boldsymbol{t}^D = B(T')g(d)\boldsymbol{D}\,, \quad d = \left(\tfrac{1}{2}\mathrm{tr}\boldsymbol{D}^2\right)^{1/2}\,, \tag{11.3}
$$

but this cannot be an exact inversion of (11.2). This representation is used for ice shelves.

where the dot is the inner product of two vectors and

$$\boldsymbol{n}_s = \frac{1}{N_s}\left(-\frac{\partial h}{\partial x}, -\frac{\partial h}{\partial y}, +1\right) \quad N_s := \left(1 + \left(\frac{\partial h}{\partial x}\right)^2 + \left(\frac{\partial h}{\partial y}\right)^2\right)^{\frac{1}{2}} \tag{11.8}$$

is the exterior unit normal vector, expressed here in Cartesian coordinates, x, y horizontal, z vertical. The accumulation–ablation rate function $a_s^{\perp}(\boldsymbol{x}, t)$ is one of the two climate input functions and describes how much volume of ice is added per unit surface area by precipitation or melting; the free surface is therefore in general non-material. The second of equations (11.7) prescribes the boundary traction to be zero, so the atmospheric pressure and wind stress components tangential to the surface are ignored. Finally, the third of equations (11.7) is a prescription of the atmospheric surface temperature as a function of position and time. The functions $a_s^{\perp}(\boldsymbol{x}, t)$ and $T_s(\boldsymbol{x}, t)$ describe the climate driving input from above into the ice mass. The other boundary for ice sheets or rumpled ice shelf regions is the *bed-rock*, described by $z = b(x, y, t)$. Here, the kinematic boundary condition takes the form

$$\frac{1}{N_b}\frac{\partial b}{\partial t} + \boldsymbol{v} \cdot \boldsymbol{n}_b = a_b^{\perp} H(T - T_M) \, , \tag{11.9}$$

$$\boldsymbol{n}_b = \frac{1}{N_b}\left(\frac{\partial b}{\partial x}, \frac{\partial b}{\partial y}, -1\right) , \quad N_b := \left(1 + \left(\frac{\partial b}{\partial x}\right)^2 + \left(\frac{\partial b}{\partial y}\right)^2\right)^{\frac{1}{2}} ,$$

in which $\boldsymbol{n}_b$ is the exterior unit normal vector to the basal surface, $a_b^{\perp} = a_b^{\perp M} + a_b^{\perp A}$ is the volume of ice per unit surface area and unit time added to the surface by melting ($a_b^{\perp M}$) or by accretion ($a_b^{\perp A}$). $H(\cdot)$ is the Heavyside step function

$$H(T - T_M) = \begin{cases} 0 \, , & \text{if } \; T < T_M \, , \\ 1 \, , & \text{if } \; T \geq T_M \, , \end{cases} \tag{11.10}$$

where T is the basal temperature; so melting is activated only when the basal temperature reaches the melting point. Furthermore,

$$\frac{\partial b}{\partial t} = \begin{cases} 0 \, , & \text{if the base is rigid, as is the case for ice shelves in rumpled regions} \, , \\ -\dfrac{1}{\tau_v}\left[b - \left(b_{\text{st}} - \dfrac{\rho}{\rho_a}H\right)\right] , & \text{if the base moves vertically} \, . \end{cases} \tag{11.11}$$

The parameterization (11.11) assumes a relaxation type response of the asthenosphere–lithosphere compound with a time lag $\tau_v \approx 3000\,[\text{yr}]$ for the isostatic bed adjustment; b_{st} is the steady state bed position for zero ice thickness H and $\rho_a \approx 3000\,[\text{kg}\,\text{m}^{-3}]$ is the mass density of the asthenosphere. The first dynamic condition comprises the no-slip/sliding law

$$
\begin{aligned}
\boldsymbol{v}_{sl} &= -\left[C_t(\| \boldsymbol{t}_{\|} \|^2, p_\perp) H(T - T_M)\right] \boldsymbol{t}_{\|} \, , \\
\boldsymbol{v}_{sl} &:= (\mathbf{1} - \boldsymbol{n}_b \otimes \boldsymbol{n}_b)(\boldsymbol{v} - \boldsymbol{v}_r) \, , \quad \boldsymbol{v}_r = \left(0, 0, \frac{\partial b}{\partial t}\right) \, , \\
\boldsymbol{t}_{\|} &:= \boldsymbol{t}\boldsymbol{n}_b + p_\perp \boldsymbol{n}_b \, , \quad p_\perp := -\boldsymbol{n}_b \cdot \boldsymbol{t}\boldsymbol{n}_b \, ,
\end{aligned}
\tag{11.12}
$$

which activates sliding when the melting temperature is reached. Thus, $\boldsymbol{v}_{sl}$ is the tangential component[3] of the difference of the velocity of the ice and the rock at the interface. $C_t(\cdot)$ is the sliding coefficient; it depends on the shear traction $\boldsymbol{t}_{\|}$ and the surface normal pressure $p_\perp$, and $\boldsymbol{v}_r$ is the velocity field of the rock at the ice rock interface. The second dynamic condition,

$$
\begin{aligned}
&\kappa(\operatorname{grad} T \cdot \boldsymbol{n}_b) - \kappa_r(\operatorname{grad} T_r \cdot \boldsymbol{n}_b) = \left[\delta - a_b^{\perp M}\left(\rho L - \frac{\rho}{\rho_w} p_\perp\right)\right] H(T - T_M) \, , \\
&T = T_r
\end{aligned}
\tag{11.13}
$$

is a Stefan-type energy jump condition that determines the melting rate $a_b^{\perp M}$ when sliding occurs. L is the latent heat of melting, ρ_w the density of water and κ_r the heat conductivity of the rock. Moreover, the temperature T_r and its surface gradient $\operatorname{grad} T_r \cdot \boldsymbol{n}_b$ follow from solving a heat conduction problem in the rock-bed and assigning the geothermal heat at a lower boundary. The quantity

$$
\delta = C_t(\cdot)(\boldsymbol{t}_{\|} \cdot \boldsymbol{t}_{\|}) \tag{11.14}
$$

is the dissipation rate due to sliding friction. The sliding coefficient is commonly parameterized as

$$
C_t(\| \boldsymbol{t}_{\|} \|^2, p_\perp) = \mathfrak{c} \, \| \boldsymbol{t}_{\|} \|^{p} \cdot p_\perp^{-q} \, , \qquad p = 3 \quad \text{or} \quad 4 \, , \quad q = 2 \, , \tag{11.15}
$$

where $\mathfrak{c}$ is a constant. Physically, (11.13) states that the heat flow from the interface into the ice minus that towards that interface from below is balanced by the heat due to sliding dissipation minus that used up by melting. The boundary condition of energy (11.13), also applies in floating regions of ice shelves. T_r is then the ocean temperature at the ice–ocean interface, $\kappa_r(\operatorname{grad} T_r \cdot \boldsymbol{n}_b)$ the heat flow from the ocean into the ice and $C(\boldsymbol{t}_{\|} \cdot \boldsymbol{t}_{\|})$ the dissipation at the ice–ocean interface due to the ocean boundary layer flow at the interface. The above field equations and boundary conditions also apply for an ice shelf, its free and basal surfaces where ice rumples occur. However, in regions where the ice is floating the boundary conditions at the ice–ocean interface require the kinematic condition (11.9) to be satisfied, where $a_b^\perp = a_b^{\perp M} + a_b^{\perp A}$ is composed of a contribution due to melting, $a_b^{\perp M}$, determinable by (11.13), and a second contribution, $a_b^{\perp A}$, due to accretion by frazil ice from the ocean; the latter is prescribed by an ocean

[3] The operator $(\mathbf{1} - \boldsymbol{n} \otimes \boldsymbol{n})$ is a projection on the tangential plane of the surface with unit normal vector $\boldsymbol{n}$.

model and considered here to be given. The traction boundary condition,

$$\begin{aligned} &\boldsymbol{t}\boldsymbol{n}_b = -p_b \boldsymbol{n}_s + \boldsymbol{t}_{\parallel} \,, \\ &p_b = \begin{cases} p_{\mathrm{sw}} \,, \\ p_{\mathrm{gr}} \,, \end{cases} \quad \boldsymbol{t}_{\parallel} = \begin{cases} 0 \,, & \text{floating region} \,, \\ \boldsymbol{\tau}_{\mathrm{gr}} \,, & \text{rumpled region} \,, \end{cases} \end{aligned} \tag{11.16}$$

differentiates between floating and rumpled regions. In the former case, the traction exerted on the ice–water interface is given by the hydrostatic sea water pressure[4]

$$p_{\mathrm{sw}} = \rho_{\mathrm{sw}} g [z_s(t) - b(x,y,t)] \,, \tag{11.17}$$

where $z = z_s(t)$ is the sea water level (given function of time) and the shear traction is ignored (note, however, that the dissipation due to viscous shearing in the ocean boundary layer is accounted for in (11.13)). Over rumpled regions the basal pressure is parameterized by

$$p_b = p_{\mathrm{sw}} + p_{\mathrm{rump}} = p_{\mathrm{sw}} + [\rho g(h-b) - \rho_{\mathrm{sw}} g(z_s - b)] = \rho g H \,. \tag{11.18}$$

The pressure due to the ice rumples, p_{rump}, equals the weight of the ice column minus the buoyancy of the submerged portion, but the total pressure is larger by that of the sea water; so together with the floating condition, (11.17) and (11.18) are the same! At a rumpled portion $\boldsymbol{\tau}_{\mathrm{gr}}$ is not prescribed but the outcome of the model.

The thermal coupling of ice sheets and ice shelves in regions where the ice is grounded is accomplished through a rock layer of approximately 5 km thickness adjacent to the grounded portion of the ice. Here the boundary value problem of heat

$$\begin{aligned} &\rho_r c_r \left(\frac{\partial T_r}{\partial t} + v_z \frac{\partial T_r}{\partial z} \right) = \nabla^2 T_r \,, && v_z = \frac{\partial b}{\partial t} \,, \\ &T_r = T \,, && \text{at} \quad z = b(x,y,t) \,, \\ &-\kappa_r \frac{\partial T_r}{\partial z} = Q_{\mathrm{geoth}} \,, && \text{at} \quad z = b_r (= \text{constant}) \,, \end{aligned} \tag{11.19}$$

is solved, where b_r is the bottom boundary of the bedrock. Notice that the horizontal motion is ignored in (11.19). This implies that for constant Q_{geoth} the steady heat flow into the ice at $z = b(x,y,t)$ is simply Q_{geoth}.

In *summary*, cold ice sheets and ice shelves are described by the field equations (11.1)–(11.6) and (11.12). Ice sheets are subject to the boundary conditions (11.7)–(11.15); $T_s, a_s^{\perp}$ and the geothermal heat Q_{geoth} are input quantities.

[4] If the ice shelf and the sea water are locally in equilibrium then the floating condition $\rho_{\mathrm{sw}} g(z_s - b) = \rho g H$ applies which, when substituted in (11.17) implies $p_{\mathrm{sw}} = \rho g H$; the sea water pressure at a point at the base equals the pressure due to the weight of the ice column above it. Ignoring vertical accelerations this result will be shown to apply in the ice shelf model deduced below.

Table 11.1. Physical constants and values of parameters pertinent to cold ice sheets and ice shelves

Quantities	Values
Density of ice, ρ	$910\,\mathrm{kg\,m^{-3}}$
Heat conductivity of ice, κ	$9.828\,\mathrm{e}^{-0.0057\,T[\mathrm{K}]}\,\mathrm{W\,m^{-1}K^{-1}}$
Specific heat of ice, c	$(146.3 + 7.253\,T[\mathrm{K}])\,\mathrm{J\,kg^{-1}K^{-1}}$
Latent heat of ice, L	$335\,\mathrm{kJ\,kg^{-1}}$
Clausius–Clapeyron gradient, β	$8.7 \cdot 10^{-4}\,\mathrm{K\,m^{-1}}$
Density of sea water, ρ_{sw}	$1000\,\mathrm{kg\,m^{-3}}$
Density of lithosphere, ρ_r	$3000\,\mathrm{kg\,m^{-3}}$
Density $\times$ specific heat of the lithosphere, $\rho_r c_r$	$2000\,\mathrm{kJ\,m^{-3}K^{-1}}$
Heat conductivity of the lithosphere, κ_r	$3\,\mathrm{W\,m^{-1}K^{-1}}$
Thickness of the upper lithosphere layer considered in this model, H_r	$5\,\mathrm{km}$
Time lag for bed adjustment, τ_V	$3000\,\mathrm{a}$
Density of the asthenosphere, ρ_a	$3300\,\mathrm{kg\,m^{-3}}$
Gravity acceleration, g	$9.81\,\mathrm{m\,s^{-2}}$
Creep response function, $f(\sigma)$	$\sigma^{(n-1)}, \quad n = 3$
Viscosity, η (dimensionless)	10^{-3}
Rate factor, $A(T)$	$A_0\,\mathrm{e}^{-Q/(RT)}$
	with $A_0 = 3.985 \cdot 10^{-13}\,\mathrm{s^{-1}\,Pa^{-3}}$
Activation energy, Q	$60\,\mathrm{kJ\,mol^{-1}}$ for $(T' < -10^{\circ}\mathrm{C})$
	$139\,\mathrm{kJ\,mol^{-1}}$ for $(T' \geq -10^{\circ}\mathrm{C})$
Universal gas constant, R	$8.314\,\mathrm{J\,mol^{-1}K^{-1}}$
Melting temperature at normal pressure, $p_N = 10^5\,\mathrm{Pa}$	$T_0 = 273.10\,\mathrm{K}$
Melting temperature, at pressure p	$T_M = T_0 - \beta p/(\rho g)$

By contrast, ice shelves must obey the boundary conditions (11.7)–(11.10) and (11.19), but (11.11) only at rumpled regions (since at floating portions $\partial b/\partial t$ is left to be determined). Furthermore, (11.12)–(11.18) must hold. Input quantities for shelves are $a_s^{\perp}$ and T_s at the free surface, $a_b^{\perp A}$, $\kappa_r(\operatorname{grad} T_r \cdot \boldsymbol{n}_b)$, $C(\boldsymbol{t}_{\|} \cdot \boldsymbol{t}_{\|})$, i.e. the accretion rate, the normal heat flow and the dissipation rate from the ocean water and/or the base of the rumpled region into the ice. These quantities must either be parameterized or evaluated by coupling the ice sheet with an atmospheric and/or ocean model. We close by collecting in Table 11.1 the physical constants that are pertinent to this model.

PART A: ICE SHEETS

11.3 Scale Analysis and Perturbation Scheme

The scalings for ice sheets and ice shelves are different from one another; we therefore separate from now on the two cases.

It will be assumed that the geometry of the ice sheet is such that its extent in the horizontal directions is much larger than in the vertical direction, i.e., the ice sheet is presumed shallow. If $[L]$ and $[H]$ are such typical length scales then $\epsilon := [H]/[L]$, called *aspect ratio*, is small, typically 10^{-2} to 10^{-3}. The non-dimensionalization of the physical quantities is now implemented by choosing their typical values as suggested by the physical problem at hand. We will write $\psi = [\psi]\tilde{\psi}$; the quantities in square brackets represent the respective typical values, and those with the tilde are the corresponding dimensionless quantities. All the typical values are chosen such that for the processes under consideration the dimensionless quantities are likely to be of order unity. This is the reason why in the list below certain scales are premultiplied by ϵ or even ϵ^2, where

$$\epsilon = [H]/[L] = [V_H]/[V_L] \,. \tag{11.20}$$

This choice for ϵ and the assumption to select the velocity ratio equal to the aspect ratio expresses the fact that the motion is predominantly horizontal[5]. We shall choose the following non-dimensionalizations:

$$\begin{aligned}
(x,y) &= [L]\,(\tilde{x},\tilde{y})\,, & (h,b) &= [H]\,(\tilde{h},\tilde{b})\,,\\
z &= [H]\,\tilde{z}\,, & (a_b^\perp, a_s^\perp) &= [V_H]\,(\tilde{a}_b^\perp,\tilde{a}_s^\perp)\,,\\
(v_x,v_y) &= [V_L]\,(\tilde{v}_x,\tilde{v}_y)\,, & A(T') &= [A]\,\tilde{A}(\tilde{\theta}')\,,\\
v_z &= [V_H]\,\tilde{v}_z\,, & f(\sigma) &= [f]\tilde{f}(\tilde{\sigma})\,,\\
t &= ([L]/[V_L])\,\tilde{t}\,, & Q_{\text{geo}}^\perp &= [Q_{\text{geo}}^\perp]\,\tilde{Q}_{\text{geo}}^\perp\,,\\
T &= T_0 + [\Delta T]\,\tilde{\theta}\,, & c(T) &= [c]\,\tilde{c}(\tilde{\theta})\,,\\
T' &= [\Delta T]\,\tilde{\theta}\,, & c_r &= [c_r]\,\tilde{c}_r\,,\\
p &= \rho g[H]\,\tilde{p}\,, & \kappa(T) &= [\kappa]\,\tilde{\kappa}(\tilde{\theta})\,,\\
(t_{xz}^D, t_{yz}^D, \sigma) &= \epsilon\rho g[H]\,(\tilde{t}_{xz}^D,\tilde{t}_{yz}^D,\tilde{\sigma})\,, & \kappa_r &= [\kappa_r]\,\tilde{\kappa}_r\,,\\
(t_{xx}^D, t_{yy}^D) &= \epsilon^2\rho g[H]\,(\tilde{t}_{xx}^D,\tilde{t}_{yy}^D)\,, & C_t(t_\perp,\ldots) &= [C_t]\,\tilde{C}_t(\tilde{t}_\perp,\ldots)\,,\\
(t_{xy}^D, t_{zz}^D) &= \epsilon^2\rho g[H]\,(\tilde{t}_{xy}^D,\tilde{t}_{zz}^D)\,, & w &= [w]\tilde{w}\,.
\end{aligned} \tag{11.21}$$

Whereas most selections of the scales are "natural", those for the deviatoric stress components are special insofar as they involve prefactors ϵ and ϵ^2 which

[5] The readers may easily show that with the choice (11.20) the continuity equation is scale invariant.

implicitly say that the horizontal shear stresses t^D_{xz} and t^D_{yz} are of order ϵ smaller than the overburden pressure and that the longitudinal stress deviator components $t^D_{xx}, t^D_{yy}, t^D_{zz}$ and t^D_{xy} are of order ϵ^2 smaller than the overburden pressure. These estimates are not obvious and have been so selected because any other choices do not allow a proper balance of the momentum equations. With the above scalings the following independent dimensionless products can be formed:

$$
\begin{aligned}
\epsilon &= \frac{[H]}{[L]} = \frac{[V_H]}{[V_L]} \sim 10^{-3}\,, & \mathsf{F} &= \frac{[V_L]^2}{g[L]} & &\sim 10^{-18}\,,\\
\mathcal{D} &= \frac{[\kappa]}{\rho[c][H][V_H]} \sim 0.34\,, & \mathcal{F}_t &= \frac{\rho g[H]^2[C_t]}{[L][V_L]} & &\sim 6\times 10^{-3}\,,\\
\alpha &= \frac{g[H]}{[c][\Delta T]} \sim 0.25\,, & \mathcal{B} &= \frac{\beta[H]}{[\Delta T]} & &\sim 0.04\,,\\
\alpha_t &= \frac{g[H]}{L[\mathcal{W}]} \sim 2.98\,, & \mathcal{D}_r &= \frac{[\kappa_r]}{\rho_r[c_r][H][V_H]} & &\sim 0.473\,,\\
\mathcal{K} &= \frac{\rho g[H]^3[A][f]}{[L][V_L]} \sim 0.38\,, & \mathcal{N}_r &= \frac{[H][Q_{\mathrm{geoth}}]}{[\kappa_r][\Delta T]} & &\sim 0.7\,,\\
\mathcal{T}_r &= \frac{\tau_V[V_H]}{[H]} \sim 0.3\,, & & [\mathcal{W}] & &\sim 0.01\,,\\
&\frac{[\kappa_r]}{[\kappa]} \sim 1.5\,, & & \frac{\rho_r}{\rho} & &\sim 3.29\,,\\
&\frac{\rho_a}{\rho} \sim 3.62\,,
\end{aligned}
\tag{11.22}
$$

where F is the Froude number, $\mathcal{D}$ the heat diffusion number, α the ratio of potential energy to internal energy for cold ice, $\mathcal{K}$ the fluidity number, α_t the ratio of potential to thermal energy, $\mathcal{F}_t$ the sliding number, $\mathcal{B}$ the Clausius–Clapeyron number, $\mathcal{D}_r$ the heat diffusion number of the lithosphere, $\mathcal{T}_r$ the time-lag number for isostatic bed adjustment, and finally $\mathcal{N}_r$ the geothermal heat number in the lithosphere. The numerical values assigned to these dimensionless products are obtained from the values of the physical constants in Table 11.1 and the typical scales listed in Table 11.2. The reader is urged, by choosing his/her own values for [ ]-variables, to acquire a feeling for the above orders of magnitude.

11.3.1 Scaled Equations

If the field equations and boundary conditions are non-dimensionalized with the scalings (11.21) then they take the following forms (tildes are everywhere omitted)

(i) Field equations in the ice

$$
\frac{\partial v_x}{\partial x} + \frac{\partial v_y}{\partial y} + \frac{\partial v_z}{\partial z} = 0\,, \tag{11.23}
$$

Table 11.2. Typical and constant values of the *ice-sheet scales* which are required, in addition to the values given in Table 11.1, for estimating the dimensionless parameters as shown in (11.22)

Quantities	Values
Vertical dimension, $[H]$	$1\,\mathrm{km} = 10^3\,\mathrm{m}$
Horizontal dimension, $[L]$	$1000\,\mathrm{km} = 10^6\,\mathrm{m}$
Vertical velocity, $[V_H]$	$0.1\,\mathrm{m\,a^{-1}}$
Horizontal velocity, $[V_L]$	$100\,\mathrm{m\,a^{-1}}$
Water content, $[w]$	$1\% = 0.01$
Temperature difference, $[\Delta T]$	$20\,\mathrm{K}$
Homologous temperature, T'	$-10\,\mathrm{K}$ to $-30\,\mathrm{K}$
Geothermal heat, $[Q_{\mathrm{geo}}]$	$0.042\,\mathrm{W\,m^{-2}}$
Density of the lithosphere, ρ_r	$3000\,\mathrm{kg\,m^{-3}}$
Heat conductivity, $[\kappa(T = 273\,\mathrm{K})]$	$9.828\,\mathrm{e}^{-0.0057\,T[\mathrm{K}]}\,\mathrm{W/(mK)} \approx 2\mathrm{W/(mK)}$
Specific heat, $[c(T = 273\,\mathrm{K})]$	$(146.3 + 7.253\,T[\mathrm{K}])\,\mathrm{J/(kgK)} \approx 2000\,\mathrm{J/(kgK)}$
Creep function, $[f(\sigma) = (\epsilon\rho g[H])^{n-1}]$	$91^2 \times 10^4\,\mathrm{kg^2\,m^{-2}\,s^{-4}}$; for $n = 3$
Rate factor, $[A(T')]$	$1.6 \times 10^{-24}\,\mathrm{Pa^{-3}s^{-1}}$; for $T' = -5^0\,\mathrm{C}$
Sliding function for temp. ice, $[C_t]$	
$= (C_{sl} \parallel t_{\parallel} \parallel^p)/(\rho g(\rho g H)^q \parallel t_{\parallel} \parallel)$	$2.12 \times 10^{-12}\,\mathrm{m^2\,s/Kg}$; for $p = 3, q = 2$
	and $\parallel t_{\parallel} \parallel \approx \epsilon\rho g[H] \approx 91 \times 10^2\,\mathrm{kg\,m^{-1}\,s^{-2}}$

$$\frac{\mathsf{F}}{\epsilon}\frac{dv_x}{dt} = -\frac{\partial p}{\partial x} + \epsilon^2\frac{\partial t^D_{xx}}{\partial x} + \epsilon^2\frac{\partial t^D_{xy}}{\partial y} + \frac{\partial t^D_{xz}}{\partial z}\,, \tag{11.24}$$

$$\frac{\mathsf{F}}{\epsilon}\frac{dv_y}{dt} = \epsilon^2\frac{\partial t^D_{xy}}{\partial x} - \frac{\partial p}{\partial y} + \epsilon^2\frac{\partial t^D_{yy}}{\partial y} + \frac{\partial t^D_{yz}}{\partial z}\,, \tag{11.25}$$

$$\mathsf{F}\epsilon\frac{dv_z}{dt} = \epsilon^2\frac{\partial t^D_{xz}}{\partial x} + \epsilon^2\frac{\partial t^D_{yz}}{\partial y} - \frac{\partial p}{\partial z} + \epsilon^2\frac{\partial t^D_{zz}}{\partial z} - 1\,. \tag{11.26}$$

$$\frac{\partial\theta}{\partial t} + v_x\frac{\partial\theta}{\partial x} + v_y\frac{\partial\theta}{\partial y} + v_z\frac{\partial\theta}{\partial z} = \frac{\mathcal{D}}{c(\theta)}\left[\epsilon^2\frac{\partial}{\partial x}\left(\kappa\frac{\partial\theta}{\partial x}\right) + \epsilon^2\frac{\partial}{\partial y}\left(\kappa\frac{\partial\theta}{\partial y}\right)\right.$$
$$\left.+\frac{\partial}{\partial z}\left(\kappa\frac{\partial\theta}{\partial z}\right)\right] + 2\frac{\alpha}{c(\theta)}\mathcal{K}A(\theta')f(\sigma)\sigma^2\,. \tag{11.27}$$

$$\frac{\partial v_x}{\partial z} + \epsilon^2 \frac{\partial v_z}{\partial x} = 2\mathcal{K} t_{xz}^D A(\theta') f(\sigma) , \tag{11.28}$$

$$\frac{\partial v_y}{\partial z} + \epsilon^2 \frac{\partial v_z}{\partial y} = 2\mathcal{K} t_{yz}^D A(\theta') f(\sigma) . \tag{11.29}$$

$$\frac{\partial v_x}{\partial x} = \mathcal{K} t_{xx}^D A(\theta') f(\sigma) , \tag{11.30}$$

$$\frac{\partial v_y}{\partial y} = \mathcal{K} t_{yy}^D A(\theta') f(\sigma) , \tag{11.31}$$

$$\frac{\partial v_z}{\partial z} = \mathcal{K} t_{zz}^D A(\theta') f(\sigma) , \tag{11.32}$$

$$\frac{\partial v_x}{\partial y} + \frac{\partial v_y}{\partial x} = 2\mathcal{K} t_{xy}^D A(\theta') f(\sigma) , \tag{11.33}$$

$$\sigma = \sqrt{(t_{zx}^D)^2 + (t_{yz}^D)^2 + \tfrac{1}{2}\epsilon^2 \left[(t_{xx}^D)^2 + (t_{yy}^D)^2 + (t_{zz}^D)^2 + 2(t_{xy}^D)^2\right]} . \tag{11.34}$$

These are, in order, the continuity equation, three components of the momentum equations, the energy equation, the six components of the stretching-stress relations (whereby those for t_{xz}^D and t_{xz}^D are separated from the others) and the expression for the effective stress σ. These equations involve ϵ and some other dimensionless products defined in (11.22) will further be simplified.

(ii) Boundary conditions at the ice surface. These apply at $z = b(x, y, t)$ and are given by (11.7) and (11.8), and their dimensionless counterparts are

$$\frac{\partial h}{\partial t} + v_x \frac{\partial h}{\partial x} + v_y \frac{\partial h}{\partial y} - v_z = \sqrt{1 + \epsilon^2 \left(\frac{\partial h}{\partial x}\right)^2 + \epsilon^2 \left(\frac{\partial h}{\partial y}\right)^2} \; a_s^{\perp} , \tag{11.35}$$

$$-\left(-p + \epsilon^2 t_{xx}^D\right) \frac{\partial h}{\partial x} - \epsilon^2 t_{xy}^D \frac{\partial h}{\partial y} + t_{xz}^D = 0 , \tag{11.36}$$

$$-\epsilon^2 t_{xy}^D \frac{\partial h}{\partial x} - \left(-p + \epsilon^2 t_{yy}^D\right) \frac{\partial h}{\partial y} + t_{yz}^D = 0 , \tag{11.37}$$

$$-\epsilon^2 t_{xz}^D \frac{\partial h}{\partial x} - \epsilon^2 t_{yz}^D \frac{\partial h}{\partial y} - p + \epsilon^2 t_{zz}^D = 0 , \tag{11.38}$$

$$\theta(x, y, z, t) = \theta_s(x, y, z, t) . \tag{11.39}$$

These comprise of the kinematic surface equation, the three components of the zero traction condition and the prescribed surface temperature.

(iii) Transition conditions at the basal surface. These conditions apply at $z = b(x, y, t)$ and involve (11.9)–(11.13). Their dimensionless counterparts read

$$
\begin{aligned}
&\frac{\partial b}{\partial t} + \frac{\partial b}{\partial x} v_x^i + \frac{\partial b}{\partial y} v_y^i - v_z^i = N_b \, a_b^{\perp} \, H(\theta - \theta_M) \,, \\
&N_b = \sqrt{1 + \epsilon^2 \left(\frac{\partial b}{\partial x}\right)^2 + \epsilon^2 \left(\frac{\partial b}{\partial y}\right)^2} \,,
\end{aligned}
\tag{11.40}
$$

$$
\theta_M = -\mathcal{B} p \,, \tag{11.41}
$$

$$
\frac{\partial b}{\partial t} = \begin{cases} 0 \,, & \text{if the base is rigid} \,, \\ -\dfrac{1}{\mathcal{T}_r}\left(b - b_{st} + \dfrac{\rho}{\rho_a} H\right) \,, & \text{if the base moves vertically by isostatic adjustment} \,. \end{cases} \tag{11.42}
$$

The worked-out forms of (11.12) are lengths but need be derived in a systematic perturbation approach. They are relegated to the Appendix and are listed there as (11.184) to (11.189). Even more cumbersome is the systematic derivation of the energy jump condition (11.13); for this reason, its dimensionless form is equally deferred to the Appendix where it is listed as (11.191).

Equations (11.40) comprise the kinematic boundary condition describing the temporal evolution of the basal surface, (11.41) is the dimensionless Clausius–Clapeyron equation, (11.42) determines the basal deformation of the bed due to the lithosphere–asthrenosphere reaction; the sliding law and energy jump condition being stated in the Appendix. All these equations are exact and hold for arbitrary values of the shallowness parameter ϵ.

(iv) Lithosphere relations. These follow by non-dimensionalizing (11.19). This yields the boundary value problem

$$
\begin{aligned}
&\frac{\partial \theta_r}{\partial t} + \frac{\partial b}{\partial z}\frac{\partial \theta_r}{\partial z} \\
&\quad = \frac{\mathcal{D}_r}{c_r}\left[\epsilon^2 \frac{\partial}{\partial x}\left(\kappa_r \frac{\partial \theta_r}{\partial x}\right) + \epsilon^2 \frac{\partial}{\partial y}\left(\kappa_r \frac{\partial \theta_r}{\partial y}\right) + \frac{\partial}{\partial y}\left(\kappa_r \frac{\partial \theta_r}{\partial z}\right)\right] \,,
\end{aligned}
\tag{11.43}
$$

$$
\theta_r = \theta \,, \qquad \text{at} \quad z = b(x, y, t) \,, \tag{11.44}
$$

$$
-\kappa_r \frac{\partial \theta_r}{\partial z} = \mathcal{N}_r Q_{\text{geoth}}^{\perp} \,, \qquad \text{at} \quad z = b_r \,. \tag{11.45}
$$

defining a parabolic initial boundary value problem with mixed boundary conditions.

This completes the formulation of the boundary value problem for ice sheets in dimensionless form. Scrutiny of all these equations shows that the aspect ratio ϵ arises only in even powers ϵ^2 and ϵ^4 except on the left-hand sides of the momentum balances (11.24)–(11.26) in combination with the Froude number. These terms will now be omitted, so that the momentum equations reduce to force balances. With $F = 10^{-18}$ and $\epsilon = 10^{-3}$ this omission is well justified as long as powers of ϵ no larger than ϵ^4 arise[6]. This is the case because no equations involve even larger powers of ϵ. We conclude with the

Formal statement 1: *Stokes flow is a justified approximation of ice sheet flows as long as formal perturbation expansions or equivalent iterative schemes are not pushed farther than to terms of $\mathcal{O}(\epsilon^4)$.*

11.3.2 Perturbation Scheme

Given the above form of the non-dimensionalized equations it is, of course tempting to seek solutions to all field variables in the form of a regular perturbation expansion, e.g.

$$F = \sum_{\nu=0}^{\infty} \epsilon^\nu F_{(\nu)} = \epsilon^0 F_{(0)} + \epsilon^1 F_{(1)} + \epsilon^2 F_{(2)} + \epsilon^3 F_{(3)} + \dots , \tag{11.46}$$

where $F_{(\nu)}$ are the ν-th approximation to F in the formal asymptotic series (11.46). The zeroth order equations, i.e., the restriction of all variables F in the expansion (11.46) to the lowest order $F_{(0)}$ is well known, has been derived or postulated several times in the literature and is known as the Shallow Ice Approximation (SIA). Full expansions to the second order are for the first time given in [1] and [2]. In these references (and in some of the literature quoted there), it is shown that this expansion would formally break down, i.e., the perturbation expansion would not be regular but singular if the creep response function $f(\sigma)$, see (11.4), would possess the property $f(0) = 0$, which is equivalent for the material to having infinite viscosity ($\eta = \infty$). This was the reason for us to postulate a flow law with finite viscosity ($\eta \neq \infty$)[7].

We will not show here the complete expansion to second order, because it is more adequate to use a perturbative-iterative procedure. We limit ourselves to illustrating the procedure to the integration of the momentum equations (11.24)–(11.26) subject to the boundary conditions (11.36)–(11.38). Substituting for the stress components expansions of the form (11.46) and collecting like powers of ϵ in the emerging equations yields the following chain of boundary value problems:

[6] More generally, if expansions are restricted to second order terms in ϵ, the left-hand sides of (11.24)–(11.26) can be ignored if $\mathsf{F} < \epsilon^3$. In this case Stokes flow is justified if approximations are restricted to $\mathcal{O}(\epsilon^2)$-terms.

[7] This conclusion can easily be inferred from (11.30) to (11.33). Indeed, if the stretchings $\partial v_x/\partial x$ etc. are known (left-hand sides of the equations) and the deviatoric stress components t^D_{xx}, etc. (right-hand sides) computed, then division by $f(\sigma)$ must be performed, which is singular whenever $f(0) = 0$

(i) The zeroth order equations are

$$-\frac{\partial p_{(0)}}{\partial x} + \frac{\partial t^D_{xz(0)}}{\partial z} = 0\,, \qquad -\frac{\partial p_{(0)}}{\partial y} + \frac{\partial t^D_{yz(0)}}{\partial z} = 0\,, \tag{11.47}$$

$$-\frac{\partial p_{(0)}}{\partial z} - 1 = 0\,, \qquad (\sigma_{(0)})^2 = (t^D_{xz(0)})^2 + (t^D_{yz(0)})^2\,. \tag{11.48}$$

subject to the boundary conditions

$$p_{(0)}(\cdot, h_{(0)}) = t^D_{xz(0)}(\cdot, h_{(0)}) = t^D_{yz(0)}(\cdot, h_{(0)}) = 0\,, \quad \text{at} \quad z = h_{(0)}(x,y,t)\,. \tag{11.49}$$

The third of these equations expresses hydrostatic equilibrium – sometimes called, since ice is involved – cryostatic equilibrium. Once $p_{(0)}$ is determined subject to the top boundary conditions, the shear stresses $t^D_{xz(0)}$ and $t^D_{yz(0)}$ may be determined by integration subject to the remaining boundary conditions. Subsequently, the effective stress $\sigma_{(0)}$ may be computed. The solutions are

$$p_{(0)} = h_{(0)} - z\,, \tag{11.50}$$

$$\left.\begin{aligned} t^D_{xz(0)} &= -\frac{\partial h_{(0)}}{\partial x}(h_{(0)} - z)\,, \\ t^D_{yz(0)} &= -\frac{\partial h_{(0)}}{\partial y}(h_{(0)} - z)\,, \end{aligned}\right\} \quad \boldsymbol{\tau} = -(h_{(0)} - z)\nabla_{\mathrm{H}} h_{(0)}\,, \tag{11.51}$$

$$\sigma_{(0)} = (h_{(0)} - z)\sqrt{\left(\frac{\partial h_{(0)}}{\partial x}\right)^2 + \left(\frac{\partial h_{(0)}}{\partial y}\right)^2}\,. \tag{11.52}$$

It should be noticed that the pressure $p_{(0)}$ and the shear stresses, $t^D_{xz(0)}$, $t^D_{yz(0)}$ could be determined without even using any information about the material behaviour of the ice. Furthermore, the zeroth order effective stress is determined by those two shear stress components *alone.*

(ii) The first order equations take the form

$$-\frac{\partial p_{(1)}}{\partial x} + \frac{\partial t^D_{xz(1)}}{\partial z} = 0\,, \quad -\frac{\partial p_{(1)}}{\partial y} + \frac{\partial t^D_{yz(1)}}{\partial z} = 0\,, \quad \frac{\partial p_{(1)}}{\partial z} = 0\,, \tag{11.53}$$

which must obey the boundary conditions

$$p_{(1)} = h_{(1)}\,, \quad t^D_{xz(1)} = -h_{(1)}\frac{\partial h_{(0)}}{\partial x}\,, \quad t^D_{yz(1)} = -h_{(1)}\frac{\partial h_{(0)}}{\partial y}\,, \tag{11.54}$$

at $z = h_{(0)}(x,y,t)$. These equations are also easily integrated; the solutions are

$$p_{(1)} = h_{(1)}(x,y,t)\,, \tag{11.55}$$

$$t^D_{xz(1)} = -\frac{\partial h_{(0)}}{\partial x}h_{(1)} - \frac{\partial h_{(1)}}{\partial x}(h_{(0)} - z)\,, \tag{11.56}$$

$$t^D_{yz(1)} = -\frac{\partial h_{(0)}}{\partial y}h_{(1)} - \frac{\partial h_{(1)}}{\partial y}(h_{(0)} - z)\,. \tag{11.57}$$

These solutions show, that the first order corrections to p, t^D_{xz}, t^D_{yz} are likewise independent of the material behaviour and that still hydrostatic conditions prevail. Indeed, by adding (11.55)–(11.57) to (11.50)–(11.52) we obtain

$$p = h(x,y,z,t) - z + \mathcal{O}(\epsilon^2)\,, \tag{11.58}$$

$$\boldsymbol{\tau} = (h(x,y,z,t) - z)\nabla_{\mathrm{H}} h_{(0)} + \mathcal{O}(\epsilon^2)\,, \tag{11.59}$$

where $h = h_{(0)} + \epsilon h_{(0)}$. The material dependence of p and τ must necessarily manifest itself in the $\mathcal{O}(\epsilon^2)$ terms. This is an interesting property of the SIA.

(iii) The second order equations can be written as

$$\frac{\partial p_{(2)}}{\partial x} - \frac{\partial t^{\mathrm{D}}_{xz(2)}}{\partial z} = \frac{\partial t^{\mathrm{D}}_{xx(0)}}{\partial x} + \frac{\partial t^{\mathrm{D}}_{xy(0)}}{\partial y}\,, \tag{11.60}$$

$$\frac{\partial p_{(2)}}{\partial y} - \frac{\partial t^{\mathrm{D}}_{yz(2)}}{\partial z} = \frac{\partial t^{\mathrm{D}}_{xy(0)}}{\partial x} + \frac{\partial t^{\mathrm{D}}_{yy(0)}}{\partial y}\,, \tag{11.61}$$

$$\frac{\partial p_{(2)}}{\partial z} = \frac{\partial t^{\mathrm{D}}_{xz(0)}}{\partial x} + \frac{\partial t^{\mathrm{D}}_{yz(0)}}{\partial y} + \frac{\partial t^{\mathrm{D}}_{zz(0)}}{\partial z}\,, \tag{11.62}$$

and are subject to the boundary conditions

$$p_{(2)} = (h_{(2)} - t^{\mathrm{D}}_{zz(0)})\,, \tag{11.63}$$

$$t^{\mathrm{D}}_{xz(2)} = -\frac{\partial h_{(0)}}{\partial x} h_{(2)} - \frac{\partial h_{(1)}}{\partial x} h_{(1)} - (t^{\mathrm{D}}_{zz(0)} - t^{\mathrm{D}}_{xx(0)})|_{h_{(0)}} \frac{\partial h_{(0)}}{\partial x} + t^{\mathrm{D}}_{xy(0)} \frac{\partial h_{(0)}}{\partial y}\,, \tag{11.64}$$

$$t^{\mathrm{D}}_{yz(2)} = -\frac{\partial h_{(0)}}{\partial y} h_{(2)} - \frac{\partial h_{(1)}}{\partial y} h_{(1)} - (t^{\mathrm{D}}_{zz(0)} - t^{\mathrm{D}}_{yy(0)})|_{h_{(0)}} \frac{\partial h_{(0)}}{\partial y} + t^{\mathrm{D}}_{xy(0)} \frac{\partial h_{(0)}}{\partial x}\,, \tag{11.65}$$

at $z = h_{(0)}(x,y,z)$. These equations contain new ingredients. First, the force balances (11.60)–(11.62) have now a nontrivial right hand side, comprising those zeroth order stress deviator components which so far were not computed but are computable when the zeroth order velocity field is known, see (11.30)–(11.33). An anologous statement holds true for the boundary conditions (11.63)–(11.65). We now momentarily regard these stress deviator terms as known. Then the above equations can be integrated but computations are no longer easy. The reader may check that the expressions

$$p_{(2)} = \tfrac{1}{2}\left(h_{(0)} - z\right)^2 \Delta_{\mathrm{H}} h_{(0)} + \left(h_{(0)} - z\right) \|\nabla_{\mathrm{H}} h_{(0)}\|^2 + h_{(2)} + t^{\mathrm{D}}_{zz(0)}\,, \tag{11.66}$$

$$\begin{aligned} t^{\mathrm{D}}_{xz(2)} = &-\frac{\partial}{\partial x}\left(\tfrac{1}{6}\left(h_{(0)} - z\right)^3 \Delta_{\mathrm{H}} h_{(0)} + \tfrac{1}{2}\left(h_{(0)} - z\right)^2 \|\nabla_{\mathrm{H}} h_{(0)}\|^2\right) \\ &-\frac{\partial h_{(0)}}{\partial x} h_{(2)} - \frac{\partial h_{(1)}}{\partial x} h_{(1)} - \frac{\partial h_{(2)}}{\partial x}\left(h_{(0)} - z\right) \\ &-\frac{\partial}{\partial x}\int_z^{h_{(0)}} \left(t^{\mathrm{D}}_{zz(0)} - t^{\mathrm{D}}_{xx(0)}\right) \mathrm{d}z' + \frac{\partial}{\partial x}\int_z^{h_{(0)}} t^{\mathrm{D}}_{xy(0)} \mathrm{d}z'\,, \end{aligned} \tag{11.67}$$

$$
\begin{aligned}
t^{\mathrm{D}}_{yz(2)} = &-\frac{\partial}{\partial y}\left(\tfrac{1}{6}\left(h_{(0)}-z\right)^3 \Delta_{\mathrm{H}} h_{(0)} + \tfrac{1}{2}\left(h_{(0)}-z\right)^2 \|\nabla_{\mathrm{H}} h_{(0)}\|^2\right) \\
&-\frac{\partial h_{(0)}}{\partial y} h_{(2)} - \frac{\partial h_{(1)}}{\partial y} h_{(1)} - \frac{\partial h_{(2)}}{\partial y}\left(h_{(0)}-z\right) \\
&-\frac{\partial}{\partial y}\int_z^{h_{(0)}}\left(t^{\mathrm{D}}_{zz(0)} - t^{\mathrm{D}}_{xx(0)}\right)\mathrm{d}z' + \frac{\partial}{\partial y}\int_z^{h_{(0)}} t^{\mathrm{D}}_{xy(0)}\,\mathrm{d}z' ,
\end{aligned}
\tag{11.68}
$$

satisfy both, the differential equations (11.60)–(11.62) and boundary conditions (11.63)–(11.65). ∇_{H} and Δ_{H} are the horizontal gradient and Laplace opetators.

We note that the above equations are only computationally useful in the form they are given, if $h_{(0)}, h_{(1)}, t^D_{xx(0)}, t^D_{yy(0)}, t^D_{zz(0)}$ and $t^D_{xy(0)}$ are known. This is not so without a full expansion, but we will see below that they form an important cornerstone in the process of generalizing the SIA.

11.3.3 Second Order Stress Formulas

If the above stress formulas for p, t^D_{xz}, t^D_{yz} are combined according to (11.46), their second order accurate expressions are obtained as follows:

$$
\begin{aligned}
p = &\left(h_{(0)} + \epsilon h_{(1)} + \epsilon^2 h_{(2)} + .. - z\right) \\
&+ \epsilon^2\left(\tfrac{1}{2}\left(h_{(0)}-z\right)^2 \Delta_{\mathrm{H}} h_{(0)} + \left(h_{(0)}-z\right)\|\nabla_{\mathrm{H}} h_{(0)}\|^2\right) + \epsilon^2 t^{\mathrm{D}}_{zz} ,
\end{aligned}
\tag{11.69}
$$

$$
\begin{aligned}
t^{\mathrm{D}}_{xz} \simeq\ & t^{\mathrm{D}}_{xz(0)} + \epsilon t^{\mathrm{D}}_{xz(1)} + \epsilon^2 t^{\mathrm{D}}_{xz(2)} + \ldots = -\frac{\partial}{\partial x}\left(h_{(0)} + \epsilon h_{(1)} + \epsilon^2 h_{(2)} + ..\right) \times \\
&\left(h_{(0)}-z\right) - \frac{\partial}{\partial x}\left(h_{(0)} + \epsilon h_{(1)} + ..\right)\left(\epsilon h_{(1)} + \ldots\right) - \epsilon^2 \frac{\partial h_{(0)}}{\partial x} h_{(2)} \\
&-\epsilon^2 \frac{\partial}{\partial x}\left(\tfrac{1}{6}(h_{(0)}-z)^3 \Delta_{\mathrm{H}} h_{(0)} + \tfrac{1}{2}(h_{(0)}-z)^2 \|\nabla_{\mathrm{H}} h_{(0)}\|^2\right) \\
&-\epsilon^2\left(\frac{\partial}{\partial x}\int_z^{h_{(0)}}\left(t^{\mathrm{D}}_{zz(0)} - t^{\mathrm{D}}_{xx(0)}\right)\mathrm{d}z' - \frac{\partial}{\partial y}\int_z^{h_{(0)}} t^{\mathrm{D}}_{xy(0)}\mathrm{d}z'\right) ,
\end{aligned}
\tag{11.70}
$$

and where t^{D}_{yz} is obtained from t^{D}_{xz} by replacing x- and y-derivatives. If the free surface level is identified with $h = h_{(0)} + \epsilon h_{(1)} + \epsilon^2 h_{(2)} + \ldots$, then the above formulas may be replaced by the more suggestive formulas

$$
\begin{aligned}
p\ &= (h-z) + \epsilon^2\left(\tfrac{1}{2}(h-z)^2 \Delta_{\mathrm{H}} h + (h-z)\|\nabla_{\mathrm{H}} h_{(0)}\|^2\right) \\
&\quad + \epsilon^2 t^{\mathrm{D}}_{zz} + \mathcal{O}(\epsilon^3) ,
\end{aligned}
\tag{11.71}
$$

$$
\begin{aligned}
t^{\mathrm{D}}_{xz} = &-\frac{\partial h}{\partial x}(h-z) - \epsilon^2 \frac{\partial}{\partial x}\left(\tfrac{1}{6}(h-z)^3 \Delta_{\mathrm{H}} h + \tfrac{1}{2}(h-z)^2 \|\nabla_{\mathrm{H}} h\|^2\right) \\
&-\epsilon^2\left(\frac{\partial}{\partial x}\int_z^h \left(t^{\mathrm{D}}_{zz} - t^{\mathrm{D}}_{xx}\right)\mathrm{d}z' - \frac{\partial}{\partial y}\int_z^h t^{\mathrm{D}}_{xy}\mathrm{d}z'\right) + \mathcal{O}(\epsilon^3) ,
\end{aligned}
\tag{11.72}
$$

and analogously,

$$t_{yz}^{\mathrm{D}} = -\frac{\partial h}{\partial y}(h-z) - \epsilon^2 \frac{\partial}{\partial y}\Big(\tfrac{1}{6}(h-z)^3 \Delta_{\mathrm{H}} h + \tfrac{1}{2}(h-z)^2 \|\nabla_{\mathrm{H}} h\|^2\Big)$$
$$-\epsilon^2 \left(\frac{\partial}{\partial y} \int_z^h \left(t_{zz}^{\mathrm{D}} - t_{yy}^{\mathrm{D}}\right) \mathrm{d}z' - \frac{\partial}{\partial x} \int_z^h t_{xy}^{\mathrm{D}} \mathrm{d}z' \right) + \mathcal{O}(\epsilon^3) . \qquad (11.73)$$

Here alterations relative to (11.69) and (11.70) have errors of $\mathcal{O}(\epsilon^3)$ or smaller, which explains the order symbols arising in (11.71)–(11.73). We emphasize that variables above are without indices $(\cdot)_{(\nu)}$, i.e., (11.71)–(11.73) are written in terms of the unperturbed fields; this is also correct with an error of $\mathcal{O}(\epsilon^3)$ or smaller.

For later use we will write these formulas as

$$p(.,z) = (h(.) - z) + p^{\mathrm{corr}}(.,z) , \qquad (11.74)$$

$$t_{xz}^{D}(.,z) = \frac{\partial h(.)}{\partial x}(z - h(.)) + t_{xz}^{D\ \mathrm{corr}}(.,z) , \qquad (11.75)$$

$$t_{yz}^{D}(.,z) = \frac{\partial h(.)}{\partial y}(z - h(.)) + t_{yz}^{D\ \mathrm{corr}}(.,z) , \qquad (11.76)$$

$$p^{\mathrm{corr}}(.,z) = \epsilon^2 \Big(\tfrac{1}{2} (h-z)^2 \Delta_{\mathrm{H}} h + (h-z) \|\nabla_{\mathrm{H}} h_{(0)}\|^2 \Big) + \epsilon^2 t_{zz}^{\mathrm{D}} , \qquad (11.77)$$

$$t_{xz}^{D\ \mathrm{corr}}(.,z) = -\epsilon^2 \frac{\partial}{\partial x}\Big(\tfrac{1}{6}(h-z)^3 \Delta_{\mathrm{H}} h + \tfrac{1}{2}(h-z)^2 \|\nabla_{\mathrm{H}} h\|^2\Big)$$
$$-\epsilon^2 \left(\frac{\partial}{\partial x} \int_z^h \left(t_{zz}^{\mathrm{D}} - t_{xx}^{\mathrm{D}}\right) \mathrm{d}z' - \frac{\partial}{\partial y} \int_z^h t_{xy}^{\mathrm{D}} \mathrm{d}z' \right) , \qquad (11.78)$$

$$t_{yz}^{D\ \mathrm{corr}}(.,z) = -\epsilon^2 \frac{\partial}{\partial y}\Big(\tfrac{1}{6}(h-z)^3 \Delta_{\mathrm{H}} h + \tfrac{1}{2}(h-z)^2 \|\nabla_{\mathrm{H}} h\|^2\Big)$$
$$-\epsilon^2 \left(\frac{\partial}{\partial y} \int_z^h \left(t_{zz}^{\mathrm{D}} - t_{yy}^{\mathrm{D}}\right) \mathrm{d}z' - \frac{\partial}{\partial x} \int_z^h t_{xy}^{\mathrm{D}} \mathrm{d}z' \right) . \qquad (11.79)$$

The above representations of the pressure and the shear stresses on horizontal planes have for the first time appeared in [2] and can be physically interpreted.

The first terms in (11.71)–(11.73) are the well known pressure and shear stress formulas of the SIA, the bracketed terms in the middle describe the influence of the variation of the surface topography and have a quadratic and linear z-dependence, respectively, are largest at the base and smallest at the free surface (namely zero); so topography affects the pressure by the mean curvature $\Delta_{\mathrm{H}} h$ and the norm of the slope $\|\nabla_{\mathrm{H}} h\|$ of the free surface. These contributions to the dominant stresses can be very accurately computed, since satellite-altimetry measurements allow a very high resolution of the surface topography. Alternatively, in time-dependent evolutions of ice sheets under climate variations accurate determination of the surface geometry is required. The terms on the second line describe the stress-deviator effects of which the determination we address below. Thus, *higher order corrections to the SIA are due to two sources, (1) topography effects, not related to material behaviour, and (2) stress deviator effects ignored at the SIA level. To call these effects simply "longitudinal stress deviator effects" is at best misleading and at worst simply wrong.*

The above correction terms are important additions beyond the SIA where the latter proves to yield results that are inaccurate; this is the case close to the ice divide and near the surface as explained already in the Introduction. On lengthscales of the total ice thickness surface topographies are smooth and so the topography effects are in most cases much smaller than the stress deviator effects, which, as we shall see, are due to the material behaviour of the ice.

11.4 Second Order Shallow Ice Approximation (SOSIA)

Basic idea of the ensuing solution scheme for the ice sheet equations is an iterative approach using the exact equations except for the momentum balance laws which are replaced by the second order stress formulas (11.71)–(11.73). This iteration procedure is very much motivated by the classical SIA and will for this reason be called SOSIA – despite small changes in the solution approach. In pursuing the analysis, the fact that the stress-stretching relationship exhibits finite viscosity at zero stretching or zero stress deviator is very significant.

Consider the field equations and boundary conditions of Subsect. (11.3.1) but replace the momentum balance equations (11.24)–(11.26) and the traction boundary conditions (11.36)–(11.38) by the stress formulas (11.71)–(11.73). This set of differential equations and boundary conditions consists of terms of $\mathcal{O}(\epsilon^0), \mathcal{O}(\epsilon^2)$ and $\mathcal{O}(\epsilon^4)$[8]. Writing formally in each equation the terms of $\mathcal{O}(\epsilon^0)$ on the left-hand side and those of $\mathcal{O}(\epsilon^2, \epsilon^4)$ on the right-hand side, and supposing that the terms on the right-hand side are known, all fields can be determined. This scheme requires initialization namely zero right-hand side, which is equivalent to the SIA, but in so doing an iteration scheme is started which improves on the SIA. For this innovative scheme, the mechanical fields are decoupled from the thermal fields and the free and basal surfaces, i.e., we initially prescribe the geometry and the temperature field.

Given these, the stress and velocity fields are determined. With them at hand, a forward step is performed to find the new geometry and the new temperature field at $t = t_0 + \Delta t$. This forward stepping in time can be used, in principle to follow ice sheet dynamics through ice ages, e.g. from inception to maximum extent to disintegration.

11.4.1 Velocity and Stress Deviator Fields

We start with the determination of the velocity field v_x, v_y, v_z. To this end, (11.28) and (11.29) are integrated over $\bar{z}$ from $\bar{z} = b$ to $\bar{z} = z$ and subject to the

[8] Actually all $\mathcal{O}(\epsilon^4)$-terms can be dropped as the stress formulas possess only $\mathcal{O}(\epsilon^2)$-accuracy.

sliding boundary conditions (11.184)–(11.190) (stated in Appendix A1). This formal integration yields

$$v_x(.,z) = 2\mathcal{K}\int_{b(.)}^{z} t^{\mathrm{D}}_{xz}(.,\bar z)\mathcal{A}(.,\bar z)f(\sigma(.,\bar z))\mathrm{d}\bar z - \epsilon^2\int_{b(.)}^{z}\frac{\partial v_z(.,\bar z)}{\partial x}\mathrm{d}\bar z + (v_\mathrm{b})_x(.)\,, \tag{11.80}$$

$$v_y(.,z) = 2\mathcal{K}\int_{b(.)}^{z} t^{\mathrm{D}}_{yz}(.,\bar z)\mathcal{A}(.,\bar z)f(\sigma(.,\bar z))\mathrm{d}\bar z - \epsilon^2\int_{b(.)}^{z}\frac{\partial v_z(.,\bar z)}{\partial y}\mathrm{d}\bar z + (v_\mathrm{b})_y(.)\,, \tag{11.81}$$

$$v_z(.,z) = -\int_{b(.)}^{z}\left(\frac{\partial v_x(.,\bar z)}{\partial x}+\frac{\partial v_y(.,\bar z)}{\partial y}\right)\mathrm{d}\bar z + (v_\mathrm{b})_z(.)\,, \tag{11.82}$$

in which $(v_\mathrm{b})_x$, $(v_\mathrm{b})_y$ and $(v_\mathrm{b})_z$ are the components of the sliding velocity at the base which can be determined from the conditions (11.184)–(11.190) and are given by

$$(v_\mathrm{b})_x(.) = v_x^+ + \frac{\mathcal{F}_t C_t}{N_\mathrm{b}}\left(p\frac{\partial b}{\partial x} + t^{\mathrm{D}}_{xz} + \frac{1}{N_\mathrm{b}^2}p\frac{\partial b}{\partial x}\right) + (v_\mathrm{sl})_x^{\mathrm{corr}} - \Delta v_x^{\mathrm{corr}}, \tag{11.83}$$

$$(v_\mathrm{b})_y(.) = v_y^+ + \frac{\mathcal{F}_t C_t}{N_\mathrm{b}}\left(p\frac{\partial b}{\partial y} + t^{\mathrm{D}}_{yz} + \frac{1}{N_\mathrm{b}^2}p\frac{\partial b}{\partial y}\right) + (v_\mathrm{sl})_y^{\mathrm{corr}} - \Delta v_y^{\mathrm{corr}}, \tag{11.84}$$

$$\begin{aligned}(v_\mathrm{b})_z(.) = \left(1-\frac{1}{N_\mathrm{b}^2}\right)^{-1}\Bigg\{ v_z^+ - \frac{1}{N_\mathrm{b}^2}\Bigg[\left((v_\mathrm{b})_x(.)-v_x^+\right)\frac{\partial b}{\partial x} + \left((v_\mathrm{b})_y(.)-v_y^+\right)\frac{\partial b}{\partial y}\\ +v_z^+\Bigg] + \mathcal{F}_t C_t\left(t^{\mathrm{D}}_{xz}\frac{\partial b}{\partial x} + t^{\mathrm{D}}_{yz}\frac{\partial b}{\partial y}\right) + (v_\mathrm{sl})_z^{\mathrm{corr}}\Bigg\}\,;\end{aligned} \tag{11.85}$$

$$N_b = \sqrt{1+\epsilon^2\left(\frac{\partial b}{\partial x}\right)^2 + \epsilon^2\left(\frac{\partial b}{\partial y}\right)^2}\,,$$

$$\sigma^2 = (t^{\mathrm{D}}_{xz})^2 + (t^{\mathrm{D}}_{yz})^2 + \underbrace{\tfrac{1}{2}\epsilon^2\left[(t^{\mathrm{D}}_{xx})^2 + (t^{\mathrm{D}}_{yy})^2 + (t^{\mathrm{D}}_{zz})^2 + 2(t^{\mathrm{D}}_{xy})^2\right]}_{\sigma^2_{\mathrm{corr}}}\,,$$

$$v_x^+ = v_y^+ = 0;\quad v_z^+ = \frac{\partial b(x,y,t)}{\partial t} = -\frac{1}{\mathcal{T}_\mathrm{a}}\left(b - b_\mathrm{st} + \frac{\rho}{\rho_\mathrm{a}}H\right)\,. \tag{11.86}$$

The expressions (11.83)–(11.86) define the sliding velocity at the base. The corrective terms $(\cdot)^{\mathrm{corr}}$ are in full length listed in the Appendix; they are boundary corrections beyond the SIA and are significant in those places where, on length scales of the ice depth the variations of the basal topography are not small. In addition, the pressure and the shear stresses on horizontal planes are given in (11.74)–(11.79), whilst t_{ii} (no sum over i, $i = x, y, z$) and t^D_{xy} are evaluated from

(11.30)–(11.33), viz.,

$$t^{\mathrm{D}}_{xx} = \frac{1}{\mathcal{K}Af(\sigma)}\frac{\partial v_x}{\partial x} \,, \tag{11.87}$$

$$t^{\mathrm{D}}_{yy} = \frac{1}{\mathcal{K}Af(\sigma)}\frac{\partial v_y}{\partial y} \,, \tag{11.88}$$

$$t^{\mathrm{D}}_{zz} = -(t^{\mathrm{D}}_{xx} + t^{\mathrm{D}}_{yy}) \,, \tag{11.89}$$

$$t^{\mathrm{D}}_{xy} = \frac{1}{2\mathcal{K}Af(\sigma)}\left(\frac{\partial v_y}{\partial x} + \frac{\partial v_x}{\partial y}\right) . \tag{11.90}$$

Notice that to $\mathcal{O}(\epsilon)$, $N_\flat = 1$ and all $(\cdot)^{\mathrm{corr}}$ variables vanish. So to zeroth or first order the velocity components may formally be evaluated as in the SIA using the effective stress evaluated by using $\sigma^2 = \left({t^{\mathrm{D}}_{xz}}^2 + {t^{\mathrm{D}}_{yz}}^2\right)$, thereby ignoring the terms of $\mathcal{O}(\epsilon^2)$. This, of course, requires the use of a finite-viscosity creep law to guarantee regularity at the free surface and ice divide[9]. To this order the integrand variables in the quadratures (11.80)–(11.82) also need to be known to order zero only. This level of computation is needed to evaluate the stress deviator components (11.87)–(11.90) in order to determine the pressure and the stresses, t^{D}_{xz} and t^{D}_{yz}, (11.74)–(11.79), to second order. With all stress components at hand the velocity components v_x, v_y, v_z can be calculated in a further iteration, now with all stress components evaluated to second order. Only in the second quadrature formulas on the right-hand sides of (11.80) and (11.81) the lower-order formulas must be substituted for the vertical velocity component v_z.

It is of course not mandatory that this scheme is initiated with $N_\flat = 1$ and all $(\cdot)^{\mathrm{corr}}$ variables set to zero. In fact it is likely advantageous to keep the $\mathcal{O}(\epsilon^2)$ terms wherever possible. However, it is presently not known whether and how much the convergence can be accalarated. For a suggested numerical procedure see [3].

11.4.2 Updating the Geometry and Temperature Field

In the last Subsect. 11.4.1 it was assumed that the geometry and temperature distribution of the ice sheet are known. At the initial time they are prescribed, later they are known from the previous computational step.

The new geometry of the ice sheet at time $t + \Delta t$ can be best determined by integrating the continuity equation (11.23) from $z = b(x, y, t)$ to $z = h(x, y, t)$ and using the kinematic boundary conditions (11.35) and (11.41). This yields

$$\frac{\partial H}{\partial t} + \frac{\partial}{\partial x}\int_{\mathrm{b}}^{h} v_x \mathrm{d}z + \frac{\partial}{\partial y}\int_{\mathrm{b}}^{h} v_y \mathrm{d}z = \sqrt{1 + \epsilon^2\left(\frac{\partial h}{\partial x}\right)^2 + \epsilon^2\left(\frac{\partial h}{\partial y}\right)^2} \cdot a_{\mathrm{s}}^{\perp}$$

---

[9] In the SIA, σ is computed using t^{D}_{xz} and t^{D}_{yz} only. Both are zero at the free surface and at all depths at the ice divide, (τ is proportional to $\nabla_H h (= 0)$ and so $\sigma^{(0)} = 0$). Thus, with $f(0) = 0$ the formulas (11.87)–(11.90) could be singular.

$$-\sqrt{1+\epsilon^2\left(\frac{\partial b}{\partial x}\right)^2+\epsilon^2\left(\frac{\partial b}{\partial y}\right)^2}\cdot a_b^{\perp} H(\theta-\theta_M)\,, \quad (11.91)$$

where $H = h - b$ and $a_b^{\perp} = -(\boldsymbol{w} - \boldsymbol{v}) \cdot \boldsymbol{n}$. Additionally, (11.42) must hold. Both equations are first order differential equations in time; using a forward Euler step, the new positions of the free surface and the ice–rock interface can be determined. The presence of the ϵ-dependent terms does not affect the integration procedure and no precautions need to be taken because of the non-SIA terms in the equations.

A similar updating is also needed for the temperature distributions in the ice sheet and the rock layer. Here, (11.27) and (11.43) must be solved subject to the boundary conditons (11.39) at the free surface, the continuity requirement (11.44) and the energy jump condition (11.45) at the ice–rock interface and the flux boundary condition (11.191) at the lower boundary of the lithosphere. It is possible to determine an SIA-like first order approximation to these equations by integrating them with the terms premultiplied by $\mathcal{O}(\epsilon^2)$ evaluated with the old temperature field and iterating on corresponding corrections. As such, the equations in their entirety are fully parabolic advection–diffusion-reaction equations for which a large body of reliable commercial software is available. Future will show which technique will win.

11.5 Some Results

The zeroth order theory has found its application in a wealth of climatological applications of the Greenland and Antarctic Ice Sheets and other sheet flows. More than fifty contributions are referenced in [2]. These essentially discuss velocity and temperature fields in an ice sheet of given geometry or follow these fields plus that of the geometric evolution through entire ice age cycles. The fallacies of these zeroth order models is known to most authors of such papers, but corrections such as those presented here, have so far hardly ever been analysed in detail. An account is again given in [2].

The complete second order model has still not been implemented in a software program; such an endevour is presently under way. However, the second order stress formulas (11.77)–(11.79) have been computed for the Greenland Ice Sheet in a computation from the past to the present using the geometry and temperature field of the zeroth order approximation. The results are discussed in [2]. Probably the most important finding in this regard is that the topography effects in (11.77)–(11.79) are very much smaller than the stress deviator effects. This is, of course, only so because the spatial step in the zeroth order surface discretization was too large to disclose any significant contributions. A more local analysis will most likely alter this. A second inference implied by such calculations clearly pointed at significant contributions of the second order stress effects close to the Summit region, as one would have expected.

To illustrate the potential of this theoretical formulation we show in Fig. 11.3 the results of a computation of the change of the geometry of the Greenland ice sheet when it is subjected to a sudden rise of surface air temperature by 10°C. The left panel illustrates the surface topography as of today and computed by integration from the past 250.000 years, the panel on the right shows the small ice cap left after 1000 years of Greenhouse scenario.

Future analyses will have to incorporate the improved model equations in ice sheet models simply because without them dating the ice at depth will be fraught with large errors and thus of little use in climate reconstructions from ice core analyses.

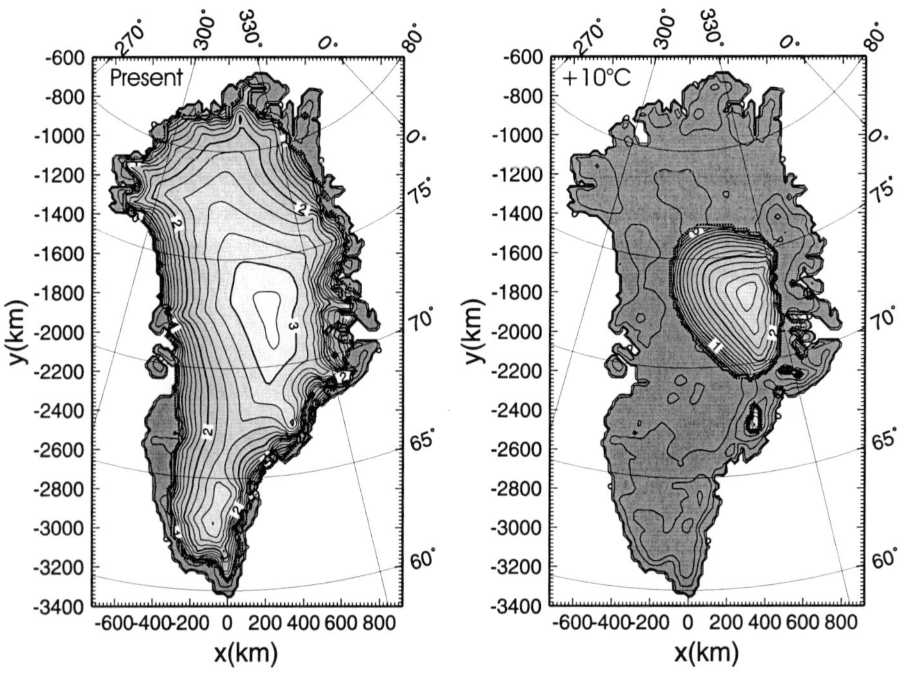

Fig. 11.3. Surface topography of the Greenland ice sheet as of today computed by integration from the past 250.000 years (*left*) and after 1000 years subjected to a rise of surface air temperature by 10°C (*right*) (Courtesy of R. Greve)

PART B: ICE SHELVES

11.6 Scale Analysis and Perturbation Scheme

As for ice sheets shallowness of the geometry and the velocity field is the outstanding feature of ice shelves. If $[L]$, $[H]$ and $[V_L]$, $[V_H]$ are typical horizontal and vertical length and velocity scales, then, as for ice sheets, we require the aspect ratio ϵ to satisfy the conditions of the shallow-shelf approximation (SSA), i.e.

$$\frac{[H]}{[L]} = \frac{[V_H]}{[V_L]} = \epsilon \ll 1 \,. \tag{11.92}$$

However, this property is the only one shared with the SIA. Since ice shelves float on the water except at the few grounding areas, the shear stresses t^D_{xz} and t^D_{yz} cannot be large, because they vanish at the free surface and are practically zero at the ice–water interface. On the other hand, the longitudinal stresses t^D_{xx}, t^D_{yy} and the shear stress t^D_{xy} are of principal importance, i.e., of $\mathcal{O}(\epsilon^0)$, because they respond to the horizontally diverging and converging flow conditions. We will assume that t^D_{xz} and t^D_{yz} are of $\mathcal{O}(\epsilon)$ smaller. The quantities subscripted in the list below as $(\cdot)_{\text{gr}}$ are scaled to account for the ice rumples due to the ocean bottom contact. These quantities are τ_{gr}, the shear traction at the "rumpled bed", and $a^{\perp}_{\text{gr}}$, the melting take there, they are absent at floating points of the ice shelf base; they are of $\mathcal{O}(\epsilon^0)$ in the SIA and should be smaller at ice rumples, so it is tempting to postulate them to be of $\mathcal{O}(\epsilon^2)$ small. By taking $a^{\perp}_{\text{gr}}$ of $\mathcal{O}(\epsilon^2)$ smaller than the ocean melting-freezing-accretion rates, we explicitly account for the fact that melting rates at the grounding portions are considerably smaller than at the floating parts. There is some indirect observational evidence for this. The shallow shelf scales, as chosen by us, and motivated in greater detail in [1], are given as follows:

$$\begin{aligned}
(x,y) &= [L](\tilde{x},\tilde{y})\,, & z &= [H]\tilde{z}\,,\\
(v_x,v_y) &= [V_L](\tilde{v}_x,\tilde{v}_y)\,, & v_z &= [V_H]\tilde{v}_z = \epsilon[V_L]\tilde{v}_z\,,\\
t = [t]\tilde{t} &= \frac{[L]}{[V_L]}\tilde{t} = \frac{[H]}{[V_H]}\tilde{t}\,, & d &= \frac{1}{[t]}\tilde{d}\,,\\
p_{\text{gr}} &= \frac{\rho_{\text{sw}}-\rho}{\rho_{\text{sw}}}\rho g[H]\tilde{p}_{\text{gr}}\,, & p &= \frac{\rho_{\text{sw}}-\rho}{\rho_{\text{sw}}}\rho g[H]\tilde{p}\,,\\
t^D_{xx} &= \frac{\rho_{\text{sw}}-\rho}{\rho_{\text{sw}}}\rho g[H]\tilde{t}^D_{xx}\,, & p_{\text{sw}} &= \frac{\rho_{\text{sw}}-\rho}{\rho_{\text{sw}}}\rho g[H]\tilde{p}_{\text{sw}}\,,\\
t^D_{yy} &= \frac{\rho_{\text{sw}}-\rho}{\rho_{\text{sw}}}\rho g[H]\tilde{t}^D_{yy}\,, & \tau_{\text{gr}} &= \epsilon^2\frac{\rho_{\text{sw}}-\rho}{\rho_{\text{sw}}}\rho g[H]\tilde{\tau}_{\text{gr}}\,,\\
t^D_{zz} &= \frac{\rho_{\text{sw}}-\rho}{\rho_{\text{sw}}}\rho g[H]\tilde{t}^D_{zz}\,, & t^D_{xz} &= \epsilon\frac{\rho_{\text{sw}}-\rho}{\rho_{\text{sw}}}\rho g[H]\tilde{t}^D_{xz}\,,
\end{aligned} \tag{11.93}$$

$$
\begin{aligned}
(t^D_{xy}, \sigma) &= \frac{\rho_{\rm sw} - \rho}{\rho_{\rm sw}} \rho g [H] (\tilde{t}^D_{xy}, \tilde{\sigma}) \,, & t^D_{yz} &= \epsilon \frac{\rho_{\rm sw} - \rho}{\rho_{\rm sw}} \rho g [H] \tilde{t}^D_{yz} \,, \\
(z_s, b) &= [H](\tilde{z}_s, \tilde{b}) \,, & \tau_{\rm oc} &= \epsilon^3 \frac{\rho_{\rm sw} - \rho}{\rho_{\rm sw}} \rho g [H] \tilde{\tau}_{\rm oc} \,, \\
a^\perp_b &= [V_H] \tilde{a}^\perp_b \,, & A(\cdot) &= [A] \tilde{A}(\cdot) \,, \\
a^\perp_{\rm gr} &= \epsilon^2 [V_H] \tilde{a}^\perp_{\rm gr} \,, & f(\sigma) &= [f] \tilde{f}(\sigma) \,, \\
B(\cdot) &= [B] \tilde{B}(\cdot) \,, & \delta_{\rm oc} &= [\delta_{\rm oc}] \tilde{\delta}_{\rm oc} \,, \\
g(d) &= [g] \tilde{g}(d) \,, & \delta_{\rm gr} &= \frac{[\delta_{\rm oc}]}{\epsilon} \tilde{\delta}_{\rm gr} \,, \\
(q^\perp_{\rm oc}, q^\perp_{\rm gr}) &= [q^\perp] (\tilde{q}^\perp_{\rm oc}, \tilde{q}^\perp_{\rm gr}) \,, & \kappa(T) &= [\kappa] \, \tilde{\kappa}(\tilde{\theta}) \,, \\
c^\perp &= [V_L] \tilde{c}^\perp \,, & \kappa_r &= [\kappa_r] \, \tilde{\kappa}_r \,, \\
T &= T_0 + [\Delta T] \, \tilde{\theta} \,, & c(T) &= [c] \, \tilde{c}(\tilde{\theta}) \,, \\
T' &= [\Delta T] \, \tilde{\theta} \,, & c_r &= [c_r] \, \tilde{c}_r \,, \\
Q^\perp_{\rm geo} &= [Q^\perp_{\rm geo}] \, \tilde{Q}^\perp_{\rm geo} \,.
\end{aligned}
\tag{11.94}
$$

Some of these scales are the same as for sheets in (11.21), but those typical of ice shelves are collected in Table 11.3. From the above scalings the independent dimensionless products

$$
\begin{aligned}
\epsilon &= \frac{[H]}{[L]} && \sim 10^{-3} \,, & \alpha &= \frac{g[H]}{[c][\Delta T]} && \sim 0.125 \,, \\
\mathcal{B} &= \frac{\beta [H]}{[\Delta T]} && \sim 0.02 \,, & \mathcal{D} &= \frac{[\kappa]}{\rho [c][H][V_H]} && \sim 0.069 \,, \\
\varrho &= \frac{\rho_{\rm sw} - \rho}{\rho_{\rm sw}} && \sim 0.09 \,, & \alpha_s &= \frac{g[H]}{L} && \sim 0.014 \,, \\
\mathcal{K}_S &= \frac{\rho g [H]^2 [A][f]}{[V_H]} && \sim 19249 \,, & \mathcal{K}'_S &= \frac{[B][V_H][t]^{(n-1)/n}}{\rho g [H]^2} && \sim 0.007 \,, \\
\mathcal{N}_{q^\perp_{\rm oc}} &= \frac{[H][q^\perp_{\rm oc}]}{[\kappa][\Delta T]} && \sim 0.75 \,, & \mathcal{N}_{\delta_{\rm oc}} &= \frac{[H][\delta_{\rm oc}]}{[\kappa][\Delta T]} && \sim 0.01 \,, \\
\mathcal{N}_{q^\perp_{\rm gr}} &= \frac{[H][q^\perp_{\rm gr}]}{[\kappa][\Delta T]} && \sim 0.75 \,, & \mathcal{N}_{\delta_{\rm gr}} &= \frac{[H][\delta_{\rm gr}]}{[\kappa][\Delta T]} && \sim 10.23 \,, \\
\gamma &= \frac{\beta_S N_0}{[S][\Delta T]} && \sim 0.003 \,, & \mathcal{S} &= \frac{L[S]}{[c][\Delta T]} && \sim 0.418 \,, \\
\mathsf{F} &= \frac{[V_L]^2}{g[L]} && \sim 10^{-15} \,,
\end{aligned}
\tag{11.95}
$$

can be formed. The assigned numerical values are so obtained by substituting the values of the physical constants and the shelf-related scales from Tables 11.1

Table 11.3. Typical and constant values of the ice-shelf scales which are required, in addition to the values given in Table 11.1, for estimating the dimensionless parameters as written in (11.95)

Quantities	Typical Values
Vertical dimension, $[H]$	$500\,\mathrm{m}$
Horizontal dimension, $[L]$	$500\,\mathrm{km} = 500 \times 10^3\,\mathrm{m}$
Vertical velocity, $[V_H]$	$1\,\mathrm{m\,a^{-1}}$
Horizontal velocity, $[V_L]$	$500\,\mathrm{m\,a^{-1}}$
Temperature difference, $[\Delta T]$	$20\,\mathrm{K}$
Homologous temperature, T'	$-10\,\mathrm{K}$ to $-20\,\mathrm{K}$
Heat flux, $[q^{\perp}_{\mathrm{oc}}] \approx [q^{\perp}_{\mathrm{gr}}]$	$0.06\,\mathrm{W\,m^{-2}}$
Density of sea water, ρ_{sw}	$1000\,\mathrm{kg\,m^{-3}}$
Heat conductivity, $[\kappa(T = 273\,\mathrm{K})]$	$9.828\,\mathrm{e}^{-0.0057\,T[\mathrm{K}]}\,\mathrm{W\,m^{-1}K^{-1}}$ $\approx 2\mathrm{W\,m^{-1}K^{-1}}$
Specific heat, $[c(T = 273\,\mathrm{K})]$	$(146.3 + 7.253\,T[\mathrm{K}])\,\mathrm{J\,kg^{-1}K^{-1}}$ $\approx 2000\,\mathrm{J\,kg^{-1}K^{-1}}$
Creep function, $[f(\sigma)]$	$(0.09 \times \rho g[H])^{n-1}\,\mathrm{kg^2\,m^{-2}\,s^{-4}}$; $n = 3$
Rate factor, $[A(T')]$	$1.6 \times 10^{-24}\,\mathrm{Pa^{-3}s^{-1}}$; for $t = -5^0\,\mathrm{C}$
Rate factor, $[B(T')]$	$([A])^{-1/3}$; for $n = 3$
Salinity correction coefficient, β_s	$1.86\,[\mathrm{K\,kg\,mol^{-1}}]$
Salt content in marine ice, $[N_0]$	$2/1000$
Brine content, $[S]$	0.05
Oceanic dissipation, $[\delta_{\mathrm{oc}}]$	$(v^+ - v^-)\tau_{\mathrm{oc}} \approx 7.9 \times 10^{-4}\,\mathrm{kg\,s^{-3}}$
Grounding dissipation, $[\delta_{\mathrm{gr}}]$	$0.79\,\mathrm{kg\,s^{-3}}$

and 11.3. They are typical of most ice shelves. Finally, we emphasize once more that the essential difference between the non-dimensionalization of the ice-sheet and ice-shelf equations is the scaling of these stresses[10]. For ice sheets all stresses are small of $\mathcal{O}(\epsilon)$ or $\mathcal{O}(\epsilon^2)$ except the (mean) pressure which is $\mathcal{O}(\epsilon^0)$. For shelves all stress components are $\mathcal{O}(\epsilon^0)$ except the shear stresses t^D_{xz}, t^D_{yz} which are $\mathcal{O}(\epsilon)$.

[10] We again urge the reader to pause here and to ask himself/herself about the form of the scalings and the orders of magnitude of the $[\cdot]$ quantities. Most critical of all these quantities are the stresses which contain a factor $\varrho := (\rho_{\mathrm{sw}} - \rho)/\rho_{\mathrm{sw}} \approx 0.09$, the question being why? At this stage of the computations this factor is not readily understandable. It may be made plausible by stating that ice shelves are floating and only the difference between the gravity force and the buoyancy force of the shelf are dynamically significant. This difference is exactly proportianal to $(\rho_{\mathrm{sw}} - \rho)/\rho_{\mathrm{sw}}$.

11.6.1 Non-Dimensionalized Ice Shelf Equations

As was done in Subsect. 11.3.1 for ice sheets, the governing equations are now put in dimensionless form by substituting the scalings (11.92) into the field equations and boundary conditions. Below we list these equations in dimensionless form, but will delete the tilde that characterizes the non-dimensionality of the variables in question.

(i) Field equations for ice shelves. These are

$$\frac{\partial v_x}{\partial x}+\frac{\partial v_y}{\partial y}+\frac{\partial v_z}{\partial z}=0\,, \tag{11.96}$$

$$\frac{1}{\epsilon}\frac{\mathsf{F}}{\varrho}\frac{dv_x}{dt}=-\frac{\partial p}{\partial x}+\frac{\partial t^D_{xx}}{\partial x}+\frac{\partial t^D_{xy}}{\partial y}+\frac{\partial t^D_{xz}}{\partial z}\,, \tag{11.97}$$

$$\frac{1}{\epsilon}\frac{\mathsf{F}}{\varrho}\frac{dv_y}{dt}=\frac{\partial t^D_{xy}}{\partial x}-\frac{\partial p}{\partial y}+\frac{\partial t^D_{yy}}{\partial y}+\frac{\partial t^D_{yz}}{\partial z}\,, \tag{11.98}$$

$$\epsilon\frac{\mathsf{F}}{\varrho}\frac{dv_z}{dt}=\epsilon^2\frac{\partial t^D_{xz}}{\partial x}+\epsilon^2\frac{\partial t^D_{yz}}{\partial y}-\frac{\partial p}{\partial z}+\frac{\partial t^D_{zz}}{\partial z}-\frac{1}{\varrho}\,, \tag{11.99}$$

$$\begin{aligned}&\frac{\partial\theta}{\partial t}+v_x\frac{\partial\theta}{\partial x}+v_y\frac{\partial\theta}{\partial y}+v_z\frac{\partial\theta}{\partial z}\\&=\frac{\mathcal{D}}{c}\left[\epsilon^2\frac{\partial}{\partial x}\left(\kappa\frac{\partial\theta}{\partial x}\right)+\epsilon^2\frac{\partial}{\partial y}\left(\kappa\frac{\partial\theta}{\partial y}\right)+\frac{\partial}{\partial z}\left(\kappa\frac{\partial\theta}{\partial z}\right)\right]+2\frac{\varrho^2\alpha}{c}\mathcal{K}_S A(\theta')f(\sigma)\sigma^2\,,\end{aligned} \tag{11.100}$$

$$\theta'=\theta+\varrho\mathcal{B}p\,, \tag{11.101}$$

$$\frac{\partial v_x}{\partial x}=\varrho\mathcal{K}_S A(\theta')f(\sigma)t^D_{xx}\,, \tag{11.102}$$

$$\frac{\partial v_y}{\partial y}=\varrho\mathcal{K}_S A(\theta')f(\sigma)t^D_{yy}\,, \tag{11.103}$$

$$\frac{\partial v_z}{\partial z}=\varrho\mathcal{K}_S A(\theta')f(\sigma)t^D_{zz}\,, \tag{11.104}$$

$$\frac{\partial v_x}{\partial y}+\frac{\partial v_y}{\partial x}=2\varrho\mathcal{K}_S A(\theta')f(\sigma)t^D_{xy}\,, \tag{11.105}$$

$$\frac{\partial v_x}{\partial z}+\epsilon^2\frac{\partial v_z}{\partial x}=2\epsilon^2\varrho\mathcal{K}_S A(\theta')f(\sigma)t^D_{xz}\,, \tag{11.106}$$

$$\frac{\partial v_y}{\partial z}+\epsilon^2\frac{\partial v_z}{\partial y}=2\epsilon^2\varrho\mathcal{K}_S A(\theta')f(\sigma)t^D_{yz}\,, \tag{11.107}$$

$$\sigma=\sqrt{\tfrac{1}{2}\left[(t^D_{xx})^2+(t^D_{yy})^2+(t^D_{zz})^2\right]+(t^D_{xy})^2+\epsilon^2\left[(t^D_{xz})^2+(t^D_{yz})^2\right]}\,. \tag{11.108}$$

These are the continuity equation, three components of the momentum equations, the energy equation, the pressure-adjusted homologous temperature, six components of the symmetric stretching-stress-deviator relation and, finally, the expression for the effective stress. These equations involve the aspect ratio ϵ and some other dimensionless products defined in (11.95). They allow the following physical interpretations: The continuity equation appears in its preserved form $\operatorname{div} \boldsymbol{v} = 0$; exact satisfaction of the balance of mass is therefore thought significant. This equally implies that the condition $\operatorname{tr} \boldsymbol{t}^D = 0$ must be satisfied at all levels of perturbation if such a perturbation is pursued, see (11.102)–(11.104). The momentum equations (11.97)–(11.99) show that acceleration terms can be ignored at least as long as terms of order 10^{-11} are not accounted for. (Note, $\mathsf{F} \approx 10^{-15}, \epsilon \approx 10^{-3} \to \mathsf{F}/(\varrho\epsilon) \approx 10^{-11}$). This is the case as long as perturbation expansions are restricted to $\mathcal{O}(\epsilon^3)$ terms. Thus we conclude with the[11]

Formal statement 2: *Stokes flow is a justified approximation of ice shelf flows as long as perturbation expansions or equivalent iterative procedures are not pushed beyond $\mathcal{O}(\epsilon^3)$ terms.*

Since our approximations will not go beyond $\mathcal{O}(\epsilon^3)$-terms we shall limit our attention to Stokes flows only. Scrutiny of the force balances (11.97)–(11.98) then shows that the dominant stress contributions are those parallel to the (x, y)-plane, and all stress components contribute. On the other hand, the vertical force balance (11.99) states that, to lowest order, the vertical normal stress gradient is balanced by the buoyancy force. This may be interpreted as a cryostatic balance, but it involves $t_{zz} = -p + t^D_{zz}$ and thus contains a materially dependent contribution. The heat equation (11.100) implies that, to lowest order, advection and strain heating must be fully accounted for, but that horizontal diffusion is small of $\mathcal{O}(\epsilon^2)$.

The material equations (11.106), (11.107) imply that $\partial v_x/\partial z$ and $\partial v_y/\partial z$ are small of $\mathcal{O}(\epsilon^2)$; this is so despite the fact that v_x and v_y themselves are of $\mathcal{O}(1)$. This implies that to $\mathcal{O}(1)$ and $\mathcal{O}(\epsilon)$ the horizontal velocity components cannot depend on the z-coordinate. A possible z-dependence only enters at the $\mathcal{O}(\epsilon^2)$-level. Any student of strength of materials learns that bending of beams or plates arises because in-plane deformations (here horizontal velocities) are nonuniform in the direction normal to the beam axis or plate plane. If such variations can be ignored or are absent the beam behaves as a rod and the plate as a membrane with internal forces/deformations being longitudinal and parallel to the plane, respectively. Thus, we have

[11] In fluid mechanics, flows which are mathematically constructed from the governing equations by omitting in the momentum equations the acceleration terms (left-hand sides in (11.97)–(11.99)) are called Stokes flows. Creeping flow is often simply said to be *Stokes flow*, however, as the above scaling analysis shows, it is an approximation and can only be fulfilled to a certain degree of approximation, if Froude numbers are sufficiently small. This fact simultaneously demonstrates the usefulness of the scaling and asymptotic approach.

Formal statement 3: *In an expansion accounting for $\mathcal{O}(\epsilon^0)$ and $\mathcal{O}(\epsilon^1)$-terms the ice shelf deforms as a membrane, or: zeroth (SSA) and first order (FOSSA) shallow shelf approximations are membrane theories. Bending effects are necessarily of $\mathcal{O}(\epsilon^2)$ or smaller.*

The reader might be interested to learn that differences in the horizontal velocities at the free surface and the ice–ocean interface have been observed in the Ross Ice Shelf, (Doake, personal communication), but systematic measurements have so far not been taken.

The dual represenations to (11.102)–(11.107) and (11.108) read

$$
\begin{aligned}
t^D_{xx} &= \frac{1}{\varrho}\mathcal{K}'_S B(\theta')g(d)\frac{\partial v_x}{\partial x}\,, && (11.109)\\
t^D_{yy} &= \frac{1}{\varrho}\mathcal{K}'_S B(\theta')g(d)\frac{\partial v_y}{\partial y}\,, hutter2 && (11.110)\\
t^D_{zz} &= \frac{1}{\varrho}\mathcal{K}'_S B(\theta')g(d)\frac{\partial v_z}{\partial z}\,, && (11.111)\\
t^D_{xy} &= \frac{1}{2\varrho}\mathcal{K}'_S B(\theta')g(d)\left(\frac{\partial v_x}{\partial y}+\frac{\partial v_y}{\partial x}\right), && (11.112)\\
\epsilon^2 t^D_{xz} &= \frac{1}{2\varrho}\mathcal{K}'_S B(\theta')g(d)\left(\frac{\partial v_x}{\partial z}+\epsilon^2\frac{\partial v_z}{\partial x}\right), && (11.113)\\
\epsilon^2 t^D_{yz} &= \frac{1}{2\varrho}\mathcal{K}'_S B(\theta')g(d)\left(\frac{\partial v_y}{\partial z}+\epsilon^2\frac{\partial v_z}{\partial y}\right), && (11.114)
\end{aligned}
$$

$$
\begin{aligned}
d = \Bigg[&\frac{1}{2}\left\{\left(\frac{\partial v_x}{\partial x}\right)^2+\left(\frac{\partial v_y}{\partial y}\right)^2+\left(\frac{\partial v_z}{\partial z}\right)^2\right\}+\frac{1}{4}\left(\frac{\partial v_x}{\partial y}+\frac{\partial v_y}{\partial x}\right)^2\\
&+\left(\frac{1}{\epsilon}\frac{\partial v_x}{\partial z}+\epsilon\frac{\partial v_z}{\partial x}\right)^2+\left(\frac{1}{\epsilon}\frac{\partial v_y}{\partial z}+\epsilon\frac{\partial v_z}{\partial y}\right)^2\Bigg]^{\frac{1}{2}}. \qquad (11.115)
\end{aligned}
$$

They also lead to the formal statement 3, and so the two terms in (11.115) premultiplied by ϵ^{-1} are actually of $\mathcal{O}(\epsilon)$. Before we list the boundary conditions in dimensionless form it is worth illustrating the differences of the ice sheet and ice shelf approximations by scrutinising the formulas of the effective stresses as listed in (11.34) and (11.124), respectively. We have

$$
\begin{aligned}
\sigma^2_{\text{sheet}} &= \left(t^D_{xz}\right)^2+\left(t^D_{yz}\right)^2+\left(\sigma^{\text{corr}}_{\text{sheet}}\right)^2\,,\\
\sigma^2_{\text{shelf}} &= \frac{1}{2}\left[\left(t^D_{xx}\right)^2+\left(t^D_{yy}\right)^2+\left(t^D_{zz}\right)^2+\left(t^D_{xy}\right)^2\right]+\left(\sigma^{\text{corr}}_{\text{shelf}}\right)^2\,,
\end{aligned}
$$

in which the corrective terms are $\mathcal{O}(\epsilon)^2$. Interestingly, these corrective terms are in one formulation the dominant terms of the other, illustrating in a nice and complementary fashion the essential differences of the SIA and SSA: Vertical shearing in the first and in-plane stresses in the second. This duality could also be observed if the second stretching invarants of the two formulations would be compared.

(ii) Boundary conditions at the free surface. These apply at $z = h(x,y,t)$ and are given by (11.7), (11.8); their dimensionless counterparts are

$$\frac{\partial h}{\partial t} + v_x \frac{\partial h}{\partial x} + v_y \frac{\partial h}{\partial y} - v_z = N_s\, a_s^{\perp}\,, \quad N_s = \sqrt{1+\epsilon^2 \left(\frac{\partial h}{\partial x}\right)^2 + \epsilon^2 \left(\frac{\partial h}{\partial y}\right)^2} \tag{11.116}$$

$$-\left(-p + t_{xx}^D\right) \frac{\partial h}{\partial x} - t_{xy}^D \frac{\partial h}{\partial y} + t_{xz}^D = 0\,, \tag{11.117}$$

$$-t_{xy}^D \frac{\partial h}{\partial x} - \left(-p + t_{yy}^D\right) \frac{\partial h}{\partial y} + t_{yz}^D = 0\,, \tag{11.118}$$

$$-\epsilon^2 t_{xz}^D \frac{\partial h}{\partial x} - \epsilon^2 t_{yz}^D \frac{\partial h}{\partial y} + \left(-p + t_{zz}^D\right) = 0\,, \tag{11.119}$$

$$\theta = \theta_s(x,y,z,t)\,. \tag{11.120}$$

As can be inferred from these equations, the shear traction conditions (11.117), (11.118) must be satisfied in their entirety at all perturbation levels. On the other hand, the surface pressure may, to lowest order, be replaced by the vertical normal stress $-p + t_{zz}^D$. The form (11.117)–(11.118) of the shear traction boundary conditions in this dimensionless form (without involving an explicit ϵ-dependence) is physically very significant. Using "naive" scaling arguments, one would expect an ϵ-factor of the terms involving $\partial h/\partial x$ and $\partial h/\partial y$. That these ϵ-factors are not present in (11.117) and (11.118) means that the shear stresses t_{xz}^D and t_{yz}^D do not vanish whenever the free surface is not horizontal. (An analogous argument also applies to (11.122) and (11.123)). These nonvanishing shear stresses are obviously due to the thickness variations of the ice shelf (which are small but always present), and they induce at $\mathcal{O}(\epsilon^2)$ the bending effects. A naive sealing would have eliminated these ab initio.

(iii) Boundary conditions at the ice-shelf base. The ice-shelf base, $z = b(x,y,t)$, is either the ice–ocean interface or, at rumpled regions, the contact surface with the rock bed. The kinematic, traction and thermal boundary conditions are

$$\frac{\partial b}{\partial t} + v_x \frac{\partial b}{\partial x} + v_y \frac{\partial b}{\partial y} - v_z = -N_b\,, \quad N_b = \sqrt{1+\epsilon^2 \left(\frac{\partial b}{\partial x}\right)^2 + \epsilon^2 \left(\frac{\partial b}{\partial y}\right)^2}\,, \tag{11.121}$$

$$-\left(-p + t_{xx}^D\right) \frac{\partial b}{\partial x} - t_{xy}^D \frac{\partial b}{\partial y} + t_{xz}^D = p_b \frac{\partial b}{\partial x} - \epsilon \tau_{\mathrm{gr}} e_x N_b\,, \tag{11.122}$$

$$-t_{xy}^D \frac{\partial b}{\partial x} - \left(-p + t_{yy}^D\right) \frac{\partial b}{\partial y} + t_{yz}^D = p_b \frac{\partial b}{\partial y} - \epsilon \tau_{\rm gr} e_y \mathcal{N}_b \,, \tag{11.123}$$

$$-\epsilon^2 t_{xz}^D \frac{\partial b}{\partial x} - \epsilon^2 t_{yz}^D \frac{\partial b}{\partial y} + \left(-p + t_{zz}^D\right) = -p_b - \epsilon^3 \tau_{\rm gr} e_z \mathcal{N}_b \,, \tag{11.124}$$

$$\theta = \theta^+ \,, \tag{11.125}$$

$$\kappa \left(\epsilon^2 \frac{\partial \theta}{\partial x} \frac{\partial b}{\partial x} + \epsilon^2 \frac{\partial \theta}{\partial y} \frac{\partial b}{\partial y} - \frac{\partial \theta}{\partial z} \right) - \mathcal{N}_b \, q_b^\perp \mathcal{N}_b$$
$$-\mathcal{N}_b \left\{ \mathcal{N}_b \, \delta + \frac{\alpha}{\alpha_S \mathcal{D}} a_b^{\perp M} \right\} H(\theta - \theta_M) = 0 \,, \tag{11.126}$$

where H is the Heaviside function and

$$p_b = \begin{cases} p_{\rm sw} = \dfrac{\rho_{\rm sw}}{\varrho \rho} (z_s - b) \,, \\ p_{\rm gr} = \dfrac{1}{\varrho} (h - b) \,, \end{cases} \qquad \tau_{\rm gr} = \begin{cases} 0 \,, & \text{floating regions} \,, \\ \tau_{\rm gr} \,, & \text{rumpled regions} \,, \end{cases} \tag{11.127}$$

$$\mathcal{N}_b, \delta_b = \begin{cases} \mathcal{N}_{q_{\rm oc}^\perp}, \delta_{\rm oc} \,, \\ \mathcal{N}_{q_{\rm gr}^\perp}, \delta_{\rm gr} \,, \end{cases} \qquad a_b^\perp = \begin{cases} a_{\rm oc}^\perp = a_b^{\perp M} + a_b^{\perp A} \,, & \text{floating regions} \,, \\ \epsilon^2 a_{\rm gr}^\perp \,, & \text{rumpled regions} \,. \end{cases} \tag{11.128}$$

We note that over rumpled regions a sliding law applies. It is seen from (11.128) that melting/freezing and accretion rates over floating regions are of $\mathcal{O}(\epsilon^0)$ whereas melting rates in grounding areas are of $\mathcal{O}(\epsilon^2)$. The shear traction conditions (11.122), (11.123) involve to lowest order all terms except the shear tractions due to sliding over the base in rumpled regions, in agreement of what was already said above. The basal pressure in (11.124), on the other hand is given to lowest order by the vertical normal stress; corrections enter only at $\mathcal{O}(\epsilon^2)$ and $\mathcal{O}(\epsilon^3)$. This is qualitatively analogous to (11.119). Finally, the energy jump condition (11.126) shows that the horizontal components of the heat flux vector do not contribute to lowest order to the heat transfer through the surface.

(iv) Ice thickness evolution equation. If the continuity equation is integrated over depth from $z = b$ to $z = h$ and the kinematic boundary conditions (11.116), (11.121) are used in the emerging equation then the following evolution equation involving h and b is obtained:

$$\frac{\partial (h-b)}{\partial t} + \frac{\partial}{\partial x} \int_b^h v_x {\rm d}z + \frac{\partial}{\partial y} \int_b^h v_y {\rm d}z = N_s \, a_s^\perp + N_b \, a_b^\perp \,, \tag{11.129}$$

with $a_b^\perp = (\boldsymbol{w} - \boldsymbol{v}) \cdot \boldsymbol{n}$.

In what follows zeroth and first order equations for shallow shelves will be presented. However, the analysis will be more detailed than for sheets because the mathematical details are far more involved.

11.6.2 Zeroth Order Ice Shelf Equations

These are obtained from (11.96)–(11.129) by omitting all terms premultiplied by a positive power of ϵ; variables will carry the index $(\cdot)_{(0)}$ to identify that they are of $\mathcal{O}(\epsilon^0)$. Furthermore, we shall for the time being not deal with the thermal problem and also impose the Stokes assumption. The force balances (11.97)–(11.99) and the material equations (11.106)–(11.107) take the forms

$$0 = -\frac{\partial p_{(0)}}{\partial x} + \frac{\partial t^D_{xx(0)}}{\partial x} + \frac{\partial t^D_{xy(0)}}{\partial y} + \frac{\partial t^D_{xz(0)}}{\partial z} , \tag{11.130}$$

$$0 = \frac{\partial t^D_{xy(0)}}{\partial x} - \frac{\partial p_{(0)}}{\partial y} + \frac{\partial t^D_{yy(0)}}{\partial y} + \frac{\partial t^D_{yz(0)}}{\partial z} , \tag{11.131}$$

$$\frac{1}{\varrho} = -\frac{\partial p_{(0)}}{\partial z} + \frac{\partial t^D_{zz(0)}}{\partial z} . \tag{11.132}$$

$$\frac{\partial v_{x(0)}}{\partial z} = 0 , \qquad \frac{\partial v_{y(0)}}{\partial z} = 0 . \tag{11.133}$$

and the continuity equation remains formally unchanged

$$\frac{\partial v_{x(0)}}{\partial x} + \frac{\partial v_{y(0)}}{\partial y} + \frac{\partial v_{z(0)}}{\partial z} = 0 . \tag{11.134}$$

The other stress-deviator-stretching relationships (11.102)–(11.105) or (11.109)–(11.112) are also unchanged (because no ϵ-term arises) and will therefore not be repeated. However, the effective stress and strain rate simplify and read, to zeroth order,

$$\sigma^2_{(0)} = (t^D_{xx(0)})^2 + (t^D_{yy(0)})^2 + t^D_{xx(0)} t^D_{yy(0)} + (t^D_{xy(0)})^2 , \tag{11.135}$$

$$d^2_{(0)} = \left(\frac{\partial v_{x(0)}}{\partial x}\right)^2 + \left(\frac{\partial v_{y(0)}}{\partial y}\right)^2 + \frac{\partial v_{x(0)}}{\partial x}\frac{\partial v_{y(0)}}{\partial y} + \frac{1}{4}\left(\frac{\partial v_{x(0)}}{\partial y} + \frac{\partial v_{y(0)}}{\partial x}\right)^2 . \tag{11.136}$$

The kinematic and traction boundary conditions (11.116)–(11.119) become at zeroth order

$$\frac{\partial h_{(0)}}{\partial t} + v_{x(0)}\frac{\partial h_{(0)}}{\partial x} + v_{y(0)}\frac{\partial h_{(0)}}{\partial y} - v_{z(0)} = a^{\perp}_{s(0)} , \tag{11.137}$$

$$\left(p_{(0)} - t^D_{xx(0)}\right)\frac{\partial h_{(0)}}{\partial x} - t^D_{xy(0)}\frac{\partial h_{(0)}}{\partial y} + t^D_{xz(0)} = 0 , \tag{11.138}$$

$$-t^D_{xy(0)}\frac{\partial h_{(0)}}{\partial x} + \left(p_{(0)} - t^D_{yy(0)}\right)\frac{\partial h_{(0)}}{\partial y} + t^D_{yz(0)} = 0 , \tag{11.139}$$

$$-p_{(0)} + t^D_{zz(0)} = 0 , \tag{11.140}$$

at $z = h_{(0)}(x,y,t)$. At the base one must differentiate between floating and rumpled regions. From (11.121) to (11.128) it is easy to deduce that to zeroth order,

$$\frac{\partial b_{(0)}}{\partial t} + v_{x(0)}\frac{\partial b_{(0)}}{\partial x} + v_{y(0)}\frac{\partial b_{(0)}}{\partial y} - v_{z(0)} = -a_b^{\perp}{}_{(0)} \,, \tag{11.141}$$

$$-\left(-p_{(0)} + t^D_{xx(0)}\right)\frac{\partial b_{(0)}}{\partial x} - t^D_{xy(0)}\frac{\partial b_{(0)}}{\partial y} + t^D_{xz(0)} = p_{b(0)}\frac{\partial b_{(0)}}{\partial x} \,, \tag{11.142}$$

$$-t^D_{xy(0)}\frac{\partial b_{(0)}}{\partial x} - \left(-p_{(0)} + t^D_{yy(0)}\right)\frac{\partial b_{(0)}}{\partial y} + t^D_{yz(0)} = p_{b(0)}\frac{\partial b_{(0)}}{\partial y} \,, \tag{11.143}$$

$$-p_{(0)} + t^D_{zz(0)} = -p_{b(0)} \,, \tag{11.144}$$

at $z = b_{(0)}(x,y,t)$, where

$$a_b^{\perp}{}_{(0)} = \begin{cases} a_b^{\perp M}{}_{(0)} + a_b^{\perp A}{}_{(0)} \,, \\ 0 \,, \end{cases} \qquad p_{b(0)} = \begin{cases} p_{\mathrm{sw}(0)} = \dfrac{\rho_{\mathrm{sw}}}{\varrho\rho}(z_s - b_{(0)}) \,, & \text{floating regions} \\ \dfrac{1}{\varrho}(h_{(0)} - b_{(0)}) \,, & \text{rumpled regions} \,. \end{cases}$$

These zeroth order equations are formally still rather complicated and do not seem to offer essential simplification when compared with the original three-dimensional equations. The relations (11.133), which state that the horizontal velocity components are functions of the z-coordinate, however suggest that thickness averaged equations, or equations that are integrated over the ice-shelf thickness may yield this simplification. The appearance of the shear stresses $t^D_{xz(0)}$ and $t^D_{yz(0)}$ in the force balances (11.130) and (11.131) may look counter-productive in this regard, but a thickness integration of these equations shows that these stresses have to be evaluated at the surfaces and can be eliminated with the aid of (11.138), (11.139), (11.142) and (11.143). These observations are indications that vertical integration is the operation which achieves the desired simplification.

Vertical integration. If we integrate the vertical force balance (11.132) over z and incorporate the boundary conditions (11.140), (11.144) – note there is one boundary condition too many, so $(h - z_s)$ and $(z_s - b)$ can be related to one another by the so-called *floating condition* – it is found that

$$t_{zz(0)}(\cdot, z) = \frac{1}{\varrho}(z - z_s) - H_{(0)} \,, \quad p_{\mathrm{sw}(0)} = \frac{1}{\varrho}H_{(0)} \,, \quad \text{floating regions} \,. \tag{11.145}$$

So the floating condition motivated in Sect. 11.2 is indeed satisfied. In rumpled regions the analogous steps yield

$$t_{zz(0)}(\cdot, z) = \frac{1}{\varrho}(h_{(0)} - z) \,, \quad \text{rumpled regions} \,, \tag{11.146}$$

implying that $t_{zz(0)}(\cdot, h_{(0)}) = 0$ and $t_{zz(0)}(\cdot, b_{(0)}) = -H_{(0)}/\varrho$, demonstrating consistency in the formulation made with ice rumples. We emphasise, that vertical integration of the vertical force balance together with the associated boundary conditions (including the floating condition) has determined $t_{zz(0)}$ as an unknown field variable.

Next, we determine the *vertical component of the velocity*, $v_{z(0)}$. Integrating (11.134) with the proviso (11.133) then yields

$$v_{z(0)}(.,z) = -\nabla_H \cdot \boldsymbol{v}^H_{(0)}(z - c_0(x,y,t)) \,, \tag{11.147}$$

where ∇_H is the horizontal gradient operator and $\boldsymbol{v}^H_{(0)} = (v_{x(0)}, v_{y(0)})$; c_0 follows best by evaluating (11.147) at $z = h_{(0)}$ and using (11.137). This yields

$$v_{z(0)}(.,z) = (\nabla_H \cdot \boldsymbol{v}^H_{(0)})(h_{(0)} - z) + \frac{\partial h_{(0)}}{\partial t} + \boldsymbol{v}_{(0)} \cdot \nabla_H h_{(0)} - a_s^{\perp}{}_{(0)} \tag{11.148}$$

which holds true for floating as well as rumple regions.

Next, the *thickness evolution equation* for the ice shelf follows from (11.129) by omitting the $\mathcal{O}(\epsilon^2)$-terms and accounting for the fact that v_x, v_y are z-independent,

$$\frac{\partial H_{(0)}}{\partial t} + \frac{\partial}{\partial x}\left(H_{(0)} v_{x(0)}\right) + \frac{\partial}{\partial y}\left(H_{(0)} v_{y(0)}\right) = a_s{}^{\perp}_{(0)} + a_b{}^{\perp}_{(0)} \,, \tag{11.149}$$

where

$$a_b{}^{\perp}_{(0)} = \begin{cases} a_b^{\perp}{}^M_{(0)} + a_b^{\perp}{}^A_{(0)} \,, & \text{floating regions} \,, \\ 0 \,, & \text{rumpled regions} \,, \end{cases} \tag{11.150}$$

which are known either from the climatological input or from interface thermodynamics.

Last, the horizontal force balance equations are integrated over the ice shelf thickness. To this end, the first step is to introduce the *membrane stress tensor* N_{ij} as the depth integral of the stress deviators t^D_{ij}

$$N_{ij} = \int_b^h t^D_{ij}\,\mathrm{d}z, \quad \Rightarrow \quad N_{ij(0)} := \int_{b_{(0)}}^{h_{(0)}} t^D_{ij(0)}\mathrm{d}z \,, \quad i,j = x,y \,. \tag{11.151}$$

In the second step the vertical normal stress $t^D_{zz(0)} = -(t^D_{xx(0)} + t^D_{yy(0)})$, given by (11.145) and (11.146) is substituted into the horizontal force balances (11.130), (11.131). The result is given in (11.152) below in Table 11.4. Third, these equations are then integrated over the ice sheet thickness from $z = b_{(0)}$ to $z = h_{(0)}$ and the definitions (11.151) are used. This leads to (11.153). The last step consists in using the constitutive relations (11.109)–(11.112) and to express $N_{ij(0)}$ in terms of thickness integrals of the rate factor B. The result is shown in (11.154) with $\bar{\nu}$ as expressed in (11.155). Notice that for a power law $g(d_{(0)}) = d_{(0)}{}^{(1-n)/n}$, $n > 1$ which is singular at $d_{(0)} = 0$. So, a finite viscosity law should be used, and we suggest to use the law (11.156). Its limit behaviours are $g(0) = \eta$ and $g(\infty) = d_{(0)}{}^{(1-n)/n}$.

Table 11.4. Zeroth order ice shelf equations (for details see main text). The repeated right-hand sides are for floating and rumpled regions, respectively

equations	floating region	rumpled region	
$2\frac{\partial t^D_{xx(0)}}{\partial x} + \frac{\partial t^D_{yy(0)}}{\partial x} + \frac{\partial t^D_{xy(0)}}{\partial y} + \frac{\partial t^D_{xz(0)}}{\partial z}$	$= \frac{\partial H_{(0)}}{\partial x}$	$= \frac{1}{\varrho}\frac{\partial h_{(0)}}{\partial x}$	(11.152)
$\frac{\partial t^D_{xx(0)}}{\partial y} + 2\frac{\partial t^D_{yy(0)}}{\partial y} + \frac{\partial t^D_{xy(0)}}{\partial x} + \frac{\partial t^D_{yz(0)}}{\partial z}$	$= \frac{\partial H_{(0)}}{\partial y}$	$= \frac{1}{\varrho}\frac{\partial h_{(0)}}{\partial y}$	
$2\frac{\partial N_{xx(0)}}{\partial x} + \frac{\partial N_{yy(0)}}{\partial x} + \frac{\partial N_{xy(0)}}{\partial y}$	$= H_{(0)}\frac{\partial H_{(0)}}{\partial x}$	$= \frac{1}{\varrho}H_{(0)}\frac{\partial h_{(0)}}{\partial x}$	
$\frac{\partial N_{xx(0)}}{\partial y} + 2\frac{\partial N_{yy(0)}}{\partial y} + \frac{\partial N_{xy(0)}}{\partial x}$	$= H_{(0)}\frac{\partial H_{(0)}}{\partial y}$	$= \frac{1}{\varrho}H_{(0)}\frac{\partial h_{(0)}}{\partial y}$	(11.153)
$2\frac{\partial}{\partial x}\left(\bar{\nu}\frac{\partial v_{x(0)}}{\partial x}\right) + \frac{\partial}{\partial x}\left(\bar{\nu}\frac{\partial v_{y(0)}}{\partial y}\right) + \frac{1}{2}\frac{\partial}{\partial y}\left[\bar{\nu}\left(\frac{\partial v_{x(0)}}{\partial y} + \frac{\partial v_{y(0)}}{\partial x}\right)\right]$	$= \varrho H_{(0)}\frac{\partial H_{(0)}}{\partial x}$	$= H_{(0)}\frac{\partial h_{(0)}}{\partial x}$	(11.154)
$\frac{\partial}{\partial y}\left(\bar{\nu}\frac{\partial v_{x(0)}}{\partial x}\right) + 2\frac{\partial}{\partial y}\left(\bar{\nu}\frac{\partial v_{y(0)}}{\partial y}\right) + \frac{1}{2}\frac{\partial}{\partial x}\left[\bar{\nu}\left(\frac{\partial v_{x(0)}}{\partial y} + \frac{\partial v_{y(0)}}{\partial x}\right)\right]$	$= \varrho H_{(0)}\frac{\partial H_{(0)}}{\partial y}$	$= H_{(0)}\frac{\partial h_{(0)}}{\partial y}$	

where

$$\bar{\nu} = \mathcal{K}'_s g(d_{(0)}) \int_{b_{(0)}}^{h_{(0)}} B(\theta')\,\mathrm{d}z, \tag{11.155}$$

with

$$g(d_{(0)}) = \frac{\eta}{1 + \eta\, {d_{(0)}}^{(n-1)/n}} \tag{11.156}$$

The governing equations of this SSA are the thickness evolution equation (11.149) and the equilibrium equations (11.154) expressed in terms of the velocities. The former is hyperbolic and driven by the accumulation rate functions $a_{s(0)}^{\perp} + a_{b(0)}^{\perp}$; in numerical computations this equation is prone to instabilities so non-oscillating central (NOC) schemes with cell limiters should be used. The latter is elliptic and driven by the horizontal thickness gradients; subject to appropriate boundary conditions these equations are solved first for a given thickness distribution, and (11.149) is solved afterwards for the new geometry. This must be done together with an updating of the temperature field (which we have not demonstrated here). Once this is done, (11.109)–(11.114) may be used to determine the local stress components, and (11.147) will yield the distribution of the vertical velocity component.

Of course, the solution of (11.154) must be constructed subject to boundary conditions appropriate for this elliptical problem. Along the grounding line the flux $\boldsymbol{v}_{(0)}H_{(0)}$ is prescribed as the flow from the inland ice, along the side boundaries one best prescribes the no-slip condition $\boldsymbol{v}_{(0)} = \mathbf{0}$ and along the ice shelf front a kinematic equation and a calving rate equation are to be prescribed.

11.6.3 First Order Mechanical Ice Shelf Equations

The above zeroth order solution is inaccurate in rumpled regions and close to side boundaries. It can be improved by constructing the first- and eventually second-order correction. This will be done now. The equations will be somewhat involved, but their derivation has also the merit to demonstrate that the asymptotic approach is correct insofar as it does not develop singularities.

The first order force balances, (11.97)–(11.99), material equations (11.106)–(11.107) and continuity equation (11.96) take the forms

$$0 = -\frac{\partial p_{(1)}}{\partial x} + \frac{\partial t^D_{xx(1)}}{\partial x} + \frac{\partial t^D_{xy(1)}}{\partial y} + \frac{\partial t^D_{xz(1)}}{\partial z}\,, \tag{11.157}$$

$$0 = \frac{\partial t^D_{xy(1)}}{\partial x} - \frac{\partial p_{(1)}}{\partial y} + \frac{\partial t^D_{yy(1)}}{\partial y} + \frac{\partial t^D_{yz(1)}}{\partial z}\,, \tag{11.158}$$

$$0 = -\frac{\partial p_{(1)}}{\partial z} + \frac{\partial t^D_{zz(1)}}{\partial z}\,, \tag{11.159}$$

$$\frac{\partial v_{x(1)}}{\partial z} = 0\,, \qquad \frac{\partial v_{y(1)}}{\partial z} = 0\,, \tag{11.160}$$

$$\frac{\partial v_{x(1)}}{\partial x} + \frac{\partial v_{y(1)}}{\partial y} + \frac{\partial v_{z(1)}}{\partial z} = 0\,, \tag{11.161}$$

whilst the first order effective stress and strain rates are given by

$$2\sigma_{(0)}\sigma_{(1)} = t^D_{xx(0)}t^D_{xx(1)} + t^D_{yy(0)}t^D_{yy(1)} + t^D_{zz(0)}t^D_{zz(1)} + t^D_{xy(0)}t^D_{xy(1)}\,, \tag{11.162}$$

$$2d_{(0)}d_{(1)} = \frac{\partial v_{x(0)}}{\partial x}\frac{\partial v_{x(1)}}{\partial x} + \frac{\partial v_{y(0)}}{\partial y}\frac{\partial v_{y(1)}}{\partial y} + \frac{\partial v_{z(0)}}{\partial z}\frac{\partial v_{z(1)}}{\partial z} + \frac{1}{2}\frac{\partial v_{x(0)}}{\partial y}\frac{\partial v_{x(1)}}{\partial y}$$
$$+\frac{1}{2}\left(\frac{\partial v_{x(0)}}{\partial y}\frac{\partial v_{y(1)}}{\partial x} + \frac{\partial v_{x(1)}}{\partial y}\frac{\partial v_{y(0)}}{\partial x}\right) + \frac{1}{2}\frac{\partial v_{y(0)}}{\partial x}\frac{\partial v_{y(1)}}{\partial x} \,. \qquad (11.163)$$

It is seen from (11.157)–(11.159) that the equilibrium equations are formally the same as those of zeroth order, see (11.130)–(11.132) except that they are here homogeneous. Equation (11.160) implies that $v_{x(1)}$ and $v_{y(1)}$ are z-independent and (11.161) requires the velocity field to be solenoidal. All this is very similar to the zeroth order model and thus we should not be surprised if similar results may emerge. The above equations suffice in explaining to the reader how the computations principally proceed. Nevertheless, for the explicit computations the stress-deviator stretching relations and the kinematic and traction boundary conditions at the free surface and the base must also be used. For the benefit of the reader these are stated in the Appendix.

The above equations together with those listed in Appendix A3 are formidably looking equations, and it does not seem possible to use them sensibly. However, it turns out that integration through depth simplifies the structure of the model considerably. Therefore, from (11.160) and (11.161) we have

$$v_{z(1)}(\cdot, z) = -\left(\nabla_H \cdot \boldsymbol{v}^H_{(1)}\right)(z + c_1(x,y,t)) \,, \qquad (11.164)$$

and if the kinematic boundary condition (11.203) in Appendix A3 is employed to determine c_1, we have

$$v_{z(1)}(\cdot, z) = -\left(\nabla_H \cdot \boldsymbol{v}^H_{(1)}\right) z + \frac{\partial h_{(1)}}{\partial t} + \nabla_H \cdot \left(\boldsymbol{v}^H_{(0)} h_{(1)} + \boldsymbol{v}^H_{(1)} h_{(0)}\right)$$
$$-a^\perp_{s\,(1)} - \frac{\partial a^\perp_{s\,(0)}}{\partial z}\Big|_{h_{(0)}} h_{(1)} \,, \qquad (11.165)$$

valid over floating as well as rumpled regions.
This equation can be combined with (11.148) to yield the vertical velocity component to first order as follows

$$v_z(\cdot, z) = (\nabla_H \cdot \boldsymbol{v}_H)(h - z) - a^\perp_s + \mathcal{O}(\epsilon^2) \,, \qquad (11.166)$$

where $v_z = v_{z(0)} + \epsilon v_{z(1)}$, etc.
Vertical integration. We integrate the vertical force balance (11.159) subject to the boundary conditions (11.206), (11.210), (11.214) of Appendix A3, (11.145) and (11.146). This yields

$$t_{zz(1)} = -p_{(1)}(.,z) + t^D_{zz(1)}(.,z) = -\frac{h_{(1)}}{\varrho} \qquad (11.167)$$

and the *floating condition*

$$\varrho b_{(1)} = (\varrho - 1) h_{(1)} \,, \qquad \text{floating region} \,. \qquad (11.168)$$

Equation (11.167) states that $t_{zz(1)}$ is z-independent, but obviously (note $\operatorname{tr} \boldsymbol{t}^D_{(1)} = 0$ is used)

$$-p_{(1)}(.,z) = -\frac{h_{(1)}}{\varrho} + t^D_{xx(1)}(.,z) + t^D_{yy(1)}(.,z) \tag{11.169}$$

is not, since $t^D_{xx(1)}$ and $t^D_{yy(1)}$ may be z-dependent; (11.168) shows that $b_{(1)}$ and $h_{(1)}$ have different signs (as they ought to). In rumpled regions (11.168) does not apply, but

$$\left.\begin{aligned} &b_{(1)} = 0\,, && \text{rigid base, or} \\ &\frac{\partial b_{(1)}}{\partial t} = -\frac{1}{\mathcal{T}_r}\left(b_{(1)} - b_{st(1)} + \frac{\rho}{\rho_a}H_{(1)}\right), && \\ & && \text{deformable base} \end{aligned}\right\} \text{over rumples}\,. \tag{11.170}$$

The next step is the substitution of the result (11.169) into the horizontal force balances (11.157)–(11.158). The result is

$$\frac{2\partial t^D_{xx(1)}}{\partial x} + \frac{\partial t^D_{yy(1)}}{\partial x} + \frac{\partial t^D_{xy(1)}}{\partial y} + \frac{\partial t^D_{xz(1)}}{\partial z} = \frac{1}{\varrho}\frac{\partial h_{(1)}}{\partial x}\,, \tag{11.171}$$

$$\frac{\partial t^D_{xx(1)}}{\partial y} + \frac{2\partial t^D_{yy(1)}}{\partial y} + \frac{\partial t^D_{xy(1)}}{\partial x} + \frac{\partial t^D_{yz(1)}}{\partial z} = \frac{1}{\varrho}\frac{\partial h_{(1)}}{\partial y}\,. \tag{11.172}$$

for both floating and rumpled regions. Finally, we now integrate these equations over the ice-shelf depth. However, this is not so straight-forward. The difficulty lies in the definition of the zeroth, first and second order membrane forces N_{ij}. To see the difficulty, consider

$$\begin{aligned} F :=& \int_b^h f\,\mathrm{d}z = \int_{b_{(0)}+\epsilon b_{(1)}+\epsilon^2 b_{(2)}+\cdots}^{h_{(0)}+\epsilon h_{(1)}+\epsilon^2 h_{(2)}+\cdots} \left(f_{(0)} + \epsilon f_{(1)} + \epsilon^2 f_{(2)} + \cdots\right)\mathrm{d}z \\ =& \cdots\cdots\cdots\cdots \\ =& \left\{\int_{b_{(0)}}^{h_{(0)}} f_{(0)}\,\mathrm{d}z\right\} + \epsilon\left\{\int_{b_{(0)}}^{h_{(0)}} f_{(1)}\,\mathrm{d}z + f_{(0)}\,|_{h_{(0)}}\,h_{(1)} - f_{(0)}\,|_{b_{(0)}}\,b_{(1)}\right\} \\ &+\epsilon^2\left\{\int_{b_{(0)}}^{h_{(0)}} f_{(2)}\,\mathrm{d}z + \left[f_{(0)}\,|_{h_{(0)}}\,h_{(2)} + \frac{1}{2}\frac{\partial f_{(0)}}{\partial z}\,|_{h_{(0)}}\,h^2_{(1)} + f_{(1)}\,|_{h_{(0)}}\,h_{(1)}\right]\right. \\ &\left. - \left[f_{(0)}\,|_{b_{(0)}}\,b_{(2)} + \frac{1}{2}\frac{\partial f_{(0)}}{\partial z}\,|_{b_{(0)}}\,b^2_{(1)} + f_{(1)}\,|_{b_{(0)}}\,b_{(1)}\right]\right\} + \cdots\cdots \\ =& F_{(0)} + \epsilon F_{(1)} + \epsilon^2 F_{(2)} + \cdots. \end{aligned} \tag{11.173}$$

It is seen that it is the terms in curly brackets which define the zeroth, first, second and higher order integrated quantities and not simply the integral terms like $\int_{b_{(0)}}^{h_{(0)}} f_{(i)}\,\mathrm{d}z$. So, in particular

$$N_{ij\,(1)} := \int\limits_{b_{(0)}}^{h_{(0)}} t^D_{ij\,(1)}\,\mathrm{d}z + t^D_{ij\,(0)}\,|_{h_{(0)}}\, h_{(1)} - t^D_{ij\,(0)}\,|_{b_{(0)}}\, b_{(1)}\,, \qquad (11.174)$$

for $(i,j) = x, y, z$. If we now integrate (11.171), (11.172) over the thickness, we do this for each term, as expanded in (11.173) and use Leibniz' rule to interchange integration with respect to z and differentiation with respect to x and y. The boundary terms that emerge in this process are simplified wherever possible by using the boundary conditions (11.204), (11.205), (11.208), (11.209) in Appendix A3. After massive calculations, what obtained is as follows:

$$2\frac{\partial N_{xx\,(1)}}{\partial x} + \frac{\partial N_{yy\,(1)}}{\partial x} + \frac{\partial N_{xy\,(1)}}{\partial y} + \left\{\begin{matrix} 0 \\ \tau^x_{\mathrm{gr}\,(0)} \end{matrix}\right\}$$

$$= \left\{\begin{array}{ll} \dfrac{\partial H_{(0)}}{\partial x}(h_{(1)} - b_{(1)}) + \dfrac{1}{\varrho} H_{(0)} \dfrac{\partial h_{(1)}}{\partial x}\,, & \text{floating region} \\ \dfrac{1}{\varrho}\dfrac{\partial h_{(0)}}{\partial x}(h_{(1)} - b_{(1)}) + \dfrac{1}{\varrho} H_{(0)} \dfrac{\partial h_{(1)}}{\partial x}\,, & \text{rumpled region} \end{array}\right\} \qquad (11.175)$$

$$2\frac{\partial N_{yy\,(1)}}{\partial y} + \frac{\partial N_{xx\,(1)}}{\partial y} + \frac{\partial N_{xy\,(1)}}{\partial x} + \left\{\begin{matrix} 0 \\ \tau^y_{\mathrm{gr}\,(0)} \end{matrix}\right\}$$

$$= \left\{\begin{array}{ll} \dfrac{\partial H_{(0)}}{\partial y}(h_{(1)} - b_{(1)}) + \dfrac{1}{\varrho} H_{(0)} \dfrac{\partial h_{(1)}}{\partial y}\,, & \text{floating region} \\ \dfrac{1}{\varrho}\dfrac{\partial h_{(0)}}{\partial y}(h_{(1)} - b_{(1)}) + \dfrac{1}{\varrho} H_{(0)} \dfrac{\partial h_{(1)}}{\partial y}\,, & \text{rumpled region} \end{array}\right\} \qquad (11.176)$$

in which $\tau^{x,y}_{\mathrm{gr}\;(0)}$ may, via a sliding law, be related to the zeroth order velocity field

$$\boldsymbol{\tau}_{\mathrm{gr}\,(0)} = C(\cdot)\boldsymbol{v}_{(0)}\,. \qquad (11.177)$$

Unfortunately, it is not convenient as it was in the case for the zeroth order equations to derive a set of partial differential equations for $\boldsymbol{v}_{(1)}$. Nevertheless, the first order membrane forces (11.174) can be related to the zeroth and first order deformation fields by substituting (11.197)–(11.202) in Appendix A3 into the definition (11.174). The result is

$$N_{(ij)(1)} = \begin{pmatrix} \frac{\partial v_{x(0)}}{\partial x} & \frac{1}{2}\left(\frac{\partial v_{x(0)}}{\partial y} + \frac{\partial v_{y(0)}}{\partial x}\right) \\ \frac{1}{2}\left(\frac{\partial v_{x(0)}}{\partial y} + \frac{\partial v_{y(0)}}{\partial x}\right) & \frac{\partial v_{y(0)}}{\partial y} \end{pmatrix} N_{(1)}$$

$$+ \Gamma \begin{pmatrix} \frac{\partial v_{x(1)}}{\partial x} & \frac{1}{2}\left(\frac{\partial v_{x(1)}}{\partial y} + \frac{\partial v_{y(1)}}{\partial x}\right) \\ \frac{1}{2}\left(\frac{\partial v_{x(1)}}{\partial y} + \frac{\partial v_{y(1)}}{\partial x}\right) & \frac{\partial v_{y(1)}}{\partial y} \end{pmatrix}, \tag{11.178}$$

where

$$N_{(1)} := \frac{\mathcal{K}_s{}'}{\varrho} \left\{ \frac{\mathrm{d}g(d_{(0)})}{\mathrm{d}d} \left[\left(\int_{b_{(0)}}^{h_{(0)}} B(\theta'_{(0)}) \mathrm{d}z \right) d_{(1)} + B(\theta'_{(0)}) \,|_{h_{(0)}}\, h_{(1)} \right.\right.$$
$$\left.\left. - B(\theta'_{(0)}) \,|_{b_{(0)}}\, b_{(1)} \right] + g(d_{(0)}) \left(\int_{b_{(0)}}^{h_{(0)}} \frac{\mathrm{d}B(\theta'_{(0)})}{\mathrm{d}\theta'} (\theta'_{(1)}) \mathrm{d}z \right) \right\}, \tag{11.179}$$

and

$$\Gamma = \frac{\mathcal{K}_s{}'}{\varrho} \int_{b_{(0)}}^{h_{(0)}} B(\theta'_{(0)})\, g(d_{(0)})\, \mathrm{d}z\,,$$

and $d_{(0)}$ and $d_{(1)}$ are given in (11.136) and (11.163), respectively.

The kinematic evolution equation for $h_{(1)}$ is obtained by applying the rule (11.173) to the kinematic equation (11.129). The $\mathcal{O}(\epsilon)$-equation is then given by

$$\frac{\partial \left(h_{(1)} - b_{(1)}\right)}{\partial t} + (v_{x(1)} + v_{y(1)})H_{(0)} + (v_{x(0)} + v_{y(0)})(h_{(1)} - b_{(1)})$$
$$= a_{s(1)}^{\perp} + \frac{\partial a_{s(0)}^{\perp}}{\partial z} \,|_{h_{(0)}}\, h_{(1)} + a_{b\,(1)}^{\perp} + \frac{\partial a_{b\,(0)}^{\perp}}{\partial z} \,|_{b_{(0)}}\, b_{(1)}\,. \tag{11.180}$$

This formally completes the presentation of the first order mechanical equations. The unknown fields are $v_{x(1)}$, $v_{y(1)}$, $h_{(1)}$ and $b_{(1)}$. If we think the expressions (11.178), (11.179) for $N_{ij\,(0)}$ formally be substituted in (11.175), (11.176), then these equations form a system of two differential equations involving $v_{x(1)}$, $v_{y(1)}$, $h_{(1)}$ and $b_{(1)}$, provided the zeroth order fields are known. A third equation is given by (11.180) and a fourth by the floating condition (11.168) or the evolution equation of the base (11.170) over rumpled regions. Integration proceeds as for the zeroth order field equations: For a given temperature distribution and a given distribution of $h_{(1)}$ and $b_{(1)}$ at a fixed time t the velocity field is determined.

Equations (11.180) and (11.168) and (11.170) are then used to step forward and determine the new geometry at $t + \Delta t$. Thus we have shown, as with the zeroth order model, that the first order mechanical equations exhibit *membrane structure*, and the integrated balances form a well posed set of equations, if the temperature field is already determined.

11.7 First Order Shallow Shelf Approximation (FOSSA)

11.7.1 General Procedure

As mentioned before, it is not thought worthwhile to also formally develop the temperature field into zeroth and first order contributions and to derive individual boundary value problems for each. It is advantageous to solve the mechanical equations as outlined in Subsects. 11.6.2 and 11.6.3, provided the temperature field and the free surface and basal geometry be given (i.e., functions $\theta(x,y,z,t)$, $h_{(0)}(x,y,t)$, $b_{(0)}(x,y,t)$, $h_{(1)}(x,y,t)$ and $b_{(1)}(x,y,t)$ are known for fixed t). With these prerequisites the zeroth order velocities $v_{x(0)}(x,y,t)$, $v_{y(0)}(x,y,t)$ and $v_{z(0)}(x,y,t)$ and then the corrections $v_{x(1)}(x,y,t)$, $v_{y(1)}(x,y,t)$ and $v_{z(1)}(x,y,t)$ are computed.

The second computational step then consists in evaluating the new geometry by stepping forward a temporal increament Δt and solving the new free surfaces $h_{(0)}$, $h_{(1)}$ and bases $b_{(0)}$, $b_{(1)}$, as equally outlined in Subsects. 11.6.2 and 11.6.3. Adding the two solutions in the sense $f = f_{(0)} + \epsilon f_{(1)}$ yields the total fields with an error of $\mathcal{O}(\epsilon^2)$ and also defines the new geometry of the ice sheet. For this new geometry the new temperature field $\theta = \theta_{(0)} + \epsilon\theta_{(1)}$ is then constructed at $t + \Delta t$, however, not by solving for $\theta_{(0)}$ and $\theta_{(1)}$ individually but for both together.

11.7.2 Determination of the Temperature Field

The idea that the temperature field is determined for the complete field $\theta = \theta_{(0)} + \epsilon\theta_{(1)}$ with an error of $\mathcal{O}(\epsilon^2)$ works, because the scaled thermal field equations and boundary conditions contain no $\mathcal{O}(\epsilon)$-terms. Thus dropping the $\mathcal{O}(\epsilon^2)$-terms in these equations leads to an equation set with an error of $\mathcal{O}(\epsilon^2)$. The equations are

$$\frac{\partial\theta}{\partial t} + v_x\frac{\partial\theta}{\partial x} + v_y\frac{\partial\theta}{\partial y} + v_z\frac{\partial\theta}{\partial z} = \frac{\mathcal{D}}{c}\frac{\partial}{\partial z}\left(\kappa\frac{\partial\theta}{\partial z}\right) + 2\frac{\varrho^2\alpha}{c}\mathcal{K}_S A(\theta')f(\sigma)\sigma^2, \quad (11.181)$$

$$\theta = \theta_s(x,y,z,t) \quad \text{on} \quad z = h(x,y,t)\,, \quad \theta = \theta^+ \quad \text{on} \quad z = b(x,y,t)\,, \quad (11.182)$$

$$-\kappa\frac{\partial\theta}{\partial z} - \mathcal{N}_b a_b^\perp - \left\{\mathcal{N}_b\delta + \frac{\alpha}{\alpha_s\mathcal{D}}a_b^{\perp M}\right\} H(\theta - \theta_M)\,. \quad (11.183)$$

These equations agree with (11.100), (11.120), (11.125) and (11.126). We emphasize that for v_x, v_y, v_z, d, h and b the total fields have to be substituted. Apart from this, the temperature field is no more difficult to solve than for the zeroth order problem.

11.8 Closure

In these notes an account has been given on the higher order accurate models of shallow ice sheets and ice shelves. The notes are brief and arguments sometimes terse – especially for ice shelves. The reason has not been a lack of understanding but the relative complexity of the subject matter[12]. Readers interested in more details should consult the references [1–3].

Structurally, the improvements of the SIA and SSA are the same. A formal perturbation expansion in terms of the aspect ratio ϵ is pursued with the mechanical equations, treating the temperature field as known. This expansion of the stress fields could be pushed to $\mathcal{O}(\epsilon^2)$ in the scaled equations for ice sheets and led to an iterative solution procedure for stress, velocity and temperature fields that is systematic and applicable *to all* geometric situations in large ice sheet dynamics. The result is the second order shallow ice approximation (SOSIA) that will certainly be developed and numerically implemented in the future.

For ice shelves the situation is more difficult. The algebraic manipulations were so complicated that so far we can only present a first order approximation with errors that are second order, $\mathcal{O}(\epsilon^2)$, in the aspect ratio. This led to the first order shallow shelf approximation (FOSSA). It was demonstrated that to this order shelves behave like membranes with horizontal velocity fields that are independent of the z-coordinate. Since observations have been made which show differences in the amount of velocity at the top and bottom of an ice shelf, bending effects must eventually come into the picture. This z-dependence of the velocity field could be accounted for by considering the $\mathcal{O}(\epsilon^2)$-terms. This is work still untouched, except for a series of first steps in [1]. It is not likely to be easy to develop these equations.

Remark: We have restricted the quoted references to the three items listed below. This does not mean that they are the only significant ones. Refs [1,2] contain more than 300 quotations relevant to the topic and we refer the reader to those for details.

Acknowledgement

We thank the editors for their criticism of an earlier manuscript.

[12] The derivation of theories of the thermomechanical response of thin sheets spans more than two centuries, starting with the Bernoulli's in the early 18th century. The second half of the 20th century brought the systematic deductions of plate equations by asymptotic methods. Ice shelves are from a physical point of view about the most complex situation one may encounter: coupling between mechanical and thermal effects including possible phase change processes. Under such conditions it would be very difficult to derive these equations by ad-hoc methods. This is the reason why this systematic derivation is necessary.

Appendix

This Appendix lists equations of the main text which are complicated and needed, but would make reading of the main text cumbersome.

Appendix A1: Sliding and energy jump conditions at the basal surface

The boundary conditions (11.12) and (11.13) when written in dimensionless form according to the scalings introduced in (11.21) take the forms

$$
\begin{aligned}
(v_{sl})_x = & -\frac{\mathcal{F}_t C_t}{N_b} H(\theta-\theta_M)\left\{ t_{xx}\frac{\partial b}{\partial x} + \epsilon^2 t^D_{xy}\frac{\partial b}{\partial y} - t^D_{xz} \right. \\
& -\frac{1}{N_b^2}\left[\epsilon^2 t_{xx}\left(\frac{\partial b}{\partial x}\right)^3 + \epsilon^2 t_{yy}\frac{\partial b}{\partial x}\left(\frac{\partial b}{\partial y}\right)^2 + t_{zz}\frac{\partial b}{\partial x}\right. \\
& \left.\left. +2\epsilon^4 t^D_{xy}\left(\frac{\partial b}{\partial x}\right)^2\frac{\partial b}{\partial y} - 2\epsilon^2 t^D_{xz}\left(\frac{\partial b}{\partial x}\right)^2 - 2\epsilon^2 t^D_{yz}\frac{\partial b}{\partial x}\frac{\partial b}{\partial y}\right]\right\}
\end{aligned} \tag{11.184}
$$

$$
\begin{aligned}
(v_{sl})_y = & -\frac{\mathcal{F}_t C_t}{N_b} H(\theta-\theta_M)\left\{ \epsilon^2 t^D_{xy}\frac{\partial b}{\partial x} + t_{yy}\frac{\partial b}{\partial y} - t^D_{yz} \right. \\
& -\frac{1}{N_b^2}\left[\epsilon^2 t_{xx}\left(\frac{\partial b}{\partial x}\right)^2\frac{\partial b}{\partial y} + \epsilon^2 t_{yy}\left(\frac{\partial b}{\partial y}\right)^3 + t_{zz}\frac{\partial b}{\partial y}\right. \\
& \left.\left. +2\epsilon^4 t^D_{xy}\frac{\partial b}{\partial x}\left(\frac{\partial b}{\partial y}\right)^2 - 2\epsilon^2 t^D_{xz}\frac{\partial b}{\partial x}\frac{\partial b}{\partial y} - 2\epsilon^2 t^D_{yz}\left(\frac{\partial b}{\partial y}\right)^2\right]\right\}
\end{aligned} \tag{11.185}
$$

$$
\begin{aligned}
(v_{sl})_z = & -\frac{\mathcal{F}_t C_t}{N_b} H(\theta-\theta_M)\left\{ t^D_{xz}\frac{\partial b}{\partial x} + t^D_{yz}\frac{\partial b}{\partial y} - \frac{t_{zz}}{\epsilon^2} \right. \\
& -\frac{1}{N_b^2}\left[-t_{xx}\left(\frac{\partial b}{\partial x}\right)^2 - t_{yy}\left(\frac{\partial b}{\partial y}\right)^2 - \frac{t_{zz}}{\epsilon^2}\right. \\
& \left.\left. -2\epsilon^2 t^D_{xy}\frac{\partial b}{\partial x}\frac{\partial b}{\partial y} + 2t^D_{xz}\frac{\partial b}{\partial x} + 2t^D_{yz}\frac{\partial b}{\partial y}\right]\right\} .
\end{aligned} \tag{11.186}
$$

$$
(v_{sl})_x = v_x - v_x^+ - \frac{\epsilon^2}{N_b^2}\left[(v_x - v_x^+)\frac{\partial b}{\partial x} + (v_y - v_y^+)\frac{\partial b}{\partial y} - (v_z - v_z^+)\right]\frac{\partial b}{\partial x} \tag{11.187}
$$

$$
(v_{sl})_y = v_y - v_y^+ - \frac{\epsilon^2}{N_b^2}\left[(v_x - v_x^+)\frac{\partial b}{\partial x} + (v_y - v_y^+)\frac{\partial b}{\partial y} - (v_z - v_z^+)\right]\frac{\partial b}{\partial y} \tag{11.188}
$$

$$
(v_{sl})_z = v_z - v_z^+ + \frac{1}{N_b^2}\left[(v_x - v_x^+)\frac{\partial b}{\partial x} + (v_y - v_y^+)\frac{\partial b}{\partial y} - (v_z - v_z^+)\right] . \tag{11.189}
$$

In these relations, $(\cdot)^+$ denotes the variable $(\cdot)$ evaluated immediately below the ice rock interface. Moreover, $t_{ii} = -p + \epsilon^2 t_{ii}^D$ (no summation over i) for $i = x, y, z$, and for our modelling of the isostatic lithosphere adjustment one has

$$v_x^+ = v_y^+ = 0\,, \quad v_z^+ = \frac{\partial b}{\partial t}\,. \tag{11.190}$$

The energy jump condition (11.13), after lengthy manipulations is written as

$$\begin{aligned}
&\kappa\left(\epsilon^2 \frac{\partial \theta}{\partial x}\frac{\partial b}{\partial x} + \epsilon^2 \frac{\partial \theta}{\partial y}\frac{\partial b}{\partial y} - \frac{\partial \theta}{\partial z}\right) - \frac{[\kappa_r]}{[\kappa]}\kappa_r\left(\epsilon^2 \frac{\partial \theta^+}{\partial x}\frac{\partial b}{\partial x} + \epsilon^2 \frac{\partial \theta^+}{\partial y}\frac{\partial b}{\partial y} - \frac{\partial \theta^+}{\partial z}\right) \\
&= \Bigg\langle -\frac{\alpha}{\mathcal{D}}\Bigg\{(v_{sl})_x \Bigg[\left(-p+\epsilon^2 t_{xx}^D\right)\frac{\partial b}{\partial x} + \epsilon^2 t_{xy}^D \frac{\partial b}{\partial y} - t_{xz}^D - \frac{1}{{N_b}^2}\times \\
&\Bigg(\epsilon^2\left(-p+\epsilon^2 t_{xx}^D\right)\left(\frac{\partial b}{\partial x}\right)^3 + \epsilon^2\left(-p+\epsilon^2 t_{yy}^D\right)\frac{\partial b}{\partial x}\left(\frac{\partial b}{\partial y}\right)^2 + \left(-p+\epsilon^2 t_{zz}^D\right)\frac{\partial b}{\partial x} \\
&+2\epsilon^4 t_{xy}^D\left(\frac{\partial b}{\partial x}\right)^2\frac{\partial b}{\partial y} - 2\epsilon^2 t_{xz}^D\left(\frac{\partial b}{\partial x}\right)^2 - 2\epsilon^2 t_{yz}^D\frac{\partial b}{\partial x}\frac{\partial b}{\partial y}\Bigg)\Bigg] + (v_{sl})_y\Bigg[\epsilon^2 t_{xy}^D\frac{\partial b}{\partial x} \\
&+\left(-p+\epsilon^2 t_{yy}^D\right)\frac{\partial b}{\partial y} - t_{yz}^D - \frac{1}{{N_b}^2}\Bigg(\epsilon^2\left(-p+\epsilon^2 t_{xx}^D\right)\left(\frac{\partial b}{\partial x}\right)^2\frac{\partial b}{\partial y} \\
&+\epsilon^2\left(-p+\epsilon^2 t_{yy}^D\right)\left(\frac{\partial b}{\partial y}\right)^3 + \left(-p+\epsilon^2 t_{zz}^D\right)\frac{\partial b}{\partial y} + 2\epsilon^4 t_{xy}^D\frac{\partial b}{\partial x}\left(\frac{\partial b}{\partial y}\right)^2 \\
&-2\epsilon^2 t_{xz}^D\frac{\partial b}{\partial x}\frac{\partial b}{\partial y} - 2\epsilon^2 t_{yz}^D\left(\frac{\partial b}{\partial y}\right)^2\Bigg)\Bigg]\Bigg\} - \frac{\alpha}{\mathcal{D}}\Bigg\{(v_{sl})_z\Bigg[\epsilon^2 t_{xz}^D\frac{\partial b}{\partial x} + \epsilon^2 t_{yz}^D\frac{\partial b}{\partial y} \\
&-\left(-p+\epsilon^2 t_{zz}^D\right) - \frac{1}{{N_b}^2}\Bigg(-\epsilon^2\left(-p+\epsilon^2 t_{xx}^D\right)\left(\frac{\partial b}{\partial x}\right)^2 - \epsilon^2\left(-p+\epsilon^2 t_{yy}^D\right)\left(\frac{\partial b}{\partial y}\right)^2 \\
&-\left(-p+\epsilon^2 t_{zz}^D\right) - 2\epsilon^4 t_{xy}^D\frac{\partial b}{\partial x}\frac{\partial b}{\partial y} + 2\epsilon^2 t_{xz}^D\frac{\partial b}{\partial x} + 2\epsilon^2 t_{yz}^D\frac{\partial b}{\partial y}\Bigg)\Bigg]\Bigg\} \\
&-N_b\frac{1}{[\mathcal{W}]}\frac{\alpha}{\alpha_t}\frac{1}{\mathcal{D}}a_b^\perp - \frac{\rho}{\rho_w}\frac{1}{N_b}\frac{\alpha}{\mathcal{D}}\Bigg\{a_b^\perp\Bigg[\epsilon^2\left(-p+\epsilon^2 t_{xx}^D\right)\left(\frac{\partial b}{\partial x}\right)^2 \\
&+2\epsilon^4 t_{xy}^D\frac{\partial b}{\partial x}\frac{\partial b}{\partial y} - 2\epsilon^2 t_{xz}^D\frac{\partial b}{\partial x} + \epsilon^2\left(-p+\epsilon^2 t_{yy}^D\right)\left(\frac{\partial b}{\partial y}\right)^2 - 2\epsilon^2 t_{yz}^D\frac{\partial b}{\partial y} \\
&+\left(-p+\epsilon^2 t_{zz}^D\right)\Bigg]\Bigg\}\Bigg\rangle H(\theta-\theta_M)\,.
\end{aligned} \tag{11.191}$$

Appendix A2: SIA corrections to the sliding law

Here we list explicit formulas for the corrections $(\cdot)^{\mathrm{corr}}$ of the sliding law arising in (11.83)–(11.86).

$$
\begin{aligned}
(v_{\mathrm{sl}})_x^{\mathrm{corr}} := -\epsilon^2 \frac{\mathcal{F}_t C_t}{N_{\mathrm{b}}} \Bigg[t_{xx}^{\mathrm{D}} \frac{\partial b}{\partial x} + t_{xy}^{\mathrm{D}} \frac{\partial b}{\partial y} - \frac{1}{N_{\mathrm{b}}^2} \Bigg\{ \left(-p + \epsilon^2 t_{xx}^{\mathrm{D}}\right) \left(\frac{\partial b}{\partial x}\right)^3 \\
+ \left(-p + \epsilon^2 t_{yy}^{\mathrm{D}}\right) \frac{\partial b}{\partial x} \left(\frac{\partial b}{\partial y}\right)^2 + t_{zz}^{\mathrm{D}} \frac{\partial b}{\partial x} + 2\epsilon^2 t_{xy}^{\mathrm{D}} \left(\frac{\partial b}{\partial x}\right)^2 \frac{\partial b}{\partial y} \\
- 2 t_{xz}^{\mathrm{D}} \left(\frac{\partial b}{\partial x}\right)^2 - 2 t_{yz}^{\mathrm{D}} \frac{\partial b}{\partial x} \frac{\partial b}{\partial y} \Bigg\} \Bigg] ,
\end{aligned}
\tag{11.192}
$$

$$
\begin{aligned}
(v_{\mathrm{sl}})_y^{\mathrm{corr}} := -\epsilon^2 \frac{\mathcal{F}_t C_t}{N_{\mathrm{b}}} \Bigg[t_{xy}^{\mathrm{D}} \frac{\partial b}{\partial x} + t_{yy}^{\mathrm{D}} \frac{\partial b}{\partial y} - \frac{1}{N_{\mathrm{b}}^2} \Bigg\{ \left(-p + \epsilon^2 t_{xx}^{\mathrm{D}}\right) \left(\frac{\partial b}{\partial x}\right)^2 \frac{\partial b}{\partial y} \\
+ \left(-p + \epsilon^2 t_{yy}^{\mathrm{D}}\right) \left(\frac{\partial b}{\partial y}\right)^3 + t_{zz}^{\mathrm{D}} \frac{\partial b}{\partial y} + 2\epsilon^2 t_{xy}^{\mathrm{D}} \frac{\partial b}{\partial x} \left(\frac{\partial b}{\partial y}\right)^2 \\
- 2 t_{xz}^{\mathrm{D}} \frac{\partial b}{\partial x} \frac{\partial b}{\partial y} - 2 t_{yz}^{\mathrm{D}} \left(\frac{\partial b}{\partial y}\right)^2 \Bigg\} \Bigg] ,
\end{aligned}
\tag{11.193}
$$

$$
\begin{aligned}
(v_{\mathrm{sl}})_z^{\mathrm{corr}} := -\epsilon^2 \mathcal{F}_t C_t \Bigg[t_{xx}^{\mathrm{D}} \left(\frac{\partial b}{\partial x}\right)^2 + t_{yy}^{\mathrm{D}} \left(\frac{\partial b}{\partial y}\right)^2 + 2 t_{xy}^{\mathrm{D}} \frac{\partial b}{\partial x} \frac{\partial b}{\partial y} \\
+ \left(\left(\frac{\partial b}{\partial x}\right)^2 + \left(\frac{\partial b}{\partial y}\right)^2\right) \Bigg\{ p \left(\left(\frac{\partial b}{\partial x}\right)^2 + \left(\frac{\partial b}{\partial y}\right)^2\right) \\
+ \frac{5}{2} t_{xz}^{\mathrm{D}} \frac{\partial b}{\partial x} + \frac{5}{2} t_{yz}^{\mathrm{D}} \frac{\partial b}{\partial y} - t_{zz}^{\mathrm{D}} \Bigg\} \Bigg] ;
\end{aligned}
\tag{11.194}
$$

$$
\begin{aligned}
\mathit{\Delta} v_x{}^{\mathrm{corr}} := -\frac{\epsilon^2}{N_b^2} \Bigg[\left(v_{x,b}(.) - v_x^+\right) \frac{\partial b}{\partial x} \\
+ \left(v_{y,b}(.) - v_y^+\right) \frac{\partial b}{\partial y} - \left(v_{z,b}(.) - v_z^+\right) \Bigg] \frac{\partial b}{\partial x} ,
\end{aligned}
\tag{11.195}
$$

$$
\begin{aligned}
\mathit{\Delta} v_y{}^{\mathrm{corr}} := -\frac{\epsilon^2}{N_b^2} \Bigg[\left(v_{x,b}(.) - v_x^+\right) \frac{\partial b}{\partial x} \\
+ \left(v_{y,b}(.) - v_y^+\right) \frac{\partial b}{\partial y} - \left(v_{z,b}(.) - v_z^+\right) \Bigg] \frac{\partial b}{\partial y} .
\end{aligned}
\tag{11.196}
$$

Appendix A3: Detailed equations for the FOSSA

We list here those equations relevant to the first order shallow shelf approximation which are too lengthy and therefore would deter from the text. Nevertheless the statements are needed for the construction of the main theory. Next, the first order stress-deviator stretching relations are

$$t^D_{xx(1)} = \frac{\mathcal{K}_S{}'}{\varrho}\left[\frac{\partial v_{x(0)}}{\partial x}\left\{B(\theta'_{(0)})\frac{\mathrm{d}g(d_{(0)})}{\mathrm{d}d}d_{(1)} + \frac{\mathrm{d}B(\theta'_{(0)})}{\mathrm{d}\theta'}g(d_{(0)})\theta'_{(1)}\right\}\right.$$
$$\left. + \frac{\partial v_{x(1)}}{\partial x}\Big(B(\theta'_{(0)})g(d_{(0)})\Big)\right], \tag{11.197}$$

$$t^D_{yy(1)} = \frac{1}{\varrho}\mathcal{K}_S{}'\left[\frac{\partial v_{y(0)}}{\partial y}\left\{B(\theta'_{(0)})\frac{\mathrm{d}g(d_{(0)})}{\mathrm{d}d}d_{(1)} + \frac{\mathrm{d}B(\theta'_{(0)})}{\mathrm{d}\theta'}g(d_{(0)})\theta'_{(1)}\right\}\right.$$
$$\left. + \frac{\partial v_{y(1)}}{\partial y}\Big(B(\theta'_{(0)})g(d_{(0)})\Big)\right], \tag{11.198}$$

$$t^D_{zz(1)} = \frac{1}{\varrho}\mathcal{K}_S{}'\left[\frac{\partial v_{z(0)}}{\partial z}\left\{B(\theta'_{(0)})\frac{\mathrm{d}g(d_{(0)})}{\mathrm{d}d}d_{(1)} + \frac{\mathrm{d}B(\theta'_{(0)})}{\mathrm{d}\theta'}g(d_{(0)})\theta'_{(1)}\right\}\right.$$
$$\left. + \frac{\partial v_{z(1)}}{\partial z}\Big(B(\theta'_{(0)})g(d_{(0)})\Big)\right], \tag{11.199}$$

$$t^D_{xy(1)} = \frac{1}{2\varrho}\mathcal{K}_S{}'\left[\left(B(\theta'_{(0)})\frac{\mathrm{d}g(d_{(0)})}{\mathrm{d}d}d_{(1)} + \frac{\mathrm{d}B(\theta'_{(0)})}{\mathrm{d}\theta'}g(d_{(0)})\theta'_{(1)}\right)\times\right.$$
$$\left.\left(\frac{\partial v_{x(0)}}{\partial y} + \frac{\partial v_{y(0)}}{\partial x}\right) + B(\theta'_{(0)})g(d_{(0)})\left(\frac{\partial v_{x(1)}}{\partial y} + \frac{\partial v_{y(1)}}{\partial x}\right)\right], \tag{11.200}$$

$$0 = \frac{1}{2\varrho}\mathcal{K}_S{}'\left[\frac{\partial v_{x(0)}}{\partial z}\left\{B(\theta'_{(0)})\frac{\mathrm{d}g(d_{(0)})}{\mathrm{d}d}d_{(1)} + \frac{\mathrm{d}B(\theta'_{(0)})}{\mathrm{d}\theta'}g(d_{(0)})\theta'_{(1)}\right\}\right.$$
$$\left. + \frac{\partial v_{x(1)}}{\partial z}\Big(B(\theta'_{(0)})g(d_{(0)})\Big)\right], \tag{11.201}$$

$$0 = \frac{1}{2\varrho}\mathcal{K}_S{}'\left[\frac{\partial v_{y(0)}}{\partial z}\left\{B(\theta'_{(0)})\frac{\mathrm{d}g(d_{(0)})}{\mathrm{d}d}d_{(1)} + \frac{\mathrm{d}B(\theta'_{(0)})}{\mathrm{d}\theta'}g(d_{(0)})\theta'_{(1)}\right\}\right.$$
$$\left. + \frac{\partial v_{y(1)}}{\partial z}\Big(B(\theta'_{(0)})g(d_{(0)})\Big)\right], \tag{11.202}$$

and the kinematic and traction boundary conditions at first order become

$$
\begin{aligned}
&\frac{\partial h_{(1)}}{\partial t} + v_{x(0)}(.,h_{(0)})\frac{\partial h_{(1)}}{\partial x} + v_{x(1)}(.,h_{(0)})\frac{\partial h_{(0)}}{\partial x} \\
&+v_{y(0)}(.,h_{(0)})\frac{\partial h_{(1)}}{\partial y} + v_{y(1)}(.,h_{(0)})\frac{\partial h_{(0)}}{\partial y} \\
&-\frac{\partial v_{z(0)}(.,h_{(0)})}{\partial z}h_{(1)} - v_{z(1)}(.,h_{(0)}) = \frac{\partial a_s^{\perp}(.,h_{(0)})}{\partial z}h_{(1)} \, ,
\end{aligned}
\tag{11.203}
$$

$$
\begin{aligned}
&\frac{\partial p_{(0)}}{\partial z}h_{(1)}\frac{\partial h_{(0)}}{\partial x} + p_{(0)}\frac{\partial h_{(1)}}{\partial x} + p_{(1)}\frac{\partial h_{(0)}}{\partial x} - \frac{\partial t^D_{xx(0)}}{\partial z}h_{(1)}\frac{\partial h_{(0)}}{\partial x} - t^D_{xx(0)}\frac{\partial h_{(1)}}{\partial x} \\
&-t^D_{xx(1)}\frac{\partial h_{(0)}}{\partial x} - \frac{\partial t^D_{xy(0)}}{\partial z}h_{(1)}\frac{\partial h_{(0)}}{\partial y} - t^D_{xy(0)}\frac{\partial h_{(1)}}{\partial y} - t^D_{xy(1)}\frac{\partial h_{(0)}}{\partial y} + \frac{\partial t^D_{xz(0)}}{\partial z}h_{(1)} \\
&+t^D_{xz(1)} = 0 \, ,
\end{aligned}
\tag{11.204}
$$

$$
\begin{aligned}
&-\frac{\partial t^D_{xy(0)}}{\partial z}h_{(1)}\frac{\partial h_{(0)}}{\partial x} - t^D_{xy(0)}\frac{\partial h_{(1)}}{\partial x} - t^D_{xy(1)}\frac{\partial h_{(0)}}{\partial x} + \frac{\partial p_{(0)}}{\partial z}h_{(1)}\frac{\partial h_{(0)}}{\partial y} + p_{(0)}\frac{\partial h_{(1)}}{\partial y} \\
&+p_{(1)}\frac{\partial h_{(0)}}{\partial y} - \frac{\partial t^D_{yy(0)}}{\partial z}h_{(1)}\frac{\partial h_{(0)}}{\partial y} - t^D_{yy(0)}\frac{\partial h_{(1)}}{\partial y} - t^D_{yy(1)}\frac{\partial h_{(0)}}{\partial y} + \frac{\partial t^D_{yz(0)}}{\partial z}h_{(1)} \\
&+t^D_{yz(1)} = 0 \, ,
\end{aligned}
\tag{11.205}
$$

$$
-\frac{\partial p_{(0)}}{\partial z}h_{(1)} - p_{(1)} + \frac{\partial t^D_{zz(0)}}{\partial z}h_{(1)} + t^D_{zz(1)} = 0 \, ,
\tag{11.206}
$$

valid at $z = h_{(0)}(x,y,t)$. At the base, one needs to differentiate between the floating and rumpled regions. From (11.121)–(11.128) the following first order expressions can be deduced:

$$
\begin{aligned}
&\frac{\partial b_{(1)}}{\partial t} + \frac{\partial v_{x(0)}(.,b_{(0)})}{\partial z}b_{(1)}\frac{\partial b_{(0)}}{\partial x} + v_{x(0)}(.,b_{(0)})\frac{\partial b_{(1)}}{\partial x} + v_{x(1)}(.,b_{(0)})\frac{\partial b_{(0)}}{\partial x} \\
&+\frac{\partial v_{y(0)}(.,b_{(0)})}{\partial z}b_{(1)}\frac{\partial b_{(0)}}{\partial y} + v_{y(0)}(.,b_{(0)})\frac{\partial b_{(1)}}{\partial y} + v_{y(1)}(.,b_{(0)})\frac{\partial b_{(0)}}{\partial y} \\
&-\frac{\partial v_{z(0)}(.,b_{(0)})}{\partial z}b_{(1)} - v_{z(1)}(.,b_{(0)}) = -\frac{\partial a^{\perp}_{b\,(0)}(.,b_{(0)})}{\partial z}b_{(1)} - a^{\perp}_{b\,(1)}(.,b_{(0)}) \, , \\
&= -a_b^{\perp}\left(\cdot, b_{(0)} + \epsilon b_{(1)}\right) \, ,
\end{aligned}
\tag{11.207}
$$

$$
\frac{\partial p_{(0)}}{\partial z}b_{(1)}\frac{\partial b_{(0)}}{\partial x} + p_{(0)}\frac{\partial b_{(1)}}{\partial x} + p_{(1)}\frac{\partial b_{(0)}}{\partial x} - \frac{\partial t^D_{xx(0)}}{\partial z}b_{(1)}\frac{\partial b_{(0)}}{\partial x} - t^D_{xx(0)}\frac{\partial b_{(1)}}{\partial x}
$$

$$
\begin{aligned}
&-t^D_{xx(1)}\frac{\partial b_{(0)}}{\partial x} - \frac{\partial t^D_{xy(0)}}{\partial z} b_{(1)}\frac{\partial b_{(0)}}{\partial y} - t^D_{xy(0)}\frac{\partial b_{(1)}}{\partial y} - t^D_{xy(1)}\frac{\partial b_{(0)}}{\partial y} + \frac{\partial t^D_{xz(0)}}{\partial z} b_{(1)} \\
&+t^D_{xz(1)} \;=\; \mathrm{RHS}_x \,,
\end{aligned}
\tag{11.208}
$$

$$
\begin{aligned}
&-\frac{\partial t^D_{xy(0)}}{\partial z} b_{(1)}\frac{\partial b_{(0)}}{\partial x} - t^D_{xy(0)}\frac{\partial b_{(1)}}{\partial x} - t^D_{xy(1)}\frac{\partial b_{(0)}}{\partial x} + \frac{\partial p_{(0)}}{\partial z} b_{(1)}\frac{\partial b_{(0)}}{\partial y} + p_{(0)}\frac{\partial b_{(1)}}{\partial y} \\
&+p_{(1)}\frac{\partial b_{(0)}}{\partial y} - \frac{\partial t^D_{yy(0)}}{\partial z} b_{(1)}\frac{\partial b_{(0)}}{\partial y} - t^D_{yy(0)}\frac{\partial b_{(1)}}{\partial y} - t^D_{yy(1)}\frac{\partial b_{(0)}}{\partial y} + \frac{\partial t^D_{yz(0)}}{\partial z} b_{(1)} \\
&+t^D_{yz(1)} \;=\; \mathrm{RHS}_y \,,
\end{aligned}
\tag{11.209}
$$

$$
-\frac{\partial p_{(0)}}{\partial z} b_{(1)} - p_{(1)} + \frac{\partial t^D_{zz(0)}}{\partial z} b_{(1)} + t^D_{zz(1)} \;=\; \mathrm{RHS}_z \,,
\tag{11.210}
$$

where

$$
a_b^\perp\left(\cdot, b_{(0)} + \epsilon b_{(1)}\right) = \begin{cases} a_b^{\perp M} + a_b^{\perp A} \,, & \text{floating region}\,, \\ 0\,, & \text{rumpled region}\,, \end{cases}
\tag{11.211}
$$

$$
\mathrm{RHS}_x = \begin{cases} \dfrac{\partial p_{\mathrm{sw}(0)}}{\partial z} b_{(1)}\dfrac{\partial b_{(0)}}{\partial x} + p_{\mathrm{sw}(0)}\dfrac{\partial b_{(1)}}{\partial x} + p_{\mathrm{sw}(1)}\dfrac{\partial b_{(0)}}{\partial x}\,, & \text{floating region}\,, \\ \dfrac{\partial p_{\mathrm{gr}(0)}}{\partial z} b_{(1)}\dfrac{\partial b_{(0)}}{\partial x} + p_{\mathrm{gr}(0)}\dfrac{\partial b_{(1)}}{\partial x} + p_{gr(1)}\dfrac{\partial b_{(0)}}{\partial x} - \left(\tau_{\mathrm{gr}(0)} e_{x(0)}\right)\,, & \\ \qquad\qquad\qquad\qquad\qquad\qquad\qquad \text{rumpled region}\,, \end{cases}
\tag{11.212}
$$

$$
\mathrm{RHS}_y = \begin{cases} \dfrac{\partial p_{\mathrm{sw}(0)}}{\partial z} b_{(1)}\dfrac{\partial b_{(0)}}{\partial y} + p_{\mathrm{sw}(0)}\dfrac{\partial b_{(1)}}{\partial y} + p_{\mathrm{sw}(1)}\dfrac{\partial b_{(0)}}{\partial y}\,, & \text{floating region}\,, \\ \dfrac{\partial p_{\mathrm{gr}(0)}}{\partial z} b_{(1)}\dfrac{\partial b_{(0)}}{\partial y} + p_{\mathrm{gr}(0)}\dfrac{\partial b_{(1)}}{\partial y} + p_{gr(1)}\dfrac{\partial b_{(0)}}{\partial y} - \left(\tau_{\mathrm{gr}(0)} e_{y(0)}\right)\,, & \\ \qquad\qquad\qquad\qquad\qquad\qquad\qquad \text{rumpled region}\,, \end{cases}
\tag{11.213}
$$

$$
\mathrm{RHS}_z = \begin{cases} -\left(\dfrac{\partial p_{\mathrm{sw}(0)}}{\partial z} b_{(1)} + p_{\mathrm{sw}(1)}\right)\,, & \text{floating region}\,, \\ -\left(\dfrac{\partial p_{\mathrm{gr}(0)}}{\partial z} b_{(1)} + p_{gr(1)}\right)\,, & \text{rumpled region}\,, \end{cases}
\tag{11.214}
$$

and

$$
\frac{\partial p_{\mathrm{sw}(0)}}{\partial z} b_{(1)} + p_{\mathrm{sw}(1)} = -\frac{\rho_{\mathrm{sw}}}{\varrho\rho} b_{(1)}\,,
\tag{11.215}
$$

valid at $z = b_{(0)}(x, y, t)$.

References

1. D.R. Baral: Asymptotic theories of large scale motion, temperature and moisture distributions in land based polythermal ice shields and in floating ice shelves: A critical review and new development. Doctoral Dissertation, Department of Mechanics, Darmstadt University of Technology (1999)
2. D.R. Baral, K. Hutter, R. Greve: Asymptotic theories of large scale motion, temperature and moisture distributions in land based polythermal ice sheets. A critical review and new development. Applied Mechanics Reviews, in press
3. D.R. Baral, K. Hutter: 'An iterative solution procedure for shallow Stokes flows. The shallow ice Approximation revisited'. In: *Advances in Cold-Region Thermal Engineering and Sciences*, ed. by K. Hutter, Y. Wang, H. Beer. Lecture Notes in Physics No 533, (Springer Verlag, Berlin, Hydelberg, New York 1999)

12 Aspects of Iceberg Deterioration and Drift

S.B. Savage

McGill University, Montreal, Quebec H3A 2K6, Canada

12.1 Introduction

During her maiden voyage from Southampton to New York, the ocean liner RMS Titanic struck an iceberg off the Newfoundland Banks and sank on April 15, 1912. Of the 2228 passengers and crew on board, only 705 survived. This tragedy generated a public outcry that subsequently provoked government action. Representatives of the world's various maritime powers signed a convention in 1914 to inaugurate an international derelict-destruction, ice observation, and ice patrol service. Today, the International Ice Patrol (IIP) is comprised of 17 member national organizations (including Belgium, Canada, Denmark, Finland, France, Germany, Greece, Italy, Japan, Netherlands, Norway, Panama, Poland, Spain, Sweden, the United Kingdom, and the United States of America). Its mission is "to monitor the extent of the iceberg danger near the Grand Banks of Newfoundland and provide limits of all known ice to the maritime community". In addition to participation in the IIP, several countries have their own independent organizations that keep track of individual iceberg positions, trajectories, size and melt decay in Northern waters. Typically, this is accomplished through the use of satellites (RADARSAT), aerial reconnaissance making use of Side-Looking Airborne Radar (SLAR) and Forward-Looking Airborne Radar (FLAR), as well as observations from commercial shipping.

Most of the icebergs that endanger the shipping lanes of the North Atlantic are generated by the tidewater glaciers off the West Coast of Greenland. Glaciers are formed from thousands of years of snowfall accumulation. New fallen snow changes over periods of several months into a form of granular snow called firn. Over several decades this is compressed into dense ice by the overburden. Ice at the bottom of a glacier can come from snow that was deposited as many as 150,000 years ago.

The glacier ice can be regarded as a fluid of an extremely high viscosity; it deforms and flows under its own weight. When the leading edge of a flowing glacier enters the sea, large slabs of ice can be weakened and then broken off by buoyant forces and the rise and fall of the tides (see Fig. 12.1). Up to 20,000 icebergs are calved in this way each year; most of these come from the 20 main glaciers between the Jacobshaven and Humboldt Glaciers. Water and wind currents then force the icebergs southward to the Atlantic Ocean. As they travel to the warmer waters, they melt and break up. Typically, an average of around 500 icebergs survive past the 48th parallel in a given year. The largest icebergs have the potential to cause the most damage to ships, but they are easy to detect by ship radar and can be avoided. On the other hand, the smaller ice pieces, having lengths of a few meters, are much more difficult to discern. In particular, this is due to the fact that they can be shielded by waves in choppy seas. Most of their mass is submerged, and an ice piece even as small as 3–5 m in length can cause very significant damage if it collides with a vessel.

Fig. 12.1. Icebergs being calved from leading edge of glacier (courtesy of Environment Canada, Canadian Ice Service)

In addition to the icebergs that are found in the North Atlantic, a very small number of icebergs originate in Alaska and in Siberia or south of Franz Joseph Land in the Barents Sea.

Icebergs are also calved from the Antarctic ice sheet as it moves towards the sea. These Antarctic icebergs are generally very much larger than their Arctic counterparts; some recent ones have been of truly enormous size. The Super Iceberg B10A, which broke off in 1992 from the Thwaites Ice Tongue in Antarctica, is approximately the size of the state of Rhode Island. The danger to mariners comes not from the iceberg itself, but from the 'smaller' kilometer sized ice pieces that break away from the edges of the main iceberg as it moves Northward and melts. An iceberg designated as B15 broke off from the Ross Ice Shelf in March 2000. It is approximately 11,000 square kilometers in area, about the size of the state of Connecticut, making it one of the largest icebergs ever observed. While some have expressed concern that these events are evidence of global warming, others have cautioned that they are probably just part of the normal process of ice shelf growth and loss. It is worth noting that the largest known Antarctic iceberg, having a length of 333 km and a width of 100 km, was observed in 1957 by the icebreaker USS Glacier.

There are many photographs and satellite images of Antarctic icebergs on the Antarctic Meteorological Research Center (AMRC) Iceberg Page of the internet web site `http://uwamrc.ssec.wisc.edu`. This web page contains a comprehensive archive of information about icebergs recently calved from the Ross Ice Shelf. Numerous photographs of Arctic icebergs are available from the Lane Gallery in Newfoundland (cf. `http://www.tidespoint.com/lanegallery/index.htm`).

Besides the hazards posed by icebergs to shipping and offshore structures such as oil platforms, icebergs can cause other problems. In shallow water, the base of the iceberg can gouge wide and deep troughs in the sea bed. In doing so, it can damage cables and pipelines resting on the sea bottom. Iceberg scour also can cause ecological damage by exterminating meiofauna such as snails, lobsterlikes, testaceans, worms, eels, etc. While the seabed ecosystem sometimes recovers quickly, thereby benefiting the diversity of the species, occasionally it may take decades to fully recuperate.

12.1.1 The Link Between Iceberg Deterioration and Drift

Before discussing various aspects of iceberg deterioration and drift, we shall attempt to provide some motivation and rational for this review. Because the icebergs, as well as the smaller ice pieces that are calved from them, pose hazards to shipping, offshore structures, underwater pipelines, and the seabed ecosystem, it is essential to have information about their size, positions, and velocities. While some information is provided by satellites, aerial reconnaissance and commercial shipping observations, these data are obtained on an intermittent and often irregular basis. Therefore, it is necessary to have forecasting models that can predict iceberg motion and deterioration. Typically, the required forecast time frame ranges from a couple of days to a week into the future.

Specifically, the kinds of things that might be expected from a forecasting model are as follows:

1. Predictions of the position and velocity of the parent icebergs. As field observations are accumulated, these predictions would be corrected or updated prior to the next forecasting period.
2. Information about the iceberg size and how it changes with time because of melting and calving of small ice pieces. Estimates of the iceberg lifetimes, i.e. the time it takes for the iceberg to deteriorate to negligible size such that it is harmless.
3. Some particulars about the calving of the smaller ice pieces from the parent iceberg. Predictions of the initial size distributions and the subsequent melting of these smaller ice pieces. It is also of interest to estimate how far the smaller ice pieces disperse from their parent iceberg before they melt

to an insignificant size. While the parent iceberg usually can be seen, it is desirable to have some idea of the size of the danger zone around the parent where there is a likelihood of encountering smaller calved pieces that are not so easily observed.

The topics covered in this chapter are presented in the context of assembling material that could be used in the formulation of operational iceberg forecasting models. The combined study of iceberg deterioration and drift involves a large number of fields, including fluid mechanics, atmospheric science, ocean engineering, dynamics, various modes of heat transfer, solid mechanics, and fracture mechanics. While the topic of drift is fairly straightforward, theoretically grounded, and easy to follow, the material on deterioration is much more ambiguous and inexact, and it encompasses a wider range of fields. Because the iceberg geometry is so irregular and is also changing with time as a result of calving, it is much more difficult to derive detailed and meaningful deterministic models to predict the various kinds of deterioration processes. In the absence of such approaches, one is forced to put together rather crude empirical models.

Let us now very briefly outline the sequence of steps involved in computing the evolution of iceberg dynamics and thermodynamics. The purpose is to point out the close coupling between iceberg deterioration and drift. One would start out with the initial size and shape of the iceberg being given. Using independent atmospheric and oceanographic forecasting models, the environmental forcing, i.e. the wind and the water currents at various levels, would be calculated. The wind and water drags, which are needed for the prediction of the iceberg dynamics, both depend upon the iceberg's shape and the cross-sectional areas associated with each water current layer. The equations of motion would then be used to predict the iceberg position and velocity at the next time step. During this time step, the melting due to the various heat transfer mechanisms, most of which depend upon the relative velocities between the iceberg and the fluids (air or water) would be determined. The melting reduces the volume of the iceberg. It has been found that iceberg shape can be correlated as a function of its total volume. For example, the ratio of the iceberg draft to its length is observed to increase as the volume decreases. One could make use of such shape correlations to approximate a new iceberg geometry after a certain amount of melting has occurred. This new iceberg geometry would be used in the determination of the forces required for calculating the iceberg dynamics during the next time step. The computations proceed in this way, cycling back and forth between the dynamics and deterioration.

12.1.2 Outline of the Chapter

The chapter begins with a review of terminology pertaining to iceberg size and shape. The various deterioration mechanisms responsible for melting, wave erosion and calving, and the equations used for their computation are then reviewed. This is followed by a discussion of iceberg lifetimes and population studies. A recent melt model for small ice pieces in the bergy bit and growler size ranges is presented. A final section deals with iceberg dynamics and drift. It reviews the governing differential equations of motion for the iceberg, and the various driving and resisting forces. Numerical schemes to integrate the differential equations are considered.

12.2 Terminology – Classification of Shape and Size

Information about iceberg populations, and their statistical distributions of mass, draft and various other linear dimensions is important for hazard assessment in connection with marine transportation and the design of offshore oil drilling and production platforms. As noted in the previous section, knowledge about the iceberg shape is essential for the calculations of iceberg drift and deterioration. The next few subsections also have the goal of familiarizing the reader with the commonly used terminology dealing with iceberg shape and size.

12.2.1 Shape

The morphology of icebergs is of interest for scientific reasons as well as in the above-mentioned practical applications. The variety and complexity of iceberg shapes makes it difficult to classify them. However, in an effort to do so, the IIP has categorized iceberg shapes into the five common forms of blocky, drydocked, domed, pinnacled and tabular as defined below (see Fig. 12.2).

Blocky – Steep precipitous sides with horizontal or flat top. Very solid berg. Length/height ratio of 2.5:1.
Drydocked – Eroded such that a large U-shaped slot is formed, with twin columns or pinnacles. Slot extends into the water line or close to it.
Domed – Large smoothed rounded top. Solid type berg.
Pinnacled – Large central spire or pyramid of one or more spires dominating shape. Less massive than dome shaped berg of similar dimensions.
Tabular – Horizontal or flat topped berg with length/height ratio of 5:1.

Hotzel and Miller [1] have reported on the frequency of observation of these various forms in the Labrador Sea from 1973 to 1978 during the months of July through October. They noted that the pinnacled iceberg was most commonly observed (36%), followed by tabular (29%), domed (16%), drydocked (15%), and blocky (4%). Using a coarser, binary classification of *tabular* and *non-tabular* icebergs, yields the result that the non-tabular icebergs are found with more than twice the frequency as the less degraded tabular forms (71% versus 29%). On the other hand, the immense icebergs found in the Antarctic are tabular in shape.

An early method to estimate the drafts of icebergs was to use typical ratios of height to draft [2]. The more detailed study of Hotzel and Miller [1] reveals the uncertainties inherent in this simple approach. They presented statistical information on distributions of various important iceberg physical dimensions such as height, length, draft, mass and width, as well as parameter ratios

Fig. 12.2. Photographs of five common iceberg shapes (courtesy of Environment Canada, Canadian Ice Service)

including width/length, draft/length, height/length, mass/length, draft/width, height/width, mass/width, height/draft, and mass/draft. To emphasize the variability of some of these data, Hotzel and Miller [1] note that while the *average* length to height ratio is about 3:1, the observed values ranged from 0.5:1 to 50:1. On the other hand, the ratio of width to length is about 0.8 with a moderate standard deviation, so that it is reasonable to use the waterline length as an effective diameter of the iceberg.

Hotzel and Miller [1] obtained power law correlations between various physical quantities. For example, the relationship between height H_I and length L was found to be

$$H_I = 0.402\, L^{0.89} \,. \tag{12.1}$$

Such functional relationships can be used to provide a very rough first estimate of various iceberg dimensions based on the simplest and most easily measured dimension which is the iceberg waterline length L. These kinds of functional relationships are specific to a particular location and their use must be tempered by a recognition of the uncertainties inherent in such equations.

12.2.2 Size Classifications of Tabular and Non-tabular Icebergs

There is a commonly used terminology associated with size classification. *Growlers* and *bergy bits* are terms used to designate two particular sizes of ice blocks. A *growler* is a mass of glacial ice that has calved from an iceberg or is the remains of an iceberg. Typically, it has a length between 3 and 10 m. The name growler comes from the 'growling' sound made when the ice piece bobs up and down in the waves. A *bergy bit* is a mass of glacial ice that is smaller than an iceberg but larger than a growler, i.e. about the size of a small cottage. Icebergs that are larger than bergy bits can be classified as small, medium, large, or very large. Crocker and Cammaert [3] have given specific size ranges associated with these classifications as shown in Table 12.1.

Table 12.1. Iceberg size classification used by Crocker and Cammaert [3]

Classification	Length (m)
Growler	5–10
Bergy Bit	10–20
Small Iceberg	20–50
Medium Iceberg	50–100
Large Iceberg	100–150
Very Large Iceberg	150+

A very similar classification scheme for both tabular and non-tabular icebergs giving more detailed descriptions of some shape characteristics was proposed somewhat earlier by the IIP for tabular and non-tabular icebergs [4]. This is illustrated in Table 12.2.

Table 12.2. Iceberg size classification used by International Ice Patrol (IIP) for tabular and non-tabular icebergs and average values of iceberg parameters [4]

Code	Type*	Size	Mass (metric tons)	Length (m)	Perimeter (m)	Above water surface area (m^2)	Subwater surface area (m^2)	Underwater side surface area (m^2)
1	2	Growler	450	10	30	100	250	180
2	2	Small	75,000	55	155	2,300	8,000	6,300
3	2	Medium	900,000	125	360	12,000	36,000	26,000
4	2	Large	5,500,000	225	650	40,000	110,000	83,000
5	1	Small	250,000	80	235	5,000	15,000	11,000
6	1	Medium	2,170,000	175	500	25,000	67,000	50,000
7	1	Large	8,230,000	260	750	54,000	150,000	112,000

* 1 = tabular
2 = non-tabular

12.2.3 Iceberg Geometry for Use in Drift and Deterioration Calculations

For the computation of iceberg drift, the IIP makes use of a model in which the water current field is divided into four layers below the water surface; from 0–20 m, 20–50 m, 50–100 m, and from 100–120 m depths. The water drag on the iceberg is calculated by summing up the forces acting on each corresponding layer of the iceberg. Each iceberg layer is subjected to a drag that depends on the average relative water velocity in each layer and a water drag coefficient that is based on a reference submerged cross-sectional area of the iceberg for each respective layer. The wind drag is calculated in a similar way using an air drag coefficient that is based on the cross-sectional area above the waterline, but using only one reference air velocity. The IIP has proposed typical values for the iceberg characteristic cross-sectional areas for different sizes, and for both tabular and non-tabular icebergs [5].

Other iceberg models consider larger numbers of thinner water layers. Smith and Donaldson [6] have provided plots of above-water and below-water cross-sections in two perpendicular vertical planes for 14 separate icebergs measured during three cruises. Iceberg mass and cross-sectional areas in the air and in 10 m layers below the waterline were determined and used in their iceberg drift calculations.

Sayed [7] analysed a large number of icebergs of various sizes and obtained correlations of the geometry as a function of iceberg size. He thus obtained a generic iceberg geometry that is statistically appropriate for a certain geographic location and whose shape changes as it melts. The underwater iceberg shape obtained from the correlations for a given size could then be sliced into 10 m layers and the cross-sectional area, volume and wetted area for each layer determined for use in drift and deterioration calculations. As the iceberg melts, the volume decreases and the new shapes and surface areas can be obtained from his correlation.

12.3 Iceberg Deterioration Mechanisms

12.3.1 Relative Importance of Various Mechanisms

El-Tahan et al. [8] have listed the major mechanisms of iceberg deterioration that were originally identified by Job [9]. They are arranged in somewhat different order below:

1. Melting of the exposed iceberg surface by solar radiation.
2. Buoyancy-induced natural convection along the submerged sides and to a lesser extent along the bottom.
3. Forced convection melting of the iceberg due to the differential velocities between the icebergs and the surrounding fluids. This kind of melting occurs because of: (a) the relative velocities between the submerged portions of the iceberg and the sea water, and (b) the relative velocities between the exposed portions of the iceberg above the waterline and the wind.
4. Waterline wave erosion and undercutting.
5. Calving of the ice overhang caused by wave erosion at the waterline.
6. Subsurface calving caused by upward buoyant forces on underwater shelves (recall that the ice is less dense than sea water), or by other melting mechanisms.
7. Convection induced by wallowing or overturning.
8. Differential melting along cracks and faults in the berg leading to further calving.
9. Fracture of the ice due to internal stresses.

It is difficult to obtain accurate estimates of the last four items because of the lack of quantitative theories, and because of uncertainties in geometry, shape effects and inhomogeneities in material and fracture affinities. Item 7 is not too significant since typically icebergs spend a relatively small fraction of their lifetime in the wallowing and overturning modes.

To give an appreciation of the relative importance of these various deterioration mechanisms, Table 12.3 reproduces some estimates given in [8] for three icebergs:

Case 1 This drydock-shaped 636,000 ton iceberg was observed in the Grand Banks area off Newfoundland [10]. The water temperatures were low (0–1 °C). It lost about 12% of its mass in 8 days.

Case 2 Robe et al. [11] reported on a tabular-shaped iceberg initially 600 m long by 300 m wide, observed in the Northeast Grand Banks.

Case 3 The data was collected at the Ogmund E-72 drilling site offshore Labrador [12] for Iceberg No. 032. The iceberg mass was reduced from 486,000 to 264,000 metric tons in 3 days.

Table 12.3. Contribution of each deterioration mechanism to volume loss (in $1000\,\mathrm{m}^3$)

Deterioration Mechanism	Iceberg #1 [10]	Iceberg #2 [11]	Iceberg #3 [12]
Solar	3.4 (5.3%)	76.2 (2.9%)	0.6 (0.3%)
Buoyant convection	0.8 (1.2%)	35.7 (1.4%)	3.4 (1.5%)
Wind convection	0.6 (0.9%)	74.8 (2.9%)	0.7 (0.3%)
Forced convection (water)	8.0 (12.4%)	643.2 (24.9%)	25.3 (11.3%)
Wave erosion	32.6 (50.6%)	1757.0 (67.9%)	146.3 (65.5%)
Wave induced calving	19.0 (29.5%)	nil	47.0 (21.0%)
Total loss	64.4 (100%)	2587.0 (100%)	233.3 (100%)

It is clear from Table 12.3 that wave erosion is the dominant deterioration mechanism, followed by wave induced calving and forced convection in the water, whereas incident solar radiation and wind erosion play relatively minor roles.

12.3.2 Equations to Predict Deterioration Mechanisms

We now briefly describe the various deterioration mechanisms and equations that have been developed to predict them.

Surface Melting Due to Solar Radiation. Monthly values of measured and interpolated solar radiation of the earth's surface have been given by De Jong [13]. For example, in the center of the Labrador Sea (60° N), the measured insolation I ranges from roughly $30\,\mathrm{cal/cm^2/day}$ in December and January to about $420\,\mathrm{cal/cm^2/day}$ in July. Because of fog and cloud cover, the ratio of radiation received to the maximum incident radiation is about 40%.

Let us estimate the rate at which the surface of the iceberg is receding because of melting. We can equate the energy absorbed by an element of the iceberg surface $\mathrm{d}A$ in a time $\mathrm{d}t$ to that needed to melt an iceberg volume element $\mathrm{d}A\,\mathrm{d}z$, where $\mathrm{d}z$ is an element normal to the surface. Thus we find

$$I(1-\alpha)\,\mathrm{d}t\,\mathrm{d}A = \Gamma \rho_I\,\mathrm{d}z\,\mathrm{d}A\,, \qquad (12.2)$$

where Γ is the latent heat of melting of ice (334 J/gm or 79.8 cal/gm), ρ_I is the mass density of ice (0.91 gm/cm^3), and α is the albedo which is defined as the ratio of reflected to incident solar radiation. The albedo α can range from as low as 0.1 for a clear, flat ice surface to as much as 0.95 for fresh, dry snow.

Hence, the velocity of melting of the iceberg due to solar radiation is given by (12.2) as

$$V_s = \frac{\mathrm{d}z}{\mathrm{d}t} = \frac{I}{\Gamma\rho_I}(1-\alpha)\,. \tag{12.3}$$

Assuming a reasonable value of $\alpha = 0.7$ [8], and a maximum value of $I =$ 420 cal/cm^2/day corresponding to the Labrador Sea in July, then (12.3) yields a melt velocity $V_s = 1.7$ cm/day, which is still relatively small.

Melting Due to Buoyant Vertical Convection. Josberger [14] has presented detailed laboratory and field studies of free convection adjacent to icebergs that is caused by both thermal and salinity gradients. As the iceberg melts, the adjacent sea water is both cooled and diluted. The diluted sea water next to the iceberg has a lower salinity, causing a positive buoyancy and a tendency to rise relative to the denser surrounding sea water. Closest to the iceberg surface, salinity effects dominate over the thermal effects. On the other hand, because the thermal diffusivity is considerably larger than the saline diffusivity, the cooled water further away from the iceberg surface has the ambient salinity. Being denser, because of the reduced temperature, it tends to sink. It turns out that salinity effects are generally more important and result in an upward flow around the iceberg except for a very small region at the very bottom of the iceberg. White et al. [15] have suggested that the saline buoyancy will dominate above the lowest meter or so of the vertical iceberg wall. The flow field observed by Josberger [14] with the small lower laminar region and the upper turbulent region is shown schematically in Fig. 12.3.

Further studies of this free convection heat transfer problem have been performed by Gebhart and Mollendorf [16], Neshyba and Josberger [17], and White et al. [15]. The empirical correlation of Neshyba and Josberger [17] is simple and provides estimates that are close to the mean of the other predictions. Their relationship for V_b, the melt rate (in m/yr) of a vertical surface due to buoyant convection was obtained by a least square parabolic curve fit to the melt rate data and is given by

$$V_b = 2.78T + 0.47T^2\,, \tag{12.4}$$

where T is the water temperature in degrees C. Here again, the melt velocities are relatively small.

Melting Due to Forced Convection. First, to get an idea of the flow conditions as defined by a range of relevant Reynolds numbers, consider icebergs having a relative drift velocity $v_r = 0.2$ m/s and characteristic waterline lengths l=10

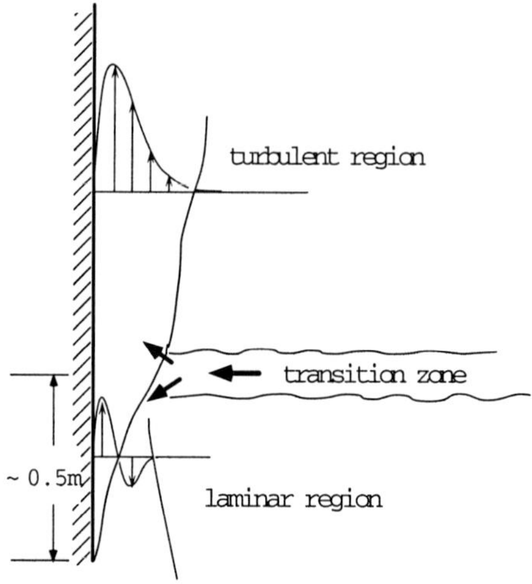

Fig. 12.3. The buoyant convection flow field adjacent to ice observed by Josberger [14]

to 100 m. Under these circumstances, the Reynolds number, $Re = \rho_w v_r l/\mu_w$, would have values ranging between approximately 10^6 and 10^7, where ρ_w and μ_w are respectively the density and viscosity of the sea water. For this Reynolds number range, the flow around a bluff body like an iceberg would involve turbulent boundary layers, flow separation and turbulent wakes. It is not presently possible to theoretically predict such complicated flows with any confidence and engineers typically rely on empirical correlations for heat transfer calculations around such bodies.

Let us begin by noting some classical results of forced convection heat transfer for flow past cylinders and spheres. The total forced convection heat flux can be expressed in terms of the average heat transfer coefficient $\overline{h}$ as

$$Q = A_s \, \overline{h} \, \Delta T \, , \tag{12.5}$$

where A_s is the surface area of the body and ΔT is the temperature difference between the free stream and the body. Hilpert [18] developed the following empirical correlation for the average Nusselt number for cylinders in cross flow

$$\overline{Nu_D} = \frac{\overline{h}D}{k} = C Re_D^m Pr^{1/3} \, , \tag{12.6}$$

where ρ, μ and k are the fluid density, viscosity and thermal conductivity, $Re_D = \rho u D/\mu$ is the Reynolds number based on cylinder diameter D and free stream velocity u, $Pr = \nu/\kappa$ is the Prandtl number, $\nu = \mu/\rho$ is the kinematic viscosity, κ is the thermal diffusivity, and C and m are parameters that depend upon Re_D. For $40,000 < Re_D < 400,000$ the exponent of Re_D was determined to be $m = 0.8$. Zhukaukas [19] has presented an equation almost identical in form to (12.6). For $2 \times 10^5 < Re_D < 10^6$, Zhukaukas [19] gives the value $m = 0.7$.

A number of similar correlations have been developed for the flow around spheres. For example, Achenbach [20] obtained the correlation

$$\overline{Nu}_D = 2 + \left(\frac{Re_D}{4} + 3 \times 10^{-4} {Re_D}^{1.6} \right)^{1/2} . \quad (12.7)$$

We can curve fit this result by an equation having a power law Reynolds number dependence like (12.6) in a Reynolds number range $Re_D \sim 10^6$. Similar to the case of the circular cylinders noted previously, one finds $m \simeq 0.8$.

Note that the surface area of a sphere with a diameter D, or a cylinder having a length equal to some multiple of its diameter, is proportional to the square of the diameter, $A_s \sim D^2$. Making use of the definition of the Nusselt number (12.6), and the dependence of the Nusselt number on D as noted in the above correlations, it is found from (12.5) that the forced convection heat flux has the following dependence on D

$$Q \sim D^{1+m} , \quad (12.8)$$

where m was found to be between 0.7 and 0.8 for Reynolds number of around 10^6.

We now focus on the forced convection problem in the context of icebergs. Strong prevailing Arctic winds cause iceberg melting in two ways. Firstly, there is the direct forced convection air melting of the exposed surface above the waterline. Secondly, the winds drive the iceberg through the water at relative velocities in the range of 10–30 cm/s resulting in forced convection melting of the submerged portion of the iceberg. In general, we can express V_f, the average surface melt rate due to forced convection, as

$$V_f = \frac{q_f}{\rho_I \Gamma} , \quad (12.9)$$

where q_f is the averaged forced convection heat flux per unit surface area (Q/A_s) given by

$$q_f = Nu\, k_f \Delta T / l , \quad (12.10)$$

ΔT is the temperature difference between the streaming fluid and the melting ice, k_f is the thermal conductivity of the fluid, l is the maximum waterline length (the straight line distance from bow to stern) and Nu is the Nusselt number. White et al. [15] suggested that the Nusselt number can be written, in a form very similar to (12.6), as

$$Nu = C Re^{0.8} Pr^{0.4} , \quad (12.11)$$

where $C = 0.058$ for tabular icebergs and $C = 0.055$ for non-tabular icebergs. The Reynolds number is defined as

$$Re = v_r l / \nu , \quad (12.12)$$

where v_r is the relative velocity between the fluid and the iceberg, and ν is the kinematic viscosity. Equations (12.9) to (12.12) can be used to compute the forced convection melting in both the air and the water by using the appropriate fluid properties. Typically, computations are based on a relative air velocity corresponding to a height of 13 m above the ocean surface, and a relative water velocity that is averaged over the draft of the iceberg. Some coupled forecasting models predict ocean currents in each of several layers beneath the surface. The CIS model, for example, considers 10 m thick layers. In such cases, one could make use of an averaged relative water velocity determined by an appropriate weighting that accounts for the iceberg surface areas corresponding to each of these layers. It is found that the surface melt rate due to forced *air convection* is comparable to the melt rate due to solar radiation and is very small compared to the forced water convection.

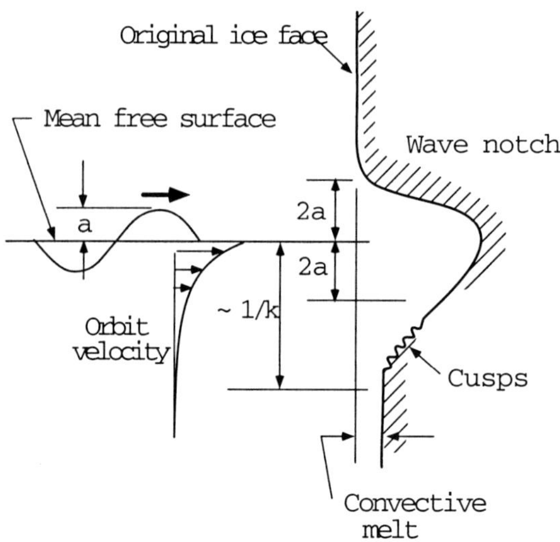

Fig. 12.4. Schematic diagram of wave erosion at the waterline and the formation of a waterline notch

Wave Erosion. Wave erosion at the waterline is the most important iceberg deterioration mechanism. The basic heat transfer mechanisms are much the same as the forced convection mode described above. What is different is the enhanced heat transfer resulting in increased melting in the region near the ocean surface because of the larger water velocities associated with the wave motion. Waves can erode a notch in the iceberg after which calving and/or fracture can occur. Figure 12.4 shows a wave train having an amplitude a, wavelength λ, and wavenumber k approaching the iceberg from the left. In deep water, the particle motions associated with the waves have the form of circular orbits whose amplitudes decay exponentially with depth as shown in Fig. 12.4. When a wave of amplitude a is reflected by a plane wall, its amplitude is doubled to $2a$ at the wall itself. So we might expect the additional melting of the iceberg due to the waves to occur within region of around $2a$ above the mean free surface of the water, and a region somewhat larger below the mean free water surface because of the exponentially decaying orbital velocities. In addition, because of the relative velocity between the mean ocean currents and the iceberg there will be forced convection melting of the iceberg as shown in Fig. 12.4. The result is a waterline notch having a shape as indicated in Fig. 12.4. An interesting feature of the lower part of the notch is the formation of ripples or cusps in the ice surface; the mechanisms responsible for their development have not been discussed in the iceberg literature.

White et al. [15] developed relationships to predict the waterline melt rates. For the case of a rough wall, the waterline melt rate per degree C is

$$V_{we} = 0.000146 \left(\frac{R}{H}\right)^{0.2} \left(\frac{H}{\tau}\right) , \tag{12.13}$$

where R is the roughness height of the ice surface, and τ and H are the mean period and height of the waves. A typical value for the roughness height R is 1 cm [15]. The waterline melt rate can be as much as 1.0 m/day/°C.

Calving of Overhanging Slabs. Section 12.3.1 has listed several iceberg deterioration mechanisms, three of which involve fracturing or calving of the parent iceberg. Two of the latter mechanisms are (a) calving as a result of thermal stresses caused by solar radiation or rolling in warm waters, and (b) breaking off of underwater shelves due to buoyancy. These two mechanisms seem to be of lesser importance and, furthermore, it is very difficult to quantify them other than by statistical means. However, the major calving mechanism is that resulting from the waterline wave erosion described in the previous subsection. Waves progressively cut rounded notches in the iceberg at the waterline. As the notches become deeper, a protruding overhang above the waterline develops as shown in Fig. 12.5. The overhanging slab acts like a cantilever beam with the downward loads arising from the weight of the overhanging ice. As is discussed in most standard textbooks on engineering solid mechanics, so-called bending stresses are developed over vertical cross-sections of the overhang. As a result of the weight of the overhang, tensile bending stresses develop on the upper portion of the cross section and compressive bending stresses occur in the lower portion. At some point the bending stresses due to the weight of the overhanging slab become so great that the ice fractures and the slab calves off.

White et al. [15] have performed extensive axisymmetric finite analyses of this problem and devised a simple expression for F_l, the critical length when failure occurs

$$F_l = 0.33 \left(37.5H + h^2\right)^{1/2} , \tag{12.14}$$

where H is the wave height and h is the thickness of the overhanging slab, both expressed in meters.

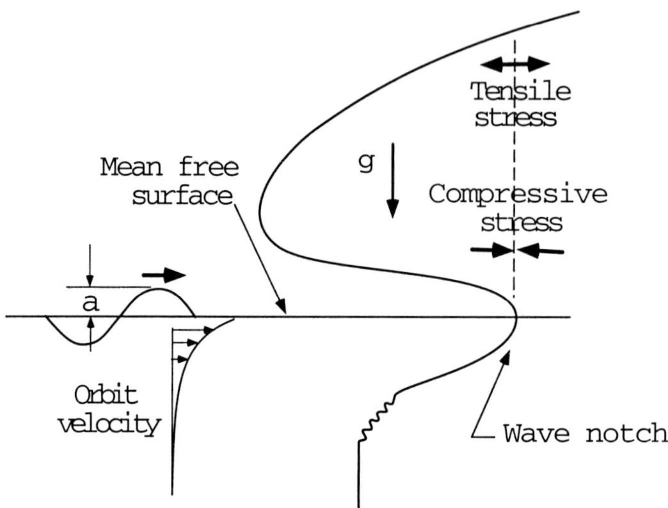

Fig. 12.5. Schematic diagram of overhanging slab above the notch caused by waterline wave erosion

For a steady wave field, t_c, the time taken to calve is given by

$$t_c = F_l/V_{\mathrm{we}} \,, \tag{12.15}$$

where V_{we} is the wave erosion velocity given by (12.13). If the wave field were unsteady, then one would integrate the expression for the waterline erosion rate (12.13) until the critical value of F_l was reached.

White et al. [15] have noted that wave erosion is significantly inhibited in the lee of waves, and hence the amount of ice lost in calving is strongly shape dependent. The iceberg shape is also quite variable and difficult to predict. As a result, the above equations for calving rates are expected to give only crude approximations.

12.3.3 Validation of Deterioration Equations

El-Tahan et al. [8] performed a study to validate the iceberg deterioration model described above in Sect. 12.3.2 that is essentially due to White et al. [15]. They simulated the deterioration of three icebergs, one in the Labrador Sea (during September) and two in the Grand Banks area (periods from April to June). These predictions were described earlier in Sect. 12.3.1 and the contributions of the various deterioration mechanisms were summarized in Table 12.3. When the sea ice melts away and the water temperatures are above $0\,^\circ$C, wave erosion and calving are the dominant deterioration mechanisms, and can account for as much as 80% of the deterioration rate. The predicted overall mass losses compared reasonably well with the observed quantities.

A more recent comparison between predicted deterioration rates and observations of two icebergs off the Canadian East coast has been made by Venkatesh, et al. [21]. They concluded that both the calving events and the deterioration of the icebergs were well modeled (within 10 to 20%) by the above approaches. This good agreement is indeed surprising in view of the uncertainties in iceberg geometry and environmental conditions, as well as the many approximations made in the analyses of the various deterioration mechanisms.

12.4 Melt Deterioration of Bergy Bits and Growlers

The previous section has dealt with the deterioration of icebergs having lengths greater than about 20 m. It is unlikely that all of the equations of White et al. [15] can be applied to determine melt rates for bergy bits and growlers whose lengths range from 20 m down to a couple of meters. These smaller ice pieces will be carried along with the wave motion to a larger extent than in the case of larger icebergs. The effect of waterline erosion, which is the major deterioration mechanism for icebergs, is expected to be different for large icebergs than for small ice pieces that heave, pitch and roll with the wave motions.

Thus, in the present section, we describe a different, simpler approach to determine small ice piece melt rates. It was initially proposed by Crocker and English [22] and was further employed by Savage et al. [23]. Let us assume that the wave enhanced heat flux Q from the ocean to the ice piece is a function of the wave height H, wave period τ, water kinematic viscosity ν, the temperature difference between the water and the melting ice ΔT, the thermal conductivity of the water k_w (0.562 W/m/°K), gravitational acceleration g, and a characteristic length of the ice piece D, i.e.

$$Q = fn\,(H, \tau, \nu, \Delta T, k_w, g, D) \ . \tag{12.16}$$

Applying the standard dimensional analysis techniques [24] to (12.16) yields four dimensionless parameters and the functional relationship:

$$\frac{Q}{k_w \Delta T H} = fn\left(\frac{H^2}{\tau\nu}, \frac{H}{g\tau^2}, \frac{D}{H}\right) , \tag{12.17}$$

where the parameter $Q/k_w \Delta T H$ is the dimensionless heat flux, $H^2/\tau\nu$ is a wave Reynolds number and D/H is an ice piece characteristic length to wave height ratio. If we consider deep water waves where the square of the wave celerity $c^2 = \lambda g/2\pi$, then it turns out that we can regard $H/g\tau^2$ as being proportional to the wave steepness. Let us further assume the particular product form of (12.17)

$$\frac{Q}{k_w \Delta T H} = \beta \left[\frac{H^2}{\tau\nu}\right]^{c_1} \left[\frac{H}{g\tau^2}\right]^{c_2} \left[\frac{D}{H}\right]^{c_3} , \tag{12.18}$$

where β, c_1, c_2, and c_3 are constants to be determined by means of fitting with some laboratory and/or field data.

Note that we can express the characteristic length D as the cube root of the volume V of the ice piece, which is merely one convenient way to define a length scale. We can also express the volume in terms of the ice piece waterline length L, thus

$$V = k_{BG} L^3 , \tag{12.19}$$

and hence, we can write

$$D = V^{1/3} = k_{BG}^{1/3} L , \tag{12.20}$$

where k_{BG} is a volume coefficient for bergy bits and growlers assumed to have a numerical value of 0.45 [23]. Using (12.18), it is possible to express the rate of change of ice mass $\dot{m}$ (in kg/s) as

$$\dot{m} = -Q/\Gamma , \tag{12.21}$$

where Γ is the latent heat of fusion (3.34×10^5 J/kg).

12.4.1 Determination of Constants in Heat Flux Equation

To make use of the dimensionless heat flux equation (12.17), the constants β, c_1, c_2, and c_3 must determined, and empirical data provides a means to do so. Unfortunately, such data is very limited. Savage et al. [23] considered laboratory tests and field investigations of Crocker and English [22]. The laboratory tests of Crocker and English were of two types. One set consisted of experiments, performed in the Ocean Engineering Basin (OEB) at the Institute for Marine Dynamics (IMD) in St. John's, Newfoundland, which measured the melt rates of fresh-water blocks subjected to regular waves. Similar series of tests were carried out in the Institute for Marine Dynamics (IMD) ballast tank; they entailed fully submerged blocks of ice that were forced harmonically under water. In the field experiments, video images of melting bergy bits and growlers were collected. These tests were carried out near Baie Verte, Newfoundland from July 19 to 22, 1994. Six ice pieces were monitored from a nearby ship and 14 melt rate measurements were obtained along with measurements of water temperature and estimates of wave heights and periods.

Multiple linear regression analyses were applied to the collected data sets and the constants β, c_1, c_2, and c_3 appearing in (12.18) were obtained. The field data showed considerably more scatter than the laboratory tests. However, after a careful examination of the small scale laboratory test results and comparisons with the field data and classical forced convection heat transfer results, Savage et al. [23] concluded that it was more appropriate to determine the constants on the basis of the field results alone. It was found that the ice pieces in the small scale laboratory tests behaved in a rather different way than in the field tests. As the ice pieces in the laboratory tests melted and became smaller, there was a greater tendency for the very small floating ice pieces to be carried along with the waves as their size decreased. Thus, the differential velocity between the water and ice pieces was reduced as melting occurred. This gave rise to a different dependence of heat flux on characteristic length scale D than was observed in the field tests. The larger ice pieces in the field tests have less of a tendency to bob up and down with the waves and are more likely to experience a washing by the waves. Considering only the Baie Verte field data, then (12.18) takes the form

$$\frac{Q}{k_w \Delta T H} = 933 \left[\frac{H^2}{\tau \nu}\right]^{0.347} \left[\frac{H}{g\tau^2}\right]^{0.171} \left[\frac{D}{H}\right]^{1.75} . \qquad (12.22)$$

12.5 Iceberg Life Expectancies

12.5.1 Computations Based on Deterioration Model

Using the deterioration models discussed previously in Sect. 12.3.1 and Sect. 12.3.2, Venkatesh and El-Tahan [4] have computed the life expectancies for each of the standard tabular and non-tabular iceberg size classifications. Calculations were performed using climatological environmental data (mean sea, sea surface and air temperatures, wave period and height, and wind speeds) for each month in the Grand Banks and in Labrador. On the basis of these computations, they made a number of conclusions corresponding to these two areas.

Grand Banks area – A growler will vanish in less than one day in any month of the year. A small non-tabular iceberg will completely disintegrate in less than three days during July to November.

Labrador Sea – During December to May, when the surface and layer mean temperatures are at or below 0 ° C, no significant melting occurs. During June to November, a growler will completely melt in less than 33 hours, while a small non-tabular iceberg needs more than five days to disintegrate completely. Also in the Labrador Sea, the life expectancy of medium and large icebergs is of the order of weeks.

The shortest life expectancy of any iceberg occurs in September for both the Grand Banks area and the Labrador Sea.

12.5.2 Empirical Model for Iceberg Life Expectancy

Venkatesh and El-Tahan [4] took the results of the above mentioned calculations for a given month and particular location, and plotted them in the form of iceberg mass versus life expectancy. On log–log scales, the graphs were approximately linear. They were then able to express the life expectancy t_l in hours in the following form

$$t_l = a\, M_0^b \,, \qquad (12.23)$$

where M_0 is the iceberg initial mass in metric tons and a and b have constant values for each month. Using linear regression analyses, Venkatesh and El-Tahan [4] determined the values of a and b for each month and location (cf. Table 12.4).

Table 12.4. Values of coefficients a and b used in iceberg life expectancy model of Venkatesh and El-Tahan [4]

Month	Grand Banks		Labrador Sea	
	a	b	a	b
January	0.15995	0.67	–	–
February	0.38725	0.67	–	–
March	0.41591	0.67	–	–
April	0.27415	0.67	–	–
May	0.15922	0.67	–	–
June	0.089125	0.67	0.831	0.6
July	0.05333	0.67	0.282	0.6
August	0.033884	0.67	0.159	0.6
September	0.019588	0.67	0.183	0.6
October	0.026546	0.67	0.349	0.6
November	0.036224	0.67	0.681	0.6
December	0.061944	0.67	–	0.6

Equation (12.23) can be used to estimate the mass of an iceberg that has a life expectancy equal to the prediction period τ_p, thus

$$m = (\tau_p/a)^{1/b} . \tag{12.24}$$

An iceberg with a mass less than or equal to m will melt during the prediction period τ_p.

We require a frequency distribution of iceberg masses in a given ensemble if we wish to calculate the percentage of icebergs with a mass less than or equal to m that are lost during the prediction period τ_p. Venkatesh and El-Tahan [4] made use of IIP data, which covered mainly the Grand Banks area, and Fenco data, which primarily covered the Labrador Sea area. Using linear regression analyses, they were able to obtain relationships between the cumulative frequency (in percent) and the mass (in metric tons) as follows:

$$f = -33 + 20\log_{10}(m) , \tag{12.25}$$

for the Grand Banks area, and

$$f = -66 + 24\log_{10}(m) , \tag{12.26}$$

for the Labrador Sea.

By using (12.24) and (12.25) or (12.26), we can determine the percentage of icebergs that will be lost in a given prediction period. For example, if we consider a prediction period of one day (24 hours), then from (12.24), we see that all icebergs with a mass of 40,670 metric tons or less will vanish in that period. Equation (12.25), for the Grand Banks area, indicates that this corresponds to 59.2 % of the icebergs considered at the start of the prediction period.

12.6 Ice Piece Size Distributions

Because of the potential hazards caused by small ice pieces in the bergy bit and growler size range, it is important to have information about the probability of encountering them. Most of them are generated when ice is calved from the larger parent iceberg and the calved ice fractures into many thousands of pieces. It is thought that the calving events, caused by waterline wave erosion and undercutting, can occur in a roughly periodic fashion. The waves create a notch at the waterline of the iceberg resulting in the formation of an overhanging cantilever slab of ice. The slab breaks off when it reaches a critical length. The time taken to develop this critical length was given previously in (12.15); it is the calving period. Thus, we can think of a continuing supply of ice pieces being generated by the calving process. The calved ice pieces melt subsequent to the calving event, and eventually disappear. Depending upon the relative velocities between the parent iceberg and the smaller ice pieces, there will be a maximum distance that the smaller ice pieces travel from the parent iceberg before they melt to a size of negligible significance. As the ice pieces are calved from the parent iceberg and disperse from it, there develops a spatial distribution of ice piece sizes that depends upon the distance from the parent iceberg. One eventual goal of an operational iceberg drift and deterioration program would be to generate charts yielding the probabilities of encountering bergy bits and growlers in regions surrounding the parent icebergs.

Although several studies [1,3,25,26] have collected data on iceberg sizes, almost all have focused on size distributions of parent icebergs, and few have included accurate measurements for the smaller (< 20 m) ice pieces in the bergy bit and growler size ranges. This is in part due to the difficulties in making size measurements of the small pieces, the very large number of small pieces that are generated in a calving event, and the earlier lack of concern about the smaller pieces. Crocker and Cammaert [3] have suggested that most existing iceberg data sets collected prior to their own work underestimated the numbers of these smaller icebergs, and some data sets contained little or no information on them.

Crocker and Cammaert [3] presented results of aerial photographic surveys conducted north of Cape Feels, Newfoundland on July 13, 1992. The large format photographs permitted identification and measurement of ice pieces as small as 5 m in length. The data set included waterline lengths of from 5 to 175 m. The measurements were plotted in the form of size distributions using 5 m bin divisions of extreme water length. These observations revealed large numbers of small ice pieces less than 10 m in length and relatively few larger pieces in the 10 to 20 m range. The absence of the 10–20 m pieces suggested that extensive fracturing occurs during the calving process when the overhanging slab (caused by wave erosion at the water line) can no longer be supported. The important conclusion of this study was that the iceberg size distribution has two distinct parts. One part is a slightly skewed distribution of parent icebergs similar in form to what was observed in most previous studies. And most importantly, there is a second, previously undocumented part, which shows an approximately exponential increase with decreasing length in the bergy bit and growler size range ($< 20\,\mathrm{m}$).

12.6.1 Size Distributions of Newly Calved Ice Pieces

In this subsection we shall briefly discuss the initial size-frequency distribution function for the small ice pieces ($< 20\,\mathrm{m}$ in length) that are generated by a calving event [27]. Crocker [28] has noted the two primary calving processes that have been observed. One is due to the breaking off and fracture of an overhanging slab that has been developed by wave erosion of the iceberg at the waterline as discussed in Sect. 12.3.2. A second calving process is that due to stresses induced by buoyant forces and or grounding [29,30]. Related calving can also occur as a result of thermal stresses caused by solar radiation or rolling in warm waters. It is very difficult to quantify these latter kinds of calving other than by statistical means. However, Crocker [28] has suggested that the distribution of ice piece sizes appears to be qualitatively similar to that which is generated during the fracture of an overhanging slab.

Analytical Expression for Initial Size Distribution Function. Savage et al. [27] proposed an analytical expression for the size distribution function and determined the constants appearing in this expression by fitting it to field data. Their results can be summarized as follows. Defining L to be the waterline length of an ice piece, the fractional number of particles with linear sizes between $L - \mathrm{d}L/2$ and $L + \mathrm{d}L/2$ can be expressed as

$$\frac{\mathrm{d}N}{N_0} = f(L)\,\mathrm{d}L\,, \tag{12.27}$$

where $f(L)$ is the distribution function and N_0 is the total number of particles. Savage et al. [27] proposed that the Weibull [31] distribution function was a suitable form for $f(L)$, and they chose a particular form that gave a good fit to the *observed* ice piece size distributions for calving icebergs and glaciers

$$f(L) = \frac{1}{2L_0}\left(\frac{L_0}{L}\right)^{1/2} \exp\left[-\left(\frac{L}{L_0}\right)^{1/2}\right]\,. \tag{12.28}$$

Savage et al. (cf. p. 167 of [27]) noted that although a power law form for the distribution function could give a reasonable fit to the middle range of the observations, it overestimated the data for both the smallest and largest ice pieces. By using the definition of the particle mean size $L_m = \int_0^\infty L\,f(L)\,\mathrm{d}L$, they determined that the characteristic length $L_0 = L_m/2$.

The total calved mass m made up of N_0 particles, having a mass density ρ_I, and a size distribution function given by (12.28), is defined by

$$m = \int_0^\infty N_0\, \rho_I\, V\, f(L)\, \mathrm{d}L \;=\; \int_0^\infty N_0\, \rho_I\, k_{BG}\, L^3\, f(L)\, \mathrm{d}L\,, \tag{12.29}$$

where we have made use of (12.19) for the volume V of an ice piece. By carrying out the integration of (12.29), we can express the mass in terms of L_0, i.e.

$$\begin{aligned} m &= \int_0^\infty N_0\, \rho_I\, k_{BG}\, \frac{L^3}{2L_0} \left(\frac{L_0}{L}\right)^{1/2} \exp\left[-\left(\frac{L}{L_0}\right)^{1/2}\right] \mathrm{d}L \\ &= 720\, N_0\, \rho_I\, k_{BG}\, L_0^3\,. \end{aligned} \tag{12.30}$$

Rearranging (12.30) yields the total number of particles N_0 in terms of the total mass m and L_0 or L_m

$$N_0 = \frac{m}{720\, \rho_I\, k_{BG}\, L_0^3} = \frac{m}{90\, \rho_I\, k_{BG}\, L_m^3}\,. \tag{12.31}$$

Correlations with Field Data. There are few field observations that provide information about the size distributions of ice pieces generated by a calving event. We consider here the field observations of Crocker [28] involving debris from recently calved icebergs, and the observations of Dowdeswell and Forsberg [32] involving 'small' ice pieces derived from tidewater glaciers (those with marine margins grounded below sea level).

Savage et al. [27] performed nonlinear least square fits to (12.28) based on these ice piece size data. They were thus able to obtain correlations for the length scale L_0 that appears in the distribution function (12.28). The iceberg calving data of Crocker [28] were derived from aerial surveys carried out in Bonavista Bay, Newfoundland in 1991 and 1992. These two data sets included 1049 and 3461 ice pieces respectively. The small ice pieces were calved from parent icebergs that ranged in length from 50 to 100 m, with a mean length of 88 m. Further details about these data are given in [27]. By using the estimates of the total calved mass for each calving event, the total numbers of ice pieces calved in a given event were determined by means of (12.31). The 1992 data sets included 9 separate calving events involving calved masses totaling 62,237 metric tons. When the events involving very small calved masses were neglected, the remaining 1991 data sets included individual calved masses of 3032 and 767 tons. Thus, while the total calved mass per event varied by an order of magnitude, the least square fits determined a value of L_0 that remained constant at approximately 0.14 m. It follows that the mean calved ice piece size $L_m = 2L_0 \simeq 0.28$ m.

This suggests that L_0 is independent of calved mass for events that involve more than some minimal calved mass, i.e. that L_0 must depend upon some characteristic length different from, for example, L_V that could be obtained by

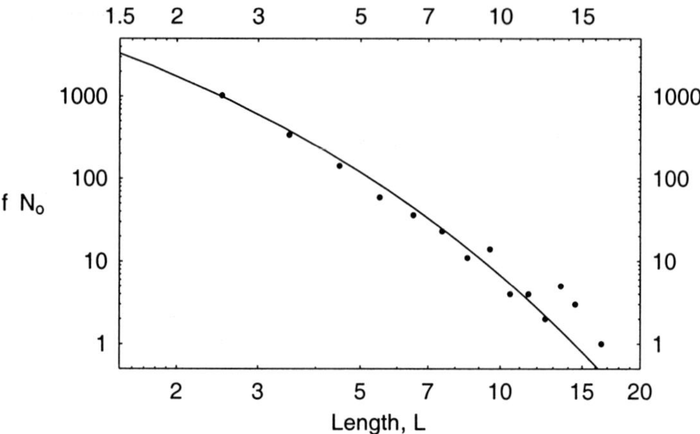

Fig. 12.6. Size-frequency distribution function for the 1992 Crocker data that include 9 separate calving events involving a total calved mass of 62,237 tons. Length scale $L_0 = 0.135\,\mathrm{m}$

taking the cube root of the calved volume. Savage et al. [27] noticed that one can form a characteristic length L_{FR} from the group

$$L_{FR} = \left[\frac{K_{Ic}}{\rho_I\, g}\right]^{2/3} , \tag{12.32}$$

where K_{Ic} is the fracture toughness (having dimensions of force/length$^{3/2}$), ρ_I is the mass density of ice, and g is the gravitational acceleration. The length L_{FR} is dependent on material properties. However, it is not dependent on a physical length scale, and thus, would be independent of calved mass.

Figure 12.6 shows a typical size-frequency distribution and compares the observed data with the curve fits corresponding to (12.28) for the 1992 data set of Crocker [28]. These data include 9 separate calving events having an average calved mass of 6915 tons per event. For the larger values of L the data points are more scattered about the curve fits. This is to be expected; the data contain only a few *large* ice pieces in a bin of pieces having lengths between L and $L + \mathrm{d}L$. On the other hand, there are of the order of 1000 ice pieces having lengths, for example, between 2 and 3 m.

Also, at first glance, the small value of the mean calved ice piece size $L_m \simeq 0.28\,\mathrm{m}$ is somewhat surprising. It transpires that this small value is a result of the extremely large numbers of very small ice pieces that arise in a calving event. Note that while the smallest ice pieces that were *measured* for the data shown in Fig. 12.6 were in the range between 2 and 3 m, enormous numbers of ice pieces as small as millimeters were present but not measured. In other field events [27], ice pieces as small as 0.15 m were observed and recorded. The size-frequency distribution function given by (12.28) was able to fit these observations quite well in the lowest size ranges. This gives some confidence in the use an equation having the form of (12.28) for estimating numbers of ice pieces in the small size ranges.

12.6.2 Evolution of Initial Calved Piece Size Distribution

The size distributions just discussed pertain to ice pieces that have just been calved. As the calved ice pieces are immersed in the warmer water, the individual pieces will melt. As a consequence, the size distribution function will evolve with time. Savage et al. [23] have determined the evolution of the distribution function by making use of the bergy bit and growler heat flux equation (12.22). We can rearrange (12.22) in the form

$$\frac{Q}{\Gamma} = \alpha\, L^{1.75}\,, \tag{12.33}$$

where Γ is the latent heat of fusion, and

$$\alpha = 933\,\frac{k_w\,\Delta T\;H}{\Gamma}\left[\frac{H^2}{\tau\nu}\right]^{0.347}\left[\frac{H}{g\tau^2}\right]^{0.171}\left[\frac{k_{BG}^{1/3}}{H}\right]^{1.75}\,. \tag{12.34}$$

Now from (12.19) the mass of an ice piece $m = \rho_I\,V = \rho_I k_{BG}\,L^3$, and thus by using (12.21), we can express the time rate of change of ice piece mass as

$$\dot{m} = 3k_{BG}\rho_I L^2\frac{\mathrm{d}L}{\mathrm{d}t} = -\frac{Q}{\Gamma} = -\alpha L^{1.75}\,, \tag{12.35}$$

or

$$\frac{\mathrm{d}L^{5/4}}{\mathrm{d}t} = -\zeta\,, \tag{12.36}$$

where

$$\zeta = \frac{5\alpha}{12k_{BG}\rho_I}\,. \tag{12.37}$$

Using the initial size distribution function (12.28) just after a calving event and the temporal variation of an individual ice piece that results from (12.36), Savage et al. [23] were able to determine the temporal evolution of the size distribution function subsequent to a calving event.

12.6.3 Dispersion of Calved Small Ice Pieces from Parent Iceberg

If we think of the well known Ekman spiral [33], it is apparent that both the magnitude and the direction of the wind driven ocean water velocity changes continuously with depth. The submerged portion of a large parent iceberg will be subjected to these depth varying currents. The effective water drag force acting on each infinitesimal horizontal layer of the iceberg will depend upon the difference between the water layer velocity and the iceberg velocity. The total effective water force acting on the iceberg mass will result from the integral of the forces acting on each differential horizontal layer of the submerged iceberg. On the other hand, the smaller calved ice pieces, whose drafts are less than that of the parent iceberg and which penetrate only an upper surface layer of the ocean, will be subject to different depth averaged velocities, and thus to total water driving forces that act in different directions.

What this means is that ice pieces which are calved from the parent iceberg will not move in unison with the parent, but will gradually move away from it and disperse, while at the same time they will melt and eventually disappear. Thus, at any given time one could expect a 'cloud' of smaller ice pieces in the neighborhood of a parent iceberg. Since these smaller pieces are diminished by melting as they drift away from the parent, there will be some maximum distance that the small ice pieces can travel relative to the parent before they become of insignificant size.

At the same time that this is going on, wave erosion can cut notches in the parent iceberg at the waterline, creating an overhanging slab that can break off when a critical sized notch has developed (cf. Sect. 12.3.2). Then, a calving event occurs and generates a new supply of small ice pieces in the neighborhood of the parent iceberg. Two processes proceed concurrently; (1) the intermittent calving that provides a source of new small ice pieces in the neighborhood of the parent iceberg, and (2) the dispersion of the small pieces away from the parent and their melting and eventual disappearance, which effectively act as a 'sink' of small ice pieces.

These processes have been implemented in an operational iceberg drift and deterioration computer program by Sayed [7] and Savage et al. [34]. This forecasting program predicts the velocities, positions and size variations of both the parent icebergs and packets of small calved ice pieces.

12.7 Iceberg Dynamics and Drift

12.7.1 Introduction

There are essentially three types of approaches used for modeling iceberg drift; these three categories have been reviewed by Marko et al. [35] who discussed the relative merits of each one. One category is the *statistical approach* that makes use of information on previous trajectories and histories to predict future iceberg positions and velocities [36–40]. The dynamics of the iceberg motion are usually ignored. A second class of methods is the *kinematic model* in which one also ignores the dynamics and estimates the drift by the use of simple empirical relationships. A simple example is the rule that icebergs move at about 2% of the wind speed [38]. The third class of methods is the *dynamical model* in which one integrates Newton's equations of motion for an iceberg of known mass that is subjected to various driving and resisting forces to obtain the iceberg's velocity and position as a function of time. Some examples of this type of model are [5,6], [41–48].

The dynamical models that yield deterministic predictions of velocities and trajectories have uncertainties associated with the various force terms in the equations of motion. These relate, for example, to inaccuracies in the air and water drag coefficients, the wave force coefficient connected with the radiation stress, the added mass coefficient, etc. The predictions of the dynamical models improve as the values of these coefficients are refined. The dynamical models also can be combined with iceberg deterioration models of various levels of sophistication (cf. Sects. 12.3, 12.4 and 12.6). For these reasons, dynamical models can be regarded as the most appropriate type for detailed forecasting. This is the model type that is discussed below. We note that the iceberg model presently used by the IIP is a dynamical model; the model used by the Canadian Ice Service (CIS) is similar to it.

12.7.2 Equations of Motion, Various Force Contributions

In this section we review the governing equations of motion and the various generic forces that arise. The basic equation governing the horizontal motion of an iceberg of mass m moving with velocity $\boldsymbol{V}$ and referred to axes at rest relative the rotating earth is based on Newton's second law, i.e.

$$m\left(\frac{\mathrm{d}\boldsymbol{V}}{\mathrm{d}t}+\boldsymbol{f}\times\boldsymbol{V}\right)=\boldsymbol{F}_a+\boldsymbol{F}_w+\boldsymbol{F}_r+\boldsymbol{F}_p+\boldsymbol{F}_{\mathrm{am}}+\boldsymbol{F}_{\mathrm{si}}\ , \tag{12.38}$$

where $\boldsymbol{f}$ is the Coriolis vector that points upward and whose magnitude is $f = 2\Omega\sin\phi$ in which $\Omega = 7.272\times10^{-5}\,\mathrm{rad/s}$ is the earth's rate of rotation and ϕ is the latitude. The terms on the right hand side of (12.38) correspond to the forces or effective forces acting on the iceberg where $\boldsymbol{F}_a$ is the wind drag force, $\boldsymbol{F}_w$ is the water drag, $\boldsymbol{F}_r$ is the force due to radiation stress associated with the waves, $\boldsymbol{F}_p$ is the pressure gradient force, $\boldsymbol{F}_{\mathrm{am}}$ is an effective force associated with the added mass of the iceberg, and $\boldsymbol{F}_{\mathrm{si}}$ is the force contribution due to interactions of the iceberg with floating sea ice. The various forces are now discussed in detail.

Wind Drag, $\boldsymbol{F}_a$. The air drag force $\boldsymbol{F}_a$ due to the wind acting on the exposed surface of the iceberg above the waterline is expressed as

$$\boldsymbol{F}_a=\frac{1}{2}\rho_a C_a A_a|\boldsymbol{u}|\boldsymbol{u}\ , \tag{12.39}$$

where $\boldsymbol{u} = (\boldsymbol{V}_a - \boldsymbol{V})$ is the relative wind velocity, $\boldsymbol{V}_a$ is the wind velocity derived from an atmospheric forecasting model, $\boldsymbol{V}$ is the iceberg drift velocity, ρ_a is the air density, and A_a is the cross-sectional area of the iceberg above the waterline and perpendicular to the relative velocity. The Reynolds number for a typical wind velocity of 10 m/s, and an iceberg length of 100 m is of the order of 10^8. Thus, the flow is certainly turbulent and a drag law like (12.39) which depends upon the square of the relative wind velocity is appropriate. The air drag coefficient for the iceberg, C_a, typically has a value of roughly 1.0. Smith [47], for example, uses a value for the wind velocity $\boldsymbol{V}_a$ measured at a height of 13 m. In most models the wind velocity $\boldsymbol{V}_a$ is assumed to be much greater than the iceberg drift velocity $\boldsymbol{V}$. As expressed in (12.39), the air drag force acts in the direction of the relative wind $\boldsymbol{u}$. This is appropriate if the iceberg shape is symmetrical about the two planes parallel to the flow direction. If the iceberg shape is unsymmetrical, then an additional force component perpendicular to the relative wind direction will arise. The iceberg would then act rather like an airfoil or a sail that generate lift forces. Because of calving, the shape of the iceberg is seldom known with any accuracy, and such transverse forces are almost always neglected.

Water Drag, $\boldsymbol{F_w}$. The water drag $\boldsymbol{F}_w$ is modeled in much the same way as the wind drag force, but since the water currents vary considerably with depth, some account must be taken of these variations. Typically, the water flow field is separated into horizontal layers, each layer having a mean relative water velocity $\boldsymbol{v}_r(i) = (\boldsymbol{V}_w(i) - \boldsymbol{V})$, where $\boldsymbol{V}_w(i)$ is the layer averaged upstream water velocity in the ith layer. The water velocities $\boldsymbol{V}_w(i)$ would be determined by an ocean current forecasting model. Summing over all the layers, the water drag force then becomes

$$\boldsymbol{F}_w = \frac{1}{2}\rho_w C_w \sum_i A_w(i)|\boldsymbol{v}_r(i)|\boldsymbol{v}_r(i) \tag{12.40}$$

where ρ_w is the density of the water, $A_w(i)$ is the cross-sectional area of the ith layer of the iceberg in a vertical plane, and C_w is the water drag coefficient again having a value of about 1.0, as consistent with turbulent flow around typical bluff bodies (recall from Sect. 12.3.2, in the subsection dealing with forced convection melting, that the Reynolds number is of the order of 10^7 for an iceberg having a length of 100 m and a relative drift velocity 0.2 m/s). Smith and Donaldson [6,46], for example, have used layer depths of 10 m. Similarly, the model of Sayed [7] also uses 10 m deep layers. The IIP model, on the other hand, uses coarser divisions of 4 unequal layer depths: from 0–20 m, 20–50 m, 50–100 m, and 100–120 m. If the draft of the iceberg is less than 100 m, then fewer than 4 layers are considered in the IIP model. Note that three-dimensional effects are not explicitly considered, and that the water drag coefficient C_w is assumed to be the same for each layer.

As a result of the water temperature and salinity variations with depth, there are associated water density variations. Moving an object through such a density stratified flow gives rise to internal wave drag, something that naval architects term the "dead water phenomenon". Stratified flow effects have generally been neglected.

Force Due to Wave Radiation Stress, $\boldsymbol{F_r}$. Longuet-Higgins and Stewart [49] formulated mean conservation laws of mass, momentum and energy for water surface wave motions imposed on variable currents and applied them in very elegant ways to a large number of problems. During the motion of a train of surface waves, there occurs an excess momentum flux associated with the wave motion that has been termed the radiation stress by [49]. It is generally acknowledged that Longuet-Higgins and Stewart coined the term "radiation stress" in the context of water surface waves [50,51]. We should note, however, that this term is also used in studies of acoustic streaming, albeit with a somewhat different meaning there.

One can think of a wave train and the excess momentum flux or radiation stress associated with it. A body exposed to the waves will experience a force depending upon the way the waves are diffracted and dissipated by the body. To appreciate this in a simple-minded way, consider the body to be replaced by a series of vertical screens in deep water. Now let us examine certain limits. First imagine that the screens are deep and that the deep water waves experience viscous dissipative as they move through the screens. If all the wave energy is absorbed (dissipated), then the momentum is also absorbed, and the force on the screens per unit width is $\rho_w g a^2/4$ where a is the amplitude of the incident waves. The amplitude a_t of the wave transmitted through to the back side of the screens vanishes. If the screens are very open, then the waves just pass through, there is no wave reflection, the transmitted wave (behind the screens) has the same amplitude as the incident wave. In this case, the force on the screens tends to zero. If on the other hand, the screens are completely closed, so as to act like a rigid vertical wall, then the incident wave energy is perfectly reflected, the momentum is all reversed, and the resulting force is doubled, i.e. equal to $\rho_w g a^2/2$. The amplitude of the reflected wave a_r is the same as that of the incident wave, and there is no transmitted wave behind the screens. For the case of partial wave reflection, we can express the force per unit width as $\rho_w g(a^2 + a_r^2 - a_t^2)/4$.

More generally, Longuet-Higgins [52] determined the radiation force on a box of width W normal to the incident wave train in water of finite depth h to be

$$\boldsymbol{F}_r = \frac{1}{4}\rho_w g W (a^2 + a_r^2 - a_t^2)\left(1 + \frac{2kh}{\sinh(2kh)}\right)\frac{\boldsymbol{V}_a}{|\boldsymbol{V}_a|}\,, \tag{12.41}$$

where $k = 2\pi/\lambda$ is the wave number, λ is the wavelength and the waves are assumed to have the same direction as the wind. For deep water, $2kh/\sinh(2kh) \to 0$, and if the waves are absorbed by a deep vertical wall, as suggested earlier the radiation force is given by

$$\boldsymbol{F}_r = \frac{1}{4}\rho_w\, g\, a^2\, W \frac{\boldsymbol{V}_a}{|\boldsymbol{V}_a|}\,. \tag{12.42}$$

For shapes such as an iceberg, the force can be less [53–56] because of wave diffraction resulting from the iceberg's three-dimensional shape and the finite draft. Commonly, the magnitude of the radiation force for icebergs is expressed as

$$|\boldsymbol{F}_r| = \frac{1}{2} C_{wf}\, \rho_w\, g\, W H^2\,, \tag{12.43}$$

where $H = 2a$ is the wave height, W is the width of the iceberg, and C_{wf} is a wave force coefficient that, in general, depends on the ratio of iceberg width to wavelength W/λ, the ratio of iceberg draft to width h_D/W, ratio of iceberg draft to water depth h_D/h, and the wave steepness H/λ. Isaacson and McTaggart [56] have noted that the wave force coefficient C_{wf} typically has values between 0 and 0.25, consistent with the analyses and experiments of Kudou [54] and Maruo [53]. In the limit of small wave heights, linear wave theory yields no dependence of C_{wf} on wave steepness H/λ.

Smith [47] has made an interesting comparison of the ratio of the radiation wave force (12.42) to the wind drag force (12.39) which is seen to be

$$\frac{\text{radiation wave force}}{\text{wind drag force}} = \frac{\rho_w \, g \, a^2}{2\rho_a \, C_a \, \overline{h} \, u^2} \,, \tag{12.44}$$

where $\overline{h}$ is the mean height of the iceberg above the waterline. Following Bigg et al. [42] one can estimate the wave amplitude as a function of wind speed by using a quadratic curve fit of wave crest to trough height $H = 0.02025 V_a^2$ from data in the marine Beaufort scale [57]. Assuming a wind speed $V_a = 10\,\mathrm{m/s}$ (much larger than the iceberg speed V), an air drag coefficient $C_a = 1.0$, and an iceberg with a mean height $\overline{h} = 20\,\mathrm{m}$, (12.44) yields that the radiation wave force is almost twice the air drag force. For higher wind speeds, the ratio of radiation wave force to air drag is much larger.

The wave radiation force has usually been neglected in most models. Bigg et al. [42] have estimated from their calculations that it *usually* contributes less than 5% of the total forces acting on the iceberg. On the other hand, Isaacson [55] has provided examples in which the wave radiation force is dominant.

Pressure Gradient Force, $\boldsymbol{F_p}$. The equation of motion for a water element can be written as

$$\frac{\mathrm{d}\boldsymbol{V}_w}{\mathrm{d}t} + \boldsymbol{f} \times \boldsymbol{V}_w = -\frac{1}{\rho_w}\nabla p + \frac{1}{\rho_w}\nabla \cdot \boldsymbol{\sigma} \,, \tag{12.45}$$

where $\mathrm{d}/\mathrm{d}t$ is the material derivative following a fluid particle, and $\boldsymbol{\sigma}$ is the stress tensor arising from viscous forces. The turbulent surface wind stresses are communicated into the fluid and generate $\boldsymbol{\sigma}$ in the interior. Let us assume that the vertical velocity (in the z-direction) is negligible, so that the velocity vector has essentially horizontal components. Furthermore, if we assume that the stress gradients in the z direction are much larger than those in the horizontal (x, y) direction, then (12.45) can be written as

$$\frac{\mathrm{d}\boldsymbol{V}_w}{\mathrm{d}t} + \boldsymbol{f} \times \boldsymbol{V}_w = -\frac{1}{\rho_w}\nabla p + \frac{1}{\rho_w}\frac{\partial \boldsymbol{\tau}}{\partial z} \,, \tag{12.46}$$

where $\boldsymbol{\tau}$ has the shear stress components $(\sigma_{xz}, \sigma_{yz})$ in the interior of the fluid. Bigg et al. [42] have suggested that the last term in (12.46), which results from the application of the surface wind stress, has a scale of the order of the surface wind stress $\boldsymbol{\tau}_s$ $(= 1.5 \times 10^{-3} \rho_a |\boldsymbol{V}_a| \boldsymbol{V}_a)$ [58], divided by the water density ρ_w and the Ekman depth. Usually the Ekman depth is defined as

$$E_k = \left(\frac{2\mu}{\rho_w f}\right)^{1/2} . \tag{12.47}$$

For the purposes of a simple order of magnitude estimate only, Bigg et al. chose, instead of (12.47), a vertical length scale corresponding to either the draft of the iceberg or 90 m, whichever is less. Substituting typical values, based on either (12.47) or the choice of Bigg et al. [42], shows that the last wind stress term in (12.46) is a small fraction of the Coriolis term $\boldsymbol{f} \times \boldsymbol{V}_w$. As noted by Bigg et al. [42], it has commonly been assumed that the ocean is in steady geostrophic balance and the pressure force per unit mass on the iceberg is simply $\boldsymbol{f} \times \boldsymbol{V}_w$. Furthermore, they suggest that the dominant term on the left hand side of (12.46) is the material derivative $\mathrm{d}\boldsymbol{V}_w/\mathrm{d}t$. It will be shown subsequently that the so-called pressure force on the iceberg depends upon $\mathrm{d}\boldsymbol{V}_w/\mathrm{d}t$. Bigg et al. [42] stated that the inclusion of the material derivative $\mathrm{d}\boldsymbol{V}_w/\mathrm{d}t$ in the determination of the pressure force was the principal factor needed to predict realistic iceberg motion.

The pressure force F_p exerted by the water on a fixed volume, such as an iceberg, can be expressed as a surface integral, which in turn may be written in terms of a volume integral by making use of Gauss' theorem, i.e.

$$\boldsymbol{F}_p = -\int\!\!\int_{Area} p\,\mathrm{d}A = -\int\!\!\int\!\!\int_{Volume} \nabla p\,\mathrm{d}\boldsymbol{r} , \tag{12.48}$$

Now following Smith and Donaldson [6], (note that there are several minor misprints in their equations), we assume that the pressure gradient ∇p is uniform horizontally in the neighborhood of the iceberg, and replace the vertical integration by a sum over a series of layers, each of volume $B(i)$

$$\boldsymbol{F}_p = -\sum_i \nabla p(i) B(i) , \tag{12.49}$$

Assuming that (12.46) can be applied to each layer and neglecting the last wind stress term, we can write

$$\boldsymbol{F}_p = \rho_w \sum_i B(i) \left(\frac{\mathrm{d}\boldsymbol{V}_w(i)}{\mathrm{d}t} + \boldsymbol{f} \times \boldsymbol{V}_w(i)\right) . \tag{12.50}$$

Since the mass of the displaced volume of water is $m_{\mathrm{d}v} = \rho_w \sum_i B(i) = m$, we can express (12.50) in terms of a volume averaged current

$$\overline{\boldsymbol{V}}_w = \frac{\sum_i B(i)\boldsymbol{V}_w(i)}{\sum_i B(i)} \simeq \frac{\sum_i A(i)^2 \boldsymbol{V}_w(i)}{\sum_i A(i)^2} , \tag{12.51}$$

where it has been assumed that the volume of each layer is proportional to the square of its measured cross-sectional area $A(i)$ in a vertical plane, since one does not normally have direct measurements of the underwater volume of each layer nor of the cross-sectional area in a horizontal plane [6]. Hence, using (12.51), we can rewrite (12.50) as

$$\boldsymbol{F}_p = m\left(\frac{\mathrm{d}\overline{\boldsymbol{V}}_w}{\mathrm{d}t} + \boldsymbol{f} \times \overline{\boldsymbol{V}}_w\right) . \tag{12.52}$$

The above expression for the pressure force differs from that of the Smith and Banke [45] model which neglected the material derivative $\mathrm{d}\overline{\boldsymbol{V}}_w/\mathrm{d}t$. In later work, which did include this term, Smith and Donaldson [6,46] indicated that while it would appear to add important physics to the model, its neglect resulted in negligible changes to the computed iceberg trajectories. They suggested that in its absence, the water drag forces act to make the iceberg velocity respond quickly to the water column. Smith [47] further qualified this in stating that while the pressure gradient term is frequently comparable to the other terms, its deletion causes the iceberg to lag behind the changes in water velocity, but the resultant motion relative to the accelerated water is reduced by changes in the water drag. The equilibrium drift rates are approached within an hour, and hence, the lag in response is so brief that it does not significantly affect the iceberg tracks.

On the other hand, as mentioned earlier, Bigg et al. [42] have forcefully argued that the material derivative $\mathrm{d}\overline{\boldsymbol{V}}_w/\mathrm{d}t$ is the key term in the expression for $\boldsymbol{F}_p$ and that only by its inclusion "could the most important iceberg zone east of Labrador and Newfoundland be reproduced." They stated that the two forces, $m\,\mathrm{d}\overline{\boldsymbol{V}}_w/\mathrm{d}t$ and the water drag $\boldsymbol{F}_w$ contribute approximately $70 \pm 15\%$ of the total forcing of the iceberg motion and are roughly in balance. It was estimated that accelerations are generally less than $10^{-7}\,\mathrm{m/s^2}$ (i.e. $1\,\mathrm{cm/s/day}$) except when grounded icebergs have just been released. Based on their own calculations, Bigg et al. [42] suggested that typically the Coriolis force and the air drag make up the remaining force balance in roughly equal parts of 15% each, whereas the wave radiation force is generally less than 5%. However, as noted earlier in the discussion of the wave radiation force, the relative sizes of the various force contributions depend very much on the environmental conditions. One should be cautious about generalizing on the basis of limited calculations.

Effective Force Associated with the Added Mass, $\boldsymbol{F}_{\mathrm{am}}$. When a solid body is accelerated in a fluid, the fluid particles are accelerated to some degree along with the body. In principle, all the fluid particles will be accelerated, but we can think of an added mass of fluid that is a weighted average of this entire mass. Thus, we can think of this added mass as being accelerated with the same value as that of the solid body [59]. For example, in an inviscid fluid, the added mass of a sphere is the mass of the fluid corresponding to half of the displaced volume of the sphere. For an iceberg we can expect a roughly similar value. Newman [59] has provided added mass coefficients for ellipsoids of revolution and other shapes. Isaacson and McTaggart [56] have discussed added mass in the context of icebergs. For example, they have considered the effects of shallow water and the presence of nearby structures. Instead of considering $m\,(\mathrm{d}\boldsymbol{V}/\mathrm{d}t + \boldsymbol{f} \times \boldsymbol{V})$ as the mass acceleration term in the momentum equation, one can write this term as $(m + m_{\mathrm{am}})\,(\mathrm{d}\boldsymbol{V}/\mathrm{d}t + \boldsymbol{f} \times \boldsymbol{V})$ where m_{am} is the added mass. Alternatively, we can express the momentum equation in the form of (12.38) with an effective force $\boldsymbol{F}_{\mathrm{am}}$ on the right hand side where

$$\boldsymbol{F}_{\mathrm{am}} = -m_{\mathrm{am}} \left(\frac{\mathrm{d}\boldsymbol{V}}{\mathrm{d}t} + \boldsymbol{f} \times \boldsymbol{V} \right) . \tag{12.53}$$

In so doing, we are making use of D'Alembert's principle, which is discussed in most dynamics textbooks. D'Alembert suggested that Newton's second law could be considered from a slightly different viewpoint in which the mass acceleration terms could be treated as effective 'inertia forces'. We note that most iceberg models have neglected added mass effects and do not contain a term like (12.53).

Smith [47] has performed calculations in which the iceberg mass was doubled and quadrupled to examine the effects of iceberg mass. It was stated that the increases in mass resulted in negligible changes in fit of the modeled track. However, these computations involved finding 'optimized' values for the air and water drag coefficients, C_a and C_w. When the mass was increased, both C_a and C_w had to be increased to obtain equivalent fits of the modeled track. For example, in the case of iceberg 83-5, his full model fit the iceberg track best when $C_a = 0.9$ and $C_w = 0.8$. When the mass was doubled, the best fit was obtained when $C_a = 1.3$ and $C_w = 1.2$. When it was quadrupled, the best fit was obtained when $C_a = 3.0$ and $C_w = 2.8$. However, this sort of 'tuning' does not seem to be entirely consistent and the significance of such results is not clear.

Force Due to Interactions with Sea Ice, $\boldsymbol{F}_{\mathrm{si}}$. Although the force contribution due to interactions of the iceberg with sea ice is sometimes mentioned as a factor in discussions of drift models [42], it is usually neglected, and the writer is unaware of any models that have explicitly taken this force into consideration for either drift or deterioration.

12.7.3 Comments Concerning Possible Errors and Uncertainties

The basic ideas behind the dynamics in the present operational iceberg model are straightforward and well established. The various contributions to the forces and effective forces acting on the iceberg have been identified. Nevertheless, there are some differing opinions concerning the relative significance of a few of the force contributions. Some of the less contentious origins of inaccuracies in existing models are mentioned below.

Drag Coefficients and Reference Areas. The wind and water drag forces depend on the fluid velocities, the drag coefficients and the pertinent reference cross-sectional areas above and below the sea surface. Frequently, the reference cross-sectional shapes are not known with any degree of certainty. They also change when the iceberg melts, and more significantly when calving and rolling of the iceberg occur. The drag coefficients for bluff bodies (such as an iceberg) at high Reynolds numbers typically have values of around 1.0 [60]. The reliable prediction of the turbulent flows around such bodies involving flow separation is not possible given the present state of the art, and one commonly relies upon empirical correlations for the drag coefficients. Banke and Smith [61] performed towing experiments on three growlers and reported water drag coefficients $C_w = 1.2 \pm 0.2$; part of this scatter was due to uncertainties about the shape and size of the bergs.

Drag coefficients are sometimes established by choosing values that give the best fit between predicted and observed iceberg trajectories. For example, Smith and Donaldson [46] found by fitting 9 iceberg track segments that the "optimum" drag coefficients were $C_a = 1.3 \pm 0.7$ and $C_w = 1.0 \pm 0.7$. The standard deviation of values in other fits of this kind are sometimes in excess of the values just noted (see also [45]). In a later study, Smith [47] suggested optimum air and water drag coefficients of 1.3 and 0.9 respectively (also see [62]). One must be cautious about such determinations since the drag coefficients are being used essentially as fitting parameters, and the particular values obtained can reflect inadequacies of the overall modeling rather than accurate assessments of the physical drag coefficients themselves. Nevertheless, the quoted values are not too different from the standard values for bluff bodies.

Uncertainties in flow forces can also be present because of sail and keel effects in which "lift" forces perpendicular to the oncoming flow are generated in addition to the usual hydrodynamic drag parallel to the flow direction. An additional effect (that is customarily neglected) is the additional drag force due to the generation of internal waves in the stratified upper water layers.

Wind and Water Currents Forecasts. The major factors that influence iceberg drift are the wind and water currents. These currents are responsible, either directly or indirectly, for all of the forces or effective forces that act on the iceberg. Clearly, inaccurate information about their magnitude, direction and variations with depth can lead to errors in predicted iceberg trajectories. It is essential to have wind and water current forecasts that are as accurate as possible.

For example, in an iceberg model currently under development [7], the Canadian Ice Service (CIS) is making use of the wind and water current forecasting components that are part of the Community Ice Ocean Model (CIOM). The CIOM is a coupled ice, ocean numerical model used for support to ice analysis and forecast operations. It uses a coupling framework shared by the Bedford Institute of Oceanography, the Maurice Lamontagne Institute and the Canadian Ice Service. The ocean component of the CIOM is POM, the Princeton Ocean Model (cf. `http://www.aos.princeton.edu/WWWPUBLIC/htdocs.pom/`). The CIOM is forced by forecast surface wind from models of the Canadian Meteorological Centre.

A tentative CIS iceberg forecasting model [7], which accounts for all of the forcing terms mentioned in Sect. 12.7.2, was able to yield quite good simulations of the iceberg trajectories measured by Smith and Donaldson [6] when it made use of the observed iceberg cross-sectional shapes and the observed water currents. This suggests that, if the model is forced by accurate predictions of the wind and water velocities, it should give reliable results. Ocean current forecasts from the particular implementation of the POM into the CIS iceberg forecasting model have not been compared with observed ocean currents, but a field project to do so is presently being planned.

12.7.4 Numerical Integration Schemes

The equations of motion based on (12.38) can be rewritten in the form of two first order differential equations

$$\frac{\mathrm{d}\boldsymbol{V}}{\mathrm{d}t} = \boldsymbol{a}(t,\boldsymbol{V}) = -\boldsymbol{f} \times \boldsymbol{V} + \frac{1}{m}(\boldsymbol{F}_a + \boldsymbol{F}_w + \boldsymbol{F}_r + \boldsymbol{F}_p + \boldsymbol{F}_{\mathrm{am}} + \boldsymbol{F}_{\mathrm{si}}) \,, \quad (12.54)$$

$$\frac{\mathrm{d}\boldsymbol{x}}{\mathrm{d}t} = \boldsymbol{V} \,, \quad (12.55)$$

where $\boldsymbol{x} = (x, y)$ is the horizontal position of the iceberg at time t. These equations can be integrated in the simplest manner by using an explicit or forward Euler method (cf. p. 550 [63]). Thus, one considers an advanced time step $t^{(j+1)} = t^{(j)} + \Delta t$, where $t^{(j)}$ is the previous time step and Δt is the time increment, and writes (12.54) and (12.55) in the following finite difference form

$$\boldsymbol{V}^{(j+1)} = \boldsymbol{V}^{(j)} + \boldsymbol{a}^{(j)}\Delta t \,, \quad (12.56)$$

$$\boldsymbol{x}^{(j+1)} = \boldsymbol{x}^{(j)} + \boldsymbol{V}^{(j)}\Delta t \,, \quad (12.57)$$

where $\boldsymbol{a}$ in (12.56) is based on the velocities $\boldsymbol{V}^{(j)}$ at the earlier time step $t^{(j)}$ and the velocities $\boldsymbol{V}^{(j)}$ in (12.57) are taken at the position $\boldsymbol{x}^{(j)}$.

For example, Smith and Donaldson [6,46], Smith [47] and Bigg et al. [42] have used this simple Euler approach. Smith and Donaldson [6,46] and Smith [47] found the appropriate time step to be 24 s. Shorter time steps gave almost the same results whereas longer time steps, greater than 100 s occasionally led to computational instabilities. Bigg et al. [42] used the same discretization as in [47], but used somewhat longer time steps of 135 s.

Although the forward Euler method is very simple, it is only first order accurate and other workers such as Mountain [5] and Sodhi and El-Tahan [48] have used more accurate fourth order Runge–Kutta approaches. The present IIP model uses a fourth order Runge–Kutta scheme with an adaptive time step. The computations start with a specified time step and if the computations are not within a predetermined error tolerance, the time steps are successively halved. The initial default time step in the current IIP model is taken as 225 s, which is considerably larger than the value of 24 s used by Smith [47] in his forward Euler model.

The experience of the Canadian Ice Service is that computations based on the IIP model sometimes become unstable. This suggests that the stability of this approach may depend upon the values of various parameters or the forcing functions. It is be desirable to have a more robust numerical scheme, one that can accommodate much larger time steps, and preferably one that is unconditionally stable.

Implicit Backward Euler Method. One alternative to the above explicit approaches is to use an implicit or semi-implicit numerical scheme. Even for nonlinear equations, such techniques are usually stable. The simplest of the implicit schemes is based on the implicit backward Euler method (cf. p. 735 [63]). Consider the nonlinear set of equations (12.54)

$$\frac{\mathrm{d}\boldsymbol{V}}{\mathrm{d}t} = \boldsymbol{a}(t, \boldsymbol{V}) \,, \tag{12.58}$$

which after implicit differencing yield

$$\boldsymbol{V}^{(j+1)} = \boldsymbol{V}^{(j)} + \boldsymbol{a}^{(j)}(t^{(j+1)}, \boldsymbol{V}^{(j+1)})\Delta t \,. \tag{12.59}$$

Equation (12.59) can be linearized to give

$$\boldsymbol{V}^{(j+1)} = \boldsymbol{V}^{(j)} + \Delta t \left[\boldsymbol{a}^{(j)}(t^{(j+1)}, \boldsymbol{V}^{(j)}) + \left. \frac{\partial \boldsymbol{a}}{\partial \boldsymbol{V}} \right|_{\boldsymbol{V}^{(j)}} \cdot \left(\boldsymbol{V}^{(j+1)} - \boldsymbol{V}^{(j)} \right) \right] \tag{12.60}$$

where $\partial \boldsymbol{a}/\partial \boldsymbol{V}$ is the matrix of partial derivatives of the right hand side of (12.58). Solving (12.60) for $\boldsymbol{V}^{(j+1)}$ we obtain

$$\boldsymbol{V}^{(j+1)} = \boldsymbol{V}^{(j)} + \Delta t \left[\mathbf{1} - \Delta t \frac{\partial \boldsymbol{a}}{\partial \boldsymbol{V}} \right]^{-1} \cdot \boldsymbol{a}^{(j)} \,. \tag{12.61}$$

Thus, at each step we must invert the matrix

$$\left[\mathbf{1} - \Delta t \frac{\partial \boldsymbol{a}}{\partial \boldsymbol{V}} \right] , \tag{12.62}$$

in order to solve (12.60) for $\boldsymbol{V}^{(j+1)}$. This result is first order accurate; Press et al. [63] discuss higher order semi-implicit methods.

Savage [69] has studied the explicit and implicit Euler approaches, and various linearly implicit Rosenbrock methods [64–67], including the Wolfbrandt [68] and modified Rosenbrock triple formulae in the context of iceberg drift. It was concluded that the simple semi-implicit Euler method was extremely robust, even for very large time steps, and provided sufficiently accurate results for time steps as large as 1000 s.

12.8 Concluding Remarks

This chapter has reviewed some of the physics and mechanics of the processes responsible for the drift and deterioration of seaborn icebergs. It has attempted to provide some background of what might be involved in the development of operational iceberg forecasting models.

The general methodology for the prediction of the various iceberg melt mechanisms and calving was first established by the work of White et al. [15]. It was found that wave erosion and calving are the dominant deterioration mechanisms, followed by forced water convection. Some of the melt models are fairly simple and relatively crude, but they can be justified in part by the existence of persistent and relatively unpredictable changes of the geometry of an individual iceberg. When calving and fracture occur, the shape of the remaining portion of parent iceberg can be altered significantly. The iceberg can pitch and roll, changing its equilibrium position in the water, consistent with its new center of gravity. Under these circumstances, it is not clear how one could handle the detailed heat transfer mechanisms in a purely deterministic way. Nevertheless, some studies have suggested that the deterioration predictions including both calving and melting are reasonably accurate (within 10 to 20%). Such good agreement is noteworthy in view of the complicated governing physical mechanisms, the simplified approaches used to handle them, a nd the complexity and variability of possible iceberg shapes. Using these models for the deterioration mechanisms, correlations of iceberg life expectancies for specific locations and given months were obtained.

The basic ideas behind the calculation of iceberg drift are also reasonably well established. The various contributions to the forces and effective forces acting on an iceberg have been identified, and explicit expressions to calculate them have been proposed. Among various researchers there are different opinions about the relative significance of a few of the force contributions. Some in question are the force due to wave radiation stress, the material derivative of the water velocity that appears in the pressure gradient force term and the effective force associated with the added mass of the iceberg. While some researchers have neglected some of these contributions in the formulation of their drift models, all of the contributions have been described in the present chapter.

There are uncertainties associated with the numerical values of the wind and water drag coefficients. Sometimes these have been determined *a posteriori* by choosing best fits of observed trajectories by predictions based on selected values of the drag coefficients. Such a fit procedure masks the possible deficiencies of the modeling of the various force contributions and can give inappropriate values for the drag coefficients. A discussion of various numerical integration schemes for the prediction of iceberg drift concluded that a simple implicit Euler was both very robust and had adequate accuracy. Details concerning the wind and water currents are probably the most important inputs to the modeling since these currents determine, either directly or indirectly, all of the forces acting on the iceberg. For iceberg forecasting models, it is apropos to couple the iceberg drift and deterioration model with state of the art ocean and atmospheric forecasting models.

Acknowledgements

I am indebted to M. Sayed, G.B. Crocker and T. Carrieres for many discussions, suggestions, and thoughtful criticisms concerning the material contained in this chapter. I am particularly grateful to A. Provenzale and N. Balmforth for their careful reading of the manuscript and many suggestions for improvements. Much of my own work was supported through contracts from the Canadian Ice Service, Environment Canada. I gratefully acknowledge the support of a Natural Sciences and Engineering Research Council (NSERC) Grant during the preparation of this chapter.

References

1. S. Hotzel, J. Miller: Annals of Glaciology **4**, 116–123 (1983)
2. R.Q. Robe: 'Height to draft ratios of icebergs'. In: *POAC 75: Proceedings of the Third International Conference on Port and Ocean Engineering under Arctic Conditions, Fairbanks, Alaska, 1975, Vol. 1*, 407–415 (1976)
3. G.B. Crocker, A.B. Cammaert: 'Measurements of bergy bit and growler populations off Canada's East Coast'. In: *Proc. of IAHR Ice Symposium, Trondheim, Norway, August 23-26, 1994*, Vol. 1, 167–176 (1994)
4. S. Venkatesh, M. El-Tahan: Cold Regions Sci. Technol. **15**, 1–11 (1988)
5. D.G. Mountain: Cold Regions Sci. Technol. **1**, 273–282 (1980)
6. S.D. Smith, N.R. Donaldson: Dynamic modeling of iceberg drift using current profiles. *Canadian Technical Report of Hydrography and Ocean Sciences No. 91, Bedford Institute of Oceanography, Dartmouth, Nova Scotia, October 1987* (1987)
7. M. Sayed: Implementation of iceberg drift and deterioration model. *Technical Report HYD-TR-049, March 2000, Canadian Hydraulics Centre, National Research Council Canada, Ottawa, Canada* 39 pp. (2000)
8. M. El-Tahan, S. Venkatesh, H. El-Tahan: J. Offshore Mech. Arctic Eng. **109**, 102–108 (1987)
9. J.C. Job: J. Glaciology **20**, 533–542 (1978)
10. R.C. Kollmeyer: Iceberg deterioration. Report No. 11, U.S. Coast Guard, Washington, D.C. 41–64 (1965)
11. R.Q. Robe, D.C. Maier, R.C. Kollmeyer: Nature **267**, 505–506 (1977)
12. Fenco Newfoundland Ltd.: Ice and environmental surveillance offshore Labrador. Operations Report for Petro–Canada Exploration Inc. (1980)
13. B. DeJong: *Net Radiation Received by a Horizontal Surface at the Earth* (Delft University Press, Groningen 1973)
14. E.G. Josberger: 'A laboratory and field study of iceberg deterioration'. In: *Proc. First Intern. Conf. on Iceberg Utilization*, ed. by A.A. Husseiny (Pergamon Press, New York 1977) pp. 245–264
15. F.M. White, M.L. Spaulding, L. Gominho: Theoretical estimates of the various mechanisms involved in iceberg deterioration in the open ocean environment. U.S. Coast Guard Report No. CG-D-62-80. 126 pp. (1980)
16. B. Gebhart, J.C. Mollendorf: J. Fluid Mech. **80**, 637–707 (1978)
17. S. Neshyba, E.G. Josberger: 'On the estimation of Antarctic iceberg melt rate'. In: *Iceberg Dynamics Symposium, St. John's, Newfoundland, June 4-5, 1979* (1979)
18. R. Hilpert: Wärmeabgabe von geheizten Drähten und Rohren. *Forsch. Geb. Ingenieurwes* **4**, 215 (1933)

19. A. Zhukaukas: 'Heat transfer from tubes in cross flow'. In: *Advances in Heat Transfer, 8*, ed. by J.P. Hartnett and T.F. Irvine, Jr., (Academic Press, New York 1972)
20. E. Achenbach: 'Heat transfer from spheres up to $Re = 6 \times 10^6$'. In: *Proc. 6th Int. Heat Transfer Conf.* Vol. 5 (Hemisphere, Washington, D.C. 1978)
21. S. Venkatesh, D.L. Murphy, G.F. Wright: Atmosphere–Ocean **32**, 469–484 (1994)
22. G.B. Crocker, G. English: Verification and implementation of a methodology for predicting bergy bit and growler populations. Contract Report prepared for Ice Services, Environment Canada, C-Core Contract 97 - C13, 46 pp. (1997)
23. S.B. Savage, G.B. Crocker, M. Sayed, T. Carrieres: J. Geophys. Res. – Oceans (accepted for publication) (2001)
24. H.L. Langhaar: *Dimensional Analysis and the Theory of Models* (Wiley, New York 1951)
25. Mobile Oil Canada, Ltd.: Hibernia Development Project, Environmental impact statement. Vol. IIIa, Biophysical assessment (1985)
26. Fenco Newfoundland Ltd.: Husky/Bow Valley East Coast Project, Environmental Data Archive, 1984–87. Vol. 1, Summary Report (1987)
27. S.B. Savage, M. Sayed, G.B. Crocker, T. Carrieres: Cold Regions Sci. Technol. **31**, 163–172 (2000)
28. G.B. Crocker: Cold Regions Sci. & Techn. **22**, 113–119 (1993)
29. D. Diemand, W. Nixon, J. Lever: 'On the splitting of icebergs – natural and induced'. In: *Proc. of the 6th Conference on Offshore Mechanics and Arctic Engineering, Houston, Texas*, Vol. 4, 379–385 (1987)
30. J. Lever, D. Bass, C. Lewis, K. Klein, D. Diemand: 'Iceberg seabed interaction events observed during the DIGS experiment'. In: *Proc. of the 8th Conference on Offshore Mechanics and Arctic Engineering, The Hague*, Vol. 4, 205–220 (1989)
31. W. Weibull: J. Applied Mech. **18**, 293–297 (1951)
32. J.A. Dowdeswell, C.F. Forsberg: Polar Research, **11**, 81–91, (1992)
33. J.J. von Schwind: *Geophysical Fluid Dynamics for Oceanographers* (Prentice-Hall, New York 1980)
34. S.B. Savage, M. Sayed, G.B. Crocker, T. Carrieres: 'Overview of a new operational iceberg prediction model'. In: *Proceedings of the 10th International Offshore and Polar Engineering Conference, ISOPE 2000, Seattle, Washington, May 28 – June 2, 2000* (2000)
35. J.R. Marko, D.B. Fissel, J.D. Miller: 'Iceberg movement prediction off the Canadian East coast'. In: *Natural and Man-Made Hazards*, ed. by M.I. El-Sabh, T.S. Murty (D. Reidel Publishing Co. 1988) pp. 435–462
36. S. de Margerie, J. Middleton, C. Garrett, S. Marquis, F. Majaess, K. Lank: An operational iceberg trajectory forecasting model for the Grand Banks. ASA Consulting Ltd., Dartmouth, Nova Scotia, Canada. Report No. 052, Environmental Studies Revolving Funds, 95 pp. (1986)
37. C.J.R. Garrett: Cold Regions Sci. Technol. **11**, 255–266 (1985)
38. C.J.R. Garrett, J. Middleton, M. Hazen, F. Majaess: Science **227**, 1333–1335 (1985)
39. H.S. Gaskill, J. Rochester: Cold Regions Sci. Technol. **8**, 223–234 (1984)
40. M. Moore: Cold Regions Sci. Technol. **14**, 263–272 (1987)
41. E.G. Banke, S.D. Smith: A hindcast study of iceberg drift on the Labrador coast. Canadian Technical Report of Hydrography and Ocean Sciences, No. 49, 161 pp. (1984)
42. G.R. Bigg, M.R. Wadley, D.P. Stevens, J.A. Johnson: Cold Regions Sci. Technol. 113–135 (1997)

43. M. El-Tahan, H.W. El-Tahan, S. Venkatesh: 'Forecast of iceberg ensemble drift'. In: *Proc. Annual Offshore Technology Conference, OTC Paper No. 4460, Houston, Texas, May 2–5, 1983*, pp. 151–158 (1983)
44. D.L. Murphy, Lt.I. Anderson: 'An evaluation of the International Ice Patrol drift model'. In: *Proc. of the Canadian East Coast Workshop on Sea Ice, January 1986*, Canadian Tech. Report of Hydrography and Ocean Sciences No. 73, G. Symonds, I.K. Peterson (1986)
45. S.D. Smith, E.G. Banke: Cold Regions Sci. Technol. **6**, 241–255 (1983)
46. S.D. Smith, N.R. Donaldson: 'Innovations in dynamic modeling of iceberg drift'. In: *Proc. IEEE, Oceans 87, Halifax, September 28 - October 1, 1987*, 5–10 (1987)
47. S.D. Smith: Cold Regions Sci. Technol. **22**, 34–45 (1993)
48. D.S. Sodhi, M. El-Tahan: Annals of Glaciology **1**, 77–82 (1980)
49. M.S. Longuet-Higgins, R.W. Stewart: Deep Sea Research **11**, 529–562 (1964)
50. O.M. Phillips: *Dynamics of the upper ocean* (Cambridge University Press, Cambridge 1966)
51. R.M. Sorensen: *Basic wave mechanics for coastal and ocean engineers* (Wiley, 1993)
52. M.S. Longuet-Higgins: Proc. R. Soc. A **352**, 463–480 (1977)
53. H. Maruo: J. Ship Research December **4**, No. 3, 1–10 (1960)
54. K. Kudou: J. Soc. Naval Arch., Japan **141**, 71–77 (1977)
55. M. Isaacson: 'Influence of wave drift force on ice mass motions.' In: *Proc. Seventh Intern. Conf. on Offshore Mechanics and Arctic Engineering, Houston, Texas, February 7–12, 1988* (ASME, New York 1988) pp. 125–130
56. M. Isaacson, K. McTaggart: Can. J. Civil Eng. **17**, 329–337 (1990)
57. Meteorological Office: *Marine Observer's Handbook, 9th ed.* (HMSO, London 1969)
58. A.E. Gill: *Atmosphere–Ocean Dynamics* (Academic Press, New York 1982)
59. J.N. Newman: *Marine Hydrodynamics* (MIT Press, Cambridge 1978)
60. S.F. Hoerner: *Fluid Dynamic Drag* (Published by the author, New York 1965)
61. E.G. Banke, S.D. Smith: 'Measurements of towing drag on small icebergs'. In: *Proc. IEEE, Oceans 74, International Conference on Engineering in the Ocean Environment, Halifax, Canada, Vol. 1*, pp. 130–132 (1974)
62. J.E. Chirivella, C.G. Miller: 'Hydrodynamics of icebergs in transit'. In: *Proc. of the First Conference on Iceberg Utilization for Freshwater Production*, ed. by A.A. Husseiny, Iowa State University, 315–333 (1978)
63. W.H. Press, B.P. Flannery, S.A. Teukolsky, W.T. Vetterling: *Numerical Recipes, The Art of Scientific Computing, 2nd Ed.*. (Cambridge Univ. Press, Cambridge 1994)
64. L.F. Shampine, M.W. Reichelt: SIAM Journal on Scientific Computing **18**, 1–22 (1997)
65. H.H. Rosenbrock: Comput. J. **5**, 329–330 (1963)
66. E. Hairer, G. Wanner: *Solving Ordinary Differential Equations II. Stiff and Differential-Algebraic Problems* (Springer-Verlag, Berlin 1991)
67. A. Sandu, J.G. Verwer, J.G. Blom, E.J. Spee, G.R. Carmichael: Benchmarking stiff ODE solvers for atmospheric chemistry problems, II: Rosenbrock solvers. Centrum voor Wiskunde en Informatica (CWI), National Institute for Mathematics and Computer Science, Netherlands, Department of Numerical Mathematics, Report NM-R9614, 20 pp. (1996)
68. T. Steihaug, A. Wolfbrandt: Math. Comp. **33**, 521–534 (1979)
69. S.B. Savage: Numerical integration schemes for iceberg drift and deterioration code. Technical Report 2000-02, Contract KM149-9-85-051 for Canadian Ice Service, Ottawa, March 31, 2000, 49 pp. (2000)

13 Snow Avalanches

C. Ancey

Cemagref, unité Erosion Torrentielle, Neige et Avalanches, Domaine Universitaire, 38402 Saint-Martin-d'Hères Cedex, France

13.1 Introduction

Over the last century, mountain ranges in Europe and North America have seen substantial development due to the increase in recreational activities, transportation, construction in high altitude areas, etc. In these mountain ranges, avalanches often threaten man's activities and life. Typical examples include recent disasters, such as the avalanche at Val d'Isère in 1970 (39 people were killed in a hostel) or the series of catastrophic avalanches throughout the Northern Alps in February 1999 (62 residents killed). The rising demand for higher safety measures has given new impetus to the development of mitigation technology and has given rise to a new scientific area entirely devoted to snow and avalanches. This paper summarises the paramount features of avalanches (formation and motion) and outlines the main approaches used for describing their movement. We do not tackle specific problems related to snow mechanics and avalanche forecasting. For more information on the subject, the reader is referred to the main textbooks published in Alpine countries [1–8].

13.1.1 A Physical Picture of Avalanches

Avalanches are rapid gravity-driven masses of snow moving down mountain slopes. With this fairly long definition, we try to characterise avalanches with respect to other snow flows. For instance, a snowdrift involves transport of snow particles, driven not by gravity but by wind. The slow slide and creep of the snow cover is driven by gravity but with a slow kinetic (typical velocities are in mm/day). Likewise, the slide of a snowpack down a roof cannot be considered an avalanche.

13.1.2 Avalanche Release

Successive snowfalls during the winter and spring accumulate to form snow cover. Depending on the weather conditions, significant changes in snow (types of crystal) occur as a result of various mechanical (creep, settlement) and thermodynamic processes (mass transfer). This induces considerable variations in its mechanical properties (cohesion, shear strength). Due to its layer structure, the snow cover is liable to internal slides between layers induced by gravity. When the shear deformation exceeds the maximum value that the layers of snow can undergo, a failure arises, usually developing first along the sliding surface, then propagating throughout the upper layers across a crack perpendicular to the downward direction. This kind of release is very frequent. In the field evidence of such failures consists of a clear fracture corresponding to the breakaway wall at the top edge of the slab and a bed surface over which the slab has slid (see Fig. 13.1). If the snow is too loose, the failure processes differ significantly from the ones governing slab release. Loose snow avalanches form near the surface. They usually start from a single point, then they spread out laterally by pushing and incorporating more snow.

Fig. 13.1. Slab avalanche released by gliding wet snow

The stability of a snow cover depends on many parameters. We can distinguish the fixed parameters related to the avalanche path and the varying parameters, generally connected to weather conditions. Fixed parameters include:

- *Mean slope.* In most cases, the average inclination of starting zones ranges from 27° to 50°. On rare occasions, avalanches can start on gentle slopes of less than 25° (e.g. *slushflow* involving wet snow with high water content), but generally the shear stress induced by gravity is not large enough to cause failure. For inclinations in excess of 45° to 50°, many slides (*sluffs*) occur during snowfalls; thus amounts of snow deposited on steep slopes are limited.
- *Roughness.* Ground surface roughness is a key factor in the anchorage of the snow cover to the ground. Dense forests, broken terrain, starting zones cut by several ridges, ground covered by large boulders generally limit the amount of snow that can be involved in the start of an avalanche. Conversely, widely-spaced forests, large and open slopes with smooth ground may favour avalanche release.

- *Shape and curvature of starting zone.* The stress distribution within the snowpack and the variation in its depth depend on the longitudinal shape of the ground. For instance, convex slopes concentrate tensile stresses and are generally associated with a significant variation in the snowcover depth, favouring snowpack instability.
- *Orientation to the sun.* The orientation of slopes with respect to the sun has a strong influence on the day-to-day stability of the snowpack. For instance, in winter, shady slopes receive little incoming radiation from the sun and conversely lose heat by long-wave radiation. It is generally observed that for these slopes, the snowpack is cold and tends to develop weak layers (faceted crystals, depth hoar). Many fatalities occur each year in such conditions. In late winter and in spring, the temperature increase enhances stability of snowpacks on shady slopes and instability on sunny slopes.

Among the varying factors intervening in avalanche release, experience clearly shows that in most cases, avalanches result from changes in weather conditions:

- *New snow.* Most of the time, snowfall is the cause of avalanches. The hazard increases significantly with the increase in the depth of new snow. For instance, an accumulation of 30 cm/day may be sufficient to cause widespread avalanching. In European mountain ranges, heavy snowfalls with a total precipitation exceeding 1 m during the previous three days may produce large avalanches, with possible extension down to the valley bottom.
- *Wind.* The wind is an additional factor which significantly influences the stability of a snowpack. Indeed it causes uneven snow redistribution (accumulation on lee slopes), accelerates snow metamorphism, forms cornices, whose collapses may trigger avalanches. On the whole, influence of the wind is very diverse, either consolidating snow (compacting and rounding snow crystals) or weakening it.
- *Rain and liquid water content.* The rain plays a complex role in snow metamorphism. Generally, for dry snow, a small increase in the liquid water content (LWC$<$ 0.5%) does not significantly affect the mechanical properties of snow. However, heavy rain induces a rapid and noticeable increase in LWC, which results in a drop in the shear stress strength. This situation leads to widespread avalanche activity (wet snow avalanches).
- *Snowpack structure.* A given snowpack results from the successive snowfalls. The stability of the resulting layer structure depends a great deal on the bonds between layers and their cohesion. For instance, heterogeneous snowpacks, made up of weak and stiff layers, are more unstable than homogeneous snowpacks.

13.1.3 Avalanche Motion

It is very common and helpful to consider two limiting cases of avalanches depending on the form of motion [7]:

- The *flowing avalanche* (avalanche coulante, Fliesslawine, valanga radente): a flowing avalanche is an avalanche with a high-density core at the bottom. Motion is dictated by the relief. The flow depth does not generally exceed a few meters (see Fig. 13.2). The typical mean velocity ranges from 5 m/s to 25 m/s. On average, the density is fairly high, generally ranging from 150 kg/m^3 to 500 kg/m^3.
- The *airborne avalanche* (avalanche en aérosol, Staublawine, valanga nubiforme): it is a very rapid flow of a snow cloud, in which most of the snow particles are suspended in the ambient air by turbulence (see Fig. 13.3). Relief has usually little influence on this aerial flow. Typically, for the flow depth, mean velocity, and mean density, the order of magnitude is 10–100 m, 50–100 m/s, 5–50 kg/m^3 respectively.

Fig. 13.2. Flowing avalanche impacting a wing-shaped structure in the Lauratet experimental site (France)

The avalanche classification proposed here only considers the form of motion and not the quality of snow. In the literature, other terms and classifications have been used. For instance, it is very frequent to see terms such as dry-snow avalanches, wet-snow avalanches, powder avalanches, etc. In many cases and probably in most cases in ordinary conditions, the motion form is directly influenced by the quality of snow in the starting zone. For instance, on a sufficiently steep slope, dry powder snow often gives rise to an airborne avalanche (in this case no confusion is possible between airborne and powder snow avalanches). However, in some cases, especially for extreme avalanches (generally involving large volumes of snow), motion is independent of the snow type. For instance, wet snow may be associated with airborne (e.g. Favrand avalanche in the Cha-

Fig. 13.3. Airborne avalanche descending a steep slope (Himalayas)

monix valley, France, on 16 May 1983). Between the two limiting cases above, there is a fairly wide variety of avalanches, which exhibit characteristics common to both airborne and flowing avalanches. Sometimes, such flows are referred to as "mixed-motion avalanches". The use of this term is often inappropriate because it should be restricted to describing complex flows for which both the dense core and the airborne play a role (from a dynamic point of view). In some cases, the dense core is covered with a snow dust cloud, made up of snow particles suspended by turbulent eddies of air resulting from the friction exerted by the air on the core. This cloud can entirely hide the high-density core, giving the appearance of an airborne avalanche, but in fact, it plays no significant role in avalanche dynamics. It should be born in mind that the mere observation of a cloud is generally not sufficient to specify the type of an avalanche. Further elements such as the features of the deposit or the destructive effects are required.

The current terminology asserts that there are two main types of motion. In this respect, mixed-motion avalanches are seen as avalanches combining aspects of both airborne and flowing avalanches, but they are not seen as a third type of avalanche. The question of a third type of avalanche has been raised by some experts during the last few years. Indeed, there is field evidence that some events did not belong either to the group of airborne or flowing avalanches. For instance, the Taconnaz avalanche (Haute-Savoie, France) on 11 February 1999 severely damaged two concrete-reinforced structures. The impact pressure was estimated at (at least) 600 kPa. The assumption of a flowing avalanche is not supported by the shape of the deposit. Current knowledge of airborne dynamics has a hard time explaining such a high impact pressure.

To conclude it should be noticed that there is currently a limited amount of data on real events. Some of the main parameters, such as the mean density in an airborne avalanche, are still unknown. Thus, many elements of our current knowledge of avalanches have a speculative basis. Today a great deal of work is underway to acquire further reliable data on avalanche dynamics. Experimental sites, such as *la Sionne* (Switzerland) or the *Lautaret pass* (France), have been developed for that purpose. However a survey of extreme past events shows that the characteristics of extreme avalanches (involving very large volumes) cannot be easily extrapolated from the features of ordinary avalanches. In this respect, the situation is not very different from the problems encountered with large rockfalls and landslides [9,57]. Many observations that hold for ordinary events no longer hold for rare events. Examples include the role of the forest, the influence of the snow type on avalanche motion, etc.

13.2 Modelling Avalanches

Avalanches are extremely complex phenomena. This complexity has led to the development of several approaches based on very different points of view. Many papers and reports have presented an overview of current models. These include the review by Hopfinger [12] as well as a comprehensive up-to-date review of all existing models edited by Harbitz [13] in the framework of an European research programme. Here we shall only outline three typical approaches: the statistical approach, the deterministic approach, and small-scale models.

13.2.1 Statistical Methods

In land-use planning (avalanche zoning), the main concern is to delineate areas subject to avalanches. Avalanche mapping generally requires either accurate knowledge of past avalanche extensions or methods for computing avalanche boundaries. To that end several statistical methods have been proposed. The two main models used throughout the world are the one developed by Lied and Bakkehøi [15] and the one developed subsequently by McClung and Lied [14]. Both attempt to predict the extension (stopping position) of the long-return period avalanche for a given avalanche path. Generally, authors have considered avalanches with a return period of approximately 100 year. All these methods rely on the correlations existing between the runout distance and some topographic parameters. They assume that the longitudinal profile of the avalanche path governs avalanche dynamics. The topographic parameters generally include the location of the top point of the starting zone (called point A) and a point B of the path profile where the local slope equals a given angle, most often 10° (this point is usually interpreted as the deceleration point of the path). The position of the stopping position (point C) is described using the angle α, which is the angle of the line joining the starting and stopping points with respect to the horizontal (see Fig. 13.4). Likewise, β is the average inclination of the avalanche path between the horizontal and the line joining the starting point A to point B.

To smooth irregularities in the natural path profile, a regular curve (e.g. a parabola) can be fitted to the longitudinal profile. Statistical methods have so far been applied to flowing avalanches. In principle, nothing precludes using them for airborne avalanches. But in this case, one is faced with the limited

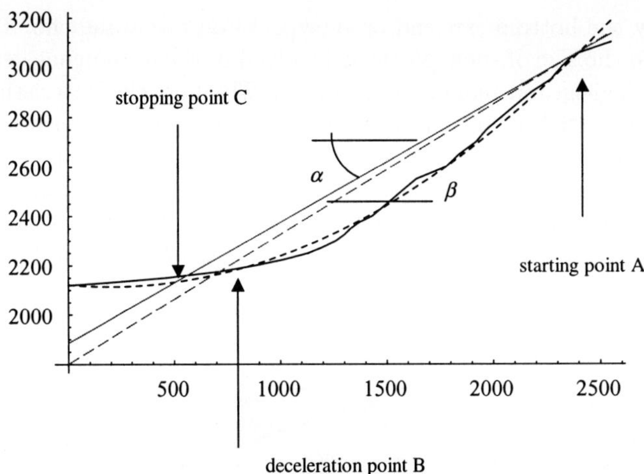

Fig. 13.4. Topographic parameters describing the profile. The dashed line represents the fitted parabola

amount of data and their poor quality (airborne avalanches are rare and the limits of their deposits are hard to delineate in the field). As an example of statistical models, we indicate the results obtained by Lied and Toppe [16]. Using regression analysis on data corresponding to the longest runout distance observed for 113 avalanche paths in western Norway, these authors have found that $\alpha = 0.96\beta - 1.7°$. The regression coefficient is fairly good ($r^2 = 0.93$) and the standard deviation is relatively small ($s = 1.4°$). Many extensions of the early model developed by Lied and Bakkehøi have been proposed over the last twenty years either to tune the model parameters to a given mountainous region or adapt the computations to other standards. For instance, subsequent work on statistical prediction of avalanche runout distance has accounted for other topographic parameters such as the inclination of the starting zone or the height difference between the starting and deposition zones. Although statistical methods have been extensively used throughout the world over the last twenty years and have given fairly reliable and objective results, many cases exist in which their estimates are wrong. Such shortcomings can be explained (at least in part) by the fact that for some avalanche paths, the dynamic behaviour of avalanches cannot be merely related or governed by topographic features.

13.2.2 Deterministic Approach (Avalanche-dynamics Models)

The deterministic approach involves quantifying the elementary mechanisms affecting the avalanche motion. Avalanches can be considered at different spatial scales (see Fig. 13.5). The larger scale, corresponding to the entire flow, leads to the simplest models. The chief parameters include the location of the gravity centre and its velocity. Mechanical behaviour is mainly reflected by the friction force

F exerted by the bottom (ground or snowpack) on the avalanche. The smallest scale, close to the size of snow particles involved in the avalanches, leads to complicated rheological and numerical problems. The flow characteristics (velocity, stress) are computed at any point of the occupied space. Intermediate models have also been developed. They benefit from being less complex than three-dimensional numerical models and yet more accurate than simple ones. Such intermediate models are generally obtained by integrating the motion equations across the flow depth in a way similar to what is done in hydraulics for shallow water equations.

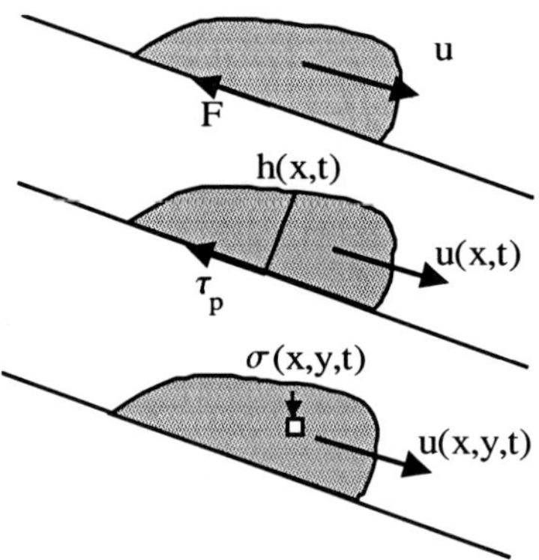

Fig. 13.5. Different spatial scales used for describing avalanches

Simple Models

Simple models have been developed for almost 80 years in order to crude estimations of avalanche features (velocity, pressure, runout distance). They are used extensively in engineering throughout the world. Despite their simplicity and approximate character, they can provide valuable results, the more so as their parameters and the computation procedures combining expert rules and scientific basis have benefited from many improvements over the last few decades.

Simple Models for Flowing Avalanches. The early models date back to the beginning of the 20th century. For the Olympic Games at Chamonix in 1924, the Swiss professor Lagotala computed the velocity of avalanches in the Favrand path [18]. His method was then extended by Voellmy , who popularised it. Since the model proposed by Voellmy, many extensions have been added. The Voellmy–Salm–Gubler (VSG) model [17] and the Perla–Cheng–McClung model [11] are probably the best-known avalanche-dynamics models used throughout the world. Here we outline the VSG model. In this model, a flowing avalanche is considered as a sliding block subject to a friction force:

$$F = mg\frac{u^2}{\xi h} + \mu\, m\, g \cos\theta\,, \tag{13.1}$$

where m denotes the avalanche mass, h its flow depth, θ the local path inclination, μ a friction coefficient related to the snow fluidity, and ξ a coefficient of dynamic friction related to path roughness. If these last two parameters cannot be measured directly, they can be adjusted from several series of past events. It is generally accepted that the friction coefficient μ only depends on the avalanche size and ranges from 0.4 (small avalanches) to 0.155 (very large avalanches) [17]. Likewise, the dynamic parameter ξ reflects the influence of the path on avalanche motion. When an avalanche runs down a wide open rough slope, ξ is close to 1000. Conversely, for avalanches moving down confined straight gullies, ξ can be taken as being equal to 400 or more. In a steady state, the velocity is directly inferred from the momentum balance equation:

$$u = \sqrt{\xi h \cos\theta\,(\tan\theta - \mu)}\,. \tag{13.2}$$

According to this equation two flow regimes can occur depending on path inclination. For $\tan\theta > \mu$, (13.2) has a real solution and a steady regime can occur. For $\tan\theta < \mu$, there is no real solution: the frictional force (13.1) outweighs the downward component of the gravitational force. It is therefore considered that the flow slows down. The point of the path for which $\tan\theta = \mu$ is called the characteristic point (point P). It plays an important role in avalanche dynamics since it separates flowing and stopping phases. In the stopping zone, we deduce from the momentum equation that the velocity decreases as follows:

$$\frac{1}{2}\frac{\mathrm{d}u^2}{\mathrm{d}x} + u^2\frac{g}{\xi h} = g\cos\theta\,(\tan\theta - \mu)\,. \tag{13.3}$$

The runout distance is easily inferred from (13.3) by assuming that at a point $x = 0$, the avalanche velocity is u_p. In practice the origin point is point P but attention must be paid in the fact that, according to (13.2), the velocity at point P should be vanishing; a specific procedure has been developed to avoid this shortcoming (see [17]). Neglecting the slope variations in the stopping zone, we find:

$$x_a = \frac{\xi h}{2g}\ln\left(1 + \frac{u_P^2}{\xi h\cos\theta\,(\mu - \tan\theta)}\right)\,. \tag{13.4}$$

This kind of model enables us to easily compute the runout distance, the maximum velocities reached by the avalanche on various segments of the path, the flow depth (by assuming that the mass flow rate is constant and given by the initial flow rate just after the release), and the impact pressure.

Simple Models for Airborne Avalanches. For airborne avalanches, simple models have been developed using the analogies with inclined thermals or starting plumes. An inclined thermal consists of the flow of a given volume of a heavy fluid into a surrounding light fluid down an inclined wall. Buoyancy is the key factor of motion. To our knowledge, the earliest model was proposed by Tochon-Danguy and Hopfinger [19], then further developments were made by Béghin and Hopfinger [20], Fukushima and Parker [21], as well as Akiyama and Ura [22]. But as for Voellmy's model, similar models were probably developed in parallel by other authors, notably Russian scientists [23]. The main difficulty encountered here is that avalanche volume increases constantly as the avalanche descends. Thus contrary to simple models developed for flowing avalanches it is necessary to consider a further equation reflecting changes in volume or mass. To that end, it is generally assumed that the avalanche volume is a half ellipsoid (three-dimensional cloud) or a half cylinder with an elliptic basis (two-dimensional cloud). Changes in volume are due to entrainment of surrounding air into the airborne avalanche and snow incorporation from the snow cover. Here, for the sake of simplicity, we only consider two-dimensional flows without snow incorporation. We further assume that the friction exerted by the ground on the cloud is negligible compared to the buoyant force.

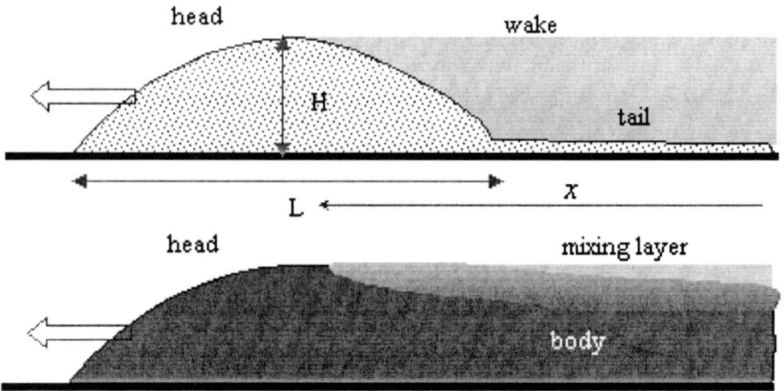

Fig. 13.6. A thermal is defined as the flow of a constant-volume flow driven by buoyancy (instantaneous release). A starting plume is a constant-supply flow (continuous release)

It is widely recognised (see [25]) that the inflow rate is proportional to a characteristic velocity (generally the mean velocity) and the surface area whatever the type of the flow (jet, plume, thermal) and the environment (uniform or stratified). Such an assumption leads to:

$$\frac{\mathrm{d}\bar{\varrho}V}{\mathrm{d}t} = \varrho_a \alpha(\theta) S\, U \,, \tag{13.5}$$

where V is the cloud volume, U its velocity (velocity of the mass centre), $\bar{\varrho}$ the mean bulk density of the "heavy" fluid, ϱ_a the density of the ambient ("light") fluid, and the surface area S (per unit width) is $k_s\sqrt{HL}$ with H the flow depth and L the flow length. We can also express the volume V (per unit width) as

κHL. We used a shape factor k_s defined by: $k_s = E(1-4k^2)/\sqrt{k}$, where $k = H/L$ and E denotes the elliptic integral function; likewise, κ is another shape factor: $\kappa = \pi/4$. In (13.5), we have also introduced $\alpha(\theta)$, which is an *entrainment coefficient* depending on the inclination θ only. This assumption needs further explanations. It is usually stated that the entrainment coefficient is a function of an overall *Richardson number*, defined here by: $Ri = g'h\cos\theta/u^2$, where we introduced the reduced gravity $g' = g\Delta\bar{\varrho}/\varrho_a$ and $\Delta\bar{\varrho} = \bar{\varrho} - \varrho_a$ is the buoyant density [24–26]. Here the overall Richardson number reflects the stabilizing effect of the density difference and the relative importance of buoyancy [24]. In the case of a gravity current with constant supply, it is observed that for a given slope, the mean velocity U reaches a constant value, insensitive to slope but depending on the buoyancy flux (per unit width) $A = g'hU$: $U \propto \sqrt[3]{A}$ [24,27]. This also means that the flow adjusts rapidly to a constant Richardson number (for a given slope). In this case, using approximate equations for the mass and momentum balances (respectively $\mathrm{d}(HU)/\mathrm{d}x = \alpha U$ and $\mathrm{d}(HU^2)/\mathrm{d}x = g'h\sin\theta$), we easily deduce that the entrainment coefficient α is a function of the Richardson number and slope: $\alpha = Ri\tan\theta$ [24]. Here, although buoyancy supply is not constant, we assume that the entrainment coefficient α depends only on the slope.

Using the fact that at any time the mean bulk density can be defined by:

$$\bar{\varrho} = \frac{\varrho_0 V_0 + \varrho_a(V - V_0)}{V} , \tag{13.6}$$

where ϱ_0 and V_0 denote the initial density and volume of the cloud, we infer the volume balance equation:

$$\kappa\frac{\mathrm{d}HL}{\mathrm{d}t} = \alpha(\theta)k_s\sqrt{HL}U . \tag{13.7}$$

In the present context, Béghin assumed that the ratio $k = H/L$ remains constant from the beginning to the collapse of the cloud. Thus, using the fact that $\mathrm{d}()/\mathrm{d}t = U\mathrm{d}()/\mathrm{d}x$, where the abscissa x refers to the downward position of the mass centre, we easily deduce from (13.7) that:

$$\frac{\mathrm{d}H}{\mathrm{d}x} = \alpha_H , \tag{13.8}$$

where $\alpha_H = \alpha(\theta)\sqrt{k}k_s/(2\kappa)$. The ambient fluid exerts two types of pressure on the cloud: a term analogous to a static pressure (Archimede's theorem), equal to $\varrho_a Vg$, and a dynamic pressure. As a first approximation, the latter term can be evaluated by considering the ambient fluid as an inviscid fluid in a irrotational flow. On the basis of this approximation, it can be shown that the force exerted by the surrounding fluid on the half cylinder is $F_{\mathrm{dyn}} = \varrho_a k_v \mathrm{d}(UV)/\mathrm{d}t$, where $k_v = 2k$ is sometimes called the *added mass coefficient* [28]. Thus the momentum balance equation can be written as:

$$\frac{\mathrm{d}\bar{\varrho}VU}{\mathrm{d}t} = \bar{\varrho}gV\sin\theta - \varrho_a gV\sin\theta - k_v\varrho_a\frac{\mathrm{d}VU}{\mathrm{d}t} , \tag{13.9}$$

or equivalently:

$$\frac{\mathrm{d}(\bar{\varrho}+k_v\varrho_a)VU}{\mathrm{d}t} = \Delta\bar{\varrho}gV\sin\theta\,. \tag{13.10}$$

The buoyant term on the right-hand side of (13.10) is constant. Indeed, using (13.6), we find that:

$$\Delta\bar{\varrho}gV\sin\theta = \Delta\bar{\varrho}_0 V_0 g\sin\theta\,, \tag{13.11}$$

with $\Delta\bar{\varrho}_0 = \bar{\varrho}_0 - \varrho_a$ the initial buoyant density. Moreover, to simplify (13.10), we can use the Boussinesq approximation, which involves neglecting the excess in density in front of the inertial terms ($\bar{\varrho} \approx \varrho_a$). Thus we infer from (13.10):

$$\frac{\mathrm{d}U^2}{\mathrm{d}x} + \frac{4}{H(x)}\alpha_H U^2 = \frac{2\beta(\theta)}{H^2(x)}\,, \tag{13.12}$$

where $\beta(\theta) = g_0' V_0 \sin\theta / [\kappa k(1+k_v)]$. After integrating (13.12), we find that the mean velocity varies as a function of the abscissa as follows:

$$U^2 = \frac{3H_0^4U_0^2 + 6\beta x H(x) + 2\beta\alpha^2x^3}{3H^4(x)}\,, \tag{13.13}$$

where (U_0, H_0) refer to the initial velocity and depth of the cloud. For large values of x, the mean velocity behaves asymptotically as: $U \propto 1/\sqrt{x}$. The velocity of the front is given by:

$$U_f = \frac{\mathrm{d}}{\mathrm{d}t}(x_f - x + x) = U + \frac{1}{2}\frac{\mathrm{d}}{\mathrm{d}t}L = U\left(1+\frac{\alpha_H}{2k}\right)\,. \tag{13.14}$$

Thus the velocity of the front is found to be proportional to the mean velocity. Asymptotically, the front position varies as:

$$U_f \approx \left(1+\frac{\alpha_H}{2k}\right)\sqrt{\frac{2\beta\alpha^2}{3\alpha_H^4}}\sqrt{\frac{1}{x}}\,, \tag{13.15}$$

or equivalently:

$$x_f \approx \left(1+\frac{\alpha_H}{2k}\right)^{2/3}\left[\frac{2}{3}\frac{\alpha^2}{\alpha_H^4}\frac{g_0' V_0\sin\theta}{\kappa k(1+k_v)}\right]^{1/3} t^{2/3}\,. \tag{13.16}$$

This result is of great interest since it is comparable to other results found using different approaches. For instance, using the von Kármán–Benjamin boundary condition at the leading edge – stating that the front motion is characterized by a constant *Froude number* $Fr = U/\sqrt{gh}$, i.e. $Fr^2 = g'/(g\,Ri)$ – Huppert and Simpson [29] developed a very simple model, sometimes called the "box model" (see also Chap. 8). They considered a two-dimensional gravity current as a series of equal cross-sectional area rectangles (of length $l(t)$ and height $h(t)$) advancing over a horizontal surface: $u = Fr\sqrt{g'h}$ and $V(t) = h(t)l(t) = V_0$ where V_0 denotes the initial volume (per unit width) of fluid (here $Fr = \sqrt{2}$ inferred from theoretical considerations using the Bernoulli equation [30]). Using $u = dl/dt$ and integrating the volume equation leads to:

$$l(t) = \left(\frac{3}{2}Fr\right)^{2/3} (g'V_0)^{1/3} t^{2/3} . \tag{13.17}$$

Comparison of (13.16) and (13.17) reveals the same asymptotic behaviour, except that in Béghin's model, the position depends on the inclination θ. This is both disturbing and comforting since these two models are based on very different approximations: Béghin's model assumes that flow is governed on average by a momentum balance while Huppert and Simpson's model states that the flow behaviour is dictated by dynamic conditions at the leading edge. Many experiments have been performed on the motion of a two-dimensional cloud over horizontal surfaces or down inclined planes (e.g. [20,27,29,31–39]). They have confirmed the theoretical trend displayed in (13.15) or (13.16). The main difference between experimental results concerns the depth increase rate α_H (ranging from 0.01 to 0.02 for $\theta = 5°$).

Many field and laboratory observations have shown the significant role played by particle sedimentation or incorporation of material from the ground into the cloud. Improvements of existing simple models have been achieved by implementing new procedures taking material entrainment into account. Research on this topic is still in process. Compared to field data, Béghin's model usually provides correct estimates of the mean front velocity (to 20%) but it may substantially underestimate the impact pressure by a factor 10. The reason why the impact pressure computed as $\varrho u^2/2$ is underestimated is not clear. Very large velocity fluctuations inside the airborne avalanche or particles clustering at the flow bottom may be responsible for very high impact pressures. Another field observation that cannot be explained by Béghin-type models is the considerable acceleration at the early stages of an aerosol; in some cases, acceleration of $6\,\mathrm{m/s^2}$ over a 40° slope has been recorded for more than 5 s. This may also be related to the controversy on reduced gravity [40]. Indeed, some authors have claimed that a flow acceleration scaling as g' is not physical and suggested the alternative g'' defined by $g'' = g(\bar{\varrho} - \varrho_a)/\bar{\varrho}$. Concerning avalanches, field data tend to show that avalanche acceleration scales as g'.

Intermediate Models (Depth-averaged Models)

Simple models can provide approximate predictions concerning runout distance, the impact pressure, or deposit thickness. However they are limited for many reasons. For instance, they are restricted to one-dimensional path profiles (the spreading of the avalanche cannot be computed) and the parameters used are fitted to past events and cannot be measured in the field or in the laboratory (rheometry), apart from airborne models if the analogy with turbidity currents is used. More refined models use depth-averaged mass and momentum equations to compute the flow characteristics. With such models, the limitations of simple models are alleviated. For instance it is possible to compute the spreading of avalanches in their runout zone or relate mechanical parameters used in the models to the rheological properties of snow. As far as we know, the early depth-averaged models were developed in the 1970s by Russian scientists (Kulikovskii, Eglit [23,41,42]) and French researchers (Pochat, Brugnot, Vila [43,44]) for flowing avalanches. For airborne avalanches, the first stage was probably the model developed by Parker, Fukushima, and Pantin [45], which, though devoted to submarine turbidity currents, contains almost all the ingredients used in subsequent models of airborne avalanches. Considerable progress in the development of numerical depth-averaged models has been made possible thanks to the increase in computer power and breakthrough in the numerical treatment of hyperbolic partial differential equation systems (see [46] for a comprehensive review on hyperbolic differential equations in physics and [47] for a practical introduction to numerical treatment).

Depth-averaged Motion Equations. Here, we shall address the issue of slightly transient flows. We focus exclusively on *gradually varied flows*, namely flows that are not far from a steady uniform state for the time interval under consideration. Moreover, we first consider flows without entrainment of the surrounding fluid and variation in density: $\varrho \approx \bar{\varrho}$. Accordingly the bulk density may be merely replaced by its mean value. In this context, the motion equations may be inferred in a way similar to the usual procedure used in hydraulics to derive the shallow water equations (or Saint–Venant equations): it involves integrating the momentum and mass balance equations over the depth. As such a method has been extensively used in hydraulics for water flow [50] as well for non-Newtonian fluids (see for instance [45,48] or [49]; see also Chap. 14) we shall briefly recall the principle and then directly provide the resulting motion equations. Let us consider the local mass balance: $\partial\varrho/\partial t + \nabla.(\varrho\mathbf{u}) = 0$. Integrating this equation over the flow depth leads to:

$$\int_0^{h(x,t)} \left(\frac{\partial u}{\partial x} + \frac{\partial v}{\partial y}\right) \mathrm{d}y = \frac{\partial}{\partial x}\int_0^h u(x,y,t)\mathrm{d}y - u(h)\frac{\partial h}{\partial x} - v(x,h,t) - v(x,0,t) \,, \tag{13.18}$$

where u and v denote the x- and y-component of the local velocity. At the free surface and the bottom, the y-component of velocity satisfies the following boundary conditions:

$$v(x,h,t) = \frac{\mathrm{d}h}{\mathrm{d}t} = \frac{\partial h}{\partial t} + u(x,h,t)\frac{\partial h}{\partial x} \,, \qquad v(x,0,t) = 0 \,. \tag{13.19}$$

We easily deduce:

$$\frac{\partial h}{\partial t} + \frac{\partial h\overline{u}}{\partial x} = 0 \,, \tag{13.20}$$

where we have introduced depth-averaged values defined as:

$$\bar{f}(x,t) = \frac{1}{h(x,t)} \int_0^{h(x,t)} f(x,y,t)\mathrm{d}y \,. \tag{13.21}$$

The same procedure is applied to the momentum balance equation: $\mathrm{d}\boldsymbol{u}/\mathrm{d}t = \rho \boldsymbol{g} + \nabla.\boldsymbol{\sigma}$, where $\boldsymbol{\sigma}$ denotes the stress tensor. Without difficulty, we can deduce the averaged momentum equation from the x-component of the momentum equation:

$$\bar{\varrho}\left(\frac{\partial h\bar{u}}{\partial t}+\frac{\partial h\overline{u^2}}{\partial x}\right)=\bar{\varrho}gh\sin\theta+\frac{\partial h\bar{\sigma}_{xx}}{\partial x}-\tau_p\,, \tag{13.22}$$

where we have introduced the bottom shear stress: $\tau_p = \sigma_{xy}(x,0,t)$. In the present form, the motion equation system (13.20)–(13.22) is not closed since the number of variables exceeds the number of equations. A common approximation involves introducing a parameter (sometimes called the Boussinesq momentum coefficient) which links the mean velocity to the mean square velocity:

$$\overline{u^2}=\frac{1}{h}\int\limits_0^h u^2(y)\,\mathrm{d}y=\alpha\bar{u}^2\,. \tag{13.23}$$

Another helpful (and common) approximation, not mentioned in the above system, concerns the computation of stress [50]. Putting ourselves in the framework of long wave approximation, we assume that longitudinal motion outweighs vertical motion: for any quantity m related to motion, we have $\partial m/\partial y \gg \partial m/\partial x$. This allows us to consider that every vertical slice of flow can be treated as if it was locally uniform. In such conditions, it is possible to infer the bottom shear stress by extrapolating its steady-state value and expressing it as a function of u and h. A point often neglected is that this method and its results are only valid for flow regimes that are not too far away from a steady-state uniform regime. In flow parts where there are significant variations in the flow depth (e.g. at the leading edge and when the flow widens or narrows substantially), corrections should be made to the first-order approximation of stress [49]. Finally, an unresolved problem concerns the nature of the front in a transient flow. The same problem has been already pointed out above in the discussion on Béghin's model and "box models". Some authors have considered it as a shock; in this case, it is included in the motion equations as a downstream boundary condition [42–44]. In contrast, authors have implicitly assumed that the front has no specific dynamic role and can be generated by the hyperbolic motion equations [51]. Other authors considered that the front may be controlled by gravity instability. For instance, numerous experiments performed on viscous and buoyant gravity currents have revealed that a shifting pattern of lobes and clefts ranges across the front due to a gravity instability [52–54].

Flowing Avalanches. The material is very concentrated in ice particles: generally the concentration ranges from 20% to 65%. The material is highly compressible (it is frequent to observe snow densities in the deposition zone three times larger than in the starting zone). This is due to the intrinsic compressibility of snow as well as dilatant behaviour when the material contains snow balls. The rheology of ice/air mixtures is rather complex: significant variations in the mixture composition are caused by minute changes in the air temperature around 0° C. This explains the considerable variability of snow consistency: granular (snow ball), loose, slush-like or pasty snow. The diversity of snow consistency, along with the size scales, makes any thorough rheometrical examination of snow involved in avalanches a tricky undertaking. To date, few experimental studies have been devoted to this topic. The authors (such as Dent [55] or Maeno and Nishimura [56]), who studied the rheological bulk behaviour of snow, have generally found that snow is a non-Newtonian viscoplastic material, which depends a great deal on density. Several constitutive equations have been proposed: Newtonian fluid, Reiner–Ericken fluid, Bingham fluid, frictional Coulombic fluid, and so on. For instance, Savage and Hutter assumed that flowing avalanches have many similarities with dry granular flows [10,48]. They have further assumed that, as a first approximation, the Coulomb law can be used to describe the bulk behaviour of flowing granular materials. Therefore they have expressed the bottom shear stress as: $\tau_p = \varrho gh \tan\delta \cos\theta$, where δ denotes a bed friction angle. Likewise, the normal mean shear stress can be written as: $\bar{\sigma}_{xx} = -k_a \varrho gh \cos\theta/2$, where the coefficient k_a is related to the earth pressure coefficient used in soil mechanics. Eventually they obtained for flows down inclined planes:

$$\frac{\partial h}{\partial t} + \frac{\partial h\bar{u}}{\partial x} = 0\,, \tag{13.24}$$

$$\frac{\partial \bar{u}}{\partial t} + \bar{u}\frac{\partial \bar{u}}{\partial x} = g\cos\theta\,(\tan\theta - \tan\delta) - k_a g\cos\theta\frac{\partial h}{\partial x}\,. \tag{13.25}$$

Laboratory tests with dry granular media have shown that such a model captures the flow features well for steep smooth inclined channels [10,57–59]. Similar models were developed using different constitutive equations. For instance, Eglit used empirical expressions for the bottom shear stress (in a form similar to (13.1)) and treated the leading edge using a specific boundary condition [42,41]. Naaim and Ancey used a Bingham constitutive equation in their model [60]. All these models must deal with the difficult problem of fitting rheological parameters. Due to the lack of relevant rheological data on snow, the parameters are usually adjusted for the runout distance to coincide with field data.

Airborne Avalanches. An airborne avalanche is a very turbulent flow of a dilute ice–particle suspension in air. It can be considered as a one-phase flow as a first approximation. Indeed, the Stokes number defined as the ratio of a characteristic time of the fluid to the relaxation time of the particles is low, implying that particles adjust quickly to changes in the air motion [61]. At the particle scale, fluid turbulence is high enough to strongly shake the mixture since the particle size is quite small. To take into account particle sedimentation, authors generally consider airborne avalanches as turbulent stratified flows. Thus, contrary to flowing avalanches, bulk behaviour is well identified in the case of airborne avalanches. The main differences between the various models proposed result from the different boundary conditions, use of the Boussinesq approximation, and the closure equations for turbulence. Parker and his co-workers [45] developed a complete depth-averaged model for turbidity currents. The motion equation set proposed by these authors is more complicated than the corresponding set for dense flows presented above, since it includes additional equations arising from the mass balance for the dispersed phase, the mean and turbulent kinetic energy balances, and the boundary conditions related to the entrainment of sediment and surrounding fluid:

$$\frac{\partial h}{\partial t} + \frac{\partial hU}{\partial x} = E_a U \ , \tag{13.26}$$

$$\frac{\partial (Ch)}{\partial t} + \frac{\partial (hUC)}{\partial x} = v_s E_s - v_s c_b \ , \tag{13.27}$$

$$\frac{\partial hU}{\partial t} + \frac{\partial hU^2}{\partial x} = RCgh \sin\theta - \frac{1}{2} Rg \frac{\partial Ch^2}{\partial x} - u_*^2 \ , \tag{13.28}$$

$$\frac{\partial hK}{\partial t} + \frac{\partial hUK}{\partial x} = \frac{1}{2} E_a U^3 + u_*^2 U - \varepsilon_0 h - \frac{1}{2} E_a URCgh - \frac{1}{2} Rghv_s \left(2C + E_s - c_b\right) \ , \tag{13.29}$$

where U is the mean velocity, h the flow depth, K the mean turbulent kinetic energy, C the mean volume concentration (ratio of particle volume to total volume), E_a a coefficient of entrainment of surrounding fluid into the current, v_s the settlement velocity, E_s a coefficient of entrainment of particles from the bed into the current, c_b the near-bed particle concentration, R the specific submerged gravity of particles (ratio of buoyant density to ambient fluid density), u_*^2 the bed shear velocity, and ε_0 the depth-averaged mean rate of dissipation of turbulent energy due to viscosity. The main physical assumption in Parker et al.'s model is that the flow is considered as one-phase from a momentum point of view but treated as two-phase concerning the mass balance. Equation (13.26) states that the total volume variation results from entrainment of surrounding fluid. In (13.27), the variation in the mean solid concentration is due to the difference between the rate of particles entrained from the bed and the sedimentation rate. Equation (13.28) is the momentum balance equation: the momentum variation results from the driving action of gravity and the resisting action of bottom shear stress; depending on the flow depth profile, the pressure gradient can contribute either to accelerate or decelerate the flow. Equation (13.29) takes into account the turbulence expenditure for the particles to stay in suspension. Turbulent energy is supplied by the boundary layers (at the flow interfaces with the surrounding fluid and the bottom). Turbulent energy is lost by viscous dissipation ($\varepsilon_0 h$ in (13.29)) as well as by mixing the flow (fourth and fifth terms in (13.29)) and maintaining the suspension against sedimentation flow mixing (last term on the right-hand side of (13.29)). Although originally devoted to submarine turbidity currents, this model has been applied to airborne avalanches, with only small modifications in the entrainment functions [21,62]. Further developments have been brought to the primary model proposed by Parker et al., notably in order to consider non-Boussinesq fluids and snow entrainment from the snow-cover [63]. To our knowledge, such models do not currently provide better results than simple models when compared to field data.

Three-dimensional Computational Models

The rapid increase in computer power has allowed researchers to integrate local motion equations directly. Compared to the depth-averaged models, the problems in the development of three-dimensional (3D) computational models mainly concern numerical treatments. For instance, the treatment of the free surface poses complicated issues. Naturally, problems linked to the constitutive equations reliable for snow are more pronounced compared to intermediate models since the entire constitutive equation must be known (not just the shear and normal stress). The development of 3D models is currently undertaken mainly for airborne avalanches generally using finite-volume codes for turbulent flows. Examples include the models by Naaim [64], Hermann [66], Schweiwiller and Hutter [65], etc.

13.2.3 Small-scale Models

A few authors have exploited the similarities between avalanches and other gravity-driven flows. For instance, Hopfinger and Tochon-Danguy used the analogy between airborne avalanches and saline density currents to perform experiments in the laboratory in a water tank [67]. In this way, examination of various aspects of airborne dynamics has been possible: effect of a dam, structure of the cloud, determination of the entrainment coefficients, etc. The chief issue raised by the analogy with density or gravity currents concerns the similarity conditions based on both the Froude (or equally the Richardson number) and Reynolds numbers [12,34,67]. Regarding flowing avalanches, authors have considered the analogy with granular flows. Various materials (ping-pong ball, sand, beads) have been used. In engineering laboratory experiments simulating flowing avalanches offer promising tools for studying practical and complicated issues, such as the deflecting action of a dam [68] or braking mounds [69]. A few scientists have conducted or are performing experiments studying snow flows down confined geometries the field [70].

References

1. A. Roch: *Neve e Valanghe* (Club Alpino Italiano, Torino 1980)
2. C. Ancey: *Guide Neige et Avalanches: Connaissances, Pratiques, Sécurité*, 2nd edn. (Edisud, Aix-en-Provence 1998)
3. W. Amman, O. Buser, U. Vollenwyder: *Lawinen* (Birhäuser, Basel 1997)
4. D.M. McClung, P.A. Schaerer: *The avalanche handbook* (The Montaineers, Seattle 1993)

5. T. Daffern: *Avalanche Safety for Skiers & Climbers* (Rocky Mountain Books, Calgary 1992)
6. W. Munter: *Lawinen, entscheiden in kritischen Situationen* (Agentur Pohl und Schellhammer, Garmisch Partenkirchen 1997)
7. R. de Quervain: *Avalanche Atlas* (Unesco, Paris 1981)
8. A.I. Mears: *Snow-avalanche hazard analysis for land-use planning and Engineering.* Bulletin **49** (Colorado Geological Survey, Denver 1992)
9. K.J. Hsü: 'Albert Heim: observations on landslides and relevance to modern interpretations'. In: *Rockslides and avalanches*, ed. By B. Voight (Elsevier, Amsterdam 1978) pp. 71–93
10. S.B. Savage: 'Flow of granular materials'. In: *Theoretical and Applied Mechanics*, ed. by P. Germain, J.-M. Piau, D. Caillerie (Elsevier, Amsterdam 1989) pp. 241–266
11. R. Perla, T.T. Cheng, D.M. Mc Clung: J. Glaciol. **26**, 197 (1980)
12. E.J. Hopfinger: Ann. Rev. Fluid Mech. **15**, 45 (1983)
13. K. Harbitz: 'A survey of computational models for snow avalanche motion'. In: *Final report of Avalanche Mapping, Model Validation and Warning Systems*, (Fourth European Framework Programme, ENV4- CT96-0258, Brussels 1999)
14. D.M. McClung, K. Lied: Cold Reg. Sci. Technol. **13**, 107 (1987)
15. K. Lied, S. Bakkehøi: J. Glaciol. **26**, 165 (1980)
16. K. Lied, R. Toppe: Ann. Glaciol. **13**, 164 (1989)
17. B. Salm, A. Burkard, H. Gubler: *Berechnung von Fliesslawinen, eine Anleitung für Praktiker mit Beispielen.* Report **47** (EISFL, Davos 1990)
18. H. Lagotala: *Etude de l'avalanche des Pélerins (Chamonix)* (Société Générale d'Imprimerie, Genève 1927)
19. J.-C. Tochon-Danguy, E.J. Hopfinger: 'Simulation of the dynamics of powder avalanches'. In: *International symposium on snow mechanics, Grindelwald, 1974*, IAHS Publication 144 (IAHS 1974) pp. 369–380
20. P. Beghin, E.J. Hopfinger, R.E. Britter: J. Fluid Mech. **107**, 407 (1981)
21. Y. Fukushima, G. Parker: J. Glaciol. **36**, 229 (1990)
22. J. Akiyama, M. Ura: J. Hydraul. Eng. ASCE **125**, 474 (1999)
23. N. Bozhinskiy, K.S. Losev: *The fundamentals of avalanche science.* Report **55** (EISLF, Davos 1998)
24. J.S. Turner: *Buoyancy effects in fluids* (C. U. P., Cambridge 1973)
25. J.S. Turner: J. Fluid Mech. **173**, 431 (1986)
26. G. Parker, M. Garcia, Y. Fukushima, W. Yu: J. Hydr. Res. **25**, 123 (1987)
27. R.E. Britter, P.F. Linden: J. Fluid Mech. **99**, 531 (1980)
28. G.K. Batchelor: *An introduction to fluid dynamics* (C. U. P., Cambridge 1967)
29. H.E. Huppert, J.E. Simpson: J. Fluid Mech. **99**, 785 (1980)
30. T.B. Benjamin: J. Fluid Mech. **31**, 209 (1968)
31. R.E. Britter, J.E. Simpson: J. Fluid Mech. **88**, 223 (1978)
32. J.W. Rottman, J.E. Simpson: J. Fluid Mech. **135**, 95 (1983)
33. J.E. Simpson, R.E. Britter: J. Fluid Mech. **94**, 477 (1979)
34. P. Beghin, X. Olagne: Cold Reg. Sci. Technol. **19**, 317 (1991)
35. Y. Fukushima, N. Hayakawa: J. Hydraul. Eng. ASCE **121**, 600 (1995)
36. M.A. Hallworth, A. Hogg, H.E. Huppert: J. Fluid Mech. **359**, 109 (1998)
37. R.T. Bonnecaze, M.A. Hallworth, H.E. Huppert, J.R. Lister: J. Fluid Mech. **294**, 93 (1995)
38. R.T. Bonnecaze, H.E. Huppert, J.R. Lister: J. Fluid Mech. **250**, 339 (1993)
39. L. Hatcher, A. Hogg, A.W. Woods: J. Fluid Mech. **416**, 297 (2000)
40. H.P. Gröbelbauer, T.K. Fanneløp, R.E. Britter: J. Fluid Mech. **250**, 669 (1993)

41. M. Eglit: 'Mathematical modeling of dense avalanches'. In: *25 years of snow avalanche research, Voss 1998*, ed. by E. Hestnes (Norwegian Geotechnical Institute, 1998) pp. 15–18
42. E.M. Eglit: 'Some mathematical models of snow avalanches'. In: *Advances in the mechanics and the flow of granular materials*, ed. by M. Shahinpoor (Trans Tech Publications, 1983 of Conference) pp. 577–588
43. G. Brugnot, R. Pochat: J. Glaciol. **27**, 77 (1981)
44. J.P. Vila: Sur la théorie et l'approximation numérique des problèmes hyperboliques non linéaires, application aux équations de Saint-Venant et à la modélisation des avalanches denses. Ph.D. Thesis, University Paris VI (1986)
45. G. Parker, Y. Fukushima, H.M. Pantin: J. Fluid Mech. **171**, 145 (1986)
46. G.B. Whitham: *Linear and nonlinear waves*, 2nd edn. (Wiley, New York 1999)
47. E.F. Toro: *Riemann solvers and numerical methods for fluid dynamics* (Springer, Berlin 1997)
48. S.B. Savage, K. Hutter: Acta Mech. **86**, 201 (1991)
49. J.-M. Piau: J. Rheol. **40**, 711 (1996)
50. V.T. Chow: *Open-channel Hydraulics* (Mc Graw Hill, New York 1959)
51. M. Wieland, J.M.N.T. Gray, K. Hutter: J. Fluid Mech. **392**, 73 (1999)
52. J.E. Simpson: J. Fluid Mech. **53**, 759 (1972)
53. C. Härtel, E. Meiburg, F. Necker: J. Fluid Mech. **418**, 189 (2000); *ibid* 213
54. D. Snyder, S. Tait: J. Fluid Mech. **369**, 1 (1998)
55. J.D. Dent, T.E. Lang: Ann. Glaciol. **4**, 42 (1983)
56. O. Maeno: 'Rheological characteristics of snow flows'. In: *International Workshop on Gravitational Mass Movements, Grenoble 1993*, ed. by L. Buisson (Cemagref 1993) pp. 209-220
57. S.B. Savage, K. Hutter: J. Fluid Mech. **199**, 177 (1989)
58. K. Hutter, T. Koch, C. Plüss, S.B. Savage: Acta Mech. **109**, 127 (1995)
59. K. Hutter, M. Siegel, S.B. Savage, Y. Nohguchi: Acta Mech. **100**, 37 (1993)
60. C. Ancey, M. Naaim: 'Modelisation of dense avalanches'. In: *Université européenne d'été sur les risques naturels, Chamonix 1992*, ed. by G. Brugnot (Cemagref, Antony 1995) pp. 173–182
61. G.K. Batchelor: 'A brief guide to two-phase flow'. In: *Theoretical and Applied Mechanics*, ed. by P. Germain, J.M. Piau, D. Caillerie (Elsevier, Amsterdam 1989) pp 27–41
62. P. Gauer: 'A model of powder snow avalanche'. In: *Les apports de la recherche scientifique à la sécurité neige, glace et avalanche, Chamonix 1995*, ed. by F. Sivardière (Cemagref, 1995), pp. 55–61
63. D. Issler: Ann. Glaciol. **26**, 253 (1998)
64. M. Naaim, I. Gurer: J. Natural Hazard **16**, 18 (1997)
65. T. Scheiwiller, K. Hutter, F. Hermann: Ann. Geophys. **5B**, 569 (1987)
66. F. Hermann, D. Issler, S. Keller: 'Numerical simulations of powder-snow avalanches and laboratory experiments in turbidity currents'. In: *International Workshop on Gravitational Mass Movements, Grenoble, 1993*, ed. by L. Buisson (Cemagref 1993) pp. 137–144
67. E.J. Hopfinger, J.-C. Tochon-Danguy: J. Glaciol. **81**, 343 (1977)
68. T. Chu: Can. Geotech. J. **32**, 285 (1995)
69. K.M. Hàkonardóttir: Retarding effects of breaking mounds–avalanches. MA dissertation, University of Bristol, Bristol (2000)
70. J.D. Dent, K.J. Burrell, D.S. Schmidt, M.Y. Louge, E.E. Adams, T.G. Jazbutis: Ann. Glaciol. **26**, 243 (1998)

14 Dense Granular Avalanches: Mathematical Description and Experimental Validation

Y.-C. Tai[1], K. Hutter[1], and J.M.N.T. Gray[2]

[1] Institute of Mechanics, Darmstadt University of Technology
Hochschulstr. 1, 64289 Darmstadt, Germany
[2] Department of Mathematics, University of Manchester, Manchester M13 9PL, UK

14.1 Introduction

Snow avalanches, landslides, rock falls and debris flows are extremely dangerous and destructive natural phenomena. The frequency of occurrence and amplitudes of these disastrous events appear to have increased in recent years perhaps due to recent climate warming. The events endanger the personal property and infra-structure in mountainous regions. For example, from the winters 1940/41 to 1987/88 more than 7000 snow avalanches occurred in Switzerland with damaged property leading to a total of 1269 deaths. In February 1999, 36 people were buried by a single avalanche in Galtür, Austria. In August 1996, a very large debris flow in middle Taiwan resulted in 51 deaths, 22 lost and an approximate property damage of more than 19 billion NT dollars (ca. 600 million US dollars) [18]. In Europe, a suddenly released debris flow in North Italy in August 1998 buried 5 German tourists on the Superhighway "Brenner–Autobahn". The topic has gained so much significance that in 1990 the United Nations declared the *International Decade for Natural Disasters Reduction (IDNDR)*; Germany has its own *Deutsches IDNDR–Komitee für Katastrophenvorbeugung e. V.* Special conferences are devoted to the theme, e.g. , the CALAR conference on Avalanches, Landslides, Rock Falls and Debris Flows (Vienna, January 2000), *INTERPRAEVENT*, annual conferences on the *protection of habitants from floods, debris flows and avalanches*, special conferences on debris flow hazard mitigation and those exclusively on *Avalanches.*

With increasing population and with the popularization of the tourism in the mountainous regions the damage equally increases, occasionally leading to excessive devastation. Reliable methods for the prevention or reduction of the effects of such disasters consist, on the one hand, in predicting the disaster itself and, on the other hand, in the determination of the likely paths of the flows, the maximum run-out distances as well as the protection against such destructive flows. They are of considerable interest to civil and environmental engineers and civil servants of municipalities responsible for the planning and development in populated mountainous regions. The Savage–Hutter theory [19,20] and its three-dimensional extension [4,15] for the gravity-driven, free-surface flow of granular material has proved to model such flows adequately and is now established as one of the leading models for this purpose. We will show in these lecture notes, how these model equations can be constructively used to describe these flows.

A successful verification of an avalanche model by laboratory experiments is a necessary requirement for it to have a chance also to be adequate in realistic situations. This second proof still needs to be completed by applying it to a real avalanche event. For the time being, we are confident on the basis that our model extends a well established model of Voellmy [26] to account for the important geometric deformations of an avalanche along its track. Entrainment of material along the avalanche track can be incorporated, and is a very significant process in real avalanches, but it is not yet incorporated in our model, because no experimental method has so far been found by which an entrainment model could be verified in the laboratory. Entrainment to and deposition from a moving granular mass are however very significant in realistic flows and constitute the "last" unsolved item in the mathematical model to be presented below.

Reviews on the subject are e.g. given in [8,9]. A further article on avalanches – mostly from the practical side – is given by Ancey (Chap. 13).

14.2 The Granular Avalanche Model of Savage & Hutter [1]

The Savage–Hutter theory [19] is a continuum theory to describe the two-dimensional motion of a finite mass avalanche over a rough inclined slope[2]. The dry cohesionless granular material is assumed to be incompressible with constant density ρ_0 throughout the entire body. During flow the body behaves as a Mohr–Coulomb plastic material at yield, which slides over a rigid basal topography. Scaling analysis isolates the physically significant terms in the governing equations and identifies those terms that can be neglected. Finally, depth integration reduces the theory by one spatial dimension.

A simple curvilinear coordinate system was introduced by Savage & Hutter [20] to enable the avalanche motion to be modelled from initiation on a steep slope to run-out on a rough curved bed. The coordinates are defined and aligned with the curved rigid basal topography, so that the local inclination angle ζ varies as a function of the downslope coordinate x.

The Savage–Hutter theory has been extended to three-dimensions by Hutter et al. [15], Greve et al. [6], Gray et al. [4], and Wieland et al. [27] for the case of

[1] This section is taken from [24].

[2] All avalanche models known to us and used in practice are essentially based on Voellmy's [26] original model, which may be interpreted as a rigid mass or hydraulic model (depending on view point). In these models, the physics is incorporated in the parameterization of the resistive forces comprising of a Coulomb and a viscous type contribution. The modern trend is to use the basic balance laws of physics and to account for the internal physics as well as the variation of the basal topography, just as attempted in the Savage–Hutter model in a very simple form. There have been attempts to model avalanches by molecular dynamics procedures, and these are successful when the number of particles or grains is small, i.e. a few hundred or thousand, as e.g. in rock falls. Possible farther reaching conceptual formulations of both the continuum and molecular dynamics concepts are given in Chap. 4 and the literature cited there.

unconfined three-dimensional flow. Hutter et al. [15] derived the leading order equations for the motion of unconfined flow on an inclined plane with constant inclination angle. Greve et al. [6] introduced a quasi one-dimensional curvilinear system to model unconfined flow on a simple chute without lateral curvature, whilst Gray et al. [4] generalized this theory to allow the flow over complex three-dimensional topography. The final three-dimensional theory is able to predict the flow over realistic topography and provide information about the maximum run-out distance in site specific applications.

In this section a brief introduction to the Savage–Hutter theory and its three-dimensional extension over realistic topography is given. Different from the procedure of the original derivation in [19], the governing equations are integrated through the depth before the procedure of scaling analysis. In accordance with the conservation laws of mass and linear momentum the governing equations are derived in conservative form as described in [4], which will be used to model granular shocks that have been observed in laboratory experiments, as they are in conservative form and therefore allow discontinuities in the physical variables to be considered.

14.2.1 Governing Equations in Conservative Form

The avalanche is treated as a material with constant density[3] ρ_0 throughout the entire avalanche body, the local differential forms of the mass and momentum conservation laws are therefore

$$\operatorname{div} \boldsymbol{v} = 0 \,, \tag{14.1a}$$

$$\rho_0 \left\{ \frac{\partial \boldsymbol{v}}{\partial t} + \operatorname{div}(\boldsymbol{v} \otimes \boldsymbol{v}) \right\} = -\operatorname{div} \boldsymbol{p} + \rho_0 \, \boldsymbol{g} \,, \tag{14.1b}$$

where $\boldsymbol{v}$ is the velocity, $\otimes$ the dyadic product, $\boldsymbol{p}$ the pressure tensor and $\boldsymbol{g}$ the gravitational acceleration.

Following [19], the body is assumed to have Mohr–Coulomb constitutive properties. This implies that yield occurs when the internal shear stress S and the normal pressure N are related by

$$|S| = N \tan\phi \,, \tag{14.2}$$

where ϕ is the so-called internal friction angle.

The body is subject to kinematic and traction boundary conditions at the free surface $F^s(\boldsymbol{x}, t) = 0$ and at the base $F^b(\boldsymbol{x}, t) = 0$ of the avalanche. The kinematic boundary conditions are

$$\frac{\partial F^s}{\partial t} + \boldsymbol{v}^s \cdot \operatorname{grad} F^s = 0 \,, \tag{14.3}$$

$$\frac{\partial F^b}{\partial t} + \boldsymbol{v}^b \cdot \operatorname{grad} F^b = 0 \,, \tag{14.4}$$

[3] Possible sizable volume changes occur at the instants of avalanche inception and settling, but not so much during motion, see [12]. This is the reason, why a density preserving model delivers good results.

where the superscripts s and b indicate the variable evaluated at the free surface and at the base, respectively. Note that for a rigid basal topography, $F^b(\boldsymbol{x}) = 0$, the kinematic boundary condition reduces to $\boldsymbol{v}^b \cdot \operatorname{grad} F^b = 0$, which implies that the avalanche slides on the basal surface without inflow or outflow, i.e. an impenetrable base.

The kinematic free surface of the avalanche is assumed to be traction free, and at the base a sliding Coulomb dry friction law[4] is applied. That is,

$$\boldsymbol{p}^s \boldsymbol{n}^s = \boldsymbol{0} , \tag{14.5}$$

$$\boldsymbol{p}^b \boldsymbol{n}^b - (\boldsymbol{n}^b \cdot \boldsymbol{p}^b \boldsymbol{n}^b)\boldsymbol{n}^b = (\boldsymbol{v}^b/|\boldsymbol{v}^b|) N^b \tan\delta , \tag{14.6}$$

where $(\boldsymbol{p}\boldsymbol{n})_i = p_{ij} n_j$, $N^b = \boldsymbol{n}^b \cdot \boldsymbol{p}^b \boldsymbol{n}^b$ indicates the normal pressure at the base of the avalanche, δ is the basal angle of friction, $\boldsymbol{n}^s$ and $\boldsymbol{n}^b$ are outward pointing normal vectors at the free surface and base, respectively,

$$\boldsymbol{n}^s = \frac{\nabla F^s}{|\nabla F^s|} , \qquad \boldsymbol{n}^b = \frac{\nabla F^b}{|\nabla F^b|} . \tag{14.7}$$

14.2.2 Curvilinear Coordinate System

An orthogonal curvilinear coordinate system, $Oxyz$, is defined by a reference surface [4], which is illustrated in Fig. 14.1a. The x axis is oriented in the downslope direction, the y axis lies in the cross slope direction to the reference surface and the z axis is normal to it. The downslope inclination angle of the reference surface ζ, to the horizontal, changes as a function of the downslope coordinate x, and there is no lateral variation in the y direction. The complex shallow basal topography is defined by its elevation $z = z^b(x, y)$ above the reference surface, as illustrated in Fig. 14.1b. The region above the reference surface $z = 0$ can be described by the coordinates xyz that is based on the metric with the squared arc length

$$\mathrm{d}s^2 = (1 - \kappa z)^2 \mathrm{d}x^2 + \mathrm{d}y^2 + \mathrm{d}z^2 , \tag{14.8}$$

where $\kappa = -\partial\zeta/\partial x$ is the curvature of the reference surface. The metric defines each point in a domain of the three-dimensional space uniquely as long as the z-coordinate is locally smaller than $1/\kappa$. In the ensuing analysis this will automatically be assumed.

In this curvilinear coordinate system the divergence of the velocity $\boldsymbol{v}$ in (14.1a) is

$$\nabla \cdot \boldsymbol{v} = \frac{\partial}{\partial x}(\psi u) + \frac{\partial v}{\partial y} + \frac{\partial w}{\partial z} - \psi^2 \kappa' z u - \psi \kappa w , \tag{14.9}$$

where $\psi = 1/(1 - \kappa z)$ and u, v, w are the physical velocity components in the x, y and z directions, respectively. $\kappa' = \partial\kappa/\partial x$ is the derivative of the curvature

[4] An elementary account on the Coulomb law is given in Chap. 4.

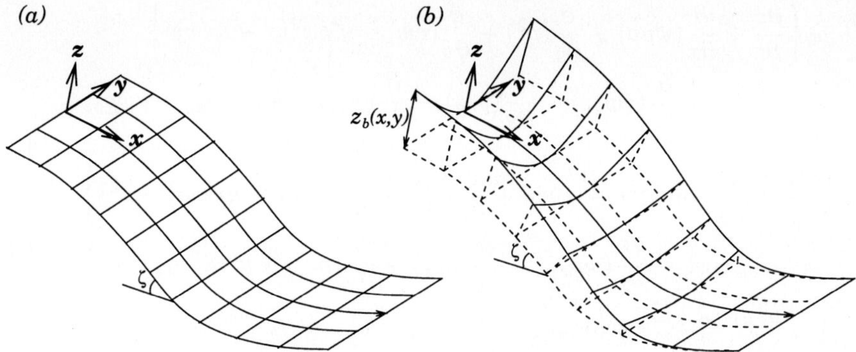

Fig. 14.1. (**a**) The curvilinear reference surface which defines the curvilinear coordinate system $Oxyz$, where the downslope inclination angle of the reference surface ζ, to the horizontal, changes as a function of the downslope coordinate x. (**b**) The shallow basal topography is defined by its height $z = z^b(x,\, y)$ above the curvilinear reference surface

with respect to the downslope coordinate x. In the curvilinear coordinate system the divergence of a second order tensor $\boldsymbol{T}$ [6,4] is expressed by

$$\begin{aligned}\nabla\cdot\boldsymbol{T} = &\left\{\frac{\partial}{\partial x}(\psi T_{xx}) + \frac{\partial T_{xy}}{\partial y} + \frac{\partial T_{xz}}{\partial z} - \psi^2\kappa' z T_{xx} - 2\psi\kappa T_{xz}\right\}\boldsymbol{e}_x\\ &+\left\{\frac{\partial}{\partial x}(\psi T_{xy}) + \frac{\partial T_{yy}}{\partial y} + \frac{\partial T_{yz}}{\partial z} - \psi^2\kappa' z T_{xy} - \psi\kappa T_{yz}\right\}\boldsymbol{e}_y\\ &+\left\{\frac{\partial}{\partial x}(\psi T_{xz}) + \frac{\partial T_{yz}}{\partial y} + \frac{\partial T_{zz}}{\partial z} - \psi^2\kappa' z T_{xz} - \psi\kappa (T_{zz} - T_{xx})\right\}\boldsymbol{e}_z\,,\end{aligned} \tag{14.10}$$

where $\boldsymbol{e}_x$, $\boldsymbol{e}_y$ and $\boldsymbol{e}_z$ are the unit vectors in the downslope, cross slope and normal directions, respectively. Furthermore, the gradient of a scalar field F in this curvilinear coordinate system is

$$\nabla F = \frac{1}{1-\kappa z}\frac{\partial F}{\partial x}\boldsymbol{e}_x + \frac{\partial F}{\partial y}\boldsymbol{e}_y + \frac{\partial F}{\partial z}\boldsymbol{e}_z\,. \tag{14.11}$$

Using (14.9) the mass balance equation (14.1a) becomes

$$\frac{\partial}{\partial x}(\psi u) + \frac{\partial v}{\partial y} + \frac{\partial w}{\partial z} - \psi^2\kappa' z u - \psi\kappa w = 0\,. \tag{14.12}$$

By virtue of (14.9) and (14.10), the downslope, cross slope and normal components of the momentum balance equations are

$$\begin{aligned}&\rho_0\left\{\frac{\partial u}{\partial t} + \frac{\partial}{\partial x}(\psi u^2) + \frac{\partial}{\partial y}(uv) + \frac{\partial}{\partial z}(uw) - \kappa' z\psi^2 u^2 - 2\kappa\psi uw\right\}\\ &= \rho_0 g\sin\zeta - \frac{\partial}{\partial x}(\psi p_{xx}) - \frac{\partial}{\partial y}(p_{xy}) - \frac{\partial}{\partial z}(p_{xz}) + \kappa' z\psi^2 p_{xx} + 2\kappa\psi p_{xz}\,,\end{aligned} \tag{14.13}$$

$$\rho_0\left\{\frac{\partial v}{\partial t}+\frac{\partial}{\partial x}(\psi uv)+\frac{\partial}{\partial y}(v^2)+\frac{\partial}{\partial z}(vw)-\kappa' z\psi^2 uv-\kappa\psi vw\right\}$$
$$= \qquad -\frac{\partial}{\partial x}(\psi p_{xy})-\frac{\partial}{\partial y}(p_{yy})-\frac{\partial}{\partial z}(p_{yz})+\kappa' z\psi^2 p_{xy}+\kappa\psi p_{yz}\,, \tag{14.14}$$

$$\begin{aligned}\rho_0&\left\{\frac{\partial w}{\partial t}+\frac{\partial}{\partial x}(\psi uw)+\frac{\partial}{\partial y}(vw)+\frac{\partial}{\partial z}(w^2)-\kappa' z\psi^2 uw-\kappa\psi(w^2-u^2)\right\}\\ &=-\rho_0 g\cos\zeta-\frac{\partial}{\partial x}(\psi p_{xz})-\frac{\partial}{\partial y}(p_{yz})-\frac{\partial}{\partial z}(p_{zz})+\kappa' z\psi^2 p_{xz}\\ &\quad+\kappa\psi(p_{zz}-p_{xx})\,,\end{aligned} \tag{14.15}$$

respectively, where p_{ij}, $i, j = x, y, z$ are the components of the pressure tensor in this curvilinear coordinate system. The free and basal surfaces are defined by their heights above the reference surface,

$$\begin{aligned}F^s(\boldsymbol{x},t)&=z-z^s(x,y,t)=0\,,\\ F^b(\boldsymbol{x},t)&=z^b(x,y,t)-z=0\,,\end{aligned} \tag{14.16}$$

which ensure the normals $\boldsymbol{n}^s$ and $\boldsymbol{n}^b$ point outwards from the avalanche body. The kinematic boundary conditions reduce to

$$-\frac{\partial z^s}{\partial t}-\psi^s u^s\frac{\partial z^s}{\partial x}-v^s\frac{\partial z^s}{\partial y}+w^s=0 \tag{14.17}$$

for the free surface and

$$\frac{\partial z^b}{\partial t}+\psi^b u^b\frac{\partial z^b}{\partial x}+v^b\frac{\partial z^b}{\partial y}-w^b=0 \tag{14.18}$$

for the base, where we recall that the superscripts s, b indicate the values at the free and basal surface, respectively. The traction condition on the free surface (14.5) yields

$$\begin{aligned}-p^s_{xx}\psi^s\frac{\partial z^s}{\partial x}-p^s_{xy}\frac{\partial z^s}{\partial y}+p^s_{xz}&=0\,,\\ -p^s_{xy}\psi^s\frac{\partial z^s}{\partial x}-p^s_{yy}\frac{\partial z^s}{\partial y}+p^s_{yz}&=0\,,\\ -p^s_{xz}\psi^s\frac{\partial z^s}{\partial x}-p^s_{yz}\frac{\partial z^s}{\partial y}+p^s_{zz}&=0\,,\end{aligned} \tag{14.19}$$

and the sliding condition at the base (14.6) becomes

$$\begin{aligned}p^b_{xx}\psi^b\frac{\partial z^b}{\partial x}+p^b_{xy}\frac{\partial z^b}{\partial y}-p^b_{xz}&=\left(\psi^b\frac{\partial z^b}{\partial x}+|\nabla F^b|\frac{u^b}{|\boldsymbol{v}^b|}\tan\delta\right)N^b\,,\\ p^b_{xy}\psi^b\frac{\partial z^b}{\partial x}+p^b_{yy}\frac{\partial z^b}{\partial y}-p^b_{yz}&=\left(\frac{\partial z^b}{\partial y}+|\nabla F^b|\frac{v^b}{|\boldsymbol{v}^b|}\tan\delta\right)N^b\,,\\ p^b_{xz}\psi^b\frac{\partial z^b}{\partial x}+p^b_{yz}\frac{\partial z^b}{\partial y}-p^b_{zz}&=-N^b\,,\end{aligned} \tag{14.20}$$

where $\psi^b = 1/(1-\kappa z^b)$. The Coulomb dry friction shear traction is related by the normal basal pressure N^b and the bed friction angle δ. Applying the definitions of the normal basal pressure $N^b = \boldsymbol{n}^b \cdot \boldsymbol{p}^b \boldsymbol{n}^b$ and the basal normal vector (14.7) yields

$$
\begin{aligned}
N^b = \frac{1}{|\nabla F^b|^2} \Bigg\{ & (\psi^b)^2 \left(\frac{\partial z^b}{\partial x}\right)^2 p^b_{xx} + 2\psi^b \left(\frac{\partial z^b}{\partial x}\right) \left(\frac{\partial z^b}{\partial y}\right) p^b_{xy} \\
& -2\psi^b \left(\frac{\partial z^b}{\partial x}\right) p^b_{xz} + \left(\frac{\partial z^b}{\partial y}\right)^2 p^b_{yy} - 2 \left(\frac{\partial z^b}{\partial y}\right) p^b_{yz} + p^b_{zz} \Bigg\} .
\end{aligned}
\tag{14.21}
$$

14.2.3 Depth Integration

The mass and momentum balance equations are integrated through the avalanche depth to simplify the problem. The avalanche thickness (depth) is the difference between the height of the free surface z^s and the height of the basal topography z^b

$$
h = z^s - z^b , \tag{14.22}
$$

and is measured normal to the reference surface. The depth integrated mean value is denoted by $\langle \cdot \rangle$ and defined by

$$
\langle f \rangle = \frac{1}{h} \int_{z^b}^{z^s} f \, \mathrm{d}z \tag{14.23}
$$

for any field quantity f.

Using Leibniz's rule and integrating the mass balance equation (14.9) through the avalanche depth subject to the kinematic boundary conditions at the free (14.17) and basal (14.18) surfaces in the curvilinear coordinate system, it follows that

$$
\frac{\partial h}{\partial t} + \frac{\partial}{\partial x}(h\langle \psi u \rangle) + \frac{\partial}{\partial y}(h\langle v \rangle) - \kappa' h \langle \psi^2 z u \rangle - \kappa h \langle \psi w \rangle = 0 . \tag{14.24}
$$

Similarly, integrating the linear momentum balance equations, (14.13), (14.14) and (14.15), through the avalanche depth and applying the kinematic as well as traction boundary conditions at both the free surface (14.17), (14.19) and the basal surface (14.18), (14.20), the depth integrated downslope, cross slope and normal components of the momentum balance are

$$
\begin{aligned}
& \rho_0 \left\{ \frac{\partial}{\partial t}(h\langle u \rangle) + \frac{\partial}{\partial x}(h\langle \psi u^2 \rangle) + \frac{\partial}{\partial y}(h\langle uv \rangle) - \kappa' h \langle \psi^2 z u^2 \rangle - 2\kappa h \langle \psi u w \rangle \right\} \\
& = \rho_0 g h \sin\zeta - \frac{\partial}{\partial x}(h\langle \psi p_{xx} \rangle) - \frac{\partial}{\partial y}(h\langle p_{xy} \rangle) \\
& \quad - \left(\psi^b \frac{\partial z^b}{\partial x} + |\nabla F^b| \frac{u^b}{|\boldsymbol{v}^b|} \tan\delta \right) N^b + \kappa' h \langle \psi^2 z p_{xx} \rangle + 2\kappa h \langle \psi p_{xz} \rangle ,
\end{aligned}
\tag{14.25}
$$

$$\rho_0\Big\{\frac{\partial}{\partial t}(h\langle v\rangle)+\frac{\partial}{\partial x}(h\langle\psi uv\rangle)+\frac{\partial}{\partial y}(h\langle v^2\rangle)-\kappa' h\langle\psi^2 zuv\rangle-\kappa h\langle\psi vw\rangle\Big\}$$
$$\begin{aligned}=\quad&-\frac{\partial}{\partial x}(h\langle\psi p_{xy}\rangle)-\frac{\partial}{\partial y}(h\langle p_{yy}\rangle)\\&-\left(\frac{\partial z^b}{\partial y}+|\nabla F^b|\frac{v^b}{|\boldsymbol{v}^b|}\tan\delta\right)N^b+\kappa' h\langle\psi^2 zp_{xy}\rangle+\kappa h\langle\psi p_{yz}\rangle\ ,\end{aligned}\tag{14.26}$$

$$\rho_0\Big\{\frac{\partial}{\partial t}(h\langle w\rangle)+\frac{\partial}{\partial x}(h\langle\psi uw\rangle)+\frac{\partial}{\partial y}(h\langle vw\rangle)-\kappa' h\langle\psi^2 zuw\rangle-\kappa h\langle\psi(w^2-u^2)\rangle\Big\}$$
$$\begin{aligned}&=-\rho_0 gh\cos\zeta-\frac{\partial}{\partial x}(h\langle\psi p_{xz}\rangle)-\frac{\partial}{\partial y}(h\langle p_{yz}\rangle)\\&\qquad+N^b+\kappa' h\langle\psi^2 zp_{xz}\rangle+\kappa h\langle\psi(p_{zz}-p_{xx})\rangle\ ,\end{aligned}\tag{14.27}$$

respectively. For details of the derivation see [4].

14.2.4 Non-Dimensionalization and Ordering

Three length scales are introduced to isolate the physically significant terms in the governing equations, a longitudinal length scale, L, a depth scale, H, and a scale for the basal curvature in the downslope direction, $1/R$. Following [19], [6] and [4], the physical variables are non-dimensionalized using the scalings

$$\begin{aligned}(x,\,y,\,z)_{\text{dim}}&=L(x,\,y,\,\varepsilon z)_{\text{non-dim}}\ ,\\(u,\,v,\,w)_{\text{dim}}&=\sqrt{gL}(u,\,v,\,\varepsilon\,w)_{\text{non-dim}}\ ,\\(p_{xx},\,p_{yy},\,p_{zz},\,N^b)_{\text{dim}}&=\rho_0 gH(p_{xx},\,p_{yy},\,p_{zz},\,N^b)_{\text{non-dim}}\ ,\\(p_{xy},\,p_{xz},\,p_{yz})_{\text{dim}}&=\rho_0 gH\mu(p_{xy},\,p_{xz},\,p_{yz})_{\text{non-dim}}\ ,\\(t)_{\text{dim}}&=\sqrt{L/g}\,(t)_{\text{non-dim}}\ ,\\(\kappa)_{\text{dim}}&=1/R\,(\kappa)_{\text{non-dim}}\ ,\end{aligned}\tag{14.28}$$

where $\varepsilon=H/L$ is the aspect ratio and μ indicates a typical magnitude of the friction coefficient, $\tan\delta_0$.

Observations of avalanches in nature and laboratory experiments suggest that they are long and thin and that the basal surfaces on which they slide often have shallow curvature. The shallowness assumption for the avalanche geometry implies that the aspect ratio of the avalanche is small,

$$\varepsilon=H/L\ll 1\ .\tag{14.29}$$

The measure of the curvature of the reference surface geometry with respect to the length of the avalanche $\lambda=L/R$ and the friction coefficient μ are assumed to be of magnitude [4]

$$\lambda=\mathcal{O}(\varepsilon^\alpha)\ ,\quad \mu=\mathcal{O}(\varepsilon^\beta)\ ,\quad 0<\alpha\ ,\ \ \beta<1\ .\tag{14.30}$$

Applying the scalings (14.28) and assumption (14.30), it follows that the depth integrated non-dimensional mass balance equation in curvilinear form is

$$\frac{\partial h}{\partial t} + \frac{\partial}{\partial x}(h\langle u\rangle) + \frac{\partial}{\partial y}(h\langle v\rangle) = 0 + \mathcal{O}(\varepsilon^{1+\alpha}) , \tag{14.31}$$

where all variables in this equation and in the remainder of this text are now non-dimensional unless stated otherwise. Using (14.28) the normal component of the momentum balance (14.15) reduces to

$$\frac{\partial p_{zz}}{\partial z} = -\cos\zeta + \mathcal{O}(\varepsilon^{\alpha}) , \tag{14.32}$$

which implies that p_{zz} varies linearly with respect to z to order ε^{α}. The normal basal pressure (14.21) gives $N_b = p^b_{zz} + \mathcal{O}(\varepsilon^{1+\beta})$ that the normal component of the non-dimensional depth integrated momentum balance (14.27) reduces to

$$p^b_{zz} = \lambda\kappa h\langle u^2\rangle + h\cos\zeta + \mathcal{O}(\varepsilon) . \tag{14.33}$$

The downslope and cross slope components are

$$\begin{aligned} &\frac{\partial}{\partial t}(h\langle u\rangle) + \frac{\partial}{\partial x}(h\langle u^2\rangle) + \frac{\partial}{\partial y}(h\langle uv\rangle) \\ &\quad = h\sin\zeta - \varepsilon\frac{\partial}{\partial x}(h\langle p_{xx}\rangle) - \varepsilon h\cos\zeta\frac{\partial z^b}{\partial x} \\ &\qquad - \frac{u^b}{|\boldsymbol{v}^b|}h\tan\delta(\cos\zeta + \lambda\kappa\langle u^2\rangle) + \mathcal{O}(\varepsilon^{1+\gamma}) , \end{aligned} \tag{14.34}$$

$$\begin{aligned} &\frac{\partial}{\partial t}(h\langle v\rangle) + \frac{\partial}{\partial x}(h\langle uv\rangle) + \frac{\partial}{\partial y}(h\langle v^2\rangle) \\ &\quad = -\varepsilon\frac{\partial}{\partial y}(h\langle p_{yy}\rangle) - \varepsilon h\cos\zeta\frac{\partial z^b}{\partial y} \\ &\qquad - \frac{v^b}{|\boldsymbol{v}^b|}h\tan\delta(\cos\zeta + \lambda\kappa\langle u^2\rangle) + \mathcal{O}(\varepsilon^{1+\gamma}) , \end{aligned} \tag{14.35}$$

respectively, in which $\gamma = \min(\alpha, \beta)$.

14.2.5 Earth Pressure Coefficients

In the original Savage–Hutter theory [19,20] the stress state of the avalanche is assumed to satisfy both the Coulomb sliding friction law and the internal yield criterion simultaneously at the base of the avalanche. In addition, since the motion is predominantly downslope it is assumed that the basal cross slope pressure, p^b_{yy}, is a principal stress and that it is equal to one of the other two principal stresses in the xz plane. The details of this analysis have been performed many times and are well documented in the literature, see e.g. [4–9,12–15,17,19,20], the most useful references probably being [6,17].

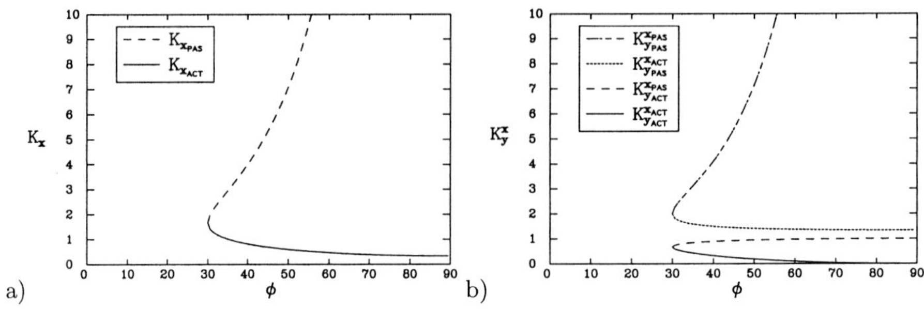

Fig. 14.2. The downslope (**a**) and cross slope (**b**) earth pressure coefficients are plotted as functions of the internal friction angle ϕ with constant bed friction angle $\delta = 30°$. The various active and passive stress states are indicated by different line styles

Under these assumptions two Mohr stress circles can be constructed that satisfy the yield criterion and the sliding law. For the basal normal pressure, p^b_{zz}, and the shear stress, $-p^b_{xz}$, the basal down slope pressure, p^b_{xx}, can therefore assume two values, one on the larger circle, $p^b_{xx} > p^b_{zz}$, and the other on the smaller circle, $p^b_{xx} < p^b_{zz}$. These downslope and normal pressures can be related by introducing the earth pressure coefficient $K^b_x = p^b_{xx}/p^b_{zz}$. Using elementary geometrical arguments K^b_x is described as a function of the internal and basal friction angles [19]

$$K^b_{x_{\text{act/pass}}} = 2\left(1 \mp \sqrt{1 - \cos^2\phi/\cos^2\delta}\right)\sec^2\phi - 1\,. \tag{14.36}$$

This is real valued provided $\delta \le \phi$. Savage & Hutter [19] made the *ad hoc* definition that the *active* state was associated with divergent motion and the *passive* state was associated with convergent motion, i.e.

$$K^b_x = \begin{cases} K^b_{x_{\text{act}}} & \partial u/\partial x \ge 0\,, \\ K^b_{x_{\text{pass}}} & \partial u/\partial x < 0\,. \end{cases} \tag{14.37}$$

The left panel in Fig. 14.2 illustrates the values of $K^b_{x_{\text{act/pass}}}$ as a function of the internal friction angle ϕ for constant basal friction angle $\delta = 30°$. When $\phi = \delta$ the active and passive earth pressure coefficients are equal, $K^b_{x_{\text{act}}} = K^b_{x_{\text{pass}}}$. For $\phi < \delta$ the earth pressure coefficients are not real valued.

As mentioned above, the basal cross slope pressure is equal to one of the other two principal stresses in the xz plane. Introducing the cross slope earth pressure coefficient at the base, $K^b_y = p^b_{yy}/p^b_{zz}$, Hutter et al. [13] showed that it is equal to

$$K^b_{y_{\text{act/pass}}} = \frac{1}{2}\left(K^b_x + 1 \mp \sqrt{(K^b_x - 1)^2 + 4\tan^2\delta}\right)\,, \tag{14.38}$$

which is not only a function of the internal and basal friction angle but also depends on the downslope earth pressure coefficient. Since there are two principal stresses for the Mohr stress circle determined by the stress state in the xz plane, there are four possible stress states for the cross slope pressure. As in the two-dimensional theory they are distinguished from one another by *ad hoc* definitions dependent upon whether the downslope and cross slope deformation is divergent or convergent [13],

$$K_y^b = \begin{cases} K_{y_{\text{act}}}^{x_{\text{act}}} & \partial u/\partial x \geq 0\,,\ \partial v/\partial y \geq 0\,, \\ K_{y_{\text{pass}}}^{x_{\text{act}}} & \partial u/\partial x \geq 0\,,\ \partial v/\partial y < 0\,, \\ K_{y_{\text{act}}}^{x_{\text{pass}}} & \partial u/\partial x < 0\,,\ \partial v/\partial y \geq 0\,, \\ K_{y_{\text{pass}}}^{x_{\text{pass}}} & \partial u/\partial x < 0\,,\ \partial v/\partial y < 0\,. \end{cases} \tag{14.39}$$

In the first of these inequalities the flow is extending in both the x- and y-directions, in the last it is contracting in the two directions. In the right panel of Fig. 14.2, K_y^b is illustrated as a function of the internal friction angle ϕ for constant $\delta = 30°$. Like the downslope earth pressure coefficient, K_y is real valued if and only if $\delta \leq \phi$ and $K_{y_{\text{act}}}^b = K_{y_{\text{pass}}}^b$ when $\delta = \phi$. In addition the earth pressure coefficients are ordered in the following way: $K_{y_{\text{act}}}^{x_{\text{act}}} \leq K_{y_{\text{act}}}^{x_{\text{pas}}} < K_{y_{\text{pas}}}^{x_{\text{act}}} \leq K_{y_{\text{pas}}}^{x_{\text{pas}}}$.

Note that the theory is not objective[5] in the xy plane, but it is a good approximation if the assumption $v^b \ll u^b$ holds, i.e. this simple representation is reasonable when the flow is chiefly downhill and the shearing in the xy plane is small in comparison with the shearing in the xz and yz planes. With these *ad hoc* definitions (14.37), (14.39), Koch et al. [17], Gray et al. [4] and Wieland et al. [27] obtained good agreement between theory and experiments.

The *ad hoc* definitions (14.37) and (14.39) define the earth pressure coefficients in two limiting states with piecewise constant values, respectively. There is a discontinuity at $\partial u/\partial x = 0$ or $\partial v/\partial y = 0$, which results in a jump in the in-plane pressure between convergent and divergent regions. If we consider the jump condition of the linear momentum [1], there must be a corresponding jump in the avalanche velocity, and/or the thickness, in order to balance the tractions on either side of the jump interface.

A regularization[6] for these two limiting stress states was proposed by Tai & Gray [21], in which the discontinuity is regularized by introducing a smooth transition between the two limiting stress states. This is illustrated in Fig. 14.3 for the downslope earth pressure coefficient. For large convergence they approach the passive stress state and for large divergence they approach the active stress state. Between these two limiting stress states there is a smooth monotonically decreasing transition, which crosses the $\partial u/\partial x = 0$ line at $K_x^b = K_{x_0}$ for the down slope component and $K_y^b = K_{y_0}$ at $\partial v/\partial y = 0$ for the cross slope component, where K_{x_0} and K_{y_0} are the downslope and cross slope coefficients with $\delta = \phi$, respectively. The regularized downslope and cross slope earth pressure coefficients are given by

[5] Objectivity refers here to invariance under rigid body rotations.

[6] There is no other reason for this regularization than to make the earth pressure coefficient a continuous function of the strain rate. A partial physical argument for its introduction can be found in [23].

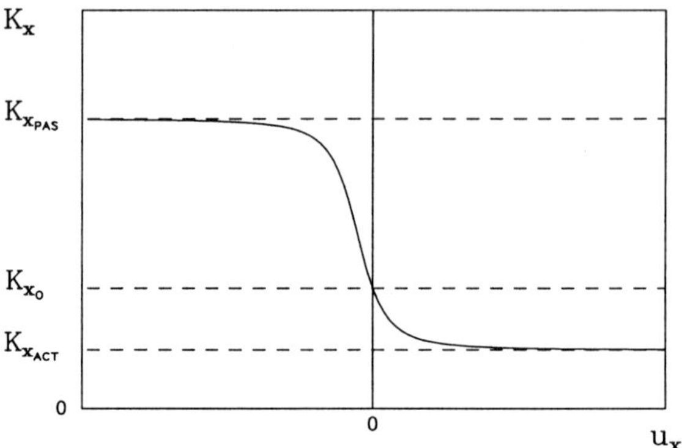

Fig. 14.3. The down slope earth pressure coefficient is regularized by introducing a smoothly varying monotonically decreasing function of the downslope divergence $\partial u/\partial x$, which approaches the limiting values, $K_{x_{\mathrm{act}}}$ and $K_{x_{\mathrm{pass}}}$, for large divergence and convergence, respectively. At $\partial u/\partial x = 0$ the downslope earth pressure coefficient equals K_{x_0}

$$
\begin{aligned}
K_x^b &= \frac{1}{2}\{(K_{\mathrm{act}}^x + K_{\mathrm{pass}}^x) + f_x(\partial u/\partial x)(K_{\mathrm{act}}^x - K_{\mathrm{pass}}^x)\}\,, \\
K_y^b &= \frac{1}{2}\{(K_{\mathrm{act}}^y + K_{\mathrm{pass}}^y) + f_y(\partial v/\partial y)(K_{\mathrm{act}}^y - K_{\mathrm{pass}}^y)\}\,,
\end{aligned}
\tag{14.40}
$$

where the regularization functions f_x and f_y are dependent on the downslope and cross slope velocity gradients $\partial u/\partial x$ and $\partial v/\partial y$, respectively. They are chosen to be the monotonically decreasing functions

$$
\begin{aligned}
f_x(\partial u/\partial x) &= (\alpha_K \partial u/\partial x - c_0^x)/\left(1 + (\alpha_K \partial u/\partial x - c_0^x)^2\right)^{1/2}\,, \\
f_y(\partial v/\partial y) &= (\alpha_K \partial v/\partial y - c_0^y)/\left(1 + (\alpha_K \partial v/\partial y - c_0^y)^2\right)^{1/2}\,,
\end{aligned}
\tag{14.41}
$$

where α_K determines the steepness of the transition. The constants c_0^x and c_0^y are chosen that $K_x\big|_{\partial u/\partial x=0} = K_{x0}$ and $K_y\big|_{\partial v/\partial y=0} = K_{y0}$, respectively[7]. Using this regularization of the earth pressure coefficients, Tai & Gray [21] demonstrated that a *necking* of the avalanche is resolved in simulating a channelized free-surface flow, in which the Wieland et al. [27] Lagrangian moving grid technique is applied. The *necking* form is observed in the transition zone when the material flows down in a channel into the horizontal flat runout zone.

14.2.6 Model Equations in Conservative Form

In the one-dimensional Savage–Hutter [19,20] theory and in the two-dimensional extensions of their theory [4,13] the downslope and cross slope pressures are

[7] The parameters K_{x0} and K_{y0} must be identified by experiment or via inverse methods, which is not easy. Tai et al. [23] design an experiment with rotating drums from which K_{x0} can directly be inferred.

assumed to vary linearly over the thickness of the avalanche. In accordance with (14.32) this implies that $K_x = K_x^b$ and $K_y = K_y^b$ throughout the avalanche depth. With the traction free assumption at the free surface it follows that the average depth integrated pressures $h\langle p_{xx}\rangle$ and $h\langle p_{yy}\rangle$ [4] are determined by

$$h\langle p_{xx}\rangle = \tfrac{1}{2}h^2 K_x^b \cos\zeta + \mathcal{O}(\varepsilon^\gamma) \,, \quad h\langle p_{yy}\rangle = \tfrac{1}{2}h^2 K_y^b \cos\zeta + \mathcal{O}(\varepsilon^\gamma) \,. \tag{14.42}$$

It is also assumed that the velocity profiles are approximately uniform through the avalanche depth, i.e. all sliding and little differential shear [19]. Thus, the basal velocities are assumed to be of the form

$$u^b = \langle u\rangle + \mathcal{O}(\varepsilon^{1+\gamma}) \,, \quad v^b = \langle v\rangle + \mathcal{O}(\varepsilon^{1+\gamma}) \,, \tag{14.43}$$

and the velocity products can be factorised [4]

$$\begin{aligned} &\langle u^2\rangle = \langle u\rangle^2 + \mathcal{O}(\varepsilon^{1+\gamma}) \,, \quad \langle uv\rangle = \langle u\rangle\langle v\rangle + \mathcal{O}(\varepsilon^{1+\gamma}) \\ &\text{and} \quad \langle v^2\rangle = \langle v\rangle^2 + \mathcal{O}(\varepsilon^{1+\gamma}) \,. \end{aligned} \tag{14.44}$$

These assumptions are supported by measurements in large scale dry snow [2] and ping-pong ball avalanches [16].

From (14.31) and (14.44) it follows that the mass balance equation reduces to order $\varepsilon^{1+\alpha}$ to

$$\frac{\partial h}{\partial t} + \frac{\partial}{\partial x}(hu) + \frac{\partial}{\partial y}(hv) = 0 \,. \tag{14.45}$$

With assumptions (14.42), (14.43) and (14.44) the depth integrated downslope (14.34) and cross slope (14.35) momentum balances yield

$$\frac{\partial}{\partial t}(hu) + \frac{\partial}{\partial x}(hu^2) + \frac{\partial}{\partial y}(huv) = hs_x - \frac{\partial}{\partial x}\left(\frac{\beta_x h^2}{2}\right) \,, \tag{14.46a}$$

$$\frac{\partial}{\partial t}(hv) + \frac{\partial}{\partial x}(huv) + \frac{\partial}{\partial y}(hv^2) = hs_y - \frac{\partial}{\partial y}\left(\frac{\beta_y h^2}{2}\right) \,, \tag{14.46b}$$

to order $\varepsilon^{1+\gamma}$, where the brackets $\langle\ \rangle$ for the mean values are dropped. The factors β_x and β_y are defined as

$$\beta_x = \varepsilon \cos\zeta K_x \quad \text{and} \quad \beta_y = \varepsilon \cos\zeta K_y \,, \tag{14.47}$$

respectively. The terms s_x and s_y represent the net driving accelerations in the downslope and cross slope directions, respectively

$$s_x = \sin\zeta - \frac{u}{|\boldsymbol{v}|}\tan\delta(\cos\zeta + \lambda\kappa u^2) - \varepsilon\cos\zeta\frac{\partial z^b}{\partial x} \,, \tag{14.48a}$$

$$s_y = \quad - \frac{v}{|\boldsymbol{v}|}\tan\delta(\cos\zeta + \lambda\kappa u^2) - \varepsilon\cos\zeta\frac{\partial z^b}{\partial y} \,, \tag{14.48b}$$

where $|\boldsymbol{v}| = (u^2 + v^2)^{1/2}$. The first term at the right-hand side of (14.48a) is due to the gravitational acceleration. It has no contribution in the lateral, y, direction. The second terms of both (14.48a) and (14.48b) indicate the dry Coulomb friction and guarantee that basal shear traction and the sliding velocity are collinear. The third terms are the contributions from the basal topography. The system of equations (14.45)–(14.46b) shall be referred to as the *two-dimensional conservative system* (2DCS) of equations.

For *smooth solutions* the mass balance can be used to simplify the convective terms in the momentum balances (14.46a), (14.46b). Providing $h \neq 0$ the mass and momentum balance equations reduce to

$$\frac{\partial h}{\partial t} + \frac{\partial}{\partial x}(hu) + \frac{\partial}{\partial y}(hv) = 0 \,, \tag{14.49a}$$

$$\frac{\partial u}{\partial t} + u\frac{\partial u}{\partial x} + v\frac{\partial u}{\partial y} = s_x - \beta_x \frac{\partial h}{\partial x} - \frac{h}{2}\frac{\partial \beta_x}{\partial x} \,, \tag{14.49b}$$

$$\frac{\partial v}{\partial t} + u\frac{\partial v}{\partial x} + v\frac{\partial v}{\partial y} = s_y - \beta_y \frac{\partial h}{\partial y} - \frac{h}{2}\frac{\partial \beta_y}{\partial y} \,, \tag{14.49c}$$

These equations and their spatially one-dimensional analogues ($\partial(\cdot)/\partial y = 0$, and (14.49c) missing) were derived earlier and numerically integrated by a Lagrangian finite difference method[8]. These cannot capture possible shocks; but they proved the model to be adequate for many avalanche tests performed in the laboratory.

14.3 Numerical Integration of the Savage–Hutter Equations

14.3.1 Standard Form of the Differential Equations and Characteristic Speeds

The two dimensional model equations (14.45)–(14.46b) can be written in general vector form

$$\frac{\partial \boldsymbol{w}}{\partial t} + \frac{\partial \boldsymbol{f}}{\partial x} + \frac{\partial \boldsymbol{g}}{\partial y} = \boldsymbol{s} \,, \tag{14.50}$$

where $\boldsymbol{w}$ denotes the vector of conservative variables, $\boldsymbol{f}$ and $\boldsymbol{g}$ represent the transport fluxes in the x- and y-directions, respectively, and $\boldsymbol{s}$ means the source

[8] The one-dimensional model was derived by Savage & Hutter [19,20] and tested against laboratory chute experiments by Greve & Hutter [5], and Hutter et al. [13]. Two-dimensional spreading was attacked by Hutter [7], Hutter et al. [13], Greve et al. [6] and Koch et al. [17] on the basis that the basal topography was flat perpendicular to the direction of steepest descent with good agreement with granular avalanches from laboratory experiments. Sidewise confinement was then incorporated in [4,27,3] with equally satisfactory agreement between model output and laboratory experiments. The chute topography in these cases was a weak parabolic channel merging into a horizontal plane.

term. They are

$$\boldsymbol{w} = \begin{pmatrix} h \\ m^x \\ m^y \end{pmatrix}, \qquad \boldsymbol{f} = \begin{pmatrix} m^x \\ (m^x)^2/h + \beta_x h^2/2 \\ m^x m^y/h \end{pmatrix},$$
$$\boldsymbol{g} = \begin{pmatrix} m^y \\ m^x m^y/h \\ (m^y)^2/h + \beta_y h^2/2 \end{pmatrix}, \quad \boldsymbol{s} = \begin{pmatrix} 0 \\ hs^x \\ hs^y \end{pmatrix}, \tag{14.51}$$

where the source terms in the momentum balance equations, s_x and s_y, are defined in (14.48a) and (14.48b), respectively and equations are written in the conservative variables h, $m^x = hu$ and $m^y = hv$. The spatially one-dimensional version of (14.50) is

$$\frac{\partial \boldsymbol{w}}{\partial t} + \frac{\partial \boldsymbol{f}}{\partial x} = \boldsymbol{s}, \tag{14.52}$$

where

$$\boldsymbol{w} = \begin{pmatrix} h \\ m^x \end{pmatrix}, \quad \boldsymbol{f} = \begin{pmatrix} m^x \\ (m^x)^2/h + \beta_x h^2/2 \end{pmatrix}, \quad \boldsymbol{s} = \begin{pmatrix} 0 \\ hs^x \end{pmatrix}. \tag{14.53}$$

It can be obtained from (14.51) by setting $\boldsymbol{g} = \boldsymbol{0}$ and ignoring in $\boldsymbol{w}$, $\boldsymbol{f}$ and $\boldsymbol{s}$ the third line.

The characteristic speeds of the system (14.50)–(14.51) can be computed by rewriting (14.50) as

$$\frac{\partial \boldsymbol{w}}{\partial t} + \begin{pmatrix} \boldsymbol{A}_x & 0 \\ 0 & \boldsymbol{A}_y \end{pmatrix} \begin{pmatrix} \dfrac{\partial \boldsymbol{w}}{\partial x} \\ \dfrac{\partial \boldsymbol{w}}{\partial y} \end{pmatrix} = \boldsymbol{s}, \tag{14.54}$$

where

$$\boldsymbol{A}_x := \frac{\partial \boldsymbol{f}}{\partial \boldsymbol{w}} = \begin{pmatrix} 0 & 1 & 0 \\ -(m^x)^2/h^2 + \beta_x h & 2m^x/h & 0 \\ -m^x m^y/h^2 & m^y/h & m^x/h \end{pmatrix},$$
$$\boldsymbol{A}_y := \frac{\partial \boldsymbol{g}}{\partial \boldsymbol{w}} = \begin{pmatrix} 0 & 0 & 1 \\ -m^x m^y/h^2 & m^y/h & m^x/h \\ -(m^y)^2/h^2 + \beta_y h & 0 & 2m^y/h \end{pmatrix}, \tag{14.55}$$

and evaluating the eigenvalues of $\boldsymbol{A}$. These follow from the characteristic equation

$$\det(\boldsymbol{A} - \lambda \boldsymbol{I}_6) = \det(\boldsymbol{A}_x - \lambda \boldsymbol{I}_3)\det(\boldsymbol{A}_y - \lambda \boldsymbol{I}_3) = 0 \tag{14.56}$$

with six solutions, given by

$$\begin{aligned} \lambda_1 &= u\,, \quad \lambda_{3,5} = m^x/h \pm \sqrt{\beta_x h}\,, \\ \lambda_2 &= v\,, \quad \lambda_{4,6} = m^y/h \pm \sqrt{\beta_y h}\,. \end{aligned} \tag{14.57}$$

The first two solutions yield as characteristic speed the particle velocity $c_p = (u^2+v^2)^{1/2}$ and as characteristic directions the streamline directions. $\lambda_{3,\dots,6}$ give rise to four different characteristic speeds

$$\begin{aligned} C^{++} &= \left(\lambda_3^2 + \lambda_4^2\right)^{1/2}\,, \quad C^{+-} = \left(\lambda_3^2 + \lambda_6^2\right)^{1/2}\,, \\ C^{-+} &= \left(\lambda_5^2 + \lambda_4^2\right)^{1/2}\,, \quad C^{--} = \left(\lambda_5^2 + \lambda_6^2\right)^{1/2} \end{aligned} \tag{14.58}$$

with four different directions; C^{++} is the fastest and C^{--} the slowest. Whenever $c_p > C^{--}$ the flow is called *supercritical*; otherwise, i.e. when $c_p < C^{--}$, it is *subcritical*. Any transition from a supercritical to a subcritical flow state is associated with a shock. This inevitably happens when a finite avalanching mass moves down a steep slope (where it reaches supercritical speeds) and is considerably decelerated (when it approaches the runout zone) and eventually approaches a subcritical speed. This transition is always accomplished by the formation of a shock front across which the avalanche depth and speed experience sudden changes from small heights and large speeds to larger heights and smaller speeds. The numerical schemes must cope with this situation.

14.3.2 Remarks on Numerical Integration

It is not the place here to present a detailed introduction into shock-capturing numerical methods. Such an overview is given in Chap. 4 of [24]. We sketch the method only and must direct the interested reader to the literature, see [24] and [25] for detail.

Let us commence by recalling that the Lagrangian integration technique applied to the SH equations faced difficulties whenever a supercritical extending (diverging) flow became subcritical and contracting. Numerical solution in the vicinity of such transitions were accompanied with high oscillations of the depth profile and velocity field which often led to instabilities unless this was properly counteracted by a sufficient amount of numerical diffusion. The regularization of the earth pressure coefficient outlined in Sect. 14.2.5 helped to improve the situation, but the difficulties encountered with shocks were thereby not resolved.

The shock-capturing numerical methods give a high resolution of shock solutions without any spurious oscillations near a discontinuity. The traditional high order accuracy methods result in unexpected oscillations near the discontinuity. The Total Variation Diminishing (TVD) method for equation systems in conservative form achieves this goal; its application to the Savage–Hutter equations allows integration across shocks without the introduction of additional numerical diffusion.

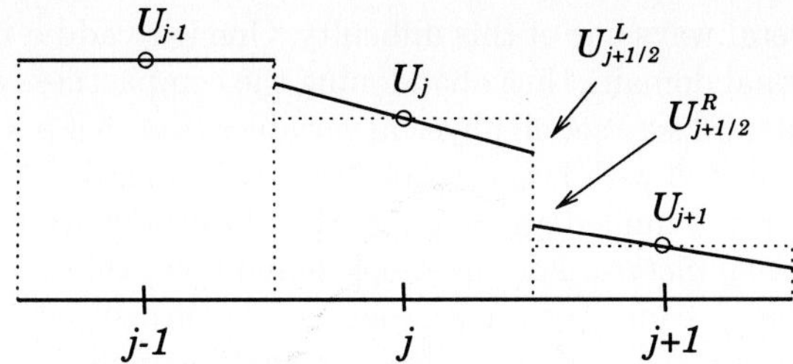

Fig. 14.4. The cell average physical value U_j (*dashed line*) and the linear piecewise cell reconstruction (*solid line*), where there are two values for each interface, e.g. $U^L_{j+1/2} = U_j + U'_j/2$ and $U^R_{j+1/2} = U_{j+1} - U'_{j+1}/2$, one from the left side cell, U_j, with the approximate derivatives U'_j, the other from the right side element, U_{j+1} with U'_{j+1}

TVD means that the sum of the variations of the variables over the whole computational domain does not increase as the time evolves. Now, the numerical schemes are designed such that they provide only the cell average values of the variable to be determined. In classical schemes of higher order approximation the numerically determined variable is continuous or even differentiable across cell boundaries. In TVD methods jump discontinuities are allowed over cell boundaries, whilst within each cell C^n-continuity may prevail, for an illustration, see Fig. 14.4.

In regions where the variations of the field variables are small no jumps are needed, but in the neighbourhood of shocks and in regions of large gradients of the field variable the cell re-constructions are such, i.e. the slopes of the variable within the cell kept so small, that possible spurious oscillations are avoided. The operators that achieve the limitation of the cell slope (just sufficient to avoid oscillations) are called *slope limiters* (and several different versions have been proposed: e.g. *Superbee, Minmod* or *Woodward*). Several schemes have been tested with the application of these three slope limiters to find the optimal scheme for smooth as well as discontinuous solutions.

There is a further numerical subtlety associated with the motion of a finite mass of granular material along an inclined plane or curved topography. The material does not occupy the whole region of topographic surface available to it but covers a region with compact support. The margin separates the regions with and without material. It can be shown that the governing equations (14.45)–(14.46b) do not admit solutions with cliffs [24] at the margin, so that margins always have vanishing avalanche height and the transition from the avalanche region to its complement is continuous. Now in an Eulerian numerical scheme with the cells fixed in space and a moving boundary problem as this one, it happens more often than not that the margin lies between the cells than exactly on cell boundaries. This is different from the Lagrangian integration technique in which the grid moves with the deforming avalanche mass and margins are always exactly traced. It is in general associated with a considerable loss of accuracy.

There are several ways out of this difficulty. One is to add a thin layer to the whole computational domain, thus abandoning the compactness of the avalanche body; a second method is to set all physical variables to zero if $h = 0$; both are not ideal and still associated with large errors close to the margin. A third method is to treat the cells in the immediate neighbourhood of the margin separately by a special *front-tracking method.* For one-dimensional flows this has been done [25], and results for the spreading of a parabolic profile turned out to be very much improved; for two-dimensional situations the method must still be developed. We now present a few computational solutions.

14.4 Examples

In this section we[9] present a number of solutions that were constructed with the shock capturing finite difference schemes developed by Tai [24]. Further results are also given in [25].

14.4.1 Similarity Solutions

For flows of a finite mass of granular materials down an inclined plane the deformation of an initially compact mass of granular materials is everywhere extending and so no shock will form in this case. The equations may then be used in the form (14.49a)–(14.49c), either in their one- or two-dimensional case. For a parabolic linear (1D) or circular (2D) initial hump at rest exact similarity solutions were constructed. These solutions ([19], 1D; [14], 1D; [11], 2D) allow determination of the asymptotic behaviour of the motion. They show that without a viscous contribution to the drag force the avalanches do not reach an asymptotic constant velocity. The parabolic profile remains preserved, but the originally circular hump becomes elliptical. Thus, the streamwise extension is larger than the cross slope extension. These exact solutions are useful, because numerical solutions obtained by other techniques can be checked against them.

It is interesting to note that such parabolic ellipses have not been observed experimentally. The profiles have rather tear drop shape [10]. These indicate that either exact initial conditions to arrive at these solutions were not realized in the experiments or the model equations – in particular the Coulomb sliding law with constant bed friction angle δ – are not adequate. The problem is still open.

14.4.2 Motion of a Granular Avalanche on an Inclined Plane Chute into the Horizontal Run-Out Zone

Shock formations are often observed when the avalanche slides from an inclined slope into the horizontal run-out zone, where the frontal part comes to rest

9 The authors acknowledge help received from S. Noelle on shock capturing integration techniques and the software for gas dynamics from A.-K. Lie which was adopted to avalanche flow by Y.-Ch. Tai [24].

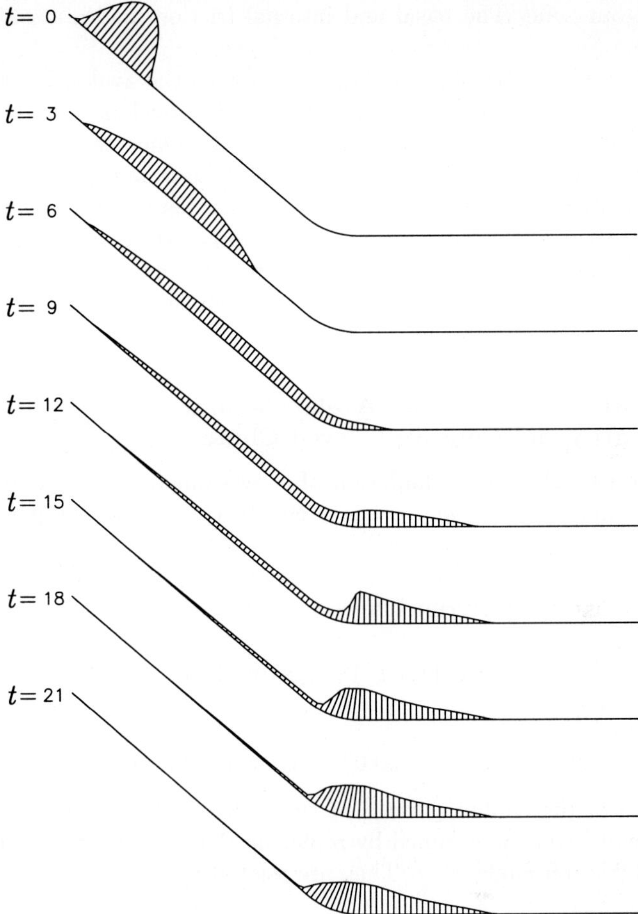

Fig. 14.5. Process of the avalanche simulated by the shock-capturing method at $t = 0, 3, 6, 9, 12, 15, 18, 21$ dimensionless time units. As the front reaches the run-out zone and comes to rest, the rear part of the avalanche accelerates further and the avalanche body contracts. Once the velocity becomes supercritical, a shock wave develops, which moves upward.

and the part of the tail still accelerates further so that its velocity becomes supercritical. A test simulation is made by the shock-capturing method.

The granular material released from a parabolic cap slides down an inclined plane chute and merges into the horizontal run-out zone. The parabolic cap is initially located at the top of the slope with a linearly increasing velocity distribution, so that the avalanche extends by maintaining its parabolic form if it slides on an infinitely long slope [19]. The inclination angle of the inclined plane is prescribed as 40°, and a transition region lies between the inclined slope and the run-out zone. The basal and internal friction angles are 35° and 38°, respectively.

Figure 14.5 illustrates the simulated process as the avalanche slides on the inclined plane into the horizontal run-out zone. The avalanche body extends on the inclined plane with a parabolic form ($t = 3$). Once the front reaches the horizontal run-out zone the basal friction brings the frontal part of the granular material to rest, but the part of the rear accelerates further. At this stage, if the velocity becomes super-critical, a shock (surge) wave is created ($t = 12$ to $t = 18$), which moves a short distance backwards as can clearly be seen (compare the humps at $t = 12$ to $t = 18$). At $t = 21$ the whole avalanche body comes to rest.

14.4.3 Motion of a Granular Avalanche in a Convex and Concave Curved Chute

In this section we show the simulation of a two-dimensional avalanche moving down in a confined convex and concave curved chute, and compare the result with one of many experiments, called here exp. 29 in [5]. The experiment was performed in a 10 cm wide chute of length greater than 400 cm. The basal surface was formed to follow a prescribed function, so that the inclination angle is given by

$$\zeta(x) = \zeta_0 e^{-0.1x} + \zeta_1 \xi/(1+\xi^8) - \zeta_2 \exp(-0.3(x+10/3)^2) \ , \qquad (14.59)$$

where

$$\xi = \tfrac{4}{15}(x-9) \ , \quad \text{and} \quad \zeta_0 = 60.0° \ , \quad \zeta_0 = 31.4° \ , \quad \zeta_2 = 37.0° \ . \qquad (14.60)$$

The influence of the confining walls of the chute on the bed friction was also considered, which was determined by replacing the bed friction angle, δ, by the *effective* bed friction angle, δ_{eff}. They are related by

$$\delta_{eff} = \delta_0 + \varepsilon k_{\text{wall}} h \ ; \qquad (14.61)$$

here ε is the aspect ratio, h is the dimensionless depth and k_{wall} the measured correction factor to account for the side wall effects in the bed friction angle, see [12].

In the simulation all parameters are assigned as in [5], where $\delta_0 = 26.5°$, $K_{\text{wall}} = 11°$, and the internal angle of friction is selected to be $\phi = 37°$. Figure 14.6 shows the computed profile of the avalanche height on the real chute geometry. Once mobile the avalanche rapidly accelerates downslope until it reaches the shallow rise in the topography. This is enough to retard the granular material until the pressure from the material behind has sufficiently accumulated to push it over the bump. The material accelerates again and when the slope angle decreases the mass comes to rest. Normally the deposit is divided into two parts on the both sides of the bump: the rear deposit and the front deposit.

Figure 14.7 shows time slices of the computed (solid) and experimentally determined (dashed) profiles of the avalanche height for exp. 29 in [5], where $\delta_0 = 26.5°$, $\phi = 37°$ and $K_{\text{wall}} = 11°$. The division of the avalanche body into two parts is well described by the simulation.

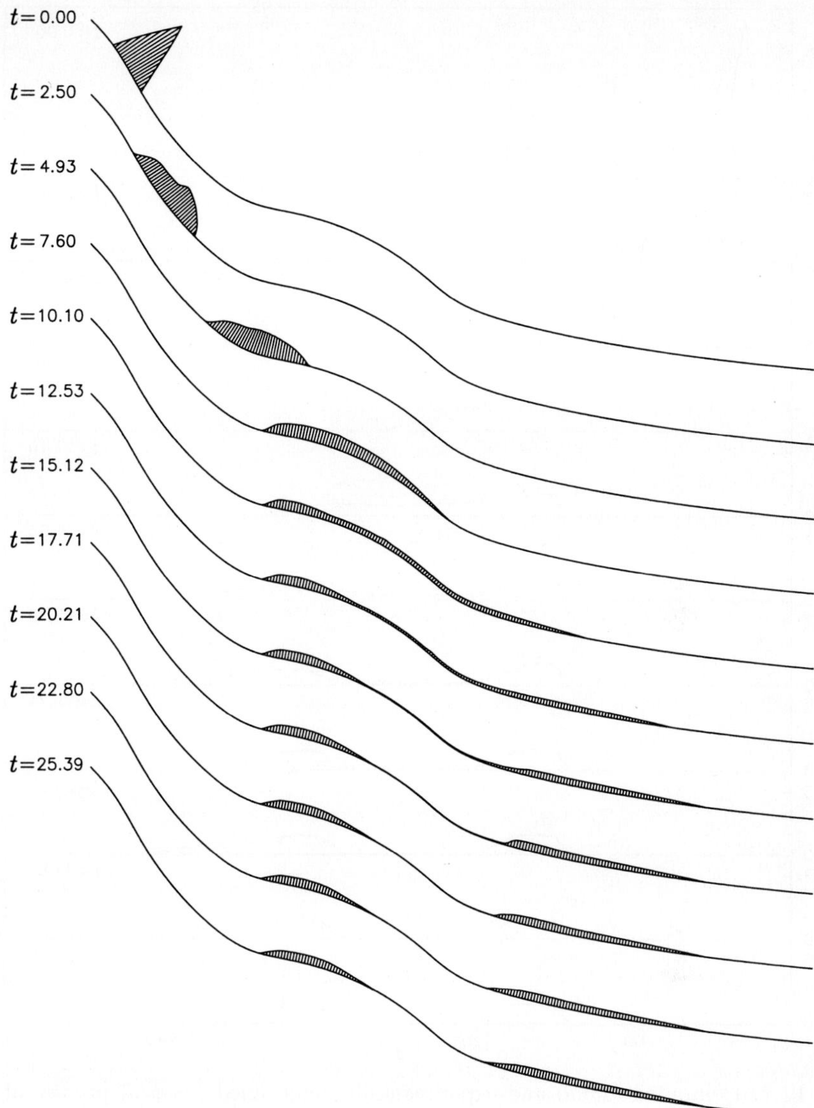

Fig. 14.6. Computed profile of the avalanche height on the real chute geometry for exp. 29 in [5]. Since compared to the length the deposited height of the avalanche is very small, the height is three times exaggerated

14.4.4 Granular Avalanche over Complex Basal Topography

In this section a simulation example on a chute with complex basal topography is presented to describe the two-dimensional shock formation. A simple reference surface is defined consisting of an inclined plane ($\zeta = 40°$) that is connected to a horizontal run-out zone ($\zeta = 0°$) by a transition zone. Superposed

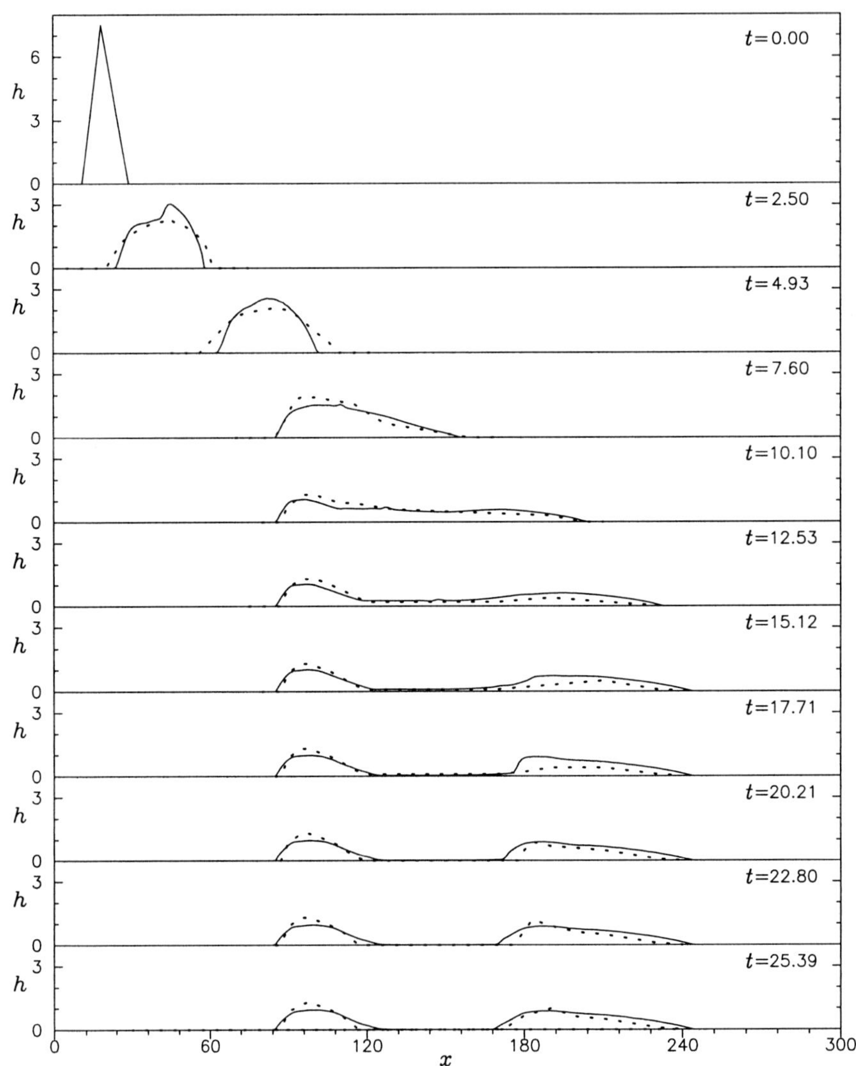

Fig. 14.7. Computed (*solid*) and experimentally determined (*dashed*) profiles of the avalanche height for exp. 29 in [5]; $\delta_0 = 26.5°$, $\phi = 37°$ and $K_{\text{wall}} = 11°$. The horizontal distance is arc length measured along the basal surface

on the inclined section of the chute is a shallow parabolic cross-slope topography, $z^b(y) = y^2/(2R)$ with $R = 110$ cm, which forms a channel that partly confines the avalanche motion. The inclined parabolic channel lies in the range $0 < x < 215$ cm and the run-out zone lies in the range $x > 245$ cm, between which a transition zone smoothly joins the two regions. At $x = 160$ cm there is a small parabolic hill with radius 15 cm and heigh 5 cm, see Fig. 14.8. In the transition zone, $215 < x < 245$, a smooth change in the topography defined by

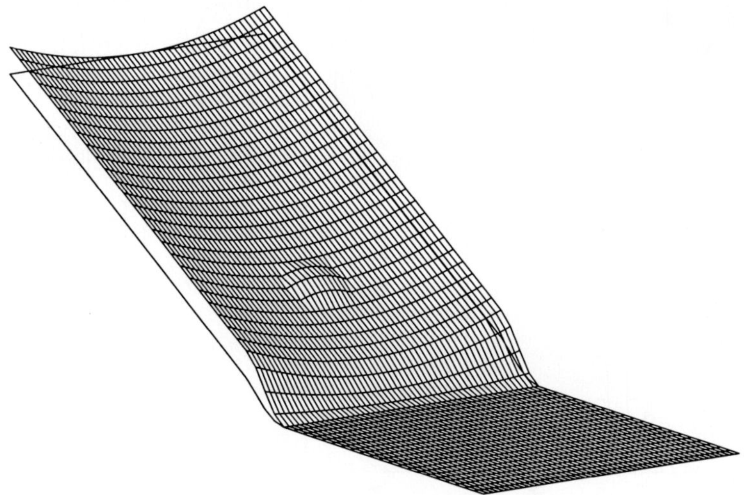

Fig. 14.8. The complex basal topography for the test problem describing the two-dimensional shock formation. A simple reference surface is defined consisting of an inclined plane that is connected to a horizontal run-out zone by a transition zone. Superposed on the reference surface is a shallow parabolic cross-slope topography, which forms a channel that partly confines the avalanche motion. The parabolic channel in restricted to the inclined range. It is connected with the horizontal run-out zone by a smooth transition zone. A small parabolic hill lies in the channel centre of the inclined portion and constitutes a partial obstruction

the inclination angle

$$\zeta(x) = \begin{cases} \zeta_0 \, , & 0 \le x \le 215 \, , \\ \zeta_0[1 - (x - 215)/40] \, , & 215 < x < 245 \, , \\ 0^\circ \, , & x \ge 245 \, , \end{cases} \tag{14.62}$$

is prescribed, where $\zeta_0 = 40^\circ$.

The simulation is performed with an internal angle of friction $\phi = 37^\circ$ and a bed friction angle $\delta = 32^\circ$. The material is suddenly released from a hemispherical shell with radius $r_0 = 32\,\mathrm{cm}$. It is so fitted to the basal chute topography, that the projection of the line of intersection onto the reference surface is approximately elliptical in shape. The major axis of the ellipse is 32 cm long and the maximum height of the cap above the reference surface is 22 cm.

Figure 14.9 shows the depth contours of the simulated results for a sequence of non-dimensional times, from the release of the material ($t = 0$) until the

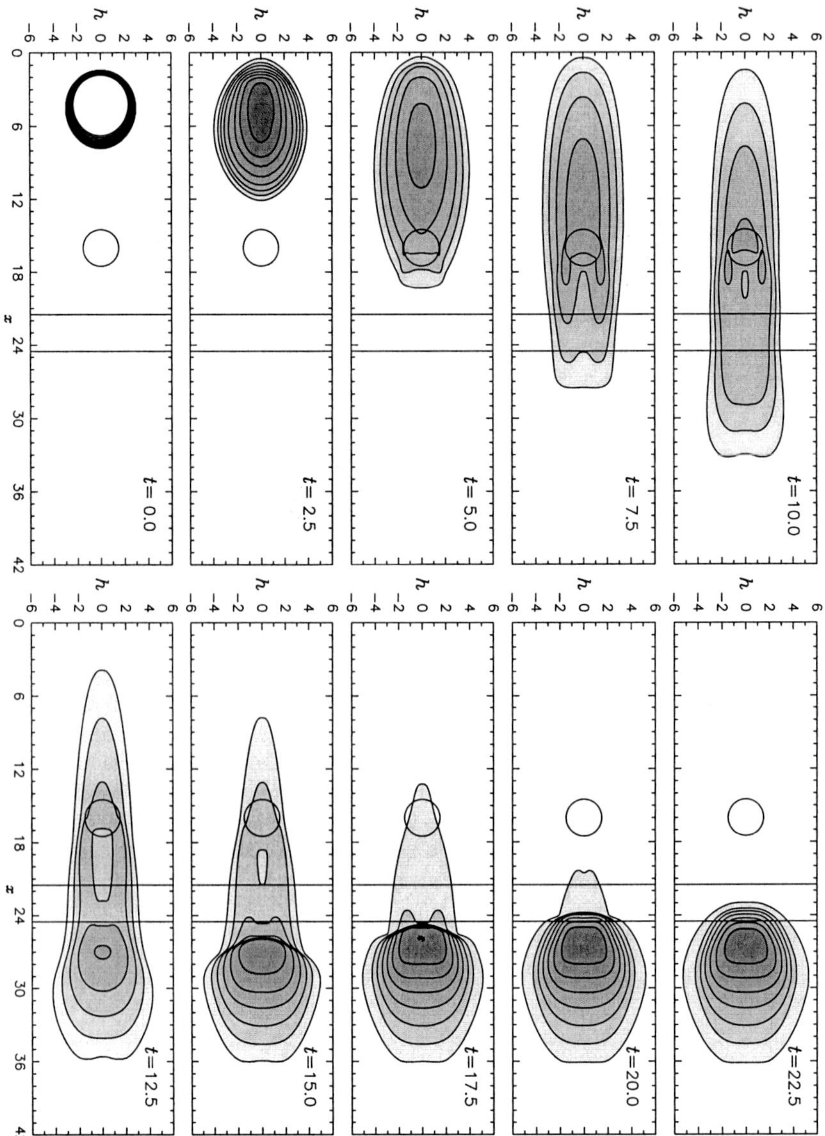

Fig. 14.9. Depth contour of the simulated results at a series of non-dimensional times, from the release of the material ($t = 0$) until the avalanche comes to rest ($t = 22.5$). The length unit is in dm

avalanche comes to rest ($t = 22.5$). Once the cap is open, the avalanche accelerates and extends, where the acceleration in the down-slope $x-$direction is obviously dominant ($t = 2.5$). Because of the back pressure the rear part of the avalanche moves slightly backwards at the initial stage of the motion. Due to the curvature in the cross-slope direction the extension in y-direction is limited in the channel region (see $t = 5.0$ to $t = 12.5$). The hill holds the material partly up ($t = 5.0$ to $t = 12.5$), but immediately below the hill and on either side of it two knolls form. Furthermore, behind the hill the reduction of inflowing mass causes a dent to form. Basically, the material accelerates until it reaches the horizontal run-out zone. With increasing basal drag the front comes to rest ($t = 7.5$ to $t = 12.5$) but the part of the tail accelerates further. In this stage the avalanche body contracts. Once the supercritical velocity becomes subcritical, a shock wave (steep surface gradient) is formed. This occurs just after the end of the transition zone at *ca.* $x = 260$ ($t = 15$). With the approaching mass from the tail, the shock wave propagates backwards ($t = 15$ to $t = 20$), i.e., as time proceeds, this shock wave propagates upstream. At $t = 22.5$ the avalanche comes to rest.

The velocities inside the avalanche body for the same times as the avalanche geometries in Fig. 14.9 are illustrated in Fig. 14.10, in which the arrows denote the direction of the velocity, and their lengths indicate the speed. The velocity of the elements with depth $h < 0.1$ cm are not shown here. Although the hill holds the material partly up and side knolls around it and a dent behind it are formed, the velocity is not strongly affected by these features ($t = 5.0$ to $t = 12.5$) and the material is obviously accelerated in the downslope direction. The front comes to rest in the run-out zone but the part of the tail accelerates further ($t = 7.5$ to $t = 12.5$). At $t = 15$ there is obviously a jump of velocity taking place at the transition zone, which corresponds to the steep surface gradient in Fig. 14.9. With the mass approaching from the tail, the jump propagates backwards ($t = 15$ to $t = 20$). At $t = 22.5$ the avalanche comes to rest.

14.5 Concluding Remarks

In this contribution a simple theoretical model due to Savage and Hutter was presented and results obtained with it were compared with experiments. It consists of depth integrated balance laws of mass and momentum of an incompressible fluid that obeys a dry friction Coulomb type constitutive relation with constant internal angle of friction. A second phenomenological parameter entering this model is the bed friction angle which measures the roughness between the granules and the bed. This model, which is based on a shallowness assumption and supposes that downhill velocities are large in comparison to cross-channel velocities, is expressed as a hyperbolic system of partial differential equations with an (earth pressure) coefficient appearing in them which, depending on the solution, may be discontinuous. Both the hyperbolicity of the equations and the discontinuity of the earth pressure coefficient pose difficulties in the integration process and may require shock capturing numerical techniques. This requires that the differential equations are formulated in conservative form and that total variation diminishing finite difference schemes are used and combined with frontal techniques. Simulations conducted for avalanches, observed in the lab-

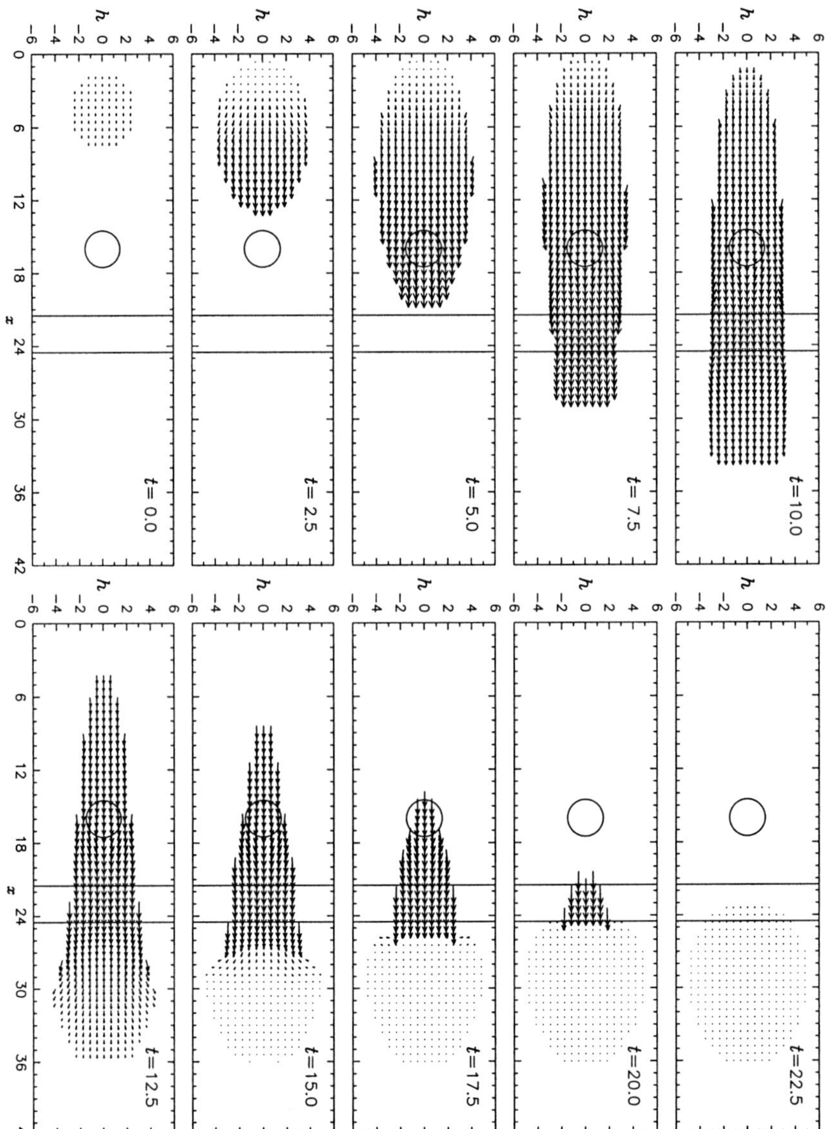

Fig. 14.10. The velocities inside the avalanche body at the same times as in Fig. 14.9 are illustrated in Fig. 14.10, in which the arrows denote the direction of the velocity and their lengths indicate the speed

oratory, show that agreement with the observations is good including in those cases when shocks are formed.

Finally we mention that the fact that the dimensionless form of the Savage–Hutter equations does not depend on any non-dimensional parameters such as

Froude, Reynolds numbers or any other π-product can constructively be used to perform laboratory experiments on rapid flow of granular materials. As a consequence, there are no scale effects in this theory, and all that must be observed in a physical model is geometric similitude and reproduction of the internal and bed friction angles. This has been done by Tai et al. [22] in a model simulation of the flow of an avalanche around a wedge that was protecting a construction. Shocks that form in such processes have also been adequately reproduced.

There remains the incorporation of entrainment/deposition processes to make the model applicable to realistic situations.

Acknowledgement

Financial support by the Deutsche Forschungsgemeinschaft through the Sonderforschungsbereich 298 is gratefully acknowledged. We thank the editors for a constructive review of the manuscript, and Y. Wang for help with the TEXing.

References

1. P. Chadwick: *Continuum Mechanics. Concise Theory and Problems* (Allen & Unwin, London 1976)
2. J.D. Dent, K.J. Burrell, D.S. Schmidt, M.Y. Louge, E.E. Adams, T.G. Jazbutis: Ann. of Glaciology, **26**, 247–252 (1998)
3. J.M.N.T. Gray, K. Hutter: Physik granularer Lawinen, Physikalische Blätter, **54** (1), 37–43 (1998)
4. J.M.N.T. Gray, M. Wieland, K. Hutter: Proc. R. Soc. London **455**, 1841–1874 (1999)
5. R. Greve, K. Hutter: Phil. Trans. R. Soc. London **A 342**, 573–604 (1993)
6. R. Greve, T. Koch, K. Hutter: Proc. R. Soc. London **A 445**, 399–413 (1994)
7. K. Hutter: Acta Mechanica (Suppl.) **1**, 167–181 (1991)
8. K. Hutter: Lawinen Dynamik – eine Übersicht. Der Maschinenschaden, **92**(5), 181–191 (1992)
9. K. Hutter: 'Avalanche Dynamics'. In: *Hydrology of Disasters*, ed. by V.P. Singh (Kluwer Academic Publ., Dordrecht–Boston–London 1996) pp. 317–394
10. K. Hutter: 'Order and disorder in granular materials'. In: *CISM, Kinetic and continuum theories of granular and porous media*, ed. by K. Hutter, K. Wilmanski (Springer, Heidelberg 1998) pp. 1–65
11. K. Hutter, R. Greve: J. Glaciology **39**, 357–372 (1993)
12. K. Hutter, T. Koch: Phil. Trans. R. Soc. London **A 334**, 93–138 (1991)
13. K. Hutter, T. Koch, C. Plüss, S.B. Savage: Acta Mechanica **109**, 127–165 (1993)
14. K. Hutter, Y. Nohguchi: Acta Mechanica **82**, 99–127 (1990)
15. K. Hutter, M. Siegel, S.B. Savage, Y. Nohguchi: Acta Mechanica **100**, 37–68 (1993)
16. S. Keller, Y. Ito, K. Nishimura: Ann. of Glaciology **26**, 259–264 (1998)
17. T. Koch, R. Greve, K. Hutter: Proc. R. Soc. London **A 445**, 415–435 (1994)
18. M.L. Lin, F.S. Jeng, C.C. Wu: Sino-Geotechnics **57**, 25–30 (1996)
19. S.B. Savage, K. Hutter: J. Fluid Mech. **199**, 177–215 (1989)
20. S.B. Savage, K. Hutter: Acta Mechanica **86**, 201–223 (1991)
21. Y.C. Tai, J.M.N.T. Gray: Annals of Glaciology, **25** 272–276 (1998)

22. Y.-C. Tai, Y. Wang, J.M.N.T. Gray, K. Hutter: 'Methods of similitude in granular avalanche flows'. In: *Advances in Cold-Region Thermal Engineering and Sciences*, ed. by K. Hutter et al. (Springer–Verlag, Berlin 1999) pp. 415–428
23. Y.-C. Tai, K. Hutter, J.M.N.T. Gray: The Chinese Journal of Mechanics **16**(2), 67–72 (2000)
24. Y.-C. Tai: Dynamics of Granular Avalanches and their Simulations with Shock-Capturing and Front-Tracking Numerical Schemes. Dissertation for Doctor Degree, Darmstadt University of Technology, Shakers Verlag, Aachen, ISBN 3-8265-7249-1, 146p (2000)
25. Y.-C. Tai, S. Noelle, J.M.N.T. Gray, K. Hutter: Shock capturing and front tracking methods for granular avalanches. J. Comput. Phys., (submitted)
26. A. Voellmy: Über die Zerstörungskraft von Lawinen, Schweizerische Bauzeitung, Jahrg. **73**, Hf 12, 159–162 (1955)
27. M. Wieland, J.M.N.T. Gray, K. Hutter: J. Fluid Mech. **392**, 73–100 (1999)

Part IV

Dirty

15 Patterns of Dirt

N.J. Balmforth[1] and A. Provenzale[2]

[1] Department of Applied Mathematics and Statistics, School of Engineering, University of California at Santa Cruz, CA 95064, USA
[2] Istituto di Cosmogeofisica, Corso Fiume 4, 10133 Torino, Italy; and ISI Foundation, V.le Settimio Severo 65, 10133 Torino, Italy

> Nor can the geomorphologist rest content, [...] until he knows *why* sand collects into dunes at all, [...] and *how* the dunes assume and maintain their own special shapes.
>
> R. A. Bagnold
> The Physics of Blown Sand and Desert Dunes [1]

15.1 Introduction

Many natural patterns appear when a simply structured equilibrium state is no longer preferred in comparison to a more complicated restructuring or rearrangement of the system. Our goal is typically a theoretical explanation or rationalization of the physical process, and invariably proceeds by way of mathematics; we formulate equations that describe the physical processes and seek to solve them in the appropriate context. Many standard techniques are available for the purpose to aid our analysis. For example, sometimes, hints about the patterns that will form can be extracted from a study of disturbances of infinitesimal amplitude, and so linear stability theory and decomposition into normal modes are our tools. Often, however, the ultimate, nonlinear mechanism of saturation is critical to selecting or shaping the forming pattern, and this cannot be revealed by linear stability analysis alone. Instead, we must advance into the nonlinear regime where we can use ideas from weakly nonlinear and dynamical systems theory complemented by numerical simulation.

In fluid flows, instabilities are common and many kinds of patterns are observed, ranging from convection cells and surface waves to meandering jets and vortices. Such instabilities arise purely from the hydrodynamics of the flow, but by transporting solid particles in suspension, they may further shape the walls or bed containing the fluid. This is one mechanism by which patterns may appear in geomorphology, but much more significant are new effects introduced by the interaction of the flow with the erodible bed over which the fluid runs. By this interaction, many new complex patterns arise.

In their full complexity, erosion and sedimentation result from the dynamical interaction between a turbulent fluid flow and the granular medium composing the bed. But we know only a little about turbulence, much less about granular media, and not much at all about their interaction. Consequently, the theoretical problems one must formulate to understand geomorphological patterns generated by sediment motion are all, at present, intractable. Instead, drastic simplifications must be tolerated, and in many instances empiricism necessarily replaces first principles. However, even though some of the basic aspects of this field are "built upon sand," in the last forty years much progress has been made, and our overall understanding of geomorphic patterns has significantly increased.

In this contribution, we briefly discuss some of these results, choosing those that are both simple and close to our hearts. We concentrate mainly on patterns that are found in association with waters running in sloping channels. Similar patterns are generated by the interaction of sea water and coastal sediments (such as sand waves, or tidal bedforms), and by the interaction of air and sand (aeolian bedforms, such as desert ripples and dunes) – and these are described in following chapters.

We embark with an introduction to the phenomenology of natural channel dynamics through a succession of images of bedforms, channel shapes and drainage networks. Our theoretical discussion then begins with a description of sediment transport and it continues with the formulation of the governing equations and an exploration of the linear stability problem for the simple case of a shallow water approximation. We pose briefly to describe an example of a purely fluid dynamical instability – the roll wave. Subsequently, we plunge into the dynamics of flows over an erodible, non-cohesive bottom, and we discuss some of the pattern-forming instabilities of the coupled fluid–dirt system. We hope that these specific examples will be of some use to start the journey into the realm of dirty flows, which continues further and deeper in chapters to follow.

15.2 Bedform Phenomenology

When a turbulent fluid flows down a channel having a non-cohesive bottom, composed by sediment such as sand that can be moved from one place to another, several things can happen.[1] A first important point concerns the bottom stress exerted by the flowing water. If the stress is large enough to lift the bottom material, then the interaction between the flow and the sediment can generate bedforms – patterns of sand on the bottom of the watercourse. These include ripples, dunes and anti-dunes (Fig. 15.1). Second, the water course itself can become disrupted or diverted by the large-scale redistribution of bed material. This creates a pattern in the shape of the watercourse, such as braids and meanders (Figs. 15.2, 15.3). Finally, on the grandest scale, there can be multiple

[1] In this discussion we consider only cases where gravity is a stabilizing factor that competes with the sediment mobilization induced by the stress exerted by the fluid. When the channel slope is large, gravity can become a destabilizing factor, and a large portion of the bottom sediment becomes entrained in the fluid. In such a situation, one speaks of a *debris flow*, a fluid composed of a mixture of water and sediment. One of the following chapters is entirely devoted to debris flows.

watercourses that compete in accumulating the overland flow, or that ally themselves and merge into wide rivers. The pattern now is the landscape itself, the complex terrain scoured by the running water of a drainage network (Fig. 15.4).

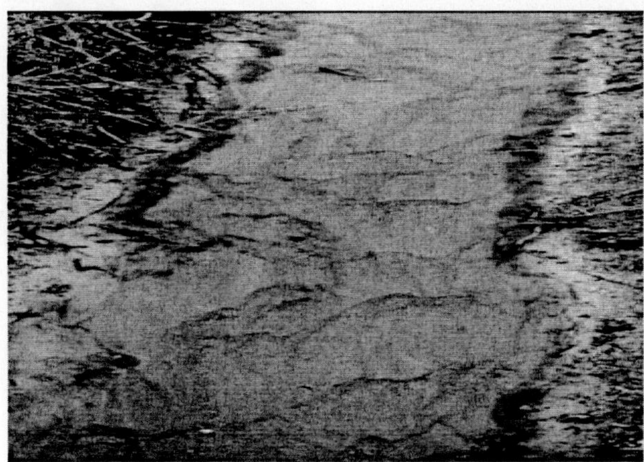

Fig. 15.1. Ripples and dunes on the bed of a small ditch in the countryside

Our main discussion will be on bedforms, and we now describe some of the phenomenology observed on the bottom of natural watercourses. However, the discussion is far from complete, and surrounds only the main types of bedform. General introductions to river geomorphology are given in [2,3]. A classic discussion of aeolian bedforms can be found in the book by Bagnold [1].

The two main quantities influencing the evolution of the bed are the bottom stress exerted by the fluid, τ_b, and the average size of the sediment, d. As we shall see later, their ratio enters an important non-dimensional quantity, the Shields number, that controls sediment transport. In channel flows, one often assumes that the bottom stress can be estimated in terms of the average flow rate, U, which is far easier to measure. This leads one to quantify sediment transport and therefore bed morphology in terms of the Froude number, $F = U/(g'H)^{1/2}$, where H is the average water depth, $g' = g\cos\phi$ is the projection of the gravitational acceleration perpendicular to the channel, and ϕ is the angle of inclination (typically, $g' \approx g$). The distribution of sediment sizes constitutes a far more complicated input to the problem. In general, very fine sediment can be mobilized by slow flows with small Froude number while larger sediment grains can only be moved by strong enough flows with large Froude number. In order to give a brief overview of bedform phenomenology, we concentrate here on the case of sand, with grains of typical size of about 0.1 mm and a fairly narrow distribution about that mean.

In this case, when the average flow speed is low and $F \ll 1$, not much happens. A flat bottom remains that way on average, and the sediment grains do not move much. At larger flow speed, the shear stress that the fluid exerts

Fig. 15.2. Meandering rivers. First two panels: Arctic coastal river. Third panel: Rio Ucayali river (Peru). These pictures are courtesy of the NASA Goddard Space Flight Center's Distributed Active Archive, http://daac.gsfc.nasa.gov/

Fig. 15.3. The braided Waimakariri river in New Zealand. Aerial photo by Bianca Federici, University of Genoa

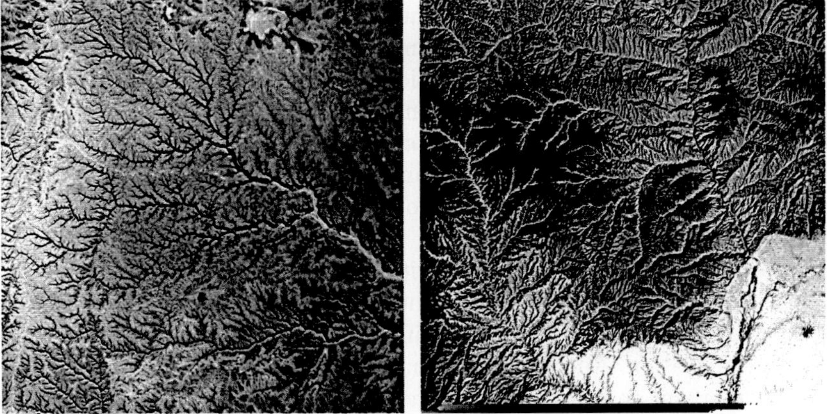

Fig. 15.4. Dendritic river networks. A network in Yemen and that creating the Huang He (Yellow) river. These pictures are courtesy of the NASA Goddard Space Flight Center's Distributed Active Archive, http://daac.gsfc.nasa.gov/

on the bottom becomes sufficient to mobilize the sediment, and the grains start to move. For $F < 1$ (the "subcritical" regime), the grains move mainly either by rolling or by taking small jumps from one location to another ("saltation"). This is called "bedload transport", and precipitates pattern formation. In this regime, two main types of bedforms are observed. The first type is a small corrugation of the bottom, with typical wavelengths of a few centimeters and amplitudes of a few millimeters. These bedforms are called ripples, and they can be seen for values of the Froude number close to incipient sediment motion. The other bedforms are called dunes, and they can be both longer and of larger amplitude than ripples. Dunes are asymmetric bedforms, with the side facing downstream (the "lee" side) being much steeper than the side facing upstream (the "stoss" side). Dunes can be either two-dimensional (i.e. with little variation in the cross-stream direction), or fully three-dimensional. The presence of dunes is also reflected in the shape of the free surface of the overlying water; the surface is typically depressed above a dune.

Both ripples and dunes move slowly downstream, at a speed (called celerity) that is much smaller than the fluid velocity. This is because erosion takes place on the stoss side of the dune, and sediment is deposited immediately after the crest, on the lee side. The speed of motion of a dune or ripple is determined by the effectiveness of bedload transport and the amount of material that must be moved (the volume of the bedform). For fixed water flux, the degree of bedload transport is roughly given by the surface area of the dune. Thus, larger dunes (or ripples) have lower bedload transport relative to volume, and so are slower than smaller dunes by a factor given roughly by the ratio of their respective lengths.

For still higher flow speeds, close to (but usually smaller than) the critical value $F = 1$, bedforms become either irregular and not well defined, or do not form at all and the bottom becomes flat again. This regime is typically associated with a change in the dominant form of sediment transport: The stress exerted by the flow is now sufficient to mobilize much of the sediment and keep it suspended in the turbulent fluid motions. Thus, rather than bouncing along, the sediment can be simply carried along within the body of the fluid, leading to the notion of "suspended load." Of course, the distinction between the two modes of transportation is not sharp, and in sediments with varying grain size one can observe the largest grains to move as bedload and the finest ones to be suspended in the fluid.

At the largest flow rates (typically, with $F > 1$, in the "supercritical" regime), a new type of bedform appears: The antidune. Antidunes are characterized by the fact that the free surface of the flowing water is approximately in phase with the bed perturbation. Another important difference is that antidunes can move either downstream or upstream (i.e. against the flowing water). For upstream-moving antidunes, erosion takes place mainly on the lee side of an antidune, and the grains are moved to the stoss side of the following antidune. Thus, although grains move downflow, the bedforms can migrate upstream. Antidunes are often associated with the presence of suspended load, which is able to carry particles between bedforms, but this is not always the case and rivers and laboratory flumes with dominant bedload transport also form antidunes.

Finally, ripples and dunes are observed on the surface of wind-blown sand. Antidunes, on the other hand, are typical of channel flows and are not observed either in wind-tunnel experiments or in the desert. This is because one cannot usually access the supercritical regime for blown dust or sand in the atmosphere: The height of the air layer is so large that the Froude number is much smaller than one. For instance, in a layer of height of 1000 m (the approximate height of the planetary boundary layer), and with a wind speed of 180 km/hr (typical of bad weather in Antarctica), one gets a Froude number as small as 0.05.

15.3 Moving a Sandy Bottom

15.3.1 Initiation of Sediment Motion

For a granular material on a slope, there is a critical angle (the "angle of repose") beyond which the material starts to avalanche downhill. The value of the angle of repose depends on the specific properties of the medium; for natural dry sand, it is about 30 degrees. In the case of a granular sediment immersed in water, the detailed mechanics of the sediment are much complicated by the presence of the overlying fluid, but the idea that the medium can become unstable and begin avalanching remains roughly the same – this is still crudely characterized by an angle of repose.

When a channel slope is less than this special angle, the bed only moves if the fluid can mobilize sediment particles. This requires the fluid to exert sufficient stress on a particle to displace it from bottom, which, as we have implied above, occurs when the flow is sufficiently strong. However, the problem of the initiation of erosion, or of the displacement of sediment grains, is an old and complicated one, that is still only partially understood, see for example [1–6], and the next chapter of the present volume.

Very crudely, for a horizontal cohesionless bottom composed by tightly packed sand grains, the bed becomes mobilized when the force exerted on a submerged particle exceeds the restoring force produced by gravity [1]. The main force exerted by the fluid on the grain is the surface drag force, F_D, that is due to the direct effect of the fluid velocity on the grain and can be obtained by integrating the stresses exerted by the fluid over the surface of the grain. The result is a force proportional to the bottom shear stress, τ_b, and to the square of the grain diameter, d: $F_D \approx \tau_b d^2$. This force competes with the downward force of gravity, $F_G = -g(\rho_s - \rho)\pi d^3/6$, where ρ_s is the density of the sediment and ρ is the fluid density. Their ratio provides the Shields number, defined as

$$\tau_* = \frac{\tau_b}{(\rho_s - \rho)gd} \,. \tag{15.1}$$

For values of the Shields number larger than a critical threshold τ_{cr}, the drag force overcomes gravity, and the sediment is set into motion. For $\tau_* < \tau_{cr}$, the drag exerted by the fluid is too low and the grains do not move. As shown by its definition, the Shields number is directly proportional to the stress exerted by the fluid but it is inversely proportional to the size of the sediment, d. This illustrates how slower flows can, in principle, mobilize finer sediments. However, the critical shear stress is also a function of particle size (see [1–6] and the following chapter).

As a final comment, we recall that the value of τ_b for a turbulent flow is a wildly variable quantity, characterized by strong spatio-temporal intermittency and huge departures from the average value. In turbulent wall flows, in particular, the dynamics is characterized by the presence of intense coherent structures, called bursts and sweeps, that violently take fluid from and toward the wall [7]. It is not yet clear what happens to sediment transport in real turbulent flows where the bottom shear stress violently fluctuates around the critical value [8]. This is matter of present research, and it is perhaps one of the basic issues to be solved in order to gain a more fundamental understanding of erosion and sedimentation.

15.3.2 Sediment Transport

Once set into motion, the sediment grains are transported by the fluid flow either as bedload or suspended load. In bedload transport, the sediment moves much more slowly than the fluid. At the moment, there is no fully satisfactory theoretical derivation of the bedload sediment flux as a function of the fluid and bottom properties. There are, however, a few empirical expressions. A well-known one is the Meyer–Peter and Muller formula [9],

$$J = \frac{q_b}{[(\rho_s/\rho - 1)gd^3]^{1/2}} = A\left[\tau_* - \tau_{cr}\right]^{3/2} \,, \tag{15.2}$$

where q_b is the modulus of the (bedload) sediment flux, J is its nondimensional counterpart, τ_* is the Shields parameter, τ_{cr} is the nondimensional critical shear stress, and A is an empirical constant; Meyer–Peter and Muller suggest $A = 8$ and $\tau_{cr} = 0.047$. Equation (15.2) only gives the value of the modulus of the bedload flux, and not its direction. To complete the formulation, the flux is usually taken in the direction of the fluid velocity, $\mathbf{u}$. On sloping bottoms, however, sediment inertia can deflect the bedload motion from the direction of the fluid flow, demanding revision of the formula.

When the fluid motion is more intense, the suspended load dominates the transport of sediment. Because the suspended particles follow the fluid in its rush downhill, the sediment is moved at approximately the same speed as the water, and fills the entire water column. In reality, suspended grains do not fully behave like fluid parcels, due to their relative inertia and finite size. Again, the forces on the suspended particles are very complex. However, all these complications are usually discarded, and the suspended grains are considered to move with the local fluid velocity, corrected to include the settling speed induced by the gravitational force acting on the grain. In this circumstance, a simple model of the sediment load is given by the conservation equation,

$$\frac{\partial C}{\partial t} + (\mathbf{u} - \mathbf{W}_s) \cdot \nabla C = \kappa \nabla^2 C \,, \tag{15.3}$$

where $C(\mathbf{x}, t)$ is the concentration, and $\mathbf{W}_s$ is the free-fall velocity for the sediment. The term on the right hand side represents an empirical diffusion and κ is the (phenomenological) eddy diffusivity of the sediment concentration field. A complicated boundary condition must be imposed on this equation at the lower boundary where there are sources and sinks of sediment due to erosion and deposition. Equation (15.3) does not have the status of an exact description, and is obtained by averaging over the random trajectories of the suspended grains. At the same time, many forces acting on the grains have to be discarded, as well as the grains' acceleration. The diffusive term in (15.3) tries to incorporate some of these effects. However, overall, the validity of the approach has not been verified in detail and the theory remains open to question – a concentration equation for particles with inertia has yet to be derived.

With prescriptions for bedload and suspended load, we are now ready to write an equation for the evolution of the bottom surface, that we denote by $\zeta(x, y, t)$. This is known as the Exner equation, and it is written as

$$(1 - \lambda_p)\frac{\partial \zeta}{\partial t} = -\nabla \cdot \mathbf{q}_b - \mathcal{E} + \mathcal{D} \,, \tag{15.4}$$

where λ_p is the sediment porosity and $\mathbf{q}_b$ is the sediment flux due to bedload transport. The terms, $\mathcal{E} - \mathcal{D}$, represent erosion and deposition of the suspended load, and can be rewritten using the vertical integral of (15.3), assuming no diffusive flux through the boundaries:

$$(1 - \lambda_p)\frac{\partial \zeta}{\partial t} = -\left(\nabla \cdot \mathbf{q}_b + \nabla \cdot \mathbf{q}_s + \frac{\partial Q_s}{\partial t}\right) \,, \tag{15.5}$$

where

$$Q_s = \int_\zeta^\eta C(x, y, z, t)\, \mathrm{d}z\ , \tag{15.6}$$

$$\mathbf{q}_s = \int_\zeta^\eta (\mathbf{u} - \mathbf{W}_s)\, C\, \mathrm{d}z\ , \tag{15.7}$$

and η is the free surface of the fluid. In most cases, the amount of suspended sediment is relatively small, and the time derivative of Q_s is usually discarded. This allows for writing the Exner equation as

$$(1 - \lambda_p)\frac{\partial \zeta}{\partial t} = -\nabla \cdot \mathbf{q}\ , \tag{15.8}$$

where $\mathbf{q} \equiv \mathbf{q}_b + \mathbf{q}_s$ is the total load. Moreover, rather than deal with the full transport equation in (15.3), many theorists revert to empirical formulae for the total load, in which case $\mathbf{q}$ becomes a known function of the fluid flow.

At this stage, we have the ingredients that are needed to study how the erodible bottom responds to the overlying fluid motion. Together with a prescription for the fluid dynamics, we are then set to explore whether bedforms can be generated as a result of a linear instability. In the next section, we specify a simple fluid model – known as the shallow water approximation – and, in this idealized framework, we couple the Exner equation to the fluid dynamics.

15.4 A Shallow World

Before attempting a mathematical description of the interaction between a fluid flow and its erodible bed, it is essential to appreciate that the flow of interest is fully turbulent: For a channel with depth $H = 1\,\mathrm{m}$, fluid velocity $U = 3\,\mathrm{ms}^{-1}$, and water viscosity $\nu \approx 10^{-6}\,\mathrm{m}^2\mathrm{s}^{-1}$, one gets a Reynolds number, $R = UH/\nu$, larger than 10^6. By all standard yardsticks (critical Reynolds numbers for pipe and channel flows in the laboratory) this places the flow in the fully developed turbulence regime. Thus, the equations we seek are in the realm of Reynolds averaged equations (approximate equations in which one takes into account mean, large-scale properties of the turbulent flow and roughly parameterizes the effect of small-scale turbulent fluctuations), and we have to cope with standard closure problems.

A simplified model that has often been used in past studies of channel and river flows is the one-dimensional shallow-water approximation for a fluid flowing down a channel with constant (and small) slope $S = \tan\phi$. The configuration is sketched in Fig. 15.5 and it is characterized by the water depth, $h(x, t)$, the vertically averaged fluid velocity, $u(x, t)$, and the elevation of the surface of the erodible bed, $\zeta(x, t)$, where t is time and x is the downstream spatial coordinate. There is no lateral variability, and the dynamics takes place only in the downstream direction. Of course, we know that the turbulent motion is indeed fully three dimensional, and that there are strong vertical motions near the bottom that make a shallowness assumption quite unrealistic. In the shallow-water theory, all of these unknowns become synthesized into eddy diffusion and drag terms in the momentum equations for the fluid, and into the erosion and deposition law for the sediment. The description can therefore only be a pale metaphor of reality. Encouraged by these considerations, we proceed.

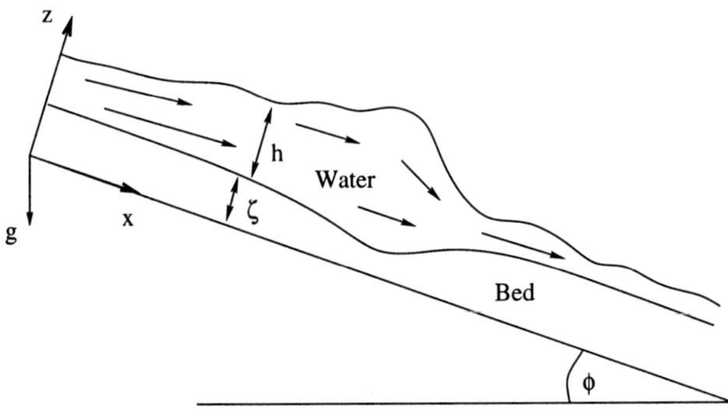

Fig. 15.5. A sketch of the configuration for a vertical slice of a fluid flowing down a channel with erodible bed

The shallow-water equations for the model can be derived as a crude approximation of the governing Navier–Stokes equations: We assume that vertical lengthscales are far smaller than horizontal scales, that the pressure distribution is therefore almost hydrostatic, and vertically average the momentum equations. This procedure is identical to that used for any kind of shallow fluid layer, but at this juncture, the approaches for inviscid or viscous, laminar or turbulent, fluid all diverge. In inviscid theory one postulates that the horizontal velocity is independent of the vertical coordinate or that flow is irrotational (either of which remain true if the initial conditions are so) and one obtains a closure of the system of equations. For very viscous fluids, one can close the equations by choosing an appropriate balance of terms in the scheme of an asymptotic expansion (this is the lubrication approximation that is used for the relatively slow motion of glaciers, lava and other geophysical fluids elsewhere in this volume). However, for turbulent fluid, neither approach is suitable and there is really no justifiable step forward. Instead, we must replace asymptotic expansion by a crude closure argument. Various options have been chosen in the past (heuristic reasoning, Reynolds averaging, von Karman's power integral method, Galerkin projection amongst others), some of which lead to the equations (see e.g. [10]),

$$\frac{\partial u}{\partial t} + u\frac{\partial u}{\partial x} = g'\left(S - \frac{\partial h}{\partial x} - \frac{\partial \zeta}{\partial x}\right) - C_f\frac{u^2}{h} + \frac{1}{h}\frac{\partial}{\partial x}\left(h\tilde{\nu}_t\frac{\partial u}{\partial x}\right) \tag{15.9}$$

and

$$\frac{\partial h}{\partial t} + \frac{\partial}{\partial x}(hu) = 0\,, \tag{15.10}$$

where $g' = g\cos\phi$, the drag term $C_f u^2/h$ represents the stress exerted on the fluid by the bed, and $\tilde{\nu}_t$ is an eddy viscosity. There are empirical estimates of the friction coefficient, C_f, used in the drag formula, which is often referred to as the Chezy formula. Equation (15.9) is also known as the Saint Venant equation.

Next, we must specify the dynamics of erosion of the bed. That is, we require an evolution equation for $\zeta(x,t)$. We use the one-dimensional form of the Exner equation:

$$(1-\lambda_p)\frac{\partial \zeta}{\partial t} + \frac{\partial q}{\partial x} = 0\,, \tag{15.11}$$

where $q(u,h,s)$ represents the sediment flux and is an empirical function of the flow speed and depth, and of the local bed slope, s (consisting of the mean slope S plus a correction due to the bed shape). We do not yet need to specify the functional dependence of q, and leave open the choice of sediment transport law. Also, at this stage we do not make a distinction between bedload and suspended load.

In the case of a flat bottom ($\zeta = 0$), (15.9) admits a family of stationary and homogeneous equilibrium solutions where the flow has constant velocity U and depth H. We set all terms containing spatial or temporal derivatives to zero, leaving

$$g'S = C_f\frac{U^2}{H}\,. \tag{15.12}$$

This is a relation between the fluid velocity and the water depth, which prescribes the equilibrium flow given the total downslope water flux, $Q = HU$. That homogeneous state owes its existence to a balance between the forcing provided by the gravitational acceleration, which pushes the fluid down the channel, and the effective friction from the stress on the bed, which brakes the fluid. In a more realistic description, effects of turbulent motion may be more varied, but in this vertically-averaged approach the effective friction is conceptually similar to the friction exerted by the air on a falling body, and the flow rolls down smoothly at a suitably defined "terminal velocity."

The properties of the homogeneous state allow for naturally introducing dimensionless variables. Different choices are possible, and here we use

$$x \to x' = x/L\,,\ \ u \to u' = u/U\,,\ \ t \to t' = tU/L\,, \tag{15.13}$$

$$(\eta,\zeta,h) \to (\eta',\zeta',h') = (\eta/H,\zeta/H,h/H)\,, \tag{15.14}$$

where L represents the horizontal scale of motion. To be consistent with the shallow-water assumption we require only that $H/L \ll 1$; a convenient choice is $L = H/S$. With these definitions, and after dropping the prime decoration on the dimensionless variables for notational ease, we obtain

$$F^2\left(\frac{\partial u}{\partial t} + u\frac{\partial u}{\partial x}\right) = 1 - \frac{u^2}{h} - \frac{\partial h}{\partial x} - \frac{\partial \zeta}{\partial x} + \frac{1}{h}\frac{\partial}{\partial x}\left(h\nu_t\frac{\partial u}{\partial x}\right)\,, \tag{15.15}$$

$$\frac{\partial h}{\partial t} + \frac{\partial}{\partial x}(hu) = 0 \quad \text{and} \quad \frac{\partial \zeta}{\partial t} + \varepsilon\frac{\partial J}{\partial x} = 0\,, \tag{15.16}$$

where the Froude number is

$$F = \frac{U}{(g'H)^{1/2}} = \frac{S}{C_f} , \tag{15.17}$$

the dimensionless eddy diffusivity is ν_t (as measured by the unit UL/F^2), ε is a dimensionless parameter given by the ratio of the characteristic sediment flux to the water flux ($\varepsilon = (1-\lambda_p)^{-1}[q]/HU$ or $(1-\lambda_p)^{-1}V_s/U$, where V_s is a characteristic sediment transport speed), $J(u,h,s)$ is the sediment transport law in dimensionless units (with characteristic values of unity), and $s = |1-\zeta_x|$ is the local slope.

Equations (15.15)–(15.16) have stationary solutions characterized by $u = h = constant$ and $\zeta = 0$. The relevant solution is that used to non-dimensionalize the problem, $(h,u,\zeta) = (1,1,0)$. A standard approach is then to explore the linear stability of the homogeneous state by setting $(h,u,\zeta) = (1,1,0)+(\hat{h},\hat{u},\hat{\zeta})$, where $(\hat{h},\hat{u},\hat{\zeta})$ are infinitesimally small. By substituting these expressions in (15.15)–(15.16) and retaining only linear terms one obtains

$$F^2\left(\frac{\partial \hat{u}}{\partial t} + \frac{\partial \hat{u}}{\partial x}\right) = -\frac{\partial \hat{h}}{\partial x} - \frac{\partial \hat{\zeta}}{\partial x} + \hat{h} - 2\hat{u} + \nu_0 \frac{\partial^2 \hat{u}}{\partial x^2} , \tag{15.18}$$

$$\frac{\partial \hat{h}}{\partial t} + \frac{\partial \hat{u}}{\partial x} + \frac{\partial \hat{h}}{\partial x} = 0 \qquad \text{and} \qquad \frac{\partial \hat{\zeta}}{\partial t} + \varepsilon\left[J_u \frac{\partial \hat{u}}{\partial x} + J_h \frac{\partial \hat{h}}{\partial x} - J_s \frac{\partial^2 \hat{\zeta}}{\partial x^2}\right] = 0 , \tag{15.19}$$

where $\nu_0 = \nu_t(u_0, h_0)$ is the value of the equilibrium eddy diffusivity, and the subscripts on J indicate the partial derivatives with respect to u, h and slope, evaluated for the equilibrium state (that is, $J_u = [\partial J/\partial u]_{u=1,h=1,s=1}$, and so on).

To proceed further, we look for normal modes with an exponential dependence on space and time: $(\hat{u},\hat{h},\hat{\zeta}) \propto \exp ik(x-ct)$, where k is the downstream wavenumber and c is the wavespeed. On substituting the form of the solution into the linear equations, and after a little algebra, we obtain the dispersion relation,

$$[2 + \nu_0 k^2 + ikF^2(1-c)](1-c) + 1 - ik = ik\varepsilon \frac{J_h - (1-c)J_u}{c + ik\varepsilon J_s} , \tag{15.20}$$

that implicitly determines c as a function of k and the parameters of the problem. Because (15.20) is a cubic, there are three types of normal modes. If $c = c_r + ic_i$ turns out to have positive imaginary part, the normal mode grows exponentially in time; for $c_i = 0$, the mode is neutrally stable. The equilibrium is stable only if $c_i < 0$ for all three modes and for all values of k. The real part of c determines the propagation speed of the mode: If $c_r > 0$, the perturbation propagates with the flow in the positive x direction, while $c_r < 0$ indicates propagation upstream. If $c_r > 1$, the perturbation propagates faster than the flow (i.e. at a speed that is larger than the fluid velocity).

A key property of the problem we are considering is that, for bedload transport, the water flows much faster than the sediment. Indeed, we often observe that low-amplitude waves on the surface of a river are brief and insignificant as far as the rearrangement of the bed is concerned. And ripples and dunes form far more gradually than the water rushes over them. In mathematical terms, this means that the parameter ε is usually small. The smallness of this parameter allows us to solve the cubic dispersion relation in an asymptotic fashion and to decode the physics that is woven into it.

First, given that $\varepsilon \ll 1$, we are tempted to simply ignore the right-hand side of (15.20). The result is a quadratic equation for c:

$$[2 + \nu_0 k^2 + ikF^2(1-c)](1-c) + 1 - ik \approx 0\,. \tag{15.21}$$

On solving this quadratic we observe that the corresponding values for $|c|$ are order one. Thus we uncover directly two of the modal solutions. These are rapidly evolving modes and correspond to water waves that are barely modified by the erodible bed. Indeed, they are the solutions that we would have found had there been no erosion of the bed at all; (15.21) is the dispersion relation for the non-erosive problem. The erosion generated by these water waves is recovered on looking at the higher-order terms in ε:

$$\zeta = \frac{[(1-c)J_u - J_h]}{(1-c)(c + ik\varepsilon J_s)}\varepsilon u\,. \tag{15.22}$$

Thus the bed perturbation remains small.

The third solution that is missing from the approximate quadratic is a slowly evolving or "sediment" mode. We uncover this solution by observing that if $c \sim \varepsilon$, we can no longer neglect the right-hand side of (15.20). Instead, we find another solution,

$$c \approx \frac{ik\varepsilon(J_h - J_u)}{3 + \nu_0 k^2 + ik(F^2 - 1)} - ik\varepsilon J_s\,. \tag{15.23}$$

This slow mode describes the diffusive-like evolution of the bed. Because $c \sim \varepsilon$, the time derivatives in the momentum and continuity equations are always small for this mode. Thus the water evolves "quasi-statically" as the bed erodes or builds up through deposition.

Each of the different kinds of modes can, in principle, be unstable. Unstable fast-modes correspond to purely hydrodynamical instabilities, or roll waves as they are often called. Unstable sediment modes, on the other hand, could create bedforms such as dunes and antidunes. Hence by exploring the linear stability of the diffusive sediment mode we may begin to rationalize bedform dynamics.

15.5 Rolling Shallowly Downhill

First we consider the purely hydrodynamic instability which occurs when the fluid velocity (equivalently, the Froude number) is large enough. As these perturbations propagate and grow, they develop into a series of fronts, or shocks, that propagate downstream – the roll waves. Photographs of roll waves in a long rectangular open channel were first published by Cornish [11].

As described above, for roll waves we can neglect the disturbance of the bed (for $\varepsilon \ll 1$). In linear theory, the waves are then described by the dispersion relation in (15.21). This equation is quadratic in c, and so there are two different solutions for c for any value of the controlling parameters and of k. One wave moves slower than the mean flow and is always decaying ($c_i < 0$). However, the other wave moves faster than the mean flow and can be unstable, depending on the values of F, ν_0 and k. The inception of instability can be detected by demanding that the wavespeed is purely real: $c = c_r$ and $c_i = 0$. Then the real and imaginary parts of the dispersion relation provide the two equations,

$$(2 + \nu_0 k^2)(c - 1) = 1 \qquad \text{and} \qquad F^2(1 - c)^2 = 1\,. \tag{15.24}$$

The latter implies that $c = 1 \pm F^{-1}$. The slower solution, $c = 1 - F^{-1}$, cannot satisfy the first relation in (15.24) for positive viscosity. However, the faster solution, $c = 1 + F^{-1}$, leads to $2 + \nu_0 k^2 = F$. A little more algebra shows further that all modes with $\nu_0 k^2 < F - 2$ are unstable. Notably, if $\nu_0 = 0$, all wavenumber are unstable for $F > 2$, which implies that the problem without eddy diffusivity is potentially ill-conceived.

From a physical perspective, the instability has a simple explanation: Consider an initial perturbation in which we rearrange the fluid so that the water depth is depressed at some point, but is elevated just downstream. Because the friction coefficient depends on the inverse of the water depth, the fluid elements with the elevated surface experience less friction; they accelerate downstream. Behind them, the fluid elements where the surface is depressed experience greater friction and slow down. As a result, the surface swell picks up speed and propagates downstream, collecting fluid in its passage and leaving behind a shallower layer.

In summary of linear theory, for $F < 2$, the homogeneous state is stable, and travelling perturbations decay. For $F > 2$, the homogeneous state becomes unstable, and small perturbations that travel faster than the flow grow exponentially. This instability is how we rationalize mathematically the emergence of roll waves. However, their finite-amplitude behaviour must be explored by other means, particularly since stability theory predicts a wide band of unstable modes when ν_0 is small and there is no way to select the characteristic separation of roll waves. To that end, steady finite-amplitude waves were constructed by Dressler [12] and Needham & Merkin [13,14], and weakly nonlinear results were obtained by Kranenburg [15] and Yu & Kevorkian [16]. These results were compared favourably to full numerical simulations of the non-erosive equations by Brook et al. [17].

Kranenburg [15] suggested an image for the behaviour of finite-amplitude roll waves: From a generic initial field, linear instability nucleates growing roll waves. These roll waves steepen into a sequence of shock-like objects. In general, because the different crests are not equally spaced, the individual shock-like roll waves experience different water depths. The waves that begin in deeper water move faster as they feel a reduced drag with the bottom. This allows a stronger roll wave to catch up with a smaller, slower moving roll wave lying immediately downstream; the two waves then collide and merge into a single wave.

The merged waves then accelerate downstream, catching up with more distant companions, leading to further collisions and mergers on a longer timescale. This "coarsening" of the roll-wave pattern continues inexorably until only a single roll wave remains with the size of the simulation domain. Thus the system displays a tendency toward generating structures on the largest possible spatial scale – an "inverse cascade."

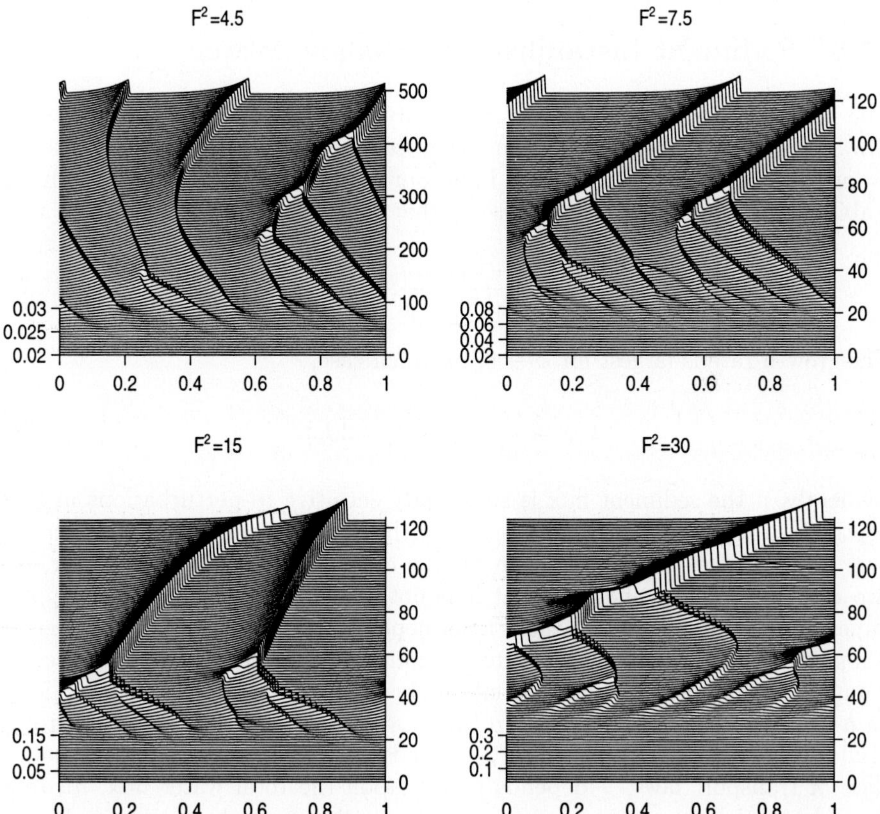

Fig. 15.6. Nonlinear evolution of the free surface for the erosionless shallow water system discussed in the text. Each panel shows a case with different Froude number. The downstream drift of the whole pattern is subtracted in each case, so that we can easily observe the coarsening dynamics of the roll-wave patterns. The system has been numerically integrated by a finite difference code with third-order predictor-corrector time stepping and second-order accuracy in the spatial derivatives. The initial condition is low-amplitude random noise. The horizontal scale indicates distance in the unit periodic box (0,1), the vertical scale on the left of the figures indicates wave amplitude, the vertical scale on the right of the figures indicates time in nondimensional units. The numerical simulations were performed by Jost von Hardenberg, University of Oxford

Figure 15.6 shows solutions obtained by numerically integrating the full hydrodynamic equations (15.15)–(15.16) without erosion ($\zeta = 0$) from an initial condition consisting of a random perturbation away from the homogeneous state. We show the results for four values of the Froude number, all larger than two. The perturbations grow and form roll waves; these waves then merge with one another and eventually generate a single roll wave with the spatial scale of the simulation box (see also [18]).

15.6 Sediment Instabilities in Shallow Water

The previous discussion focusses on hydrodynamic instabilities, the roll waves. However, it is the slow, diffusive modes of sediment redistribution that are relevant to the creation of bedforms. From our approximate solution of the linear stability problem in (15.23), we observe that these modes have the growth rate,

$$kc_i \approx \varepsilon k^2 \left[\frac{(3+\nu_0 k^2)(J_h - J_u)}{(3+\nu_0 k^2)^2 + k^2(F^2-1)^2} - J_s \right] . \qquad (15.25)$$

The growth rate is largest for $k \to 0$, in which case,

$$kc_i \to \frac{\varepsilon k^2}{3}(J_h - J_u - 3J_s) . \qquad (15.26)$$

Evidently, if the sediment flux is sufficiently sensitive to perturbations in fluid depth ($J_h > J_u + 3J_s$), sediment modes with long wavelength (small wavenumber) are unstable. This mechanism for the creation of bedforms, however, has two drawbacks which indicate that it is not a tenable explanation for ripples, dunes and anti-dunes. First, there is no dependence on the Froude number, in complete contrast with the bedform phenomenology described earlier. Second, many of the commonly used sediment flux laws have at most a weak dependence on h and thus give no instability in this way. For example, the Meyer–Peter & Muller formula (15.2) implies that J depends only on u (so $J_h = 0$); in another class of transport laws, J depends purely upon the total water flux, hu (and $J_u = J_h$). Either way, it is impossible for the sediment mode to be unstable.

We illustrate further with a specific example: Consider the explicit form for the total sediment load given by Coleman & Fenton [19] which summarizes the Engelund and Hansen formula [20] and is fit to the experiments of Jain & Kennedy [21]:

$$q = \frac{0.05U^2(HS)^{3/2}}{d(\rho_s/\rho - 1)^2 g^{1/2}} u^2 (h|1-\zeta_x|)^{3/2} , \qquad (15.27)$$

where ρ_s/ρ is the ratio of sediment to water density, and d is a characteristic sediment particle size. Thence,

$$\varepsilon = \frac{0.05UH^{1/2}S^{3/2}}{d(1-\lambda_p)(\rho_s/\rho - 1)^2 g^{1/2}} . \qquad (15.28)$$

For one of Jain & Kennedy's experiments, Coleman & Fenton quote $\rho_s = 2.65\rho$, $d = 2.5\times10^{-4}$ m, $H = 0.0771$ m, $U = 0.3847$ m/s and $S = 0.00267$, which implies that $\varepsilon \approx 2.5 \times 10^{-4}/(1-\lambda_p)$, and is very small as we remarked earlier. Given this data, we may also independently estimate the drag coefficient,

$$C_f = \frac{gHS}{U^2} \approx 1.4 \times 10^{-2}\;; \tag{15.29}$$

this is consistent with values typically quoted in the literature. Also $Re = UH/\nu \approx 3 \times 10^4$, so the flow is presumably fully turbulent. For this transport law, we find that $J_u = 2$, $J_h = 3/2$ and $J_s = 3/2$, which implies that the sediment mode is stable. This is confirmed by solving the eigenvalue equation, as shown in Fig. 15.7. Importantly, because of the bottom friction, the three types of modes are all well separated on the spectral plane for all values of k. This is unlike the problem considered by Coleman & Fenton, and as a result there is no chance of an instability (for these parameter values) occuring due to mode interaction.

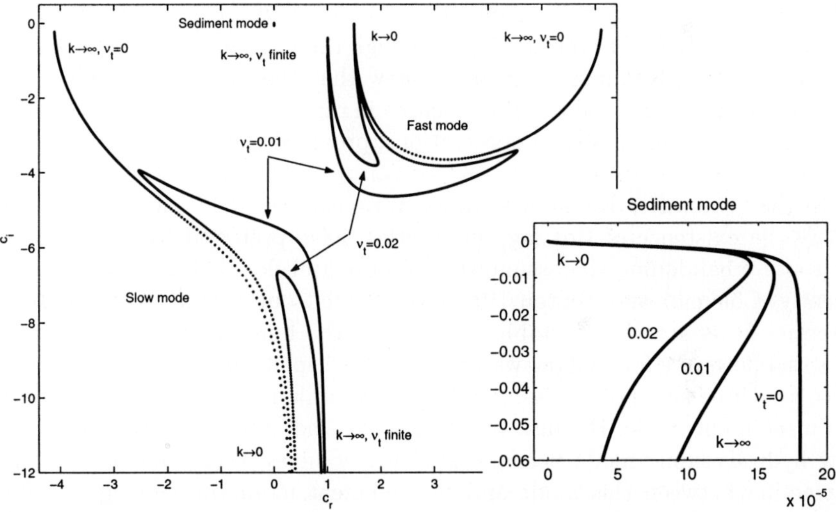

Fig. 15.7. Solution of the dispersion relation. The values of c are shown for the different modes with k ranging from 0.05 to 125, for three values of ν_0 (0, 0.01 and 0.02). Somewhat curiously, the large k limit of the fast mode is $c \to 1$ for finite ν_0. $F^2 = 0.2$, $J_u = 2$, $J_h = J_s = 3/2$ and $\varepsilon = 2.5 \times 10^{-4}$

Numerical simulations of the full equations (15.15)–(15.16) including erosion also fail to show any instability. For $F < 2$, one finds that there is no instability; for $F > 2$, the main effect of the transport of sediment is to modify roll waves and their dynamics slightly, but not qualitatively. There is some rearrangement of the bed, with the roll wave dredging sediment with it as it propagates.

15.7 Lagging Behind

Given that our simple model fails to incorporate the necessary physics of bedform pattern formation, we must next ask what is missing. Many effects have been left out. One possibility is that the shallow water theory does not contain sufficient information about the vertical structure of the instability to capture the physics. Another option is that we have been using a bad parameterization for the erosion law.

Vertical structure can be taken into account in another way if we consider irrotational motions in inviscid fluid. This approach provides an alternative to the shallow-water theory that keeps the mathematical analysis relatively simple. In this circumstance, the addition of rearrangements of the bed through the Exner equation constitutes a straightforward generalization of classical theory of water waves, and can be used to try to detect bedform instability. Unfortunately, Kennedy [22] found no instabilities in this problem. To resolve the issue, he proposed that, if the bed perturbation lagged behind the fluid perturbation by an angle $k\delta$, instability would occur provided that δ lay within certain ranges. The physics of the phase lag was not worked out in detail, and is evidently not present in the Exner equation. In this case, we write the sediment flux as $J \to J[u(x+\delta,t), h(x+\delta,t), \zeta(x,t)]$. Although the introduction of this phase lag is rather ad hoc, Kennedy went on to show that the modified stability theory predicted bedform instabilities that occured in ranges of wavenumber and Froude number that roughly matched observations, provided the phase lag was suitably chosen.

In the years after Kennedy's works, various researchers have attempted to justify the existence of this lag, while other have preferred to follow a different route, abandoning the assumption of potential flow [23–25]. Actually, very recently, Coleman and Fenton [19] have pointed out that it is not true that potential flows are always stable. They solve the linear stability equations for potential flows. As for shallow-water flows, the dispersion relation is a cubic and there are three types of modes: Two hydrodynamic modes (one slow, one fast), and a sediment mode. Because the fluid is inviscid, when the wavespeed of the slow hydrodynamic mode becomes sufficiently small, there can be a resonant interaction between this mode and the sediment mode, promoting instability. Some rationalization for bedforms in laboratory experiments is offered, but the theory does not give a definitive explanation of bedforms in general.

Mode interactions do not occur in shallow water theory because of the importance of the bottom drag (which separates all the eigenvalues, see Fig. 15.7). This highlights a key difference between the two approaches, even though both, in principle, apply to high Reynolds number fluids. The potential flow approach neglects turbulent dissipation altogether, and treats the dynamics as though it were purely inviscid. On the other hand, the shallow-water (or, better, Saint–Venant) approach admits the existence of dissipation and turbulence and crudely parameterizes their effects. These are two quite different approaches that can be relevant in different limits of the fluid problem. For example, it is often argued that inviscid fluid dynamics is relevant to phenomena in high Reynolds number flows when the timescales of interest are relatively short. In this circumstance, there is no time for a turbulent cascade to channel energy down to the dissipative scales. At the other extreme, for relatively slowly evolving phenomena (compared to the typical turn-over times of eddies), the turbulent motions may be in some instantaneous equilibrium state in which energy continually cascades to the dissipation scale to create an effective friction. In this limit, the braking action of turbulent motions are clearly essential, and simple parameterizations of the drag may be useful. This is the Saint-Venant regime. For bedform processes, the evolutionary timescales are relatively long, which suggests that the Saint–Venant approach may be the more plausible of the two. (An even simpler argument is that, over the timescales of interest and on a slope, inviscid fluid would continually accelerate, precluding a steady downflowing equilibrium.)

As in potential flow theory, we can explore the effect of an artificial phase lag in the shallow-water analysis. Following Kennedy, we simply adopt the modified sediment transport law, and omit a discussion of the difficult matter of its physical origin. The normal-mode analysis proceeds as before except that we must replace J_u and J_h by $J_u \mathrm{e}^{\mathrm{i}k\delta}$ and $J_h \mathrm{e}^{\mathrm{i}k\delta}$. We then obtain a dispersion relation like (15.20), and the eigenvalue of the sediment mode becomes

$$c + \mathrm{i}k\varepsilon J_s \approx \frac{\mathrm{i}k\varepsilon(J_h - J_u)}{\sqrt{(3+\nu_0 k^2)^2 + k^2(F^2-1)}} \mathrm{e}^{\mathrm{i}(\Delta + k\delta)} , \tag{15.30}$$

where

$$\tan \Delta = \frac{k(F^2-1)}{3+\nu_0 k^2} . \tag{15.31}$$

Hence, the growth rate takes the form,

$$kc_i \approx \frac{\varepsilon k^2 (J_h - J_u)}{\sqrt{(3+\nu_0 k^2)^2 + k^2(F^2-1)}} \cos(\Delta + k\delta) - \varepsilon k^2 J_s . \tag{15.32}$$

For instability,

$$(J_h - J_u)\cos(\Delta + k\delta) > J_s \sqrt{(3+\nu_0 k^2)^2 + k^2(F^2-1)} . \tag{15.33}$$

Because of the phase lag, it is no longer necessary that $J_h > J_u$ for instability (although the instability condition cannot be satisfied unless $|(J_h - J_u)| > 3J_s$). Curiously, if $J_s = 0$, the flow is always unstable, and the growth rate is an oscillating function of wavenumber with the limit, $kc_i \to (\varepsilon/\nu_0)(J_h - J_u)\cos k\delta$ as $k \to \infty$. For small ν_0, the higher wavenumber modes are even the most unstable. Neither feature would seem particularly desirable from a physical perspective. The theory is to some degree saved by the slope-dependence of the sediment flux: By using a transport law with $J_s \neq 0$ we can eliminate many of the ranges of unstable wavenumber, and reduce the wavenumbers of the most unstable modes. Nevertheless, the problem of wavelength selection still appears particularly severe in this model, a feature also brought out by Coleman & Fenton for potential flow with a phase lag. Thus, whereas an arbitrary phase lag can permit unstable sediment modes where none previously existed, it can also create just as many problems as it cures.

The above results indicate that a better treatment of the turbulent flow and transport is needed in order to detect bed instabilities. Along these lines, shortly after Kennedy's work, several authors [23–25] studied the linear instability of flows with erodible beds, treating turbulence by means of a simple eddy diffusion, and sediment transport using empirical formulae. By directly considering the stability of the vertical flow profile of the steady solution, these explorations avoid any crude vertical averaging of the sort that goes into the Saint-Venant model. However, the price one pays for this generalization is that one must then solve differential equations for the vertical structure of each normal mode together with complications regarding how one deals with the movable lower boundary. Engelund [23] considered both bedload and suspended load transport and predicted linear instabilities with the lengthscales of dunes. However, finite growth rates were again predicted for bed waves of infinitesimal wavelength. This problem was cured by Fredsoe [24] who incorporated a slope-dependence into the sediment flow law (as we exploited $J_s \neq 0$ in the Saint-Venant model). Further detailed computations were performed by Richards [25], who found two ranges of maximally unstable wavelengths that he associated with ripples and dunes.

These results suggest that one can explain the formation of certain types of bedforms if one introduces a more detailed parameterization of turbulence. However, though plausible, there is no guarantee that such parameterization is any more accurate that the physics captured in our shallow-water model – turbulence is the big unknown ingredient in all of the theory. Moreover, for particle saltation in water, Wiberg & Smith [26] suggest that the dependence of the sediment transport on local bed slope, although present, may be too small to have an appreciable effect on bed instabilities (see also [27]). Thus, all is by no means solved here.

15.8 Landscaping

Up to this point, we have considered only one-dimensional situations. However, real bedforms are two-dimensional. In addition, there are other phenomena – such as bars, braids, meanders and channelization – that demand a two-dimensional description.

The model we explore here is the two-dimensional version of the Saint Venant picture. The setup is now a sloping plane in which x points down the slope and y across the slope. The governing equations are

$$F^2(u_t + uu_x + vu_y) = 1 - \frac{\sqrt{u^2+v^2}}{h}u - (h_x + \zeta_x) + \frac{1}{h}[\partial_x(h\nu_t u_x) + \partial_y(h\nu_t u_y)] , \tag{15.34}$$

$$F^2(v_t+uv_x+vv_y) = -\frac{\sqrt{u^2+v^2}}{h}v-(h_y+\zeta_y)+\frac{1}{h}[\partial_x(h\nu_t v_x)+\partial_y(h\nu_t v_y)] , \tag{15.35}$$

$$h_t + (hu)_x + (hv)_y = 0 \qquad \text{and} \qquad \zeta_t + \varepsilon \nabla \cdot \mathbf{j} = 0 \, , \tag{15.36}$$

where the sediment flux,

$$\mathbf{j} = \frac{(u, v)}{\sqrt{u^2 + v^2}} J \left(\sqrt{u^2 + v^2}, h, \sqrt{(1 - \zeta_x)^2 + \zeta_y^2} \right) \, , \tag{15.37}$$

is assumed to lie in the same direction as the water flow, and to depend only on water speed, depth and local slope. More complicated expressions can be introduced.

15.8.1 Two-dimensional Instabilities

Again there is a basic state with $u = h = 1$ and $\zeta = v = 0$. In linear theory, we explore perturbations to this state of infinitesimal amplitude: Let $(h, u, v, \zeta) \to (1, 1, 0, 0) + \epsilon(\tilde{h}, \tilde{u}, \tilde{v}, \tilde{\zeta}) \exp[ik(x - ct) + ily]$. Then a little algebra provides the dispersion relation,

$$(2 + \Gamma)[(1 + \Gamma)(1 - c) - \kappa^2] + (1 - ik)[1 + \Gamma + (2 + \Gamma)\kappa^2] = ik\varepsilon\mathcal{F} \tag{15.38}$$

where $\Gamma = \nu_0(k^2 + l^2) + ikF^2(1 - c)$, $\kappa = l/k$ and

$$\mathcal{F} = \frac{J_h[1 + \Gamma + \kappa^2(2 + \Gamma)] - J_u[(1 + \Gamma)(1 - c) - \kappa^2] - \kappa^2[(2 + \Gamma)(1 - c) + 1]}{c + ik\varepsilon J_s} \, . \tag{15.39}$$

For $\varepsilon \ll 1$, the left-hand side of (15.38) provides the simpler dispersion relation of the two-dimensional erosionless problem. It is a cubic equation, reflecting how a third hydrodynamic mode now appears. However, as before, when $\varepsilon \ll 1$, the hydrodynamic modes are all much faster than the sediment mode, which has the approximate eigenvalue,

$$c + ik\varepsilon J_s = ik\varepsilon \frac{J_h(1 + \Gamma + 2\kappa^2 + \kappa^2\Gamma) - J_u(1 + \Gamma - \kappa^2) - \kappa^2(3 + \Gamma)}{(3 + \Gamma)(1 + \Gamma) - ik(1 + \Gamma + 2\kappa^2 + \kappa^2\Gamma)} \, , \tag{15.40}$$

with $\Gamma \approx \nu_0(k^2 + l^2) + ikF^2$.

A number of different types of instabilities are encoded in this formula, and manifest themselves in different ways depending on the sediment transport formula that is employed. For example, using the sediment transport relation (15.27) and the values quoted by Coleman & Fenton, we derive the results shown in Fig. 15.8. Although perturbations with purely streamwise structure are stable, those with lateral structure are unstable. The stability theory also predicts that these objects should migrate upstream (see panel (a) of Fig. 15.8). The maximal growth rates occur for values of l of around unity, with the corresponding values for k being smaller by a factor of three or so ($\kappa \approx 3$). Thus these instabilities would be somewhat longer than they were wide (length measured downstream).

Related instabilities with other transport laws have been used to rationalize the appearance of classes of two-dimensional bedforms, namely, bars and braids in rivers [28–30]. Alternating bars are sedimentary features in rivers with roughly

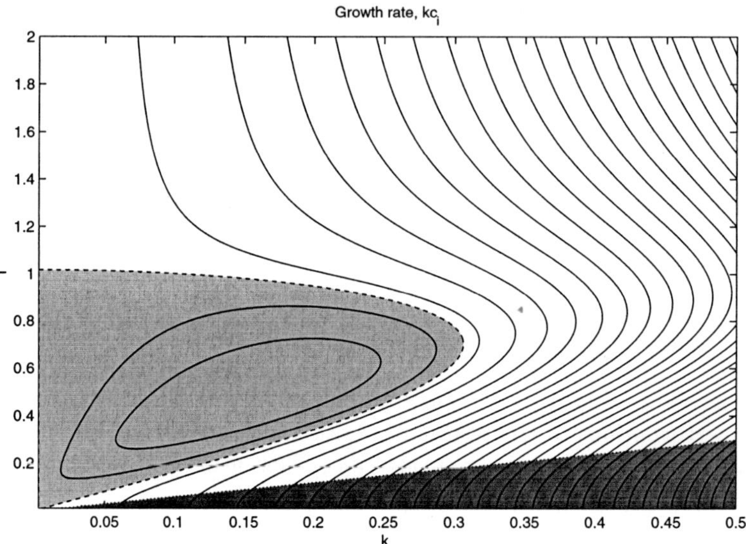

Fig. 15.8. Contours of constant growth rate for the sediment mode eigenvalue predicted for the sediment transport law quoted by Coleman & Fenton (with $\nu_0 = 0.001$, $F^2 = 0.2$, $J_u = 2$, $J_h = J_s = 3/2$ and $\varepsilon = 2.5 \times 10^{-4}$). The unstable region is lightly shaded. The dark region shows the eigenvalues with $c_r > 0$

regular spacing in which bed elevation alternates from one side of the river to the other on proceeding downstream. If we assume that water flows in a straight channel and the banks simply quantize the transverse wavenumber, then we may apply the preceding linear stability theory to the study of this problem. In this way, alternate bars could be understood as resulting from the growth of unstable modes whose wavelength matched the channel width. Braids, on the other hand, would be interpreted as having smaller wavelengths than the channel width, and so more bed elevations would occur at any given streamwise position. Note that a commonly accepted philosophy is that once bars form, meandering of the whole watercourse can begin [28,29]. Hence the prediction of bar instability further offers the beginning of an explanation of river meanders [30]. Various experiments and observations offer support for these theoretical rationalizations [29].

Given the freedom to vary the parameters of the problem, we can tune the system so that the instability is gradually turned off and we arrive at a marginally stable state. At that point, we can exploit asymptotic tools to derive a weakly nonlinear theory for the growth of the instability just beyond onset (see Chap. 1). If we consider flows that are strictly periodic in both x and y, we then arrive at a Landau equation, as did Colombini et al [31] in a related problem. Alternatively, if we allow perturbations to develop spatially downstream, and allow long variations in x, the result is a complex Ginzburg–Landau equation

(Chap. 1; [32]). This suggests that bar patterns on long length scales need not be periodic.

15.8.2 Channelization

A particularly interesting limit of the stability problem is that with $F^2 \to 0$, which corresponds to slow overland flow on a sloping plane. As considered by Smith & Bretherton [33] and Izumi and Parker [34], linear instability of such flow could be responsible for the initiation of channelization and the formation of watercourses. Moreover, as the water continues to collect in the seeded streams and flows downhill, a river network may be created through nonlinear pattern dynamics [35]. Alternative views to this basis of landscape patterns are provided by cellular automata models [36] and stochastic models [37].

For $F^2 = 0$, $\Gamma \approx \nu_0(k^2 + l^2)$ is real, and so

$$c_i + k\varepsilon J_s =$$

$$k\varepsilon \frac{(3+\Gamma)(1+\Gamma)[J_h(1+\Gamma+2\kappa^2+\kappa^2\Gamma) - J_u(1+\Gamma-\kappa^2) - \kappa^2(3+\Gamma)]}{(3+\Gamma)^2(1+\Gamma)^2 + k^2(1+\Gamma+2\kappa^2+\kappa^2\Gamma)^2} . \tag{15.41}$$

In the original work by Smith & Bretherton, J depended only on the water flux, so $J_u = J_h$, and turbulent viscosity was neglected. Thence,

$$kc_i = \varepsilon k^2 \left[\frac{9l^2(J_u - 1)}{9k^2 + (k^2 + 2l^2)^2} - J_s \right] . \tag{15.42}$$

Hence, for sufficiently large J_u, the system is unstable over a range of wavenumbers. Actually, the present approximation is not completely equivalent to the original Smith & Bretherton model, which was not derived from a shallow-water system but was written down as a phenomenological model. The Smith–Bretherton system predicts a growth rate that diverges with l if $J_u > 1$, a fact that requires the introduction of an external regularization [38]. Here, this ultraviolet catastrophe is avoided and the most unstable wavenumbers are finite even with vanishing eddy viscosity.

15.9 Conclusion

Hopefully, the topics discussed in this chapter can serve as an introduction to the complicated world of dirty flows. We have seen that it is possible to generate linear bed instabilities that can be associated with incipient ripples, dunes, antidunes, bars, braids, or entire drainage networks. The whole picture, however, resides heavily upon the empirical parameterizations that are adopted for turbulence and the bed evolution. Changing the parameterizations can significantly alter the results. Clearly, this is worrying especially because there are no solid reasons, based on first principles, to prefer one formulation over another.

Indeed, simple shallow-water models with empirical transport laws offer explanations for two-dimensional erosional features in rivers (bars and braids), but they seem unable to explain the simpler, one-dimensional structures that form on the bed (ripples, dunes and anti-dunes). It seems a little far-fetched to believe that theoretical modelling could be successful in the one context and not the other. Nevertheless, this pessimism should be tempered by the fact that there are countless examples in engineering in which empirical parameterizations of turbulence have been very effective.

In conclusion, there is still room for new approaches and for developing deeper theoretical insight in bedform evolution. A better understanding of erosion and deposition in turbulent flows is clearly the crucial ingredient for describing the instability of a flat erodible bed. Even when linear instability is more completely understood, on the other hand, we will still need to understand what happens next – that is, the nonlinear evolution of the bedforms and their further instabilities to secondary perturbations.

Acknowledgements

We are grateful to Andrew Fowler and Giovanni Seminara for critically reading the manuscript.

References

1. R.A. Bagnold: *The Physics of Blown Sand and Desert Dunes* (Chapman & Hall, London 1941)
2. L.B. Leopold, M.G. Wolman, J.P. Miller: *Fluvial Processes in Geomoprhology* (Dover, New York 1964)
3. D. Knighton: *Fluvial Forms and Processes* (Arnold, London 1998)
4. M. Selim Yalin: *Mechanics of Sediment Transport* (Pergamon Press, Oxford 1972)
5. N.M. Coleman: 'A theoretical and experimental study of drag and lift forces acting on a sphere resting on a hypothetical stream bed'. In: *Proc. 12th Cong. I.A.H.R.*, paper C22 (Fort Collins 1967)
6. S. Ikeda: J. Hydraul. Div. ASCE **108**(HY1), 95–114 (1982)
7. P. Holmes, J.L. Lumley, G. Berkooz: *Turbulence, Coherent Structures, Dynamical Systems and Symmetries* (Cambridge Un. Press, Cambridge 1996)
8. J.M. Nelson, R.L. Shrieve, S.R. McLean, T.G. Drake: Water Resour. Res. **31**(8), 2071–2086 (1995)
9. E. Meyer-Peter, R. Muller: 'Formulas for bed-load transport'. In: *Int. Assoc. Hydraul. Res.*, 2nd meeting, Stockholm, pp.39–64 (1948)
10. J.J. Stoker: *Water waves* (Interscience Publishers, New York 1957)
11. V. Cornish: *Ocean Waves and Kindred Geophysical Phenomena* (Cambridge Un. Press, Cambridge 1934)
12. R.F. Dressler: Commun. Pure Applied Maths **2**, 149–194 (1949)
13. D.J. Needham, J.H. Merkin: Proc. R. Soc. London A **394**, 259–278 (1984)
14. J.H. Merkin, D.J. Needham: Proc. R. Soc. London A **405**, 103–116 (1986)
15. C. Kranenburg: J. Fluid Mech. **245**, 249–261 (1992)
16. J. Yu, J. Kevorkian: J. Fluid Mech. **243**, 575–594 (1992)

17. B.S. Brook, S.A.E.G. Falle, T.J. Pedley: J. Fluid Mech. **396**, 223–256 (1999)
18. H.-C. Chang, E.A. Demekhin, E. Kalaidin: Phys. Fluids **12**, 2268–2278 (2000)
19. S.E. Coleman, J.D. Fenton: J. Fluid Mech. **418**, 101–117 (2000)
20. V.A. Vanoni: *Sedimentation Engineering. Manuals and Reports of Engineering Practice.* Vol. 54, ASCE (1975)
21. S.C. Jain, J.F. Kennedy: 'The growth of sand waves'. In: *Proc. Intl. Symp. on Stochastic Hydr.*, pp. 449–471 (Pittsburgh University Press, Pittsburgh 1971); J. Fluid Mech. **63**, 301–314 (1974)
22. J.F. Kennedy: J. Fluid Mech. **16**, 521–544 (1963); Ann. Rev. Fluid Mech. **1**, 147–168 (1969)
23. F. Engelund: J. Fluid Mech. **42**, 225–244 (1970)
24. J. Fredsoe: J. Fluid Mech. **64**, 1–16 (1974)
25. K.J. Richards: J. Fluid Mech. **99**, 597–618 (1980)
26. P.L. Wiberg, J.D. Smith: J. Geophys. Res. **90**, 7341–7354 (1985)
27. S.R. McLean: Earth-Science Reviews **29**, 131–144 (1990)
28. J. Fredsoe: J. Fluid Mech. **84**, 609–624 (1978)
29. G. Parker: J. Fluid Mech. **76**, 457–480 (1976)
30. P. Blondeaux, G. Seminara: J. Fluid Mech. **157**, 449–470 (1985)
31. M. Colombini, G. Seminara, M. Tubino: J. Fluid Mech. **181**, 213–232 (1987)
32. R. Schielen, A. Doelman, H.E. de Swart: J. Fluid Mech. **252**, 325–356 (1993)
33. T.R. Smith, F.P. Bretherton: Water Resour. Res. **8**, 1506–1529 (1972)
34. N. Izumi, G. Parker: J. Fluid Mech. **283**, 341–363 (1995)
35. T.R. Smith, B. Birnir, G.E. Merchant: Comput. Geosci. **23**, 811–822 and 823–849 (1997)
36. I. Rodriguez-Iturbe, A. Rinaldo: *Fractal River Basins: Chance and Self-Organization* (Cambridge Univ. Press, Cambridge 1997)
37. P.S. Dodds, D. Rothman: Ann. Rev. Earth Planet. Sci. 28, 571–610 (2000)
38. D.S. Loewenherz: J. Geophys. Res. **96B**, 8453–8464 (1991)

16 Invitation to Sediment Transport

G. Seminara

Dipartimento di Ingegneria Ambientale, Università di Genova, Via Montallegro 1, 16145 Genova, IT

16.1 Formulation

As stated in [12] the mechanical system analyzed in morphodynamics '....consists of a low concentration two phase mixture of water and sediment particles subject to a free surface flow bounded by a granular medium. Flow of water also occurs very slowly through the interstices of the granular medium: however such a weak filtration process may be safely ignored. The interface between flowing mixture and granular medium can move as a result of a continuous exchange of sediment particles. The *general problem of morphodynamics may then be stated as that of determining the motion of the above interface for given boundary and initial conditions for the flowing mixture and the granular medium*'.

The classical mathematical formulation of the latter problem is based on the assumption that the presence of the solid phase does not affect significantly the motion of the fluid phase, a reasonable assumption when the concentration of solid particles is sufficiently low, as it is typical of fluvial and tidal environments, except within a thin layer close to the bed. The theoretical scheme then adopts the *continuity and Navier–Stokes equations* (in suitably averaged form) *for the fluid phase*, coupled to *mass and momentum conservation equations for the solid phase* along with *appropriate boundary conditions.* The formulation is completed by deriving an *evolution equation for the free boundary* consisting of the interface between the flowing mixture and the granular medium.

16.2 Mass Conservation of the Solid Phase

Let the motion of the flowing mixture be referred to a fixed cartesian reference frame (x_1, x_2, x_3) with x_3 vertical coordinate pointing upwards. The region occupied by the *flowing mixture* is defined by the following relationship (Fig. 16.1):

$$\eta(x_1, x_2, t) < x_3 < H(x_1, x_2, t) \ , \tag{16.1}$$

where t is time, $\eta(x_1, x_2, t)$ and $H(x_1, x_2, t)$ are the elevations of the bed interface and free surface, respectively. The granular medium fills the region $x_3 < \eta$.

Mass conservation of the solid phase imposes the following continuity requirement:

$$\frac{\partial c}{\partial t} + \nabla \cdot \mathbf{q}_s = 0 \tag{16.2}$$

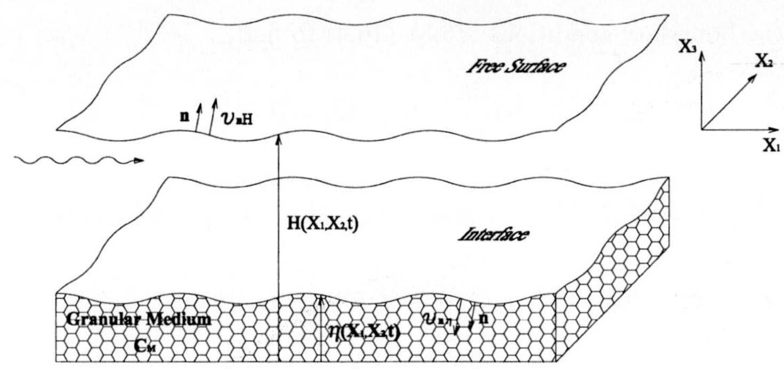

Fig. 16.1. Sketch of the system

where $c(x_1, x_2, x_3, t)$ and $\mathbf{q_s}(x_1, x_2, x_3, t)$ are volume concentration and volume flux of sediments, respectively.

The boundary conditions associated with (16.2) must express the following constraints:

i) no net flux of sediments enters or leaves the flow region through the free surface, hence

$$(\mathbf{q}_s \cdot \mathbf{n} - c\, v_{nH}) = 0 \qquad (x_3 = H) \tag{16.3}$$

where v_{nH} is the normal component of the velocity of the free surface and $\mathbf{n}$ is the outward normal unit vector;

ii) the motion of the bed interface leads to a net flux of sediments entering or leaving the flow region, due to the difference between the near bed concentration of the flowing mixture and the packing concentration c_M of the granular medium bounding the flow region, hence

$$\mathbf{q}_s \cdot \mathbf{n} = E = [v_{n\eta}(c - c_M)] \qquad (x_3 = \eta) \tag{16.4}$$

where E is the entrained flux of sediments and $v_{n\eta}$ is the normal component of the speed of the bed interface.

Also note that in natural flows the motion of the flowing mixture is nearly invariably turbulent, hence both c and $\mathbf{q}_s$ are fluctuating quantities amenable to the classical Reynolds decomposition.

In order to make any progress with the latter formulation appropriate equations interpreting the motion of sediments are needed. They are presented in Sect. 16.4, while, in the next section, we show that from (16.2) one immediately derives an evolution equation for the bed interface.

16.3 Evolution Equation of the Bed Interface

Let us perform a Reynolds average of (16.2), integrate it over depth, use Leibniz's rule to account for the spatial and temporal dependence of the boundaries and employ the boundary conditions (16.3)–(16.4) to find:

$$c_M \frac{\partial \eta}{\partial t} + \nabla_H \cdot \mathbf{Q}_s = 0 \tag{16.5}$$

where ∇_H is the gradient operator in two dimensions ($\equiv (\partial/\partial x_1, \partial/\partial x_2)$) and $\mathbf{Q}_s$ is the depth integrated volumetric sediment discharge, defined in the form:

$$\mathbf{Q}_s = (Q_{s1}, Q_{s2}) = \int_\eta^H \langle (q_{s1}, q_{s2}) \rangle \mathrm{d}x_3 \ , \tag{16.6}$$

with $\langle \cdot \rangle$ denoting Reynolds average. Equation (16.5) plays the role of an evolution equation for the bed interface.

A one-dimensional form of the latter, first derived by Exner [4], is readily obtained by integrating (16.5) in the lateral direction x_2. Further, stipulating that no flux of sediment can enter or leave the stream through the banks, one finds:

$$c_M b \frac{\partial \overline{\eta}}{\partial t} + \frac{\partial}{\partial s}(b\overline{Q_s}) = 0 \ , \tag{16.7}$$

where b is the width of the active portion of the bed interface assumed to undergo spatial (but not temporal) variations, s is a longitudinal coordinate (which replaces the cartesian coordinate x_1) and an overbar denotes average in the lateral direction between the banks at $x_2 = \pm b/2$, hence:

$$(b\overline{\eta}, b\overline{Q_s}) = \int_{-b/2}^{b/2} (\eta, Q_s) \mathrm{d}x_2 \ . \tag{16.8}$$

The possible presence of a net sediment flux entering or leaving the stream through the sidewalls may be incorporated in a 2D context by assigning the vector $\mathbf{Q}_s$ at the side walls. In a 1D context the latter effect would lead to adding to the right hand side of (16.7) an additional contribution, q_{sl}, namely the given value of the lateral sediment flux per unit width.

Both the 2D and the 1D equations (16.5) and (16.7) require boundary conditions. The choice of the appropriate form of the latter conditions requires an analysis of the mathematical nature of the governing equations. The interested reader is referred to [16] for the 1D case. The 2D case has yet to be thoroughly investigated.

In order to complete the formulation we need closure relationships for the vector $\mathbf{Q}_s$ or for the scalar quantity $\overline{Q_s}$ in the 2D or 1D contexts, respectively.

16.4 Motion of the Solid Phase

Providing such closure relationships requires the knowledge of the instantaneous quantities q_{s1} and q_{s2}, which should then be Reynolds averaged, depth integrated (see (16.6)) to determine the vector $\mathbf{Q}_s$ and finally integrated in the lateral direction (see (16.8)) to evaluate the scalar quantity $\overline{Q_s}$. This is a formidable task which still awaits the development of fully satisfactory, theoretically well founded tools. In spite of such relative uncertainty a few fairly well established results of semiempirical nature have been widely employed in the literature. They refer to the simplest flow conditions, namely uniform free surface flow on a homogeneous cohesionless bed, a configuration that by definition is in *equilibrium*, i.e. such that the elevation of the bed interface keeps constant in time. Let us summarize the main features of such results.

16.4.1 Incipient Transport

Threshold hydrodynamic conditions exist below which no bed particle is set in motion. Such conditions have been expressed in terms of a relationship between a *critical value* Θ_c of the dimensionless form of the average shear stress acting on the bed Θ (called *Shields stress*, [14]) and a *particle Reynolds number* R_p, defined as follows:

$$\Theta = \frac{u_*^2}{\Delta g d_s}, \qquad R_p = \frac{\sqrt{\Delta g d_s^3}}{\nu}, \tag{16.9}$$

where u_* is the average friction velocity, Δ reads $(\varrho_s/\varrho - 1)$ with ϱ and ϱ_s water and particle density, respectively, d_s is particle diameter, ν is kinematic viscosity and g is gravity. Such relationship was first plotted by Shields [14] and is known as Shields curve.

Simple theoretical derivations of Shields curve have been proposed by Coleman [3] and Ikeda [6] among others.

16.4.2 Bedload Transport

For values of Θ exceeding Θ_c but lower than a second threshold value Θ_s to be defined below, particles are intermittently entrained by the flow, either individually or collectively, and move close to the bed, mostly saltating or rolling but also occasionally sliding. Particles eventually come to rest to be entrained again after some time. In this mode of sediment transport, which keeps confined within a layer of thickness ranging about few grain diameters and is called *bedload transport*, particles have a *distinct dynamics* driven by, but different from, fluid motion. Recent detailed investigations have clarified that the agent of particle motion is the spatially and temporally intermittent generation of turbulent eruptions in the near wall region. More precisely, events responsible for bedload transport are those called *sweeps* in the turbulent literature [9].

Under *uniform conditions* theoretical approaches (e.g. [1]) as well as laboratory observations (e.g. [8]) suggest that the average bedload flux:

- is aligned with the average bottom stress $\boldsymbol{\tau}$;
- has an intensity which, once suitably made dimensionless, is found to be a monotonically increasing function of the excess Shields stress $(\Theta - \Theta_c)$ and of the particle Reynolds number R_p.

In other words, we may write:

$$\hat{\mathbf{Q}}_{sb} = \frac{\mathbf{Q}_{sb}}{\sqrt{\Delta g d_s^3}} = \Phi[(\Theta - \Theta_c), R_p] \frac{\boldsymbol{\tau}}{|\boldsymbol{\tau}|} \, . \tag{16.10}$$

Various relationships have been proposed in the literature to quantify the function Φ (e.g. [8,10]). They provide estimates which can be considered as qualitatively reliable for practical purposes.

In order to make use of the latter result in the context of investigations on morphodynamical patterns, a crucial extension is needed such to describe bedload transport on a weakly sloping topography. This is readily accomplished as described in [96]. Assuming that the dimensionless quantity $\nabla_H \eta$ is small enough to allow for a linearized treatment of the effect of local slope on sediment transport, on pure dimensional ground one derives the following relationship:

$$\hat{\mathbf{Q}}_{sb} = \Phi(\Theta - \Theta_c, R_p) \left(\frac{\boldsymbol{\tau}}{|\boldsymbol{\tau}|} + \mathbf{G} \cdot \nabla_H \eta \right) \, , \tag{16.11}$$

where the bedload function Φ must be evaluated in terms of the local value of the Shields stress and $\mathbf{G}$ is a (2×2) matrix whose elements are in general functions of the dimensionless parameters θ_c, θ and R_p. Physically (16.11) simply states that the direction of bedload transport deviates from the direction of the local average bottom stress by an amount which is linearly related to the local value of the gradient of bed elevation.

Theoretical [11] and experimental [15] works suggest the following estimates for the elements of the matrix $\mathbf{G}$:

$$G_{ss} = -\frac{\Theta_c}{\mu \Phi} \frac{\mathrm{d}\Phi}{\mathrm{d}\Theta} \, , \tag{16.12}$$

$$G_{sn} = G_{ns} = 0 \, , \tag{16.13}$$

$$G_{nn} = -\frac{r}{\Theta^m} \, , \tag{16.14}$$

where n is a lateral coordinate, orthogonal to the s-longitudinal coordinate, μ is the friction coefficient of the mixture, r and m are empirical constants ranging about 0.56 and 0.5 respectively.

The equation of motion of the solid phase appropriate in the context of 2D models of the dynamics of the bed interface (see (16.5)) is (16.11). It is restricted to the case of slowly varying bed topographies, hence it fails close to sharp fronts where the weakly sloping assumption does not apply. The reader interested in extensions of the above treatment to the case of *arbitrarily sloping beds* is referred to [7] and the later developments in [13].

The equation of sediment motion employed in the context of one-dimensional models of the dynamics of the bed interface (see (16.8)) in dimensionless form reads:

$$\overline{\hat{Q}_{sb}} = \frac{\overline{Q}_{sb}}{\sqrt{\Delta g d_s^3}} = \Phi[(\overline{\Theta} - \Theta_c); R_p] \tag{16.15}$$

where $\overline{\Theta}$ is the cross sectionally averaged value of the Shields stress. Note that the latter formulation would be strictly valid (i.e. it would be obtained performing a cross sectional average of (16.10)) if the bedload function Φ were linear in the excess Shields stress. Since the latter function is non linear, the above approximation is increasingly less satisfactory the less uniform is the lateral distribution of Shields stress in the cross section. Also note that lateral averaging cancels the dependence of bedload transport on lateral slope, while its dependence on longitudinal slope is usually neglected in the context of 1D formulations applying to very slowly varying configurations.

Are the above *equilibrium* formulations appropriate to patterns forming under unsteady or spatially varying conditions, as typical of tidal environments? *Non equilibrium effects* may in principle lead to a phase lag in the response of sediment transport to changing hydrodynamic conditions. This is a subject which will require attention in the near future. Current models commonly neglect non equilibrium effects and employ the equilibrium formulations (16.11) and (16.13) for the equations of motion of the solid phase under bedload dominated conditions.

16.4.3 Transport in Suspension

As the local instantaneous value of the Shields stress Θ exceeds a second threshold value Θ_s, depending on the particle Reynolds number, a second mode of transport, namely *transport in suspension*, is observed to coexist with bedload transport. Particles are individually or collectively entrained by the flow, driven by near wall ejection events, and are 'nearly passively' transported by the fluid in the outer flow region until they return to the bed under the effect of their excess weight. In other words, transport in suspension differs from bedload transport because sediment particles are able to escape the near wall flow barrier and their dynamics is not significantly distinct from that of fluid particles except for their tendency to settle. The threshold condition for entrainment in suspension can be expressed in the form:

$$\Theta > \Theta_s = N(R_p) . \tag{16.16}$$

Various semiempirical estimates have been proposed for the function N (e.g. [2,10]) which attains values of order one. The sediment flux per unit width in this regime can be written in the form:

$$\mathbf{Q}_s = \mathbf{Q}_{sb} + \mathbf{Q}_{ss} , \tag{16.17}$$

where $\mathbf{Q}_{sb}$ is the bedload flux per unit width defined by (16.11) and $\mathbf{Q}_{ss}$ is the flux of suspended sediment per unit width. In order to evaluate the latter quantity we note that the local value of the average flux of suspended sediments is defined in the form:

$$\mathbf{q}_{ss} = \langle c\, \mathbf{v}_p \rangle = \langle c\, (\mathbf{v} - W_s\, \hat{\mathbf{x}}_3) \rangle , \tag{16.18}$$

with $\mathbf{v}$ and $\mathbf{v}_p$ local and instantaneous values of the fluid and particle velocities respectively, W_s settling speed of sediment particles and $\hat{x}_3$ unit vector in the vertical direction. Note that (16.16) is based on the assumption that sediment particles are 'nearly passive' tracers which are advected by the flow except for their tendency to settle, accounted for through the second term in the right hand side of (16.16). The latter scheme is reasonable provided the particle suspension is sufficiently diluted and grains are small enough for their presence not to affect the fine structure of turbulence. Such conditions are usually fairly well satisfied in fluvial and tidal environments. Performing a Reynolds decomposition we write:

$$(c, \mathbf{v}) = (C, \mathbf{V}) + (c', \mathbf{v}') \, . \tag{16.19}$$

Recalling definition (16.6) we then find:

$$\mathbf{Q}_{ss} = \int_{\eta}^{H} [C(V_1, V_2) + \langle c'(v_1', v_2') \rangle] \, dx_3 \, . \tag{16.20}$$

In order to make any progress with thc evaluation of $\mathbf{Q}_{ss}$ we then need to evaluate the spatial and temporal distribution of the mean concentration C as well as the turbulent fluxes $\langle c' v_j' \rangle$ $(j = 1, 2)$. This is achieved by substituting from (16.16) into (16.2), performing the Reynolds decomposition (16.17) and thus deriving a convection-diffusion equation for C which reads:

$$\frac{\partial C}{\partial t} + V_j \frac{\partial C}{\partial x_j} - W_s \frac{\partial C}{\partial x_3} = \frac{\partial \langle -c' v_j' \rangle}{\partial x_j} \, . \tag{16.21}$$

The latter equation poses a closure problem for the turbulent fluxes. The classical approach employed in the engineering literature is to employ a diffusive type of closure by analogy with the semiempirical closures employed for Reynolds stresses. The theoretical foundation of such an approach is fairly weak and may find some justification only in the slowly varying character of the flow fields typically encountered in morphodynamics, which suggests that the turbulent structure may be considered in quasi equilibrium with the local and instantaneous conditions. Using the so called diffusion approximation one writes:

$$\langle -c' v_j' \rangle = D_T \frac{\partial C}{\partial x_j} \tag{16.22}$$

with D_T turbulent diffusivity, a quantity which several experimental investigations suggest to attain values slightly different from the corresponding values of the eddy viscosity. Suitable estimates for D_T must then be associated with (16.20). In order to complete the formulation of the convection–diffusion equation, appropriate boundary conditions are needed. The condition that the net flux of suspended sediment through the free surface must vanish instantaneously, (16.3), is readily averaged over turbulence and it gives:

$$\langle c' \mathbf{v}' \cdot \mathbf{n} \rangle = C W_s \qquad (x_3 = H) \, . \tag{16.23}$$

At the bed interface an entrainment condition of the type (16.4) is enforced with the entrained flux E given in the form:

$$E = W_s (C - C_{eq}) \qquad (x_3 = \eta) \, , \tag{16.24}$$

where C_{eq} is the average concentration established under equilibrium conditions at some conventional small distance from the bed. Various empirical or semiempirical relationships have been proposed in the literature to estimate the latter quantity as a function of the local and instantaneous value of Shields stress and particle Reynolds number (e.g. [10]). Note that (16.22) correctly predicts that under equilibrium conditions, i.e. for uniform flows, the net flux entrained by the stream vanishes, i.e. entrainment and deposition balance exactly. The knowledge of the distribution of concentration at some initial cross section or, alternatively, some periodicity condition, along with the condition of vanishing flux of suspended sediment through impermeable boundaries and the knowledge of the state at an initial time complete the three-dimensional formulation of transport of suspended sediment. Depth averaged two-dimensional formulations and cross sectionally averaged one-dimensional formulations can be derived from the three-dimensional formulation outlined above. The interested reader is referred to [5].

16.5 Conclusive Remarks

The mathematical problem of morphodynamics can be formulated at various levels of complexity on the basis of the conservation equations and semiempirical inputs briefly outlined in the previous sections.

It will appear in the following chapters that, in spite of the relatively weak foundation of the above approach, a variety of patterns observed in different sedimentary environments of the earth can be satisfactorily investigated using the above formulation. Appropriate averaged forms of the latter apply in various contexts depending on the spatial scale of the morphodynamic pattern to be investigated.

In particular, *large scale patterns* characterized by a 'longitudinal' scale vastly exceeding the typical depth and width of the flow are suitable interpreted by means of a 1D model. This is the case of the longitudinal profile of rivers and tidal channels.

Mesoscale patterns, like the so called bars, which are repetitive forms typically observed in rivers and estuaries as well as in coastal regions, are conveniently treated by means of the 2D version of the above formulation.

More detailed 3D models are needed when treating the formation and development of patterns characterized by *smaller scales*, of the order of flow depth (dunes and antidunes in rivers, sandwaves in coastal areas) of even smaller (fluvial and coastal ripples, sand ridges).

Some of the patterns mentioned above will be examined in the next chapters.

Acknowledgments

This work represents an outcome of the work initiated under the umbrella of the Italian National Research Program 'Morfodinamica fluviale e costiera', coordinated by G. Seminara and cofunded by various Italian Universities with further financial support provided by the Italian Ministry of Scientific Research. Such work is continuing in the context of the new National Research Program 'Idrodinamica e morfodinamica di ambienti a marea' coordinated by A. Rinaldo and of the Research project 'Analisi e monitoraggio dei processi morfologici nel sistema lagunare' funded by CORILA.

References

1. K. Ashida, M. Michiue: Proc. Jpn. Soc. Civ. Eng. **206**, 59–69 (1972)
2. R.A. Bagnold: Philos. Trans. R. Soc. London A **249**, 235–297 (1956)
3. N.L. Coleman: 'A theoretical and experimental study of drag and lift forces acting on a sphere resting on a hypotetical stream-bed'. In: *Proc. 12th Cong. I.A.H.R* paper C22 (Fort Collins 1967)
4. F.M. Exner: *Sitzber Akad. Wiss* 165–180 (1925)
5. G. Galappatti, G.B. Vreugdenhil: J. Hydr. Res. IAHR **23**(4), 359–377 (1985)
6. S. Ikeda: J. Hydraul. Div. ASCE, **108**(HY1), 95–114 (1982)
7. A. Kovacs, G. Parker: J. Fluid Mech. **267**, 153–183 (1994)
8. E. Meyer-Peter, R. Müller: 'Formulas for bedload transport'. In: *III Conf. Of Internat. Ass. of Hydraul. Res., Stockolm, Sweden* (1948)
9. J.M. Nelson, R.L. Shreve, S.R. McLean, T.G. Drake: Water Resour. Res. **31**(8), 2071–2086 (1995)
10. L.C. van Rijn: J. Hydr. Engng. ASCE **110**(11), 1613–1641 (1984)
11. M. Sekine, H. Kikkawa: J. Hydraul. Eng. ASCE **118**(4), 536–558 (1992)
12. G. Seminara: Meccanica **33**, 59–99 (1998)
13. G. Seminara, L. Solari, G. Parker: Water Resour. Res., submitted for publication (2001)
14. A. Shields: *Mitteil. Preuss. Versuchanst. Wasser. Erd. Shiffbau*, Berlin, n. 26 (1936)
15. A.M. Talmon, N. Struiksma, C.L.M. van Mierlo: J. Hydr. Res. IAHR **33**(4), 495–517 (1995)
16. M. de Vries: 'Riverbed variations. Aggradation and degradation'. In: *IAHR Seminar* (New Delhi 1969)

17 Types of Aeolian Sand Dunes and Their Formation

H. Tsoar

Department of Geography and Environmental Development Ben-Gurion University of the Negev Beer-Sheva 84105, Israel

17.1 Introduction

The accumulation of windblown sand creates sand dunes which are one of nature's most dynamics and intriguing phenomena. Sand dunes are found in most climates of the world as coastal dunes and in some arid regions. Grains of sand between 0.062 and 2.0 mm in diameter are not cohesive and therefore are easily carried by the wind. Paradoxically, finer grains of silt and clay (< 0.050 mm) are cohesive and can resist wind erosion. This property of sand is reflected in the wind threshold speed curve for sand transport (Fig. 17.1) explaining why dune sand, in most cases, is composed of fine particles between 0.125–0.250 mm.

While sand dunes, mobile and immobile, are found in almost all climates, from humid Europe [1,2], boreal Alaska [3], and Central Canada [4] to semiarid [5] and arid areas [6], more than 99% of all sand dunes are located in deserts. Less than 1% are located in humid climates and along some of the world's coastlines [7]. Coastal dunes are known to be relatively young, no older than 6,500 years, which is when the sea reached its present level after the last rapid postglacial rise. The common characteristics for all dunes, in all world climates, are that their formation indicates an abundant supply of sand-sized sediment, strong sand-moving winds, and conditions favoring sedimentation of the sand. Most of the world's sand dunes were active during the period 20,000 to 15,000 BP, known as the last glacial maximum, when wind power was much higher than in present-day wind storms [8,9].

17.2 Wind Power

Wind should not only be above the threshold velocity (Fig. 17.1) to initiate sand transport, but should also have a certain drift potential to prevent plants from growing in the sand and stabilizing it. The accepted method of quantifying wind power is by referring to the drift potential (DP) of the wind, which is based on the sand transport equation [10] in which the sand flux is directly proportional to the cube of the wind:

$$q = K u_*^2 (u_* - u_{*t}) , \tag{17.1}$$

where u_* is the shear velocity of the wind, u_{*t} is the threshold shear velocity and K depends upon variables such as grain size, sand sorting and air density. Since K characterizes variables that have little variation from one dune field to

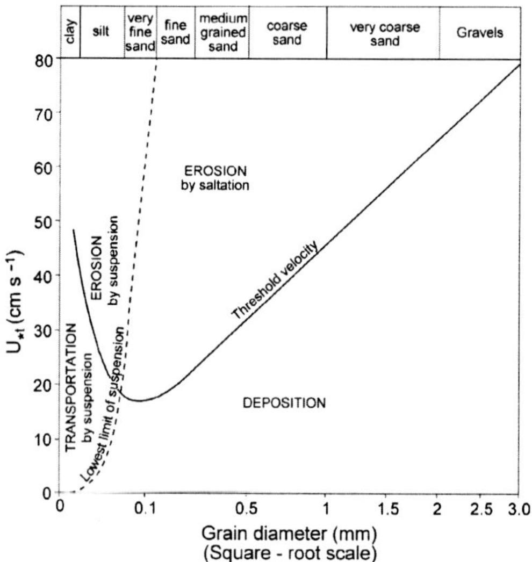

Fig. 17.1. Threshold friction velocity (U_{*t}) curve for quartz grains of different diameters (*solid line*). The broken line separates saltation from suspension (after [10])

another, the drift potential (DP) in vector units can be calculated by simplifying (17.1) and referring to the cube of the wind velocity above the threshold speed [11]:

$$DP = \sum q = \frac{u^2 (u - u_t)}{100} t \,, \tag{17.2}$$

where q is the rate of sand drift, u is the wind velocity (in knots), u_t is the threshold velocity (in knots) and t is the amount of time the wind blew above the threshold (in %). An index of wind direction variability is illustrated by the ratio between the resultant drift potential and the drift potential (RDP/DP), where values close to one indicate narrow unidirectional drift potential, and values close to zero indicate multidirectional drift potential.

The average yearly DP forms sand roses that indicate the relative potential transport of sand from various directions. An example is demonstrated by Fig. 17.2, which shows sand roses in three sites along the coastal plain of the southeastern Mediterranean. The total yearly average DP and the ratio RDP/DP can explain mobility and stability of sand dunes mostly because the limiting factor for vegetation on dune sand is wind erosion [12,13]. When RDP/DP is low, wind energy is distributed on more than one slope of the dune and the energy exerted on each slope is lower. For that reason sand dunes with high rates of directional variability are covered by vegetation on their slopes (as in some star dunes, Fig. 17.3) while under the same DP and low rates of directional variability, the dunes are bare of vegetation.

Sand dunes in areas where the annual average rainfall is ≥ 50 mm are unvegetated and mobile under the conditions in which $DP > 1000$ and RDP/DP is

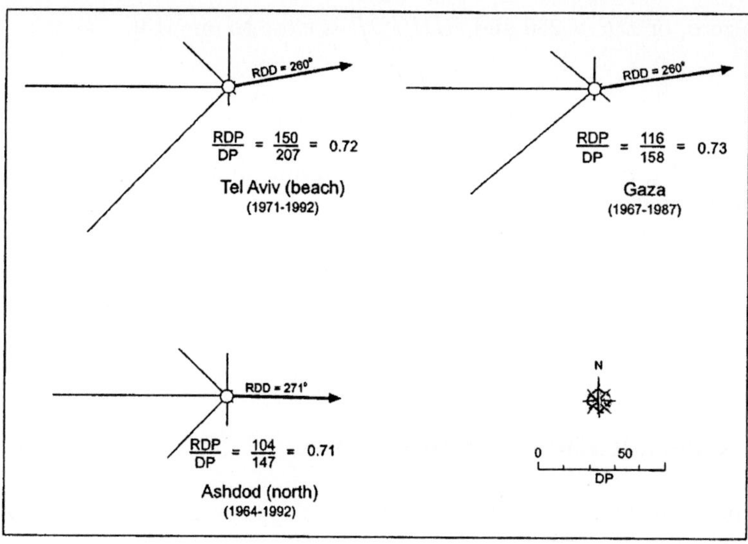

Fig. 17.2. Three sand roses of the drift potential (DP) vectors for three stations along the southeastern Mediterranean coast. Note the ratio of the total DP and the resultant drift potential (RDP), which is a parameter for the wind direction variability

Fig. 17.3. Star dunes in Gran Desierto, Mexico, where their slopes are covered by vegetation because of low value of RDP/DP (about 0.21 for the nearest meteorological station), in spite of the low amount of rainfall (about 70 mm on annual average). Data from [90]

close to zero, or $DP > 250$ and RDP/DP is close to one [13]:

$$\frac{DP}{1000 - (750\,RDP/DP)} > 1\,. \tag{17.3}$$

Table 17.1 gives the DP and RDP in various dune sites of the world. The table shows that rainfall is not as decisive a factor for dune mobilization or stabilization as is the DP. In the Negev Desert dunes are fully stabilized where the average yearly rainfall is 90 mm. The dunes in Sinai (east of Al-Ismailiya) and the coastal dunes in Gaza are fully active due to human impact [14]. Dunes that were artificially stabilized, such as along the Netherlands coast, will become active once the vegetation is destroyed.

17.3 Classification of Sand Dunes

The tremendous variety of sand dunes makes their classification a difficult task. Three main factors (two climatic and one sedimentary) influence the piling of sand into dunes with particular shapes:

1. Wind magnitude (above the threshold velocity), direction, and frequency.
2. Vegetation cover.
3. Grain size.

In addition, other factors – obstructions to wind flow, climatic changes manifested by dramatic change in wind direction, velocity and frequency of storms, sand availability, thickness of sand cover, and sudden removal of vegetation cover – can affect dune morphology. Because dunes are bed-forms in which a great deal of energy has been invested, daily or seasonal changes in wind direction do not easily reshape them. Therefore dune shape is the manifestation of a long-term average of wind conditions.

Distinction of sand dunes into simple (basic), compound and complex forms was suggested by McKee [15]. Simple dunes consist of individual dune forms which are spatially separate from nearby dunes. Compound dunes consist of two or more dunes of the *same* type which have coalesced or are superimposed. Complex dunes consist of two or more *different* types of simple dunes which have coalesced or are superimposed. Complex and compound are in most cases megadunes and abound in most of the world's great sand seas. Simple sand dunes are small in most cases, with wavelengths (shortest distance from one dune crest to the other) of 10 to 500 m.

Three general types of active sand dunes are classified by movement:

1. *Migrating dunes:* the whole dune body advances with little or no change in shape and dimension. Transverse and barchan dunes are the most representative specimens.
2. *Elongating dunes:* the dunes elongate and become extended in length with time. Linear dunes are the most representative specimens.

Table 17.1. Drift potential (DP), directional variability of the wind (RDP/DP), and rainfall of several dune field sites. * approximate value; N/A not available

Location, Country	DP	RDP/DP	Average yearly rainfall (mm)	Dunes status
Al-Ismailiya, Egypt (western Sinai)	62	0.47	50	Fully active
Nizzana, Israel (Negev Desert)	108	0.70	90	Fixed (fully vegetated)
Gaza, Palestine (coast)	158	0.73	400	Active (partly vegetated)
Ashdod, Israel (coast)	147	0.71	500	Semi-active (partly vegetated)
Upington, South Africa (Kalahari Desert)	560	0.66	183	Stabilized linear dunes
Port Elizabeth South Africa (coast)	951	0.49	660	Fully active with no vegetation
Newport, Oregon USA (coast)	2000*	N/A	1750	Fully active with no vegetation
Luderitz, Namibia (coast)	2300	0.85	< 100	Fully active with no vegetation
Ijmuiden The Netherlands (beach)	3999	0.51	768	Fully stabilized by vegetation, active when vegetation is destroyed

3. *Accumulating dunes:* the dunes have little or no net advance or elongation. Star dunes best represent this type.

These three types are distinguished by wind direction variability (RDP/DP) [16–19]. Migrating dunes are formed by a wind regime that is unimodal or close to a unimodal direction ($RDP/DP \geq 0.6$). The wind directions of elongating dunes are bimodal when the two modes are 90°–70° apart ($0.8 > RDP/DP > 0.5$). Accumulating dunes are formed under bimodal or multimodal wind directions when the two main modes form an obtuse angle that is about 180° ($0 < RDP/DP < 0.4$).

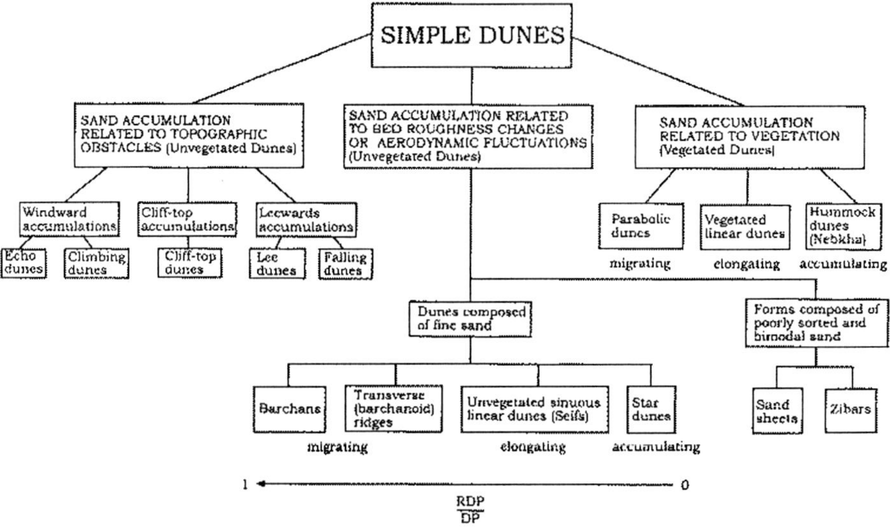

Fig. 17.4. Classification of major dune types based on dune genesis and wind directional variability (after [10])

The above classification along with a classification based on dune genesis and processes [10] is shown in Fig. 17.4. Dunes are segregated into three different categories of sand accumulation:

1. Accumulation as a consequence of topographic barriers (mostly cliffs) interfering with airflow. These dunes are not vegetated.
2. Accumulation of sand in areas of open terrain due to changes of bed roughness or aerodynamic fluctuations. These dunes are not vegetated.
3. Accumulation due to vegetation that determines dune formation and shape. All dunes in this category are vegetated. If vegetation is destroyed by human impact the reaction is either transformation to a category 2 dune type, or the formation of superimposed dunelets on top of the main dune.

The second category, unvegetated dunes that are self-accumulated, is widespread mostly in arid lands. This category is subdivided according to the grain size into sand dunes with bimodal and unimodal sand-grains. The fine sand mode of the bimodal sand dunes is within the range of the mean size of unimodal desert dune sand (0.125–0.250 mm). All dunes composed of bimodal coarse sand have a moderate aspect ratio of $h/L < 0.3$ (where h is the hill height and L is the horizontal distance from the hilltop to the point where the elevation is half its maximum). Dunes composed of unimodal, well-sorted fine sand display slip-faces, pronounced crests and a much higher aspect ratio ($1.3 > h/L > 0.3$). Despite the upslope increase in wind shear stress, the effect of gravity ($mg \sin\theta$) on grains, which is dependent upon the weight of the grain (mg) and the slope angle (θ), is stronger [20]. Bimodal sand with one coarse mode can only form *sand-sheets* and *sand-strips*. Coarse sand-sheets (also known as *zibars*) are the most common types of aeolian depositional surfaces in deserts, covering an area of 1,520,000 km^2 [21].

17.4 Dunes Accumulated and Controlled by Topographic Barriers

Topographic obstacles such as cliffs, buttes, boulders or shrubs act as baffles and induce separation of airflow into zones of acceleration and deceleration, thus producing local changes in direction and enhanced atmospheric turbulence.

When the airflow approaches the front of an isolated obstacle, such as a boulder, butte, mesa, cliff or mountain, it slows down suddenly, causing a build-up of pressure against the obstructing face [22]. The affected streamlines are forced to separate from the surface. Some of them rise and flow upwards with increasing wind velocity over the obstacle, while others make a loop and create a windward reverse-flow eddy (Fig. 17.5). The shape of the obstacle causes the flow in the windward eddy to spiral and sweep around the obstacle for some distance downwind, thus producing a three-dimensional horseshoe vortex in which the helicoidal vortex, around the windward and lateral sides of the obstacle, causes erosion [23]. The two vortices trail downwind, leeward of the obstruction, and fade out (Fig. 17.5). Unlike self-accumulated dunes, dunes that are related to

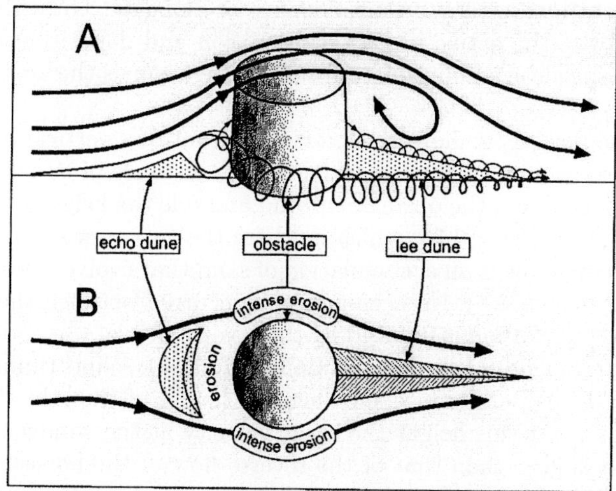

Fig. 17.5. A diagram showing the development of a horseshoe vortex in front of and around an obstruction and the resultant sand deposition. (**a**): side view; (**b**): plan view (after [10])

Fig. 17.6. Echo dune developed in front of a cliff. The extremities of the dune develop into climbing dunes and are the only places where sand leaves the dune and climbs the cliff

obstacles are static, i.e. they do not advance or elongate. Once they achieve a steady state form, the active processes of erosion and deposition do not affect their stability since the amount of sand they receive equals the amount they lose [24].

The reverse flow of the horseshoe vortex in front of a vertical obstacle (such as a cliff) causes a drop in wind magnitude at a distance of $d/h = 3.3$ (d is the horizontal distance from the obstacle upwind and h is the height of the obstacle) and a minimum at $d/h = 0.75$, which is where the two opposite flow directions meet [24]. The outcome is an accumulation of sand that evolves into a static *echo dune* (Figs. 17.5 and 17.6). No accumulation occurred between the distances of $0 < d/h < 0.3$, but erosion is induced by the reverse flow of the horseshoe vortex and on the lateral sides of the obstruction (Fig. 17.5). Wind tunnel simulation shows that a steady state is reached when the height of the echo dune is about $0.3\,h$ to $0.4\,h$ [24]. At this height the shear velocity at the windward side of the crest is slightly higher than that of the reverse flow of the horseshoe vortex on the lee side of the crest. Sand that is added to the dune from the windward side will move onto the trough between dune and obstacle and be carried by the horseshoe vortex to the rear side of the obstacle where some of it will accumulate in the 'shadow' of the obstacle, between the two horseshoe vortices, to form a *lee dune* (Fig. 17.5).

Lee dunes are best developed by a nearly unidirectional wind regime. At a distance where the topographic obstacle is no longer effective as a barrier, they tend to break up downwind into individual *barchans,* which are the preferred dune form in an open area unidirectional wind regime. The size of the lee dune is directly related to the size of the obstacle; the dune can be small, a few centimetres high and extending a few dozen centimetres downwind behind small shrubs or boulders, or it can attain a height of hundreds of metres and extend several kilometres downwind [23]. The term *longitudinal*, applied to lee dunes [25,26], originates from the resemblance of big lee dunes to *seif* dunes. However, according to the classification used in this paper (Fig. 17.4), seif dunes are not related to obstacles.

When the brink of the cliff is straight, without any projection, there is no convergence of flow on the lee side of the cliff but a great abatement of wind velocity. The result is, literally, *falling dunes* that in some cases mix with scree (talus) deposits [27,28].

Sand is normally incapable of climbing slopes [20]. Wind that encounters a cliff is diverted at the foot of the cliff to flow parallel to the cliff front. When the wind impinges obliquely upon the cliff, the separation eddy turns into a helical roll vortex that moves along the cliff front, causing sand transport along the foot of the cliff and preventing sand from climbing it [29]. Despite the above, aeolian sand is known to climb slopes, forming *climbing dunes.* This happens when a bell-shaped slope narrows, creating a funneling effect in which wind-carried sand is forced to climb the cliff, or where the helical roll vortex climbs the cliff through drainage channels (Fig. 17.6). Leeward of the slope crest there is a great abatement in wind velocity, resulting in *cliff-top dunes* [30,31].

17.5 Self-accumulated Dunes

Sand has the propensity of self-accumulation into mounds, or dunes, in the absence of topographic obstacles and vegetation. This tendency is due to the fact that the change from a rough to a smooth surface, i.e. from gravel to a sand patch, will cause a sharp drop in shear velocity, leading to sand deposition [32]. This process is only effective when strong, sand-laden winds are able to carry sand over a rough surface and then allow it to accumulate on the sand patch. Under gentler winds the sand is trapped over the rough surface so that a sand patch would be eroded and extended down-wind [33].

A different explanation is given by the *wave-form theory.* It speculates that a wave-like movement in the air, initiated by an irregularity in the bed, brings about variations in surface shear stress, causing an increase followed by a decrease in the sand transport rate [34]. This means that there are alternating transverse or longitudinal zones of erosion and deposition under which a bed-form shape of ridges and troughs starts developing and builds up until the wind velocity at the surface of the new mound is sufficient to remove as much sand as is deposited. In this case a steady state is formed, similar to a ripple formation [1], [34–36].

17.5.1 The Steady-state Dune Profile

The presence of the rudimentary aeolian bed-forms produces a number of modifications in the airflow of the atmospheric boundary layer, as both wind shear velocity and turbulence structure change when the wind blows over any mound of sand. Any such change in wind velocity has a significant influence on surface shear stress and, hence, on sand transport rates and dune morphology. On every hillock patch there is an increase in surface shear stress up the windward slope toward the crest and a decrease on the lee side [37–40]. Sand dunes, as a dynamic geomorphic system, have the attribute of negative feedback, self-regularity bedform which subjects them to periodic changes in energy (wind velocity and direction) and material (sand). Any alteration in wind and sand supply produces a change in dune morphology by forming a negative feedback which regulates the effect of that change, and brings the dune to a new state of balance whereby input and output of material and energy are equalized. This condition of equilibrium is known as a steady state [41]. A simple dune can be considered to be in a steady state when its shape and size do not change while the dune is advancing (the self-preservation criterion), i.e. when the rate of advance of all parts of the dune is constant [42,43]. A dune that is pushed out of steady state by a change in wind direction or alteration in sand supply produces negative feedback which adjusts itself to the effect of those modifications and restores a new state of balance. A final steady state does not occur because of continuing small changes in wind direction and magnitude. Therefore, all sand dunes are actually in a time-independent quasi-steady state, with little change in their configuration [44].

The first rudimentary aeolian nascent bed-form is not in a steady state; it is a result of the harmonious interrelation of wind velocity, direction and sand transport that brings the bed-form into a steady state where the rate of sand transport increases on the windward slope toward the crest in such a way as to ensure a constant and steady rate of advance at all points on the windward slope. The dune shape changes the wind shear stress above it. The change in wind speed above the dune is essential for the increased rate of sand transport on the windward side, which maintains the dune in a steady state. The change in wind velocity over the dune, measured at a particular height (z) above ground level, can be expressed as the *speed-up ratio* (A_z) [45]:

$$A_z = \frac{\bar{U}_2}{\bar{U}_1} \tag{17.4}$$

where $\bar{U}_2$ is the mean wind velocity at height z above the dunes and $\bar{U}_1$ is the mean wind velocity at the same height above a flat surface. The rate in which the speed-up ratio changes over the dune depends upon the dimensions of the dune and its shape [41]. A dune attains a steady-state shape when the speed-up ratio increases the wind shear stress in such a way that sand is eroded and carried along the windward slope while adhering to the steady-state principle (no change in shape while the dune is advancing). This rate of advance (c) depends upon the rate of erosion $\mathrm{d}q/\mathrm{d}x$ (where q is sand-transport rate per unit width and x is the longitudinal coordinate along the windward slope) and on the declination of the slope ($\tan\alpha$). In a two-dimensional configuration the above can be formulated as [42]:

$$\frac{\mathrm{d}q}{\mathrm{d}x} = \gamma c \tan\alpha \tag{17.5}$$

where γ is the specific weight of the sand in bulk. The rate of erosion is defined by the wind velocity as specified by the rate of increase in speed-up ratio, the latter depending upon the shape of the slope (Fig. 17.7). Two-dimensional pro-

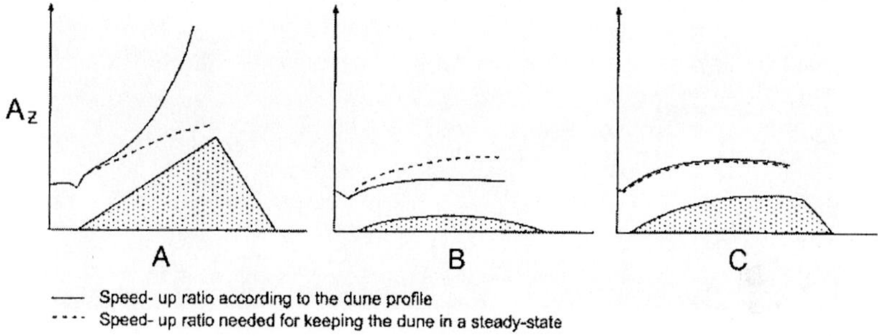

Fig. 17.7. The rate of change of speed-up ratio (A_z) over three different windward slope shapes. The solid line presents the speed-up ratio formed due to the shape of the slope. The broken line presents the speed-up ratio that would keep such a shape in steady state condition. (**a**) A uniform slope. (**b**) A small convex shape. (**c**) A bigger convex shape (for explanation see text)

file analysis [41] and three-dimensional modeling [46–48] indicate that a dune's steady-state profile in a unidirectional wind regime is a low convex slope (boat-shaped) similar to the form of a barchan dune (Figs 17.7(C) and 17.8). Three different shapes of dune profiles are shown in Fig. 17.7 with the corresponding speed-up ratio for each profile (solid line). Also shown in Fig. 17.7 is the profile of the theoretical speed-up ratio (broken line) needed for maintaining a steady-state dune shape.

When a dune's windward slope near the crest is steeper than that of a steady-state dune (Fig. 17.7(A)) the speed-up ratio on the upper windward slope attains a greater magnitude than needed to carry all the sand previously eroded from the windward slope. Consequently, sand will be eroded from the upper part of the windward slope and the crest will be lowered until a steady state is reached (Fig. 17.7(C)). A self-regulatory process can occur in the opposite direction. A bed-form can have a conformation that is flatter than that of the steady state (Fig. 17.7(B)). In that case the rate of increase of the speed-up ratio towards the crest is insufficient to prevent sand from being deposited on the windward slope; consequently the dune will grow in height until a steady state is reached (Fig. 17.7(C)). This is a self-regulatory process whereby sand deposition changes

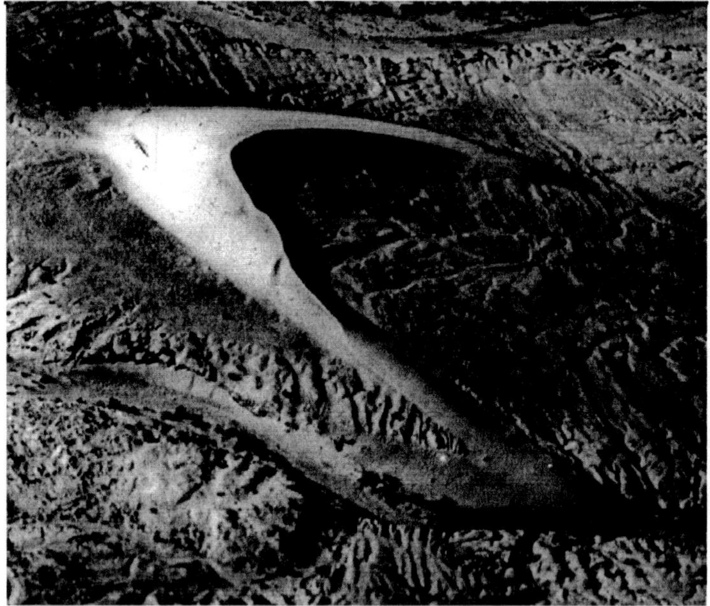

Fig. 17.8. Oblique air view of a barchan dune. Note the convex windward slope (the left side of the dune), the steep lee slope (slip-face) where avalanches occur, and the two horns typical to barchans

the bed-form shape, which, in turn, increases the wind velocity, and gradually a steady-state form is achieved. The processes of sand accumulation and self-regulation are very short, acting until the dune achieves a steady state. Therefore, most dunes in the field fall into the categories of steady state or quasi-steady state.

Sand eroded on the windward side is deposited on the upper lee side, causing oversteepening until it reaches the critical *angle of internal friction*, which is approximately 35°. At this angle the upper lee slope is not steady and failure occurs by the formation of a series of avalanche 'tongues' (Fig. 17.9) that reduce the slope angle to 32°–33° – the *angle of repose* [10].

The profile and processes described above are typical of transverse and barchan dunes, which are *migrating dunes.* They advance by erosion on the windward slope at a rate expressed in (17.5), and deposition on the lee slope, followed by avalanche that forms a *slip-face* (Fig. 17.9).

17.5.2 Transverse and Barchan Dunes

Transverse and barchan dunes are the same dune type, migrating according to the same unidirectional-wind mechanism. It is the amount of sand available for aeolian transport that causes the difference between the two. Barchans are isolated mounds of sand, formed in limited sand supply areas which overlie coarse

Fig. 17.9. A field of barchans in the Namibian desert looking upwind. Note the avalanche tongues on the slip-face

sand or non-sandy surfaces (Figs 17.8 and 17.9); a single long transverse dune is built of many barchans that have coalesced into one long dune.

In three dimensions, the wind climbing the barchan diverges a bit from the crest towards the flanks, thus increasing the speed on the barchan sides which advance more quickly than the crest and form the typical crescentic shape (Figs 17.8 and 17.9).

The rate of advance of barchan and transverse dunes according to Bagnold [42] is in direct proportion to the rate of sand transport (q) and in inverse proportion to the specific weight of the sand in bulk (γ) and the slip face height (h):

$$c = \frac{q}{\gamma h} \,. \tag{17.6}$$

Results from measurements of the barchans' rate of advance show that the link between dune height and rate of advance is not linear. Since wind increases with height, q would not be the same for dunes with dissimilar heights. High dunes will experience relatively higher sand transport than low dunes. The data from barchan displacement in Peru [49] give this exponential relation:

$$c = 33.6\,\mathrm{e}^{-0.19h} \tag{17.7}$$

and data from Sinai [50] give another exponential relation:

$$c = 8.7\,\mathrm{e}^{-0.26h} \,. \tag{17.8}$$

According to (17.6, 17.7, and 17.8), a field of barchans will arrange themselves with the smallest dunes in front and the biggest in the back.

Most of the sand of the barchan is circulated by saltation on the windward slope and avalanching on the slip-face, where it is trapped until resurfacing again on the windward slope as the dune advances one dune length. However, a barchan loses sand through horns which are devoid of slip-faces (Figs. 17.8 and 17.9). The sand that escapes through the horns of the dune is used to create another barchan downwind (Fig. 17.10). Every barchan should be in a steady state wherein the amount of sand it loses through the horns is nearly equal to the amount of sand it gains from behind.

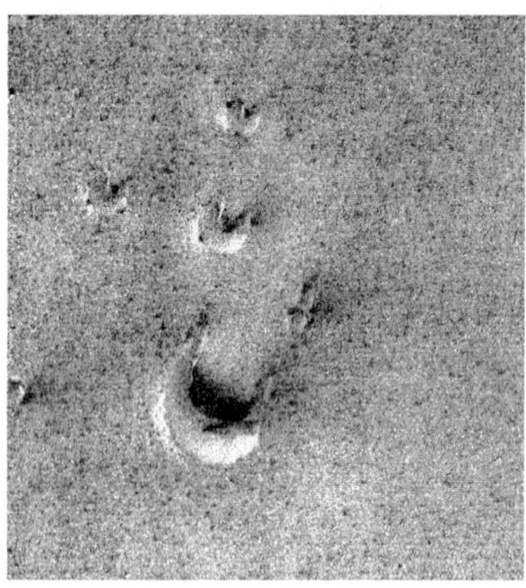

Fig. 17.10. A field of barchans on Mars (Near 76.7°N, 254.0°W). Note how the horns of one barchan serve as a source of sand for another barchan downwind. Sub-frame of MOC image SP2-45205 acquired on 26 July 1998. Area shown is approximately 2.4 km by 2.5 km and pixel sizes are approximately 3.3 meters per pixel. By courtesy of NASA/JPL/Malin Space Science Systems, San Diego

17.5.3 Linear Seif Dunes

A seif, an elongated dune type, is formed under bidirectional wind regimes beating the dune obliquely. Seifs are completely devoid of vegetation and possess a triangular profile with a sharp crest, which explains the term *seif* (an Arabic word for sword). Another typical characteristic of seifs is the tortuosity of their crest-lines, with their intermittent peaks and saddles (Fig. 17.11).

From its primary formation, the seif dune is affected by wind flows coming obliquely from both sides of its slopes, meeting the dune crest at an acute angle of attack and separating over the crest line. Each wind is diverted along the lee slope, after reattachment of the separated flow, to blow parallel to the crest-line in a down-dune direction. This process is referred to as the *flow diversion model* [51,52].

Fig. 17.11. Oblique aerial picture of seif dunes. Note the dunes' tortuosity and the sharp crest-line

It follows that two different processes act upon the lee slope of a dune. If the wind encounters the slope at a right angle, as is the case with barchan and transverse dunes, the flow will separate from the dune brink and create a separation bubble followed by an abrupt drop in wind velocity. Hence, deposition and formation of slip-face are the main processes acting upon the lee slope (Fig. 17.9). When, due to dune tortuosity, the wind direction is at an angle of 30°–40° to the crest-line, the flow separates obliquely to the crest-line and the reattachment flow on the surface of the lee slope is deflected to a direction parallel to the crest-line at a magnitude that is above the threshold speed [51]. Therefore, there is less deposition and more erosion and transport of sand in this segment of the lee slope (Fig. 17.12). The changing angle of attack explains why seif dunes meander. There are parts of the dune where angle of attack is acute and others where it is around 90° because of dune meandering (Fig. 17.12). Sand that is eroded and transported across the windward slope will not be deposited on the lee slope when the angle of attack is acute enough to create a strong diverted flow on the lee slope. This sand will be deposited on the lee slope when the dune meanders and, as a result, heightens the angle of attack. This process is exerted on the dune by bidirectional side-winds. Erosion by the wind from one side is offset by deposition of sand on the other side of the dune (Fig. 17.12). As was previously mentioned, only strong winds cause sand accumulation on desert surfaces. For that reason barchan dunes can also form by a bidirectional wind regime with one dominant strong wind direction (e.g. the W and SW sand transport winds in Fig.17.2). After the barchan is built up, the second, gentler

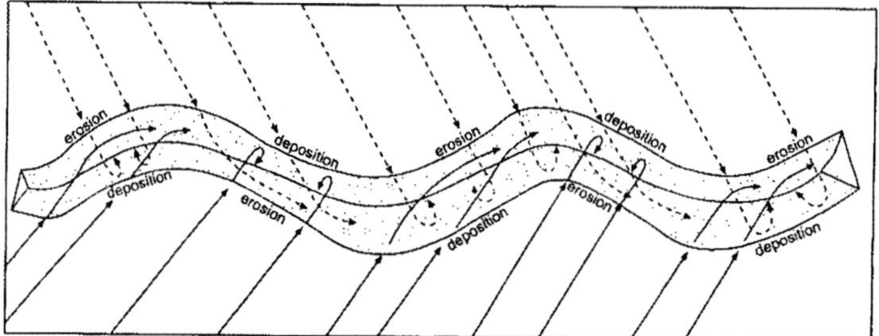

Fig. 17.12. Schematic sketch of a seif dune that is under a bidirectional wind regime (one wind direction is shown by the solid line and the other by the broken line). Note that erosion occurs on the lee side when the wind flow is diverted to flow parallel to the crest-line, and deposition when the wind encounters the crest-line perpendicularly

wind direction (e.g. the N and NW sand transport winds in Fig. 17.2) starts to affect the dune. In that case there are two main wind directions that encounter one horn of the barchan at an acute angle from both sides. Figure 17.13 is an aerial photograph of crescentic transverse dunes that are affected by a bidirectional wind regime. The strongest wind is from SW and the gentler direction is from NW. The southern horns of these transverse dunes are oriented obliquely between these two main wind directions, resulting in along-horn sand movements that elongate the horns and turn them into seif dunes (Fig. 17.13).

When seifs are formed from barchan or transverse dunes, two different aeolian bedforms are linked together in one dune system [53]. The barchan or transverse dune advances while the seif elongates. It is obvious from Fig. 17.13 that the rate of elongation is faster than the rate of advance, so it is only a matter of time until the seif dunes are the dominant dune type in this field.

17.5.4 Hybrid Dunes

In rare cases when the wind regime is bidirectional with opposing directions, the reversing wind regime will re-form a transverse dune with a complete reversing profile [42,54]. When the wind alternates from two opposite directions, the dune formed has the straight, linear, triangular shape of a *reversing dune* (Fig. 17.14). The mechanism of advance of unvegetated sand dunes can be either the transverse (barchan) mechanism, in which sand is eroded from the windward slope and deposited on the lee slope, or the linear seif mechanism, where there is also a considerable along-dune sand transport on the lee side by winds encountering the dune at an acute angle from both sides. However, in some areas the wind regime can be bidirectional where one direction is oblique to the crest line and the other is perpendicular to it. Cooper [55] noticed that Oregon coastal dunes are under such a wind regime. He termed these dunes, which resemble both transverse and linear seif dunes, *oblique ridges*. It seems more appropriate [4,56] to classify the Oregon dunes and other similar forms as *hybrid dunes*. Some see hybrid dunes

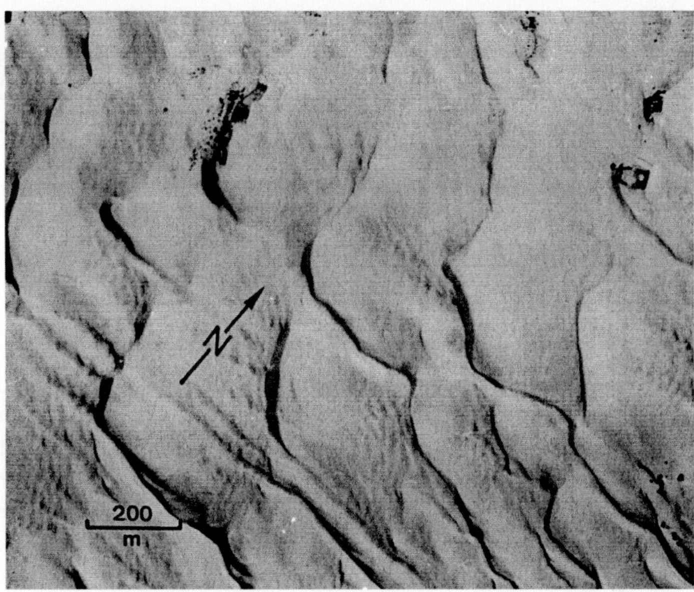

Fig. 17.13. Aerial photograph of crescent-shaped transverse dunes showing how the southern horns are turning into seif dunes

Fig. 17.14. Oblique aerial view of reversing dunes

as transverse [57] while others see them as linear dunes [58,59], although they exhibit both migrating and elongating attributes in one dune type.

Hybrid dunes can move sideways if the perpendicular wind is strong enough to form a slip-face that reaches the plinth of the dune. Pure seif dunes have small slip-faces on the lee slope undergoing erosion (Fig. 17.11) and are therefore deprived of any lateral migration [51].

The rate of elongation of the hybrid dunes is less than the rate of elongation of seif dunes because a greater proportion of the sand transport is perpendicular to the dune axis rather than parallel to it. However, hybrid dunes have a greater volume of sand per length and are therefore high dunes [4,57,58]. The confusion between hybrid and longitudinal (linear) dunes led some researchers to conclude that linear dunes migrate laterally [59–62].

17.5.5 Star Dunes

The bedforms that characterize accumulating dunes are the largest known. Star dunes are the most widespread type of accumulating dune, with sinuous arms radiating from a central, pyramid-shaped peak (Figs 17.3 and 17.15). Star dunes are formed by a wind regime with high directional variability ($RDP/DP < 0.4$) and for that reason are found in high desert latitudes where there are marked seasonal changes in wind direction [63]. Observations made in some sand seas indicate that star dunes originate as reversing or hybrid dunes in cases where sand is transported to the dune from several directions and adds to its bulk. Secondary wind directions create secondary arms that are perpendicular to the main arm of the reversing dune [63,64]. The secondary flow becomes effective once the dune increases its height and becomes exposed to winds that are below the threshold at lower elevations. Approximately 11% of all desert dunes are accumulating star dunes, and they constitute about 5% of the aeolian depositional surfaces [21].

17.6 Vegetated Dunes

Vegetation can grow on sand dunes in arid areas with less than 100 mm of annual average rainfall. The limitation of vegetation on dune sand is, first of all, human impact. The most dominant natural limitation is wind power (17.3) (Table 17.1). Rainfall is a limiting factor only where the annual average is very low (< 50 mm).

Paradoxically, vegetated surfaces cause steeper velocity gradients, and thus greater shear stress, than unvegetated surfaces. This is because vegetation causes a greater friction effect which, in turn, causes greater drag on the flow. The increased stress is not usually transferred to the ground surface, and is therefore ineffective in entraining sand. Full vegetation cover precludes aeolian entrainment but a partially vegetated canopy can only curtail particle entrainment by the wind to a certain degree [65]. Hence, dune activity can occur in the presence of vegetation.

There are several typical vegetated dune types in arid and humid areas (Fig. 17.4). Isolated clumps of vegetation act as sand traps and thus lead to the formation of *nebkhas* (coppice dunes) – hummocks that can reach up to 30 m high and 100 m across; variations in shape depend upon the shape of the canopy. They are considered to be static bedforms which change in shape as the

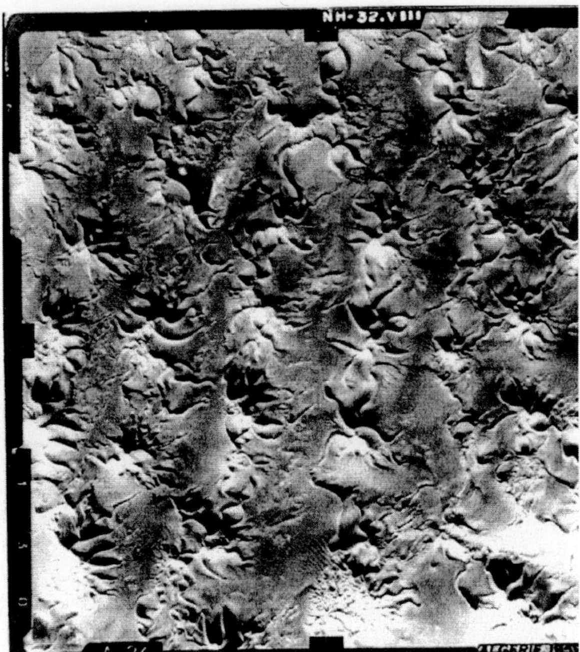

Fig. 17.15. Aerial photograph of star dunes. Note the radiating arms from the dune central peak. The main arms are from left to right and the secondary are perpendicular to them

vegetation changes with time. Experiments made by Hesp [66] demonstrated that dunes formed by the effect of isolated plants are similar to those formed by sand accumulation due to obstacles (Fig. 17.5).

17.6.1 Vegetated-linear Dunes

This dune type belongs to the group of elongating sand dunes found in many deserts of the world (Australian deserts, the Kalahari, Indian deserts and the Negev). Vegetated-linear dunes are low with rounded profiles. They range in height from a few metres up to dozens of metres. Vegetation covers them, sometimes entirely, and sometimes abundantly on the plinth and lower slopes but very sparse or absent on the crest. Those that are fully covered by vegetation have become partly or wholly stabilized. Vegetated-linear dunes may run in parallel for scores of kilometres (Fig. 17.16). An exclusive attribute of vegetated-linear dunes is the tendency for two adjacent dunes to converge and continue as a single ridge. Convergence is in the form of a Y-junction (the tuning fork shape; Fig. 17.16) commonly open to formative winds [67]. Vegetated-linear dunes are distinguished from seif dune by: 1) coextension along the strongest dominant wind direction; 2) the cover of vegetation; 3) straight alignment with no tortuosity; and 4) Y-junctions. However, when vegetation is removed, the creation of sec-

Fig. 17.16. Aerial photograph of vegetated linear dunes. These linear dunes elongate in the direction of the dominant, strong wind. Note the straight alignment of the dune (different from the meandering seif of Fig. 17.11) and the Y-junctions, which indicate wind direction from left to right

ondary, superimposed transverse dunelets with slip-faces facing downwind may change the normal low shape and rounded profile of the vegetated-linear dune (Fig. 17.17). Destruction of vegetation on vegetated-linear dunes that change their azimuth of alignment on the order of 16°–25° occurs when they converge to form a Y-junction, causing the formation of seifs (Fig. 17.17). As stated before, seif dunes form and develop under bidirectional wind regimes. Therefore, after the destruction of vegetation, the transformation takes place in those areas (Y-junctions) where vegetated-linear dunes became obliquely aligned to the strongest dominant wind.

It can be concluded from the above that vegetated-linear dunes undoubtedly owe their form and development to the vegetation cover – an important factor in the mechanism of their formation. It is worthwhile to stress that the vegetated-linear dunes in the southern and eastern Simpson Desert are located leeward of mounds that are adjacent to playas [68]. Some southwest Kalahari linear dunes also originate from pan-fringing lunette dunes [69]. In northeastern Arizona they are formed downwind of a protrusion in the cliff [70].

17.6.2 Parabolic Dunes

Parabolic dunes are mostly found in humid and cold areas. These dunes are U-shaped (parabolic) with the arms pointing upwind (Fig. 17.18). Parabolic dunes can be active and transgressive or fully stabilised and inactive. Most work on parabolic dunes and their formation was done in humid areas. However, parabolic

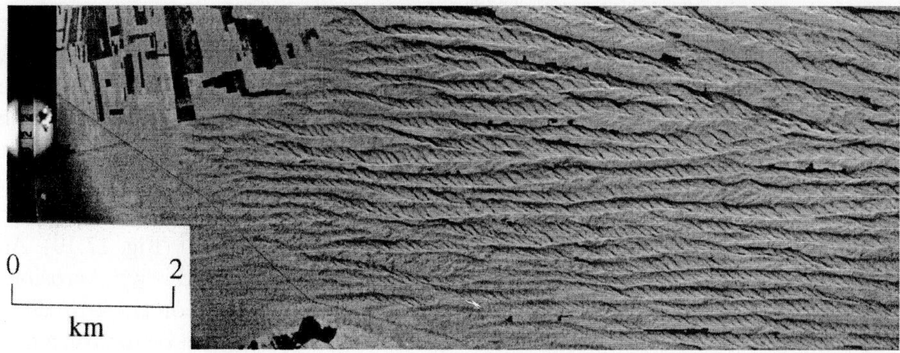

Fig. 17.17. Aerial view of linear dunes that have reacted to destruction of vegetation by grazing and trampling. Note the braided pattern of small dunelets that formed after the vegetation was destroyed. Seif dunes in the upper part of the photograph resulted from a change of 16° to 25° from the linear dune alignment because of the Y-junction

Fig. 17.18. Oblique aerial view of a parabolic dune. The wind is from right to left pushing the apex of the dune forward and leaving behind the two tails

dunes are also formed in arid and semiarid areas where vegetation is present, such as in the Jufara Desert of Arabia [71], the semiarid areas of Arizona [72], the Thar Desert of India [73] and the Kalahari desert in South Africa [5].

The mechanism of parabolic dune formation in coastal humid areas is due to the fact that vegetation is more easily established at the base of the dune near the water table. Vegetation or dampness along the lower sides of the dune retards sand motion and both are considered to be anchors. Vegetation in parabolic dune formation is said to protect the less mobile arms against wind action, thereby allowing the central part to advance downwind [74–78]. In this way the advancing apex leaves behind trailing ridges that elongate and turn the dune into a hairpin form [76]. Some see the U-shaped dune as a further development of a spot blowout [79,80].

It was recently found that barchan and transverse dunes turned into parabolic dunes in areas where the human impact was reduced or curtailed (Fig. 17.19). As mentioned above, the limiting factor for vegetation on dune sand is wind erosion. Accordingly, vegetation should be able to germinate and sprout on those areas of the dune that have little or no erosion. The rate of sand erosion or deposition is proportional to the tangent of the angle of inclination of the dune surface (17.5). According to the profile of barchan or transverse dunes (Fig. 17.8) erosion on the windward slope of the dune diminishes gradually toward the crest, which is an area of neither erosion nor deposition. Hence, once human impact stops, vegetation will recover on the barchan crest, thereby starting the process of transformation of these dunes into parabolic ones [81] (Fig. 17.19). The pioneer plant in this process of recovery is *Ammophila arenaria*, which is the hardiest shrub in shifting sand areas, able to withstand transport and burial by sand-baring erosion [82,83].

The dynamics and steady state of barchan or transverse dunes are disturbed once vegetation clutches at the dune crest. Some of the sand that is eroded from the windward side is trapped on the crest by clumps of *Ammophila arenaria* and is not deposited on the lee side. Sand deposition on the crest gradually changes the profile of the windward side of the dune from convex to concave (Fig. 17.19). The rate of wind erosion on the windward side of parabolic dunes increases because the airflow tends to compress when encountering the concave slope, both vertically and horizontally, and the velocity gradient above the dune increases [84]. Such a flow over the concave parabolic dune is characterized by funneling which strengthens the bed scour. Once vegetation is established, the dune will advance with sand eroded from the concave, windward slope, becoming trapped by the vegetation on the crest and the lee slope. The strong bed scour on the upper windward slope undercuts the shrubs and exposes their roots, thus forming a knife-edge shape at the inner apex of the dune, which is supported by the exposed roots [76]. The knife-edge shape divides the windward erosional slope from the vegetated depositional face. Where undermining breaks the edge, a wind channel may cut through [74]. Hence, the parabolic dune advances by undermining the frontal row of vegetation on the windward part of the crest. This last mechanism differs from that presented in the theory based on the anchoring of trailing arms by vegetation and the relative high advance forward of the central apex.

17.6.3 Foredunes

Foredunes are the most commonly found vegetated sand ridges on sandy backshores where pioneer vegetation can grow and trap aeolian sand (Fig. 17.20).

Fig. 17.19. Parabolic dunes (looking downwind) formed from transverse and barchan dunes a few years after the human impact was significantly reduced. The picture shows the bare windward slope of the dune, which is being eroded. Vegetation has recovered preeminently on the crest where there is little erosion

Foredunes develop into continuous vegetated ridges, which lie parallel to coastlines exposed to onshore wind energy. The foredune is the only dune type that involves the exchange of sand with the beach. Other coastal dunes are mostly transgressive types (barchan, transverse, and parabolic dunes), formed when sand is transferred inland where foredunes are absent, or through blowouts (wind-excavated gaps through which sand is transport landward) in the foredune ridges [85].

Two types of foredunes are distinguished by Hesp [86] – incipient and established. The incipient foredunes are newly developing dunes formed by the trapping of sand in pioneer plant seedlings (mostly *Ammophila arenaria*). Incipient foredunes are small (less than 2 meters high) and may be seasonal if formed in annual plants. Established foredunes develop from incipient foredunes when other vegetation species, generally woody plants, colonize the foredune. They can reach heights of up to 30–35 m but in most cases are less than 20 m [87]. Foredunes are undermined by storm waves, a process followed by some avalanching and retreat of the dunes' seaward slope. Between eroding storms, sand returns to the dune slopes in a recovery cycle [88].

Established and densely vegetated foredunes with no blowouts can obstruct the transmission of sand inland. Foredunes were formed on the west coast of North America after the introduction of *Ammophila arenaria* more than 100 years ago. The establishment of the foredunes cut off the sand nourishment to the coastal dunes. As a result, the sand surface leeward of the foredunes deflated to the level of the water table [82,89]. Erosion of the foredunes, which

Fig. 17.20. Foredune about 5m high formed on the backshore where pioneer vegetation can thrive

is commonly known to begin when vegetation is disrupted by human activities (trampling, traffic, fire or for pasturing livestock) may form blowouts.

References

1. I. Högbom: Geogr. Annlr **5**, 113 (1923)
2. B. Izmailow: Pr. Geog. **43**, 39 (1976)
3. J.W.A. Dijkmans, E.A. Koster: Geogr. Annlr **72A**, 93 (1990)
4. M.A. Carson, P.A. MacLean: Canad. J. Earth Sci. **23**, 1794 (1986)
5. P.G. Eriksson, N. Nixon, C.P. Snyman, et al.: J. Arid Environ. **16**, 111 (1989)
6. I.G. Wilson: Sed. Geol. **10**, 77 (1973)
7. J.A. Klijn: Catena (Suppl.) **18**, 1 (1990)
8. B. Thom, P. Hesp, E. Bryant: Palaeo. Palaeo. Palaeo. **111**, 229 (1994)
9. M. Sarnthein: Nature **272**, 43 (1978)
10. K. Pye, H. Tsoar: *Aeolian Sand and Sand Dunes* (Unwin Hyman, London 1990)
11. S.G. Fryberger: *A Study of Global Sand Seas*, ed. by E.D. McKee (Prof. Pap. U.S. Geol. Surv., Washington 1979), **1052**, p. 137
12. H. Tsoar: Aarhus Geosci. **7**, 79 (1997)
13. H. Tsoar, W. Illenberger: J. Arid Land Studies **7**S, 265 (1998)
14. H. Tsoar, J.T. Møller: *Aeolian Geomorphology*, ed. by W.G. Nickling (Allen and Unwin, Boston 1986) p. 75

15. E.D. McKee: *A Study of Global Sand Seas*, ed. by E.D. McKee (Prof. Pap. U.S. Geol. Surv, Washington, D.C. 1979), **1052**, p. 1
16. R.J. Wasson, R. Hyde: Nature **304**, 337 (1983)
17. D.G.S. Thomas: J. Arid Environ. **22**, 31 (1992)
18. N. Lancaster: *Geomorphology of Desert Dunes* (Routledge, London 1995)
19. D.S.G. Thomas: *Arid Zone Geomorphology*, ed. by D.S.G. Thomas (Wiley, Chichester 1997), p. 373
20. B. White, H. Tsoar: Geomorphology **22**, 159 (1998)
21. S.G. Fryberger, A.S. Goudie: Prog. Phys. Geog. **5**, 420 (1981)
22. K.K. Bofah, W. Ahmed: *Sand Transport and Desertification in Arid Lands*, ed. by F. El-Baz, I.A. El-Tayeb, M.H.A. Hassan (World Scientific, Singapore 1990), p. 389
23. R. Greeley: *Aeolian Geomorphology*, ed. by W.G. Nickling (Allen and Unwin., Boston 1986), p. 195
24. H. Tsoar: *Eolian Sediments and Processes*, ed. by M.E. Brookfield, T.S. Ahlbrandt (Elsevier, Amsterdam 1983), p. 247
25. R.S.U. Smith: *Aeolian Features of Southern California: A Comparative Planetary Geology Guide Book*, ed. by R. Greeley, M.B. Womer, R.P. Papson, P.D. Spudis (Arizona State Univ., College of the Desert and NASA – Ames Res. Center 1978) p. 66
26. R.S.U. Smith: *Reference Handbook on the Deserts of North America*, ed. by G.L. Bender (Greenwood Press, Westport 1982), p. 481
27. G.B. Hellström, R.A. Lubke: J. Coast. Res. **9**, 647 (1993)
28. N. Lancaster: J. Arid Environ. **27**, 113 (1994)
29. H. Tsoar, D. Blumberg: Acta Mech. (suppl.) **2**, 131 (1991)
30. B. Hétu: Earth Surf. Proc. Landf. **17**, 95 (1992)
31. H. Tsoar, B. White, E. Berman: Lands. Urban Plan. **34**, 171 (1996)
32. R. Greeley, J.D. Iversen: *Wind as a Geological Process* (Cambridge Univ. Press, Cambridge 1985)
33. R.A. Bagnold: Geogr. J. **89**, 409 (1937)
34. I.G. Wilson: Sedimentology **19**, 173 (1972)
35. R.L. Folk: Sedimentology **23**, 649 (1976)
36. A. Warren: *Process in Geomorphology*, ed. by C. Embleton, J. Thornes (Edward Arnold, London 1979), p. 325
37. P.S. Jackson, J.C.R. Hunt: Q. J. R. Meteorol. Soc. **101**, 929 (1975)
38. P.J. Mason, R.I. Sykes: Quart. J. R. Met. Soc. **105**, 383 (1979)
39. J.R. Pearse, D. Lindley, D.C. Stevenson: Boundary-Layer Meteorol. **21**, 77 (1981)
40. J.J. Finnigan, M.R. Raupach, E.F. Bradley, et al.: Boundary Layer Meteorol. **50**, 277 (1990)
41. H. Tsoar: Geogr. Annlr **67**, 47 (1985)
42. R.A. Bagnold: *The physics of blown sand and desert dunes* (Methuen, London 1941)
43. A.D. Howard, J.B. Morton, M. Gad-el-Hak, et al.: Sedimentology **25**, 307 (1978)
44. W.B. Bull: Bull. Geol. Soc. Am **86**, 1489 (1975)
45. A.J. Bowen, D. Lindley: Boundary-Layer Meteorol. **12**, 259 (1977)
46. A.D. Howard, J.L. Walmsley. In: *Proc. Int. Workshop on the Physics of Blown Sand*, ed. by O.E. Barndorff-Nielsen, J.T. Moller, K.R. Rasmussen, B.B. Willetts (Dept. Theoretical Statistics, Universty of Aarhus 1985), **8**, p. 377
47. F.K. Wipperman, G. Gross: Boundary-Layer Meteorol. **36**, 319 (1986)
48. H. Momiji, R. Carretero-Gonzalez, S.R. Bishop, et al.: Earth Surf. Process. Landf. **25**, 905 (2000)

49. H.J. Finkel: J. Geol **67**, 614 (1959)
50. H. Tsoar: Z. Geomorph. Supp. Bd. **20**, 41 (1974)
51. H. Tsoar: Sedimentology **30**, 567 (1983)
52. H. Tsoar, D.H. Yaalon: Sed. Geol. **36**, 25 (1983)
53. H. Tsoar: Z. Geomorph **28**, 99 (1984)
54. J.R. Burkinshaw, W.K. Illenberger, I.C. Rust: *The Dynamics and Environmental Context of Aeolian Sedimentary Systems*, ed. by K. Pye (Geological Society, Bath 1993), **72**, p. 25
55. W.S. Cooper: *Coastal Sand Dunes of Oregon and Washington* (Geol. Soc. Am. Mem. 72, New York 1958)
56. M.A. Carson, P.A. MacLean: Bull. Geol. Soc. Am. **96**, 409 (1985)
57. R.E. Hunter, B.M. Richmond, T.R. Alpha: Bull. Geol. Soc. Am. **94**, 1450 (1983)
58. I. Livingstone: Geography **73**, 105 (1988)
59. C.S. Bristow, S.D. Bailey, N. Lancaster: Nature **406**, 56 (2000)
60. D.M. Rubin, R.E. Hunter: Sedimentology **32**, 147 (1985)
61. P. Hesp, R. Hyde, V. Hesp, et al.: Earth Surf. Proc. Landf. **14**, 447 (1989)
62. D.M. Rubin: Earth Surf. Proc. Landf. **15**, 1 (1990)
63. I.N. Lancaster: Sedimentology **36**, 273 (1989)
64. I.N. Lancaster: Prog. Phys. Geogr. **13**, 67 (1989)
65. S.A. Wolfe, W.G. Nickling: Prog. Phys. Geog. **17**, 50 (1993)
66. P.A. Hesp: J. Sed. Petrol. **51**, 101 (1981)
67. D.S.G. Thomas: Z. Geomorph. **30**, 231 (1986)
68. C.R. Twidale: Die Erde **112**, 231 (1981)
69. D.S.G. Thomas, H.E. Martin: J. Sed. Petrol. **57**, 572 (1987)
70. F.A. Melton: J. Geol. **48**, 113 (1940)
71. D. Anton, P. Vincent: J. Arid Environ. **11**, 187 (1986)
72. J.T. Hack: Geogr. Rev. **31**, 240 (1941)
73. R.J. Wasson, S.N. Rajaguru, V.N. Misra, et al.: Z. Geomorph. Supp. Bd. **45**, 117 (1983)
74. S.Y. Landsberg: Geogr. J. **122**, 176 (1956)
75. P. David: GEOS **18**, 12 (1979)
76. K. Pye: Geogr. Annlr. **64**A, 213 (1982)
77. C.H. Thompson, A.W. Moore: (CSIRO, Division of Soils, 1984), p. 1
78. I. Livingstone, A. Warren: *Aeolian Geomorphology: An Introduction* (Longman, Harlow 1996)
79. P.A. Hesp, R. Hyde: Sedimentology **43**, 505 (1996)
80. J.S. Olson, E. van der Maarel: *Perspectives in Coastal Dune Management*, ed. by F. Van Der Meulen, P.D. Jungerius, J. Visser (SPB Academic Publishing, The Hague 1989), p. 3
81. H. Tsoar, D. Blumberg: *Symposium on Process Geomorphology*, Netherlands Centre for Geo-ecological Research (ICG, University of Amsterdam 1999), p. 14
82. A.M. Wiedemann, A. Pickart: Landsc. Urban Plan. **34**, 287 (1996)
83. A.B. Arun, K.R. Beena, N.S. Raviraja, et al.: Curr. Sci. **77**, 19 (1999)
84. P.P. David: Geol. Assoc. Canada and Geol. Soc. Am. Annual Meeting **3**, 385 (1978)
85. N.P. Psuty: Proc. R. Soc. Edin. **96**b, 289 (1989)
86. P.A. Hesp: *Coastal Geomorphology in Australia*, ed. by B.G. Thom (Academic Press, London 1984) p. 69
87. P.A. Hesp, *Handbook of Beach and Shoreface Morphodynamics*, ed. by A.D. Short (John Wiley & Sons, Chichester 1999) p. 145

88. R.W.G. Carter, P.A. Hesp, K.F. Nordstrom: *Coastal Dunes Form and Process*, ed. by K.F. Nordstrom, N. Psuty, B. Carter (Wiley, Chichester England, 1990) p. 217
89. A.M. Wiedemann: J. Coastal Res. – Special Issue **26**, 45 (1998)
90. N. Lancaster, R. Greeley, P.R. Christensen: Earth Surf. Proc. Landf. **12**, 277 (1987)

18 Dunes and Drumlins

A.C. Fowler

Mathematical Institute, Oxford University, 24-29 St Giles', Oxford OX1 3LB

18.1 Introduction

18.1.1 Dunes

Dunes are landforms which occur when a turbulent fluid flow occurs above an erodible substrate. The most obvious example occurs in deserts, where the wind blows sand into a wide variety of different shapes (see Chap. 17 for many illustration of such dunes). Linear dunes, or 'seifs', are ridges which form parallel to the prevailing wind direction, while transverse dunes are ridges perpendicular to the wind. A variety of other shapes can occur, amongst them star dunes and barchan dunes.

Dunes also occur under rivers, for similar reasons. Because the flow in this situation is uni-directional, such exotica as star dunes do not occur. On the other hand, when the flow is rapid enough, *anti-dunes* occur; these are associated with waves at the water surface which are in phase with the underlying bed forms.

Dunes occur due to an instability which arises through a coupling between the bed transport rate and the overlying flow. In rivers and deserts, the bed material is transported (in rivers as *bedload*) through the imposition of a wind or water driven shear stress. If a perturbation in the bed elevation occurs, then the increased roughness alters the bed shear stress, and hence the bed transport rate. It is this feedback which causes the instability. As we shall see, the instability relies crucially on the fact that the perturbed shear stress is out of phase with the perturbed bed form.

18.1.2 Drumlins

Drumlins are small oval hills. They occur in swarms in regions which were formerly covered by ice sheets (in the last ice age). For example, much of the northern part of Ireland is covered by drumlins. They are typically formed of subglacial *till*, which is a dispersion of coarse, angular rock fragments in a matrix of finer grained material. Drumlins have typical dimensions of 100–1000 metres in length, and 10–50 metres elevation. As in the case for dunes, drumlins come in many different forms. In particular, analogues to various different dune types exist. Rogen moraine consists of ridges aligned perpendicular to the former ice flow, while glacial flutes and mega-flutes are lineations parallel to the ice flow. The three-dimensional drumlin forms themselves also have varying styles, such as spindle drumlins and barchanoid drumlins, and it has been suggested that they occur through the action of massive subglacial floods, which erode the bedforms by analogy with dunes. While this is not inconceivable, it does require subglacial floods on a scale more massive than is commonly thought possible.

In this paper we show how an erosional theory for dune formation can be provided, and we will also show how the analogous theory for ice sheet flow can predict drumlin-forming instabilities, despite the vast disparity in Reynolds number. Notes and references to some of the literature can be found at the end of the paper.

18.2 Dunes

The basic geometry of the system is shown in Fig. 18.1. The water surface is $z = \eta$, where z is a coordinate normal to the mean bed slope, and the bed is $z = s$. For simplicity we consider only two-dimensional motions, so that $s = s(x,t)$, $\eta = \eta(x,t)$. The water depth is thus

$$h = \eta - s\,, \tag{18.1}$$

and the basic equation which describes the evolution of s is the *Exner equation*

$$(1-n)\frac{\partial s}{\partial t} + \frac{\partial q}{\partial x} = 0\,. \tag{18.2}$$

Here, n is the porosity of the bed and q is the bedload transport, usually written as a prescribed function of the mean basal shear stress τ; a typical example is the Meyer–Peter and Müller [23] law

$$q = q(\tau) = C(\tau - \tau_c)_+^{3/2}\,, \tag{18.3}$$

where $[x]_+ = \max(x, 0)$, and τ_c is a yield stress, called the *Shields stress* (after Shields [30]); bedload transport only occurs for stresses above this value.

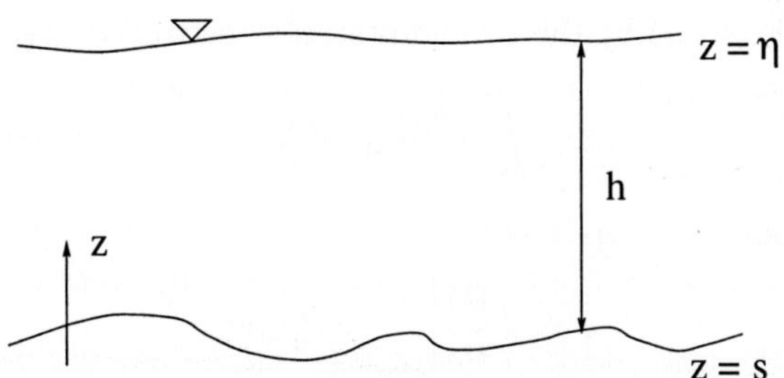

Fig. 18.1. Geometry for dune model. Water of depth h flows over an erodible bed $z = s$

The bedload transport q is an increasing function of τ, which itself depends on the mean flow velocity $\bar{u}$, for example we can take

$$\tau = f\rho_w\bar{u}^2\,, \tag{18.4}$$

where ρ_w is the density of water, and f is a dimensionless friction factor having a typical value of 0.05. It varies somewhat with flow speed and bed roughness, but can be taken as constant in an initial study.

18.2.1 St. Venant Equations

The equations (18.1)–(18.4) provide four equations for the six variables $s, q, \tau, \bar{u}$, h, η, and two further equations are necessary to complete the set. These arise from mass and momentum conservation, and an attractive possibility is to use the St. Venant equations to express these.

The St. Venant equations are the classical averaged equations which are used to describe turbulent river flow. They can be derived by taking cross-sectional averages (or, for a wide channel with no cross-stream variation of depth, depth averages) of the Navier–Stokes equations. For a two-dimensional velocity field $(u, 0, w)$ down a slope S, using depth averages, we obtain

$$\frac{\partial h}{\partial t} + \frac{\partial}{\partial x}\int_s^\eta u\,\mathrm{d}z = 0,$$

$$\frac{\partial}{\partial t}\int_s^\eta u\,\mathrm{d}z + \frac{\partial}{\partial x}\int_s^\eta u^2\,\mathrm{d}z = gh(S-\eta_x) - \frac{\mu}{\rho_w}\frac{\partial u}{\partial z}\bigg|_{z=s}, \tag{18.5}$$

using in addition the assumption of a shallow flow, so that the pressure is approximately hydrostatic,

$$p \approx \rho_w g(\eta - z), \tag{18.6}$$

and neglecting longitudinal stress terms.

The average velocity $\bar{u}$ is defined by

$$\bar{u} = \frac{1}{h}\int_s^\eta u\,\mathrm{d}z, \tag{18.7}$$

and the system is closed by the two additional constitutive assumptions, that

$$\int_s^\eta u^2\mathrm{d}z = h\bar{u}^2, \tag{18.8}$$

and that the basal shear stress is

$$\tau = \mu\,\frac{\partial u}{\partial z}\bigg|_{z=s} = f\rho_w\bar{u}^2. \tag{18.9}$$

These conditions, particularly the latter, can be derived providing some suitable assumptions about the form of the turbulent shear flow near the boundary are made. The classical St. Venant equations are then (now writing $\bar{u} = u$)

$$\begin{aligned}\frac{\partial h}{\partial t} + \frac{\partial}{\partial x}(hu) &= 0, \\ \frac{\partial u}{\partial t} + u\frac{\partial u}{\partial x} &= g(S-\eta_x) - \frac{fu^2}{h}.\end{aligned} \tag{18.10}$$

Suppose the volume flux (per unit width) of the river is Q_0, and is prescribed. We non-dimensionalise the St. Venant equations using length scales h_0, velocity scale u_0, time scale t_0 and bedload transport scale q_0, chosen so that

$$u_0 h_0 = Q_0 \,, \quad gS = \frac{f u_0^2}{h_0} \,, \quad t_0 = \frac{h_0^2}{q_0(1-n)} \,, \tag{18.11}$$

where q_0 is chosen so that $q \sim q_0$ when $\tau \sim f\rho_w u_0^2$. The resulting dimensionless Exner–St. Venant model is

$$\begin{aligned} &\frac{\partial s}{\partial t} + \frac{\partial q}{\partial x} = 0 \,, \\ &\varepsilon\frac{\partial h}{\partial t} + \frac{\partial}{\partial x}(uh) = 0 \,, \\ &F^2\left(\varepsilon\frac{\partial u}{\partial t} + u\frac{\partial u}{\partial x}\right) = -\eta_x + \delta\left(1 - \frac{u^2}{h}\right) , \end{aligned} \tag{18.12}$$

and the parameters are defined by

$$\varepsilon = \frac{(1-n)q_0}{Q_0} \,, \quad F = \frac{u_0}{\sqrt{gh_0}} \,, \quad \delta = S \,. \tag{18.13}$$

Typical values are $\varepsilon \sim 10^{-2}$, $\delta \sim 10^{-3}$, $F < 1$ (if we restrict our attention to dunes), and the simplifying assumptions $\varepsilon \to 0$, $\delta \to 0$ lead to

$$\begin{aligned} &uh = 1 \,, \\ &\tfrac{1}{2}F^2u^2 + \eta = \tfrac{1}{2}F^2 + 1 \,, \end{aligned} \tag{18.14}$$

referring to a basic (scaled) state $u = h = 1$, $s = 0$. These provide the extra two equations to complete the model. Note that the specific assumptions $\varepsilon \ll 1$, $\delta \ll 1$ (both realistic) obviate the necessity to specify (18.8) and (18.9). However (18.14) does require the water flow to be slowly varying (and specifically, $s \ll 1$ or $\partial s/\partial x \ll 1$).

Elimination of h and η yields $s = s(u)$,

$$s = 1 - \frac{1}{u} + \tfrac{1}{2}F^2(1-u^2) \,, \tag{18.15}$$

and since also $q = q(\tau) = q(u)$, then $q = q(s)$, and the Exner equation is simply

$$\frac{\partial s}{\partial t} + q'(s)\frac{\partial s}{\partial x} = 0 \,, \tag{18.16}$$

or equivalently

$$\frac{\partial q}{\partial t} + \frac{1}{s'(q)}\frac{\partial q}{\partial x} = 0 \,. \tag{18.17}$$

For example, the (scaled) choice $q = \tau^{3/2} = u^3$ leads to

$$v(q) = \frac{1}{s'(q)} = \frac{3q^{4/3}}{1 - F^2 q} \,. \tag{18.18}$$

When $F < 1$, the wave speed $v(q)$ is an increasing function of q (at $q = 1$), so that shocks form and propagate downstream. Dunes are indeed shocks (and do propagate downstream); also

$$\frac{\mathrm{d}s}{\mathrm{d}\eta} = \frac{F^2 - h^3}{F^2} , \tag{18.19}$$

so that for $F < 1$, $\mathrm{d}s/\mathrm{d}\eta < 0$ at $h = 1$, and the water surface is out of phase with the bed, as is the case for dunes.

18.2.2 Instability

Of course, (18.16) does not predict instability, it simply evolves prescribed perturbations into dunes. The key to instability lies in the observation that the presence of a perturbation of the bed causes a disturbance to the flow, and this is not manifested in the St. Venant model. However, it is not in fact the shallow water assumptions (18.14) that are at fault, but rather the prescription of the shear stress in (18.9). From (18.15) $u = u(s)$, and $\mathrm{d}u/\mathrm{d}s = u^2/(1 - F^2u^3)$; at $u = 1$, $\mathrm{d}u/\mathrm{d}s = 1/(1 - F^2) > 0$ for $F < 1$: u is exactly in phase with s. Hence also τ is exactly in phase with s.

This is not realistic, because the presence of the bump perturbs the flow, and intuitively we expect that the shear stress will be greatest on the upstream face of a bump. One simple way to represent this is to replace the dimensionless version of (18.9), $\tau = u^2$, by

$$\tau = u^2|_{x+l} , \tag{18.20}$$

so that the shear stress leads the velocity (and thus the bed). This idea was introduced by Kennedy [19], and leads to instability, since with

$$q = q[s(x + l, t)] , \tag{18.21}$$

linearisation of the Exner equation via

$$q = 1 + \bar{q}\mathrm{e}^{\mathrm{i}kx+\sigma t} , \quad s = \bar{s}\mathrm{e}^{\mathrm{i}kx+\sigma t} , \tag{18.22}$$

yields

$$\bar{q} = q'(0)\mathrm{e}^{\mathrm{i}kl}\bar{s} , \quad \sigma\bar{s} + \mathrm{i}k\bar{q} = 0 , \tag{18.23}$$

and thus

$$\sigma = kq'(0)(\sin kl - \mathrm{i}\cos kl) , \tag{18.24}$$

and instability for $l > 0$. The model (18.21) is not actually very good ($\sigma \sim k$ at $k \to \infty$ indicating ill-posedness), but it does point out the modification which needs to be made.

18.2.3 The Orr–Sommerfeld Model

In order to compute the effect of the bed on the shear stress, we must consider the structure of the shear flow above the bed. The simplest model for the turbulent flow is one in which there is an eddy viscosity, which we take to be μ_T, and independent of position (for simplicity). The governing equations for the two-dimensional time-averaged velocity field $(u, 0, w)$ (note u is not the depth-averaged mean, but reverts to its original meaning) are

$$
\begin{aligned}
u_t + uu_x + wu_z &= -\frac{1}{\rho_w}\frac{\partial p}{\partial x} + \nu_T \nabla^2 u + gS \,, \\
w_t + uw_x + ww_z &= -\frac{1}{\rho_w}\frac{\partial p}{\partial z} + \nu_T \nabla^2 w - g \,, \\
u_x + w_z &= 0 \,;
\end{aligned}
\tag{18.25}
$$

here $\nu_T = \mu_T/\rho_w$ is the kinematic eddy viscosity. If the downstream slope is $S = \sin\alpha$, then g in the second equation is an approximation for $g\cos\alpha$, but the difference is slight.

Boundary conditions are those of zero stress at the top surface, and no slip at the base (modifications may be necessary for more realistic eddy viscosity models). The object is to calculate the basal shear stress

$$
\tau \approx \mu_T \left.\frac{\partial u}{\partial z}\right|_{z=s}
\tag{18.26}
$$

for s small but non-zero (whence the approximation in (18.26) is valid). We take as a reference point the supposition that $\tau = f\rho_w \bar{u}^2$ when $s = 0$; this will provide us with a consistent (flow-dependent) definition of μ_T.

In the uniform state where $s = 0$, we find

$$
u = \frac{gS}{\nu_T}\left(hz - \frac{1}{2}z^2\right) ,
\tag{18.27}
$$

so that

$$
\bar{u} = \frac{gS}{3\nu_T} h^2 ,
\tag{18.28}
$$

and these are consistent with $\tau = f\rho_w \bar{u}^2$ providing we have

$$
\nu_T = \varepsilon_T \bar{u} h ,
\tag{18.29}
$$

and we choose

$$
\varepsilon_T = f/3 .
\tag{18.30}
$$

Thus, we suppose ν_T is defined by (18.29) and (18.30), and this reproduces (18.9) for a uniform flow.

Now we want to modify the solution to find an expression for τ when $s \neq 0$. First we scale the model as before, and scale $p - \rho g(h_0 - z)$ with $\rho_w u_0^2$. Neglecting the small time derivatives as before ($\varepsilon \to 0$), we have

$$\begin{aligned} uu_x + wu_z &= -p_x + \frac{1}{R}\nabla^2 u + \frac{S}{F^2} , \\ uw_x + ww_z &= -p_z + \frac{1}{R}\nabla^2 w , \\ u_x + w_z &= 0 , \end{aligned} \tag{18.31}$$

and the new parameter is the turbulent Reynolds number

$$R = \frac{u_0 h_0}{\nu_T} = \frac{1}{\varepsilon_T} . \tag{18.32}$$

Although in general, $\bar{u}$ and h may differ (when $s \neq 0$) from u_0 and h_0, we note that $\bar{u}h = u_0 h_0$, and thus it is consistent to define the eddy viscosity $\nu_T = \varepsilon_T Q_0$, so that for a given discharge it is constant, whether s is constant or not.

We specifically choose u_0 to be the mean steady velocity (even if $s \neq 0$), so that the dimensionless uniform velocity profile (when $s = 0$) is

$$U(z) = 3\left(z - \frac{1}{2}z^2\right) . \tag{18.33}$$

Now we write

$$(u, w) = (U(z) + \psi_z, -\psi_x) , \tag{18.34}$$

where ψ is small (as s is), and thus the model is linearised: we find

$$U\nabla^2\psi_x - U''\psi_x = \frac{1}{R}\nabla^4\psi . \tag{18.35}$$

The condition of zero normal stress at the surface becomes, approximately,

$$\eta \approx 1 - F^2 p|_{z=1} , \tag{18.36}$$

and if we conveniently suppose $F^2 \ll 1$, then we can take $\eta \equiv 1$ in the perturbed flow. The linearised boundary conditions for the flow are then

$$\begin{aligned} \psi = \psi_{zz} = 0 \quad &\text{at} \quad z = 1 , \\ \psi = 0 , \quad \psi_z = -U_0' s \quad &\text{at} \quad z = 0 , \end{aligned} \tag{18.37}$$

where $U_0' = U'(0) = 3$.

If we can solve this problem, then the dimensional basal shear stress is

$$\tau = \rho_w \varepsilon_T u_0^2 U_0' \left(1 + s\frac{U_0''}{U_0'} + \frac{1}{U_0'}\psi_{zz}|_{z=0}\right) , \tag{18.38}$$

and since $f = 3\varepsilon_T = \varepsilon_T U_0'$, $u_0 = \bar{u}$, this is

$$\tau = f\rho_w \bar{u}^2 \left(1 + \frac{sU_0''}{U_0'} + \frac{1}{U_0'}\psi_{zz}|_0\right) . \tag{18.39}$$

We write (time dependence is implicit)

$$
\begin{aligned}
s &= \int_{-\infty}^{\infty} \hat{s}(k) e^{ikx}\, dk\,, \\
\psi &= -U_0' \int_{-\infty}^{\infty} \hat{s} e^{ikx} \Psi(z,k)\, dk,
\end{aligned}
\tag{18.40}
$$

so that Ψ satisfies

$$
U(\Psi'' - k^2\Psi) - U''\Psi = \frac{1}{ikR}(\Psi^{iv} - 2k^2\Psi'' + k^4\Psi)\,, \tag{18.41}
$$

with

$$
\begin{aligned}
\Psi &= \Psi'' = 0 && \text{on } z = 1\,, \\
\Psi &= 0\,, \quad \Psi' = 1 && \text{on } z = 0\,.
\end{aligned}
\tag{18.42}
$$

Engelund [8] and Smith [31] solved this problem numerically, incorporating it into a linearised model of the Exner equation, and finding instability to occur, essentially because of the phase shift of the shear stress. However, since the parameter $1/R = f/3$ is relatively small (e.g. $1/R = 0.02$ if $f = 0.06$), an alternative approach is to derive an asymptotic solution based on the limit $R \gg 1$.

This can be done using an analysis pioneered by Bill Reid, and expounded in the book by Drazin and Reid [6]. The neglect of the terms of $O(1/R)$ in (18.41) gives a second order equation which can be solved by Frobenius's method to give power series solutions in z. Near $z = 0$, there is a boundary layer of (complex) thickness $(ikRU_0')^{-1/3}$, in which the approximating equations can be solved in terms of a class of generalised Airy functions introduced by Reid [25], defined explicitly by contour integral representations, whose asymptotic form far from $z = 0$ can be explicitly computed. Matching of the two expansions can be carried out, and it is found that

$$
\Psi''(0) \sim -3(ikRU_0')^{1/3}\mathrm{Ai}(0) + O(1) \tag{18.43}
$$

for $k > 0$, where $i^{1/3} = e^{i\pi/6}$ (and $\Psi''(0,-k) = \overline{\Psi''(0,k)}$, where the overbar denotes the complex conjugate).

Now the basal stress is, from (18.39),

$$
\tau = f\rho_w \bar{u}^2 \left[1 - s - \int_{-\infty}^{\infty} e^{ikx} \hat{s}(k) \Psi''(0,k)\, dk\right], \tag{18.44}
$$

and by use of the convolution theorem, this is

$$
\tau = f\rho_w \bar{u}^2 \left[1 - s + \int_{-\infty}^{\infty} K(x-\xi) \frac{\partial s}{\partial \xi}\, d\xi\right], \tag{18.45}
$$

where the kernel $K(x)$ is

$$
K(x) = -\frac{1}{2\pi} \int_{-\infty}^{\infty} \frac{\Psi''(0,k)}{ik} e^{ikx}\, dk\,. \tag{18.46}
$$

Writing $c = 3(RU_0')^{1/3}\mathrm{Ai}(0)$ in (18.43), we have

$$K(x) = \frac{c}{\pi}\int_0^\infty \cos(kx - \tfrac{\pi}{3})\frac{\mathrm{d}k}{k^{2/3}}\,, \tag{18.47}$$

and this can be evaluated to give

$$\begin{aligned} K(x) &= \frac{\mu}{x^{1/3}}\,, \quad x > 0\,, \\ &= 0\,, \quad x < 0\,, \end{aligned} \tag{18.48}$$

where

$$\mu = \frac{3^{2/3}R^{1/3}}{\Gamma(\frac{2}{3})^2} \approx 1.98R^{1/3}\,. \tag{18.49}$$

Thus (18.45) is

$$\tau = f\rho_w\bar{u}^2\left[1 - s + \mu\int_0^\infty \xi^{-1/3}\frac{\partial s}{\partial x}(x-\xi,t)\,\mathrm{d}\xi\right]: \tag{18.50}$$

the shear stress is corrected by a weighting which increases its value upstream of maxima of s, consistent with our previous heuristic expectations. For stability purposes, note that with

$$K(x) = \int_{-\infty}^\infty \hat{K}(k)\mathrm{e}^{\mathrm{i}kx}\mathrm{d}k, \tag{18.51}$$

then

$$\hat{K} = -\frac{\Psi''(0,k)}{2\pi\mathrm{i}k} = \frac{c}{2\pi k^{2/3}}\mathrm{e}^{-\mathrm{i}\pi/3}\,, \quad k > 0\,. \tag{18.52}$$

18.2.4 Orr–Sommerfeld–Exner–St. Venant Model

As explained previously, as long as s is small, the shallow water approximation (18.14) applies, and thus $s \approx s(u)$. In fact, when F is small, then $\eta \approx 1$, $h \approx 1-s$ and thus

$$u \approx \frac{1}{1-s}\,. \tag{18.53}$$

The dimensionless Exner model thus becomes

$$\begin{aligned} \frac{\partial s}{\partial t} + \frac{\partial q}{\partial x} &= 0\,, \\ \tau \approx \frac{1}{1-s} + \frac{1}{(1-s)^2}&\int_{-\infty}^\infty K(x-\xi)\frac{\partial s}{\partial \xi}(\xi,t)\,\mathrm{d}\xi\,, \end{aligned} \tag{18.54}$$

where $q = q(\tau)$. We write $s = \hat{s}\mathrm{e}^{\mathrm{i}kx+\sigma t}$, $\tau = 1 + \hat{\tau}\mathrm{e}^{\mathrm{i}kx+\sigma t}$, and then a linearisation of (18.54) yields

$$\begin{aligned} \sigma\hat{s} + \mathrm{i}kq'(1)\hat{\tau} &= 0\,, \\ \hat{\tau} &= \hat{s} + 2\pi\mathrm{i}k\hat{K}\hat{s}\,, \end{aligned} \tag{18.55}$$

whence (for $k > 0$)

$$\begin{aligned}\sigma &= 2\pi k^2 q'(1)\hat{K} - \mathrm{i}kq'(1)\\ &= cq'(1)k^{4/3}\mathrm{e}^{-\mathrm{i}\pi/3} - \mathrm{i}kq'(1)\,.\end{aligned}\tag{18.56}$$

The growth rate is thus

$$\mathrm{Re}\,\sigma = \tfrac{1}{2}cq'(1)k^{4/3}\,,\tag{18.57}$$

which is positive, denoting instability; the wave speed is

$$-\frac{\mathrm{i}\sigma}{k} = q'(1)\left(1 + \frac{\sqrt{3}}{2}ck^{1/3}\right)\,,\tag{18.58}$$

and waves propagate downstream.

18.2.5 Well-posedness

This model is also ill-posed because of the rapid growth of short wavelength disturbances. The remedy here is to account for the local slope of the bed on the mobility. For a particle of grain diameter D_s, the buoyancy-induced stress τ_p acting in the x direction on the particle is approximately $\tau_p = -\Delta\rho g D_s \partial s/\partial x$, where $\Delta\rho = \rho_s - \rho_w$, ρ_s is sediment density, for small s. Thus the effective driving stress for bedload transport is $\tau + \tau_p$, and we should take q as a function of $\tau + \tau_p$. Equivalently we add τ_p to the definition of τ, and when this is non-dimensionalised, we replace $(18.54)_2$ (the notation $(a)_b$ indicates the b-th equation of the equation set (a)) by

$$\tau = \frac{1}{1-s} + \frac{\mu}{(1-s)^2}\int_0^\infty \xi^{-1/3}\frac{\partial s}{\partial x}(x-\xi,t)\,\mathrm{d}\xi - \beta\frac{\partial s}{\partial x}\,,\tag{18.59}$$

where

$$\beta = \frac{\Delta\rho g D_s}{f\rho_w u_0^2} = \frac{\Delta\rho D_s}{\rho_w S h_0}\,.\tag{18.60}$$

Typical values of $D_s = 1\,\mathrm{mm}$, $h_0 \gtrsim 1\,\mathrm{m}$, $S \gtrsim 10^{-3}$, give values $\beta \sim O(1)$. The extra term is diffusive, and the growth rate (18.57) is modified as

$$\mathrm{Re}\,\sigma = q'(1)\left(\tfrac{1}{2}ck^{4/3} - \beta k^2\right)\,,\tag{18.61}$$

and high wave number disturbances decay.

18.2.6 The Canonical Dune Equation

The linear integral correction to the stress is computed on the basis that s is small compared to the depth, and also uses the fact that R is large. On the face of it, this implies that only the linearisation of (18.59) gives a self-consistent approximation. However, it is possible to argue that some nonlinear terms in (18.59) may be included, at least formally, in certain circumstances. Suppose the amplitude of variations in s is of order $s_0 \ll 1$ and varies on a length scale $L \gg 1$. Then if $s_0 \sim \mu/L \sim \beta/L$, a self consistent approximation correct to $O(s_0^2)$ is

$$\tau \approx 1 + s + s^2 + \mu \int_0^\infty \xi^{-1/3} \frac{\partial s}{\partial x}(x-\xi,t)\,\mathrm{d}\xi - \beta \frac{\partial s}{\partial x}\,, \tag{18.62}$$

and if we write

$$q(\tau) \approx q(1) + q'(1)(\tau - 1) + \frac{q''(1)}{2}(\tau-1)^2 + O(\tau-1)^3\,, \tag{18.63}$$

and define the moving spatial coordinate

$$X = x - q'(1)t\,, \tag{18.64}$$

then we find that, correct to terms of $O(s^3)$,

$$\frac{\partial s}{\partial t} + \frac{\partial}{\partial X}\left[\{q'(1) + \tfrac{1}{2}q''(1)\}s^2 + \mu \int_0^\infty \xi^{-1/3} \frac{\partial s}{\partial X}(X-\xi,t)\,\mathrm{d}\xi - \beta \frac{\partial s}{\partial X}\right] = 0\,, \tag{18.65}$$

and by a suitable rescaling of s, t and X we obtain the canonical equation

$$\frac{\partial s}{\partial t} + \frac{\partial}{\partial X}\left[\tfrac{1}{2}s^2 + \int_0^\infty \xi^{-1/3} \frac{\partial s}{\partial X}(X-\xi,t)\,\mathrm{d}\xi - \frac{\partial s}{\partial X}\right] = 0\,. \tag{18.66}$$

We propose this equation as a first canonical equation for the study of nonlinear dune formation. It bears comparison to the Kuramoto–Sivaskinsky equation, and we may hope that the properties of its solutions may be dune-like.

18.2.7 Caveats

The principal feature of real dunes which we have neglected, but which it would be essential to include in future models, is that of boundary layer separation at the dune crest. Essentially, we may expect growth of unstable perturbations to lead to shock formation (smoothed by the diffusion term), but in fact when the bed slope reaches the angle of repose, spontaneous slip occurs, and it is a familiar feature of desert dunes that there is separation at the resulting slope discontinuity.

It may in fact still be possible to model the separated flow in the same way, except that in the lee of the dune a constant pressure (or vorticity) cavity exists, whose extent is unknown *a priori*, but this is a more difficult problem to address.

18.3 Drumlins

The basic geometry of the model for drumlin formation is shown in Fig. 18.2; it is similar to the fluvial dune geometry of Fig. 18.1. We suppose the ice flows over a layer of till lying above an impermeable basement. Typical ice sheet thicknesses are on the order of kilometres, and it is commonly the case that the basal ice reaches the melting point, due to geothermal heat input, together with the insulating effect of the ice cover. In this situation, basal meltwater is produced and the underlying till will become deformable if its pore pressure is high enough (within about a bar of the overburden pressure). The resulting ice motion may then be almost entirely due to deformation of the till, which can be thought of as acting like a power-law viscous fluid, with an effective viscosity which decreases as the pore water pressure increases. This is the situation we study. The vertical coordinate is z, the ice-till interface is $z = s$, and the top surface is denoted by $z = \eta$, thus the ice depth is $h = \eta - s$.

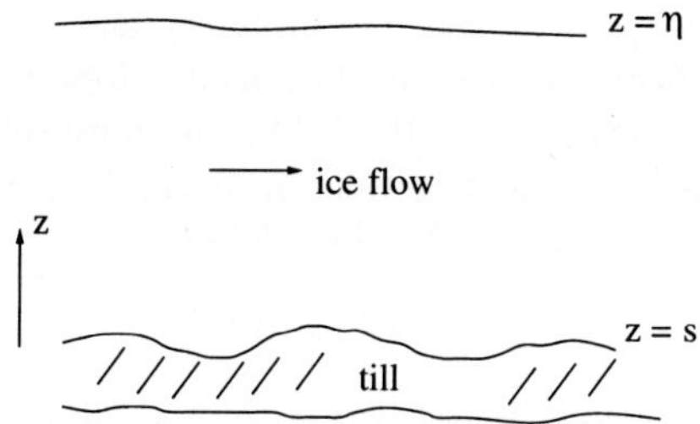

Fig. 18.2. Geometry for the drumlin model. Ice flows over a layer of deformable till

In the uniform state, h and s are constant, and a uniform shear flow exists in the ice. There are two features (other than scale) which distinguish the ice-till flow problem from the water-sediment flow problem. The first is that the flow of till is thought to depend both on basal stress (as for bedload) and on the effective pressure N, defined as the difference between overburden pressure and the interstitial pore water pressure within the sediments. Without this dependence, the instability does not occur. The second is that the Reynolds number for ice flow is essentially zero. Despite these differences, the problems are similar in structure. In particular, it is essential to the instability to take account of the effect of a perturbed bed on the basal stress.

18.3.1 The Hindmarsh Model

The basic model is due to Hindmarsh [16,17], and the present formulation is due to Fowler [13]. For a slow, two-dimensional flow in $s < z < \eta$, we define a stream function ψ and the reduced pressure Π via

$$p = p_a + \rho_i g(\eta - z) + \Pi \,, \tag{18.67}$$

where p_a is atmospheric pressure and ρ_i is ice density. Then Stokes's equations are

$$\begin{aligned} \rho_i g \eta_x + \Pi_x &= \mu \nabla^2 \psi_z \,, \\ \Pi_z &= -\mu \nabla^2 \psi_x \,, \end{aligned} \tag{18.68}$$

where μ is the viscosity of ice, which we take to be constant, and we require conditions of stress continuity at $z = \eta$, and stress and velocity continuity at $z = s$. (Velocity continuity implies no flow of ice into the till, and also implies that the ice does not slip over the till. This is reasonable, though it may be inaccurate, but there is little information on which to base a description of such slip.) The normal and tangential stresses at $z = s$ are, respectively,

$$\begin{aligned} \tau_{nn} &= -\frac{2\mu}{1+s_x^2}[(1-s_x^2)\psi_{xz} + s_x(\psi_{zz}-\psi_{xx})] , \\ \tau &= \frac{\mu}{1+s_x^2}[(1-s_x^2)(\psi_{zz}-\psi_{xx}) - 4s_x\psi_{xz}] , \end{aligned} \tag{18.69}$$

and similar expressions apply at $z = \eta$ (simply replace s by η).

The flow in an ice sheet is driven by the surface slope η_x; η varies on a length scale of order 1000 km, so that $\eta_x \sim 10^{-3}$. On the more relevant drumlin length scale of $\lesssim 1000$ m, it is convenient to take η_x as small and constant, but also to take η as approximately constant. We thus define

$$-\eta_x = \delta \sim 10^{-3} \tag{18.70}$$

to be constant in (18.68), but we solve the resulting model equations assuming the top surface η is constant.

These equations then have a uniform solution in which $\Pi = 0$, $s = 0$, $\eta = \bar{h}$, and the velocity and shear stress at the ice-till interface are $\bar{u}$ and $\bar{\tau}$, respectively. This solution is (with zero shear stress at the surface)

$$\psi = \bar{u}z + \frac{\bar{\tau}}{2\mu}\left(z^2 - \frac{z^3}{3\bar{h}}\right) , \tag{18.71}$$

and the basal shear stress is

$$\bar{\tau} = \rho_i g \delta \bar{h} . \tag{18.72}$$

In allowing uniform η, we are in fact letting $\delta \to 0$ while keeping $\bar{\tau}$ finite, an approximation of Boussinesq type.

Next, consider the zero normal stress condition at the surface. Under a perturbation of the flow, the perturbation to the surface $\Delta\eta$ is given from (18.67) by

$$\Delta\eta = -\frac{(\Pi - \tau_{nn})}{\rho_i g} , \tag{18.73}$$

and we can suppose $\Pi - \tau_{nn} \lesssim \mu u/l$, where u is ice velocity and l is the drumlin length scale. Then

$$\Delta\eta \lesssim \frac{\mu u \delta \bar{h}}{l\bar{\tau}} , \tag{18.74}$$

and with $\bar{\tau} \sim \mu u/\bar{h}$, then $\Delta\eta/\bar{h} \sim \delta\bar{h}/l \sim \delta$. Thus we can consistently neglect variations of η under flow perturbations (just as we did for dunes when the Froude number was small).[1]

[1] This observation is due to Christian Schoof.

Now we perturb the flow, supposing s to be small relative to the depth, but not zero, by writing

$$\psi = \bar{u}z + \frac{\bar{\tau}}{2\mu}\left(z^2 - \frac{z^3}{3\bar{h}}\right) + \Psi\,, \tag{18.75}$$

so that Ψ satisfies

$$\begin{aligned} \Pi_x &= \mu\nabla^2\Psi_z\,, \\ \Pi_z &= -\mu\nabla^2\Psi_x\,, \end{aligned} \tag{18.76}$$

with

$$\Psi_x = 0\,,\quad \Psi_{zz} = 0 \quad \text{on} \quad z = \bar{h}\,, \tag{18.77}$$

and, correct to terms of $O(s)$, (18.69) gives

$$\begin{aligned} \tau_{nn} &\approx -2\mu\Psi_{xz} - 2\bar{\tau}s_x\,, \\ \tau &\approx \bar{\tau}(1 - s/\bar{h}) + \mu(\Psi_{zz} - \Psi_{xx})\,, \end{aligned} \tag{18.78}$$

which can be taken to be evaluated on $z = 0$ rather than $z = s$.

If the horizontal till velocity at $z = s$ is u, then the velocity continuity conditions are

$$-\psi_x = s_t + us_x\,,\quad \psi_z = u \quad \text{at} \quad z = s\,, \tag{18.79}$$

and, under linearisation about $z = 0$, these imply

$$\begin{aligned} u &= \bar{u} + \frac{\bar{\tau}s}{2\mu} + \Psi_z\,, \\ -\Psi_x &= s_t + \bar{u}s_x\,, \end{aligned} \tag{18.80}$$

at $z = 0$.

The ice flow problem thus reduces to the solution of the biharmonic equation

$$\nabla^4\Psi = 0\,, \tag{18.81}$$

together with the two conditions in (18.77), and the four conditions in (18.78) and (18.80). Only two of these latter four are necessary, and so the solution gives two extra relations, which can be taken to be those in (18.78), i.e. we obtain τ_{nn} and τ as linear functionals of u and s. The model is now completed by relating u to τ, τ_{nn}, and s through the dynamics of the deforming till. When this is done, we can eliminate u to find τ and τ_{nn} in terms of s, and hence the till flux q can be written in terms of s, and an Exner equation will provide a model for s.

We also need to specify boundary conditions as $x \to \pm\infty$, and if we suppose $s \to 0$ there, then we would have $\Psi \to 0$. This implies that we can replace (18.77) with the conserved flux condition, thus

$$\Psi = 0\,,\quad \Psi_{zz} = 0 \quad \text{on} \quad z = \bar{h}\,. \tag{18.82}$$

The solution of the problem is now completed by consideration of the till flux.

18.3.2 Till Rheology and Flow

This has been considered in some detail by Fowler [13]. The till is shallow, and the velocity and flux can be determined explicitly. The principal feature of the till flow is that it should increase with applied shear stress and also with pore water pressure. A simple (empirical) choice is of the form

$$\frac{\partial v}{\partial z} = A \exp(\alpha\tau/p_e) \,, \tag{18.83}$$

where v is horizontal velocity and p_e is the effective pressure within the till. This has some basis in experimental measurements on deformation in clays and in till, and is consistent with observed plastic-like behaviour when α is large. The effective pressure increases with depth from the ice-till interface due to hydrostatic effects:

$$p_e = N + (1-n)\Delta\rho_{\mathrm{sw}} g(s-z) \,, \tag{18.84}$$

where N is the value at $z = s$, n is the till porosity, $\Delta\rho_{\mathrm{sw}} = \rho_s - \rho_w$, ρ_s is sediment density, ρ_w is water density, and g is gravity. The shear stress is taken as constant, while the interfacial effective pressure itself is given by

$$N = \bar{N} + \Delta\rho_{\mathrm{wi}} g s + \Pi - \tau_{nn} \,, \tag{18.85}$$

evaluated on $z = s$, or (since linearised) on $z = 0$. Here $\bar{N}$ is a reference value which is supposed fixed, and corresponds to the effective pressure which we suppose is determined by the subglacial drainage characteristics.

The upshot of these assumptions is that the till flux is (approximately)

$$q = A^* \zeta^{*2} [1 - (1+X)\mathrm{e}^{-X}] \,, \tag{18.86}$$

while the ice-till interface velocity is

$$u = A^* \zeta^* (1 - \mathrm{e}^{-X}) \,, \tag{18.87}$$

and the parameters are

$$X = s_0/\zeta^* \,, \quad \zeta^* = \frac{N^2}{\alpha r \tau} \,, \quad A^* = A \exp(\alpha\tau/N) \,, \tag{18.88}$$

where s_0 is the till thickness, and we can take $s_0 = \bar{s} + s$, where $\bar{s}$ is the (uniform) till depth in the undisturbed flow; also the parameter r is defined by

$$r = \Delta\rho_{\mathrm{sw}} g (1-n) \approx 0.1 \ \mathrm{bar\,m^{-1}} \,. \tag{18.89}$$

These expressions give $u = u(s, \tau, N)$, $q = q(s, \tau, N)$, as shown in Fig. 18.3. From (18.78), (18.80) and (18.85), we can write

$$u = L(s, \tau - \bar{\tau}) \,, \quad N - \bar{N} = M(s, \tau - \bar{\tau}) \,, \tag{18.90}$$

where L and M are bilinear (integral) operators. Thus the ice-till velocity is

$$u[s, \tau, \bar{N} + M\{s, \tau - \bar{\tau}\}] = L\{s, \tau - \bar{\tau}\} \,, \tag{18.91}$$

whence $\tau = P[s]$ and thus $q = q[s]$. We then evolve s via the Exner-type equation

$$\frac{\partial s}{\partial t} + \frac{\partial q}{\partial x} = 0 \,. \tag{18.92}$$

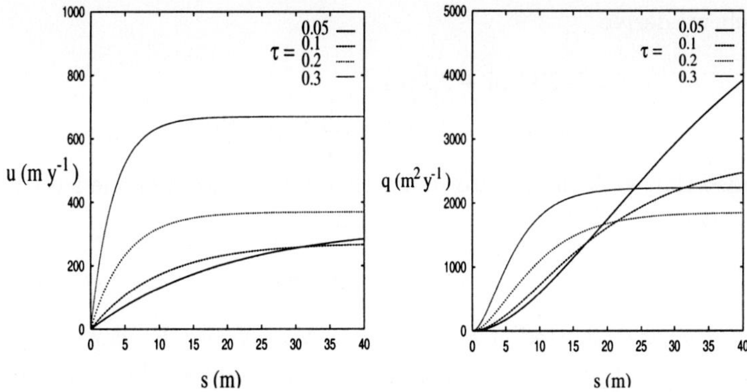

Fig. 18.3. Profiles of $u(s,\tau)$ and $q(s,\tau)$ calculated using (18.87) and (18.86). Graphs are shown for $\tau = 0.05$, 0.1, 0.2, and 0.3 bars. The constants are taken as $\alpha = 10$, $\Delta\rho_{\rm sw}g(1-n) = 0.1\,\text{bar m}^{-1}$, $N = 1\,\text{bar}$, $A = 10\ \text{y}^{-1}$

18.3.3 Fourier Integral Solution

We can write a solution of (18.76), (18.78), (18.80) and (18.82) in terms of Fourier integrals. It is convenient to suppose that $\bar{h} = \infty$, corresponding to a limit in which $\bar{h}/l \gg 1$. This is in fact marginal, since we expect $l \sim 100$–$1000\,\text{m}$, and $\bar{h} \sim 1000\,\text{m}$, but simplifies the algebra without (apparently) seriously compromising the physics. We can then write the solution in the form

$$
\begin{aligned}
\Psi &= \int_{-\infty}^{\infty} (a+bz)\mathrm{e}^{-|k|z}\mathrm{e}^{\mathrm{i}kx}\,\mathrm{d}k,\\
\Pi &= -\int_{-\infty}^{\infty} 2\mu \mathrm{i}kb\mathrm{e}^{-|k|z}\mathrm{e}^{\mathrm{i}kx}\,\mathrm{d}k\ ,\\
u-\bar{u} &= \int_{-\infty}^{\infty} \tilde{u}\mathrm{e}^{\mathrm{i}kx}\,\mathrm{d}k\ ,\\
\tau-\bar{\tau} &= \int_{\infty}^{\infty} \tilde{\tau}\mathrm{e}^{\mathrm{i}kx}\,\mathrm{d}k,\\
s &= \int_{-\infty}^{\infty} \tilde{s}\mathrm{e}^{\mathrm{i}kx}\,\mathrm{d}k,\\
N &= \bar{N} + \int_{\infty}^{\infty} \tilde{N}\mathrm{e}^{\mathrm{i}kx}\,\mathrm{d}k\ ,
\end{aligned}
\tag{18.93}
$$

and the boundary conditions imply

$$
\begin{aligned}
b-|k|a &= \tilde{u} - \frac{\bar{\tau}\tilde{s}}{\mu}\ ,\\
-\mathrm{i}ka &= \tilde{s}_t + \mathrm{i}k\bar{u}\tilde{s},\\
\tilde{\tau} &= -2|k|u(b-|k|a)\ ,\\
\tilde{N} &= -2\mu \mathrm{i}k|k|a + 2\mathrm{i}k\bar{\tau}\tilde{s} + \Delta\rho_{\rm wi}g\tilde{s}\ ,
\end{aligned}
\tag{18.94}
$$

from which we derive

$$\begin{aligned}\tilde{\tau} &= -2|k|\mu\tilde{u} + 2|k|\bar{\tau}\tilde{s} \, ,\\ \tilde{N} &= 2\mu|k|(\tilde{s}_t + \mathrm{i}k\bar{u}\tilde{s}) + (\Delta\rho_{\mathrm{wi}}g + 2\mathrm{i}k\bar{\tau})\tilde{s} \, .\end{aligned} \tag{18.95}$$

Suppose also that $\tilde{s}$, $\tilde{N}$, and $\tilde{\tau}$ are small; then linearisation of the ice-till velocity $u(s, \tau, N)$ gives the relation

$$\tilde{u} = u_s\tilde{s} + u_\tau\tilde{\tau} + u_N\tilde{N} \, , \tag{18.96}$$

where $u_s = \partial u/\partial s$ evaluated at $s = 0$, $\tau = \bar{\tau}$, $N = \bar{N}$, etc. Fowler [13] omitted to include the second term in $(18.95)_1$; he also argued that the last term in $(18.95)_2$ was negligible, on the basis that, typically, $\Delta\rho_{\mathrm{wi}}g, 2k\bar{\tau} \ll 2\mu k^2\bar{u}$, for values of interest when $k^{-1} \sim 100\,\mathrm{m}$, and we do the same here.

From $(18.95)_1$ and (18.96), we have

$$\tilde{\tau} = -\frac{2|k|\mu}{1 + 2|k|\mu u_\tau}\left[\left(u_s - \frac{\bar{\tau}}{\mu}\right)\tilde{s} + u_N\tilde{N}\right] \, . \tag{18.97}$$

The $\bar{\tau}/\mu$ term (which arises from the second term in $(18.95)_1$) is negligible if $\bar{\tau} \ll \mu\partial u/\partial s$. Consulting Fig. 18.3, we see that a typical range of $\partial u/\partial s$ is 10–100 y^{-1} for $s < 10\,\mathrm{m}$. Then if $\mu = 6\,\mathrm{bar\,y}$, $\mu\partial u/\partial s \gtrsim 60\,\mathrm{bar}$, and it is safe to neglect this term. Thus we derive the approximations

$$\begin{aligned}\tilde{N} &= 2\mu|k|(\tilde{s}_t + \mathrm{i}k\bar{u}\tilde{s}) \, ,\\ \tilde{\tau} &= -\frac{2|k|\mu}{1 + 2|k|\mu u_\tau}(u_s\tilde{s} + u_N\tilde{N}) \, ,\\ \tilde{u} &= \frac{u_s\tilde{s} + u_N\tilde{N}}{1 + 2|k|\mu u_\tau} \, .\end{aligned} \tag{18.98}$$

18.3.4 Linear Stability

If we now linearise the till-flux expression (18.86) and the Exner equation (18.92), we obtain two further relations

$$\begin{aligned}\tilde{q} &= q_s\tilde{s} + q_\tau\tilde{\tau} + q_N\tilde{N},\\ \tilde{s}_t &+ \mathrm{i}k\tilde{q} = 0 \, ,\end{aligned} \tag{18.99}$$

and from these we can derive the growth rate in the form

$$\tilde{s}_t/\tilde{s} = \rho - \mathrm{i}kc \, , \tag{18.100}$$

where ρ is the growth rate and c is the wave speed. Explicit expressions are given by Fowler [13], for example the growth rate is

$$\rho = \frac{2\mu k^2|k|\Delta_1\Delta_2}{(1 + 2\mu|k|u_\tau)^2 + 4\mu^2k^4\Delta_2^2} \, , \tag{18.101}$$

where

$$\Delta_1 = (1 + 2\mu|k|u_\tau)(\bar{u} - q_s) + 2\mu|k|q_\tau u_s \,,$$
$$\Delta_2 = q_N + 2\mu|k|[u_\tau q_N - q_\tau u_N] \,. \tag{18.102}$$

It is clear from these formulae that it is essential for instability that the flow law for till depend on N, for otherwise $\Delta_2 = 0$ and thus stability is neutral.

Further simplification is possible, using the anticipated fact that $k^{-1} \sim 100\,\mathrm{m}$, whence typically $2\mu|k|u_\tau \gg 1$. In fact, if we define (cf. (18.88))

$$X = \xi Y \,, \quad \xi = \frac{r\bar{s}}{N} \,, \quad Y = \frac{\alpha\tau}{N} \,, \quad K = \frac{2\mu\alpha A|k|}{r} \,, \tag{18.103}$$

then we find that $K \gg 1$, and thence (for $X, Y = O(1)$)

$$\Delta_1 \approx \frac{A^*\zeta^* W(X) K \mathrm{e}^Y [Y - F(X)]}{Y^2} \,,$$
$$\Delta_2 \approx \frac{qK\mathrm{e}^Y J(X)}{NY^2} \,, \tag{18.104}$$

where the functions W, J and F are positive, and

$$F(X) = \frac{1 - 2X\mathrm{e}^{-X} - \mathrm{e}^{-2X}}{1 - (1 + X)\mathrm{e}^{-X}} \,; \tag{18.105}$$

F increases monotonely from 0 at $X = 0$ to 1 as $X \to \infty$. Roughly, $F \approx 1 - \mathrm{e}^{0.7X}$. We suppose typical values $\bar{s} \sim 10\,\mathrm{m}$, $N \sim 1\,\mathrm{bar}$, $\alpha = 10$, $\tau = 0.1\,\mathrm{bar}$, so that $X, Y = O(1)$, and instability occurs (since $\Delta_2 > 0$) if

$$Y > F(X) = F(\xi Y) \,. \tag{18.106}$$

(We also require $X > O(1/\sqrt{K})$.) A delineation of the instability region is shown in Fig. 18.4. We see that instability generally occurs for $\xi, Y = O(1)$, and thus in general for low effective pressures.

With the same approximation, that $K \gg 1$, we find that the growth rate can be written in the approximate form

$$\rho \approx \left(\frac{AN}{2\mu}\right)^{1/2} \left[\frac{D|k/k^*|^3}{B^2 + C^2(k/k^*)^4}\right] \,, \tag{18.107}$$

where

$$k^* = \frac{r}{(2\mu AN)^{1/2}} \,, \tag{18.108}$$

and B, C, D are $O(1)$ functions of X and Y. This indicates that the preferred wavelength of growth is $O(k^{*-1})$, and the growth time scale is $(2\mu/AN)^{1/2}$. We find unstable wavelengths in the range 100–1000 m, and the growth time is of order 1 year. Of course, the model is only two-dimensional, but we might expect three-dimensional instability also, perhaps arising as a secondary instability on the primary (Rogen moraine) ridge forms.

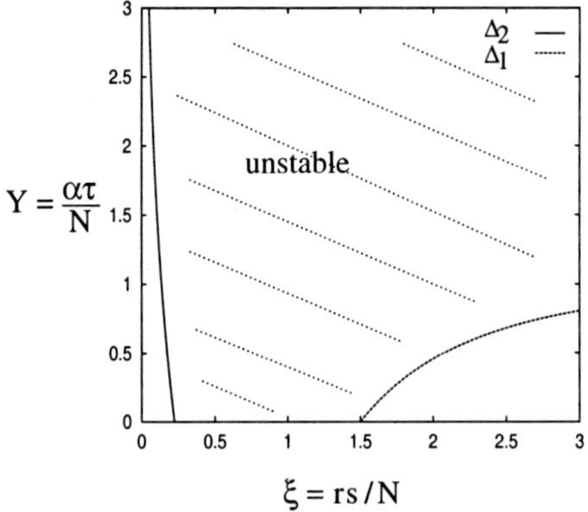

Fig. 18.4. Instability region in Y, ξ space, bounded by the curves $\Delta_1 = 0$, $\Delta_2 = 0$ (see (18.102)). A value of $K = 120$ (see (18.103)) has been used

18.3.5 A Nonlinear Model

With the approximation $2|k|\mu u_\tau \gg 1$, (18.98) may be written in the approximate form

$$\begin{aligned} \tilde{N} &= 2\mu|k|(\tilde{s}_t + ik\bar{u}\tilde{s}) \ , \\ \tilde{\tau} &\approx -\frac{(u_s\tilde{s} + u_N\tilde{N})}{u_\tau} \ , \\ \tilde{u} &\approx \frac{(u_s\tilde{s} + u_N\tilde{N})}{2|k|\mu u_\tau} \ , \end{aligned} \tag{18.109}$$

and $\tilde{N}$ and $\tilde{\tau}$ can be explicitly inverted to obtain

$$N - \bar{N} = -\frac{2\mu}{\pi} \fint_{-\infty}^{\infty} \frac{\partial a}{\partial \xi} \frac{d\xi}{\xi - x} \ , \tag{18.110}$$

where the barred integral indicates that the principal value is taken, a is given by

$$\frac{\partial s}{\partial t} + \bar{u}\frac{\partial s}{\partial x} = a \ , \tag{18.111}$$

and

$$\tau - \bar{\tau} = -\frac{(u_s s + u_N N)}{u_\tau} \ . \tag{18.112}$$

The principal restriction used to obtain these formulae is that $s \ll k^{-1}$ (and $s \ll \bar{h}$), while the assumption that perturbations in τ, s and N are small is only used in linearising $u(s, \tau, N)$. However, the fact that the resulting perturbation in u is approximately zero suggests that the restriction to small perturbations is not essential. This suggests that the formulae (18.110) and (18.112) can be used in the Exner model, but retaining the full nonlinear prescription for $q(s, \tau, N)$; in this way we obtain a nonlinear evolution equation for s.

It is appealing to seek the weakly nonlinear form of this by expanding q for small s out to quadratic terms in s, just as we did for the dune model. We retain only linear terms in $\tau - \bar{\tau}$ and $N - \bar{N}$; the result of this is

$$\frac{\partial s}{\partial t} + \left(q_s - \frac{q_\tau u_s}{u_\tau} \right) \frac{\partial s}{\partial x} + \frac{\partial}{\partial x} \left[\tfrac{1}{2} q_{ss} s^2 - \frac{2\mu (q_N u_\tau - q_\tau u_N)}{u_\tau} H(a_x) \right] = 0 \,, \quad (18.113)$$

where $H(g)$ is the Hilbert transform

$$H(g) = \frac{1}{\pi} \fint_{-\infty}^{\infty} \frac{g(\xi)\,d\xi}{\xi - x} \quad (18.114)$$

(whose Fourier transform $\widehat{H(g)} = -\pi \hat{g}\ \mathrm{sgn}k$), and a is given by (18.111).

Define the parameters

$$\begin{aligned} f &= \bar{u} + \frac{q_\tau u_s}{u_\tau} - q_s \,, \\ G &= 2\mu (u_\tau q_N - u_N q_\tau)/u_\tau \,. \end{aligned} \quad (18.115)$$

With $2|k|\mu u_\tau \gg 1$, then (18.102) implies

$$\Delta_1 \approx 2\mu |k| u_\tau f \,, \quad \Delta_2 \approx |k| u_\tau G \,. \quad (18.116)$$

We define

$$Z = x - \bar{u}t \,; \quad (18.117)$$

then in the Z frame moving with the interfacial velocity, $a = \partial s/\partial t$, and the equation (18.113) for s can be written in the form

$$\frac{\partial s}{\partial t} - f \frac{\partial s}{\partial Z} + \frac{\partial}{\partial Z} \left[\tfrac{1}{2} q_{ss} s^2 - GH \left(\frac{\partial^2 s}{\partial Z \partial t} \right) \right] = 0 \,, \quad (18.118)$$

and this is our candidate canonical nonlinear evolution equation for drumlins (or more properly, their two-dimensional version – Rogen moraine).

Finally, it is convenient to write the model in non-dimensional form. We use the length and time scales suggested by the stability results, namely

$$l = \frac{(2\mu A\bar{N})^{1/2}}{r} \,, \quad [t] = \left(\frac{2\mu}{A\bar{N}} \right)^{1/2} \,, \quad (18.119)$$

and thus the velocity scale is

$$[u] = \frac{l}{[t]} = \frac{A\bar{N}}{r} \,. \quad (18.120)$$

The scale for s is taken as $\bar{s}$. Following [13], we find

$$f = [u] v(X,Y) \,, \quad G = l^2 \gamma(X,Y) \,, \quad (18.121)$$

where the $O(1)$ functions v and γ are defined by

$$v = \frac{e^Y[Y - F(X)]}{Y[(Y-1)U + XU']}, \quad \gamma = \frac{L(X)e^Y}{Y^2[(Y-1)U + XU']}, \tag{18.122}$$

the subsidiary functions being

$$\begin{aligned} U &= 1 - e^{-X}, \\ W &= 1 - (1+X)e^{-X}, \\ L &= UW + X(U'W - UW'). \end{aligned} \tag{18.123}$$

Recall that, from (18.103), $X = \alpha r\bar{s}\bar{\tau}/\bar{N}^2$, $Y = \alpha\bar{\tau}/\bar{N}$. In terms of the scaled variables, the dimensionless version of (18.118) is

$$\frac{\partial s}{\partial t} - v\frac{\partial s}{\partial Z} + \frac{\partial}{\partial Z}\left[\tfrac{1}{2}\beta s^2 - \gamma H\left(\frac{\partial^2 s}{\partial Z \partial t}\right)\right] = 0, \tag{18.124}$$

where the parameter β is defined by

$$\beta = \frac{\bar{s}q_{ss}}{[u]}. \tag{18.125}$$

Using the expression (18.86) for q we find

$$q_{ss} = \frac{q}{s^2}\left[\left(\frac{XW'}{W}\right)^2 - \left(\frac{XW'}{W}\right) + X\left(\frac{XW'}{W}\right)'\right], \tag{18.126}$$

and thus

$$\beta = \frac{We^Y}{XY}\left[\left(\frac{XW'}{W}\right)^2 - \left(\frac{XW'}{W}\right) + X\left(\frac{XW'}{W}\right)'\right], \tag{18.127}$$

and is indeed $O(1)$. A Fourier transform of the linearisation of (18.124) gives a growth rate of

$$\sigma = \frac{-ikv + k^2|k|v\gamma}{1 + k^4\gamma^2}, \tag{18.128}$$

which, since $v \propto \Delta_1$ and $\gamma \propto \Delta_2$, reproduces (18.102). Instability occurs if $v\gamma > 0$; and with $\gamma > 0$, this is essentially $v > 0$: the resulting waveforms move forwards relative to the ice flow. (18.124) closely resembles the integrable Benjamin–Ono equation $s_t + ss_x = H(s_{xx})$ [24].

18.4 Discussion

The canonical equation for both dune and drumlin evolution follows from the Exner equation

$$\frac{\partial s}{\partial t} + \frac{\partial q}{\partial x} = 0, \tag{18.129}$$

together with a suitable prescription for q. In the case of fluvial dunes, $q = q(\tau)$, and the basal shear stress depends on the mean flow speed u and the bed elevation s. When s is small, the stress can be calculated by linearisation of a suitable turbulent flow model over a flat bed. When the turbulent Reynolds number is reasonably large (which is typically the case), then an explicit asymptotic approximation for τ can be determined, in the form of a Fourier convolution of a kernel function with the bed slope. In this way we derive (18.54), which is a nonlinear model for bed evolution. The essence of this model is captured by the reduced form (18.65), or (18.66).

Ideally, one wants an extension of the model to three dimensions, although fluvial dunes are essentially two-dimensional features. Such a generalisation is the Exner equation

$$\frac{\partial s}{\partial t} + \boldsymbol{\nabla}.\mathbf{q} = 0 , \tag{18.130}$$

together with $\mathbf{q} = \mathbf{q}(\boldsymbol{\tau})$, where the basal stress vector must now be computed from the Orr–Sommerfeld equation. It seems this should be relatively straightforward to do, simply involving a double Fourier transform.

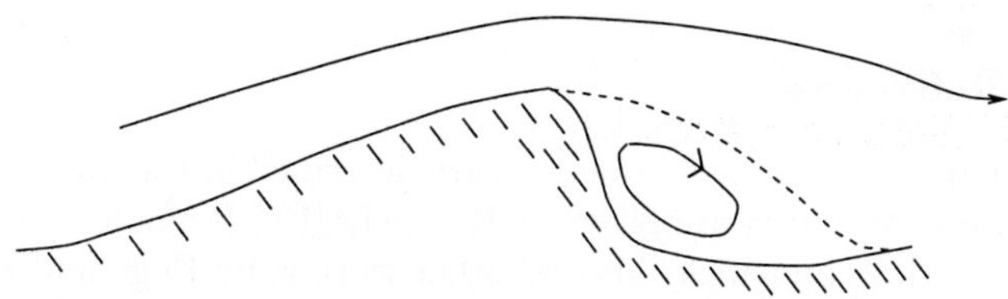

Fig. 18.5. Separation in the lee of a dune

A more important extension is to allow for boundary layer separation in the lee of dunes. Within the present framework, this could be done by supposing, for example, that the recirculating wake is a region of constant pressure or vorticity (see Fig. 18.5). The analysis is the same, but now s is unknown in the wake, whereas the pressure (or vorticity) is prescribed there. Presumably we can write the pressure as a linear functional of the bed, so that we would gain an extra equation to solve for s in the wake. It remains to be seen whether this strategy is feasible.

For drumlins, the extension to three dimensions is more essential, but seems equally straightforward. In addition, inclusion of a finite depth is straightforward, though perhaps messy. There is also a nice analogy with the formation of wakes, because the generation of bedform is likely to cause cavities to occur, and we can imagine that sediment deposition in such cavities may be one way in which layered stratigraphy is formed in drumlins, through successive fluvial deposition events. Cavity formation also complicates the prescription of the shear stress, although in this case the problem becomes one of Hilbert type, and it is possible to solve this [11].

Apart from such developments in the mathematical model, the equations (18.66):

$$\frac{\partial s}{\partial t} + \frac{\partial}{\partial X}\left[\tfrac{1}{2}s^2 + \int_0^\infty \xi^{-1/3}\frac{\partial s}{\partial X}(X-\xi,t)\,d\xi - \frac{\partial s}{\partial X}\right] = 0\,, \tag{18.131}$$

and (18.124):

$$\frac{\partial s}{\partial t} - \frac{\partial s}{\partial Z} + \frac{\partial}{\partial Z}\left[\tfrac{1}{2}s^2 - H\left(\frac{\partial^2 s}{\partial Z \partial t}\right)\right] = 0 \tag{18.132}$$

(where we can put $\beta = \gamma = v = 1$ by appropriate rescaling of s, Z and t), are interesting nonlinear evolution equations, whose study in the context of dynamical systems is of interest in its own right. In particular (18.131) bears comparison with the Kuramoto–Sivashinsky equation, while (18.132) similarly resembles the Benjamin–Ono equation. While we may expcct the dune equation (18.131) to provide a coherent model with the stabilising diffusion term, it is less clear that the drumlin model will be. In this context, it may be worth noting that there is a further stabilising term which could be included, since till will also creep down pressure gradients.

Notes and References

Dunes. Principles of sediment transport are described in the book by Allen [1]. Theories of dune formation are given by Kennedy [19], Reynolds [26], Engelund [8] and Smith [31]. A review of this and other work is by Engelund and Fredsøe [9]. Subsequently, theoretical development has been hindered by the necessity of solving the Orr–Sommerfeld equation numerically, and this has precluded the development of nonlinear theories other than through direct numerical computation.

A description of desert dunes can be found in [14] and in Chap. 17. Recently, theoretical models similar to the nonlinear models proposed here have been advanced by Herrmann and co-workers to explain the form of barchan dunes [22]; they use an integral correction for the bed shear stress proposed by Jackson and Hunt [18]. The correction is similar to that used in (18.131), but the convolution kernel is of Cauchy type. The mechanics of the resulting instability is very similar though.

River flow. The hydraulics and processes of river flow are described in the books by Chow [4], Richards [28] and Knighton [21]. An account which is aimed at applied mathematicians is in the book by Fowler [12], and there are also the classic accounts of Stoker [32] and Whitham [36].

Drumlins. The literature on drumlins is substantial; their formation has been debated for well over one hundred years. However, the debate has been largely geological, and the dynamical concept of an instability is virtually absent. Amongst early authors, the papers of Davis [5], Upham [35] and Tarr [34] may be mentioned, as well the seminal paper of Kinahan and Close [20] – the last not easily accessible, but a copy may be obtained from The Royal Irish Academy, Dawson Street, Dublin. Useful early reviews are by Charlesworth [3] and Gravenor [15], and later developments can be followed in [7] and [33]. The erosional and flood theories are expounded by Boulton [2] and Shaw [29], for example. The instability theory derives from Hindmarsh [16]. A voluminous bibliography is that of Everett [10].

Acknowledgements

Thanks are due to Neil Balmforth and Antonello Provenzale for hosting an inspirational summer school on geomorphological fluid dynamics. I thank Emanuele Schiavi and Christian Schoof for their suggestions and scrutiny.

References

1. J.R.L. Allen: *Principles of physical sedimentology* (Chapman and Hall, London 1985)
2. G.S. Boulton: 'A theory of drumlin formation by subglacial sediment deformation'. In: *Drumlin Symposium*, ed. by J. Menzies, J. Rose (Balkema, Rotterdam 1987) pp. 25–80
3. J.K. Charlesworth: *The geology of Ireland: an introduction* (Oliver and Boyd, Edinburgh 1966)
4. V.T. Chow: *Open-channel hydraulics* (McGraw-Hill, New York 1959)
5. W.M. Davis: Amer. J. Sci. **23**, 407–416 (1884)
6. P.G. Drazin, W.H. Reid: *Hydrodynamic stability* (CUP, Cambridge 1981)
7. C. Embleton, C.A.M. King: *Glacial geomorphology* (Edward Arnold, London 1975)
8. F. Engelund: J. Fluid Mech. **42**, 225–244 (1970)
9. F. Engelund, J. Fredsøe: Ann. Rev. Fluid Mech. **14**, 13–37 (1982)
10. W. Everett: An analysis of the literature on drumlins and related streamlined forms. M. Phil. thesis, University of London, pp. 491 (1987)
11. A.C. Fowler: Proc. R. Soc. Lond. **A407**, 147–170 (1986)
12. A.C. Fowler: *Mathematical models in the applied sciences* (C. U. P., Cambridge 1997)
13. A.C. Fowler: 'An instability mechanism for drumlin formation'. In: *Deformation of glacial materials*, ed. by A. Maltman, M.J. Hambrey, B. Hubbard, Spec. Pub. Geol. Soc. **176**, 307–319 (2000)
14. A. Goudie: *The nature of the environment*, 3rd edn. (Blackwell, Oxford 1993)
15. C.P. Gravenor: Amer. J. Sci. **251**, 674–681 (1953)
16. R.C.A. Hindmarsh: J. Glaciol. **44**, 285–292 (1998)
17. R.C.A. Hindmarsh: J. Glaciol. **44**, 293–314 (1998)
18. P.S. Jackson, J.C.R. Hunt: Quart. J. R. Met. Soc. **101**, 929–955 (1975)
19. J.F. Kennedy: J. Fluid Mech. **16**, 521–544 (1963)
20. G.H. Kinahan, M.H. Close: *The general glaciation of Iar-Connaught and its neighbourhood, in the counties of Galway and Mayo* (Hodges, Foster and Co., Dublin 1872) 20 pp.
21. D. Knighton: *Fluvial forms and processes* (Arnold, London 1998)

22. K. Kroy, G. Sauermann, H.J. Herrmann: A minimal model for sand dunes. Preprint (2001)
23. E. Meyer-Peter, R. Müller: 'Formulas for bed-load transport'. In: *Proc. Int. Assoc. Hydraul. Res., 3rd annual conference, Stockholm* (1948) pp. 39–64
24. J. Ockendon, S. Howison, A. Lacey, A. Movchan: *Applied partial differential equations* (O. U. P., Oxford 1999)
25. W.H. Reid: Stud. Appl. Math. **51**, 341–368 (1972)
26. A.J. Reynolds: J. Fluid Mech. **22**, 113–133 (1965)
27. K.J. Richards: J. Fluid Mech. **99**, 597–618 (1980)
28. K. Richards: *Rivers. Form and process in alluvial channels* (Methuen, London 1982)
29. J. Shaw: J. Glaciol. **29**, 461–479 (1983)
30. A. Shields: Anwendung der Aehnlichkeitsmechanik und der turbulenzforschung auf die geschiebebewegung. Mitteilung der Preussischen versuchsanstalt für Wasserbau und Schiffbau, Heft 26, Berlin 1936
31. J.D. Smith: J. Geophys. Res. **75**, 5928–5940 (1970)
32. J.J. Stoker: *Water waves: the mathematical theory with applications* (Interscience, New York 1957)
33. D.A. Sugden, B.S. John: *Glaciers and landscape* (Edward Arnold, London 1976)
34. R.S. Tarr: Amer. Geol. **13**, 393–407 (1894)
35. W. Upham: Amer. Geol. **10**, 339–362 (1892)
36. G.B. Whitham: *Linear and nonlinear waves* (Wiley, New York 1974)

19 Estuarine Patterns: An Introduction to Their Morphology and Mechanics

G. Seminara[1], S. Lanzoni[2], M. Bolla Pittaluga[1], and L. Solari[3]

[1] Dipartimento di Ingegneria Ambientale, Università di Genova, Via Montallegro 1, 16145 Genova, IT
[2] Dipartimento di Ingegneria Idraulica, Marittima e Geotecnica, Università di Padova, Via Loredan 20, 35131 Padova, IT
[3] Dipartimento di Ingegneria Civile, Università di Firenze, Via S. Marta 3, 50139 Firenze, IT

19.1 Introduction

The aim of the present review is to expose the average reader (assumed to have little previous knowledge of morphodynamics) to an overview of some aspects of estuarine morphodynamics tackled from a mechanical perspective. It will appear that the tidal analogue of several phenomena which are extensively discussed and fairly well understood in the fluvial literature, still await to be fully explored. When writing reviews of this kind, authors are usually confronted with a dilemma: either treating superficially a large number of aspects of the problem or discussing fewer topics in greater depth. In the present case we have chosen the former alternative, which seems appropriate to the non specialistic audience assumed above, though space limitation will not allow a systematic treatment of the subject. The reader interested in achieving a more advanced understanding is referred to the extensive literature quoted in the paper.

Part One: Morphology

19.2 Large-Scale Tidal Patterns

We define as *large-scale tidal patterns* those coastal features which are characterized by spatial scales of the order of the distance reached by the hydrodynamic and morphodynamic effects of tide propagation. Two main large scale tidal patterns will be discussed below, namely *estuaries* and tidal landforms, often called *wetlands*.

19.2.1 Estuaries

Definition

The definition of estuaries has been the subject of considerable debate in the geomorphological community [77].

Fig. 19.1. Estuary of the Heuningnes River (Western coast of South Africa)

A sufficiently comprehensive definition which seems generally acceptable has been proposed by Perillo [77]: *An estuary is a semienclosed coastal body of water that extends to the effective limit of tidal influence, within which sea water entering from one or more free connections with the open sea, or any other saline coastal body of water, is significantly diluted with fresh water derived from land drainage and can sustain euryhaline biological species for either part or the whole of their life cycle.*

Genetic Classification of Estuaries

Estuaries have been classified according to their geological origin by Pritchard [80–82], who identifies four classes: drowned river valleys, bar-built estuaries, fjords and tectonic estuaries.

- *Drowned river valleys*

Such estuaries originated from the flooding of Pleistocene–Holocene river valleys following the sea level rise by about 100–130 m during the Flandrian transgression (roughly 15000–18000 years ago). Such estuaries are widespread throughout the world and may be found along coastlines with relatively wide coastal plains (the type called *coastal plain estuaries* by Pritchard, [80], of which a notable example is provided by Chesapeake Bay along the eastern coast of U.S.A., Fig. 19.10) as well as along mountain and cliffy coasts (the type called rias [84], of which notable examples are found along the northern coast of the Iberian peninsula).

These estuaries are typically characterized by a funnel shape, depths of the order of 10 m, increasing towards the mouth, V-shaped cross sections with aspect ratios ranging from 10–100 in rias to thousands in coastal plain estuaries (see sketch in Fig. 19.2).

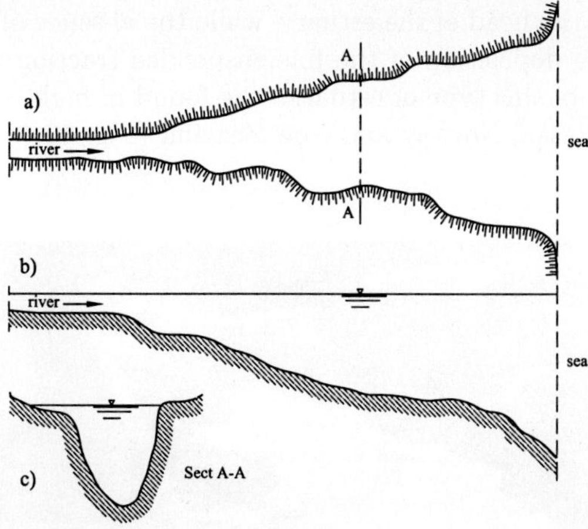

Fig. 19.2. Sketch of a drowned river valley estuary

• *Fjords*

Fjords have been formed by the drowning of river valleys at high latitudes covered by glacial troughs. They are typically extremely deep (several hundred meters), U-shaped and bounded by steep rock walls. A shallow sill (as shallow as 4 m at some locations along the Norwegian coast, up to 150 m along the coast of British Columbia) is typically present at the mouth of the estuary and constrains the tidal exchange between the estuary and the sea (Fig. 19.3). Sediments are

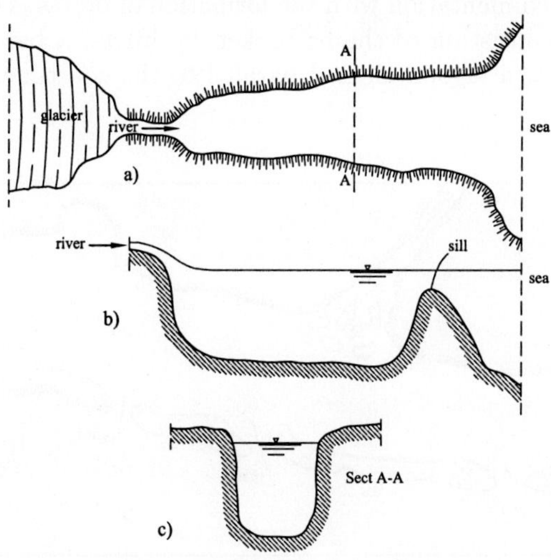

Fig. 19.3. Sketch of a fjord

typically coarse at the head of the estuary, while the absence of significant water circulation leads to deposition of the fine suspended fraction along the muddy bottom. Examples of this type of estuaries are found in high latitude mountainous regions, like Alaska, Norway and New Zealand (Fig. 19.4).

Fig. 19.4. Aereal picture of a fjord in Alaska

- *Bar-built estuaries*

The mouth of rivers debouching into seas characterized by small tidal range or undergoing rapid sedimentation with the formation of deltas is often surrounded by sand barriers, consisting of chains broken by inlets; a bar-built estuary, in the form of a bay or a lagoon aligned parallel to the coastline, then forms (see sketch in Fig. 19.5).

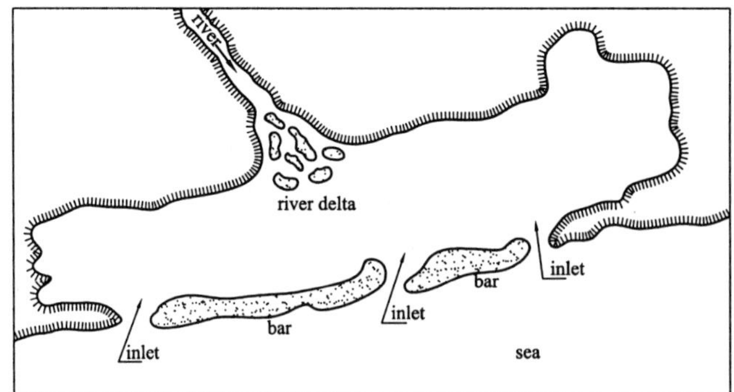

Fig. 19.5. Sketch of a bar-built estuary

Typically, such offshore estuaries are shallow, except close to the inlets kept deeper by the erosive action of the tidal currents. They are strongly influenced by littoral transport and wind action which plays a stirring role comparable with that associated with tidal motion. The embayment is most commonly bounded by tidal flats and salt marshes, dissected by highly sinuous tidal creeks. An important example of bar-built estuary is the lagoon of Venice (Fig. 19.6).

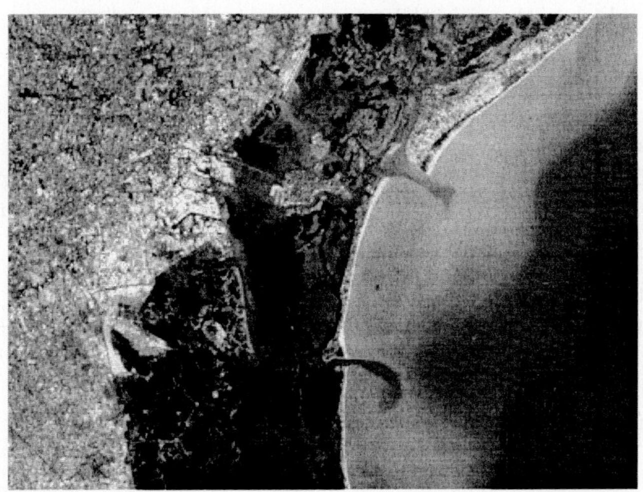

Fig. 19.6. The Lido and Malamocco inlets of the Venice Lagoon

- *Tectonically formed estuaries*

In this category Pritchard [80] includes all the estuaries which cannot be clearly recognized as belonging to any of the other three classes, and in particular '... coastal indentures formed by faulting or by local subsidence, and having an excess supply of freshwater inflow....San Francisco Bay is an example of this group of estuaries.'

Tidal Range, Estuarine Morphology and Mixing

While the genetic classification discussed above helps us tracing the origin of more recent estuarine processes, it does not allow us to associate unique morphological characteristics to each class of estuaries.

Hayes [45] and Davies [23] pointed out that a major controlling factor of estuarine morphology is *tidal range*. The intensity of tidally driven transport processes is indeed determined by the intensity of tidal currents which are increasing functions of the tidal range. The latter authors have then proposed a classification whereby estuaries are grouped according to their tidal range in three classes.

• *Micro-tidal* estuaries

When the tidal range does not exceed about 2 m the estuary is conventionally described as *microtidal.* Hayes [45] has pointed out the existence of a strong correlation between the occurrence of such tidal range and the observed formation of barrier islands and sand spits, which are associated with the formation of *bar-built estuaries*, as pointed out in Sect. 19.2.1. The latter features result from the dominant landward effect of freshwater which leads to the formation of a delta at the mouth of the estuary, along with the dominant seaward effect of wind action which leads to the delta being enclosed by sand structures broken by inlets (Fig. 19.5). Microtidal estuaries are generally wide (typical width 10 Km) and shallow (typical depth 1 m).

• *Mesotidal* estuaries

When the tidal range falls between 2–4 m, the estuary is conventionally described as *mesotidal.* These are the most common estuaries whose morphology is characterized by the following main features:

- the upstream reaches are usually strongly meandering;
- the estuary mouth displays the formation of a composite delta (Fig. 19.7), which has a seaward side (called ebb-tide delta) and a landward side (called flood-tide delta) [13];
- the estuary is bordered by tidal flats and salt marshes.

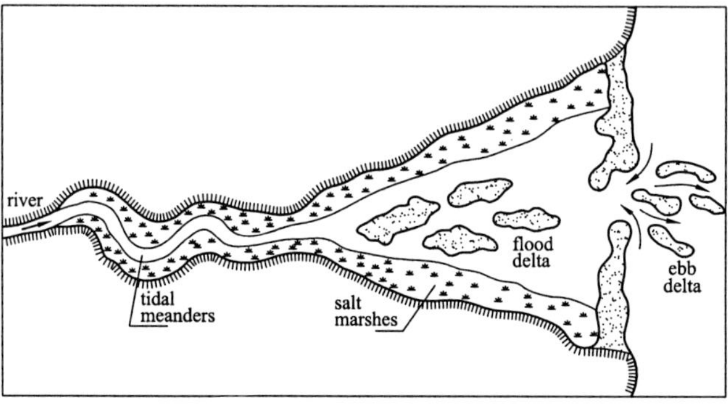

Fig. 19.7. Sketch of a mesotidal estuary with ebb-tide and flood-tide deltas formed at the mouth

• *Macrotidal* estuaries

When the tidal range exceeds about 4 m, then the estuary is conventionally described as *macrotidal.* The morphology of such estuaries is determined by the dominant effect of strong tidal currents, which may be felt at distances of hundreds of kilometers from the mouth. In this case, the mesotidal deltas are replaced by a pattern of so called ‘elongated tidal bars’ (see Sect. 19.3.3) and the alignment of the estuary is weakly meandering and strongly tapered displaying a typical funnel shape (Figs. 19.8, 19.9).

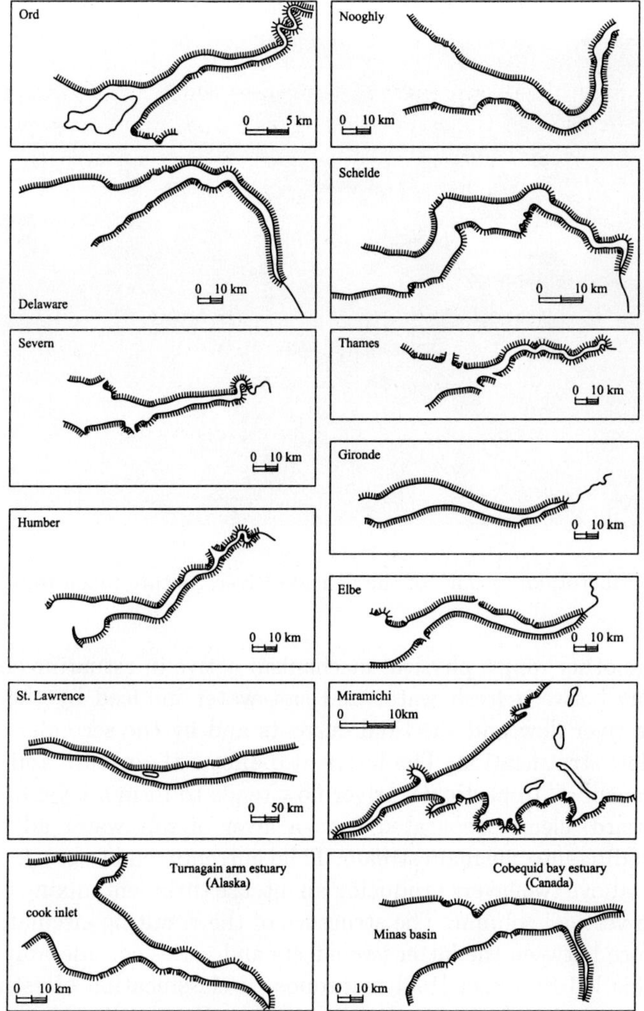

Fig. 19.8. Typical funnel shape of several macrotidal estuaries

The classification of estuaries based on tidal range is probably the most significant. However, the geomorphological definition, which relies on the actual physical value of tidal range, should more conveniently be formulated in terms of the ratio between the volume of saline water exchanged between the sea and the estuary in half a tidal cycle (the so called *tidal prism*) and the volume of fresh water discharged by the river into the estuary in the same period. 'Macro-', 'meso-' and 'micro-tidal' estuaries then correspond to the latter ratio being 'much larger than', 'of order' and 'much smaller than' one, respectively. This modified criterion also allows to classify estuaries according to the degree of stratification of the flowing stream.

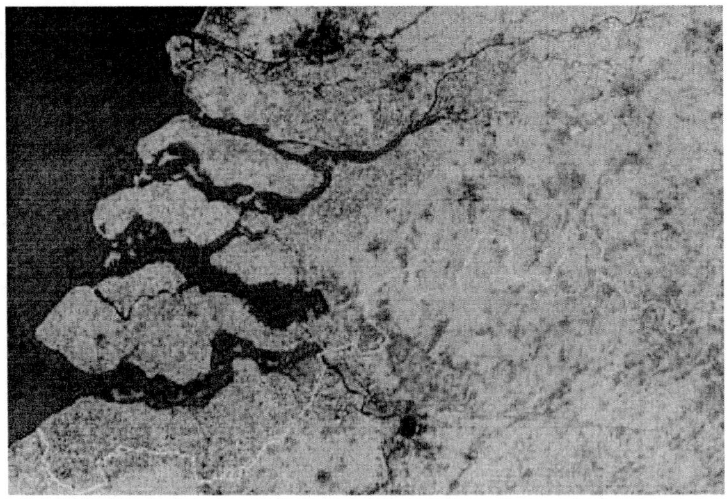

Fig. 19.9. Macrotidal estuary of the Schelde River, picture taken from a satellite

In fact, another major physical mechanism active in estuarine environments is the *mixing* between fresh water and salt water induced by the interaction between the river flow and the tidal currents and by the secondary circulation driven by flow stratification. The latter also affects the nature and intensity of sediment transport. In particular, river flow tends to form a layer of fresh water flowing seaward, clearly separated from a layer of salt water adjacent to the bottom intruding into the main stream. Tidal currents tend to break the interface between the above two layers producing an intense turbulent mixing through part or all of the vertical column. The structure of the resulting circulation depends on the balance between the latter two effects and varies considerably in different estuaries: this led Stommel [108] to propose a classification of estuaries based on the characteristics of the estuarine circulation. Essentially, Stommel identifies two limiting cases and an intermediate one.

• The first limiting case: *salt wedge estuaries.*

This is the case when fresh water flow dominates, i.e. the estuary is *microtidal.* The salt water forms a wedge intruding into the river and separated from the upper layer of fresh water by a sharp interface, which would be horizontal in the absence of friction. The effect of friction leads to a weak slope of the interface in the seaward direction, which, for a stationary salt wedge, drives a weak landward directed residual current of sea water. The mixing of salt water and fresh water is basically localized at the interface of the salt wedge and is relatively weak.

The position of the salt wedge is dependent on the river flow, the penetration of the salt wedge inland increasing for lower discharge.

The flow field within the salt wedge is usually too weak to drive a significant bedload transport but some 'residual' suspended load can indeed be carried landward through the residual current described above. Nevertheless, the fresh water flowing seaward above the salt wedge layer is actually the major source of sediment transport, which may occur in the form of both bedload and suspended load upstream to the tip of the salt wedge: downstream, the coarser bed load fraction is left behind while the suspended fraction is carried by the fresh water flow. As a result, typical morphological features of such estuaries are the formation of a bar consisting of the coarser fraction at the tip of the salt wedge layer [115] and the formation of a delta where the suspended load carried by the river is subject to deposition depending also on the intensities of littoral and tidal currents [119]. As noted by Dyer [31], unlike almost all other tidal environments, grain size increases landward in such estuaries.

A major example of salt wedge estuary characterized by the formation of a typical delta and of a strong deposit (which can reach values of the order of a meter per week) at the tip of the salt wedge layer is the Mississippi.

- The second limiting case: *well mixed estuaries.*

This is the case when tidal flow dominates, i.e. the estuary is *macrotidal.* Under such conditions tidal currents drive an intense vertical mixing such that variations of salinity throughout the water column are quite small. Of course significant longitudinal variations of salinity, which increases from the head to the mouth of the estuary, are driven by the progressive effect of tidal mixing.

Lateral variations of salinity are also induced when the width to depth ratio of the channel is sufficiently large for Coriolis force to give rise to a lateral slope of the free surface able to drive a significant secondary circulation: under these conditions salt and fresh water are found to flow in opposite directions in distinct portions of the channel adjacent to the bank located on their right side (in the northern hemisphere). This secondary circulation also drives a distinct transport of upland sediment by the fresh water flow and of marine sediments by the salt water flow.

- The intermediate case: *partially mixed estuaries.*

An intermediate behaviour between the two extreme cases treated above is typical of (mesotidal) estuaries characterized by tidal currents able to induce vertical mixing such to give rise to a distinct vertical variation of salinity, increasing from the free surface to the bottom. Two layers of flow can still be distinctly recognized, with the upper layer flowing seaward and the lower layer flowing landward, though no sharp interface exists. As the ratio between the fresh water discharge and the tidally driven discharge varies, the degree of stratification experienced by the flowing stream may vary significantly. Seasonal variations of the degree of stratification are also experienced by this type of estuary.

Note that the net entrainment of salt water by the upper layer is such that the flow discharge in this layer may be an order of magnitude larger than the river discharge: hence, mixing further enhances the ability of tidal currents to transport marine sediments both as bed load and as suspended load, a behaviour in contrast with that of salt wedge estuaries, which are dominated by river sediments.

In fact, a striking feature of partially mixed estuaries is the existence of a peak in the spatial distribution of sediment concentration (the so called *turbidity maximum*), roughly located where the salt water intrusion ceases. A classical example of the occurrence of such a peak is the so called 'Barking Mud Reaches', located a few miles from the mouth of the Thames estuary in Great Britain [52].

A further feature of this type of estuaries is the occurrence of a grain sorting process whereby larger particles are found close to the mouth, while finer particles are progressively selected upstream [30]. In other words, grain size decreases landward, in contrast with the observed behaviour of salt wedge estuaries.

Important examples of partially mixed estuaries are the Thames [52] and the Chesapeake Bay (Fig. 19.10), in particular the James River extensively investigated by Pritchard [80].

19.2.2 Wetlands

Tidal flats and salt marshes

The land bordering the incised portion of estuaries is periodically or intermittently subject to inundation or exposition. The lateral portions of most estuaries (typically tide-dominated estuaries and the inner portions of microtidal estuaries and deltas) are then characterized by a sequence of fairly flat environments whose morphology and sedimentology depends on the frequency of inundation-exposition they experience.

Geomorphologists (see the reviews [6] and [69]) have proposed various classifications of such environments. A gross distinction, which is generally agreed, may be made between tidal flats and salt marshes.

Tidal flats (Fig. 19.11) are those areas which belong to one of the following classes:

- areas which are *intermittently inundated* as their elevation falls between mean high water spring tides and mean high water neap tides;
- areas which are *inundated by every tide* as their elevation falls between mean high water neap tides and mean low water neap tides;
- areas which are *intermittently exposed* as their elevation falls below mean low water neap tides.

The latter three regions (respectively called *higher, middle* and *lower tidal flats*) define what is called the *intertidal* region [58,117], bounded by the upper *supratidal* and by the lower *subtidal* regions.

Salt marshes are 'environments high in the intertidal zone where a generally muddy substrate supports varied and normally dense stands of halophytic plants' [5].

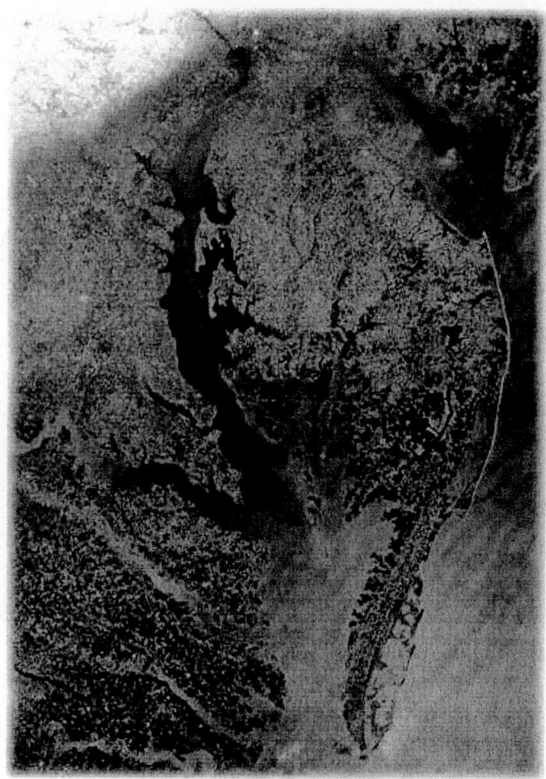

Fig. 19.10. A picture of Chesapeake Bay taken from the space shuttle

Fig. 19.11. Partially submerged tidal flat

Both the tidal flats and the vegetated areas occupied by salt marshes are fed by intricate networks of *tidal creeks.* In salt marshes creeks are flooded at each high tide, while the surface of the marsh itself is inundated only during the highest spring tides (Fig. 19.12).

Fig. 19.12. The salt marshes of Saaftinge (Westerschelde, Netherlands) with the network of tidal creeks

Sandy versus Muddy Tidal Flats

As pointed out by Amos [6], tidal flats contain sub-environments characterized by distinct slopes, grain size range and floral–faunal diversity. From the analysis of several tidal flats Amos suggests that they be distinguished into two groups, depending on the value of the mean inorganic suspended sediment concentration: *sandy tidal flats*, characterized by a fairly low value of concentration (say less than 1 g/l) and *muddy tidal flats*, characterized by larger values of concentration (say larger than 1 g/l).

In sandy tidal flats the higher, middle and lower tidal flats are, respectively, muddy, mixed sandy- muddy and sandy (Fig. 19.13). Channels are largely composed of sand and exhibit the presence of meso- and small-scale bed patterns. On the contrary, in muddy tidal flats, sandy regions are lacking while muddy and mixed regions dominate. Furthermore, biological diversity is much lower.

Examples of muddy tidal flats are found along nearly 50% of the coast of China [116] and at the mouths of the Orinoco and La Plata rivers. In all these cases mudflats are fed by huge amounts of suspended silt and clay discharged by the rivers into the ocean.

Examples of sandy tidal flats are found along the Dutch [79,109] and English [35] coasts of the North Sea. Sandy tidal flats are also typically found in the Venice Lagoon.

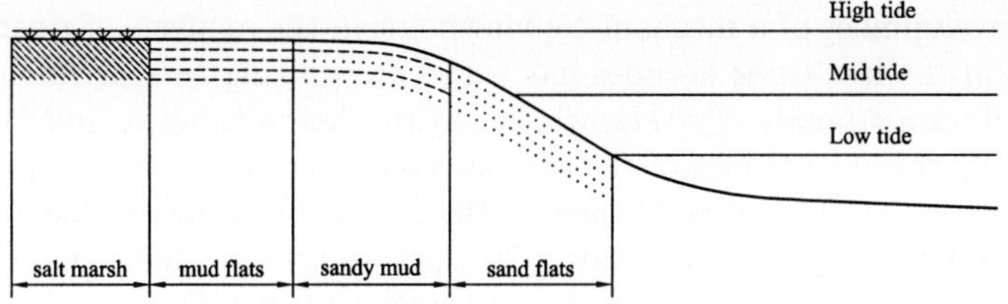

Fig. 19.13. Sketch of a sandy tidal flat

Elements of Salt Marsh Morphology

As pointed out by Pethick [78] 'the typical mature marsh profile is convexo–concave; the convex seaward margin giving way to a flat central section and a steeper concave landward edge'.

Creeks have typically a funnel shape with exponentially decreasing channel width and a meandering planimetric pattern. Typically they deepen in time [38] and migrate, at a decreasing rate the more densely vegetated is the neighbouring area [118]. A systematic analysis of the geometry of tidal creeks has been recently reported by Marani et al. [70]. Note that, in spite of the apparent similarity between tidal and fluvial networks, the oscillatory character of tidal flow and the presence of tidal flats determines a distinct function of tidal creeks which, rather than playing the role of drainage channels, behave essentially as small tidal estuaries [72,78]. The tidal control on creek morphology is also suggested by Adams [1], who points out that density and tortuosity of the creek network depends on the tidal range as well as on sediment size, with moderate tidal ranges and finer sediments being associated with more complex patterns.

The *ecology* of salt marshes, which also plays an important role in marsh dynamics, is outside the scope of the present review (see [1]).

19.3 Mesoscale Patterns

19.3.1 Bars

We define as *mesoscale estuarine patterns* those patterns which scale with channel width. In the fluvial literature such patterns are called *bars* and have been the subject of extensive investigations stem from the fundamental works of Leopold and Wolman [67] and Kinoshita [57]. Since the latter contributions, it has been recognized that three distinct channel morphologies, namely *straight, meandering* and *braiding*, can be distinguished. Straight natural rivers, the exception rather than the rule, do not exhibit the presence of mesoscale forms. Meandering rivers are typically characterized by sediment deposition at the inner bends (*point bars*) and scour (*pools*) at the outer bends, arranged in fairly regular sequences. In a braided river a highly dynamic multichannel pattern is formed with bars and pools forming a multiple row array, continuously rearranged in response to flood events.

The development of a mechanistic viewpoint in the analysis of river morphodynamics in the last three decades has led to recognizing *bars as the basic morphodynamic constituents of rivers*, which control their altimetric and planimetric evolution. Extensive reviews of the mechanics of bars, in meandering as well as braiding rivers, have been published in the recent literature [96,98,101,114] to which we refer the interested reader. Here, it suffices to point out that, from a mechanistic viewpoint, bars can be distinguished into *free* and *forced* [98], a classification based on the mechanism underlying their formation. Let us clarify this point.

19.3.2 Free Bars

Free bars arise spontaneously as a result of a *bottom instability*: they are migrating bedforms observed both in relatively narrow channels, in the form of alternate sequences of riffles and pools with diagonal fronts (*alternate bars*) and in wide channels where they form multiple row sequences (*multiple row bars*). The observed wavelength of such bedforms is of the order of six–ten times the channel width and the migration speed ranges about metres/day. Alternate bars have also been observed in estuaries by Barwis [9] for tidal creeks in South Carolina, and by Dalrymple et al. [20,21] in the Cobequid Bay-Salmon River estuary. As pointed out by Dalrymple and Rhodes [19] a spectrum of bar shapes is found depending on channel sinuosity. No information is reported about the possible migrating character of such forms. The theory of Seminara and Tubino [99,100], which will be outlined in Sect. 19.6, suggests that, due to the oscillatory character of the basic tidal flow, in the absence of residual effects, estuarine free bars are non migrating features. However, Dalrymple et al. [20,21] report that alternate bars are asymmetric in the direction of the local net transport, an observation which suggests the possibility that bars may exhibit a weak migration speed. The wavelength of estuarine alternate bars measured in the works quoted above have been plotted versus channel width by Dalrymple and Rhodes [19] (Fig. 19.14). In spite of the oscillatory character of the basic state and of the dominant suspended load prevailing in tidal environments, the dependence of wavelength on channel width seems to conform to the relationship valid for fluvial alternate bars under bedload dominated conditions. Little information is available about the shape of estuarine alternate bars.

Fewer observations of multiple row bars are available. Zaitlin [120] and Dalrymple et al. [20] describe multiple braid bars in the Cobequid Bay-Salmon River estuary. The shape of such bars appears to be fairly elongated (with lengths and widths ranging 150–1500 m and 50–200 m respectively) and asymmetric, with low bar amplitude (0.3–1.5 m) and small slopes of the lee face ($< 5° - 10°$).

No laboratory observations of estuarine free (either alternate or multiple braid) bars are known to the present authors. An extensive research appears to be needed in this area.

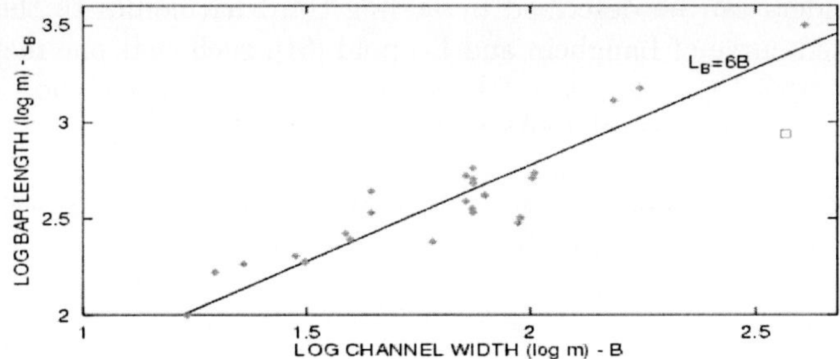

Fig. 19.14. The wavelength of estuarine alternate bars observed by Barwis [9] in tidal creeks of South Carolina and by Zaitlin [120] in the Salmon River estuary, Bay of Fundy, is plotted versus channel width (from [19]). Note that data concerning both free bars in straight or weakly sinuous channels and forced (*point*) bars in sinuous channels are included

19.3.3 Forced Bars

A second class of bars is associated with forcing effects acting on the erodible channel bottom.

The Forcing Effect of Curvature: Point Bars and Tidal Meanders

An important forcing effect is curvature of channel axis, which leads to the establishment of a secondary flow able to let sediment transport deviate from the longitudinal direction. As a result, sediments are found to accumulate in the inner bends and be eroded from the outer bends, leading to the classical point bar-pool pattern of meandering channels. Such pattern is self-sustaining as bank erodibility leads to a progressive shifting of the outer bend with the eroded material depositing along the inner banks, hence reinforcing the meandering process. Such mechanism, which has been widely investigated in the fluvial case (see [50]), has recently been extended to the tidal case [104,106] and will be discussed in Sect. 19.5.

Point bars have been observed in tidal environments by Barwis [9] and Zaitlin [120] (see also [20,21]). Note that, unlike the fluvial case where the unidirectional character of the flow leads to an asymmetric configuration of the point bar, which lags ahead of channel curvature, in the tidal case, the flow being oscillatory, the point bar structure has distinct flood and ebb components which may lead to a symmetric or weakly asymmetric pattern depending on the degree of asymmetry of the basic flow field.

The *shape of meanders* is the result of the planimetric development of the channel. In the fluvial case the latter arises from a combination of outer bank erosion and inner bend deposition, which progressively leads to lateral shifting of the channel axis and meander migration, keeping the channel width essentially constant. Kinoshita [57] has suggested that the shape of fully developed fluvial meanders can be described by adding third harmonics to the so called sine generated curve of Langbein and Leopold [61], such that one may write:

$$c(s) = c_0(\cos \lambda s + c_F \cos 3\lambda s + c_S \sin 3\lambda s) \tag{19.1}$$

with c_F and c_S fattening and skewing coefficients respectively. Figure 19.15 clarifies the meaning of such coefficients.

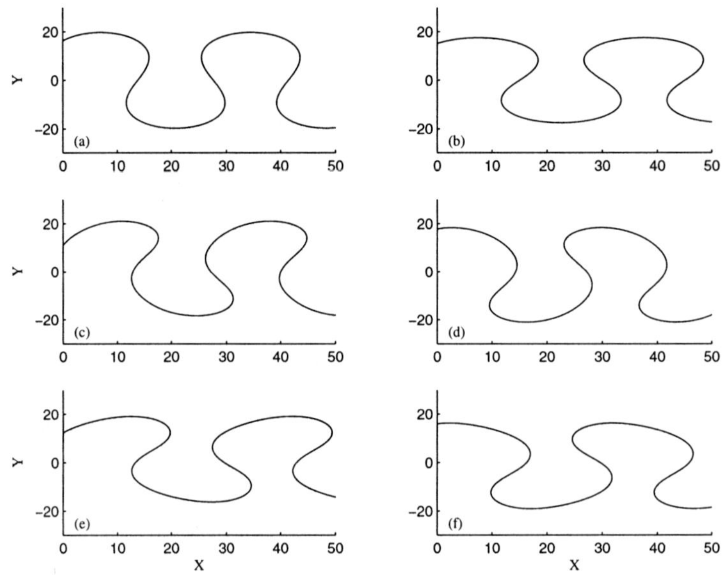

Fig. 19.15. The role of fattening and skewing coefficients in determining the shape of fully developed fluvial meanders according to Kinoshita's curve (i.e. (19.1)). The various curves have been obtained by setting $c_0 = 0.12$ and **(a)** $c_F = c_S = 0$; **(b)** $c_F = -0.4$, $c_S = 0$; **(c)** $c_F = 0$, $c_S = 0.4$; **(d)** $c_F = 0$, $c_S = -0.4$; **(e)** $c_F = -0.4$, $c_S = 0.4$; **(f)** $c_F = -0.4$, $c_S = -0.4$

In fact, (19.1) involves three contributions, a fundamental component (the sine generated curve) which generates symmetric shapes, a 'fattening' contribution which tends to widen the meander lobes while keeping the shape symmetric and a 'skewing' contribution which distorts the meander shape with the skewing being directed either downstream or upstream, depending on the coefficient c_S being positive or negative.

Meandering is also an ubiquitous feature of *tidal* environments (Fig. 19.16), though a systematic investigation of the shape of tidal meanders has only recently been performed [70]. An interesting result of the latter analysis is the observation that, unlike fluvial meanders, tidal meanders do not exhibit a distinct preferential direction of skewing of meander shape. Moreover, the mathematical representation of the shape of tidal meanders can be given a form similar to (19.1) with the inclusion of second harmonics.

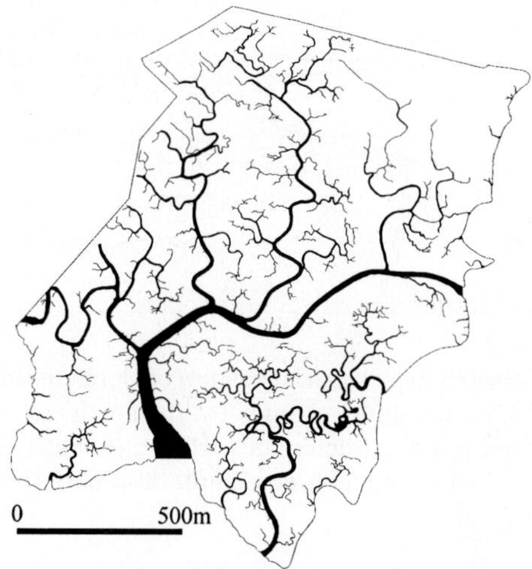

Fig. 19.16. Meandering is an ubiquitous feature of tidal environments: the figure represents a portion of the tidal channel network extracted from the Venice Lagoon (from [37])

Elongated Tidal Bars

Near the mouth of macrotidal (and sometimes mesotidal) estuaries a peculiar type of bar has been observed by various authors: the main feature of such, so called *elongated tidal bars*, is their large size. According to Dalrymple and Rhodes [19], typical bar lengths are of the order of 1–15 km with corresponding widths ranging between 0.2–4 Km and amplitudes which can reach 20 m. Bars are inclined at a small angle relative to the dominant tidal flow and are usually asymmetric with the steeper side in the direction of the dominant sediment transport.

In smoothly converging estuaries bars are typically organized in chains dissected by diagonal channels (*swatchways*, [87]) and separating ebb- from flood-dominant channels. This is the case e.g. of Cobequid Bay, Bay of Fundy, sketched in Fig. 19.17, as well as of Thames estuary [87] and Gironde estuary [110]. In other cases, where the estuary mouth exhibits a local constriction with a submerged sill, the bar pattern is more complex and apparently less organized with interconnected flood- and ebb-channels separated by zig-zag bars. This is the case e.g. of Moreton Bay [46] and of Chesapeake Bay [68].

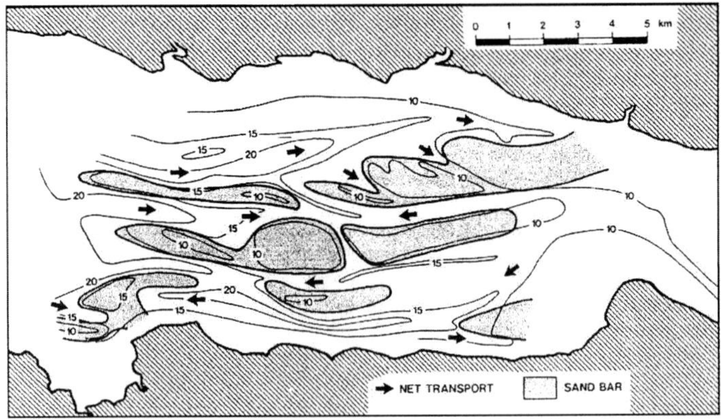

Fig. 19.17. Sketch depicting the pattern of elongate tidal bars at Cobequid Bay, Bay of Fundy. Note that the upper channel (to the north) as well as the western portion of the central channel are flood-dominant, while the south channel and the eastern portion of the central channel are ebb-dominant. Also note the presence of several swatchways (reproduced from [20])

Delta-like Patterns

A further forcing effect arises from abrupt changes of geometry which give rise to delta-like patterns. Information on the main characteristics exhibited by delta-like patterns can be found in [53].

19.4 Small-Scale Patterns

The word *small-scale* refers to bedforms which scale on flow depth: the most important of such bedforms are called *dunes* by analogy with the corresponding bedforms observed in the fluvial environment, though various alternative names have been proposed in the geomorphological literature (e.g. *sandwaves or megaripples*). A systematic description of the geomorphology of estuarine dunes is given in [19].

Dunes have been observed in several estuaries, e.g. at the mouth of tide dominated estuaries where they form on elongated tidal bars, in the tidal inlets and tidal channels of coastal plain wave-dominated estuaries, in the lower portions of tidal flats and sandy tidal channels of lagoons.

Geomorphic classification of dunes [8] have been based on three main criteria, namely size (*small, medium, large* and *very large plan form shape*, 2D and 3D) and possible presence of smaller superimposed dunes (*simple* versus *compound* dunes). A few examples of estuarine dune patterns are represented in Fig. 19.18, reproduced from [19].

As mentioned above, the spatial scale of dunes is flow depth H. Allen [2], reports two plots where dune wavelength L and dune amplitude A are correlated with flow depth: data refer to either fluvial or tidal dunes and, in spite of

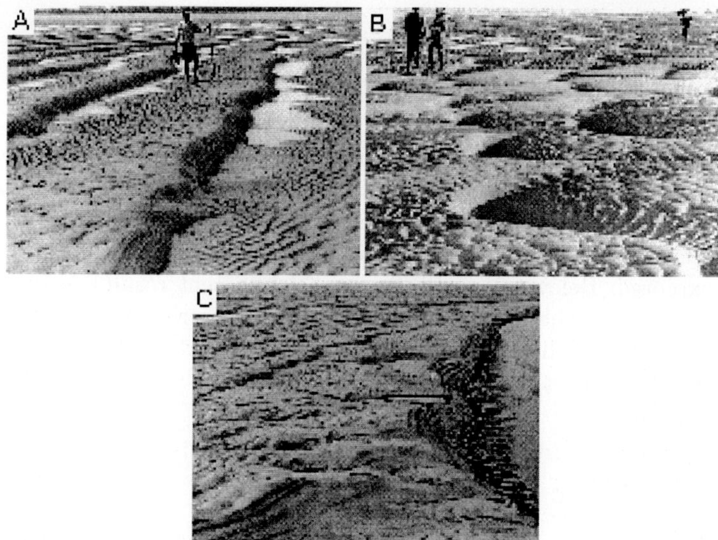

Fig. 19.18. A few examples of dune patterns observed in tidal environments. (**a**) Small, simple nearly 2D dunes observed in the Gironde estuary (France) superimposed over elongate tidal bars [4]. (**b**) Small 3D simple dunes observed in the ebb-tidal delta of North Inlet, South Carolina [9]. (**c**) 2D oblique medium simple dunes superimposed over large 2D compound dunes formed over elongate tidal bars observed in Coquebid Bay, Bay of Fundy [20]

considerable scatter, they confirm the simple observation that large dunes are observed only in subtidal channels and tidal inlets while small-medium dunes form on intertidal flats. The scatter of data in such plots suggests that the ratios A/H and L/H do not keep constant. In fact, theoretical investigations on the mechanism of dune formation (see Sect. 19.6) and laboratory observations [44,85] of fully developed dunes in steady currents suggest that the latter ratios depend on the excess Shields stress, i.e. a dimensionless form of the excess averaged stress acting on the cohesionless bottom, the word 'excess' referring to the threshold of sediment motion.

The role played by such parameter is confirmed by the observation that, as the flow speed increases, for given values of other flow parameters, dune amplitude increases up to a maximum and then decreases till dunes disappear being replaced by a flat bed or antidunes. Field observations [90] confirm that dune amplitude may decrease proceeding from the estuary mouth landward despite an increase in water depth due to a decrease in water speed.

Less clear is the observed dependence of dune amplitude and dune wavelength on grain size in spite of several laboratory investigations (see references in [19]).

Part Two: Mechanics

19.5 Morphodynamic Equilibrium

The issue of whether morphodynamic patterns may achieve equilibrium conditions, at least on time scales of decades is, needless to say, of major importance for all aspects related to the management of tidal environments. In spite of its relevance, the problem has only been partially investigated and still awaits to be fully explored. Below, we outline some fairly recent results concerning the equilibrium of tidal channels and tidal inlets. In describing theoretical developments, below we will often refer to equations introduced in the companion paper (Chap. 16). Furthermore note that hereafter a star apex will denote a dimensional quantity.

19.5.1 Equilibrium of Tidal Channels

Equilibrium Profile of Straight Convergent Tidal Channels

Let us examine the issue of the possible existence of a long term longitudinal equilibrium or quasi-equilibrium profile in estuaries and tidal channels. We consider the case of tide-dominated well-mixed estuaries. A mechanistic analysis of this problem can be performed employing a one-dimensional model. A glance at the one dimensional version of the evolution equation of the bed interface (16.7) immediately suggests that, in the absence of lateral exchange of sediments between tidal channels and adjacent tidal flats, in order to achieve equilibrium in an averaged sense, i.e. such that the average bed elevation in a tidal cycle does not vary, the net sediment flux in a tidal cycle must vanish. The reader should note that the latter condition does not imply that the instantaneous sediment flux must vanish. The field evidence analyzed by Friedrichs [40] does not contradict the latter statement. Several factors contribute to determine the sediment balance in a tidal channel, namely:

i) channel geometry, i.e. the degree of convergence of the channel induced both by its funnel planimetric shape and by the sloping character of the bed and the degree of channel sinuosity, which controls the hydrodynamics of tide propagation along the channel;
ii) the harmonic content of the tidal oscillation acting at the inlet;
iii) the fluvial transport of sediments discharged by the river at the upstream end of the estuary;
iv) the lateral exchange of sediments between the channel and the adjacent tidal flats.

A sound theoretical approach can be constructed by isolating each of the latter factors in the context of idealized models reproducing only part of the actual process. This research line has been recently pursued theoretically by Lanzoni and Seminara [62,63] and experimentally by Bolla Pittaluga et al. [12].

In the latter contributions only the factor i) was examined, considering a straight, convergent channel close at one end and connected at the other end with a tidal sea characterized by a tidal oscillation consisting of a single harmonic. The possible presence of tidal flats was ignored.

Theory

Under the latter conditions, the mathematical problem reduces to solving the evolution equation of the bed interface (16.7), starting from some initial bottom profile. This requires to evaluate the net flux of sediments at each instant and at each cross section. The latter problem is clearly coupled to the hydrodynamical problem, i.e (16.7) must be coupled to the one-dimensional continuity and momentum equations of the fluid phase. Let us then consider a tidal channel of length L_e^* and average flow depth at the channel inlet D_0^*. We denote by B^* the effective width of a rectangular cross section equivalent to the actual cross section of the channel and assume:

$$B^* = B_0^* \exp\left(-\frac{x^*}{L_b^*}\right) , \tag{19.2}$$

having denoted by L_b^* the so called convergence length and by x^* a longitudinal coordinate. Denoting by t^*, D^*, H^*, U^* and C time, flow depth, free surface elevation, cross-sectionally averaged flow speed and cross-sectionally averaged conductance respectively, the governing equations in dimensionless form read:

$$\frac{1}{\epsilon} D_{,t} + \mathcal{F}\left(UD\right)_{,x} - \mathcal{K} UD = 0 , \tag{19.3}$$

$$\mathcal{S} U_{,t} + \epsilon \mathcal{S} \mathcal{F} U U_{,x} + \frac{1}{\epsilon} H_{,x} + \mathcal{R} \frac{U|U|}{\mathcal{C}^2 D} = 0 , \tag{19.4}$$

In (19.3)–(19.4) the following dimensionless quantities have been introduced:

$$t = \omega^* t^* , \quad x = \frac{x^*}{L_0^*} , \quad (D, H) = \frac{(D^*, H^*)}{D_0^*} , \quad U = \frac{U^*}{U_0^*} , \quad \mathcal{C} = \frac{C}{C_0} , \tag{19.5}$$

and the following scales have been employed:
L_0^*: longitudinal spatial scale determined by the dominant dynamic balance;
D_0^*: characteristic flow depth, say the flow depth at the channel inlet;
U_0^*: characteristic flow speed, say the maximum flow speed at the channel inlet;
C_0: characteristic flow conductance, say the mean flow conductance at the channel inlet;
ω^*: angular frequency of the tidal wave.

Moreover, the following dimensionless parameters arise in (19.3)–(19.4):

$$\epsilon = \frac{a_0^*}{D_0^*} , \quad \mathcal{F} = \frac{1}{\epsilon} \frac{U_0^*}{\omega^* L_0^*} , \quad \mathcal{K} = \frac{1}{\epsilon} \frac{U_0^*}{\omega^* L_b^*} , \tag{19.6}$$

$$\mathcal{S} = \frac{F_0^2}{\epsilon} \frac{\omega^* L_0^*}{U_0^*} , \quad \mathcal{R} = \frac{F_0^2}{\epsilon} \frac{L_0^*}{C_0^2 D_0^*} , \tag{19.7}$$

where F_0 is a characteristic Froude number constructed with the characteristic scales U_0^* and D_0^*, and a_0^* is a characteristic amplitude of the tidal wave. The parameter $\mathcal{K}$ is a dimensionless measure of the degree of channel convergence; the parameter $\mathcal{R}$ measures the ratio between friction and gravity while $\mathcal{S}$ measures the ratio between local inertia and gravity; finally $\mathcal{F}$ is a dimensionless measure of the ratio between the contribution to mass balance associated with oscillations of the free surface and the contribution due to spatial variations of flow depth and flow speed. In order to close (16.7) we need to evaluate the cross sectionally averaged sediment flux $\overline{Q_s^*}$. The bedload component of $\overline{Q_s^*}$ is readily obtained from (16.15). The suspended load component is obtained from the one dimensional form of (16.20), namely:

$$\overline{Q_{ss}^*} = \sqrt{\Delta g d_s^{*3}}\,\overline{Q_{ss}} = \int_{-b^*/2}^{b^*/2} \mathrm{d}x_2^* \int_{\eta^*}^{H^*} \left(CU^* + \langle c' u^{*\prime}\rangle\right) \mathrm{d}x_3^* \,. \tag{19.8}$$

Equation (19.8) shows that, in order to evaluate $\overline{Q}_{ss}$, one would strictly need to evaluate the spatial and temporal distribution of the local concentration of suspended sediment averaged over turbulence C. However a simple scaling argument [42,63] shows that, at the leading order of approximation in an expansion in powers of suitable parameters, C can be simply evaluated using relationships established for the case of transport in suspension in uniform turbulent free surface flows, i.e. resorting to the classical Rouse [88] solution. In fact, let us write the convection–diffusion equation (16.21) in dimensionless form for the case of plane flows (i.e. $V^* = V_2^* = 0$). Setting $(x^*, z^*) = (x_1^*, x_3^*)$, $(U^*, W^*) = (V_1^*, V_3^*)$, employing the scaling (19.5) and introducing the further dimensionless quantities

$$z = \frac{z^*}{D_0^*}\,, \qquad W = \frac{W^*}{W_s^*}\,, \qquad \mathcal{D}_T = \frac{D_T^*}{u_{*0} D_0^*}\,, \tag{19.9}$$

one finds:

$$\delta_1 \frac{\partial C}{\partial t} + \delta_2 \langle U\rangle \frac{\partial C}{\partial x} + \delta_3 \langle W\rangle \frac{\partial C}{\partial z} - k Z_0 \frac{\partial C}{\partial z} = \frac{\partial}{\partial z}\left(\mathcal{D}_T \frac{\partial C}{\partial z}\right) + \delta_4 \frac{\partial}{\partial x}\left(\mathcal{D}_T \frac{\partial C}{\partial x}\right)\,, \tag{19.10}$$

where:

$$\delta_1 = \frac{\omega^* D_0^*}{u_{*0}}\,, \qquad \delta_2 = \frac{U_0^* D_0^*}{u_{*0} L_0^*}\,, \qquad \delta_3 = \frac{W_s^*}{u_{*0}}\,, \qquad \delta_4 = \frac{D_0^{*2}}{L_0^{*2}}\,, \tag{19.11}$$

having denoted by k the von Karman constant, by Z_0 $(= W_s^*/k u_{*0})$ the so called *Rouse number*, which controls the vertical distribution of concentration, and by u_{*0} a typical value of the friction velocity. Simple estimates immediately suggest that the parameters Z_0 and D_T are O(1) quantities, while the parameters δ_1 and δ_4 are definitely small: in fact, with ω^*, D_0^* and u_{*0} ranging about $1.4\cdot 10^{-4}\,\mathrm{s}^{-1}$, 10 m and 5 cm/s respectively, one finds that δ_1 ranges about 0.028. On the other hand, with L_0^* of the order of kilometers δ_4 is of order 10^{-3}. Finally, since the vertical velocity associated with tide propagation may reach, at most, a fraction of mm/s, the parameter δ_3 is also at most of order 10^{-2}. More delicate is the role of the longitudinal advective term: in fact with U_0^* ranging about 1 m/s and with D_0^*/L_0^* of order 10^{-2} the parameter δ_2 would range about 0.2. The above arguments suggest that the solution for C can be expanded in powers of the small parameter δ_2 possibly including the effects of vertical convection, longitudinal turbulent diffusion and local variations of mean concentration once suitable relationships are established among the sizes of the various small parameters appearing in (19.10). At the leading order of approximation (19.10) reduces to the classical uniform balance between vertical depositing flux associated with the settling speed and mean upward vertical flux associated with turbulent diffusion. Lanzoni and Seminara [63] have restricted their analysis to such approximation, which allowed them to calculate the mean total sediment flux by means of classical relationships (e.g. [34,85]) established for uniform free surface flows, evaluated in terms of the local instantaneous conditions determined at each cross section by tide propagation. We refer the reader to the above authors for details of the numerical procedure employed to solve continuity and momentum equations. In this respect, we note that two difficulties arise in performing the above calculations: firstly, the transient evolution of the bed profile starting from arbitrary initial conditions may give rise to the formation of fairly sharp fronts which must appropriately be dealt with in the numerical scheme; secondly, at the landward end, the front reflects leading to the emergence of the bottom with the consequent formation of a wetting and drying portion of the computational domain. Both features were conveniently treated by Lanzoni and Seminara [63]. In particular, their results suggest that a long term longitudinal equilibrium or quasi-equilibrium profile in tide-dominated, well-mixed estuaries may indeed be reached on time scales of the order of hundreds of years. More precisely, given a tidal input at the mouth of the tidal channel, a width distribution of the channel and the sediment size, it is shown that the longitudinal bed profile of the channel evolves from an arbitrary initial condition towards some equilibrium configuration characterized by an upward concavity increasing as the channel convergence increases. The final equilibrium configuration of the bed profile is characterized by a length determined by the emergence of the bed in the inner portion of the channel. Furthermore, the depth at the inlet section is uniquely determined by the final equilibrium state. Finally, the relationship between tidal prism and cross-sectional area in each section of the channel is found to evolve from its initial arbitrary distribution towards a linear relationship of Jarrett's type (see Sect. 19.5.2).

The problem treated in [63] has also been analyzed theoretically by Schuttelaars and de Swart [94,95]. The approach employed by the latter authors differs from the approach of Lanzoni and Seminara [63] in several respects. Space does not allow to discuss various assumptions employed by the above authors which we feel to be questionable. The interested reader is referred to [63].

Experiments

Experiments were carried out in the laboratory of the Department of Environmental Engineering of the University of Genova (Italy), on a large indoor platform. A straight rectangular channel, closed at one end and connected to a basin representing the sea at the other end, was built over the platform. The apparatus for tidal wave generation, installed in the basin, consisted of a cylinder controlled by an oleodynamic mechanism driven by a control system which could generate the desired law of motion. The flume was filled with crushed hazelnut shells characterised by a density of 1480 Kg/m^3 and median grain size $d^*_{50} = 0.31$ mm. Sediment were chosen light enough to be entrained into suspension throughout most of the tidal cycle with the values of friction velocity typically generated in the experiments. The scaling rules employed to achieve similarity between the process reproduced in the laboratory and the actual natural process, are discussed in [12].

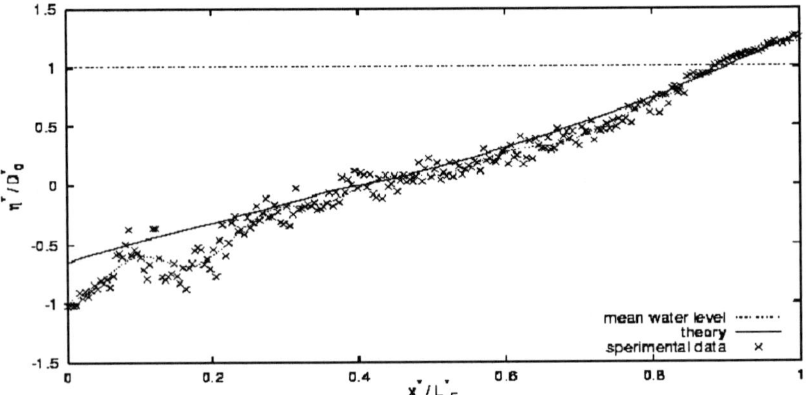

Fig. 19.19. Comparison between the observed equilibrium profile in the experiments of Bolla et al.[12] and the theoretical prediction of Lanzoni and Seminara [63]

Preliminary experiments, reported in the latter work were characterised by an initial mean flow depth $D^*_0 = 0.082$ m constant throughout the channel, dimensionless tidal amplitude at the mouth $\epsilon = 0.32$ and tidal period $T^* = 180$. Starting from a flat bottom configuration, sediments were found to be scoured by the tidal motion in the seaward portion of the channel, driven landward and deposited in the inner part of the channel. Comparison between the bed profiles at 50 h and 100 h suggested that the latter pattern was quite close to equilibrium. Note that the weak upward concavity of the final bottom profile is consistent with field observations concerning both tidal estuaries and coastal lagoons and with the theoretical findings of Lanzoni and Seminara [63].

A detailed comparison between observations and theoretical predictions has been recently pursued [12] and is shown in Fig. 19.19. It appears that the time scale of the evolutionary process observed in the experiments is samewhat faster that the one predicted theoretically, a feature possibly related to the inability of the model to reproduce accurately the actual bedforms present in nature (see Sect. 19.6.2). However, the equilibrium profile eventually reached shows a very good agreement with theoretical predictions.

Equilibrium Cross Section: Why are Tidal Channels Convergent?

The theoretical and experimental results discussed above clearly show that the flow depth of tide dominated estuaries decreases landward. This finding is directly relevant to the understanding of the spatial variations of width of tidal channels. In fact, though the problem of equilibrium of channel cross section in tidal environments has not been tackled yet, both the theoretical knowledge established in the fluvial literature [49,75,76,113] and the empirical evidence concerning the equilibrium cross section of rivers [66,93] give us a clue to postulate possible mechanisms for the observed development of a funnel shape in tide dominated estuaries. The well known observations reported by Leopold and Maddock [66] suggest that an empirical correlation can be established between river width and a power of the mean annual discharge with exponent typically ranging about 0.5 (0.45 according to Schumm, [93]). Later, Schumm [93] has extended the latter observations showing that the proportionality constant of the latter relationship depends on the average percentage of silt and clay present in the material composing banks and bed. High percentages of silt and clay lead to cross sections which are comparatively narrower than those typical of quasi cohesionless channels. An attempt to interpret the latter observations has been proposed by Parker [76] and later corrected by Ikeda and Izumi [49]. The scheme employed in these contributions ignores the effects of bank cohesion. Under these conditions and assuming sediment to be dominantly transported as suspended load, Parker's analysis [76] suggests that equilibrium is achieved when a balance is established between the sediment flux driven by turbulent diffusion from the central region towards the banks and the flux of sediment transported in the opposite direction as bedload driven by gravity acting on particles rolling down the banks. Ikeda and Izumi's [49] approach leads to the so called *regime relationships*. In particular, a relationship is found between channel width and formative water discharge Q^*, in the form:

$$\frac{B^*}{d_s^*} = \left(\frac{Q^*}{\sqrt{g d_s^*}} \right)^m , \tag{19.12}$$

with d_s^* grain diameter taken to be uniform and m exponent dependent on the dimensionless settling speed,

$$\widehat{w} = \frac{W_s^*}{\sqrt{\Delta g d_s^*}} . \tag{19.13}$$

Though the laboratory observations reported by Ikeda and Izumi [49] appear to support the above results, however the exponent m estimated from their figure ranges about 0.9, a value significantly larger than those suggested by field evidence.

The role of bank cohesion as a controlling factor of the equilibrium cross section of rivers has been given some attention by Osman and Thorne [74], Thorne and Osman [112] and by Darby and Thorne [22]. Essentially, the latter authors assume that channel widening is associated with bank collapse and examine various mechanisms which may induce the latter phenomenon. In particular, in the case of plane slip failure, typical of steep banks, a classical result of Taylor [111] and Spangler and Handy [107] suggests that a critical bank height H_c^* exists, above which uniformly cohesive banks collapse along a plane slip surface. Such critical height reads:

$$H_c^* = \frac{4c(\sin\theta\cos\phi)}{\gamma[1-\cos(\theta-\phi)]} , \tag{19.14}$$

where c is bank cohesion, θ is the angle that the bank surface forms with the horizontal, γ and ϕ are the specific weight and the friction angle of bank material, respectively. Various other mechanisms may affect bank failure: collapse may occur along curved surfaces (rotational slip failure), may be of cantilever type in composite banks characterized by finer more resistant material in the upper layers, may be influenced by drying and wetting cycles associated with the propagation of floods [86] and by the possible occurrence of piping induced by the presence of tree roots or small tunnels excavated by animals underground. The process may indeed be complex enough to defeat theoretical attempts to describe it in detail. However, the main implication of the 'cohesive' approach suggests that channels would continue widening until the formative discharge is carried by the stream with flow depths not exceeding the threshold value for bank collapse. It would be worth exploring whether both the 'cohesionless mechanism' of Parker [76] and the 'cohesive mechanism' of Osman and Thorne [74], may play a role in the establishment of an equilibrium cross section of sandy rivers through an intermediate mechanism, whereby the stability of the upper and steeper part of banks is dominated by bank cohesion, while the bank foot is somewhat 'sheltered' by the formation of submerged cohesionless deposits (often observed in nature), a sort of shallower submerged banks driven by a mechanism of the type proposed by Parker [76]. While the empirical regime relationships proposed by Leopold and Maddock [66] for rivers have been extended to tidal channels [72], to our knowledge no attempt at extending to the tidal environments the theoretical interpretations mentioned above have been proposed so far.

Equilibrium Topography of Tidal Meanders

As mentioned in part one, meandering channels are a common feature of both fluvial and estuarine environments.

The problem of meandering of alluvial rivers has been widely investigated (see [50]). In particular we know that river meanders are typically characterized by the formation of a sequence of so called *point bars*, with depositional regions typically located at the inner bank close to bend apexes and pools located at the outer bank. These alternate sequences of deposits and scours propagate typically (though not invariably) downstream at a very slow rate, of the order of metres per year, as a result of the planimetric evolution of the meandering pattern.

Since the pioneering contribution of Rozovskij [89], it has been known that the point bar-pool pattern is maintained through the development of secondary flows which act on sediment particles. In constant curvature channels a *centrifugally driven* secondary flow is established, which arises from the imbalance between two transverse forces acting on fluid particles: on one hand the force associated with the lateral pressure gradient induced by the development of a lateral slope of the free surface; on the other hand an apparent centrifugal force increasing in the vertical direction driven by streamline curvature. Sufficiently far from the bed entrance for flow and bed topography to be uniform in the longitudinal direction and in the absence of coexisting free bars, the resulting secondary flow is constrained by continuity to have a vanishing depth average and is directed inwards close to the bed where the pressure gradient exceeds the centrifugal force and outwards close to the free surface. Furthermore, sediment continuity constrains the lateral component of sediment transport to vanish. Two lateral forces act on bedload particles: an inward drag force associated with the secondary flow and an outward directed lateral component of particle weight. Equilibrium is achieved for a lateral slope increasing outwards, giving rise to deposition at the inner bank and scour at the outer bank.

In the case of meandering channels, flow continuity forces the development of an additional, *topographically driven*, component of the secondary flow which has a non vanishing depth average and transfers longitudinal momentum from each pool to the next one. Furthermore, sediment continuity forces a lateral component of bed load transport, which gives rise to an additional contribution to the lateral slope. The balance between centrifugal and topographic effects give rise to a bar-pool pattern which displays a phase lag relative to curvature, depending on meander wavenumber for given flow and sediment characteristics.

Tidal meandering channels exhibit distinct novel features with respect to the fluvial case, namely the oscillating character of the basic flow and the changing character of sediment transport throughout the tidal cycle. Due to the oscillatory character of the basic flow, which reverses its direction at each half cycle, the bar-pool pattern, unlike in the fluvial case, oscillates in time. It is worth noticing that, as one may expect, the pattern of secondary flow is not sensitive to the direction of the basic flow. The character of sediment transport changes throughout the tidal cycle, with periods of slack water characterized by the absence of any entrainment and the presence of deposition of previously suspended sediments, followed by periods where bedload transport occurs and eventually by a stage dominated by transport in suspension. Such variability does affect the lateral as well the longitudinal fluxes of sediments, hence the amplitude and the phase lag of the bar-pool pattern. However, note that the general features of the latter do not change qualitatively.

The study of bottom topography in meandering tidal channels has been recently tackled by Solari et al. [106] who considered channels formed by inerodible banks and cohesionless bottom. The sediment size distribution was taken to be uniform and the grain diameter small enough to be suspended by turbulence throughout most of the tidal cycle. Curvature of the longitudinal channel axis was taken to follow the so called 'sine-generated curve' of Langbein and Leopold [61] (recall (19.1)). Recent observational evidence, illustrated in [106], suggests that meander wavelength scales with the local channel width; therefore, at the meander scale and at the leading order of approximation, width variations may be safely neglected. Furthermore, tidal meandering channels are characterized by sufficiently large aspect ratio, though smaller than that typically observed in the fluvial case, and fairly low curvature ratio ν (defined as ratios between channel width and radius of curvature at the bend apex). Such features allow one to formulate a 3D model of flow and bed topography in meandering tidal channels along the lines of the classical framework developed in the fluvial context, where the role of side walls is ignored and the flow and bed topography are treated as slightly perturbed with respect to the basic configuration.

We refer the reader to [106] for details of the analysis. Here it suffices to state that, taking advantage of the weakly meandering character of the channel, the solution has been expanded in powers of the small parameter ν. At the leading order of approximation, one finds the basic solution, parametrically dependent on time, for a tidal wave propagating in a long, weakly meandering channel, i.e. at the meander scale, the basic flow structure is spatially uniform at least at leading order. At the first order in ν, the curvature induced perturbation of the flow field and bottom topography is obtained. The dimensionless parameters which play a role in the analysis are: width ratio β, particle Reynolds number R_p, curvature ratio ν and relative roughness d_s. Typical values applying to Venice lagoon have been adopted.

Results have been obtained assuming 'quasi-equilibrium', i.e. assuming that the bottom configuration adapts instantaneously to changes of the flow field throughout the tidal cycle. However, at a more careful examination, it turns out that the latter assumption is unduly severe and can be readily removed [105]. The latter authors have shown that, starting from an initial configuration characterized by a flat bed with a uniform flow depth along the channel, the system reaches an equilibrium topographic pattern such that throughout a tidal cycle the bottom displays relatively low amplitude oscillations around a mean level which does no longer change in time. It is found that the system attains 'quasi-equilibrium' conditions only when two limiting cases are approached; namely when the tidal wave is so 'long' that the system is able to adapt instantaneously to the changing flow configuration or when transport in suspension is extremely 'strong'.

Figure 19.20 shows the patterns of deposition and scour at the positive ($t = 0$) and negative ($t = \pi$) peaks of the tidal cycle (the arrow indicates the direction of the basic flow), respectively. Note the symmetrical position of the point bar-pool with respect to the bend apex; due to the periodicity of the basic flow throughout

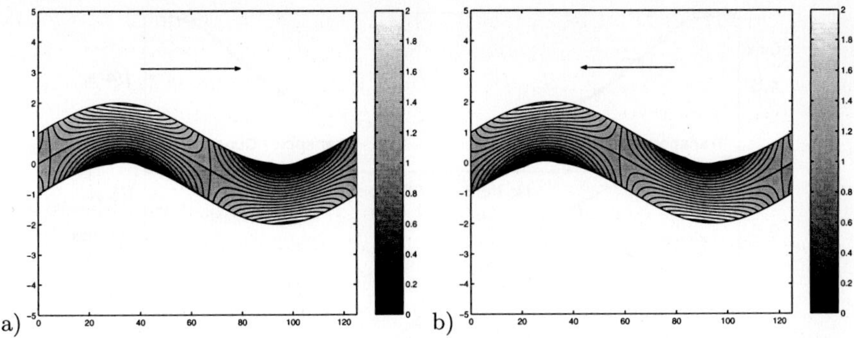

Fig. 19.20. The pattern of scour and deposition at the positive ($t = 0$) (**a**) and negative ($t = \pi$) (**b**) peak of the tidal cycle. ($\beta = 6$, $\lambda = 0.05$, $R_p = 4$, $\theta_0 = 0.6$, $d_s = 2 \cdot 10^{-5}$, dune-covered bed)

the tidal cycle, the point bar-pool pattern migrates alternatively forward and backward in a symmetric fashion: in other words no net bar migration in a cycle is present.

During the tidal cycle the instantaneous Shields number varies in time from zero to some maximum value at the tidal peak. As the Shields number θ increases, sediments are transported at first as bed load; for larger values of θ, suspension becomes an appreciable fraction of the total transport while dunes appear on the bottom surface. The oscillatory character and the intensity of the point bar are shown in Fig. 19.21 which shows position and amplitude of the location where occurs maximum scour during half a tidal cycle. Note that $\Psi_{D_{\max}}$ denotes the phase lag (in radians) of the location of the maximum flow depth $D_{\max}$ relative to the bend apex. It appears that the location of the maximum scour oscillates in time with maximum displacement of the order of a fraction of a radiant, hence a small value relative to meander wavelength. It appears that, as the intensity of the basic flow decreases, the point bar migrates upstream decreasing its amplitude.

Approaching basic flow reversal, the Shields number reaches some threshold value below which part of the channel cross section becomes inactive. In other words, starting at the inner bend, the flow can be so weak that the Shields number is not high enough to allow any sediment transport. Due to the above mechanism the present analysis fails for Shields stresses ranging about 2–3 times θ_{cr}, which, in the fluvial case, corresponds to the minimum Shields stress for the occurrence of sediment transport throughout the whole cross section [97].

19.5.2 Equilibrium of Tidal Inlets

The problem of maintaining tidal inlets in equilibrium is of major importance for its obvious implications on navigation and, in general, on the morphodynamics evolution of estuaries and lagoons.

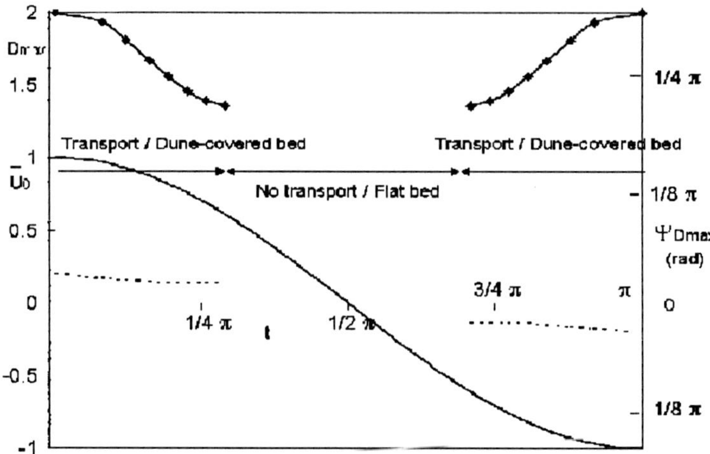

Fig. 19.21. Temporal variations of the position and amplitude of the maximum flow depth during half a tidal cycle. The continuous line describes the temporal distribution of the basic flow field $|\overline{\mathrm{U}}_0|$, the dashed lines indicate the phase lag between the cross section where the maximum flow depth is located and the cross section at the bend apex ($\Psi_{D_{\max}}$), the bold continuous lines denote the intensity of the maximum dimensionlessflow depth ($D_{\max}$). ($\beta = 6$, $\lambda = 0.05$, $R_p = 4$, $\theta_0 = 0.6$, $d_s = 2 \cdot 10^{-5}$, dune-covered bed)

Tidal inlets on littoral drift shores attain typically quasi-equilibrium conditions characterized by relatively small geometric changes affecting their location, planform, cross sectional area and shape [14]. This quasi-equilibrium configuration depends on the balance among the effects of the littoral drift carried to the entrance by flood currents, of the ebb tidal currents and of the wave action. Several studies have been devoted to investigate the stability of tidal inlets, the word stability referring to the ability of the cross section to maintain its shape. Most of these studies have proposed empirical relationships relating the tidal prism to the cross sectional area of the channel crossing the barrier connecting the ocean with the bay or lagoon (gorge) [54,73]. Others relate the degree of stability of the tidal inlet to the ratio of the tidal prism to forces by waves causing littoral drift, i.e. the material transport to the tidal entrance [15]. A review of such investigations can be found in [14]. Here, we briefly recall the very popular empirical relationships proposed by O'Brien [73] and Jarrett [54]. For tidal inlets with a semi-diurnal tide range, O'Brien [73] proposes an empirical relationship between the tidal prism P^* corresponding to the semi-diurnal spring tidal range and the minimum cross sectional area $A^*_{\min}$ characterizing the entrance channel (gorge) at M.S.L.:

$$A^*_{\min} = 4.96 \cdot 10^{-4} P^{*0.85} \tag{19.15}$$

with $A^*_{\min}$ in square feet and P^* in cubic feet. A more comprehensive study of the relationship between the tidal prism and the inlet area was carried out by Jarrett [54] on the basis of data collected along the Atlantic and Pacific coasts and concerning un-jetted, single-jetted and two-jetted inlets. Jarrett [54] observed that un-jetted and single-jetted inlets on different coasts exhibit a different relationship owing to differences in the tidal and wave characteristics. Jarrett's [54] analysis does not account for the length of jetties. The regression curve for all the data reads:

$$A^*_{\min} = 5.74 \cdot 10^{-5} P^{*0.95} \tag{19.16}$$

with $A^*_{\min}$ in square feet and P^* in cubic feet. Jarrett's relationship differs considerably from that of O'Brien's [73]. Indeed, a general relationship can hardly been proposed on purely empirical ground. In fact, the inlet geometry depends strongly on the morphology of the seaward bar which is controlled by the complex balance between the ebb and flood tidal currents, the sediment transport due to the wave induced littoral drift, and the flux of wave energy into the entrance [14].

The complexity of the problem is confirmed even under the idealized conditions examined in the laboratory experiments carried out by Bolla Pittaluga et al. [12]. In fact, even in the absence of significant wave action the flow field around the mouth of the channel is highly asymmetric throughout the tidal cycle. At the initial stage of the experiment, with the 'sea' bottom flat, the near inlet flow field is nearly irrotational during the flood phase and behaves like an unsteady turbulent jet during the ebb phase. The hydrodynamics of such flow configuration was thoroughly investigated theoretically and experimentally by Arato et al. [7]. Flood flow was indeed modelled as irrotational while the ebb flow was modelled as inviscid and rotational employing well established vortex shedding techniques from the sharp edges of the tidal inlet. The outcome of the analysis was quite interesting: vortex shedding leads to the formation of a pair of counterrotating vortices which abandons the generation area as a result of the velocity that each of the vortex induces on the other. The asymmetry of the flow field seems to imply that a net flux of sediments may be imported into (or abandon) the estuary at each tidal cycle.

However, the problem turns out to be more complex as the evolution of bed topography drives a modification of the hydrodynamics until a final equilibrium is reached. Whether the latter is 'static' in nature, i.e. associated with the establishment of threshold conditions for sediment motion throughout the channel cross section (as assumed by Marchi, [71]), or dynamic, i.e. such that sediment transport is still present at equilibrium (as found by Lanzoni and Seminara, [63] throughout the tidal channel) is still unclear. The pattern of bottom topography in the near inlet region at the initial and final stages is reported in Fig. 19.22, while the cross sections plotted in Fig. 19.23 suggest that the eroding action of the jet flow excavates a submerged channel in the sea bottom, with depth decreasing in the seaward direction.

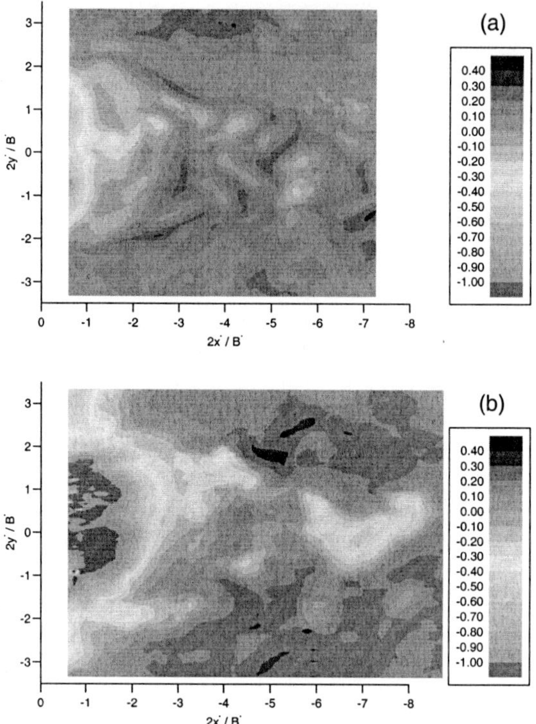

Fig. 19.22. Near inlet region bottom topography after (**a**) 1 h and (**b**) 100 h. The channel inlet is between $2y^*/B^* = -1$ and $2y^*/B^* = 1$ at the transverse cross section $2x^*/B^* = 0$. The origin of the longitudinal landward directed axis x^* is in correspondence of the channel inlet. The origin of the transverse axis y^* is set in the mid of the channel inlet

19.6 Morphodynamic Stability

19.6.1 The Formation of Tidal Free Bars

It is well known from the fluvial literature that free bars are mesoscale forms which develop spontaneously due to a bottom instability. The formation of free bars in 'infinitely' long tidal channels has been recently investigated by Seminara and Tubino [99,100]. The problem was tackled studying the altimetric instability of a cohesionless bottom in a weakly convergent channel with a rectangular cross section of width $2B^*$ and non-erodible banks, subject to the propagation of a tidal wave.

As a first step the analysis neglects various features such as the presence of tidal flats flanking the main channels. The bed was taken to be cohesionless and the sediment homogeneous. Furthermore, the channel was considered long enough to allow neglecting the effects of end conditions on the process of bar

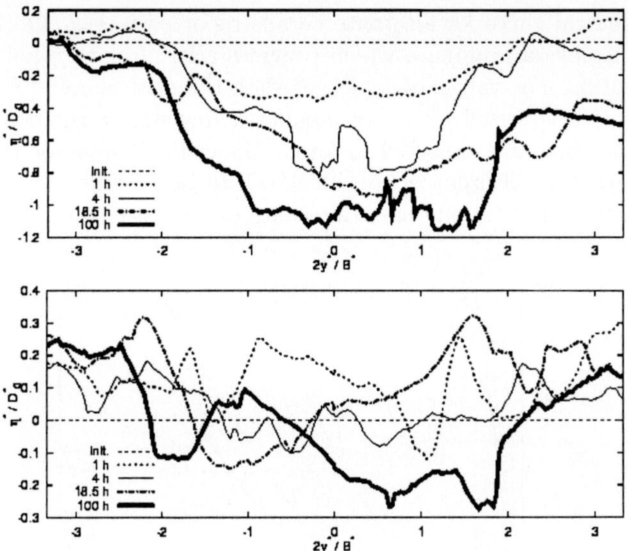

Fig. 19.23. Cross section of the near inlet region at (*top*) $2x^*/B^* = -1$ and (*bottom*) $2x^*/B^* = -3$

formation. At the scale of bars the role played by channel convergence and spatial variations of the tidal wave propagating along the channel turns out to be negligible at the leading order of approximation; hence bars feel the tidal wave as an oscillatory longitudinally uniform flow associated with a horizontal configuration of the free surface which oscillates in time. Furthermore, local inertia is found to be negligible at leading order. Perturbations of the flow field and of sediment transport induced by the formation of free bars have been investigated by means of a three dimensional model. We refer the reader to [100] for details; it suffices here to state that the formulation of the problem followed the lines discussed in Sect. 19.5. The latter problem was set at the basis of a linear stability analysis whereby the growth of infinitesimal perturbations of the bed interface, in the form of normal modes scaling on channel width, was determined. Instability turns out to be of Mathieu type: in other words a net growth (or decay) of perturbations in a tidal cycle is found to superimpose on a purely oscillatory response, in analogy with the classical behaviour of solutions of Mathieu equation. Results have been obtained in the case of a tidal wave whose cross sectionally averaged velocity consists of a single harmonic. No subharmonic or ultraharmonic response was found. The stability analysis provides a dispersion relationship for the net growth rate in a tidal cicle and the migration speed of bars as functions of dimentionless bar wavenumber λ ($\lambda = \lambda^* B^*$), width ratio β_0 ($\beta_0 = B^*/D_0^*$), Shields parameter Θ, particle Reynolds number R_p and relative roughness d_s ($d_s = d_s^*/D_0^*$). Such dispersion relationship allows one to determine neutral stability curves (or surfaces) in the space of relevant parameters. An example of a neutral curve for alternate bars is reported in Fig. 19.24. Note that such curve displays a minimum which provides critical or threshold conditions for bar formation. For values of β_0 larger than the critical value β_c, any linear perturbation characterized by a wavenumber falling within the neutral curve is unstable, hence bars are expected to form. The critical value for the formation of multiple row bars of order m can be shown to be m times the critical value for the formation of order one bars (alternate bars).

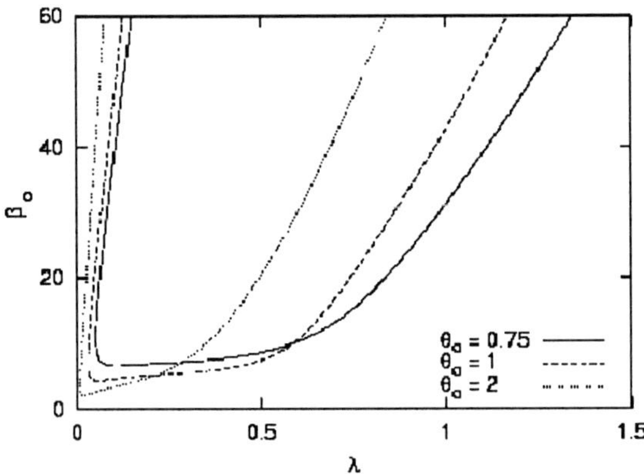

Fig. 19.24. Marginal stability curves for different values of peak Shields stress ($R_p = 4$, $d_s = 2 \cdot 10^{-5}$)

Results also show that, since the basic flow is periodic with zero mean, tidal bars, unlike river bars, do not exhibit a net migration over a tidal cycle. In other words during a tidal cycle, bars migrate alternatively forward and backward in a symmetric fashion. The role played by transport in suspension is quite significant: it turns out that the critical value of the aspect ratio β_c above which alternate bars form tends to vanish when the Shields stress attains a value so large that transport in suspension becomes dominant. Also note that, when the flow is characterized by a high value of the Shields stress, several unstable modes may be simultaneously excited for relatively low values of the aspect ratio.

Such feature indicates that the resulting bar pattern is likely to arise from a complex nonlinear competition among different unstable modes. A comparison between the neutral curves reported in Fig. 19.24 and those obtained by Colombini et al. [18] for fluvial channels with dominant bed load transport, suggests that suspension leads to a significant reduction of the width of the unstable region which is gradually shifted towards smaller values of λ. Such behaviour indicates on one hand that suspension plays a stabilizing role at sufficiently large wavenumbers, on the other hand that a longer straight reach is required to allow for the development of free bars when the suspension is dominant. The bar wavenumber characterized by the maximum growth rate as a function of the aspect ratio is plotted in Fig. 19.25 for various values of the peak Shields stress

Θ_0 and given values of the particle Reynolds number R_p and relative roughness number d_s. Note that the bar wavelengths range between 8–30 times the channel width for values of the peak Shields stress and of the aspect ratio falling in the ranges 0.6–2 and 5–30 respectively.

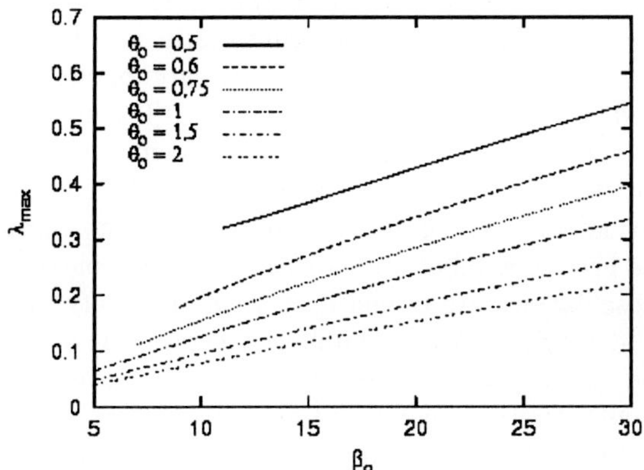

Fig. 19.25. The wavenumber $\lambda_{\max}$ characterized by maximum growth is plotted versus the width ratio β_0 for different values of the peak Shields stress Θ_0 ($R_p = 4$, $d_s = 2 \cdot 10^{-5}$)

Various developments of the above work remain to be pursued, in particular the role of channel convergence (likely to induce a slowly varying character of bars in the longitudinal direction), the non linear evolution of tidal free bars and their interaction with forced bars induced by channel curvature discussed in Sect. 19.5.1. The role of a finite length of the channel has been given some attention by Schuttelaars and de Swart [95], who claim that bar modes scaling on channel length may be unstable. A discussion of the framework on which the latter work is based is given by in [100]. The role of nonlinearity has been widely investigated in the fluvial context [18,17,91].

19.6.2 The Formation of Dunes in Tidal Channels

The formation of dunes in erodible channels is a classical topic, which has been thoroughly investigated in the fluvial literature, since the pioneering work of Kennedy [55], recently revisited by Coleman and Fenton [16]. The linear aspects of the problem of dune formation under steady conditions appear to be relatively well understood: instability of small amplitude perturbations of bottom topography are driven by a phase lag between sediment transport and bottom topography, due to several competing effects, some of which play a destabilizing role (friction and particle inertia) while others tend to damp dune formation (gravitational effects on sediment transport and suspended load). The reader interested in such features is referred to [33]. Less understood are the non linear aspects of dune development. In fact, the latter soon leads to flow separation, a feature which severely modifies the structure of the flow field, as discussed by Smith and Mc Lean [103]. It has been argued that the occurrence of separation in the early stages of dune formation is a major controlling factor of wavelength selection: in other words, due to flow separation, finite amplitude effects would dominate even in the initial stage of dune development. No attempt to examine how the latter picture is modified when dunes form in tidal channels is known to the present authors. However some qualitative arguments can be put forward to gain some understanding of the latter process. The major novel aspect to be accomodated when dealing with tidal environments is the oscillatory character of the basic flow, which, at the dune scale, can be considered as spatially uniform. Oscillations drive a continuous variations of the characteristics of the basic flow, whose Froude number oscillates around some mean value. Associated with the latter oscillations are oscillations of the intensity and nature of sediment transport: at very low Froude numbers no transport occurs, except for the deposition of sediments possibly present in suspension; as the Froude number increases, sediment is initially entrained as bedload and later both as bedload and as suspended load. Correspondingly, the stability characteristics of the basic tidal flow vary throughout the tidal cycle: the instantaneous flow, treated as steady, would be stable close to slack water, later dunes would tend to form with growth rate increasing initially as the tidal velocity increases and then decreasing for larger instantaneous Froude numbers. Close to the peak of tidal velocity, the stabilizing effect of suspended load tends to damp the amplitude of dunes. The latter qualitative arguments suggest the following three possible scenarios for dune formation in tidal channels.

i) The time scale of dune growth T_d^* is much smaller than the tidal period T^*: in this case, dunes form and disappear during each tidal cycle with characteristics in equilibrium with the local and instantaneous properties of the basic tidal flow.
ii) The time scale of dune growth is much larger than the tidal period: in this case, dunes may form as a result of their net growth in a sequence of tidal cycles.
iii) The time scale of dune growth is of the order of the tidal period: in this case, dunes will form but their properties will be out of phase relative to the basic state and will oscillate during each tidal cycle.

In order to ascertain which case is appropriate to typical tidal conditions, one needs an estimate for the dune growth rate as a function of the relevant flow parameters, namely the instantaneous Froude number F and the relative roughness d_s. Figure 19.26 shows such estimate (Colombini, personal communication) for two typical tidal configurations: a prototype configuration characterized by a value of d_s equal to 3×10^{-5} with relative density of the sediments ϱ_s/ϱ equal to 2.65 and the configuration employed in the laboratory observations discussed in Sect. 19.5.1, characterized by a value of d_s ranging about 3×10^{-3} with relative density 1.48. The quantity plotted in Fig. 19.26 is the dimensionless growth rate

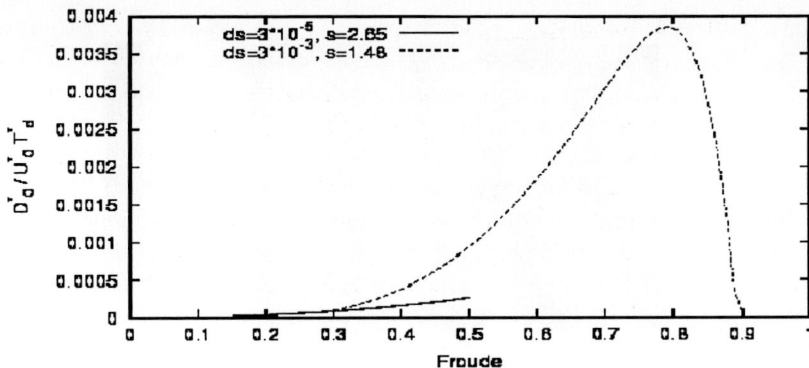

Fig. 19.26. Dimensionless growth rate of dunes characterized by dimensionless wavenumber scaled by flow depth equal one for any given (steady) Froude number

of dunes characterized by a value equal to one of the dimensionless wavenumber scaled by flow depth for any given (steady) Froude number. The growth rate is scaled by the ratio between flow depth D_0^* and cross sectionally averaged speed U_0^*: hence, the ratio T_d^*/T^* can be estimated by multiplying the quantity plotted in Fig. 19.26 by the parameter $(T^* U_0^*/D_0^*)$. The latter attains typical peak values ranging about $2 \cdot 10^{-4}$ in the prototype and 10^{-3} in our laboratory model. Using the latter estimates, Fig. 19.26 suggests that the ratio T_d^*/T^* is strongly dependent on the peak Froude number F. Typical values of F in the prototype range about 0.1 and the ratio T_d^*/T^* is much larger than one, hence the second scenario described above appears to apply under natural conditions. The unavoidable distortion introduced when attempting to reproduce the actual phenomenon in the laboratory leads to larger values of F, (say 0.4), with corresponding values of the ratio T_d^*/T^* of order one. Hence, in the laboratory, the third scenario described above is appropriate. In fact, formation of dunes along the channel was observed since the very beginning of the experiment, displaying a quasi 2D pattern, wavelengths ranging between 2 and 5 times the local flow depth and amplitudes ranging roughly about 0.1 times the local flow depth (Fig. 19.27). The damping effect of suspended load on dune development was clearly detected, since during the flood phase dunes were washed out to form again during the ebb phase which was characterized by lower values of suspended sediment transport. Dunes migrated upstream during the flood phase and downstream during the ebb phase, displaying a net upstream migration in a tidal cycle due to the asymmetry of the velocity field. The flow was indeed flood dominated. The quasi two-dimensional dune pattern was replaced by complex three-dimensional patterns all along the channel after the first hours of experiment. In the later stages of the process, when the channel profile has attained quasi equilibrium conditions, dunes undergo very weak modifications throughout each tidal cycle. In other words, the picture emerging from the experiments suggests that dunes are essentially imprinted in the channel at some initial stage.

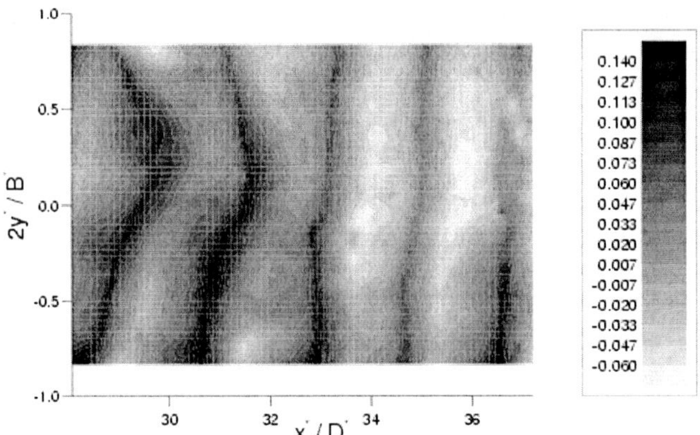

Fig. 19.27. Quasi two-dimensional dune pattern observed during the experiments of Bolla Pittaluga et al. [12] throughout a reach of the tidal channel. Bottom elevation and longitudinal coordinate (x^*) are scaled by the average local flow depth. The transverse coordinate y^* is scaled by half-channel width. Note that dune wavelenght ranges about three times the local flow depth

Theoretical attempts to explain dune formation under tide dominated conditions have been proposed by Hulscher et al. [48], Hulscher [47] and Gerkema [43]. Though the type of applications that the latter authors had in mind referred to the tidally driven formation of sandwaves and sandbanks in coastal regions, however in the absence of sea waves the problem turns out to be quite similar to that governing dune formation in tidal channels. The contributions mentioned above, which have been recently further developed by Komarova and Hulscher and by Komarova and Newell [59,60], suggest that dune formation under natural conditions indeed falls in regime ii). Moreover, dune growth is claimed to be driven by the formation of steady recirculatory cells associated with tide propagation over a wavy bottom similar to those found by Blondeaux [11] in the case of ripple formation under waves. However, a recent contribution of Besio et al. [10] questions various assumptions employed in all the latter works and suggests that the problem of dune formation under tide dominated conditions will still require attention in the near future.

19.7 Morphodynamic Evolution

Predicting the short- and long-term morphodynamic evolution of estuaries is the key problem of tidal morphodynamics. A traditional approach employed in the engineering literature has been to employ empirical relationships between the various quantities characterizing the estuarine geometry (say the cross sectional area of tidal inlets, the cross sectional area of estuarine channels, the area of the tidal flat-salt marshes) and the tidal prism [36,54]: the morphodynamic development of the system is then interpreted as a tendency to move from an equilibrium configuration to another. A second approach, more recently employed by engineers, attempts at constructing simplified models of the whole estuarine system, based on modeling assumptions providing zero order descriptions of the processes occurring in each portion of the system [26–28]. While the latter approaches play the important role to attempt providing qualitative answers to urgent practical questions, parallel to the latter, a so called 'reductionist approach' is needed, focusing on a detailed understanding of the various mechanisms whereby departures from equilibrium may arise in each part of the estuarine environment. The latter research line is still at its infancy. Below, we briefly outline some suggestions for future research.

19.7.1 Evolution of Tidal Channels

The process whereby an *equilibrium bed profile* may be reached in a straight tidal channel has been discussed in Sect. 19.5.1. Departures from the equilibrium discussed above may originate from a net exchange of sediments through the tidal inlet, possibly forced by littoral currents or by anthropic effects (like the construction of jetties at the inlets of Venice lagoon). The perturbing effects of a net sediment load carried by a river debouching into the estuary will also require attention: can an equilibrium profile of the channel still be reached and maintained? A similar question arises when the tidal channel is bounded by tidal flats: in fact, a net exchange of sediments between the channel and the neighbouring tidal flats-salt marshes is induced by various mechanisms discussed below.

Tidal channels undergo also a *planimetric development* somehow similar to that observed in the fluvial environments: as discussed in Sect. 19.5.1, tidal channels develop often a meandering pattern. The mechanism which controls such planimetric development is associated with the role of bank erosion. Fluvial meanders shift laterally their axis keeping their width fairly constant, through a process, suggested by field observations, whereby the material eroded at the outer banks is deposited at the inner bends. The latter process has been widely investigated in a series of papers of Parker's school and, more recently, in [102]. In particular, in the latter paper, the authors derive a planimetric evolution equation of meandering rivers in the form of a non linear integro-differential equation, its integro-differential character accounting for the history of the deformation of the channel axis, while nonlinearity arises from the kinematics of the deformation process. Seminara et al. [102] were able to show that periodic solutions of the latter equation exist, which display all the features of the fluvial meandering process, in particular:

- Meanders develop a Kinoshita shape, which turns out to be intrinsically embedded into the cubic non linearity of the planimetric evolution equation;
- Meanders migrate typically downstream (though upstream migration turns out to be possible under appropriate conditions) with a migration rate which decreases monotonically from incipient formation to cutoff;
- Meander amplitude increases with a rate of the order of metres per year, which increases up to a maximum and then decreases slowly;

- Neck cutoff invariably occurs at the late stage of meander development;
- No equilibrium solution (i.e. meanders of permanent form) may exist.

Further interesting features emerge when meander development is allowed to evolve from an arbitrary initial condition, the solution displaying then the typical characteristics of the solutions of the Ginzburg–Landau equation. Furthermore, the formation of compound loops of the type typically observed in nature, arises from the calculations.

The tidal analogue of the latter process could be readily investigated and, indeed, Solari et al. [106] have extended the linear, so called 'bend stability', analysis to the tidal case. However, along with the 'cohesionless' mechanism outlined above, one may be tempted to envisage an alternative 'cohesive' mechanism of development of tidal meanders: in other words, the formation of tidal meanders might be embedded in the purely erosive process whereby an initial incision in a cohesive tidal flat, widens tending to reach its equilibrium width. A slight longitudinal perturbation of the widening process might lead to the development of a periodic width perturbation essentially equivalent to the development of a meandering pattern. The analysis of such mechanism will require attention in the near future: note that the feasibility of the latter might also explain the occurrence of second harmonics in the Fourier representation of the channel axis, as well as the apparent tendency of tidal meanders to reach an equilibrium configuration in contrast with the behaviour observed in the fluvial case (most tidal meanders of Venice lagoon have been practically inactive for decades).

19.7.2 Evolution of Tidal Flats and Salt Marshes

The morphodynamic development of tidal flats and salt marshes is strictly connected with the dual problem concerning the effects of tidal flats and salt marshes on the morphodynamic equilibrium of tidal channels.

In fact, various mechanisms contribute to determine a net exchange of sediments between channels and tidal flats.

A major mechanism is associated with the role of vegetation. In fact, marsh grass is known to induce deposition of inorganic sediments, associated with the strong deceleration of the flowing stream as it expands over the flats and the trapping effect of marsh grass. Deposition rates have been documented [32] which may reach values of the order of five times that of unvegetated adjacent flats. A second mechanism of marsh accretion is driven by the production of organic matter.

Further contributions to the exchange of sediments are associated with the asymmetric character of the velocity and concentration fields in the tidal flat region during the flood and ebb phases (some aspects of this mechanism were discussed by Schijf and Schönfeld [92] and by Dronkers [29]) and to the instability of the shear layer forming at the boundary between the faster channel flow and the slower flow in the tidal flats. The latter mechanism has been widely investigated in the fluvial context [51] but is totally unexplored in the tidal case.

Tidal range also play an important role. In macrotidal marshes, tides typically provide an accretionary contribution to marsh evolution, sufficiently strong to counteract the effects of sea level rise, leading to the establishment of conditions of dynamic equilibrium [38]. In mesotidal and, more evidently, in microtidal marshes the stirring action of wind and waves during storm events are more effective in delivering sediments to the marsh [64,65].

Marsh degradation may be driven by various competing effects. While it is unlikely that tide induced flow over healthy marshes may ever lead to net erosion [39], marsh regression is typically driven by the collapse of the banks of tidal channels fringing the marshes due to tidal currents undermining the marsh at bar pools [3] or to the direct attack of wind waves [25].

Salt marsh grasses like Spartina Alterniflora do not stand waterlogging, i.e. excessive submergence [24,83], which may be driven by sea level rise, subsidence or reduced sediment supply. Note that all the latter mechanisms have been active in the recent development of Venice lagoon.

Finally various anthropic effects associated with clam fishing, outboard navigation, dredging as well as organic and minerogenic inputs may affect marsh degradation.

The picture emerging from the above discussion (but see also the recent review [41]) suggests that marshes are systems in delicate dynamic equilibrium: vertical accretion reduces the period of submergence, which in turns reduces inorganic accretion. However, such dynamic equilibrium may be disrupted by effects like an accelerated sea level rise, an enhanced subsidence or a strong reduction in sediment supply: vegetated flats respond quite rapidly, sometimes catastrofically, to such variations of the forcing mechanisms.

Incorporating all the above effects into comprehensive models of the evolution of the whole estuarine system will require an intense interdisciplinary effort which is likely to keep the next generation of engineers and scientists quite busy!

Acknowledgments

This work represents an outcome of the work initiated under the umbrella of the Italian National Research Program 'Morfodinamica Fluviale e Costiera', coordinated by G. Seminara and cofunded by various Italian Universities with further financial support provided by the Italian Ministry of Scientific Research. Such work is continuing in the context of the new National Research Program 'Idrodinamica e morfodinamica di ambienti a marea', coordinated by A. Rinaldo, and of the Research project 'Analisi e monitoraggio dei processi morfologici nel sistema lagunare', funded by CORILA.

References

1. P. Adams: *Saltmarsh Ecology* (Cambridge University Press, Cambridge 1990)
2. J.R.L. Allen: Sediment. Geol. **26**, 281–328 (1980)
3. J.R.L. Allen: Sediment. Geol. **113**(3–4), 211–223 (1997)

4. J.R.L. Allen, A. Deresseguier, A. Klingebiel: C. R. Acad. Sci. Paris, Ser. D, **269**, 2167–2169 (1969)
5. J.R.L. Allen, K. Pye: 'Coastal saltmarshes: their nature and importance'. In: *Saltmarshes: Morphodynamics, conservation and engineering significance*, ed. by J.R.L. Allen, K. Pye (Cambridge University Press, Cambridge 1992) pp. 1–18
6. C.L. Amos: 'Siliciclastic tidal flats'. In: *Geomorphology and sedimentology of estuaries*, Chapt. 10, ed. by G.M.E. Perillo (Elsevier, Amsterdam 1995)
7. E. Arato, P. Blondeaux, B. De Bernardinis, G. Seminara, L. Stagi: 'Sul meccanismo del ricambio lagunare: modello teorico e primi rilievi sperimentali'. In: *Atti Conv. di Studi 'Laguna, fiumi, lidi: cinque secoli di gestione delle acque nelle Venezie', 11-6, Venezia, 1983*
8. G.M. Ashley: J. Sedim. Petrol. **60**, 160–172 (1990)
9. J.H. Barwis: 'Sedimentology of some South Carolina tidal-creek point bars, and a comparison with their fluvial counterparts'. In: *Fluvial Sedimentology*, ed. by A.D. Miall (Can. Soc. Petrol. Geol. Mem. 5, 1978) pp. 129–160
10. G. Besio, P. Blondeaux, P. Frisina: A contribution to the study of tidally generated sand waves. In preparation (2001)
11. P. Blondeaux: J. Fluid Mech. **218**, 1–17 (1990)
12. M. Bolla Pittaluga, N. Tambroni, C. Zucca, L. Solari, G. Seminara: 'Long term morphodynamic equilibrium of tidal channels: preliminary laboratory observations'. In: *IAHR Symposium on River, Coastal and Estuarine Morphodynamics, Obihiro, Japan, 10-14 September, 2001*
13. J.C. Boothroyd: 'Mesotidal inlets and estuaries'. In: *Coastal Sedimentary Environments*, ed. by R.A. Davis (Springer–Verlag, New York 1978)
14. P. Bruun: *Stability of tidal inlets: Theory and engineering* (Elsevier, Amsterdam 1978)
15. P. Bruun, F. Gerritsen: *Stability of coastal inlets* (North Holland Publ. Co., Elsevier, Amsterdam 1960)
16. S.E. Coleman, S.E. Fenton: J. Fluid Mech. **418**, 101–117 (2000)
17. M. Colombini, M. Tubino: 'Finite amplitude free-bars: a fully nonlinear spectral solution'. In: *Euromech 262, Sand Transport in rivers, Estuaries and the Sea*, ed. by Soulsby, Bettes (Balkema, Rotterdam 1991)
18. M. Colombini, G. Seminara, M. Tubino: J. Fluid Mech. **181**, 213–232 (1987)
19. R.W. Dalrymple, R.N. Rhodes: 'Estuarine dunes and bars'. In: *Geomorphology and Sedimentology of Estuaries*, ed. by G.M.E. Perillo (Elsevier, Amsterdam 1995)
20. R.W. Dalrymple, R.J. Knight, B.A. Zaitlin, G.V. Middleton: Sedimentology **37**, 577–612 (1990)
21. R.W. Dalrymple, B.A. Zaitlin, R. Boyd: J. Sedim. Petrol. **62**, 1130–1146 (1992)
22. S.E. Darby, C.R. Thorne: J. Hydraul. Eng. **122**(8), 443–454 (1996)
23. J.L. Davies: Z. Geomorph. **8**, 127–42 (1964)
24. J.W. Day, J. Rybczyk, F. Scarton, A. Rismondo, D. Are, G. Cecconi: Estuarine, Coastal and Shelf Science **49** (1999)
25. D.J. De Jong, Z. De Jong, J.P.M. Mulder: Hydrobiologia **282/283**, 303–316 (1994)
26. G. Di Silvio: 'Interaction between marshes, channels and shoals in a tidal lagoons'. In: *IAHR Symposium on River, Coastal and Estuarine Morphodynamics, Genova, 6-10 September 1999, vol. I*, pp. 695–704
27. G. Di Silvio, G. Barusolo, L. Sutto: 'Competing driving factors in estuarine landscape'. In: *IAHR Symposium on River, Coastal and Estuarine Morphodynamics, Japan, 10-14 September* (2001)

28. Dongeren van, H. de Vriend: Coastal Engineering **22**, 287–310 (1994)
29. J. Dronkers: *16th Coastal Engineering Conference, vol.3* (A.S.C.E., New York 1978)
30. K.R. Dyer: 'Sedimentation in estuaries'. In: *The estuarine environment*, ed. by R.S.K. Barnes, J. Green (Appl. Science Publ., London 1972)
31. K.R. Dyer: *Estuaries: a physical introduction* (Wiley, New York 1973)
32. D. Eisma, K.S. Dijkema: 'The influence of salt marsh vegetation on sedimentation'. In: *Inertial Deposits*, ed. by D. Eisma (CRC Press 1997)
33. F. Engelund, J. Fredsøe: Ann. Rev. Fluid Mech. **14**, 13–37 (1982)
34. F. Engelund, E. Hansen: *A monograph on sediment transport in alluvial streams* (Danish Technical Press, Copenhagen 1967)
35. G. Evans: Q. J. Geol. Soc. Lond. **121**, 209–41 (1965)
36. W.D. Eysink: *22 Coastal Engineering Conference, vol.2* (A.S.C.E., New York 1990)
37. S. Fagherazzi, A. Adami, S. Lanzoni, M. Marani, A. Rinaldo, A. Bortoluzzi, W.E. Dietrich: Water Resour. Res. **35**(12), 3891–3904 (1999)
38. J.R. French, T. Spencer: U.K. Mar. Geol. **110**, 315–331 (1993)
39. R.W. Frey, P.B. Basan: 'Coastal salt marshes'. In: *Coastal Sedimentary Environments*, ed. R.A. Davis (Springer–Verlag, New York 1985)
40. C.T. Friedrichs: J. Coast. Res. **4**, 1062–1074 (1995)
41. C.T. Friedrichs, J.E. Perry: J. Coast. Res. **27**, 6–36 (2001)
42. G. Galappatti, G.B. Vreugdenhil: J. Hydr. Res. IAHR **23**(4), 359–377 (1985)
43. T. Gerkema: J. Fluid Mech. **417**, 303–322 (2000)
44. H.P. Guy, D.B. Simons, E.V. Richardson: *U.S. Geol. Survey Prof. Paper 252* 462-I, 96 (1966)
45. M.O. Hayes: 'Morphology of sand accumulation in estuaries'. In: *Estuarine Research, vol. II*, ed. by L. Cronin (Academic Press, New York 1975)
46. P.T. Harris, M.R. Jones: Geol. Mag. **125**, 31–49 (1988)
47. S.J.M.H. Hulscher: J. Geophys. Res. **101**(C9), 20727–20744 (1996)
48. S.J.M.H. Hulscher, H. de Swart, H. de Vriend: Cont. Shelf. Res. **13**(11), 1183–1204 (1993)
49. S. Ikeda, N. Izumi: Water Resour. Res. **27**(9), 2429–2438 (1991)
50. S. Ikeda, G. Parker, eds: *River meandering*, AGU Water Res. Mon. 12, pp. 267–320 (1989)
51. S. Ikeda, T. Sano: 'LES simulation of flow and suspended sediment transport in two-stage channels'. In: *IAHR Symposium on River, Coastal and Estuarine Morphodynamics, Genova, 6-10 September 1999, vol. I* pp. 131–140
52. C. Inglis, F. Allen: 'The regimen of the Thames estuary as affected by currents, salinities and river flow'. In: *Proc. Instn. Civ. Engrs. 7, 1957* pp. 827–68
53. S.V. Isla: 'Coastal lagoons'. In: *Geomorphology and sedimentology of estuary*, Chap. 9 (Elsevier, Amsterdam 1995)
54. J.T. Jarrett: *Tidal prism.ilet are relationship*. U.S. Army Corps Coastal Engineering Research Center, G.I.T.I. Report 3, (Vicksburg, Mississippi 1976)
55. J.F. Kennedy: J. Fluid Mech. **16**, 521–544 (1963)
56. F.J.T. Kestner: Geogr. J. **128**, 457–478 (1962)
57. R. Kinoshita: *An investigation of channel deformation of the Ishikari River*, Technical Report, Natural Resources Division (Ministry of Science and Technology of Japan 1961)
58. G. Klein, H. de Vries: 'Intertidal flats and intertidal sand bodies'. In: *Coastal sedimentary environments*, ed. by R.A. Davis (Springer–Verlag, New York 1985) pp. 187–224

59. N.L. Komarova, S.J.M.H. Hulscher: J. Fluid Mech. **413**, 219–246 (2000)
60. N.L. Komarova, A.C. Newell: J. Fluid Mech. **415**, 285–312 (2000)
61. W.B. Langbein, L.B. Leopold: Amer. J. of Science, 262 (1964)
62. S. Lanzoni, G. Seminara: J. Geoph. Res. **103**C13, 30793–30812 (1998)
63. S. Lanzoni, G. Seminara: Long term evolution and morphodynamic equilibrium of tidal channels. Under revision in the J. Geoph. Res., (2001)
64. L.A. Leonard, A.C. Hine, M.E Luther: J. Coastal Res. **11**(2), 322–336 (1995)
65. L.A. Leonard, A.C. Hine, M.E. Luther, M.E. Stumpf, E.E. Wright: Est. Coastal and Shelf Sc. **41**(2), 225–248 (1995)
66. L.B. Leopold, T. Jr. Maddock: *U.S. Geol. Survey Prof. Paper 252* **57** (1953)
67. L.B. Leopold, M.G. Wolman: *U.S. Geol. Survey, Prof. Paper 282-B* (1957)
68. J.C. Ludwick: Geol. Soc. Am. Bull. **85**, 717–726 (1974)
69. J.L. Luternauer, R.J. Atkins, A.I. Moody, H.F.L. Williams, J.W. Gibson: In: *Geomorphology and sedimentology of estuaries*, Chapt. 11, ed. by G.M.E. Perillo (Elsevier, Amsterdam 1995)
70. M. Marani, S. Lanzoni, D. Zandolin, A. Rinaldo, G. Seminara: Submitted for publication in Water Resour. Res. (2001)
71. E. Marchi: *Rend. Accademia Nazionale dei Lincei, Roma* Serie IX, 1, 137–150 (1990)
72. R. Myrick, L. Leopold: *U.S. Geol. Surv. Prof. Pap. 422–13* (1963)
73. M.P. O'Brien: J. Waterw. Harbour Div. ASCE **95**, 43–52 (1969)
74. A.M. Osman, C.R. Thorne: J. Hydr. Enging. ASCE **114**(2), 134–150 (1988)
75. G. Parker: J. Fluid Mech. **89** part 1, 109–125 (1978)
76. G. Parker: J. Fluid Mech. **89** part 2, 127–146 (1978)
77. G.M.E. Perillo: 'Definitions and geomorphologic classifications of estuaries'. In: *Geomorphology and sedimentology of estuaries*, Chap. 2, ed. by G.M.E. Perillo (Elsevier, Amsterdam 1995)
78. J. Pethick: *An introduction to coastal geomorphology* (E. Arnold, London 1984)
79. H. Postma: J. Sea Res. **1**, 148–90 (1961)
80. D. Pritchard: Advances in Geophysics **1**, 243–80 (1952)
81. D. Pritchard: *Lectures on estuarine oceanography*, ed. by B. Kinsman (J. Hopkins Univ. 1960) 154 pp.
82. D. Pritchard: 'What is an estuary: physical viewpoint'. In: *Estuaries*, ed. by G.H. Lauff (American Association for the Advancement of Science. Publ. N. 83, 1967)
83. A.C. Redfield: Ecolog. Monogr. **42**(2), 201–237 (1972)
84. F. Richthofen von: *Führer für Forschungreisende* (Oppenheim, Berlin 1886) (rias: pp.308–310)
85. L.C. Rijn van: J. Hydr. Engng. ASCE **110**(11), 1613–1641 (1984)
86. M. Rinaldi, N. Casagli: Geomorphology **26**, 253–277 (1999)
87. A.H.W. Robinson: Geography **45**, 183–199 (1960)
88. H. Rouse: Trans. A.S.C.E. **102**, (1937)
89. I.L. Rozowskij: *Flow of water in bends of open channels.* Kiew: Acad. Sci. Ukranian SSR. (1957)
90. G.G. Salsman, W.H. Tolbert, R.G. Villars: U.K. Mar. Geol. **4**, 11–19 (1966)
91. R. Schielen, A. Doelman, H.E. De Swart: J. Fluid Mech. **252**, 325–336 (1993)
92. J.B. Schijf, J.C. Schönfeld: 'Theoretical considerations on the motion of salt and fresh water'. In: *Proc. Minn. Int. Hydraul. Conv., Minneapolis, 321* (1953)
93. S.A. Schumm: *U.S. Geol. Survey Prof. Paper 352-B*, 17–30 (1960)
94. H.M. Schuttelaars, H. de Swart: Eur. Jour. Mech. B **15**, 55–80 (1996)

95. H.M. Schuttelaars, H. de Swart: J. Geoph. Res. **105** C10, 24105–24118 (2000)
96. G. Seminara: Meccanica **33**, 59–99 (1998)
97. G. Seminara, L. Solari: Water Resour. Res. **34**(6), 1585–1598 (1998)
98. G. Seminara, M. Tubino: 'Alternate bars and meandering: free, forced and mixed interactions'. In: *River Meandering*, ed. by S. Ikeda, G. Parker (AGU Water Res. Mon. 12, 1989) pp. 267–320.
99. G. Seminara, M. Tubino: 'On the formation of estuarine free bars'. In: *Physics of Estuaries and Coastal Seas*, ed. by J. Dronkers, M. Scheffers (A.A. Balkema 1998) pp. 345–353
100. G. Seminara, M. Tubino: J. Fluid Mech. (2001) (In press)
101. G. Seminara, M. Tubino, C. Paola: 'The morphodynamics of braiding rivers: experimental and theoretical results on unit processes'. In: *Gravel-Bed Rivers, New Zealand, September 2000*
102. G. Seminara, G. Zolezzi, M. Tubino, D. Zardi: J. Fluid Mech. (2001) (In press)
103. J.D. Smith, S.R. McLean: J. Geophys. Res. **82**(129), 1735–1746 (1977)
104. L. Solari, G. Seminara: 'Tidal meaders'. In: *IAHR Symposium on River Coastal and Estuarine Morphodynamics, Genova, 6-10 September 1999* pp. 629–639
105. L. Solari, M. Toffolon: 'Equilibrium bottom topography in tidal meandering channel: preliminary results'. In: *IAHR-RCEM Symposium 2001, Japan*
106. L. Solari, G. Seminara, S. Lanzoni, M. Marani, A. Rinaldo: Submitted for publication to J. Fluid Mech. (2001)
107. M.G. Spangler, R.L. Handy: *Soil engineering* (Intext Educational, New York 1973)
108. H. Stommel: Sewage Ind. Wastes **25**, 1065–1071 (1953)
109. L.M.J.U. Straaten vaan, P.H.H. Kuenen: Geol. Mijn. **19**, 329–354 (1957)
110. J.P. Tastet, H. Fenies, G.P. Allen: Bull. Inst. Géol. Bassin d'Aquitaine **39**, 165–184 (1986)
111. D.W. Taylor: *Fundamentals of soil mechanics* (J. Wiley and Sons, New York 1948)
112. C.R. Thorne, A.M. Osman: J. Hydr. Enging. ASCE **114**(2), 151–172 (1988)
113. M. Tubino, M. Colombini: 'Correnti uniformi a superficie libera e sezione lentamente variabile'. In: *XXIII Convegno di Idraulica e Costruzioni Idrauliche, Firenze, 31 Agosto-4 Settembre, 1992, vol. 3* pag. D375
114. M. Tubino, R. Repetto, G. Zolezzi: J. Hydr. Res. (2000)
115. J. Veen van: *Eb-en vloedschaarsystemen in de Nederlandse getijwateren.* Waddensymposium, Tijdschr. Kon. Ned Aardijksk. Genoot., 1950 pp 43–65.
116. Y. Wang: Can. J. Fish. Aquatic Sci. **40**, 160–71 (1983)
117. B.C. Wang, D. Eisma: Neth. J. Sea Res. **25**, 377–90 (1990)
118. R.J. Weimer, J.D. Howard, D.R. Lindsay: 'Tidal flats and associated tidal channels'. In: *Sandstone Deposition Environments*, ed. by P.A. Scholle, D. Spearing (Am. Assoc. Pet. Geol. 1981)
119. L.D. Wright: 'River Deltas'. In: *Coastal sedimentary environments*, ed. by R.A. Davis (Springer–Verlag, New York 1978)
120. B.A. Zaitlin: Sedimentology of the Cobequid Bay-Salmon River estuary, Bay of Fundy, Canada. Ph. D. Thesis. Queen's Univ., Kingston, Ont., 391 pp. (unpublished) (1987)

20 Longshore Bars and Bragg Resonance

C.C. Mei[1], T. Hara[2], and J. Yu[3]

[1] Department of Civil & Environmental Engineering, Massachusetts Institute of Technology, Cambridge MA, 02139,
[2] Graduate School of Oceanography, University of Rhode Island, Kingston, RI, 02881
[3] Division of Earth and Ocean Sciences, Duke University, Box 90227, Durham, NC 27708-0227

20.1 Introduction

Longshore bars are often found on many gently sloping beaches of large lakes, bays and sea coasts. A beautiful example can be seen in Fig. 20.1 which gives the aerial view of the Escambia Bay in Florida. Several other typical observations are summarized in Table 20.1. In contrast to bars found in rivers where the flows are

Table 20.1. Sample data of observed sand bars

Site	Beach slope	Number of bars	Bar wavelength (m)
Lake Michigan [1]	0.0072-0.012	3-4	38-321
Cape Cod [2]	0.0014-0.0029	6-8	40-105
Alaskan Artic [3]	0.0041-0.0057	4-5	141-479
Chesapeake Bay [4]	0.0017-0.0052	4-17	12-70

essentially unidirectional and characterized by very long time scales (see Chap. 15), coastal bars are usually the products of waves. Of scientific interests are the detailed physics of their generation by waves, as well as their influence on the propagation of waves.

To prepare for Normandy landing by Allied Forces near the end of the second World War, Keulegan [6][1] conducted the first laboratory research on the formation of sand bars on beaches. He sent sinusoidal waves on a plane beach covered by sand, and found that the rolling crests of breaking waves excavate sand particles from the originally plane beach, bring them into suspension, then deposit them slightly offshore. After many wave periods a bar is formed along the breaker line.

Laboratory studies have continued at a slow pace since the second world war. Herbich, Murphy & Van Weele [7] found, in a laboratory wave flume with a wave-maker at one end and a steep wall at the other, that sand bars are

[1] The report was written in 1944 and declassified in 1948.

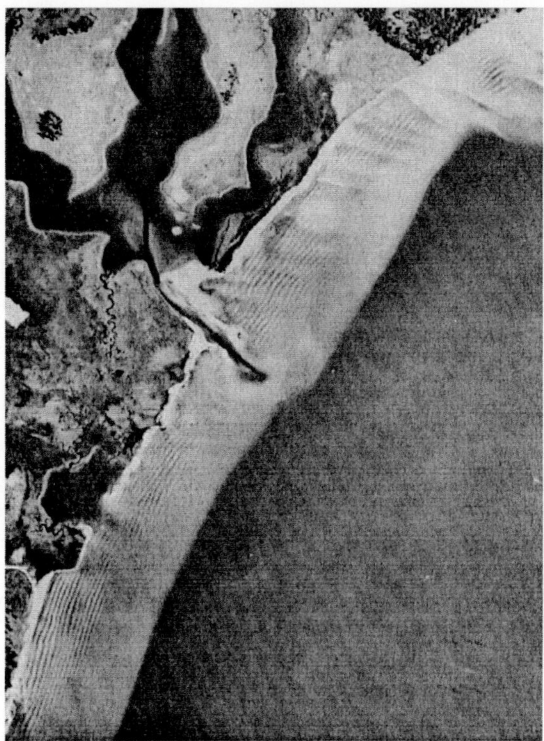

Fig. 20.1. Aerial view of submarine sandbars in Escambia Bay, Florida, USA. From [5]

formed at the wavelength roughly equal to one half of the incident wave length. To explain this special ratio considerable attention has been paid to the fluid dynamics of induced streaming in the boundary layer of an oscillating flow. As first explained for sound waves by Rayleigh, the time-averaged Reynolds stresses in the viscous boundary layer created by an oscillatory flow can induce steady streaming inside the boundary layer. Extension of Rayleigh's theory to water waves has been made by Longuet-Higgins [8]. If a two dimensional (long-crested) partially reflected wave of small amplitude exists above a horizontal rigid bed at depth h $(-\infty < x < \infty, z = -h)$, there is a viscous boundary layer of the thickness of the order $\delta = \sqrt{2\nu/\omega}$ where ν and ω denote the kinematic viscosity and frequency respectively. If U_o is the amplitude of the inviscid flow velocity just outside the boundary layer, then the first-order horizontal velocity inside the viscous layer is given by Stokes' solution:

$$u_1 = \mathrm{Re}\left[U_o\left(1 - \mathrm{e}^{-(1-\mathrm{i})\xi}\right)\mathrm{e}^{-\mathrm{i}\omega t}\right], \quad \text{where } \xi = \frac{z+h}{\delta}\,. \tag{20.1}$$

At the second order, time averaging of the horizontal momentum equation gives

$$-\nu\frac{\partial^2 \overline{u}_2}{\partial z^2} = \overline{U_I\frac{\partial U_I}{\partial x}} - \frac{\partial \overline{u_1 u_1}}{\partial x} - \frac{\partial \overline{u_1 w_1}}{\partial z}\,. \tag{20.2}$$

Thus Reynolds stresses from the convective inertia of wave fluctuations drive a mean shear stress across the boundary layer, leading to a steady Eulerian streaming with the horizontal component $\overline{u}_2$. The associated drift velocity of a fluid particle $\overline{u}_L$ is the sum of $\overline{u}_2$ and the Stokes drift

$$\overline{u}_L = \overline{u}_2 + \overline{\left(\int u_1 \, \mathrm{d}t\right) \frac{\partial u_1}{\partial x}} \, . \tag{20.3}$$

The result is also known as the *Lagrangian mass transport* velocity. For long-crested partially reflected waves over a horizontal seabed, the amplitude of the first order velocity just outside the boundary layer is

$$U_o = \frac{-\omega A}{\sinh kh} \left(\mathrm{e}^{-\mathrm{i}kx} - R\mathrm{e}^{\mathrm{i}kx}\right) \, , \tag{20.4}$$

where A and RA denote respectively the free surface amplitude of the incident and the reflected waves, with R being the reflection coefficient. Longuet-Higgins [8] found the horizontal component of the mean drift to be

$$\begin{aligned} \overline{u}_L = & \frac{k\omega A^2}{4\sinh^2 kh} \left[\left(1-R^2\right)\left(8\mathrm{e}^{-\xi}\cos\xi - 3\mathrm{e}^{-2\xi} - 5\right)\right. \\ & \left. + \; 2R\sin 2kx \left(8\mathrm{e}^{-\xi}\sin\xi + 3\mathrm{e}^{-2\xi} - 3\right)\right] \, . \end{aligned} \tag{20.5}$$

His attention was focussed more on the mass transport at the outer edge of the boundary layer for its possible implication on the suspended sediments. Hunt & Johns [9] extended the theory to bi-directional gravity waves and to tides where the effects due to earth rotation are important. Carter, Liu & Mei [10] used this theory to analyze the current in the boundary layer under a partially reflected waves whose envelope has periodic maxima (antinodes) and minima (nodes) at half wave-length intervals. If the reflection coefficient is greater than 0.414, the current in the boundary layer form closed cells. Very close to the bed, $\xi = (z+h)/\delta \ll 1$, $\bar{u}_L$ converges towards parallel lines beneath the envelope nodes. Near the upper edge of the boundary layer, u_L converges instead toward parallel lines beneath the antinodes. Heavy or large sand grains rolling or sliding on the bottom would tend to accumulate beneath the nodal lines, while light and small grains suspended at the top of the boundary layer would gather under the antinodal lines. Since the adjacent nodes and antinodes are separated at half-wavelength intervals, the bar crests should be separated likewise. With a monolayer of sand sparcely laid on an smooth bed surface Carter et al. found that visual observations support these theoretical predictions qualitatively. The mean circulation in the boundary layer may also explain the observed phenomenon of sediment sorting, i.e. fine sand is more likely found around the bar troughs and coarse sand near the bar crests.

Since on gentle beaches wind waves lose most of the energy by breaking, reflection must be weak. Why are bars frequently found near the shore? One possible reason is that there are steep seawalls at the end of the gentle beaches, another is that the breaking point bars found by Keulegan may be quite large.

Still another possible reason is that bars formed at half wavelength intervals can increase the reflection coefficient by Bragg resonance. Though well-known in optics and solid-state physics, this resonance was first demonstrated for water waves in laboratory experiments by Heathershaw [11] who installed on the horizontal bottom of a wave flume 10 rigid bars with amplitudes much smaller than the water depth. By tuning the incident waves he found that up to 60% of the incident wave energy can be reflected. Thus, a few bars created by past storms can produce larger reflection in the next storm, and induce more and higher bars by mass transport. This experimental demonstration has stimulated many theoretical studies on the effects of rigid bars on waves ([12–19]) as well as a number of small laboratory studies on sand bar formation ([20–22]).

Given the multitude of factors affecting the coastal environment (intensity and time variability of the wave climate, current, sand size distribution, bathymetric variations, etc.), there are of course other mechanisms contributing to the interaction of fluid flow and sand bars. In experiments for long waves of finite amplitude in a shallow tank, Bozar-Karakiewicz et al. [23] made extensive observations of sand bar formation on horizontal and sloping sandy beds for very long times of up to several days. Their focus was on the nonlinear effects of harmonic generation in long waves and the associated transient evolution of sand bars. In their bed profiles, half wave-length bars are prominent at the early stage before waves change significantly by nonlinearity. A sample record is plotted in Fig. 20.2. Later development was strongly modified by the effects of harmonic generation in waves [5,24,25]. The half-wavelength bars were then replaced by much longer bars with periods comparable to the beat length of the first and second harmonics. Theoretical models accounting for suspended sediments only have been pursued by Karakiewicz, Bona and associates [26,27].

In this review we shall limit our attention to progresses made on the interaction of waves and bars where Bragg resonance plays a central role. We begin with the effects of rigid bars on waves, including the linear aspect of strong reflection and the second-order nonlinear effects of setdown waves. Lastly, recent work on the mutual influence of waves and bars will be discussed.

20.2 Linear Bragg Resonance by Rigid Bars

Motivated by the experiments of Heathershaw, Davies [12] gave a linearized theory for waves scattered by a rigid bed with m periods of sinusoids on an otherwise horizontal bottom:

$$b(x) = D \cos \ell x, \quad 0 < x < 2m\pi/\ell \,, \tag{20.6}$$

where $b(x)$ denotes the bar height about the mean depth $z = -h$, and D the bar amplitude. For bars of amplitudes much smaller than the depth, the scattering is weak in general. The leading-order velocity potential is simply that of the incident wave:

$$\phi_I = -\frac{igA}{2\omega} \frac{\cosh k(z+h)}{\cosh kh} \mathrm{e}^{ikx - i\omega t} + c.c. \,, \tag{20.7}$$

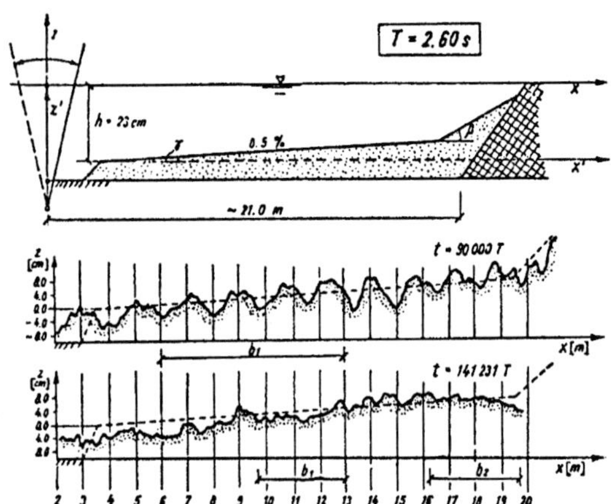

Fig. 20.2. Bed development over an initially sloping sandy beach (*top figure*). The incident wave period is $T = 2.6\,\mathrm{s}$. At $t = 90,000T$ periodic bars are developed at half-wavelength spacings (approximately 2 m, see middle figure). At $t = 141,231T$ bars are replaced by longer bedforms due to nonlinear changes of waves (*bottom figure*). From [23]

where x, z are the horizontal and vertical coordinates with origin in the mean sea surface, the wave number k and frequency ω are related by the dispersion relation,

$$\omega^2 = gk \tan kh \; . \tag{20.8}$$

At the next order in kD, the reflected wave on the side $x < 0$ is

$$-\frac{igAR}{2\omega} \frac{\cosh k(z+h)}{\cosh kh} \mathrm{e}^{-\mathrm{i}kx - \mathrm{i}\omega t} + c.c. \; , \tag{20.9}$$

where the reflection coefficient is:

$$R = \left(\frac{4k^2 D/\ell}{\sinh 2kh + 2kh} \right) \frac{\sin(2km\pi/\ell)}{(2k/\ell)^2 - 1} \; . \tag{20.10}$$

This result indicates that if the condition for Bragg resonance $2k = \ell$ is satisfied, i.e. the bar wavelength is half of the surface wavelength, the reflection coefficient becomes

$$R \to \frac{2kD}{\sinh 2kh + 2kh} \frac{m\pi}{2} \tag{20.11}$$

and is unbounded with increasing m. Mathematically the reason for this resonance can be found from the approximate boundary condition at $z = -h$

$$\left. \frac{\partial \phi}{\partial z} \right|_{z=-h} \approx -\frac{\partial}{\partial x} \left(b \frac{\partial \phi_I}{\partial x} \right) \; . \tag{20.12}$$

When the Bragg condition is met,

$$b = \frac{D}{2}\left(\mathrm{e}^{2\mathrm{i}kx} + c.c.\right) ,$$

then

$$\left.\frac{\partial \phi}{\partial z}\right|_{z=-h} \approx \frac{\mathrm{i}kDgkA}{4\omega \cosh kh}\left(-\mathrm{e}^{-\mathrm{i}kx-\mathrm{i}\omega t} + 3\mathrm{e}^{3\mathrm{i}kx-\mathrm{i}\omega t} + c.c.\right) .$$

The first forcing term is a natural mode representing a left-going wave, and must induce resonance of reflection. This is a special case of triad resonance where quadratic products of the incident wave $(\pm\omega, \mp k\ )$, and the wavy bottom $(0, \pm 2k)$, gives rise to reflected waves of the same wave number and frequency $(\pm\omega, \pm k)$. Physically, wave crests reflected by each bar differ in phase by integral multiples of 2π, i.e. $2n\pi$, hence they reinforce one another; reflection is resonant.

This physical problem has two sharply contrasting length scales: the bar wavelength, and the much greater total extent of the bar field (or the length scale of resonant growth). Mei [13] developed a theory uniformly valid near resonance for narrow-banded waves over nearly periodic bars on a gentle beach with depth contours parallel to the shoreline. He employed the WKB method well-known in the theory of wave refraction by using slow coordinates $x_1, t_1 = \epsilon(x, t)$, where $\epsilon \ll 1$ is a small parameter characterizing all of the following: the ratio of bar wavelength to the total domain of bars, the narrowness of the frequency band, and the amount of detuning from resonance. At the leading order both incident and reflected waves are allowed

$$\phi = \left(\varphi_0^+ + \mathrm{i}\epsilon\varphi_1^+ \ldots\right)\mathrm{e}^{\mathrm{i}S_+/\epsilon} + \left(\varphi_0^- + \mathrm{i}\epsilon\varphi_1^- \ldots\right)\mathrm{e}^{\mathrm{i}S_-/\epsilon} , \tag{20.13}$$

where φ_n^+ and φ_n^- are the amplitude functions at the n–th order in ϵ, and are functions of z and the slow coordinates x_1, y_1. The phase functions of the incident and reflected waves are represented by

$$S_+ = \int \alpha(x_1)\, \mathrm{d}x_1 + \beta y_1 - \omega t_1 , \quad S_- = -\int \alpha(x_1)\, \mathrm{d}x_1 + \beta y_1 - \omega t_1 \tag{20.14}$$

respectively. The local wave number vectors representing incident and reflected waves are

$$\mathbf{k}_+(x_1) = (\alpha(x_1), \beta) , \quad \mathbf{k}_-(x_1) = (-\alpha(x_1), \beta) ,$$

and the local angle of incidence θ is

$$\tan\theta(x_1) = \frac{\beta}{\alpha(x_1)} . \tag{20.15}$$

The bars are assumed to be parallel to the depth contours

$$b = \frac{1}{2}D(x_1, y_1)\left[\exp\left(2\mathrm{i}\int \alpha \mathrm{d}x_1\right) + \exp\left(-2\mathrm{i}\int \alpha \mathrm{d}x_1\right)\right]$$

and in resonance with the incident wave. Note that $D(x_1, y_1)$ is now allowed to vary with the slow coordinates.

By assuming that the surface waves and bars are comparably gentle, and imposing the condition that at the order $O(\epsilon)$ the problem for $\varphi_1^{\pm}$ must be solvable, the evolution equations coupling the envelopes of the incident and reflected waves (A and B) are derived,

$$\frac{\partial A}{\partial t_1} + C_{g_x}\frac{\partial A}{\partial x_1} + C_{g_y}\frac{\partial A}{\partial y_1} + \frac{\partial C_{g_x}}{\partial x_1}\frac{A}{2} = -\mathrm{i}\Omega_o \cos 2\theta B \;, \tag{20.16}$$

$$\frac{\partial B}{\partial t_1} - C_{g_x}\frac{\partial B}{\partial x_1} + C_{g_y}\frac{\partial B}{\partial y_1} - \frac{\partial C_{g_x}}{\partial x_1}\frac{B}{2} = -\mathrm{i}\Omega_o \cos 2\theta A \;. \tag{20.17}$$

The coupling coefficient

$$\Omega_o = \frac{gkD}{4\omega \cosh^2 kh} \tag{20.18}$$

is proportional to the bar amplitude, and

$$\mathbf{C}_g^{\pm} = \left(C_{g_x}, C_{g_y}\right) = \frac{\omega}{2k}\left(1 + \frac{2kh}{\sinh 2kh}\right)\frac{\mathbf{k}^{\pm}}{k} \tag{20.19}$$

are the group velocities of the incident and reflected waves. In the limit of constant mean depth and normal incidence, $\theta = \partial/\partial y_1 = 0$, the two equations can be combined to give the Klein–Gordon equation

$$\frac{\partial^2}{\partial t_1^2}\begin{pmatrix} A \\ B \end{pmatrix} - C_g^2 \frac{\partial^2}{\partial x_1^2}\begin{pmatrix} A \\ B \end{pmatrix} + \Omega_o^2 \begin{pmatrix} A \\ B \end{pmatrix} = 0 \;, \tag{20.20}$$

well-known in modern physics. It is easy to see that the envelopes are dispersive waves on the long scale.

As analytical examples for demonstrating the physics, we first discuss the simple case of a finite patch of bars $0 < x_1 < L$ on an otherwise horizontal seabed. The bar amplitude D is constant. A train of slightly detuned incident waves of prescribed amplitude arrives from $x_1 \sim -\infty$. Reflected waves are generated over the bars and propagate back towards $x_1 \sim -\infty$. In general, some transmitted waves can pass the end at $x_1 = L$ towards $x \sim \infty$.

We distinguish two cases in the following subsections.

20.2.1 No Reflection from $x > L$

Consider first the case where there is no reflection from $x_1 \sim \infty$, simulating crudely an idealized beach which absorbs all the incident wave energy. Let the envelope of the incident waves be

$$A = A_o \exp[\mathrm{i}K(x_1 - C_g t_1)] \;,$$

where $\Omega \equiv KC_g$ with $\epsilon\Omega, \epsilon K = O(\epsilon)$ corresponding to small detuning from perfect resonance. Then the governing equation in the region covered by bars is of the form

$$\frac{\mathrm{d}^2}{\mathrm{d}x_1^2}\begin{pmatrix} A \\ B \end{pmatrix} + \frac{1}{C_g^2}\left(\Omega^2 - \Omega_o^2\right)\begin{pmatrix} A \\ B \end{pmatrix} = 0 \;. \tag{20.21}$$

The solution changes character depending on the sign of $\Omega - \Omega_o$. Specifically the solution is monotonic in x_1 if the detuning is below cutoff, $\Omega < \Omega_o$, and is oscillatory in x_1 if the detuning is above cutoff. This difference is also present in the reflection coefficient $R(x_1) \equiv B(x_1)/A(x_1)$. For subcritical detuning ($0 \le \Omega < \Omega_o$), the reflection coefficient is monotonic in space,

$$|R|^2 = \frac{\sinh^2\left\{\frac{\Omega_o L}{C_g}\left[1-\frac{\Omega^2}{\omega_o^2}\right]^{1/2}\left(1-\frac{x_1}{L}\right)\right\}}{\cosh^2\left\{\frac{\Omega_o L}{C_g}\left[1-\frac{\Omega^2}{\Omega_o^2}\right]^{1/2}\right\}-\frac{\Omega^2}{\Omega_o^2}}\,. \tag{20.22}$$

For supercritical detuning ($\Omega > \Omega_o$), R is oscillatory

$$|R|^2 = \frac{\sin^2\left\{\frac{\Omega_o L}{C_g}\left[\frac{\Omega^2}{\omega_o^2}-1\right]^{1/2}\left(1-\frac{x_1}{L}\right)\right\}}{-\cos^2\left\{\frac{\Omega_o L}{C_g}\left[\frac{\Omega^2}{\omega_o^2}-1\right]^{1/2}\right\}+\frac{\Omega^2}{\Omega_o^2}}\,. \tag{20.23}$$

The limit of perfect tuning ($\Omega = 0$) is given by:

$$|R|^2 = \frac{\sinh^2\left(\frac{\Omega_o L}{C_g}\left(1-\frac{x_1}{L}\right)\right)}{\cosh^2\frac{\Omega_o L}{C_g}}\,. \tag{20.24}$$

As shown in Fig. 20.3, this theory checks very well with the experiments by Heathershaw whose incident waves are precisely tuned to Bragg resonance with the bars.

For imperfect tuning, envelope dispersion and cutoff frequency are the distinctive features of the theoretical results. These have been further studied theoretically and experimentally by Hara & Mei [14]. A typical prediction is shown in Fig. 20.4 in which the symbols for time τ and space x are the dimensionless equivalents of x_1 and t_1. An initial wave packet with a bell-shaped envelope was sent to a zone of periodic bars, distributed in the dimensionless region of $0 < x < 1$. At $\tau = 0$ the incident wave packet first arrives at the edge $x = 0$. At $\tau = 3\pi$, envelopes of both transmitted and reflected waves have split into multiple groups showing dispersion. At $\tau = 6\pi$ dispersion of $|A|$ and $|B|$ continues.

To confirm the predicted dispersion of envelopes, wave-packet experiments were performed in a long tank with a wave-maker at the end $x = -9.1$ m In the stretch $0 < x < 12$ m the bed is covered with periodic bars of plexiglass construction. The end $x = 12$ m is closed by a vertical and perfectly reflecting wall. Figure 20.5 shows the comparison between prediction and measurement of an incident wave packet for $\Omega = \Omega_o$. The time records of incident and reflected waves at three stations are shown in the figure. Up to its arrival at the edge $x = 0$ m, the incident wave packet has a bell-shaped envelope. At the mid-point of the zone of bars, $x = 6$ m, dispersion into two groups is evident. Dispersion increases as the waves strike and are turned back by the reflecting wall at $x = 12$ m.

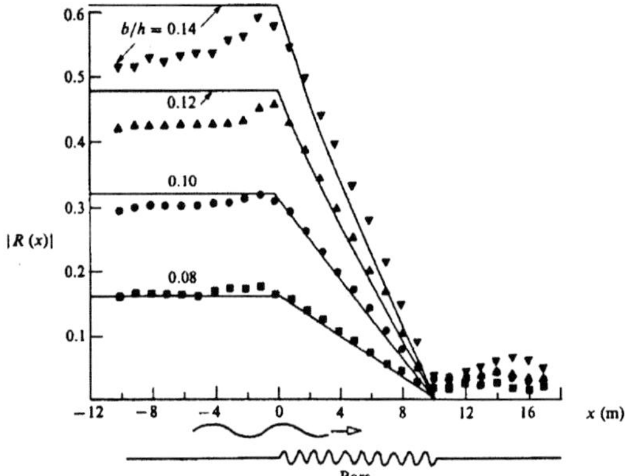

Fig. 20.3. Comparison between theory (20.24) and experiments by Heathershaw [11] for perfectly tuned waves. From [13]

To check the predicted cutoff frequency for steady incident waves, experiments were performed for a much long time with very small wave amplitude so as to avoid nonlinear effects of side-band instability. The linearized theory of Mei was further improved to include $O(kA)$, $O(kD)$ as well as $O(k^2D^2)$ terms. Not surprisingly the evolution equation for the wave envelopes include higher order dispersion:

$$\frac{\partial}{\partial t_1}\begin{pmatrix} A \\ B \end{pmatrix} \pm C_g \frac{\partial}{\partial x_1}\begin{pmatrix} A \\ B \end{pmatrix} + \mathrm{i}\Omega_o \begin{pmatrix} A \\ B \end{pmatrix} = \epsilon \left[\mathrm{i}p \frac{\partial^2}{\partial x_1^2}\begin{pmatrix} A \\ B \end{pmatrix} \mp q \frac{\partial}{\partial x_1}\begin{pmatrix} B \\ A \end{pmatrix} + \mathrm{i}r \begin{pmatrix} A \\ B \end{pmatrix} \mp s \begin{pmatrix} B \\ A \end{pmatrix} \right], \tag{20.25}$$

where p, q, r are functions of kh. With these corrections the agreement between theory and measurements is very good quantitatively, as shown in Fig. 20.6. The change from monotonic to oscillatory behaviour when Ω crosses the threshold Ω_o is indeed verified.

For further discussions on oblique incidence and on the mean slope of seabed, see [13] and [15].

20.2.2 Finite Reflection from $x_1 > L$

In the preceding subsection, both the incident and reflected waves decrease monotonically toward the transmission end of the finite bar batch, when $\Omega < \Omega_o$. Based on this result, it has been suggested that a bar patch well tuned to the incident waves can serve as a breakwater for protecting the shoreline [15,28,29]. Field attempts have so far not been successful.

On natural beaches, finite reflection from the shore is inevitable. In a numerical study, Kirby & Anton [30] considered a vertical wall at some distance

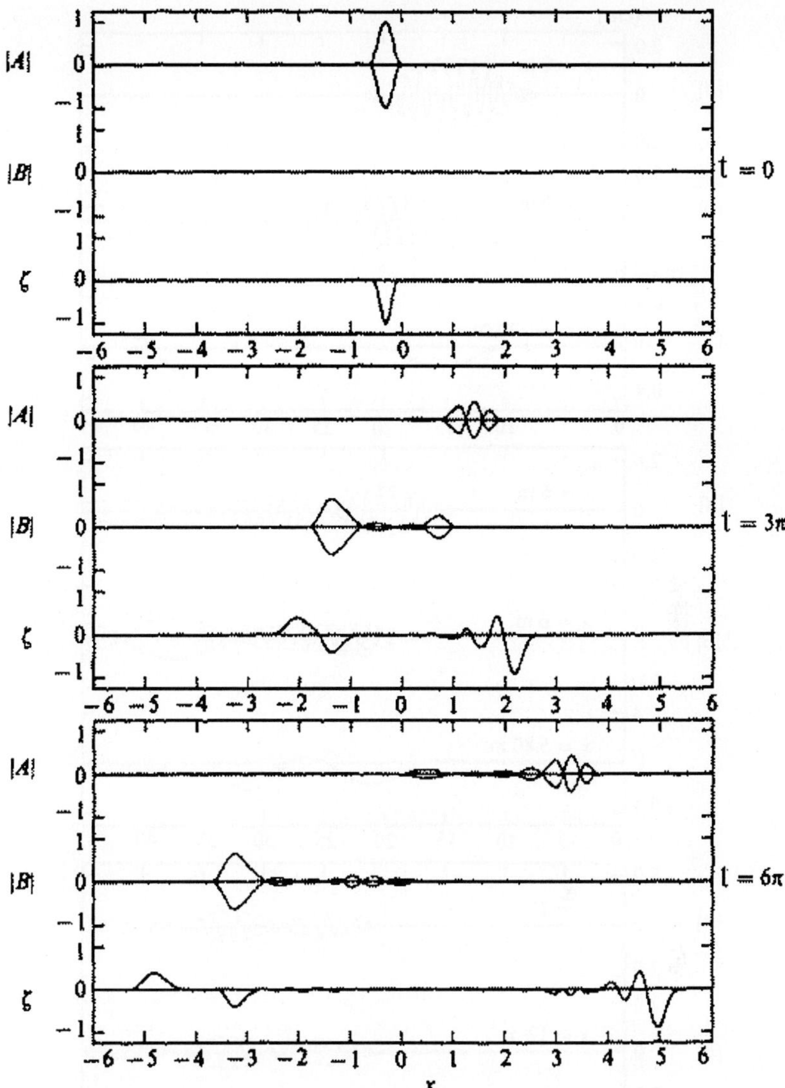

Fig. 20.4. Predicted envelopes of the incident ($|A|$) and reflected ($|B|$) waves and the induced long waves (ζ). Bars are distributed in $0 < x < 1$. In this figure reproduced from [14], x corresponds to x_1/L and t to $t_1 C_g/L$ in this article

downwave of a patch of several rigid bars, and found that near Bragg resonance the amplitude of the free-surface oscillations at the wall can vary between 1 to 3.6 times the amplitude of the incident waves, depending on the distance between the bars and the wall region between the end of the bar patch and the wall.

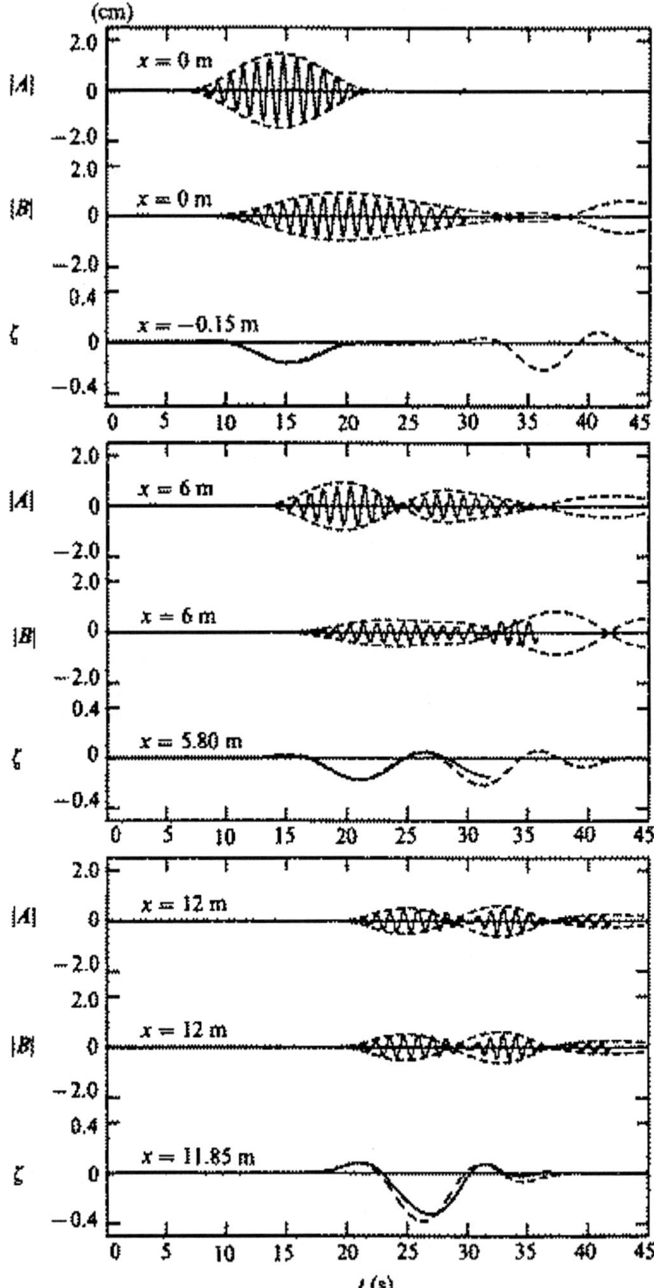

Fig. 20.5. Comparison between theory and experiments for wave packets. Predicted envelopes are shown by dashed curves, and measured free surface by solid curves. From [14]

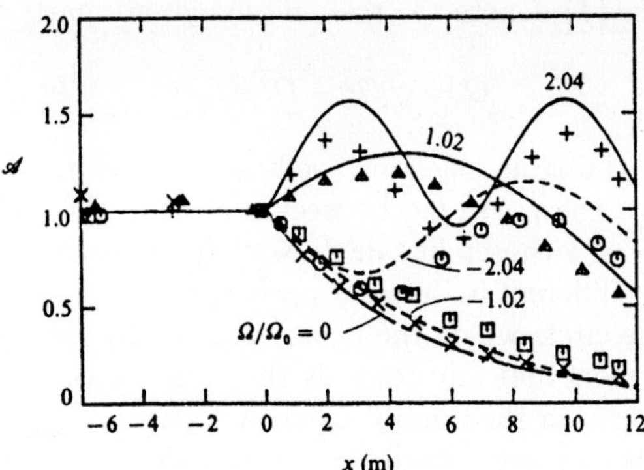

Fig. 20.6. Comparison between theory and experiments on cutoff frequency. Measured wave heights at antinodes and the corresponding values of Ω/Ω_o are + : 2.04; △ : 1.02; squares: −1.02; ∘ : −2.04. From [14]

Suppose that a finite patch of bars is present not too far from a shore along which breaking does not completely destroy all the incident wave energy. Some waves must be reflected toward the bar patch. Yu & Mei [31] carried out an explicit analysis by allowing finite reflection coefficient R_L at the transmission end of a finite patch of bars in $0 < x_1 < L$, again over a constant mean depth. The boundary condition becomes:

$$\frac{\partial A}{\partial t_1} + C_g \frac{\partial A}{\partial x_1} = \frac{\mathrm{i}D\omega k R_L}{2\sinh 2kh} A, \qquad \text{at} \quad x_1 = L\,.$$

For an incident wave of the form

$$A = A_0 \exp[\mathrm{i}(Kx_1 - \Omega t_1)] \qquad \text{at} \quad x_1 \le 0\,,$$

from side $x_1 < 0$, it can first be shown that the energy flux rate at any given station is the same:

$$C_g(|A|^2 - |B|^2) = \text{constant}\,, \tag{20.26}$$

although $|A|^2$ and $|B|^2$ vary in x_1 in general. Thus $|A|$ and $|B|$ depend on x_1 similarly. While the explicit formulas can be given, it suffices to summarize the qualitative behaviour of A and B on x_1 in the complex plane of $\widetilde{R}_L$, as shown in Fig. 20.7, where

$$\widetilde{R}_L = |R_L| \exp[-\mathrm{i}(\theta_{RL} + \theta_D)]\,,$$

and θ_{RL} and θ_D are the phases of R_L and D respectively. The unit circle represents the maximum shore reflection $|R_L| = 1$ with all possible phase angles $0 \le \theta_{RL} < 2\pi$. The second circle given by

$$\left(\mathrm{Re}\,(\widetilde{R}_L) + \frac{\Omega}{\Omega_0}\right)^2 + \left(\mathrm{Im}\,(\widetilde{R}_L) - \frac{QC_g}{\Omega_0}\coth 2QL\right)^2 = \left(\frac{QC_g/\Omega_0}{\sinh 2QL}\right)^2, \tag{20.27}$$

where Re (f) and Im (f) denote the real and imaginary parts of f, and

$$Q = \sqrt{\Omega_o^2 - \Omega^2}/C_g \ .$$

Since $|R_L| \leq 1$, all attainable reflection coefficients lie within the unit circle. The angle θ_{RL} represents the phase lag between a crest of the incident wave and a bar crest at $x_1 = L$, at a chosen instant t. Now the shoreward variation of the A or B is qualitatively different in different parts of the unit circle. In Region I, i.e. the lower half of the circle where the phase angle of R_L lies between π and 2π, A and B decrease monotonically towards the transmission end, implying that bars protect the shore. In the doubly hatched region (II), the opposite is true, i.e. wave amplitudes increase toward the transmission end, hence the shore line is adversely affected by the bars. When the complex reflection coefficient is in the singly hatched region III, the wave amplitudes first decrease then increase toward the transmission end. Since for fixed bars the phase of R_L is determined by the uncontrollable phase of the incident waves, artificially constructed bars are not a viable alternative to conventional breakwaters.

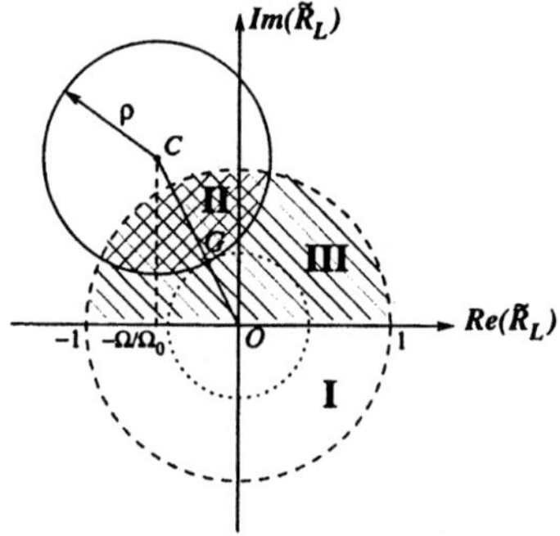

Fig. 20.7. Distinguishing behaviour of $|A|$ and $|B|$ as x increases shoreward (from 0 to L). In Region I (semicirle in the lower half plane), $|A|, |B|$ decrease monotonically. In Region II (doubly hatched), $|A|, |B|$ increase monotonically. In Region III (singly hatched), $|A|, |B|$ first decrease and then increase

20.3 Long Waves Generated by Short Waves Scattered by Bars

It is known for smooth seabeds that, due to nonlinear interactions, gentle short waves with a narrow frequency band can generate long waves (setdown). Though second-order in steepness, these long waves can reach greater depths than their parents, hence may have important effects on the evolution of sand bars. Recent experiments by Dulou, Belzons & Rey [32] are begining to shed light on these effects. For rigid bars, Mei [13] and Hara & Mei [14] have shown that the free surface displacement ζ of the long wave is governed by the forced wave equation

$$\frac{\partial^2 \zeta}{\partial t_1^2} - g\nabla_1 \cdot (h\nabla_1\zeta) = g\nabla_1 \cdot \left(h\nabla_1 \left[\frac{k(|A|^2+|B|^2)}{2\sinh 2kh}\right]\right)$$
$$-\frac{g}{2\omega}\frac{\partial}{\partial t_1}\left[\alpha\frac{\partial}{\partial x_1}\left(|A|^2-|B|^2\right)+\beta\frac{\partial}{\partial y_1}\left(|A|^2+|B|^2\right)\right], \quad (20.28)$$

where $\nabla_1 \equiv (\partial/\partial x_1, \partial/\partial y_1)$ and α and β are the components of the local incident wavenumber vector $\mathbf{k}$. The forcing terms are associated with the radiation stresses similar to Reynolds stresses in turbulence, hence long-wave generation by short waves is somewhat analogous to sound generation by turbulent fluctuations. Sample predictions of long waves are also shown in Fig. 20.4 in normalized slow variables. Before the wave packet reaches the bars located in the dimensionless range $0 < x < 1$, there is only one long wave which travels at the group velocity of the short waves, hence is bound to the wave packet; this is the classical set-down, also known as the *bound* or *forced* long wave. After reflection by the bars, two kinds of long waves, bound and *free*, can be seen. As the response to forcing by the local radiation stresses, the set-down long waves are the inhomogeneous solutions to (20.28). The free long waves correspond to the homogeneous solutions to (20.28), and travel at the higher speed $\sqrt{gh}$ in both directions, hence ahead of the wave packets. These two long waves are particularly evident at $\tau = 6\pi$ where the free long wave on the reflection side has outrun the reflected wavepacket and the set-down wave. Experimental confirmation of the long waves can also be seen in Fig. 20.5.

When the bars are sufficiently large or numerous, the first-order short waves can be totally reflected. However the second-order free long waves can still be radiated into the shadow. While the short wave is dominant on the incidence side, the free long wave is dominant on the transmission side. Thus bars can alter the wave spectrum of the near shore region[2]. The lee side of a region with bars may be free of short choppy seas, but is not necessarily a safe haven for fishing vessels since long waves may resonate the mooring system whose natural frequency is usually much lower than that of the wind waves.

We now turn to the interaction of waves with a sandy bed.

20.4 Sand-Bar Formation Dominated by Bedload

While the theory of mass transport in waves over a rigid bed suggests certain qualitative trends in the formation of sand bars, recent laboratory experiments [20] on a sandy bed have revealed a more complex physics. In particular bars can form even when the reflection coefficient is well below 0.414 predicted by Carter et al. [10].

[2] A wide zone of bars can be likened to a thick sound-transforming wall. While a soprano sings on one side, a baritone is heard on the other.

In nature there can be two modes of sediment transport. Suspended load prevails when the particles are fine and waves are strong, while bedload dominates when particles are coarse and waves are weak. In nature both sand sizes and wave intensities vary over a wide range depending on the bathymetry and seasons, hence a complete theory ought to account for both modes of sediment motion. This task is so far unfulfilled. In [26] and [27] only suspension is included. In [33], only bedload is taken into account.

In the rest of this article we sketch the work of Yu & Mei [33] which is aimed at coarse sand and/or weak waves in the shoaling zone where the typical wave lengths of sand bars and surface waves are comparable to the water depth. They assumed monochromatic gravity waves propagating over water of constant mean depth h. Let A_0 denote the typical free-surface wave amplitude and k_b and D denote the typical wavenumber and amplitude of sand bars. The typical horizontal amplitude of wave oscillations just above the bed,

$$A_b = \frac{A_0}{\sinh kh} , \tag{20.29}$$

is taken as the scale of orbital motion. Appropriate for the shoaling zone we take

$$kh = O(1) , \qquad \frac{k_b}{k} = O(1) . \tag{20.30}$$

Both surface waves and sand bars are assumed to have gentle slopes characterized by the small parameter ϵ,

$$\epsilon \equiv A_b k = O(Dk_b) \ll 1 . \tag{20.31}$$

In the field the bar surface is sometimes covered by ripples, which are much smaller in amplitudes, and grow to full size much faster than the bars. Consequently, ripples will be treated as fixed roughness in a turbulent boundary layer above the seabed. For simplicity we adopt the Boussinesq model of constant eddy viscosity ν whose value is estimated from the wave characteristics and sand diameter through an empirical procedure known in the coastal engineering literature (see [33] for details). To have some quantitative ideas, for incident wave of the same period $T = 8\,\mathrm{s}$, but two sets of amplitudes and depths $A_0 = 20\,\mathrm{cm}$, $h = 8\,\mathrm{m}$ and $A_0 = 25\,\mathrm{cm}$, $h = 7\,\mathrm{m}$, and sand diameter of $d = 0.3\,\mathrm{mm}$, the empirical estimate gives the following eddy viscosities: $\nu = (2.63, 4.75)\,\mathrm{cm}^2/\mathrm{s}$. The corresponding boundary layer thickness is of the order $\delta = \sqrt{2\nu/\omega} = (2.59, 3.48)\,\mathrm{cm}$, which is much smaller than the typical wavelength. In the perturbation theory we take specifically

$$k\delta = O(\epsilon^2) , \tag{20.32}$$

implying that

$$O\left(\frac{A_b}{\delta}\right) = O\left(\frac{D}{\delta}\right) = O(\epsilon^{-1}) . \tag{20.33}$$

The typical sand diameter d is assumed to be much smaller than the thickness of the boundary layer,

$$d/\delta = O(\epsilon) \ll 1 . \tag{20.34}$$

Under these assumptions we first derive the relation between the bar evolution and the flow properties in the water above, by employing an empirical law of sediment transport. The flow above is then examined theoretically to complete the analytical framework. The evolution equations for sand bars and waves are finally solved numerically.

20.5 Laws of Bedload Transport

Ignoring suspended load, the kinematic law of mass conservation reads

$$(1-n)\frac{\partial b'}{\partial t'} + \frac{\partial q'}{\partial x'} = 0\,, \tag{20.35}$$

where q' is the volume discharge rate of the bedload, b' is the mean bar height (after averaging over the small ripples) measured from the mean position of the bottom, and n is the bed porosity. Primes are used here to distinguish variables with physical dimensions. For spatially uniform but time-periodic flows over a plane bed of sand, Sleath [34] has given an empirical relation between the sediment discharge rate and the Shields parameter which depends solely on the local bed shear stress. With uneven bed surface the discharge must be affected by the local slope, for gravity tends to pull down sand grains. We follow an idea due to Fredsøe [35] for sand bars in steady river flows, and modify the Shields parameter by adding a term proportional to the bed slope. Specifically, Sleath's formula is first assumed to hold locally for spatially varying flow,

$$q'(x',t') = \frac{8}{3}Q'_s(x')\left\{\frac{\Theta(kx',\omega t' + \mathrm{d}\varphi)}{\widehat{\Theta}(kx')}\right\}^2 \operatorname{sgn}\left[\Theta(kx',\omega t' + \mathrm{d}\varphi)\right]\,, \tag{20.36}$$

where $\Theta(kx',\omega t' + \mathrm{d}\varphi)$ is the local Shields parameter, $\widehat{\Theta}(kx')$ denotes the local maximum of Θ within a wave period, $\Delta\varphi$ an empirical phase lag which is immaterial later after time averaging, and

$$\frac{Q'_s(x')}{\sqrt{(s-1)gd^3}} = \begin{cases} C_s\left[\widehat{\Theta}(kx') - \Theta_c\right]^{1.5} & \widehat{\Theta}(kx') > \Theta_c \\ 0 & \widehat{\Theta}(kx') \le \Theta_c\,. \end{cases} \tag{20.37}$$

Second, we redefine the Shields parameter by combining the effects of bed shear stress and bed slope,

$$\Theta(kx',\omega t' + \mathrm{d}\varphi) = \frac{\tau'_b(x',t' + \mathrm{d}\varphi/\omega)}{\rho(s-1)gd} - \beta\frac{\partial b'}{\partial x'}\,. \tag{20.38}$$

When (20.36) is substituted into (20.35), an evolution equation for h' can be obtained. Based on a heuristic argument Fredsøe [35] suggests that $\beta = O(\Theta_c/\tan\phi_s)$, where ϕ_s is the angle of repose. From the Shields diagram (e.g. [36]), Θ_c is about 0.05 for medium size sand and $\phi_s = O(30°)$ typically, $\beta \sim$

$O(0.1)$. The introduction of the term $\beta\partial b'/\partial x'$ is crucial, since it gives rise to a second derivative of b' in x', and leads to a forced diffusion equation for b', where the diffusivity is proportional to gravity. The mechanics of bar formation is therefore different from that of ripples whose growth is a consequence of instability which must initiated by a small perturbation [37]. In contrast, sand bars can grow from a perfectly plane bed by the nonuniformity of fluid shear stress which is present for all but the purely progressive waves of uniform amplitude. Since no comprehensive data is yet available, the numerical value of β can only be chosen to fit observational data. Physically this slope term is reasonable and has been employed in modeling small ripples in oscillatory flows [37–39] as well as large sand bars due to tides [40].

From (20.36) and (20.37), the magnitude of the transport rate is

$$q' \sim \sqrt{(s-1)gd^3}\,O(\widehat{\Theta}^{1.5}) \,. \tag{20.39}$$

Due to the gentle bar slope, the modified Shields parameter (20.38) is dominated by the first term, i.e. the plane-bed Shields parameter, whose order of magnitude can be estimated as:

$$O(\widehat{\Theta}) \sim \frac{A_b\omega\nu}{(s-1)gd\delta} \equiv \Theta_0 \,. \tag{20.40}$$

Introducing the following dimensionless variables without primes,

$$q' = \frac{8C_s}{3}\sqrt{(s-1)gd^3}\,\Theta_0^{1.5} q \,, \quad b' = A_b b \,, \quad t' = \frac{t}{\omega} \,, \quad x' = \frac{x}{k} \,, \tag{20.41}$$

(20.35) can be written in the dimensionless form

$$(1-n)\frac{a}{\Theta_0^{1.5}}\frac{\partial b}{\partial t} + \frac{\partial q}{\partial x} = 0 \,, \tag{20.42}$$

where a is another dimensionless parameter,

$$a = \frac{3}{8C_s}\frac{A_b\omega}{k\sqrt{(s-1)gd^3}} \,. \tag{20.43}$$

The ratio $a/\Theta_0^{1.5}$ is the dimensionless time scale normalized by ω^{-1}. Under the assumptions made at the beginning of this section, $\Theta_0 = O(1)$, $a = O(\epsilon^{-3.5})$. The ratio $a/\Theta_0^{1.5}$ is very large, hence at the leading order b does not vary significantly over a few wave periods. Denoting by $(\bar{b}, \bar{q})$ the period-averages of (b, q),

$$\bar{b} = \frac{\omega}{2\pi}\int_t^{t+2\pi/\omega} b\,\mathrm{d}t \,, \quad \bar{q} = \frac{\omega}{2\pi}\int_t^{t+2\pi/\omega} q\,\mathrm{d}t \,, \tag{20.44}$$

we get from (20.42) that

$$(1-n)\frac{a}{\Theta_0^{1.5}}\frac{\partial \bar{b}}{\partial t} + \frac{\partial \bar{q}}{\partial x} = 0 \,. \tag{20.45}$$

The part of q that oscillates at the wave frequency contributes to a small correction to b.

As a preliminary step, let us derive the form of the bar evolution equation in terms of the stream function ψ' in the boundary layer, details of which will be left to the next section. First we introduce a boundary-conforming, non-orthogonal coordinate system (x', η') with $\eta' = z' - \bar{b}'$ measured from the bar surface, and next the following normalization,

$$\eta = \eta'/\delta\ , \quad \psi = \psi'/A_b\omega\delta\ . \tag{20.46}$$

Let us also redefine $\epsilon \equiv A_b k$ which is assumed to be of the same order of magnitude as the small ratios in the WKB analysis of Sect. 20.2. Upon expanding the normalized stream function in powers of ϵ, $\psi = \psi_0 + \epsilon\psi_1 + \cdots$, the instantaneous modified Shields parameter (20.38) can be approximated by:

$$\Theta = \Theta_0 \frac{\partial^2 \psi_0}{\partial \eta^2} + \epsilon \left(\Theta_0 \frac{\partial^2 \psi_1}{\partial \eta^2} + \beta \frac{\partial h}{\partial x} \right) + O(\epsilon^2) \qquad \text{on} \quad \eta = 0\ . \tag{20.47}$$

Expanding the discharge rate similarly $q = q_0 + \epsilon q_1 + \epsilon^2 q_2 + \cdots$, and taking the time average, we get from (20.36)

$$\bar{q}_0 = 0\ , \quad \bar{q} = \epsilon \bar{q}_1\ , \tag{20.48}$$

with

$$\bar{q}_1 = 2Q_{s0} \frac{\overline{|\psi_{0,\eta\eta}|\,(\psi_{1,\eta\eta} - h_x\beta/\Theta_0)}}{\left(\widehat{\psi_{0,\eta\eta}}\right)^2}\ , \tag{20.49}$$

where

$$Q_{s0} = \begin{cases} \left(\widehat{\psi_{0,\eta\eta}} - \dfrac{\Theta_c}{\Theta_0}\right)^{1.5} & \widehat{\psi_{0,\eta\eta}} > \Theta_c/\Theta_0 \\ 0 & \widehat{\psi_{0,\eta\eta}} \le \Theta_c/\Theta_0 \end{cases} \tag{20.50}$$

is, in dimensionless form, the leading order approximation of Q_s' in (20.37). Here $\widehat{\psi_{0,\eta\eta}}$ denotes the maximun amplitude of $\psi_{0,\eta\eta}$ within a wave period.

After substituting (20.48) into (20.45), and renormalizing time,

$$\bar{t} = \frac{\epsilon \Theta_0^{1/2}}{a} t\ , \tag{20.51}$$

the normalized evolution equation for the bar height $\bar{b}$ is obtained to the leading order,

$$\frac{\partial \bar{b}}{\partial \bar{t}} - \frac{\partial}{\partial x}\left(D_\nu \frac{\partial \bar{b}}{\partial x} \right) = -\frac{\partial q_\tau}{\partial x}\ , \tag{20.52}$$

where

$$D_\nu = \frac{2\beta}{1-n} Q_{s0} \frac{\overline{|\psi_{0,\eta\eta}|}}{\left(\widehat{\psi_{0,\eta\eta}}\right)^2} \tag{20.53}$$

is the diffusivity and

$$q_T = \frac{2\Theta_0}{1-n} Q_{s0} \frac{\overline{|\psi_{0,\eta\eta}|\,\psi_{1,\eta\eta}}}{\left(\widehat{\psi_{0,\eta\eta}}\right)^2} \tag{20.54}$$

is the integrated forcing. Equation (20.52) is the forced diffusion equation, as anticipated. Note that both the source term and the diffusivity are affected by the wave-induced bed stresses, and depend on the local boundary layer flow, to be sketched next.

20.6 Fluid Flow

In most of the flow above the bed, the inviscid approximation suffices and the flow is describable by a potential theory. Inside the bottom boundary layer the flow is rotational. We shall first treat the two regions separately and then require them to be asymptotically matched.

20.6.1 The Potential Core

In the inviscid core the flow field can be described by a velocity potential Φ', $(u', w') = \nabla' \Phi'$. Since attention from now on will be focussed on the bed, the coordinates are shifted so that the mean bed surface coincides with $z' = 0$. Allowing for slow modulations of the surface waves due to either narrow-bandness or Bragg resonance, we can perform a perturbation analysis similar to that in [13]. Only the results need to be cited here, in terms of the dimensionless variables defined before in (20.41) and (20.46), plus the following,

$$\Phi = k\Phi'/A_b\omega\ , \quad z = kz'\ , \tag{20.55}$$

where ζ denotes the free surface displacement.

At the leading order $O(\epsilon^0)$, the bottom is flat. The linearized solution is

$$\Phi_0 = -\frac{\mathrm{i}}{2} \cosh z \left(A\mathrm{e}^{\mathrm{i}x} - B\mathrm{e}^{-\mathrm{i}x}\right) \mathrm{e}^{-\mathrm{i}t} + c.c. \tag{20.56}$$

Here the amplitudes A and B of the incident and reflected waves are normalized by A_b.

At the next order $O(\epsilon)$, the solution for Φ_1 is the superposition of three time harmonics $\mathrm{e}^{\pm imt}$ with $m = 0, 1, 2$, i.e.

$$\Phi_1 = \Phi_1^{[0]} + \Phi_1^{[1]} + \Phi_1^{[2]}\ . \tag{20.57}$$

We quote the result for $\Phi_1^{[2]}$ which will be needed to determine the boundary layer flow in the next section,

$$\Phi_1^{[2]} = -\frac{3\mathrm{i}}{16} \frac{\cosh 2z}{\sinh^2 kh} \left(A^2 \mathrm{e}^{2\mathrm{i}x} + B^2 \mathrm{e}^{-2\mathrm{i}x}\right) \mathrm{e}^{-2\mathrm{i}t} + c.c. \tag{20.58}$$

For the first-harmonic component, slow variations are allowed in space for resonant growth and in time for narrow frequency-band, so that A and B depend on the slow coordinates $x_1 = \epsilon x, t_1 = \epsilon t$. Now Φ_0 is the homogeneous solution to a linear boundary value problem, and $\Phi_1^{[1]}$ can be shown to be governed by the same boundary value problem with inhomogeneous forcing and boundary conditions. Fredholm alternative imposes a condition of solvability for $\Phi_1^{[1]}$ and leads to the evolution equations for the amplitudes A and B:

$$\frac{\partial A}{\partial t_1} + \frac{1}{2}\left(1 + \frac{2kh}{\sinh 2kh}\right)\frac{\partial A}{\partial x_1} = \frac{\mathrm{i}D_1 B}{2\sinh 2kh}\,, \tag{20.59}$$

$$\frac{\partial B}{\partial t_1} - \frac{1}{2}\left(1 + \frac{2kh}{\sinh 2kh}\right)\frac{\partial B}{\partial x_1} = \frac{\mathrm{i}D_1^* A}{2\sinh 2kh}\,, \tag{20.60}$$

where D_1 and its complex conjugate D_1^* are the first harmonic amplitudes of the bar profile

$$b = \frac{1}{2}\sum_{m=0}^{\infty}\left(D_m \mathrm{e}^{2\mathrm{i}mx} + D_m^* \mathrm{e}^{-2\mathrm{i}mx}\right) \qquad m \in N\,. \tag{20.61}$$

Since the phase difference between the incident wave and the bars is yet unknown, we allow D_m to be complex. The two equations (20.59) and (20.60) are formally the same as those obtained by Mei [13] for rigid bars. However the first harmonic amplitude D_1 is now unknown a priori and is a part of the solution.

20.6.2 The Boundary Layer

Inside the boundary layer, the flow is rotational and the dimensionless vorticity equation for the stream function ψ is, in the non-orthogonal coordinates (x, η),

$$\nabla^2\psi_t - \epsilon\frac{\partial(\psi, \nabla^2\psi)}{\partial(x, \eta)} = \frac{1}{2}\nabla^2\nabla^2\psi\,, \tag{20.62}$$

where

$$\nabla^2 = \left(k\delta\frac{\partial}{\partial x} - \epsilon\frac{\partial h}{\partial x}\frac{\partial}{\partial \eta}\right)^2 + \frac{\partial^2}{\partial \eta^2}\,. \tag{20.63}$$

From either observations in the laboratory or order estimates based on bedload transport formulas, the thickness of moving sand is no more than a few grain diameters, and much less than the boundary layer thickness (cf. (20.34)). Hence the no-slip boundary condition can be approximately applied at the bed surface $\eta = 0$. Specifically, we can deduce that [33],

$$\frac{\partial\psi}{\partial x} = O(\epsilon^4) \qquad \text{on} \quad \eta = 0\,, \tag{20.64}$$

$$\frac{\partial\psi^{[1]}}{\partial\eta} = O(\epsilon)\,, \quad \frac{\partial\psi^{[0]}}{\partial\eta} = \frac{\partial\psi^{[2]}}{\partial\eta} = O(\epsilon^2)\,, \qquad \text{on} \quad \eta = 0\,, \tag{20.65}$$

where the superscripts in brackets indicate the time-harmonics.

As boundary conditions at the upper edge of the boundary layer, ψ must join smoothly with the stream function $\Psi(x, z, t)$ of the inviscid flow. Now the inviscid stream function can be found at various orders as the harmonic conjugate of the corresponding velocity potential Φ by the Cauchy-Riemann condition. Thus, at the leading order the conjugate of (20.56) is

$$\Psi_0 = \frac{1}{2}\sinh z\left(A\mathrm{e}^{\mathrm{i}x} + B\mathrm{e}^{-\mathrm{i}x}\right)\mathrm{e}^{-\mathrm{i}t} + c.c. \tag{20.66}$$

At the next order the conjugate of the second-harmonic part (20.58) is

$$\Psi_1^{[2]} = \frac{3}{16}\frac{\sinh 2z}{\sinh^2 kh}\left(A^2 e^{2ix} - B^2 e^{-2ix}\right)e^{-2it} + c.c. \tag{20.67}$$

The limiting form of these two expressions at $z \to 0$ must be matched to the boundary layer solution as $\eta \to \infty$.

To solve the boundary layer problem we assume a multi-scale expansion

$$\psi = \psi_0(z, x, x_1, t, t_1, \bar{t}\,) + \epsilon\psi_1(z, x, x_1, t, t_1, \bar{t}\,) + \cdots \tag{20.68}$$

and deduce from (20.62), (20.64) and (20.65) the perturbation problems at the first two orders. By a straightforward analysis, the solution at leading order $O(\epsilon^0)$ is just the Stokes solution in the new plane (x, η),

$$\psi_0 = \frac{1}{2}\left[\eta - \frac{1+\mathrm{i}}{2}\left(1 - \mathrm{e}^{-(1-\mathrm{i})\eta}\right)\right]\left(A\mathrm{e}^{\mathrm{i}x} + B\mathrm{e}^{-\mathrm{i}x}\right)\mathrm{e}^{-\mathrm{i}t} + c.c. \tag{20.69}$$

At the second order $O(\epsilon)$, ψ_1 is the sum of the zeroth, first and second harmonics in time. The solutions, when expressed in the (x, η) coordinates, are formally the same as the classical results of Longuet-Higgins [8]. Only the zeroth and second harmonics are needed for use in (20.54),

$$\begin{aligned}\psi_1^{[0]} = &\left\{\left(\frac{1}{4}\eta + \frac{3-5\mathrm{i}}{8}\right)\mathrm{e}^{-(1+\mathrm{i})\eta} - \frac{1+\mathrm{i}}{8}\left(\mathrm{e}^{-(1-\mathrm{i})\eta} + \frac{1}{2}\mathrm{e}^{-2\eta}\right) + \frac{3-3\mathrm{i}}{8}\eta\right.\\ &\left.-\frac{3-13\mathrm{i}}{8}\right\}\left(|A|^2 - |B|^2 + AB^*\mathrm{e}^{2\mathrm{i}x} - A^*B\mathrm{e}^{-2\mathrm{i}x}\right) + c.c.\end{aligned} \tag{20.70}$$

$$\begin{aligned}\psi_1^{[2]} = &\left\{\frac{1+\mathrm{i}}{2\sqrt{2}}\left(\frac{3}{8\sinh^2(kh)} + \frac{1}{4}\right)\left(\mathrm{e}^{-\sqrt{2}(1-\mathrm{i})\eta} - 1\right) + \frac{1}{4}\eta e^{-(1-\mathrm{i})\eta}\right.\\ &\left.+\frac{3}{8\sinh^2(kh)}\eta\right\}\left(A^2\mathrm{e}^{2\mathrm{i}x} - B^2\mathrm{e}^{-2\mathrm{i}x}\right)\mathrm{e}^{-2\mathrm{i}t} + c.c.\end{aligned} \tag{20.71}$$

The terms $\psi_1^{[0]}$ and $\psi_1^{[2]}$ are respectively the wave-induced steady streaming and the second harmonic field in the boundary layer. Note that the bottom shear stresses associated with these two components, $\psi_{1,\eta\eta}^{[0]}$ and $\psi_{1,\eta\eta}^{[2]}$ on $\eta = 0$, do not depend explicitly on the bar height b.

In the numerical examples to be discussed, we limit to case where the first-order wave is monochromatic so that A and B do not depend on t_1. Equations

(20.59) and (20.60) are then ordinary differential equations in x_1, with $\bar{t}$ as a parameter through D_1. At each new time step in $\bar{t}$, we use the known values of A, B to compute ψ_0, $\psi_1^{[0]}$ and $\psi_1^{[2]}$. The diffusivity (20.53) and forcing term (20.54) can then be computed. The bar height $\bar{b}(x, \bar{t})$ is then solved for one time step in $\bar{t}$, subject to the requirement that $\bar{b}$ be periodic over the wave period with zero mean. By Fourier decomposition we get the first spatial harmonic D_1, which is then used to solve for the new wave amplitudes from (20.59) and (20.60) as functions of the long spatial scale x_1. The numerical procedure is then repeated for the next time step for $\bar{b}$, etc.

20.7 Properties of Bar Evolution Equation

Much insight can be gained before the numerical solution of the bar evolution equation. The local evolution of the sand bar at a fixed x_1 is controled by the local amplitudes of the incident and reflected waves $A(x_1)$ and $B(x_1)$, which define the local reflection coefficient $R(x_1) = B/A$ due to the bars themselves. In general the forcing term $(-\partial q_T/\partial x)$ and the diffusivity in the diffusion equation (20.52) are both periodic on the short scale x with the period π, i.e. one half of the surface wavelength. Thus, the wave-induced bottom stress tends to build up sand bars with spacings equal to half of the water wavelength. On the other hand by pulling particles down the slope, gravity limits this buildup through the diffusion term with the coefficient β. The two counter-acting mechanisms are both influenced by the local bed stress via $Q_{s0}/\widehat{\psi_{0,\eta\eta}}$. In the limit of a purely progressive wave ($R = 0$), $\widehat{\psi_{0,\eta\eta}}$ and Q_{s0} reduce to constants. Consequently, the diffusivity D_ν is uniform in x and the forcing $(-\partial q_T/\partial x)$ vanishes everywhere. Without reflection, sand bars cannot be formed from a flat bed.

Omitting the details including the empirical estimates of the ripple height and the wave friction factors, we show in Fig. 20.8 the forcing $(-\partial \bar{q}_T/\partial x)$ and diffusivity D_ν as functions of x within half of the surface wavelength between two adjacent minima (or nodes) of the envelope.

As shown in Fig. 20.8(a), the forcing term $(-\partial q_T/\partial x)$ is non-zero except for purely progressive waves with $R = 0$, therefore bars can be generated as long as there is some reflection, whether or not the mean current in the boundary layer is cellular. The hydrodynamic threshold $R = 0.414$ predicted by Carter et al. [10] for a rigid bed no longer applies to a sandy bed; this is consistent with the laboratory finding of O'Hare & Davies [20]. For finite reflection coefficient R the wave envelope has spatially periodic nodes (minima) and antinodes (maxima). Under a node, the horizontal orbital velocity is the greatest near the bed; the forcing is positive for all finite $|R|$, hence causes deposition of sand and forms bar crests. Under an antinode, the horizontal orbital velocity is the smallest; the forcing is negative, hence causes erosion and forms bar troughs.

From Fig. 20.8(b), $D_\nu(x)$ is symmetrical in x with respect to a envelope antinode for any finite $|R|$. The maximum of D_ν occurs at a node and the minimum at an antinode. This is attributable to the fact that the tangential

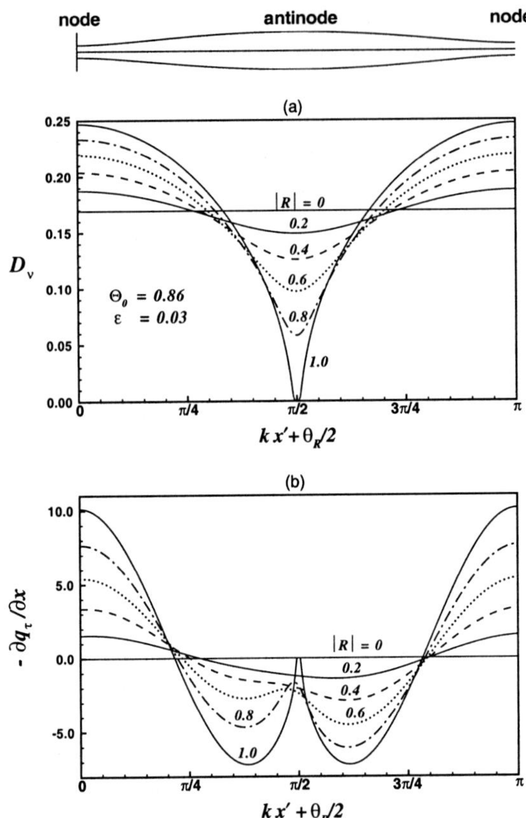

Fig. 20.8. Diffusivity and forcing term of the diffusion equation, within half a local wavelength

velocity above, and the shear stress on, the bed is the largest under a wave node (near a bar crest) and smallest under an antinode (near a bar trough).

From these two figures we conclude that gravity counteracts the forcing due to the bed shear and tends to limit the growth of bars. In principle, an equilibrium state [33] can be achieved when the total transport rate becomes uniform within a bar wavelength.

20.8 Numerical Simulation of a Laboratory Experiment

Computations have been carried out by solving the coupled equations for the wave amplitudes, the stream function in the boundary layer, and the sand bar height. In one class of problems the bed surface is initially flat; the reflection at the far end of the transmission side is specified. In another class, there are a finite number of sand bars in the range $0 < x_1 < L$. Incident waves are assumed to arrive from $x = -\infty$ and transmitted to $x_1 \sim \infty$ without further reflection. For details of these solutions, references can be made to [33] and [41].

As a check of the present theory, Yu & Mei simulated the experiments of Herbich et al. [7] from a fairly large flume (length= 20.6 m, width=depth= 61 cm). Steady monochromatic waves are generated at one end and reflected by a steep sea wall at the other. A layer of sand, initial flat and 12.7 cm thick, was placed on the bottom of the tank for a distance of 11.28 m in front of the sea wall. The mean diameter of the sand on the bed was $d = 0.4826$ mm. Half-wavelength sand bars, with ripples superimposed on them, were found to form along the bed. Only the spatially averaged depth of scour, measured from the mean bed position to each bar trough, was reported in [7] as a function of time.

Numerical simulations of the bar evolution under waves have been performed for three tests reported in [7]. For each test, β is so adjusted that the predicted bar heights at late stages (i.e. close to steady state) agree with the data. Figure 20.9 shows the comparison of the averaged depth of the bar troughs throughout the entire course of the evolution, for Test 1. The data include results for three different wall inclinations 45°, 67.5° and 90°, all of which should give complete reflection. That they fall nearly onto a single curve suggests that the phase of the end-wall reflection θ_{RL} at $x_1 = L$ is not important to the bar height. With β close to the value used by Fredsøe for river bars, the agreement between the predictions and the data is fairly good during the entire course of transient evolution.

In other numerical experiments, Yu & Mei [33] have examined the formation of sandbars on a initially flat horizontal bed due to the reflection of monochromatic incident waves. The rate of growth and the ultimate bar size and form depend on the incident wave characteristics, the reflection coefficient and sand size. Naturally, large reflection leads to higher bars spaced at exactly one half of the wavelength. There is no minimum threshold on R for bars to form.

20.9 Further Remarks on Sand Bars

20.9.1 On Scaling in Laboratory Experiments

There are several other laboratory experiments in relatively small flumes with low reflection, aiming at greater details of bar formation. One outstanding feature of these simulations is the exagerated prominence of ripples. O'Hare and Davies [20,21] used a flume 10 m long and 0.3 m wide and 0.45 m deep. The water depth was about 15 cm above the erodible bed. For sand with diameter $d = 0.08$ mm the observed ripples were typically 2 cm in height (crest to trough), while the bars were only $3 \sim 4$ cm. For ballotini grains with $d = 0.11$ mm, the observed ripple height was 0.5 cm and bar height $1 \sim 2$ cm. In Rey et al. [22], the flume was only 4.7 m long and 0.39 m wide, and the sediment size was $d = 0.08$ mm. The water depth was 4.75 cm above the sand layer. The observed ripple height was 0.23 cm, and bar height was $0.5 \sim 0.7$ cm. Thus, in both experiments the ripples were quite prominent compared with the sand bars. In contrast, acoustic

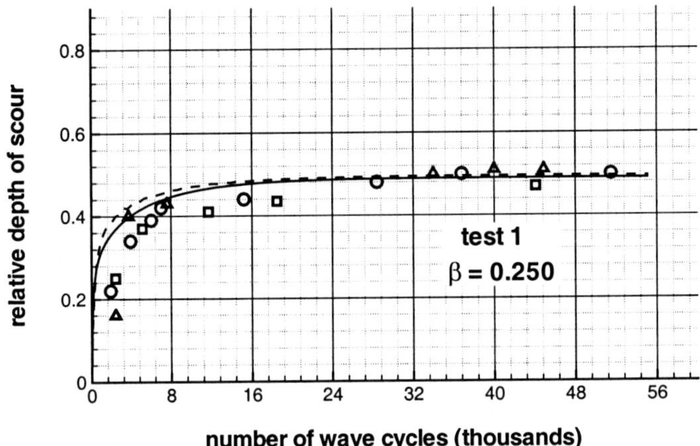

Fig. 20.9. The averaged depth of bar trough (normalized by the surface wave height) as a function of time. (*Solid curve*): the predicted height of bar crest above the mean bed position. (*Dashed curve*): the predicted depth of bar trough below the mean bed position

sounding records by Dolan [4] of natural bars in Chesapeake Bay do not show such prominence of ripples.

We give below the reasons [41] that in a laboratory experiment dynamical similarity can easily be achieved for ripples, but not for bars, especially those generated by the bedload. The implication is that accurate laboratory simulation of sand bars, hence definitive comparison between theory and measurement, requires that experiments be performed in large flumes.

The bar evolution equation (20.52) depends on three parameters: a, Θ_0 and ϵ. To simulate nature in the laboratory, it is necessary that

$$\frac{(a)_m}{(a)_p} = \frac{(\Theta_0)_m}{(\Theta_0)_p} = \frac{(\epsilon)_m}{(\epsilon)_p} = 1\,, \tag{20.1}$$

where the subscripts m and p stand for model and prototype, respectively. From the similarity of wave steepness ϵ, we have $(A_b)_m/(A_b)_p = k_p/k_m$. It then follows that

$$\frac{(a)_m}{(a)_p} = \left(\frac{k_m}{k_p}\right)^{-3/2} \left[\frac{(\tanh kh)_m}{(\tanh kh)_p}\right]^{1/2} \left(\frac{[(s-1)gd^3]_m}{[(s-1)gd^3]_p}\right)^{-1/2} . \tag{20.2}$$

The dispersion relation has been used. The factor $\tanh(kh)$ can be made the same for both the prototype and the model. Let natural sand and water be used in the laboratory experiment. From (20.2), we must require

$$(kd)_m \simeq (kd)_p \tag{20.3}$$

in order to have the same a. For 1–to–50 ratio of length scales, this would mean $d_m/d_p \sim 1/50$, which is difficult to achieve with sand.

Moreover,

$$\frac{(\Theta_0)_m}{(\Theta_0)_p} = \left(\frac{k_m}{k_p}\right)^{-1/4} \left[\frac{(\tanh kh)_m}{(\tanh kh)_p}\right]^{3/4} \left(\frac{\nu_m}{\nu_p}\right)^{1/2} \left(\frac{[(s-1)gd]_m}{[(s-1)gd]_p}\right)^{-1} . \tag{20.4}$$

In order to have the same Θ_0, we must have

$$\frac{\nu_m}{\nu_p} = \left(\frac{k_m}{k_p}\right)^{1/2} \left(\frac{[(s-1)gd]_m}{[(s-1)gd]_p}\right)^{2} . \tag{20.5}$$

This implies that $\nu_m/\nu_p = (k_m/k_p)^{1/2} \sim 7$, if natural sands are used. Thus the flow in a laboratory experiment needs to be more turbulent than that in the field. This is certainly unlikely. Thus sand bars are not easily simulated in a water tank with ordinary sand.

Ripple formation is quite different. Based on a bedload model, the linear instability analyses of Blondeaux [37] and Mei & Yu [39] have revealed that the parameters controlling the dynamics of ripples are the particle Froude number,

$$F_d = \frac{A_b \omega}{\sqrt{(s-1)gd}} , \tag{20.6}$$

and the ratio of local wave orbital radius to ripple wavelength $A_b k_r$, where k_r is the ripple wavenumber. In particular the most unstable ripple wavenumber is determined by an eigenvalue condition which is a relation between $A_b k$ and F_d. To simulate the most unstable ripple wavelength it suffices to require only $(F_d)_m = (F_d)_p$, which is not a severe constraint by using natural sand and water in the laboratory.

20.9.2 Future Challenges on Sand-bar Theories

Since the crucial term representing gravity-induced diffusion depends on the coefficient β of which we know only the order of magnitude, it would be highly desirable to perform experiments to measure β in oscillatory flows over a sloping sandy surface. Such experiments can be conducted in a U-tube with an inclined section and oscillating pistons at two ends.

On a sloping beach, the waves must intensify while propagating towards the shore. A greater fraction of sand is expected to be resuspended. Rational treatment of both bedload and suspended load is therefore necessary. In addition, small scale ripples, whose presence is only empirically accounted for here through the eddy viscosity, deserves theoretical studies in their own right. Expected to grow much faster than the bars, their length and amplitude, as well as their migration, must depend on their position on the bars. Indeed their migration may contribute to the bed load transport on the bar surface.

For sufficiently steep waves, nonlinear dynamics of waves should introduce unsteadiness of much longer time and spatial scales. In particular the infragravity waves, generated through nonlinearity by narrow-banded short waves, should have profound influence on, and be altered by, the long-time evolution of the sandy beds. The Bragg resonance mechanism and the 1–to–2 wavelength ratio will likely appear in only one of several stages of bar evolution, as suggested by the experiments of Dulou et al. [32]. In addition, it is necessary to pursue the effects of complex waves such as randomness and diffraction by large scale bathymetric variations.

Finally the mechanics of sand bars depends of course on our understanding of the sediment dynamics in waves. Thus far practical theories must all rely on empirical formulas relating the bed shear to the sediment discharge. Heuristically, these relations are largely based on real or perceived similarity with steady unidirectional flows. It is well-known in soil mechanics that persistent oscillations due to earthquakes [42] or ocean waves [43,44] can cause monotonic increase of the mean fluid pressure in the pores, hence reduce the contact between grains and liquefy the wet soil. Understanding this highly nonlinear process requires the collaboration between fluid and soil dynamicists. Also, as waves pass over a saturated sand, hydrodynamic pressures of alternating signs exert on the mud-line. For sufficiently steep waves, the passage of wave troughs can conceivably induce large enough instantaneous pore pressure gradient in the upward direction and cause periodic fluidization. Thus the transient vertical pressure gradient can likely be as important as the horizontal bed shear stress in resuspending sand particles. Subsequent transport of suspensions demands a transient analysis of two-phase flows involving turbulence. While these would involve more complex theories, perhaps similar to gas-induced fluidization extensively studied in chemical engineering, they may offer some chance for bringing the dynamics of sediments in waves to a less empirical level.

Acknowledgements

We gratefully acknowledge the recent grants by U.S. National Science Foundation (Grant CTS 9634120 directed by Drs. Roger Arndt, John Foss and Chuan F. Chen), and Office of Naval Research (Grant N00014-92-J-1754, directed by Dr.Thomas Swean), as well as earlier grants from these two agencies since 1985.

References

1. J.H. Saylor, E.B. Hands: *Proc. 12th Conf. Coastal Eng.* **22**, 839 (1970)
2. P. Nielsen: Technical University of Denmark, Institute of Hydrodynamcis and Hydraulic Engineering. Series paper No. 20, 160pp (1979)
3. A.D. Short: J. Geol. **83**, 209–211 (1975)
4. J.T. Dolan: Wave mechanisms for the formation of multiple longshore bars with emphasis on the Chesapeake Bay. Master Thesis, Dept. of CE, Univ. of Florida (1983)
5. J. Lau, B. Travis: J. Geophys. Res. **78**, 4489 (1973)
6. G.B. Keulegan: Beach Erosion Board Tech. Memo. Rep. 3, US Army Corps Engr. (1948). Reprinted in: *Spits and Bars*, ed. by M. L. Schwarz (Dowden, Hutchinson & Ross, 1972)

7. J.B. Herbich, H.D. Murphy, B. Van Weele: *Coastal Engineering, Santa Barbara Specialty Conference, ASCE*, 705 (1965)
8. M.S. Longuet-Higgins: Phil. Trans. Roy. Soc. A **245**, 535 (1953)
9. J.N. Hunt, B. Johns: Tellus **15**, 341 (1963)
10. T.G. Carter, P.L-F. Liu, C.C. Mei: J. Waterways, Harbors, Coastal Eng. Div. ASCE **99**, 165 (1973)
11. A.D. Heathershaw: Nature **296**, 343 (1982)
12. A.G. Davies, A.D. Heathershaw: J. Fluid Mech. **144**, 419 (1984)
13. C.C. Mei: J. Fluid Mech. **152**, 315 (1985)
14. T. Hara, C.C. Mei: J. Fluid Mech. **178**, 221 (1987)
15. C.C. Mei, T. Hara, M. Naciri: J. Fluid Mech. **186**, 147 (1988)
16. J.T. Kirby: J. Fluid Mech. **162**, 171 (1986)
17. P.L.-F. Liu: J. Fluid Mech. **179**, 371 (1987)
18. M. Belzons, V. Rey: E. Guazzelli, Europhysics Letters **16 (2)**, 189 (1991)
19. S.B. Yoon, P.L.-F. Liu: J. Fluid Mech. **180**, 451 (1987)
20. T.J. O'Hare, A.G. Davies: J. Coastal Res. **6.3**, 531 (1990)
21. T.J. O'Hare, A.G. Davies: Continental Shelf Research **13**, 1149 (1993)
22. V. Rey, A.G. Davies, M. Belzons: J. of Coastal Research **11. 4**, 1180 (1995)
23. B. Boczar-Karakiewicz, B. Paplinska, J. Winiecki: Polska Akademia Nauk, Inst. Budownictwa Wodnego Gdansk, Rozpr. Hydrotech. **43**, 111 (1981)
24. C.C. Mei, Ü. Ünlüata: *Waves on Beaches*, ed. by R.E. Meyer (Academic, New York 1972) p. 181
25. J. Lau, A. Barcilon: J. Phys. Oceanogr. **2**, 405 (1972)
26. B. Boczar-Karakiewicz, J.L. Bona, D.L. Cohen: *Dynamical Problems in Continuum Physics*, Vol. **4**, ed. by J.L. Bona (IMA Volume in Mathematics and its Application, Springer, 1987) p. 131
27. J.M. Restrepo, J.L. Bona: Nonlinearity **8**, 781 (1995)
28. J.A. Baillard, J.W. DeVries, J.T. Kirby, R.T. Guza: *Proc. 22nd Int. Conference on Coastal Engineering*, Delft, 757 (1990)
29. J.A. Baillard, J.W. DeVries, J.T. Kirby: J. Waterway Port Coastal Ocean Engng. **118**, 62 (1992)
30. J.T. Kirby, J.P. Anton: *Proc. 22nd Int. Conference on Coastal Engineering.* Delft, 757 (1990)
31. J. Yu, C.C. Mei: J. Fluid Mech. **404**, 251 (2000)
32. C. Dulou, M. Belzons, V. Rey: J. Geophys. Res. **105**, 19745 (2000)
33. J. Yu, C.C. Mei: J. Fluid Mech. **416**, 315 (2000)
34. J.F.A. Sleath: J. Waterway Port Coastal Ocean Eng. Div. ASCE **104(WW3)**, 291–307 (1978)
35. J. Fredsøe: J. Fluid Mech. **64**, 1 (1974)
36. J.F.A. Sleath: *Seabed Mechanics* (Wiley, New York 1984) p. 260
37. P. Blondeaux: J. Fluid Mech. **218**, 1 (1990)
38. G. Vittori, P. Blondeaux: J. Fluid Mech. **248**, 19 (1990)
39. C.C. Mei, J. Yu: Phys. Fluids **9**, 1606 (1997)
40. N.L. Komarova, A.C. Newell: J. Fluid Mech. **415**, 285 (2000)
41. J. Yu: Generation of sand ripples and sand bars by surface waves. Ph.D Thesis, Department of Civil and Environmental Engineering, Massachusetts Institute of Technology, Cambridge, MA (1999)
42. H.B. Seed: 'Liquefaction Problems in Geotechnical Engineering'. In: *Amer. Soc. Civ. Engrs. National Convention Proc., Philadelphia, Sep 27-Oct 1,* (1976)
43. K. Ishihara, A. Yamazaki: Soil Found **24**(3), 85 (1984)
44. M. Foda, S.Y. Tsang: J. Geophys. Res. **99**, 463 (1994)

21 Debris Flows and Related Phenomena

C. Ancey

Cemagref, unité Erosion Torrentielle, Neige et Avalanches, Domaine Universitaire, 38402 Saint-Martin-d'Hères Cedex, France

21.1 Introduction

Torrential floods are a major natural hazard, claiming thousands of lives and millions of dollars in lost property each year in almost all mountain areas on the Earth. After a catastrophic eruption of Mount St. Helen in the USA in May 1980, water from melting snow, torrential rains from the eruption cloud, and water displaced from Spirit Lake mixed with deposited ash and debris to produce very large debris flows and cause extensive damage and loss of life [1]. During the 1985 eruption of Nevado del Ruiz in Colombia, more than 20,000 people perished when a large debris flow triggered by the rapid melting of snow and ice at the volcano summit, swept through the town of Armero [2]. In 1991, the eruption of Pinatubo volcano in the Philippines disperses more than 5 cubic kilometres of volcanic ash into surrounding valleys. Much of that sediment has subsequently been mobilised as debris flows by typhoon rains and has devastated more than 300 square kilometres of agricultural land. Even, in European countries, recent events that torrential floods may have very destructive effects (Sarno and Quindici in southern Italy in May 1998, where approximately 200 people were killed).

The catastrophic character of these floods in mountainous watersheds is a consequence of significant transport of materials associated with water flows. Two limiting flow regimes can be distinguished. *Bed load* and *suspension* refer to dilute transport of sediments within water. This means that water is the main agent in the flow dynamics and that the particle concentration does not exceed a few percent. Such flows are typically two-phase flows. In contrast, *debris flows* are mass movements of concentrated slurries of water, fine solids, rocks and boulders. As a first approximation, debris flows can be treated as one-phase flows and their flow properties can be studied using classical rheological methods. The study of debris flows is a very exciting albeit immature science, made up of disparate elements borrowed from geomorphology, geology, hydrology, soil mechanics, and fluid mechanics. The purpose of this chapter is to provide an introduction to physical aspects of debris flows, with specific attention directed to their rheological features. Despite attempts to provide a coherent view on the topic, coverage is incomplete and the reader is referred to a series of papers and books. Three books are particularly commendable [3–5]. Some review papers provide interesting overviews, introducing the newcomers to the field to the main concepts [6–8]. The background material in rheology can be found in Chaps. 2 and 3.

21.2 A Typology of Torrential Flows

21.2.1 The Watershed as a Complex Physical System

The notion of torrent refers to a steep stream, typically in a mountainous context. According to a few authors, a stream can be referred to as a *torrent* as soon as its mean slope exceeds 6%. For bed slopes ranging from 1% to 6%, it is called a *torrential river.* For bed slopes lower than 1%, it can be merely called a *river.* In addition to the slope, the sediment supply is generally considered as another key ingredient in torrential watersheds. Depending on the nature of the soil and relief, slopes can provide a large quantity of poorly sorted solid materials to torrents. Supplied materials have sizes ranging typically from 1 μm to 10 m. The situation is very different from the one encountered for streams on a plain, where bed material is much finer and sorted (typically 1 μm to 10 cm) since it generally results from transport that occurred during previous floods. Finally, one of the chief ingredients of torrential watersheds is water. Due to the small dimensions of torrential watersheds (typically from 0.1 km^2 to 100 km^2) and the steep slopes, floods are sudden, short, and violent. The flood regime differs significantly from plain floods, which are characterised by slower kinetics and smoother variations with time. Figure 21.1 depicts a typical watershed. The upper part is generally degraded and submitted to erosion to a more or less large extent. It supplies water and sediment to the floods. Below this basin, the torrent enters a gorge, sometimes with very abrupt flanks depending on the nature of the soil. Then the torrent discharges onto the alluvial fan. The slope transition between the gorge and the alluvial provides interesting information on bed equilibrium. Generally, a watershed with an abundant supply of sediment and intense bed load transport in the past is characterised by a smooth transition from channel to fan.

For plain rivers, sediment transport results from the action of water: water entrains materials either by pushing them along the bed (bed load transport) or by keeping them in suspension as a result of turbulence (suspension). In a torrential context, as soon as the bed inclination is sufficiently high, gravity has a more pronounced role on sediment transport. Therefore, on the one hand, bed load transport is more intense and on the other hand, a new mode of transport arises: debris flow. We can define them as follows:

- *Debris flows* are highly concentrated mixtures of sediments and water, flowing as a single-phase system. Debris flows look like mudslides and landslides except that their velocity and the distances they travel are much larger. It is worth noticing that in the literature there are many terms used to refer to slides and/or debris flows, which is a source of confusion.
- *Bed load transport* involves transportation of sediment by water. Coarse particles (sand, gravel, and boulders) roll and slide in a thin layer near the bed (called the bed layer). Generally fine particles (silts and clays) are brought into suspension as a result of water turbulence. The system is typically made up of two distinct phases: liquid phase (i.e. water) and dispersed (solid) phase.

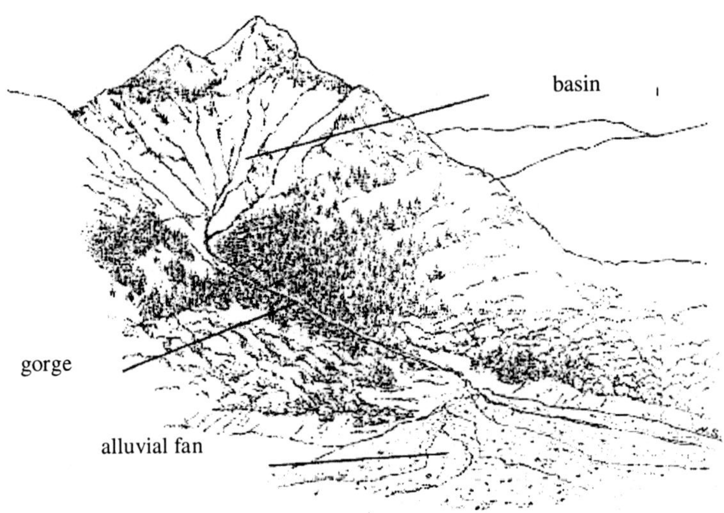

Fig. 21.1. A typical watershed (courtesy of Nicole Sardat)

21.2.2 Types of Transport

In the laboratory, it is possible to simulate torrential phenomena using an inclined channel with a mobile bed made up of sand and gravel. Figures 21.2 and 21.3 show two very different situations that can be observed when the channel slope is increased by only a few percent. Figure 21.2 corresponds to a slope of 17%. At high discharges, fine particles are in suspension while the coarsest particles are pushed down to the bed. In this photograph the largest particles are stationary and significantly affect water flow. The two phases (solid and liquid) are well separated and water flows much faster than solid particles. When the inclination exceeds a critical value (approximately 20%), significant sudden changes can be observed: a transition from a two-phase flow to a single-phase flow occurs very quickly. The mixture takes on the appearance of a "viscous" homogeneous fluid flowing down the bed. Figure 21.3 (slope of 27%) illustrates such a transition and the resulting mass movement. Most laboratory experiments conducted with water flows on erodible beds have shown that the bed inclination θ is a key factor in sediment transport dynamics [9–13]. On the whole it has been observed that:

- $\theta < 20\%$: at sufficiently high water discharges, water flow induces intense bed load transport near the bed. As a first approximation, the water and solid discharges (respectively q_w and q_s) are linearly linked: $q_s \approx 8.2\,\theta^2 q_w$ (this relationship is an overly simplified expression of discharge obtained by Smart and Jaeggi [11] or Rickenmann [10]). Three layers can be distinguished: the bed made up of stationary particles (that can be eroded), the (active) bed layer in which sediment of all sizes is set in motion (rolling and sliding), and the water layer, where fine particles are in suspension or in saltation. In two-phase flows of this type the solid concentration (ratio of solid volume to total volume) does not exceed 30%.

- $\theta > 20\%$: at sufficiently high water discharges, bed load transport is unstable. It changes into a dense single-phase flow. The solid concentration is very high, ranging from 50% to 90% depending on the size distribution of particles. Such flows simulated in the laboratory correspond to debris flows in the field.

Fig. 21.2. Small-scale simulation of bed load transport in the laboratory. The solid and liquid phases are distinct (water was coloured with fluoresceine). The typical flow depth in these experiments was 1 cm

Fig. 21.3. Small-scale simulation of a debris flow in the laboratory. The solid and liquid phases are mixed

In the laboratory, the transition from bed load transport to debris flow is reflected by a discontinuity in the solid concentration. It is suspected that such a discontinuity still exists in the field, at least in the Alps, but the underlying mechanisms are unknown. It is worth noticing that in the field, debris flows can also form from landslides [7]. In this case, the transformation mechanisms are similar to soil liquefaction processes (rapid creep of saturated soils). In the following, we will tackle the problem of debris flows, which are intrinsic to mountain torrents and steep slopes. Other chapters in this book deal with bed load transport.

21.3 Initiation, Motion, Effects of Debris Flow

21.3.1 Initiation

The torrential activity of a watershed depends on many parameters. Debris flows are common in some areas and uncommon in others. In areas prone to debris flow formation, their frequency also varies. In some watersheds, several debris flows occur each year while for other torrents, they are rare. Conditions for initiation of most debris flows usually include:

- *Steep slopes.* In the Alps, slopes in excess of 70% are liable to surface erosion (sediment transport induced by runoff) and landslides (soil failure leading to large masses of saturated materials coming loose).
- *Abundant supply of unconsolidated materials.* Debris flows originate either from the simultaneous contributions of many material sources or from a single source (landslides):
  - Slow and continuous erosive processes on slopes in the drainage basin form deposits of materials in the torrent bed. Such deposits can be subsequently mobilised during intense floods and then transform into debris flows. In this case, debris flow originate as a slurry, primarily of water and fine particles, which erodes its channel and grows in size. Presumably instabilities in the bed load transport (such as those observed in the laboratory) arise and enable debris flow initiation. Usually the volume produced every year by erosion over the whole drainage basin is small and thus the amount of sediments that can be involved by a single debris flow is limited ($< 10^5\,\mathrm{m}^3$). In the field, the absence of failure surfaces and the presence of rills in the drainage basin are generally evidence that a debris flow has picked up coarse materials from the bed.
  - Old ill-consolidated deposits (moraines, massive rockfall deposit, etc.) can mobilise into landslides to form debris flows. In this case, the volume of materials involved can be very large ($> 10^5\,\mathrm{m}^3$) depending on the total volume made available by the source. Likewise, certain soils (e.g. gypsum) are very liable to landslides and can supply materials to debris flows. Presumably, initiation is due to a combination of several mechanisms: rapid creep deformation, increase in pore pressure, increase in load, erosion at the foot of the landsliding mass, etc. In the field, the

presence of a failure surface can clearly serve to identify the source of material.

- *Large source of moisture.* Most of debris flows occur during or after heavy and sustained rainfalls. In some cases, snowmelt can be sufficient to form debris flows. (There are many other ways in which water can be provided for the formation of debris flows: thawing soil, sudden drainage of lakes, dam break, etc., but these are much less frequent.) A high liquid water content seems to be a necessary condition for the soil to be saturated, which cause: intense surface runoff, and an increase in the pore–water pressure (presumably leading to Coulomb slope failure).
- *Sparse vegetation.* Vegetation plays a role by intercepting rainfall (limitation of runoff) and increasing soil cohesion (root anchorage). Vegetation reduces the initiation potential to a certain extent but does not completely inhibit formation of debris flows. Many observations have shown that debris flows also occur in forested areas.

21.3.2 Motion

On the whole, debris flows are typically characterised by three phases, that can change with time (see Fig. 21.4):

- At the leading edge, a granular front or snout contains the largest concentration of big rocks. Boulders seem to be pushed and rolled by the body of the debris flow. The front is usually higher than the rest of the flow. In some cases no front is observed because either it has been overtaken by the body (very frequent when the debris flow spreads onto the alluvial fan), or the materials are well sorted and no significant variation in the bulk composition can be detected.
- Behind the front, the body has the appearance of a more fluid flow of a rock and mud mixture. Usually, the debris flow body is not in a steady state but presents unsteady surges. It can transport blocks of any size. Many authors have reported that boulders of relatively small size seem to float at the free surface while blocks of a few meters in size move merely by being overturned by the debris flow. The morphological characteristics of the debris flow are diverse depending on debris characteristics (size distribution, concentration, mineralogy) and channel geometry (slope, shape, sinuosity, width). Debris flows velocity varies very widely but, on the whole, ranges from 1 m/s to 10 m/s [14]. The fastest debris flows are reported to move at more than 20 m/s [14]. Flowing debris can resemble wet concrete, dirty water, or granular material but whatever the debris characteristics and appearance, viscosity is much higher than for water. Most of the time, debris flows move in a completely laminar fashion, but they can also display minor turbulence (or be highly turbulent).
- In the tail, the solid concentration decreases significantly and the flow looks like a turbulent muddy water flow.

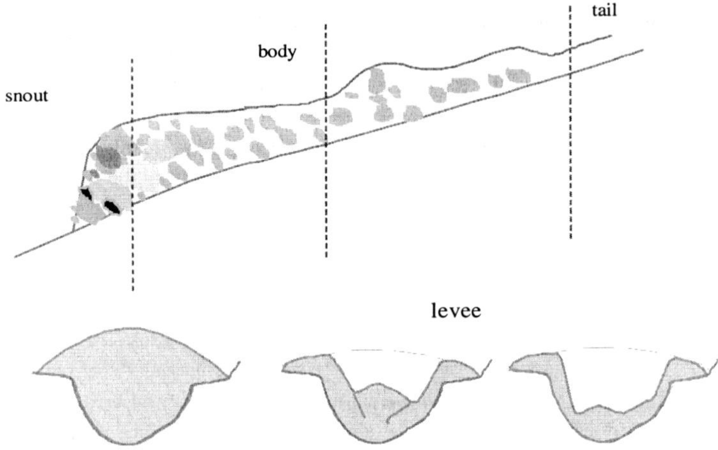

Fig. 21.4. Idealised representations of a debris flow (longitudinal profile and cross section). The different sections correspond to the dashed lines of the upper panel. Adapted from [15]

21.3.3 Deposition and Effects

The distance that a debris flow can travel depends a great deal on the mechanical characteristics of the debris as well as the total volume, channel geometry and bed inclination. For instance it is generally observed that a debris flow moving over a flat tilted plane thins by spreading laterally and stops suddenly, seemingly when the thickness reaches a critical value. In contrast, if the debris flow is channelized, it may travel quite a long distance over gentle slopes. In European alpine countries, debris flows (of sufficient volume) generally begin to decelerate when the slope ranges from 10% to 25%. For some torrents, (e.g. Illgraben in Switerland and Boscodon in France), debris flows can propagate over gentle slopes (of less than 5%). In volcanic soil areas, it has been also demonstrated that lahars (debris flows involving water–ash mixtures) can propagate over very slight slopes (less than 1%) [14].

For some debris flows, constant deposition occurs all along the channel and forms levees on the lateral boundaries of the torrent. Depending on the size distribution of the materials involved in the debris flow, a levee can have various shapes. In most cases, the cross section reveals a curved profile and, when the deposit is dry, it is characterised by strong cohesion. In other cases, the cross section has a straight free surface and even when it is dry, the deposit displays minor cohesion and looks like a sand or gravel heap. Formation of levees is not systematic. Many observers have noticed that, after a debris flow has passed through a channel, the channel bottom and sides have been swept clean of debris.

The alluvial fan is the preferential area for debris-flow deposition owing to the decrease in bed slope and widening of the channel. The slope decrease usually leads to the sudden stopping of the granular front and increase in the flow depth for the body. In many cases, debris flows overflow the channel banks and spread as broad lobes on the alluvial fan. As for levees, the morphological features of lobes vary widely. For instance, the longitudinal profile of a lobe margin can be curved (parabola-shaped), straight and tilted, or step-shaped. In the latter case, the deposit looks like an alluvial deposit. Although they move at low velocities on the alluvial fan, debris flows can impact or bury structures.

21.4 Debris Flow Classification

The diversity in the morphological features of debris flows provides evidence of different families with specific bulk behaviour. Several classifications have been proposed in the last few years. To date, there is no agreement on the chief characteristics on which classification should rely. Therefore, some classifications are based on the size distribution of materials involved, others only consider the mode of release, etc. Here we are mainly interested in the manner in which a debris flow propagates and therefore we suggest using a classification based on bulk mechanical behaviour. We shall therefore consider three families:

- *Muddy debris flow.* The transported material is usually characterised by a wide particle-size distribution. It is sufficiently rich in clay-like materials for the matrix to have a muddy consistency and lubricate contact between coarse particles. Most of the time, bulk behaviour is typically viscoplastic. That means that the material exhibits both plastic and viscous properties [3,16–20]. When the stress level is low, the material behaves as a solid body, but when the stress level exceeds of a critical value (yield stress), it flows as a fluid does. This yield stress confers specific properties to the material. For instance, when a given volume of material is released and spreads down a tilted flat plane, the flow depth decreases regularly. When the flow depth reaches a critical value (depending on the yield stress and the plane inclination), the driving shear stress is lower that the yield stress and the flow stops abruptly. In most cases, the yield stress ranges from 0.5 kPa to 15 kPa. Muddy debris flows can usually propagate over slopes greater than 5%. The limits of deposits are sharp and well delineated. Boulders and gravel are randomly distributed in a finer-grained cohesive matrix. Muddy debris flows are very frequent in the Alps.
- *Granular debris flow.* Although the size distribution is wide, the material is poor in fine (clay-like) particles. Bulk behaviour is expected to be frictional-collisional [23–26]: it is mainly governed by collisions and friction between coarse particles. Energy dissipation is usually much larger for granular debris flows than for muddy debris flows and thus, granular debris flows require steep slopes ($>$ 15%) to flow. Presumably, as for very large rockfalls, a granular debris flow involving a very large amount of materials may travel large distances over more gentle slopes. In the field, deposits are easily recognised by the irregular chaotic surface. Deposits are generally graded, with coarser debris forming mass deposits and finer debris transported downstream (due to drainage).

- *Lahar-like debris flow.* The particle-size distribution is narrow and the material contains only a small proportion of clay-like materials. This type of debris flow is typical of volcanic soil areas (soils made up of fine ash), but it can be observed on other terrain (e.g. gypsum, loess) [21]. Bulk behaviour is expected to be frictional/viscous: at low shear velocities, particles are in sustained frictional contact and bulk behaviour may be described using a Coulomb frictional equation. At high shear velocity, due to dilatancy and increased fluid inertia, contacts between coarse grains are lubricated by the interstitial fluid [27]. In the laboratory, such materials exhibit very surprising properties: at rest, they look like fine soil (silts) but once they have been stirred up, they liquefy suddenly and can flow nearly as Newtonian fluids. Contrary to muddy debris flows, the yield stress is low and therefore, lahars can move over gentle slopes of less than 1%. Deposits are very thin and flat and look like alluvial deposits.

21.5 Modelling Debris Flows

There are many similarities between debris flows and flowing avalanches. Both are rapid gravity-driven flows of dense materials down mountain slopes. Thus, approaches similar to those developed for modelling avalanches have been proposed (see Chap. 13).

21.5.1 Statistical Approach

A few authors have attempted to relate the runout distance (and other debris flow characteristics) to the watershed features. Extensive work performed by Swiss scientists and engineers has led to different equations [5,28,29]. For instance, using regression analysis, Zimmermann and co-workers found that the most significant variable in the runout distance was the surface of the watershed S (in km^2) [5]: $\alpha = 0.2\,S^{-0.26}$, where α is the angle between the line joining the top of the starting zone to the stopping point with respect to the horizontal. Such an equation differs from the ones inferred for avalanches. Indeed, in this latter case, it has been found that the angle α mainly depends on the angle β corresponding to a path characteristic (see Chap. 13). This might mean that, contrary to avalanches, the runout distance of debris flows is less influenced by channel geometry and probably depends a great deal on the sediment volume. Indeed, in the above equation, the watershed-surface dependence does reflect a debris-flow volume dependence since most of the time debris involved in debris flows result from erosion of the drainage basin, thus it is expected to depend on S. The correlation existing between the runout distance and the debris-flow volume has been further demonstrated by Rickenmann [28]. Using data from 82 events, Rickenmann inferred the following statistical equation: $L = 350\,V^{0.25}$, where L is the maximum distance (in m) that a debris flow of volume V (in m^3) can travel. Likewise, he found that for a muddy debris flow, the peak discharge Q_p (in m^3/s) can be estimated at: $Q_p = 0.0225\,V^{0.8}$, while Mizuyama found that, for a granular debris flow, it can be estimated at: $Q_p = 0.135\,V^{0.78}$. It should be noted that the peak discharge is much higher for granular debris flows than for muddy debris flows.

21.5.2 Deterministic Approach

The distinctions in the spatial scale and model complexity that have been put forward for avalanches, are still valid here. As for avalanches, Voellmy's model and depth-averaged mass and momentum equations have been proposed.

Empirical Model: The PCM Model Adapted to Debris Flow

Zimmermann has adapted the Perla–Cheng–McClung avalanche-dynamics model to compute characteristics of debris flow [5]. He assumed that a debris flow can be approximated by the motion of a solid block of mass M, subject to a frictional force including two contributions:

- a Coulombic frictional contribution (ground/debris flow): $F_C = \mu M g \cos\theta$,
- a dynamic drag: $F_D = Dv^2$,

where μ and D are two parameters, θ is the bed slope. The momentum equation in the downstream direction can then be expressed as follows:

$$\frac{1}{2}\frac{\mathrm{d}v}{\mathrm{d}x} = g(\sin\theta - \mu\cos\theta) - \frac{D}{M}v^2 \,. \tag{21.1}$$

Debris flow terrain is represented by a centreline profile stretching from the top of the starting zone to the end of the runout zone. The profile must be subdivided into several segments, which are sufficiently short for the slope to be considered constant. The length of segment i is L_i. At the end of the ith segment, the velocity (v_i^f) depends on the initial velocity (v_i^d) at the top of this segment:

$$v_i^f = \sqrt{a\frac{M}{D}(1 - \mathrm{e}^{\beta}) + (v_i^d)^2\mathrm{e}^{\beta}} \,, \tag{21.2}$$

where we have introduced $a = g(\sin\theta - \mu\cos\theta)$ and $\beta = -2L_i/(M/D)$. If the debris flow stops within segment i, then the runout distance (from the beginning of the ith segment) is given by:

$$x_s = \frac{1}{2}\frac{M}{D}\ln\left(1 - \frac{(v_i^d)^2}{aM/D}\right) \,. \tag{21.3}$$

Velocity at the bottom of a segment, v_i^f, is used to compute velocity, v_{i+1}^d, at the top of the next segment:

$$v_{i+1}^d = \cos(\theta_i - \theta_{i+1})v_i^f \,. \tag{21.4}$$

This computation is repeated downslope until the block stops. The values of the two parameters μ and D have been adjusted to 49 events that occurred in the Swiss Alps. Although this sample size may be considered too small to draw reliable correlations, it can provide helpful trends: μ and D are independent of the volume V and μ depends on the watershed surface. According to Zimmermann and co-workers [5], this dependence might suggest that the runoff over the drainage basin affects the solid concentration, and the frictional coefficient μ. They proposed two correlations:

- Lower value: $\mu = 0.18\, S^{-0.30}$.
- Upper value: $\mu = 0.13\, S^{-0.35}$.
- The mass-to-drag ratio M/D depends a great deal on size distribution. They found:
  - fine material (clays, fewer blocks): $20 \geq M/D \geq 60$ (average: 40),
  - fine-grained and coarse materials (clays, boulders): $80 \geq M/D \geq 180$ (average: 130),
  - granular materials (sand and gravel): $40 \geq M/D \geq 100$ (average: 70).
- No channelling effect was observed.
- No influence of the starting type (single or multiple sources) was detected regarding the friction coefficients.

Depth-averaged Models

The principles of depth-averaged models have been specified in Chaps. 13 and 22. The chief approximation is to consider the material involved in a debris flow as a homogeneous fluid. Thereby its behaviour can be described using a constitutive equation. Unlike snow, many experiments on debris have been conducted to gain insight into the rheology of these materials [3,20]. In a steady state, the shear stress may be written as follows:

$$\tau = f(\dot{\gamma}, \zeta)\,, \tag{21.5}$$

where τ denotes the shear stress, $\dot{\gamma}$ is the shear rate, and ζ refers to a group of mechanical parameters on which bulk behaviour depends (for instance, this group can include the solid concentration ϕ or other parameters pertaining to the microstructure). Depending on the kind of debris flows, several constitutive equations have been proposed to describe debris flows (cf. also Chaps. 3, 4, 7 and 22):

- For lahar-like debris flows, bulk behaviour may be described using a Newtonian constitutive equation as a first approximation: $\tau = \mu(\phi)\dot{\gamma}$. Viscosity is usually very high, with typical values close to 10^3 Pa s [21].
- For muddy debris flows, bulk behaviour is usually best described using a Bingham or Herschel–Bulkley model: $\tau = \tau_c(\phi) + K(\phi)\dot{\gamma}^n$, where τ_c is the yield stress, K a parameter, n a shear-thinning index ($n \leq 1$, $n = 1$ corresponding to the Bingham model). For a flow to occur ($\dot{\gamma} > 0$), the shear stress must exceed the yield stress τ_c. In the Alps, bulk yield stress values range from 0.5 kPa to 15 kPa and the ratio τ_c/K lies usually in the range 3–10 [22].

- For granular debris flows, bulk behaviour is expected to be frictional and/or collisional. To date there is no unanimity concerning the constitutive equation suitable to describe frictional–collisional flows. Different models have been proposed with very different stress generation mechanisms: collisional Bagnold models [23], collisional kinetic models [26], collisional–frictional constitutive equations [24], models based on pore-pressure effects [7], etc. Further developments on granular flows down steep slopes can be found in Chap. 4.

Once the constitutive equation has been determined, it is possible to compute some characteristics, notably the ones related to a steady state flow. Indeed, the stress distributions are known for steady uniform flows independently of the constitutive equation (cf. Chap. 3). For instance, the shear stress distribution is given by the following equation: $\tau = \varrho g(h - y)\sin\theta$, where ϱ is the bulk density and $(h - y)$ is the depth with respect to the free surface. Comparing this expression to (21.5) and after integration, we can deduce the velocity field. Further integration leads to the discharge equation. For instance, in the case of a Herschel–Bulkley fluid, we obtain:

$$y \le h_c \Rightarrow u(y) = \frac{1}{p}\sqrt[n]{\frac{\varrho g \sin\theta}{K}}\left[h_c^p - (h_c - y)^p\right] , \tag{21.6}$$

$$y \ge h_c \Rightarrow u(y) = \frac{1}{p}\sqrt[n]{\frac{\varrho g \sin\theta}{K}} h_c^p , \tag{21.7}$$

where we introduce: $p = 1/n + 1$ and $h_c = h - \tau_c/(\varrho g \sin\theta)$. Near the free surface, a non-sheared zone ("plug flow"), characterised by a constant velocity, is observed. An expression of this sort is meaningful provided that the flow depth exceeds a critical value: $h > \tau_c/(\varrho g \sin\theta)$. Thus the existence of a yield stress implies the existence of a critical flow depth, under which no steady uniform flow is possible and the existence of critical slope $\sin\theta_c = \tau_c/(\varrho g h)$. After integration, the discharge is found to be:

$$q = \frac{1}{p}\sqrt[n]{\frac{\varrho g \sin\theta}{K}} h_c^p \left(h - \frac{1}{p+1} h_c\right) . \tag{21.8}$$

The discharge is a strongly non-linear function of the flow depth. As $n \approx 0.3$, this means that small changes in the flow depth can cause large variations in the flow rate. All these computations can be extended to gradually varying flows (i.e. slightly non-uniform and unsteady). In the case of a Herschel–Bulkley fluid flowing down an inclined infinite plane, the resulting motion equation set is (see Chap. 22 for a more complete introduction):

$$\frac{\partial h}{\partial t} + \frac{\partial h\bar{u}}{\partial x} = 0 , \tag{21.9}$$

$$\frac{\partial h\bar{u}}{\partial t} + \frac{\partial h\bar{u}^2}{\partial x} = g\sin\theta - \frac{\tau_p}{\varrho} - gh\cos\theta\frac{\partial h}{\partial x} , \tag{21.10}$$

where τ_p is the bottom shear stress:

$$\tau_p = K \left(\frac{1+p}{p}\right)^n \frac{\bar{u}^n}{h_c^{n+1}\left[(1+p)h - h_c\right]^n} . \qquad (21.11)$$

Generally the motion equation set must be solved numerically since there is no analytical solution apart from the one shown for the steady-state regime. For instance, this can be done using finite-volume numerical models developed for solving hyperbolic differential equations [30,31]. They are now used in engineering problems when accurate results on a complex topography are needed. Approximate analytical or quasi-analytical solutions have also been proposed, notably by Hunt [32,33] and more recently by Huang and Garcia [35,34]. But, compared to experimental data (see Fig. 21.5), such approximations provide correct results at large times (when the flow fairly achieves a steady uniform regime) but fail to capture the flow features at any time. On explanation for partial agreement is that these models neglect the influence of normal stresses [36].

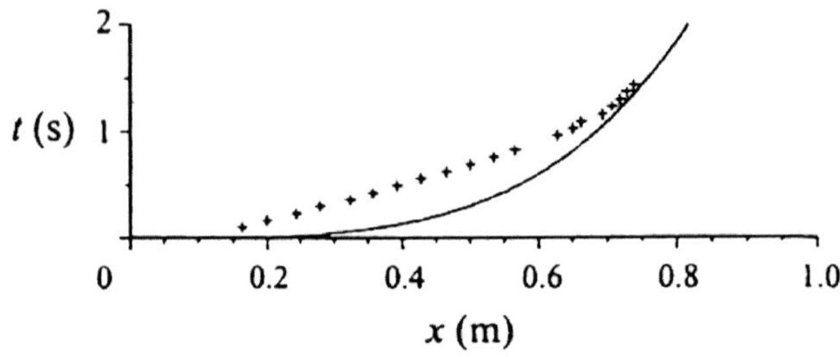

Fig. 21.5. Comparison of computed and measured position of the leading edge of a mud flow. Experiments were carried out in a tank tilted at 11° with kaolinite suspensions (solid concentration $\phi_k = 13.05\%$, released volume $24.7\,\mathrm{cm}^2$). Adapted from [34]. (Courtesy of M.H. Garcia)

In addition to providing the flow characteristics for gradually varied flows, the motion equation set may be used to investigate other interesting properties. For instance, Coussot used a depth-averaged model to demonstrate that free surface flows of Herschel–Bulkley fluids are unstable when the Froude number $Fr = \bar{u}/\sqrt{gh\cos\theta}$ exceeds a critical value (approximately 0.1 in the present context) [3]. When $Fr > 0.1$, roll waves propagate along the free surface (a similar point of view is presented in Chap. 22). This phenomenon may explain the presence of surges described by most observers for muddy debris flows. Another problem of great interest is the shape of deposit. The longitudinal profile of lobes and levees can be computed using a set of equations similar to (21.9)–(21.10) (a more general set of equations of motion is required to take two-dimensional or three-dimensional spreading into account). For instance in the case of a lobe stopped over a plane inclined at θ, it can be shown that the longitudinal profile $h(x)$ is given by:

$$\frac{\varrho g \sin^2\theta}{\tau_c \cos\theta} x = -\frac{\varrho g \sin\theta}{\tau_c} h - \ln\left(1 - \frac{\varrho g \sin\theta}{\tau_c} h\right) . \tag{21.12}$$

Equation (21.12) can be used to determine the yield stress in the field. If appropriate, the motion equation set can be cast in a dimensionless form, which yields three dimensionless number. In addition to the Froude number, we introduce a dimensionless shear stress and a generalized Reynolds number [3,34]:

$$G = \frac{\varrho g h \sin\theta}{\tau_c} , \qquad Re = \frac{\varrho \bar{u}^2}{K}\left(\frac{h}{\bar{u}}\right)^n , \tag{21.13}$$

which characterise bulk behaviour for Herschel–Bulkley fluids. These three dimensionless numbers can be used to simulate debris flows with small-scale models.

The Rheological Behaviours of Natural Slurries

On the whole, we can consider as a first approximation that soil–water mixtures involved in debris flows behave as homogeneous fluids. The solid concentration, particle-size distribution and shear rate are the key ingredients in the behaviour of natural suspensions:

- The solid volume concentration ϕ (ratio of solid volume to total volume) usually ranges from 50% to nearly 90%. The upper bound is imposed by geometrical constraints on grains. Indeed, when the solid concentration comes closer to the maximum solid concentration (0.635 for monosized suspensions, much larger for polydisperse suspensions), grain motion is increasingly impeded. Close to the maximum concentration, the material can no longer be sheared without fracturing. The lower bound reflects the minimal amount of particles required for all the particles to be in suspension. If the concentration is too low, the coarsest particles rapidly settle and the mixture can no longer be considered as a homogeneous suspension.
- When the particle-size distribution is great, typically ranging from 0.1 μm to 1 cm, interaction between particles and the surrounding fluid takes various forms [37]. For relatively small shear rates, the finest particles are generally very sensitive to Brownian motion effects or colloidal forces while coarse particles experience frictional or collisional contacts or hydrodynamic forces. As a result, bulk behaviour exhibits either plastic, frictional, or particle-inertia properties.

Microstructural theories and dimensional analysis are useful in outlining the different flow regimes and predicting flow behaviour. We begin the description of natural slurries with suspensions consisting of fine particles, then we examine how bulk behaviour is changed when the coarse-particle fraction is increased (i.e. when the particle size range is widened).

Natural suspensions of fine particles (of diameter d less than 1 μm) are usually colloidal suspensions, made-up of weakly-aggregated flocs in water. They generally exhibit viscoplastic behaviour, with sometimes time-dependent properties (thixotropy) when the solid concentration in fine particles ϕ_k is sufficiently high for particles to interact via surface forces (van der Waals attractive forces, electrostatic repulsive forces, etc.). Usually for active clays such as bentonite in pure water, concentrations as low as 0.1% are sufficient to cause the appearance of a yield stress, but for natural clay suspensions, a concentration of a few percent is required. In the opposite case, when the solid fraction is too low, the behaviour is Newtonian. A basic explanation for the existence of yield stress in polydisperse colloidal suspensions is provided by the mean-field theory of Zhou et al. [38]. These authors proposed a model for particles governed by van der Waals attractive forces. The input values of the model were Hamaker's constant A, the coordination number, the mean particle diameter d, and an interparticle separation parameter h_0, which must be fitted from experimental data. They found that the maximum yield stress can be written as:

$$\tau_k = K\left(\frac{\phi_k}{1-\phi_k}\right)^c \frac{1}{d^2}, \tag{21.14}$$

where $K = 3.1Ab/(24\pi h_0)$, and b and c are two parameters to be fitted from experimental data. They proposed the following explanation for the variation in yield stress with increasing solid concentration. A weakly flocculated dispersion may be seen as a series of weakly interconnected aggregates (flocs) made up of strongly interacting particles. At low solid concentrations, yielding results from the breakdown of the weak links between flocs. At high solid concentrations, yielding is a consequence of the rupture of interparticle bonds and resistance to the deformation of networks. This means that a critical solid concentration ϕ_{crit} separating the two domains should exist. When $\phi_k < \phi_{\mathrm{crit}}$, structural effects due to weak links between flocs prevails over those due to geometric resistance and the yield stress varies with a solid concentration such as $\tau_k \approx K\phi_k^c/d^2$. This effect is included in (21.14) since it can be derived from (21.14) by taking a series expansion to the chief order at $\phi_k = 0$. When $\phi_k > \phi_{\mathrm{crit}}$, the geometric resistance becomes more pronounced, resulting in a much higher dependence on the solid concentration $\tau_k \approx K\phi_k^{c'}/d^2$, with $c' > c$. Zhou et al. [38] considered that from a microstructural point of view, the geometric resistance enhancement is reflected by the increase in particle contacts. Assuming that the coordination number is given by Rumpf's expression ($C_N = 3.1/(1-\phi_k)$), they arrived at the conclusion that the yield stress may be scaled as a power function of $\phi_k/(1-\phi_k)$. The series expansion at $\phi_k = 0$ implies that the exponent must be c. Moreover, their experiments with alumina suspensions showed that the critical solid concentration ranged from 0.26 to 0.44 and depended on the particle diameter. As a typical application of Zhou et al.'s theory, we have reported experimental data obtained with kaolin dispersions on Fig. 21.6. It can be seen that the curve provided by (21.7) fits experimental data over a wide range of solid concentrations.

Various physical explanations for viscoplastic behaviour have been proposed. Potanin et al. developed a phenomenological fractal model to determine bulk behaviour of weakly aggregated dispersions, assuming that particles form aggregates which in turn are connected to form a network [42,43]. Thus they interpreted bulk yield stress as a consequence of chain break-up due to thermal fluctuations and rupture under compressive force. Another conceptual model inspired from glassy dynamics has been proposed by Sollich and co-workers [44,45]. They showed that the bulk mechanical properties can be related to the internal structure (described in terms of the particle energy distribution). To date such models are able to mimic bulk behaviour over a wide range of flow conditions but cannot specify the effects of particle size, size distribution, or solid concentration on the yield stress of a particulate fluid. Consequently, the flow behaviour of fine colloidal particle suspensions is usually described using the empirical Bingham or Herschel–Bulkley constitutive equation (see Chap. 2), whose rheological parameters τ_k and K are functions of the solid fraction ϕ_k while n is almost independent of the solid concentration ($n = 1$ for a Bingham fluid). The generic simple-shear expression is $\tau = \tau_k(\phi_k) + K(\phi_k)\dot{\gamma}^n$. Other empirical relationships, such as the Casson equation ($\sqrt{\tau} = \sqrt{\tau_k} + \sqrt{K\dot{\gamma}}$), are not usual. Another empirical approximation for simple shear flows of colloidal particles involves considering Krieger and Dougherty's relationship for computing bulk viscosity η_{eq} (see Chap. 3):

$$\eta_{\mathrm{eq}} = \frac{\tau}{\dot{\gamma}} = \eta\left(1 - \frac{\phi}{\phi_m}\right)^{-[\eta]\phi_m}, \tag{21.15}$$

where η is the water viscosity and $[\eta_{\mathrm{eq}}] = \lim_{\phi\to 0}(\eta_{\mathrm{eq}} - \eta)/(\eta\phi)$ is called the intrinsic viscosity. To reproduce the viscoplastic behaviour, it is assumed that the maximum solid fraction ranges from a lower value ϕ_0 to an upper bound ϕ_∞ depending on the shear stress [39–41].

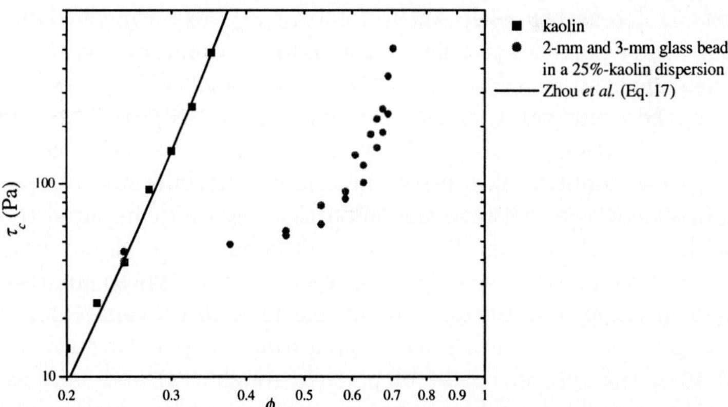

Fig. 21.6. Variation of the yield stress with solid concentration for kaolin–water dispersion and suspensions of glass beads in a kaolin dispersion. On the abscissa axis, ϕ_t denotes the total solid concentration ($\phi_t = \phi_c + \phi_k(1 - \phi_c)$). Adapted from [49]

When the particle size distribution is widened, the coarsest particles can no longer be considered as colloidal. As pointed out by Sengun and Probstein in their investigations of the viscosity of coal slurries [46–48], a useful approximation is to consider such mixtures as suspensions of force-free particles in a colloidal dispersion. As it is the interstitial phase, the dispersion resulting from the mixing of fine particles and water imparts most of its rheological properties to the entire suspension (see Chap. 3). Secondly, the coarse fraction is assumed to act independently from the fine fraction and enhance bulk viscosity. Experiments on the viscosity of coal slurries performed by Sengun and Probstein [46–48] confirmed the reliability of this concept. A typical example is provided in Fig. 21.6: we added glass beads to a water–kaolin suspension; adding a small amount of beads did not change the bulk stress significantly. Likewise, in their investigations of the behaviour of sand particle suspensions in a natural mud dispersion, Coussot and Piau also found that bulk behaviour was dictated by the fine fraction [16]. The force-free particle assumption is reliable provided the coarse particles are not too heavy (otherwise they settle), that is, they are borne by the surrounding colloidal suspension. This can be expressed in terms of dimensionless numbers by the condition: $N < 1$ where $N = \varrho' g a/\tau_k$ denotes the ratio of the buoyant force to the yield stress $\tau_k(\phi_k)$, a is the radius of coarse particles, ϱ' is the buoyant density ($\varrho' = \phi_c\left[\varrho_c - (\phi_k\varrho_k + \varrho_0 - \phi_k\varrho_0)\right]$, with ϱ_c the coarse-particle density, ϱ_k the fine-particle density, and ϱ_0 the water density).

When more and more coarse particles are added to a colloidal suspension, coarse particle motion is increasingly impeded and they begin to interact with each other. For instance for solid concentrations in the coarse fraction ϕ_c exceeding 0.35, Sengun and Probstein observed a significant change in bulk behaviour, that they ascribed to non-uniformity in shear rate distribution within the bulk due to squeezing effects between coarse particles [46–48]. Likewise, Coussot and Piau's tests [16] together with Ancey and Jorrot's experiments [49] revealed that adding coarse particles to a dispersion induced an increase in the bulk yield stress. When the solid concentration in the coarse fraction approached its maximum value, the yield stress tended towards infinity (see Fig. 21.6). When the coarse particle fraction ϕ_c comes closer to the maximum value, a network of particles in close contact takes place throughout the bulk and stresses resulting from direct contacts between coarse particles prevail compared to colloidal stresses within the dispersion. Thus, the bulk behaviour is chiefly governed by interactions between coarse particles. Two main contact types can arise depending on the suspension composition and flow features: *direct contact* for which the particle surfaces meet (i.e. the distance separating the particle surfaces is equal to or less than the typical height of particle roughness) and *indirect contact* for which there exists a fluid film between particle surfaces. In the former case, contacts between particles are generally frictional (i.e. they can be described using the Coulomb law) [50], giving rise to frictional (possibly collisional) bulk behaviour. Granular debris flows belong to this category. For the latter case, contacts are lubricated and the interstitial fluid still imparts most of its rheological characteristics to the bulk:

- If the interstitial fluid is Newtonian with viscosity η, the (squeezing) lubrication force $\mathbf{f}$ between two particles of radius a separated by a distance εa (with $\varepsilon \ll 1$) is proportional to their relative velocity U: $|\mathbf{f}| = 3\pi\eta a U/(2\varepsilon)$. To evaluate the strength of the squeezing effect, we can define a dimensionless number Γ by dividing the squeezing force by the buoyant force experienced by a test particle. In very concentrated suspensions, a network of particles in direct contact occurs, the gravity force is transmitted through the different layers such that at a depth h, a particle experiences an average "effective" normal stress $\varrho' g h$. In this case, the corresponding dimensionless number is:

$$\Gamma = \frac{9}{4}\frac{a}{\varepsilon}\frac{\eta\dot{\gamma}}{\varrho' g h} . \tag{21.16}$$

  Several experiments have shown that Γ is the relevant dimensionless number in the behaviour of concentrated suspensions when particles come in close contact. For instance, Acrivos and co-workers [51] have demonstrated that the viscous resuspension of an initially settled bed of particles is controlled by Γ. Likewise Ancey and Coussot [27] have shown that Γ could scale the flow curves for suspensions of heavy particles within a Newtonian fluid. Figure 21.7 shows such a scaling for a suspension of glass-beads within a Newtonian fluid (water–glycerol solution). A transition in the bulk behaviour can be observed at a critical value Γ ranging from 10^{-3} to 10^{-1} depending on the interstitial fluid viscosity. When $\Gamma \ll 1$, bulk behaviour is typically frictional, namely the shear stress is independent of the shear rate and varies linearly with the normal stress. Conversely when $\Gamma \gg 1$, bulk behaviour is typically Newtonian, namely the shear stress is proportional to the shear rate. Such a transition may explain the amazing behaviour and mobility of lahar-like debris flows.
- For non-Newtonian interstitial fluids, no analytical expression of the squeezing force is available. However, under the condition $N < 1$, it is expected that bulk behaviour is dictated by the interstitial fluid. This explains that, even loaded with a large amount of boulders (giving the impression that the debris is granular), a muddy debris flow behaves as a viscoplastic fluid.

Using microstructural arguments, we have shown that the different types of debris flow behaviour reflect the microstructure properties. Dimensionless groups can also be used to delimit the different flow regimes. The main difficulty in extrapolating these dimensionless numbers to field data is that we have considered truly bimodal suspensions. Natural suspensions are characterised by a continuous gradation in particle size and the cut-off between colloidal and coarse particles, ranging from 4 to 50 μm depending on the authors, is not a fixed value. It results that, in practice, inferring the debris flow type from the material composition is reserved to a limited number of cases, for which the role of different particle-size classes can be determined with sufficient accuracy.

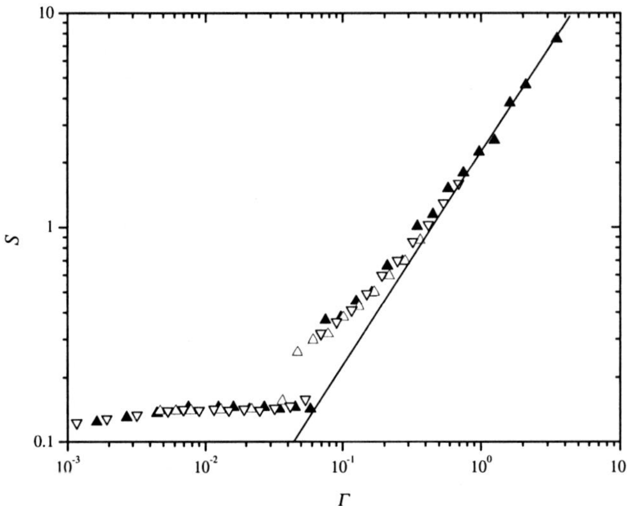

Fig. 21.7. Variation of the dimensionless shear stress $S = \tau/(\varrho g h)$ (with h the flow depth) as a function of the dimensionless number Γ. The drawn line has a slope of 1 ($S \propto \Gamma$). Experiments performed with 1-mm glass beads in a water–glycerin solution. Adapted from [27]

References

1. K.M. Scott: 'Origins, behavior, and sedimentology of lahars and lahars-runout flows in the Toutle-Cowlitz river system'. Report 1447-A (U.S. Geological Survey, 1988)
2. B. Voight: J. Volcan. Geotherm. Res. **44**, 349 (1990)
3. P. Coussot: *Mud Flow Rheology and Dynamics* (Balkema, Rotterdam 1997)
4. D. Brunsden, D.B. Prior: *Slope Instability* (John Wiley & Sons, New York 1984)
5. M. Zimmermann, P. Mani, P. Gamma, P. Gsteiger, O. Heiniger, G. Hunizker: *Murganggefahr und Klimaänderung – ein GIS-basierter Ansatz* (VDF, Zürich 1997)
6. P. Coussot, M. Meunier: Earth Sci. Rev. **3-4**, 209 (1996)
7. R.M. Iverson: Rev. Geophys. **35**, 245 (1997)
8. T. Takahashi: Ann. Rev. Fluid Mech. **13**, 57 (1981)
9. V. Koulinski: Étude de la formation d'un lit torrentiel par confrontation d'essais sur modèle réduit et d'observations de terrain. Ph.D. Thesis, University Joseph Fourier, Grenoble (1993)
10. D. Rickenmann: J. Hydraul. Eng. ASCE **117**, 1419 (1992)
11. G.M. Smart, M.N.R. Jaeggi: Sedimenttransport in steilen Gerinnen. Mitteilungen 64 (Versuchanstalt für Wasserbau, Hydrologie und Glaziologie, Zürich 1983)
12. C. Tognacca: Beitrag zur Untersuchung der Entstehungsmechanismen von Murgängen. Ph.D. Thesis, Eidgenössischen Technischen Hochschule Zürich, Zürich (1999)
13. S. Lanzoni: Meccanica di miscugli solido–liquido in regime granulo inerziale. Ph.D. Thesis, University of Padova, Padova (1993)

14. J.J. Major: Experimental studies of deposition of debris flows: process, characteristics of deposits, and effects of pore–fluid pressure. Ph.D. Thesis, University of Washington, Washington (1996)
15. A.M. Johnson, J.R. Rodine: 'Debris flow'. In: *Slope Instability*, ed. by D. Brundsen, D.B. Prior (John Wiley & Sons, New York 1984) pp. 257–361
16. P. Coussot, J.-M. Piau: Can. Geotech. J. **32**, 263 (1995)
17. P. Coussot, D. Laigle, M. Arratano, A. Deganutti, L. Marchi: J. Hydraul. Eng. ASCE **124**, 865 (1998)
18. P. Coussot, S. Proust, C. Ancey: J. Non-Newtonian Fluid Mech. **66**, 55 (1996)
19. J.J. Major, T.C. Pierson: Water Resou. Res. **28**, 841 (1992)
20. C. J. Phillips, T.R.H. Davies: Geomorphology **4**, 101 (1991)
21. Z. Wan, Z. Wang: *Hypercontrated flow* (Balkema, Rotterdam 1994)
22. P. Coussot: *Les laves torrentielles, connaissances pratiques à l'usage du praticien* (in French) (Cemagref, Antony 1996)
23. T. Takahashi: *Debris flow* (Balkema, Rotterdam 1991)
24. C. Ancey: Rhéologie des écoulements granulaires en cisaillement simple, application aux laves torrentielles granulaires. Ph.D. Thesis, Ecole Centrale de Paris, Paris (1997)
25. C.-L. Chen: Rev. Eng. Geology **7**, 13 (1987)
26. J.T. Jenkins, E. Askari: 'Hydraulic theory for a debris flow supported on a collisional shear layer'. In: *International Workshop on Floods and Inundations related to Large Earth Movements, Trent 1994*, IAHR (IAHR, 1994) pp. 6
27. C. Ancey, P. Coussot: C. R. Acad. Sci., ser. B **327**, 515 (1999)
28. D. Rickenmann: Natural Hazards **19**, 47 (1999)
29. D. Rickenmann: Schweizer Ingenieur und Architekt **48**, 1104 (1996)
30. D. Laigle, P. Coussot: J. Hydraul. Eng. ASCE **123**, 617 (1997)
31. C.C. Mei, M. Yuhi: J. Fluid Mech. **431**, 135 (2001)
32. B. Hunt: J. Hydraul. Eng. ASCE **110**, 1053 (1983)
33. B. Hunt: J. Hydraul. Eng. ASCE **120**, 1350 (1994)
34. X. Huang, M.H. Garcia: J. Fluid Mech. **374**, 305 (1998)
35. X. Huang, M.H. Garcia: J. Hydraul. Eng. ASCE **123**, 986 (1997)
36. J.M. Piau: J. Rheol. **40**, 711 (1996)
37. P. Coussot, C. Ancey: Phys. Rev. E **59**, 4445 (1999)
38. Z. Zhou, M.J. Solomon, P.J. Scales, D.V. Boger: J. Rheol. **43**, 651 (1999)
39. C.R. Wildemuth, M.C. Williams: Rheol. Acta **24**, 75 (1985)
40. C.R. Wildemuth, M.C. Williams: Rheol. Acta **23**, 627 (1984)
41. S. Mansoutre, P. Colombet, H. Van Damme: Cement Concrete Res. **29**, 1441 (1999)
42. A.A. Potanin, R. De Rooi, D. Van den Ende, J. Mellema: J. Chem. Phys. **102**, 5845 (1995)
43. A.A. Potanin, W.B. Russel: Phys. Rev. E **53**, 3702 (1996)
44. P. Sollich: Phys. Rev. E **58**, 738 (1998)
45. P. Sollich, F. Lequeux, P. Hébraud, M.E. Cates: Phys. Rev. Lett. **78**, 2020 (1997)
46. M.Z. Sengun, R.F. Probstein: Rheol. Acta **28**, 382 (1989)
47. M.Z. Sengun, R.F. Probstein: Rheol. Acta **28**, 394 (1989)
48. R.F. Probstein, M.Z. Sengun, T.-C. Tseng: J. Rheol. **38**, 811 (1994)
49. C. Ancey, H. Jorrot: J. Rheol. **45**, 297 (2001)
50. C. Ancey, P. Coussot, P. Evesque: J. Rheol. **43**, 1673 (1999)
51. A. Acrivos, R. Mauri, X. Fan: Int. J. Multiphase Flow **19**, 797 (1993)

22 Mud Flow – Slow and Fast

C.C. Mei[1], K.-F. Liu[2], and M. Yuhi[3]

[1] Department of Civil & Environmental Engineering, Massachusetts Institute of Technology, Cambridge, MA, 02139, USA
[2] Department of Civil Engineering, National Taiwan University, Taipei, Taiwan, Republic of China
[3] Department of Civil Engineering, Kanazawa University, Kanazawa, Ishikawa, 920-8667, Japan

22.1 Introduction

Heavy and persistent rainfalls in mountainous areas can loosen the hillslope and induce mud flows which can move stones, boulders and even trees, with destructive power on their path. In China where 70% of the land surface is covered by mountains, debris flows due to landslides or rainfalls affect over 18.6% of the nation. Over 10,000 debris flow ravines have been identified; hundreds of lives are lost every year [1]. While accurate assessment is still pending, mud flows caused by Hurricane Mitch in 1998 have incurred devastating floods in Central America. In Honduras alone more than 6000 people perished. Half of the nation's infrastructures were damaged.

Mud flows can also be the result of volcanic eruption. Near the volcano, lava and pyroclastic flows dominate. Further downstream solid particles become smaller and can mix with river or lake water, rainfall, melting snow or ice, or eroded soil, resulting in hyperconcentrated mud mixed with rocks. The muddy debris can travel at high speeds over tens of miles down the hill slopes and devastate entire communities. In 1985 the catastrophic eruption of Nevado del Ruiz in Colombia resulted in mud flows which took the life of 23,000 inhabitants in the town of Amero [2]. During the eruption of Mt. Pinatubo in Phillipnes in 1991, one cubic mile of volcanic ash and rock fragments fell on the mountain slopes. Seasonal rain in the following months washed down much of the loose deposits, causing damage to 100,000 villages. These catastrophes have been vividly recorded in the film documentary by Lyons [3].

Rivers running through loessial land carry a large amount of clay suspensions. In the Yellow River of China, the clay content can reach 50 percent by volume at low waters, causing siltation and floods throughout the Chinese history. Taming the Yellow River has been the challenge to Chinese river engineers for centuries. When the suspended clay particles are carried to an estuary, they coagulate due to increased salinity to form flocs and aggregates, and sink to the seabed as fluid mud. Hence coastal geologists and hydraulic engineers in many countries have also made extensive studies on the mechanical properties and behavior of fluid mud [4–12]. From most of these studies it can be concluded that for laminar mud flows the Bingham plastic model is a practical approximation. Specifically, the constitutive relation for the simple laminar shear flow reads

$$\mu \frac{\partial u}{\partial z} = \begin{cases} 0\,, & \text{if} \quad \tau_{xz} < \tau_o \\ \tau_{xz} - \tau_o\,, & \text{if} \quad \tau > \tau_o\,. \end{cases} \tag{22.1}$$

See a more general discussion in Chap. 2. In muddy water, both the yield stress τ_o and the Bingham viscosity μ depend not only on the clay concentration but also on the chemical composition as well as the salinity in water. Sample data, for typical shearing rates in the range from 50 to 500 1/sec, have been collected by Mei & Liu [47] as shown in Fig. 22.1. While both the yield stress and the Bingham viscosity increase with the clay concentration, the wide scatter of measured results reflect the complex dependency on other factors. In particular the difference caused by salinity can be seen by comparing curves (6) and (7).

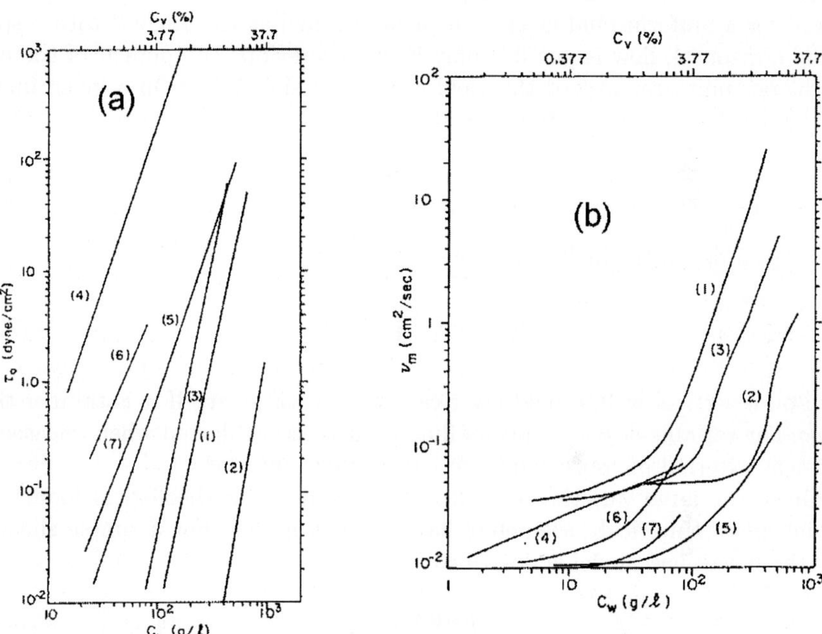

Fig. 22.1. Dependence of Bingham fluid properties on clay concentrations for various mud samples. C_v: concentration by volume; C_w: concentration by weight. Fig. (**a**): yield stress τ_o. Fig. (**b**): kinematic Bingham viscosity ν. Curve (1): Provins clay [6]; (2): powdered limestone [6]; (3): Kaolinite [6]; (4): Bentonite [7]; (5): Kaolinite [7]; (6): White River clay in salt water [8]; (7): White River clay in tap water [8]. From [47]

Fluid mud at high enough concentration ($C_v > 10\%$) has a viscosity μ hundreds times that of pure water, and a yield stress as large as $O(100)$ dynes/cm^2. As a consequence, the flow can remain laminar even if the flow velocity is as high as a few meters per second [13,14]. In Chinese literature the accepted criterion for transition from laminar to turbulent flow is that the effective Reynolds number Re_{eff} exceeds the usual threshold of 2100, where

$$\frac{1}{Re_{\text{eff}}} = \frac{1}{Re_\tau} + \frac{1}{Re_\mu}, \tag{22.2}$$

with $Re_\tau = 8\rho U^2/\tau_o$ and $Re_\mu = 4\rho Uh/\mu$ being the Reynolds numbers associated with the yield stress and the viscosity respectively, in an open-channel of mud depth h [7]. In one mud flood in Jiang Chia Ravine, Yunan Province, China, Li et al. [15] found $\rho = 2.13\,\text{g/cm}^3$, $\mu/\rho = 15\,\text{cm}^2$ sec, and $\tau_o = 2000\,\text{dynes/cm}^2$. The maximum flow speed and depth were $U = 8\,\text{m/sec}$ and $h = 1\,\text{m}$. Equation (22.2) then gives $Re_{\text{eff}} = 571$ which falls well within the laminar domain.

Because of the yield stress, a variety of features unknown in Newtonian fluids, appear even in the simplest situations. For example, a uniform layer of fluid mud can remain stationary on an incline, if the depth or the slope is sufficiently small. Indeed for a uniform mud layer on a plane inclined at the angle θ with respect to the horizontal, flow is possible only if the down-slope component of the mud weight per unit area exceed the yield stress, $\rho gh \sin\theta > \tau_o$. On a given incline ($\theta$) the depth must exceed the critical depth h_c:

$$h_c = \frac{\tau_o}{\rho g \sin\theta} \,. \tag{22.3}$$

For a given depth the inclination must exceed the angle of repose:

$$\theta_c = \sin^{-1}\left(\frac{\tau_o}{\rho g h}\right) \,. \tag{22.4}$$

Even on a vertical wall, a mud layer can still be stationary if it is thinner than $\tau_o/\rho g$. For variable depths, a pile of fluid mud need not flatten when released. A transient external pressure can leave a permanent imprint.

Given the large varieties of mud in the world, other rheological models involving more empirical parameters have been proposed. For a simple shearing flow the power-law model is described by

$$\mu_n \left(\frac{\partial u}{\partial z}\right)^n = \tau_{xz} \,, \tag{22.5}$$

and the Herschel–Bulkley model by

$$\mu_n \left(\frac{\partial u}{\partial z}\right)^n = \begin{cases} 0 \,, & \text{if} \quad \tau_{xz} < \tau \\ \tau_{xz} - \tau_o \,, & \text{if} \quad \tau_{xz} > \tau_o \,, \end{cases} \tag{22.6}$$

which combines the features of Bingham and power-law models. For more comprehensive surveys, see [16,17], and Chap. 2.

All the non-Newtonian models are highly nonlinear and analytical studies are difficult in general. Fortunately in most geomorphological flows the mud depth is often much smaller than the length scale in the flow direction, $h/L = \epsilon \ll 1$. Thus the long-wave approximation is possible. Specifically the pressure is nearly hydrostatic and convective inertia is of the order $O(\epsilon R_{\text{eff}})$. Further approximation can be made if the flow is either very slow so that convective inertia can be neglected, or so fast that the boundary layer approximation is appropriate. In this review we restrict our discussions to approximate theories for laminar flows characterized by either $R_{\text{eff}} = O(1)$ [18–20] or $R_{\text{eff}} = O(\epsilon^{-1}) \gg 1$ [19,21,31]. Focus will be directed only to physical implications; mathematical justification of the approximate equations can be found in [22,23] and in Chaps. 2 and 11.

22.2 One-Dimensional Slow Flows

In this section we describe theories for slow flows in a thin layer on an inclined plane. Due either to the gentle slope of the initial profile, the small layer thickness, or to very high viscosity, the longitudinal velocity is so small that the fluid inertia is negligible and the lubrication approximation applies. The lateral extent is assumed to be so large that the flow is essentially two-dimensional. After stating the approximate equations, we first discuss the threshold profiles where a moving mud mass comes to rest. Stationary waves on an incline are then derived. Finally a simple example of the transient collapse of a mud pile is described.

22.2.1 The Lubrication Approximation

Consider a single layer of fluid flowing down a plane ($z = 0$) inclined at the angle θ with respect to the horizontal. Let the x axis coincide with the plane bed and be directed downward. Under the long-wave approximation the fluid pressure is hydrostatic,

$$p = \rho g (h - z) \cos \theta \,, \tag{22.7}$$

where $h(x,t)$ is the mud depth. The fluid velocity is essentially in the x direction, i.e. $u \gg w$. For slow enough flows $R_{\text{eff}} = O(1)$, convective inertia is of the order $\epsilon R_{\text{eff}} \ll 1$, so that the lubrication approximation applies,

$$0 = \rho g \sin \theta - \frac{\partial p}{\partial x} + \frac{\partial \tau}{\partial z} \,, \qquad 0 \le z \le h \,. \tag{22.8}$$

Consequently the shear stress increases linearly with depth,

$$\tau = (h - z) \left[\rho g \cos \theta \left(\tan \theta - \frac{\partial h}{\partial x} \right) \right] . \tag{22.9}$$

The shear stress at the bed $z = 0$ is

$$\tau_b = \rho g h \cos \theta \left(\tan \theta - \frac{\partial h}{\partial x} \right) . \tag{22.10}$$

Clearly if $|\tau_b| > \tau_o > 0$, i.e.

$$\frac{\tau_o}{h} < \left| \frac{\rho g}{\mu} \cos \theta \left(\tan \theta - \frac{\partial h}{\partial x} \right) \right| \,, \tag{22.11}$$

the fluid moves. The flow direction is downward if $\tau_b > 0$ and upward if $\tau_b < 0$. If however $|\tau_b| < \tau_o$, or

$$\frac{\tau_o}{h} > \left| \frac{\rho g}{\mu} \cos\theta \left(\tan\theta - \frac{\partial h}{\partial x} \right) \right| , \tag{22.12}$$

the mud does not move at all. When $\tau_b = \tau_o$ (or $-\tau_o$), the mud is at the threshold of downward (or upward) flow.

Because the magnitude of the shear stress decreases from $|\tau_b|$ at the bottom to zero on the free surface, there is a yield surface at some intermediate depth $z = h_o \le h$ where $|\tau| = \tau_o$, if (22.11) is satisfied, i.e. if $|\tau_b| > \tau_o$. Below the yield surface, $0 < z < h_o$, there is shearing. The longitudinal velocity varies parabolically as

$$u = \frac{\rho g}{\mu} \cos\theta \left(\tan\theta - \frac{\partial h}{\partial x} \right) \left(\frac{1}{2} z^2 - h_o z \right) , \qquad 0 < z < h_o . \tag{22.13}$$

Above the yield surface there is a layer of plug flow within which $u = u_p$ is independent of z,

$$u_p = \frac{h_o^2}{2\mu} \left[\rho g \cos\theta \left(\tan\theta - \frac{\partial h}{\partial x} \right) \right] , \qquad h_o < z < h . \tag{22.14}$$

On the yield surface, $\tau = \pm\tau_o$ so that

$$\pm\tau_o = (h - h_o) \left[\rho g \cos\theta \left(\tan\theta - \frac{\partial h}{\partial x} \right) \right] , \tag{22.15}$$

where the upper (lower) sign corresponds to downward (upward) flow. The total volume flux at any station is

$$q = \int_0^{h_o} u \,\mathrm{d}z + u_p (h - h_o) = -\frac{1}{6\mu} \left[\rho g \left(\tan\theta - \frac{\partial h}{\partial x} \right) \right] h_o^2 (3h - h_o) . \tag{22.16}$$

Conservation of mass in the entire fluid layer requires

$$\frac{\partial h}{\partial t} + \frac{\partial q}{\partial x} = 0 . \tag{22.17}$$

Equations (22.14), (22.15) and (22.16) are the governing equations for the three unknowns $h(x,t), h_o(x,t)$ and $q(x,t)$.

Let us introduce the following set of scales:

$$[z] , \quad [h] , \quad [h_o] = \overline{h} , \quad [x] = \overline{h} \cot\theta , \quad [t] = \frac{\mu \cot\theta}{\rho g \overline{h} \sin\theta} , \tag{22.18}$$

$$[u] = \frac{\rho g \overline{h} \sin\theta}{\mu} \overline{h} , \quad [q] = \frac{\rho g \overline{h} \sin\theta}{\mu} \overline{h}^2 ,$$

where $\overline{h}$ is some characteristic depth. In this section we shall select $\overline{h} = h_c$. In dimensionless variables (without changing symbols), (22.14), (22.15) and (22.16) become

$$u_p = \frac{h_o^2}{2}\left(1 - \frac{\partial h}{\partial x}\right) , \tag{22.19}$$

$$1 - \frac{\partial h}{\partial x} = \pm \frac{1}{h - h_o} , \tag{22.20}$$

and

$$q = \frac{1}{6} h_o^2 (3h - h_o)\left(1 - \frac{\partial h}{\partial x}\right) . \tag{22.21}$$

Flow exists only if $h_o > 0$, or

$$h\left(1 - \frac{\partial h}{\partial x}\right) > 1 , \quad \text{downward flow ;} \quad h\left(1 - \frac{\partial h}{\partial x}\right) < -1, \quad \text{upward flow .} \tag{22.22}$$

A single equation can be written for the depth $h(x,t)$ of the moving mud,

$$\frac{\partial h}{\partial t} = \frac{1}{3}\left[h^3 \pm \left(1 - \frac{\partial h}{\partial x}\right)^{-3}\right]\left(\frac{\partial^2 h}{\partial x^2}\right) + h\frac{\partial h}{\partial x}\left[h\frac{\partial h}{\partial x} \mp 1\right] . \tag{22.23}$$

With an initial condition on $h(x,0)$, computations can be carried out by a Crank–Nicolson scheme.

If, on the other hand,

$$-1 < h\left(1 - \frac{\partial h}{\partial x}\right) < 1 , \tag{22.24}$$

mud remains stationary. At the threshold,

$$h\left(1 - \frac{\partial h}{\partial x}\right) = \pm 1 , \tag{22.25}$$

mud stops from downward (upward) motion along the inclined plane.

22.2.2 Profiles of Final Deposit

Because of the finite yield stress, an initial pile of mud released on an infinite plane does not ultimately collapse to zero thickness. A nonuniform profile can exist as the final state of static equilibrium where the mud bottom is the yield surface ($h_o = 0$).

In the special case of a horizontal bottom, the threshold profile satisfies the differential equation (22.15) with $h_o = \theta = 0$. Upon integration, the dimensionless surface height is

$$h - h_* = \left[\mp 2(x - x_*)\right]^{1/2} , \tag{22.26}$$

where $h = h_*$ at $x = x_*$. The upper (or lower) sign corresponds to a parabolic head facing the right (or left). For any finite bed slope, we integrate the dimensionless equation (22.25) with the equality sign. A downward threshold profile is given by

$$h - h_* + \log \frac{h-1}{h_*-1} = x - x_* \,. \qquad (22.27)$$

Three cases can be distinguished: $h_* = 1$, $h_* < 1$, and $h_* > 1$. The first case, $h_* = 1$, corresponds to the trivial limit of uniform critical depth $h = 1$ (curve (a) in Fig. 22.2). For $0 < h_* < 1$, the depth is everywhere less than the critical uniform depth and approaches unity as $x \to -\infty$. Curve (b) in Fig. 22.2 is an example for $h_* = 0$ at $x = x_* = 0$, corresponding to the downward front of a mud layer on a dry bed, i.e. a perfectly rigid and nonerodible bed. Though the slope is infinite at the front, this mathematical shortcoming is only of local significance. If $h_* > 1$, then the depth increases from $h(-\infty) = 1$ to $h(\infty) \to \infty$. At $x \to +\infty$ the mud surface is horizontal. A typical profile for $h_* = 2$ at $x = x_* = 0$ is plotted in Fig. 22.2 as curve (c), which corresponds to the mud surface connecting a uniform stationary layer on the left bank and a mud reservoir, when the draining has completely stopped.

Similarly, the upward threshold profile is given by

$$h - h_* - \log \frac{1+h}{1+h_*} = x - x_* \,, \qquad (22.28)$$

corresponding to the upper sign in (22.25). The depth increases monotonically from 0 at $x = x_* - h_* + \log(1 + h_*)$ to ∞ at $x \to \infty$, as shown by curve (d) for $h_* = 0$ at $x_* = 0$ in Fig. 22.2. This corresponds to a mud reservoir at the threshold of rising along a sloping bank.

Consider next the spreading of a two-dimensional mud pile of finite volume V per unit length, released on a wet bed, i.e. a stagnant mud layer of uniform depth $h_\infty < 1$. After a transient stage of flowing, the mud pile must settle to final equilibrium. Let x_+ and x_- denote the upstream and downstream edges of the pile respectively. The final profile is given on the front (downward) side by

$$h - h_\infty + \log \frac{h-1}{h_\infty - 1} = x - x_+ \,, \qquad (22.29)$$

and on the back (upward) side by

$$h - h_\infty - \log \frac{1+h}{1+h_\infty} = x - x_- \,. \qquad (22.30)$$

Let the maximum height h_m be at the point x_m. Substituting $h = h_m$ and $x = x_m$ in the two equations above and taking the difference, we get a relation between h_m and the final length $L \equiv x_+ - x_-$,

$$-(x_+ - x_-) = -L = \log \frac{1 - {h_m}^2}{1 - {h_\infty}^2} \,, \qquad (22.31)$$

which can be inverted to give

$$h_m = \left[1 - (1 - h_\infty^2)\mathrm{e}^{-L}\right]^{1/2} . \tag{22.32}$$

In particular, if the mud pile is released on a dry bed, $h_\infty = 0$, we then have

$$h_m = \left(1 - \mathrm{e}^{-L}\right)^{1/2} . \tag{22.33}$$

From (22.27) and (22.28), the final volume above $z = h_\infty$ can be easily calculated equated to the initial volume V, yielding

$$V = -2h_m + \log \frac{1 + h_m}{1 - h_m} . \tag{22.34}$$

With the help of (22.32) in general, or (22.33) for a dry bed, the final length of the pile L is related uniquely to the initial volume V, with h_∞ as a parameter.

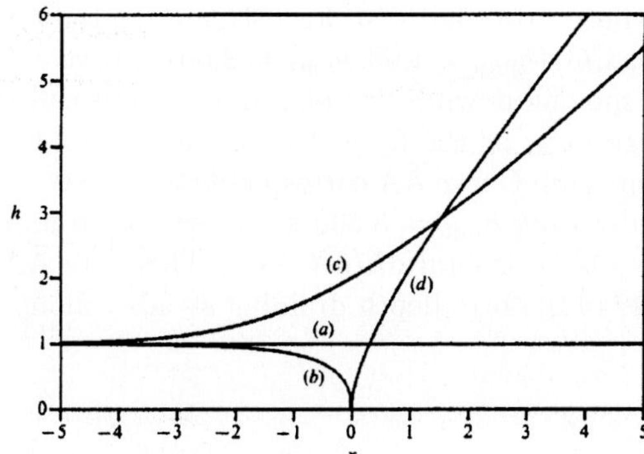

Fig. 22.2. Final profiles when mud no longer flows, $V = \infty$. Curve (**a**): a uniform layer of infinite length, (**b**): head of a uniform layer, (**c**): a mud sea that stops draining from a slope, (**d**): a mud sea that stops rising along a slope. From [18]

22.2.3 Stationary Waves

A wave is called stationary if the profile appears steady in the coordinate advancing at a constant speed C. The mathematical task is to find

$$h = h(X) , \quad h_o = h_o(X) , \quad \text{and} \quad q = q(X) , \tag{22.35}$$

as functions of

$$X \equiv x - Ct , \tag{22.36}$$

where C is to be found as a part of the solution. Assuming (22.35), we get by integrating the mass conservation law

$$q = C(h - h_e) , \tag{22.37}$$

where q is the flux measured in the fixed coordinate system, and $h_e \geq 0$ is an integration constant. Equation (22.21) becomes

$$C(h - h_e) = \frac{1}{6} h_o^2 (3h - h_o) \left(1 - \frac{dh}{dX}\right) . \tag{22.38}$$

For given C and h_e in (22.38) and (22.25), h_o can be eliminated to give a first-order nonlinear ordinary differential equation for h. The fixed points correspond to the uniform state far upstream or far downstream, whose depths $h_{\max}$ or $h_{\min}$ can be algebraically related to C and h_e.

A variety of waves has been discussed by Liu & Mei [18]. It can be shown that for waves propagating down the incline, $C > 0$, each $h_e > 1$ corresponds to a profile which connects $h_{\max}$ far upstream ($X \to -\infty$) to $h_{\min}$ far downstream ($X \to \infty$). Sample profiles are shown in Fig. (22.3). Curves $\bar{\text{B}}$B and $\bar{\text{C}}$C are typical profiles whose upstream and downstream depths are both finite. In particular we have for $\bar{\text{C}}$C: ($h_{\min} = 1.60, h_{\max} = 2.66$). Curve $\bar{\text{A}}$A′ is the limiting case of a wave front moving down a dry bed and is often called a gravity current. Despite the large slope at the front, the predicted profile is confirmed by laboratory experiments [18]. Curve $\bar{\text{A}}$A corresponds to $h_e = 0 < 1$ and the depth increases monotonically from $h_{\min} = 3.566$ at $X \to -\infty$ to infinity at $X \to \infty$ where the free surface is horizontal $dh/dX \to 0$. Therefore $\bar{\text{A}}$A represents the surface of a mud layer of uniform depth draining steadily along the plane slope into a mud sea.

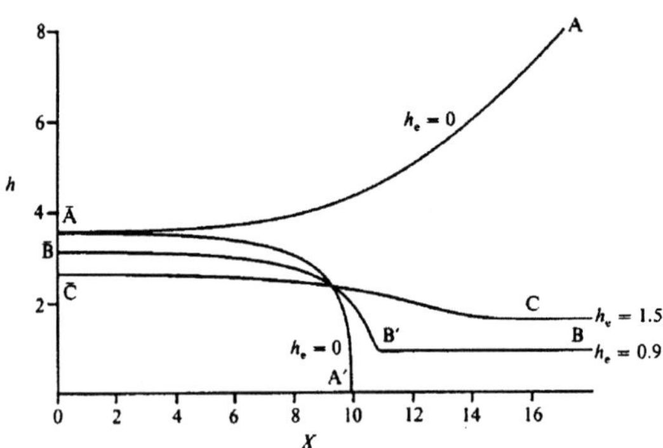

Fig. 22.3. Profiles of stationary waves at downward speed $C = 2.5$. From [18]

22.2.4 Transient Collapse of a Finite Mass

The fate of a mud pile released on a slope depends on the state of the slope, which can be a layer of mud either stagnant or flowing. Let us compare in

Fig. 22.4 the numerical results for three different starting states with uniform depths $h_\infty = 1.5, 1$ and 0.5. In physical dimensions, these depths are respectively $1.5\,h_c, h_c$ and $0.5\,h_c$, corresponding to layers that are flowing, at the threshold of motion, and stagnant, for $t < 0$.

Before its release, the initial profile of the mud pile is assumed to be a right triangle with the upstream face against a frictionless vertical wall. Over the deep layer of flowing mud ($h_\infty = 1.5$) the front of the fresh mud pile extends downward indefinitely until the entire profile flattens out eventually. Over the stagnant layer at the critical depth, the fresh pile also flattens to zero thickness, though at a slower rate. Over the thin and stagnant layer ($h_\infty = 0.5$), the pile comes to rest after a finite time. The final profile is shown by the dashed curve, the front of which matches (22.27).

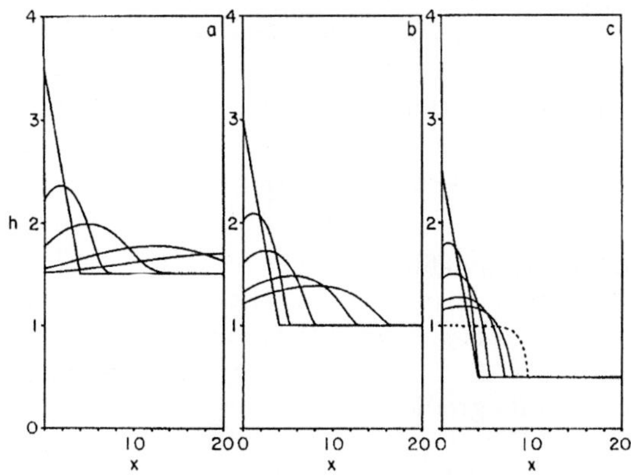

Fig. 22.4. Collapse of a mud pile on an otherwise uniform mud layer. (**a**): Supercritical initial depth with $h_\infty = 1.5$. (**b**): Critical initial depth with $h_\infty = 1.0$. (**c**): Subcritical depth with $h_\infty = 0.5$. In order of peak heights, the snapshots in each figure are for $t = 0, 0.2, 1, 4$, and 8. In (**c**) the dashed curve corresponds to $t = \infty$. From [48]

22.3 Two-Dimensional Slow Flows in a Wide Channel

To extend the theories of the previous section to mountain streams, it is necessary to include at least two additional facets of nature: finite channel width and high speed of flow. In this section we limit our attention to the first. If the channel width is comparable to the depth, the two-dimensional variation in the cross-section must be fully accounted for. This is not an easy task even for the problem of uniform flows in an open-channel of semi circular cross section, as errors have been made in earlier literature [13]. Mei & Yuhi [20] have recently studied a simpler problem for slow flows of Bingham mud in a wide and shallow channel along a constant slope. The scales in the lateral (y) direction and the longitudinal (x) directions are assumed to be comparable, while both of which are much greater than the depth (z). The lubrication approximation can be extended to facilitate the analysis.

22.3.1 The Lubrication Approximation

Let the channel axis coincide with the x axis and the bed be rigid and prescribed by $z = H(y)$ which is symmetric about the $x - z$ plane. The two-dimensional extension of the lubrication approximation of Sect. 22.2.1 can be integrated with respect to z to give the shear stresses

$$\tau_{xz} = \rho g \left(\sin\theta - \cos\theta \frac{\partial h}{\partial x} \right) (h - z) \, , \tag{22.39}$$

$$\tau_{yz} = -\rho g \cos\theta \frac{\partial h}{\partial y} (h - z) \, . \tag{22.40}$$

The dominant part of the total shear stress is

$$\tau = \sqrt{\tau_{xz}^2 + \tau_{yz}^2} \, . \tag{22.41}$$

The yield surface $h_o(x, y, t)$ is defined by

$$\tau(x, y, h_o(x, y, t), t) = \tau_o \, , \tag{22.42}$$

which relates h_o to h and its gradient:

$$\left(\sin\theta - \cos\theta \frac{\partial h}{\partial x} \right)^2 + \left(\cos\theta \frac{\partial h}{\partial y} \right)^2 = \frac{\tau_o}{\rho g} (h - h_o) \, . \tag{22.43}$$

In the shear layer below the yield surface, $\tau > \tau_o$ and the velocity components are

$$u(x, y, t) = \frac{\rho g}{\mu} \left(\sin\theta - \cos\theta \frac{\partial h}{\partial x} \right) \left[h_o z - \frac{z^2}{2} - h_o H + \frac{H^2}{2} \right] \, , \tag{22.44}$$

$$v(x, y, t) = \frac{\rho g}{\mu} \left(- \cos\theta \frac{\partial h}{\partial y} \right) \left[h_o z - \frac{z^2}{2} - h_o H + \frac{H^2}{2} \right] \, . \tag{22.45}$$

Above the yield surface, $h_o < z < h$, $\tau < \tau_o$; the plug flow[1] velocities are

$$u_p(x, y, t) = \frac{\rho g}{\mu} \left(\sin\theta - \cos\theta \frac{\partial h}{\partial x} \right) \left[\frac{h_o^2}{2} - h_o H + \frac{H^2}{2} \right] \, , \tag{22.46}$$

[1] The term *plug flow* is used here in an approximate sense and emphasizes the depth-wise uniformity of u and v. The velocities in the plug flow region are not strictly constant but varies with y slowly because of the relatively large channel width.

$$v_p(x,y,t) = \frac{\rho g}{\mu}\left(-\cos\theta\frac{\partial h}{\partial y}\right)\left[\frac{h_o^2}{2} - h_o H + \frac{H^2}{2}\right] . \tag{22.47}$$

If however, $\tau < \tau_o$ at the bed $z = H(y)$, then $h_o = H$ and there is no flow at all.

We restrict our discussion to downward flows only and use the dimensionless variables defined in (22.19), where the characteristic depth $\overline{h}$ is chosen to be the upstream maximum depth D. In normalized form, (22.43) provides the first relation between the yield surface h_0 and h:

$$(h - h_0)\left[\left(1 - \frac{\partial h}{\partial x}\right)^2 + \left(\frac{\partial h}{\partial y}\right)^2\right]^{1/2} = \frac{h_c}{D} \equiv \alpha , \tag{22.48}$$

where h_c is the threshold depth defined in (22.3). The parameter α is the Bingham number; it is the ratio of the (dimensional) critical depth h_c defined in (22.3) to the characteristic depth D, and represents the ratio of the yield stress to the bottom shear stress of a uniform flow of depth D. For a uniform flow to exist, it is necessary that $h > \alpha$. The second relation between h and h_o follows from the depth-integrated law of mass conservation,

$$\frac{\partial h}{\partial t} + \frac{\partial}{\partial x}\left[\left(1 - \frac{\partial h}{\partial x}\right)F\right] + \frac{\partial}{\partial y}\left(-\frac{\partial h}{\partial y}F\right) = 0 , \tag{22.49}$$

where

$$F(h, h_o, H) = \frac{1}{6}(3h - h_o - 2H)(h_o - H)^2 . \tag{22.50}$$

With proper initial data, $h(x,t)$ can be solved as a free boundary problem numerically. We discuss below uniform flows and transient evolution of a finite mass released from a reservoir. Discussions on stationary waves can be found in [20].

22.3.2 Steady Uniform Flow

Analytical solutions are possible for steady uniform flows with $\partial/\partial t = \partial/\partial x = 0$. It follows from (22.49) that h and h_0 are independent of y, and that $h = h_S$ and $h_o = h_{oS}$ are constants everywhere in the flow region. The flow is then confined in an effective width $2B$ whose value can be determined from (22.48) by requiring that

$$h_{oS} = H_B \equiv H(\pm B) = h_S - \alpha . \tag{22.51}$$

For a given bed profile $H(y)$ and α, the half flow width B is less than the maximum half width B_M of the channel; the bed stress is too weak for $|y| > B$ where the mud depth is too small. We shall assume for simplicity that the transverse uniformity extends to the channel banks so that $h = h_S$ up to $|y| = B_M$ where $h_S = H(B_M)$.

For any $H(y)$, the longitudinal velocities in the shear flow and plug flow zones are given by (22.44) and (22.46) respectively with $\partial h/\partial x = 0$ and $h \to h_S$ and

$h_0 \to h_{oS}$. The transverse velocity $v = v_p = 0$ vanishes everywhere. The flux per unit width at any y is

$$q(y) = \frac{1}{6}(3h_S - h_{oS} - 2H)(h_{oS} - H)^2 = \frac{1}{6}(2h_S + \alpha - 2H)(h_S - \alpha - H)^2 \ . \quad (22.52)$$

For a prescribed total discharge Q, conservation of mass requires

$$Q = \frac{1}{3}\int_0^B (2h_S + \alpha - 2H)(h_S - \alpha - H)^2 \mathrm{d}y \ . \quad (22.53)$$

Explicit results have been worked out for two types of channel cross section. In the first, case the cross section is of the power-law class

$$H = m|y|^n \ . \quad (22.54)$$

The parameter m is a measure of the bank steepness, with $m = 0$ corresponding to a flat bed of infinite width. The power n represents the channel smoothness at the center line, with $n = 1$ being the limiting case of a triangular cross section. We plot in Fig. 22.5 the typical velocity profiles for $n = 1$ and 2. The plug and shear flow regions are clearly seen above and beneath the yield surface respectively. The longitudinal velocity is the greatest along the center plane. Note that the horizontal shear rate $\partial u/\partial y = O(D/L) \ll 1$ and the corresponding shear stress is very small, $\tau_{xy} = O(D/L)^2$. The component τ_{xz} dominates the total stress in (22.41) and defines approximately the plug zone. The half width of the flow region, B, is determined from (22.51),

$$B = \left(\frac{h_S - \alpha}{m}\right)^{1/n} \ . \quad (22.55)$$

It can also be shown that

$$\begin{aligned} Q &= \frac{4n^3m^3}{(n+1)(2n+1)(3n+1)} B^{3n+1} + \frac{2n^2m^2\alpha}{(n+1)(2n+1)} B^{2n+1} \\ &= \left[\frac{4n^3(h_S - \alpha)^3}{(n+1)(2n+1)(3n+1)} + \frac{2n^2(h_S - \alpha)^2\alpha}{(n+1)(2n+1)}\right]\left(\frac{h_S - \alpha}{m}\right)^{1/n} , \end{aligned} \quad (22.56)$$

which is a function of n, α and m. The results for a semi-elliptic cross-section are given in [20].

22.3.3 Transient Spreading After Dam Collapse

Consider a parabolic channel ($n = 2$) with a mud reservoir to the left ($x < 0$) of a dam at $x = 0$. Initially the mud in the reservoir is at rest and occupies a finite length of the channel so that the horizontal free surface height is described by

$$h(x,t) = 1 + x \ , \quad my^2 - 1 \le x \le 0 \ , \quad t < 0 \ . \quad (22.57)$$

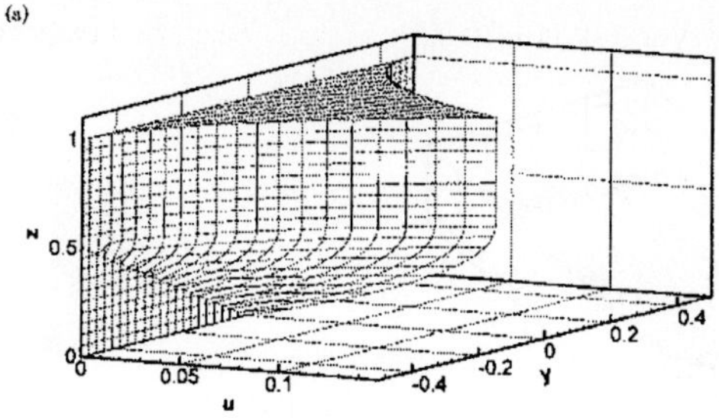

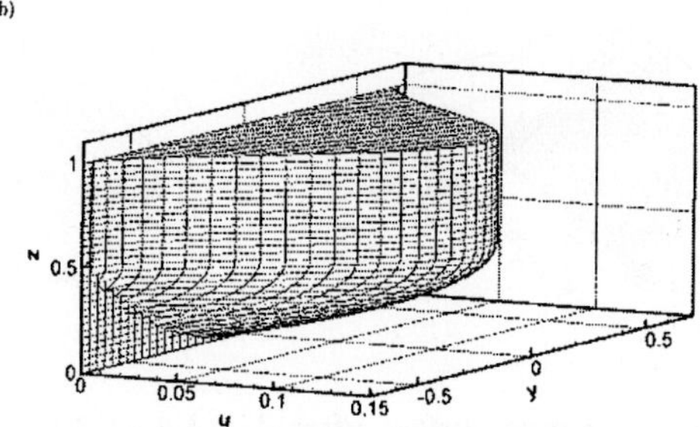

Fig. 22.5. Velocity distribution in channel of polynomial cross-section given by $H = m|y|^n$. Results shown are for $m = 1, \alpha = 0.5, h_S = 1$. Panel (**a**): n=1 (triangular channel); (**b**): n=2 (parabolic channel). From [20]

Elsewhere ($x > 0$ and $x < my^2 - 1$), the bed is initially dry. Note that the initial normalized slope of the free surface in the reservoir is unity ($\partial h/\partial x = 1$) in the present coordinate system; in physical coordinates the corresponding free surface is horizontal. At $t = 0$ the dam disappears and the reservoir mud is released suddenly, and moves downstream until the final state of static equilibrium.

Calculations have been reported in [20] for three parabolic channels with different bank steepnesses $m = 0.5, 1, 2$, and different mud plasticities $\alpha = 0.2$ to 0.9. On the centreplane ($y = 0$), symmetry is assumed.

Typical evolution of the free surface is displayed in Fig. 22.6 for $\alpha = 0.3$ and $m = 1$. Just after the dam break, the movement of the fluid is significant. The front spreads downstream and forms a fan. The free surface is convex upward in the front part and concave upward in the rear. Thus fluid is emptied from the rear to fill the advancing front. The central part of the front elongates gradually and forms a tongue. For sufficiently large t, the fluid pile comes to rest. In [20] the final extent is shown to increase as the yield stress measure α decreases.

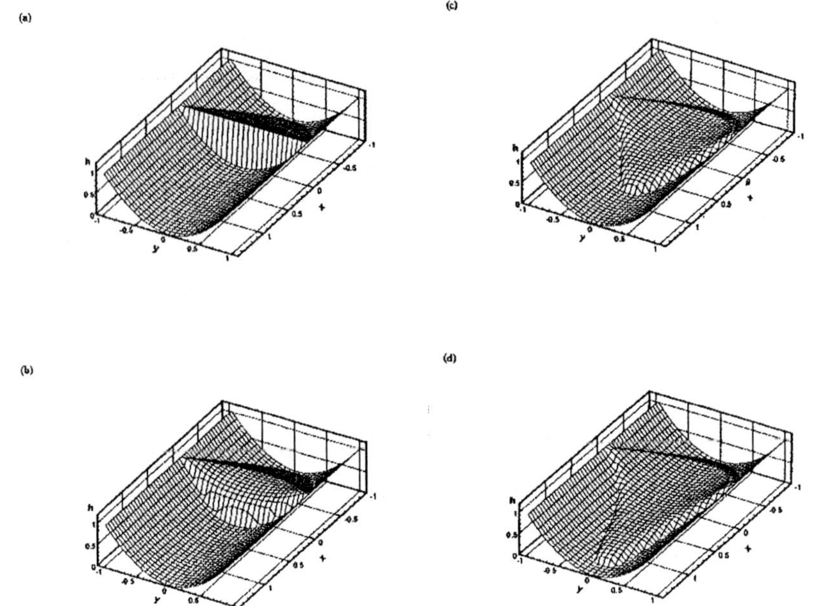

Fig. 22.6. Evolution of the mud surface after dam-break, for $\alpha = 0.3, n = 2, m = 1$. (**a**) t=0, (**b**) t=5, (**c**) t=50, (**d**) t=10,000 (final deposit). From [20]

22.4 Surges in High-Speed Flows

Highly concentrated mud flowing at moderately high speeds often develops intermittent surges [7,25,26]. Each surge has a sharp front which can be locally turbulent, but the main body is mostly laminar and tapers off to a narrow tail, to be followed by the next surge. A sample surge from Southwest China is shown in Fig. 22.7.

The number of surges in one mud flood ranges from tens to hundreds; the longest such event occured in Jiang Jia ravine, Yunan Province, China, and lasted 82 hours [1]. A sample record of the river surface at a station on Black River (a tributary of Yellow River) is shown in Fig. 22.8 [25,27]. The highest concentration of solid at $t = 0.5\,\mathrm{hr}$ was $1{,}000\,\mathrm{kg/m^3}$.

In turbulent flows of clear water, periodic surges are known as roll waves which have been modeled theoretically as one-dimensional periodic shocks (hydraulic jumps) [32]. So far two rheological models have been applied to periodic surges in flowing mud. Ng & Mei [31] examined the power-law fluid, and found from the linearized instability theory that wavy disturbances of all wavelengths are unstable, but the growth rate increases monotonically with the wave number. Hence there is no mode of finite wavelength which is the most unstable. An ad hoc criterion similar to Dressler's has to be introduced in order to determine the period of shocks. Based on the Bingham-plastic model, Liu & Mei [21] showed that for a sufficiently high yield stress, the most unstable wave length exists for

Fig. 22.7. Typical laminar surges in Southwest China. From [1]

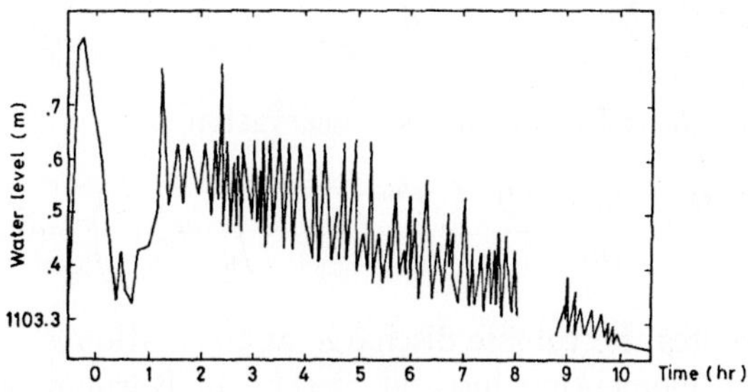

Fig. 22.8. Intermittent surges recorded at Lanshi Po, Black River, China. From [27]

a finite wavenumber and leads to periodic shocks in the nonlinear stage. This gives more evidence that the yield stress is the most important feature for proper modelling of the non-Newtonian behavior of mud. The following is a synopsis of their theory. Other related references are [33] and [34].

22.4.1 Boundary Layer Approximation and Depth-averaging

The long-wave approximation is again applied here, but with additional account of convective inertia.

Assume in general that the bottom stress is sufficiently great, so that there is a shear flow region below the yield surface at $z = h_o(x,t)$ with $0 < h_o < h$. Above this yield surface, $h_o < z < h$, the dominant strain rate must be zero and the velocity profile uniform in depth. Since the longitudinal velocity u_p is not a function of z, the shear stress in the plug flow layer is linear in z, the corresponding momentum equation is therefore

$$\frac{\partial u_p}{\partial t} + u_p \frac{\partial u_p}{\partial x} = g\left(\sin\theta - \cos\theta \frac{\partial h}{\partial x}\right) - \frac{\tau_o \mathrm{sgn}(u_p)}{\rho(h - h_o)} . \tag{22.58}$$

Below the yield surface, $0 < z < h_o$, there is velocity shear, and the longitudinal momentum equation is

$$\frac{\partial u}{\partial t} + u\frac{\partial u}{\partial x} + w\frac{\partial u}{\partial z} = g\left(\sin\theta - \frac{\partial h}{\partial x}\cos\theta\right) + \nu\frac{\partial^2 u}{\partial z^2} , \tag{22.59}$$

where $\nu = \mu/\rho$. Mass conservation in the shear flow requires

$$\frac{\partial u}{\partial x} + \frac{\partial w}{\partial z} = 0 . \tag{22.60}$$

Aside from the usual kinematic and dynamic boundary conditions on the free surface $z = h$ and the bottom $z = 0$, we must require that on the yield surface, $z = h_o$, the velocities must be continuous, as are the shear stresses, implying in turn

$$\frac{\partial u}{\partial z} = 0 . \tag{22.61}$$

The depth-integrated law of mass conservation is

$$\frac{\partial h}{\partial t} + \frac{\partial q}{\partial x} = \frac{\partial h}{\partial t} + \frac{\partial}{\partial x}\left(\int_0^{h_o} u \, \mathrm{d}z + \int_{h_o}^{h} u_p \, \mathrm{d}z\right) = 0 , \tag{22.62}$$

where $q(x,t)$ denotes the volume discharge at the station x.

Following the momentum integral method of Kármán, we shall assume a velocity profile and integrate the momentum equations. For the shear layer, we assume the velocity profile to satisfy conditions on the yield surface and the bed,

$$u = u_p\left(\frac{2z}{h_o} - \frac{z^2}{h_o^2}\right) , \qquad 0 \le z \le h_o . \tag{22.63}$$

The total flow rate can now be expressed as

$$q = u_p\left(h - \frac{1}{3}h_o\right) . \tag{22.64}$$

Substituting (22.63) into (22.59) and integrating over the shear layer, we obtain:

$$\begin{aligned} &\frac{2}{3}h_o\frac{\partial u_p}{\partial t} - \frac{1}{3}u_p\frac{\partial h_o}{\partial t} + \frac{2}{5}h_o u_p\frac{\partial u_p}{\partial x} - \frac{2}{15}u_p^2\frac{\partial h_o}{\partial x} \\ &= g h_o\left(\sin\theta - \frac{\partial h}{\partial x}\cos\theta\right) - \mu\frac{2u_p}{\rho h_o} . \end{aligned} \tag{22.65}$$

Use has been made of the fact that for any monotonic profile

$$\mathrm{sgn}(u_p) = \mathrm{sgn}\left(\lim_{z \to h_o}\frac{\partial u}{\partial z}\right) .$$

With (22.64), (22.62), (22.58) and (22.65) give a hyperbolic system of three partial differential equations for the three unknowns u_p, h, and h_o.

At the threshold of motion, the bottom stress equals the yield stress τ_o. Thus fluid moves only if gravity force and pressure gradient together exceed the yield stress, i.e.

$$\left|\rho g h\left(\sin\theta - \frac{\partial h}{\partial x}\cos\theta\right)\right| > \tau_o \,. \tag{22.66}$$

In this section we use $\overline{h}$ to denote the mean depth and define the Bingham number as

$$\alpha = \frac{h_c}{\overline{h}} = \frac{\tau_0}{\rho g \overline{h}\sin\theta} \,. \tag{22.67}$$

This is also the ratio of the yield stress to the bottom stress of a uniform flow of depth $\overline{h}$. For such a flow to exist, it is necessary that $\alpha < 1$. The Newtonian limit corresponds to $\alpha = 0$.

Using dimensionless variables normalized by the scales shown in (22.19), the normalized velocity profile is still of the form (22.63). The normalized law of mass conservation is

$$\frac{\partial h}{\partial t} + \frac{\partial}{\partial x}\left[u_p\left(h - \frac{h_o}{3}\right)\right] = 0 \,. \tag{22.68}$$

The normalized momentum equations are

$$\beta\left(\frac{\partial u_p}{\partial t} + u_p\frac{\partial u_p}{\partial x}\right) = 1 - \frac{\partial h}{\partial x} - \alpha\frac{\operatorname{sgn}(u_p)}{h - h_o} \tag{22.69}$$

for the upper plug layer, and

$$\beta\left(\frac{2}{3}h_o\frac{\partial u_p}{\partial t} - \frac{1}{3}u_p\frac{\partial h_o}{\partial t} + \frac{2}{5}h_o u_p\frac{u_p}{\partial x} - \frac{2}{15}u_p^2\frac{\partial h_o}{\partial x}\right)$$
$$= \left(1 - \frac{\partial h}{\partial x}\right)h_o - 2\frac{u_p}{h_o}\operatorname{sgn}(u_p) \tag{22.70}$$

for the lower shear layer, where the dimensionless parameter β is defined by

$$\beta = \frac{\overline{u}\overline{h}}{\nu}\tan\theta = \frac{\rho^2 g\overline{h}^3\sin\theta\tan\theta}{\mu^2} \,, \tag{22.71}$$

which depends only on material properties and the bottom slope. To fix ideas, estimates of β for Provins Bay mud are listed in the following table.

In subsequent analysis, u_p is always positive so that $\operatorname{sgn} u_p = 1$.

22.4.2 Linearized Instability of a Uniform Flow

Consider a steady uniform flow with constant depth. The corresponding shear layer depth h_o and velocity u_p are readily obtained from (22.69) and (22.70),

$$h = 1 \,, \qquad h_o = 1 - \alpha \,, \qquad u_p = \frac{1}{2}(1 - \alpha)^2 \,. \tag{22.72}$$

Table 22.1. Estimated β for Provins Bay and various bottom slopes

C	τ_o (dyne/ cm^2)	ν (cm^2/s)	β ($\theta = 1°$)	β ($\theta = 0.5°$)	β ($\theta = 0.1°$)	R_{eff}
5%	0.11	3.41	3213	802	32	5818
10%	7.85	26.4	54	13.4	0.54	706
15%	55.2	87.2	4.9	1.22	0.05	198
20%	219	204	0.9	0.23	0.009	77

Whenever the shear layer exists, $\alpha < 1$ from (22.72) so that h_o is positive. We now introduce an infinitesimal disturbance, distinguished by primes,

$$h = 1 + \epsilon h' , \qquad h_o = (1-\alpha) + \epsilon h_o' , \qquad u_p = \frac{1}{2}(1-\alpha)^2 + \epsilon u_p' , \tag{22.73}$$

where $\epsilon \ll 1$, and periodic waves for the disturbances,

$$\{h', h_o', u_p'\} = (\hat{h}, \hat{h}_o, \hat{u}_p) \mathrm{e}^{\mathrm{i}(kx-\omega t)} . \tag{22.74}$$

The corresponding perturbations in the depth of the yield surface and the plug-flow velocity $\hat{h}_o$ and $\hat{u}_p$ are, in terms of $\hat{h}$:

$$\hat{h}_o = \frac{3\beta[(1-\alpha)^2 k - 2\omega]^2 - 4(2+\alpha)k^2 - 4\mathrm{i}(2+\alpha)k}{(1-\alpha)^2\beta k[(1-\alpha)^2 k - 2\omega] - 4\mathrm{i}(2+\alpha)k}\hat{h}, \tag{22.75}$$

$$\hat{u}_p = \frac{-(1-\alpha)^2 k^2 + 2\mathrm{i}[(1-\alpha)^2 k - 3\omega]}{(1-\alpha)^2\beta k[(1-\alpha)^2 k/2 - \omega] - 2\mathrm{i}(2+\alpha)k}\hat{h} . \tag{22.76}$$

These will be used later.

From the linearized equations, we find the dispersion (eigenvalue) relation between ω and k:

$$\begin{aligned}
&\frac{\beta^2}{6}(1-\alpha)\omega^3 + \left[-\frac{\beta^2}{5}(1-\alpha)^3 k + \mathrm{i}\frac{2}{3}\frac{1+2\alpha}{\alpha(1-\alpha)}\beta\right]\omega^2 \\
&+\left\{\frac{\beta k^2}{120}(1-\alpha)\left[9\beta(1-\alpha)^4 - 20\right] - \frac{2}{\alpha(1-\alpha)^2} - \frac{\mathrm{i}\beta k}{30}\left(\frac{1-\alpha}{\alpha}\right)(16+49\alpha)\right\}\omega \\
&+\left\{\frac{2}{\alpha(1-\alpha)}k - \frac{\beta}{360}k^3(1-\alpha)^3\left[3(1-\alpha)^4\beta - 2(13-7\alpha)\right]\right. \\
&\left.+\mathrm{i}k^2\left[\frac{4}{45}\frac{\beta}{\alpha}(1-\alpha)^3(1+5\alpha) - \frac{2}{3}\frac{\alpha^2+\alpha+1}{\alpha(1-\alpha)}\right]\right\} = 0 .
\end{aligned} \tag{22.77}$$

This condition is a complex polynomial equation of the third degree for ω.

In [21] the numerical solution for the eigenvalue conditions as well as various analytical limits are studied. In particular, it is shown that, for relatively small

α, there is a minimum unstable wave number k_c below which all longer waves are stable, and above which the growth rate increases monotonically with the wavenumber. The second feature is common to Newtonian and power-law fluids, see Fig. 22.9 for $\alpha = 1/10$. However, if α is sufficiently large (more Bingham-plastic) and β sufficiently high (high speed), there is a maximum unstable wave number k_c above which all shorter waves are stable. The growth rate is the greatest for certain intermediate k between 0 and k_c, as shown in Fig. 22.10 for $\alpha = 1/2$. These results are qualitatively consistent with the field observations [27]. In Fig. 22.11 the stability boundaries are shown in the parametric plane of α vs. β. To the right of curve BIC, the longest wave $k = 0$ is unstable. To the left of AID, the shortest wave $k \to \infty$ is unstable. Thus waves of all lengths are stable to the left of AIC and unstable to the right of BID. In the wedge CID, long waves with $0 < k < k_c$ are unstable. In the wedge AIB, short waves with $k_c < k < \infty$ are unstable.

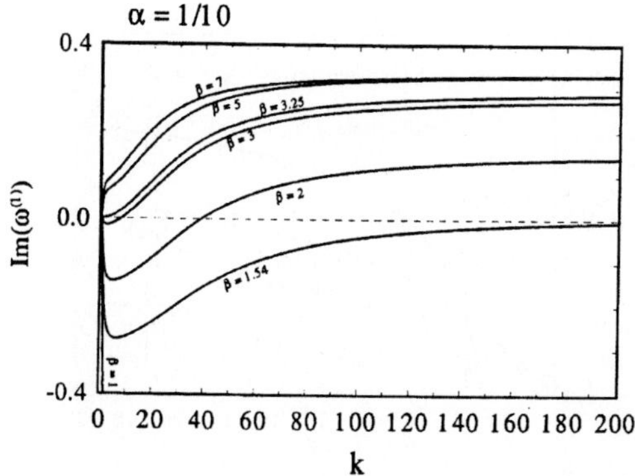

Fig. 22.9. Imaginary part of ω as a function of the wavenumber k and β, for low yield-stress fluid with $\alpha = 1/10$. From [21]

In the study of roll waves in turbulent flows in an open channel, the hydraulic approximation leads to a set of depth-averaged equations very similar to the limit of (22.62) and (22.58) at $h_o = 0$. By a linearized instability analysis it is known that all waves are unstable as long as the Froude number exceeds certain threshold, and that the growth rate increases monotonically with the wavenumber or frequency. Thus the linearized theory does not give any information on the prefered wave length in the nonlinear stage, similar to the case of small Bingham number α here. Chang et al. [28][2] have recently applied a mathematical model due to Needam & Merkin [29] by adding a diffusion term in the longitudinal momentum equation. With this fictitious term a prefered wave frequency exists. In numerical simulations of the experiments by Brock [30], they

[2] We thank Prof. Neil Balmforth for this reference.

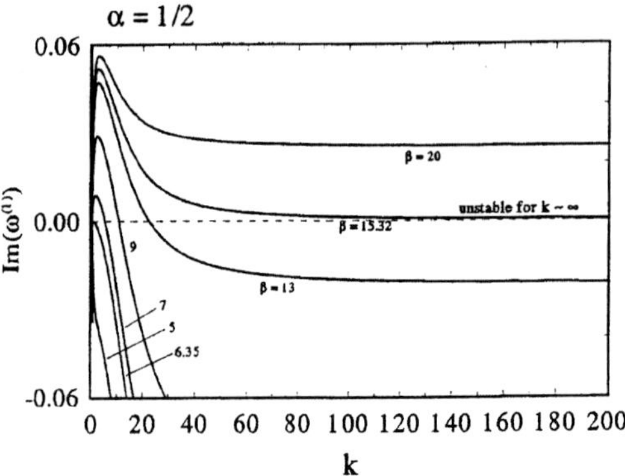

Fig. 22.10. Imaginary part of ω as a function of the wavenumber k and β, for high yield-stress fluid with $\alpha = 1/2$. From [21]

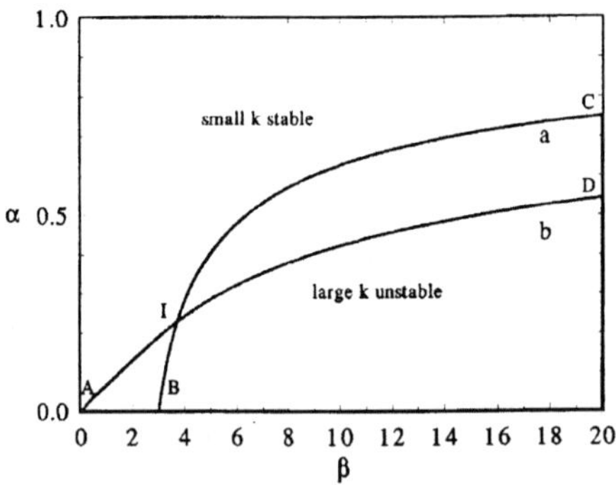

Fig. 22.11. Stability boundaries in the plane of α and β. To the right of curve (**a**), the longest wave ($k = 0$) is unstable. To the right of curve (**b**), the shortest waves ($k \to \infty$) are unstable. From [21]

adjusted the diffusivity to fit the data upstream where waves began to grow, and found encouraging agreement on the development and *coarsening* of roll waves downstream, i.e. the linear increase of shock amplitudes and wavelength with the distance away from the inlet. Mathematically speaking, fitting by choosing an artificial diffusivity differs little from choosing an initial wavelength. For the mud problem we discuss below several examples based on the numerical solutions of (22.68)-(22.70) for the nonlinear stage of surge development, including roll waves due to upstream disturbances.

22.4.3 Roll Waves by Numerical Computation

(i) Spatially Periodic Roll Waves

It is natural to expect that from initially small disturbances of different wavelengths, the one with the fastest growth will dominate the wavelength of the nonlinear roll wave at the end. As was shown before, such prefered wavenumber exists for sufficiently large α and β. With an initial perturbation corresponding to the most unstable mode, the three hyperbolic equations (22.68-22.70) governing the three unknown h, h_o and u_p are solved in [19] and [21] by the upwind method of finite differences. The amplitude of the initial disturbance is chosen to be 1% of the initial depth. The initially perturbed shear layer depth and initial velocity are determined by (22.75) and (22.76) respectively. Periodic boundary conditions are imposed.

Figure 22.12 gives the surface profile within one wavelength for $\beta = 27$, $\alpha = 0.3$ which corresponds to either large Reynolds number or highly non-Newtonian fluid. The most unstable mode occurs at $k = 1.2$, according to the linearized instability theory. As the amplitude grows in time, the wave front also steepens. At approximately $t = 48$, a shock is formed at the front. The shock amplitude gradually increases to its maximum and approaches a steady amplitude after $t = 64$.

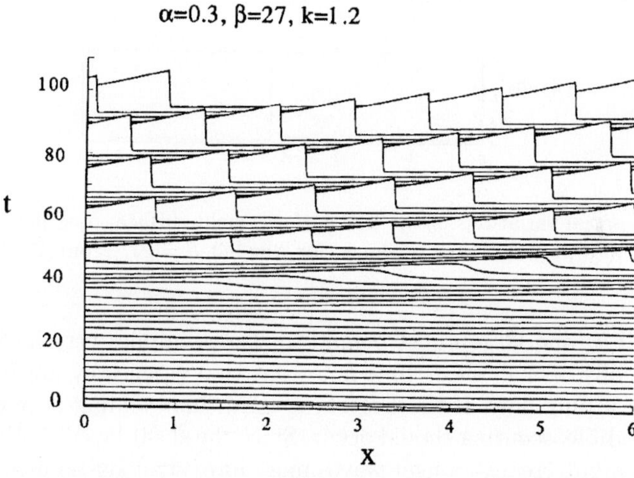

Fig. 22.12. Nonlinear development of the unstable disturbance for $\alpha = 0.3, \beta = 27$ within one wave length. An infinitesimal and periodic disturbance of the most unstable wave number $k = 1.2$ is chosen at $t = 0$. A steady shock is reached at $t > 64$. From [21]

The dimensionless propagation speed of the initial disturbance (0.25 here) has been confirmed with the linear instability theory. As the shock amplitude increases, the shock speed increases. At the steady state, the shock propagates at a higher speed C_s ($= 0.34$). As the shock grows, mass accumulates just behind the shock and is lost ahead of the next shock. Therefore, the mud depth and mud velocity u_p decrease, but the shear layer depth increases across the shock. Figure 22.13 gives the snapshots of h, u_p, h_o and the bottom stress τ_b at t=90 which is in the steady state. Since both h and u_p decrease and h_o increases across the shock.

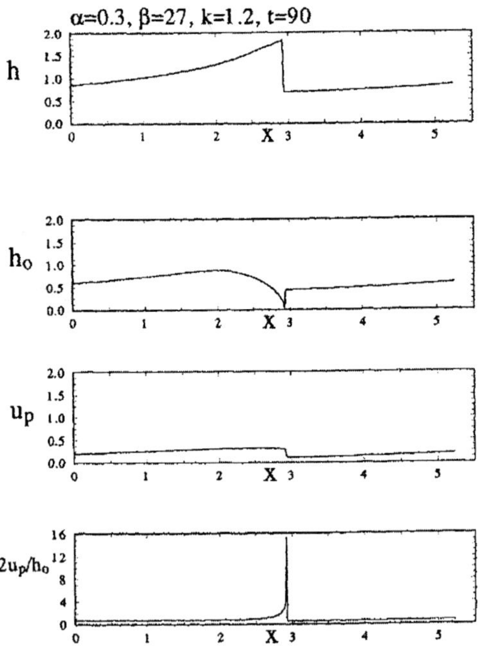

Fig. 22.13. Spatial variations of the free surface, yield surface, plug flow velocity and bottom shear in a period of steady shock, for $\alpha = 0.3, \beta = 27$. From [21]

The small h_o and large u just behind the shock implies very large bed shear there. Physically a small shear zone h_o means that essentially locally the whole mud layer is a plug flow. If the bottom is rigid, mud sliding is implied. If the bottom is erodible, scouring should occur. Since the shear layer depth h_o is found to decrease as k decreases, longer waves must have stronger scouring capacity.

When the shear layer depth h_o approaches zero, (22.70) becomes singular through the last term representing bottom friction, and computation cannot proceed towards the steady state. This numerical difficulty was avoided by applying the biviscous model of Bingham fluid with a large but finite viscosity in the plug zone. This removes the singularity and allows computation to proceed for all time, as explained in [21].

(ii) Roll Waves due to Periodic Inflow

Field studies by Davies [14] suggest that roll waves in a river nearly always occur at a distance downstream of a junction with a tributary. To simulate this phenomenon, Liu [19] introduced a small periodic source q_{in} to the right of (22.62) along a prescribed stretch upstream. The source strength (flux rate) is

$$q_{\text{in}} = \begin{cases} Q_{\text{in}} \left[1 + \sin\left(2\pi t/T - \pi/2\right)\right] , & |x - x_o| < W \\ 0 , & |x - x_o| > W , \end{cases} \tag{22.78}$$

where Q_{in} denotes the maximum discharge and T the period. Figure 22.14 shows a sample result for $\beta = 27, \alpha = 0.33$ so that a progressive wave of any wavenumber would be unstable from the linearized theory. Here the maximum source strength is $Q_{\text{in}} = 2.7$ and period $T = 3\pi/4$. The source length is taken to be $W = 0.03$ and the center of the influx is at $x_o = 5/3$. At first, small disturbances grow as they are convected downstream. After sufficiently long time shocks form at the front. As time increases, the leading shocks grow higher, while more shocks emerge from behind. The region between two successive shocks is a depression which lengthens with time. The growth and lengthening are approximately linear both in space and time. These features are qualitatively similar to the observations of Brock [30] in turbulent channel flows of clear water, and numerically simulated by Chang et al. [28] with a fictitious diffusion. Liu [19] also studied other values of β, Q_{in} and T. Generally shocks develop sooner for larger β and influx rates. Large T leads to large separation between shocks and greater shock amplitudes.

(iii) Roll Waves due to Sudden Addition of Mud

As a contrast to Sect. 22.2.4 and Sect. 22.3.3, let us examine the effects of a sudden mud addition to a fast uniform flow. Liu & Mei [21] gave some computed results for a case of high Bingham number $\alpha = 0.3$. In a uniform flow with $\beta = 27$, a triangular pile of fresh mud is added at initially

$$h(x, 0) = \begin{cases} 1 + A\left(1 + |x - x_o|/W\right) , & |x - x_o| < W \\ 1 , & |x - x_o| < W . \end{cases} \tag{22.79}$$

The resulting free surface is seen in Fig. 22.15. The initial crest first grows in height, leading to a steep front. The shock amplitude increases in time. Since the velocity increases downstream from the uniform-flow value towards the shock, mud accumulates at the front. To conserve mass a depression is formed at the rear. When the depression is sufficiently low, the local flow ceases (at about $t = 50$). The first shock is then separated from the rear and reaches a steady state. At the rear a new front is pushed down the slope by the original uniform flow from behind. Instability induces the growth of a new hump which turns into a new shock. Again a depression is formed in the rear. The process is repeated so that a few more shocks are generated, each of which diminishes in size.

For the same $\alpha = 0.3$ but smaller β, there is no unstable mode from the linearized theory. Inertia is no longer large enough; numerical computations show that the additional mud pile flattens out slowly. For more details see [19].

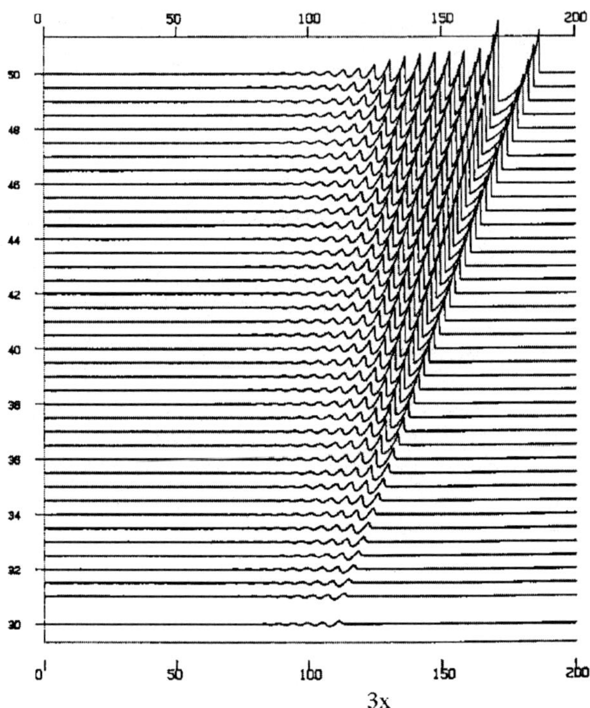

Fig. 22.14. Nonlinear development of the mud surface due to time-periodic influx upstream. The mud parameters are $\alpha = 0.3, \beta = 27$ so that waves of all lengths are linearly unstable. Reproduced from [19] with x, t renormalized according to the scales of this article.

22.5 Other Related Works

In a recent work, Huang & Garcia [35] found an analytical approximation for the one-dimensional initial value problem of the collapse of a mud pile released from a finite area, also for moderate Reynolds numbers so that inertia is ineffective. The Herschel–Bulkley model was assumed. It was shown that over the main body of the mud pile, the gentle surface slope permits the omission of the highest derivative. The resulting approximation is a hyperbolic equation of the kinematic wave type, which can be treated by the method of characteristics. A discontinuous front evolves at the front, near which the stationary wave solution serves as a local remedy and rounds up the front. This method is an efficient alternative to numerical computations.

If mud is released on a plane slope from a narrow outlet, the spreading is two dimensional (x, y). Numerical and laboratory modelling of the transient mud flow from the narrow opening of a reservoir have been reported by Coussot et al. [16,36] for Herschel–Bulkley fluids in slow flows. Extensions to fast flows has been reported by Laigle [37]. A part of the study in [36] is to predict the

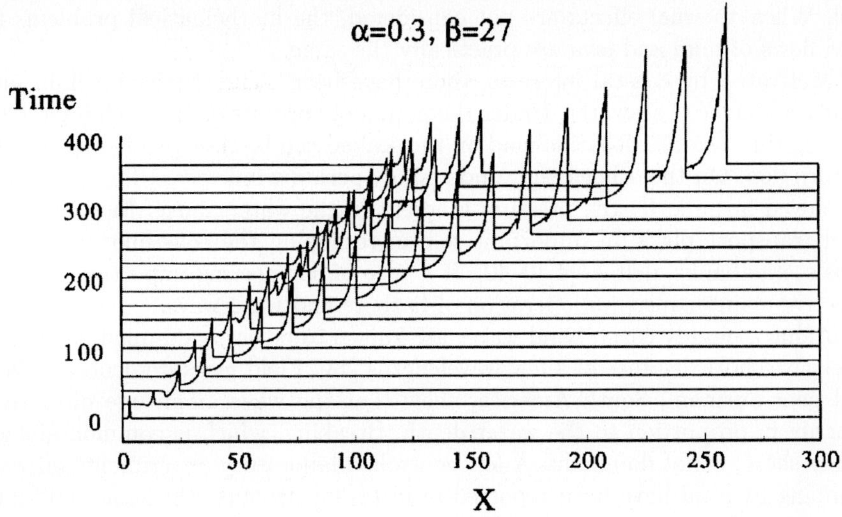

Fig. 22.15. Nonlinear development of the mud surface due to the addition of a mud pile upstream. The mud parameters are $\alpha = 0.3, \beta = 27$ so that waves of all lengths are linearly unstable. From [21]

shape and the extent of the final deposit, when flow comes to a halt. Similar to Sect. 22.2.2, they attempted to solve for the static mud depth $h(x, y)$ from the threshold condition (22.43) with $h_o = 0$ for a prescribed mud volume V, and found it necessary to add an ad hoc condition. Denoting the curve representing the outer rim of the mud pile by $R(x, y) = 0$, they impose the condition that at every point along the rim,

$$\frac{\partial R}{\partial y} = \tan\phi \frac{\partial R}{\partial x}, \tag{22.80}$$

where ϕ is the polar angle of the point. This additional constraint reduces the partial differential equation (of eikonal type) for h to an ordinary differential equation which can be solved. However, the assumption (22.80) has so far eluded convincing justification. The mathematical problem appears to be ill-posed[3]. Physically, if two elongated piles are released on a slope, the initial orientations of their axes must affect the final shape at static equilibrium. Thus the final pile shape cannot be uniquely determined by prescribing the initial volume alone. The proper answer must be found by the solution of an initial value problem as in Sect. 22.3.3.

The lubrication approximation and Bingham–plastic and Herschel–Bulkley models have been used to model the slow spreading of lava domes on a horizontal plane, as discussed in [38], and in Chap. 7. With the assumption of radial

[3] It can be shown for a pile symmetrical about the x axis, that one cannot solve (22.43) for $h(x, y)$ as an initial value problem with the initial data $h(0, y)$, because the initial line $x = 0$ is a characteristic curve.

[39]. When thermal effects are not considered, the mathematical problems for slow flows of mud and lava are practically the same.

Motivated by coastal interests, there have been some studies on fluid mud beneath a layer of seawater. Under the action of currents or tides, characterized by long time scales, estuarine mud on the seabed can be resuspended by interfacial friction. On the other hand, under short-period wind waves, fluid mud may be moved in bulk, by the pressure gradient in the water above. In either case, mud transport plays an important role in reshaping the coastline by forming shoals, mudbanks, deltas, [4,40–42]. It is also of engineering importance for the effective maintenance and operation of harbors and waterways.

Along a muddy coast, wind waves are known to be easily damped out within a relative short distance of a few wavelengths [43]. Field measurements by Wells [41] near Surinam, South America, show that the wave amplitude diminishes roughly in proportion to the water depth. Breaking, which is common along a sandy shore, is not dominant. A few controlled laboratory experiments on wave damping by mud have been reported in [6,44–46]. Because the same fluid-mud may respond very differently under forcings of significantly different magnitudes, dynamic similarity between the laboratory and nature is not easy to achieve.

In a series of papers Liu & Mei have examined the effects of muddy seabed on infinitesimal long waves shoaling in a shallow sea. In the simplest case the mud layer is supposed to be very thin compared to the depth of clear water above [47]. A consequence is that the vertical motion of the water/mud interface is quite small. When interfacial friction is also ignored, the local effect of mud motion on waves is then negligible. On the other hand the surface wave gives rise to a horizontal pressure gradient which drives the mud flow below. For sufficiently short wave periods (high frequency) the shear layer at the bottom of mud is very thin. Most of the mud then moves as a plug flow, subject to a Coulomb friction on the bottom which is also the yield surface. Energy for overcoming this Coulomb friction is supplied by the wave which is then attenuated over many wavelengths. In [48] the interfacial friction modelled as a quadratic function of the velocity jump was included. Finally in [49] the mud layer is allowed to be as deep as the overlying water. Interfacial wave can therefore exist. Both the surface wave and the interfacial wave are assumed to be of finite amplitudes and treated by the Airy approximation. The resulting long wave equations coupling water above and the plug flow below form a fourth-order hyperbolic system. Again the shear layer is negligibly thin and affects the plug flow as a Coulomb friction. When mud moves, waves are attenuated slowly by supplying energy to overcome friction. The spatial attenuation is found to be spatially linear instead of exponential, consistent with the report by Wells from field observations. A stage can be reached when waves becomes so weak that mud ceases to move. Waves can then resume propagation and steepening by nonlinearity (convective inertia) without attenuation. Once the surface wave front becomes sufficiently steep, mud moves again and damping returns. Therefore the dynamical coupling between water and mud is an intermittent process.

22.6 Future Challenges

To simulate nature more closely, a few immediate extensions of theories summarized in this article should be made. Fast flows in shallow open channels of fine width can be treated by modifying Sect. 22.3 for Bingham and other rheological models. Channels with width comparable to depth present more difficulty, as the problem is fully two dimensional in the cross section.

In order to enable quantitative modelling of the entire phenomenon of mud induced catastrophies, it is necessary to include the initiation of mud flow. The relevant scientific issues can of course be dauntingly complex and varied, and future progress demands the cooperation of several disciplines beyond the usual realm of fluid mechanics. There are areas where joint effort with other branches of applied mechanics (e.g. hydrology, soil mechanics, chemical engineering, granular mechanics) can be fruitful.

Take for example mud slides and flows due to rainfalls. There are then two main stages: (i) infiltration of rainwater into the dry top soil on a hill slope, leading to the loss of static equilibrium hence mud-slide; (ii) mud flows down a hillslope along channels.

In Stage (i), persistent and heavy rainfall causes the spatial distribution of soil moisture to change with time. The prediction of vertical infiltration is a well-known problem in the hydrology of unsaturated seepage flow (see e.g. [50,51]). Accompanying the seepage process, the geostatic stresses in soil on a hillslope change with time. In soil plasticity theory, there are ways of predicting soil slope instability according to the failure criterion of Coulomb which depends on the cohesive strength and friction angle [52]. However, these two parameters depends strongly on the soil moisture; empirical knowledge on their dependence is very meager at present. Further progress would likely require progresses in soil chemistry.

As for Stage (ii), once the mud moves, its dynamics depends on the land surface down the hill slope. If the land is dry, mud moves down as moist debris where granular collision is a central feature. If mud slides into a body of water, one needs reliable rheological models for the mixtures of water, cohesive fine particles and stones of different sizes. The erosion of channel banks and resuspension of soil particles by flowing mud must be an even more difficult problem than the erosion of river beds by flowing water. Mechanisms of soil fluidization by transient shear and pressure gradient in the muddy water must be understood.

Our mechanistic understanding of the most common material on earth has only begun.

Acknowledgements

This review is based on research carried out with the generous supported for many years by US Office of Naval Research (Ocean Engineering Program, Contract No. 00014-89-J-3128) and U.S. National Science Foundation (Natural Hazards Program, Grant BCS 9112748).

References

1. Z.C. Kang: *Debris Flow Hazards and Their Control in China.* Scientific Press, Beijing, PRC (1996)
2. B. McDowell: National Geographic **169(5)**, 640 (1968)
3. B. Lyons: *Out of the Inferno*, Savage Earth video series, Thirteen/WNET and Granda Televison (1998)
4. E. Allersma: 'Mud in estuaries and along coasts'. In: *International Symposium on River Sedimentation*, Beijing, China. Also Pubs. 270, Delft Hydraulics Laboratory, (1982)
5. A.E.J. Bryant, D.J.A. Williams: 'Rheology of cohesive suspensions'. In: *Industrialized Embayments and their Environmental Problems*, ed. by M.B. Collins et al. (Pergamon Press, 1980)
6. P.C. Migniot: La Houille Blanche **7**, 591 (1968)
7. Z.H. Wan: 'Bed material movement in hyperconcentrated flow'. Tech. Univ. Denmark Series Paper No. 31, Inst. of Hydrodynamics and Hydraulic Eng. (1982)
8. R.B. Krone: A study of rheologic properties of estuarial sediments, University of California Hydraulic Engineering Lab. and Sanitary Research Lab., Berkeley, Ser. Rep. No 63-8 (1963)
9. J.S. O'Brien, P.Y. Julien: J. Hydr. Eng., ASCE **114**(8), 877 (1988)
10. N. Qian, Z. H. Wan: *A Critical Review of the Research on the Hyperconcentrated Flow in China.* International Research and Training Centre on Erosion and Sedimentation, Beijing, China (1986)
11. G. Verreet, J. Berlamont: 'Rheology and non-Newtonian behavior of sea and estuarine mud'. In: *Encyclopedia of Fluid Mechanics*, **7** **N**. Cheremissinoff(ed.) Gulf Publ. Co. (1987)
12. K. Yano, A. Daido: Annals of Disaster Prevention Research Institute, Kyoto University, Kyoto, Japan, **7**, 340 (1965)
13. A.M. Johnson: *Physical processes in geology* , Freeman, Cooper & Co. (1970)
14. T.R.H. Davies: Acta Mechanica **63**, 161 (1985)
15. J. Li, J. Yuan, C. Bi, D. Luo: Zeit. Geomorph. N.F. **27(3)**, 325 (1983)
16. P. Coussot: *Mudflow Rheology and Dynamics*, IAHR Monograph, A. A. Balkema (1997)
17. Z.H. Wan, Z.Y. Wang: *Hyperconcentrated Flow*, IAHR Monograph, A. A. Balkema (1994)
18. K.F. Liu, C.C. Mei: J. Fluid Mech. **207**, 505–529 (1989)
19. K.F. Liu: Dynamics of a shallow layer of fluid mud. Ph.D. Thesis, Civil Engineering Department, Mass. Inst. Tech. (1990)
20. C.C. Mei, M. Yuhi: J. Fluid Mech. **431**, 135 (2001)
21. K.F. Liu, C.C. Mei: Phys. of Fluids **6**, 2577 (1994)
22. K.F. Liu, C.C. Mei, Phys. of Fluids **A 2**, 30 (1990)
23. K. F. Liu, C.C. Mei: J. Eng. Sci. **31**, 145 (1993)
24. N.J. Balmforth, R.V. Craster: J. Non-Newtonian Fluid Mech. **84**, 65 (1999)
25. F. Englund, Z.H. Wan: 'Instability of hypeconcentrated flow'. Report 255, Technical University of Denmark, (1982)
26. M. Hikida: 'Field observation of roll-waves in debris flow'. In: *Symposium Proceedings of Hydraulics/Hydrology of Arid Land*, ASCE, San Diego, 410 (1990)
27. N. Qian, Z.H. Wan: *Dynamics of Sediments* (in Chinese), Science Press, Beijing, China, (1986)
28. H-C. Chang, E.A. Demekhin, E. Kalaidin: Phys. of Fluids **12**, 2268 (2000)

29. D.J. Needam, J.H. Merkin: Proc. Roy. Soc. Lond. **A 394**, 259 (1984)
30. R.R. Brock: Am. Soc. Civ. Engr. **Hy4**, 1401 (1969)
31. C.O. Ng, C.C. Mei: J. Fluid Mech. **263**, 151 (1994)
32. R.F. Dressler: Comm. Pure & Appl. Math **2**, 149 (1949)
33. T. Kajiuchi, A. Saito: J. Chem. Eng. Japan **17**(1), 34, (1984)
34. P.Y. Julien, D.M. Harteley: J. Hydr. Res. **24**, 5 (1986)
35. X. Huang, M.H. Garcia: J. Fluid Mech. **374**, 305 (1998)
36. P. Coussot, S. Proust, C. Ancey: J. Non-Newtonian Fluid Mech. **66**, 55 (1996)
37. D. Laigle: 'A two-dimensional model for the study of debris flow spreading on a torrent debris fan'. In: *Debris-flow: Hazards Mitigation: Mechanics, Prediction and Assessment*, ASCE, 123 (1997)
38. N.J. Balmforth, A.S. Burbidge, R.V. Craster, J. Salzig, A. Shen: J. Fluid Mech. **403**, 37 (1999)
39. R.W. Griffiths: Ann. Rev. Fluid Mech. **32**, 477 (2000)
40. R.B. Krone: A field study of flocculation as a factor in estuarial shoaling processes, Tech. Rep. 19, Corps of Engineers, U.S. Army (1972)
41. J.T. Wells: Shallow-water waves and fluid-mud dynamics, Coast of Sarinam, South America, Tech. Rep. 157, Coastal Studies Inst., Louisiana State University (1978)
42. M.P. Leeder: *Sedimentology: Process and Product*, George Allen & Unwin Ltd., United Kingdom (1982)
43. G.Z. Forristall, A.M. Reece: J. Geophys. Res. **90(C2)**, 3367 (1985)
44. H.G. Gade: J. Mar. Res. **16**, 61 (1958)
45. A.J. Mehta, P.Y. Maa: Continental Shelf Research **7**, 1268 (1987)
46. T. Nagai, T. Yamamoto, L. Figeroa: L. Cont. Shelf Res. **5**, 521 (1986)
47. C.C. Mei, K.F. Liu: J. Geophys. Res. **92**, 14581 (1987)
48. K.F. Liu, C.C. Mei: J. Coastal Research **5**, 139 (1989)
49. K.F. Liu, C.C. Mei: J. Eng. Sci. **31**, 125 (1993)
50. J.Y. Parlange: Soil Sci. **114**, 1 (1972)
51. R. Phillip: Soil Sci. **83**, 345 (1957)
52. V.V. Sokolovskii: *Statics of Granular Media* (Pergamon, 1965) pp. 270

Index

bedform
– ripples 1, 15, 16, 19, 20
– bars 15, 19, 20
– dunes 1, 15, 17, 18, 19
– anti-dunes 15
bedload 15, 16, 19, 20
Bingham fluid/model 2, 3, 4, 6, 7, 13, 21, 22
bubbles 2, 8
Cahn-Hilliard equation 1
characteristics 14
climate 9, 10, 11, 17
concentration equations 2, 3, 15, 16
convection (thermal, forced) 1, 5, 12
coulomb friction 4, 13, 14
deposition 8, 15, 16, 19
depth-averaged models 1, 8, 13, 14, 15, 16, 18, 19, 21, 22
dimensional analysis 1, 3, 7, 11, 14, 16, 20, 22
eddy viscosity 1, 15, 16, 18, 19
Exner equation 1, 12, 15, 16, 18, 19
experiments water 8, 19, 20,
– kaolin slurry 2, 3, 6, 7, 19, 22
– fluvial 15, 16, 19, 20
– wax 5, 6
– viscous fluids 5, 6
– granular media 4, 14
flow on inclined planes 3, 4, 6, 7, 9, 22
fracture 7, 11, 12
generalized newtonian fluid 2, 4
Ginburg-Landau equation 1, 15
Glen's law 2, 9, 11, 18
gravity currents 8, 13, 14, 21
Herschel-Bulkley fluid/model 2, 3, 4, 7, 21, 22
kaolin slurry 2, 3, 6, 7, 19, 22
kinetic theory 3, 4
Landau equation 1, 15
levees 6, 7
linear instability 1, 5, 6, 15, 18, 19, 22
lubrication/shallow-layer theory 1, 6, 7, 9, 11, 15, 18, 21, 22
Maxwell model 2
meanders 15, 19
molecular dynamics 4
Oldroyd and Jaumann derivatives 2
plumes 1, 5, 6, 8
rheometers 2, 3
roll waves 4, 15, 22
shear thinning 2, 7
Shields stress 15, 16, 18, 20
similarity solutions 7, 8, 14
solidification/melting 2, 5, 6, 7, 8, 9, 11, 12, 18
solitons and solitary waves 1
statistical methods 13, 19
stress tensor 2, 3, 4, 11, 14
suspension 2, 3, 4, 15, 16, 19
temperature-dependent viscosity 2, 5, 6, 7, 8, 11
turbulence 1, 5, 8, 15, 16, 17, 19, 21
viscoelasticity 2
von Mises criterion 3
water waves 1, 12, 20
weakly nonlinear theory 1, 15, 18, 19, 20
wind 1, 7, 12, 17
yield stress 2, 3, 4, 6, 7, 13, 21, 22

Lecture Notes in Physics

For information about Vols. 1–543
please contact your bookseller or Springer-Verlag

Vol. 544: T. Brandes (Ed.), Low-Dimensional Systems. Interactions and Transport Properties. Proceedings, 1999. VIII, 219 pages. 2000

Vol. 545: J. Klamut, B. W. Veal, B. M. Dabrowski, P. W. Klamut, M. Kazimierski (Eds.), New Developments in High-Temperature Superconductivity. Proceedings, 1998. VIII, 275 pages. 2000.

Vol. 546: G. Grindhammer, B. A. Kniehl, G. Kramer (Eds.), New Trends in HERA Physics 1999. Proceedings, 1999. XIV, 460 pages. 2000.

Vol. 547: D. Reguera, G. Platero, L. L. Bonilla, J. M. Rubí(Eds.), Statistical and Dynamical Aspects of Mesoscopic Systems. Proceedings, 1999. XII, 357 pages. 2000.

Vol. 548: D. Lemke, M. Stickel, K. Wilke (Eds.), ISO Surveys of a Dusty Universe. Proceedings, 1999. XIV, 432 pages. 2000.

Vol. 549: C. Egbers, G. Pfister (Eds.), Physics of Rotating Fluids. Selected Topics, 1999. XVIII, 437 pages. 2000.

Vol. 550: M. Planat (Ed.), Noise, Oscillators and Algebraic Randomness. Proceedings, 1999. VIII, 417 pages. 2000.

Vol. 551: B. Brogliato (Ed.), Impacts in Mechanical Systems. Analysis and Modelling. Lectures, 1999. IX, 273 pages. 2000.

Vol. 552: Z. Chen, R. E. Ewing, Z.-C. Shi (Eds.), Numerical Treatment of Multiphase Flows in Porous Media. Proceedings, 1999. XXI, 445 pages. 2000.

Vol. 553: J.-P. Rozelot, L. Klein, J.-C. Vial Eds.), Transport of Energy Conversion in the Heliosphere. Proceedings, 1998. IX, 214 pages. 2000.

Vol. 554: K. R. Mecke, D. Stoyan (Eds.), Statistical Physics and Spatial Statistics. The Art of Analyzing and Modeling Spatial Structures and Pattern Formation. Proceedings, 1999. XII, 415 pages. 2000.

Vol. 555: A. Maurel, P. Petitjeans (Eds.), Vortex Structure and Dynamics. Proceedings, 1999. XII, 319 pages. 2000.

Vol. 556: D. Page, J. G. Hirsch (Eds.), From the Sun to the Great Attractor. X, 330 pages. 2000.

Vol. 557: J. A. Freund, T. Pöschel (Eds.), Stochastic Processes in Physics, Chemistry, and Biology. X, 330 pages. 2000.

Vol. 558: P. Breitenlohner, D. Maison (Eds.), Quantum Field Theory. Proceedings, 1998. VIII, 323 pages. 2000

Vol. 559: H.-P. Breuer, F. Petruccione (Eds.), Relativistic Quantum Measurement and Decoherence. Proceedings, 1999. X, 140 pages. 2000.

Vol. 560: S. Abe, Y. Okamoto (Eds.), Nonextensive Statistical Mechanics and Its Applications. IX, 272 pages. 2001.

Vol. 561: H. J. Carmichael, R. J. Glauber, M. O. Scully (Eds.), Directions in Quantum Optics. XVII, 369 pages. 2001.

Vol. 562: C. Lämmerzahl, C. W. F. Everitt, F. W. Hehl (Eds.), Gyros, Clocks, Interferometers...: Testing Relativistic Gravity in Space. XVII,507 pages. 2001.

Vol. 563: F. C. Lázaro, M. J. Arévalo (Eds.), Binary Stars. Selected Topics on Observations and Physical Processes. 1999.IX, 327 pages. 2001.

Vol. 564: T. Pöschel, S. Luding (Eds.), Granular Gases. VIII, 457 pages. 2001.

Vol. 565: E. Beaurepaire, F. Scheurer, G. Krill, J.-P. Kappler (Eds.), Magnetism and Synchrotron Radiation. XIV, 388 pages. 2001.

Vol. 566: J. L. Lumley (Ed.), Fluid Mechanics and the Environment: Dynamical Approaches. VIII, 412 pages. 2001.

Vol. 567: D. Reguera, L. L. Bonilla, J. M. Rubí (Eds.), Coherent Structures in Complex Systems. IX, 465 pages. 2001.

Vol. 568: P. A. Vermeer, S. Diebels, W. Ehlers, H. J. Herrmann, S. Luding, E. Ramm (Eds.), Continuous and Discontinuous Modelling of Cohesive-Frictional Materials. XIV, 307 pages. 2001.

Vol. 569: M. Ziese, M. J. Thornton (Eds.), Spin Electronics. XVII, 493 pages. 2001.

Vol. 570: S. G. Karshenboim, F. S. Pavone, F. Bassani, M. Inguscio, T. W. Hänsch (Eds.), The Hydrogen Atom: Precision Physics of Simple Atomic Systems. XXIII, 293 pages. 2001.

Vol. 571: C. F. Barenghi, R. J. Donnelly, W. F. Vinen (Eds.), Quantized Vortex Dynamics and Superfluid Turbulence. XXII, 455 pages. 2001.

Vol. 572: H. Latal, W. Schweiger (Eds.), Methods of Quantization. XI, 224 pages. 2001.

Vol. 573: H. M. J. Boffin, D. Steeghs, J. Cuypers (Eds.), Astrotomography. XX, 434 pages. 2001.

Vol. 574: J. Bricmont, D. Dürr, M. C. Galavotti, G. Ghirardi, F. Petruccione, N. Zanghi (Eds.), Chance in Physics. XI, 288 pages. 2001.

Vol. 575: M. Orszag, J. C. Retamal (Eds.), Modern Challenges in Quantum Optics. XXIII, 405 pages. 2001.

Vol. 576: M. Lemoine, G. Sigl (Eds.), Physics and Astrophysics of Ultra-High-Energy Cosmic Rays. X, 327 pages. 2001.

Vol. 577: I. P. Williams, N. Thomas (Eds.), Solar and Extra-Solar Planetary Systems. XVIII, 255 pages. 2001.

Vol. 578: D. Blaschke, N. K. Glendenning, A. Sedrakian (Eds.), Physics of Neutron Star Interiors. XI, 509 pages. 2001.

Vol. 579: R. Haug, H. Schoeller (Eds.), Interacting Electrons in Nanostructures. X, 227 pages. 2001.

Vol. 580: K. Baberschke, M. Donath, W. Nolting (Eds.), Band-Ferromagnetism: Ground-State and Finite-Temperature Phenomena.IX, 394 pages. 2001.

Vol.581: J. M. Arias, M. Lozano (Eds.), An Advanced Course in Modern Nuclear Physics. XI, 346 pages. 2001.

Vol.582: N. J. Balmforth, A. Provenzale (Eds.), Geomorphological Fluid Mechanics. X, 579 pages. 2001.

Monographs

For information about Vols. 1–27
please contact your bookseller or Springer-Verlag

Vol. m 28: O. Piguet, S. P. Sorella, Algebraic Renormalization. IX, 134 pages. 1995.

Vol. m 29: C. Bendjaballah, Introduction to Photon Communication. VII, 193 pages. 1995.

Vol. m 30: A. J. Greer, W. J. Kossler, Low Magnetic Fields in Anisotropic Superconductors. VII, 161 pages. 1995.

Vol. m 31 (Corr. Second Printing): P. Busch, M. Grabowski, P.J. Lahti, Operational Quantum Physics. XII, 230 pages. 1997.

Vol. m 32: L. de Broglie, Diverses questions de mécanique et de thermodynamique classiques et relativistes. XII, 198 pages. 1995.

Vol. m 33: R. Alkofer, H. Reinhardt, Chiral Quark Dynamics. VIII, 115 pages. 1995.

Vol. m 34: R. Jost, Das Märchen vom Elfenbeinernen Turm. VIII, 286 pages. 1995.

Vol. m 35: E. Elizalde, Ten Physical Applications of Spectral Zeta Functions. XIV, 224 pages. 1995.

Vol. m 36: G. Dunne, Self-Dual Chern-Simons Theories. X, 217 pages. 1995.

Vol. m 37: S. Childress, A.D. Gilbert, Stretch, Twist, Fold: The Fast Dynamo. XI, 406 pages. 1995.

Vol. m 38: J. González, M. A. Martín-Delgado, G. Sierra, A. H. Vozmediano, Quantum Electron Liquids and High-Tc Superconductivity. X, 299 pages. 1995.

Vol. m 39: L. Pittner, Algebraic Foundations of Non-Com-mutative Differential Geometry and Quantum Groups. XII, 469 pages. 1996.

Vol. m 40: H.-J. Borchers, Translation Group and Particle Representations in Quantum Field Theory. VII, 131 pages. 1996.

Vol. m 41: B. K. Chakrabarti, A. Dutta, P. Sen, Quantum Ising Phases and Transitions in Transverse Ising Models. X, 204 pages. 1996.

Vol. m 42: P. Bouwknegt, J. McCarthy, K. Pilch, The W3 Algebra. Modules, Semi-infinite Cohomology and BV Algebras. XI, 204 pages. 1996.

Vol. m 43: M. Schottenloher, A Mathematical Introduction to Conformal Field Theory. VIII, 142 pages. 1997.

Vol. m 44: A. Bach, Indistinguishable Classical Particles. VIII, 157 pages. 1997.

Vol. m 45: M. Ferrari, V. T. Granik, A. Imam, J. C. Nadeau (Eds.), Advances in Doublet Mechanics. XVI, 214 pages. 1997.

Vol. m 46: M. Camenzind, Les noyaux actifs de galaxies. XVIII, 218 pages. 1997.

Vol. m 47: L. M. Zubov, Nonlinear Theory of Dislocations and Disclinations in Elastic Body. VI, 205 pages. 1997.

Vol. m 48: P. Kopietz, Bosonization of Interacting Fermions in Arbitrary Dimensions. XII, 259 pages. 1997.

Vol. m 49: M. Zak, J. B. Zbilut, R. E. Meyers, From Instability to Intelligence. Complexity and Predictability in Nonlinear Dynamics. XIV, 552 pages. 1997.

Vol. m 50: J. Ambjørn, M. Carfora, A. Marzuoli, The Geometry of Dynamical Triangulations. VI, 197 pages. 1997.

Vol. m 51: G. Landi, An Introduction to Noncommutative Spaces and Their Geometries. XI, 200 pages. 1997.

Vol. m 52: M. Hénon, Generating Families in the Restricted Three-Body Problem. XI, 278 pages. 1997.

Vol. m 53: M. Gad-el-Hak, A. Pollard, J.-P. Bonnet (Eds.), Flow Control. Fundamentals and Practices. XII, 527 pages. 1998.

Vol. m 54: Y. Suzuki, K. Varga, Stochastic Variational Approach to Quantum-Mechanical Few-Body Problems. XIV, 324 pages. 1998.

Vol. m 55: F. Busse, S. C. Müller, Evolution of Spontaneous Structures in Dissipative Continuous Systems. X, 559 pages. 1998.

Vol. m 56: R. Haussmann, Self-consistent Quantum Field Theory and Bosonization for Strongly Correlated Electron Systems. VIII, 173 pages. 1999.

Vol. m 57: G. Cicogna, G. Gaeta, Symmetry and Perturbation Theory in Nonlinear Dynamics. XI, 208 pages. 1999.

Vol. m 58: J. Daillant, A. Gibaud (Eds.), X-Ray and Neutron Reflectivity: Principles and Applications. XVIII, 331 pages. 1999.

Vol. m 59: M. Kriele, Spacetime. Foundations of General Relativity and Differential Geometry. XV, 432 pages. 1999.

Vol. m 60: J. T. Londergan, J. P. Carini, D. P. Murdock, Binding and Scattering in Two-Dimensional Systems. Applications to Quantum Wires, Waveguides and Photonic Crystals. X, 222 pages. 1999.

Vol. m 61: V. Perlick, Ray Optics, Fermat's Principle, and Applications to General Relativity. X, 220 pages. 2000.

Vol. m 62: J. Berger, J. Rubinstein, Connectivity and Superconductivity. XI, 246 pages. 2000.

Vol. m 63: R. J. Szabo, Ray Optics, Equivariant Cohomology and Localization of Path Integrals. XII, 315 pages. 2000.

Vol. m 64: I. G. Avramidi, Heat Kernel and Quantum Gravity. X, 143 pages. 2000.

Vol. m 65: M. Hénon, Generating Families in the Restricted Three-Body Problem. Quantitative Study of Bifurcations. XII, 301 pages. 2001.

Vol. m 66: F. Calogero, Classical Many-Body Problems Amenable to Exact Treatments. XIX, 749 pages. 2001.

Vol. m 67: A. S. Holevo, Statistical Structure of Quantum Theory. IX, 159 pages. 2001.

Vol. m 68: N. Polonsky, Supersymmetry: Structure and Phenomena. Extensions of the Standard Model. XV, 169 pages. 2001.

Vol. m 69: W. Staude, Laser-Strophometry. High-Resolution Techniques for Velocity Gradient Measurements in Fluid Flows. XV, 178 pages. 2001.

Vol. m 70: P. T. Chruściel, J. Jezierski, J. Kijowski, Hamiltonian Field Theory in the Radiating Regime. VI, 172 pages. 2002.